STOICHIOMETRY AND PROCESS CALCULATIONS

SECOND EDITION

K.V. Narayanan
Former Professor and Head
Department of Chemical Engineering
Former Principal
Government Engineering College
Thrissur, Kerala

B. Lakshmikutty
Former Professor and Head
Department of Chemical Engineering
Government Engineering College
Thrissur, Kerala

PHI Learning Private Limited

Delhi-110092
2025

In fond memory of ***Shri Asoke K. Ghosh*** *(October 1942 – February 2024), Founder Chairman and Managing Director of PHI Learning, whose vision endlessly inspires.*

The Legacy Continues . . . .

Published by Pushpita Ghosh, PHI Learning Private Limited, Rimjhim House, 111, Patparganj Industrial Estate, Delhi-110092 and Printed by Mudrak, C-55, Sector 65, Noida, U.P.-201301

₹1095.00

STOICHIOMETRY AND PROCESS CALCULATIONS, Second Edition
K.V. Narayanan and B. Lakshmikutty

ISBN-978-81-203-5289-6 (Print Book)
ISBN-978-93-90544-51-6 (e-Book)

The export rights of the book are vested solely with the publisher.

Contents

Preface

Ten years is a long period to revise a book dealing with subjects that undergo very fast changes on a day-to-day basis. A subject like "Stoichiometry and Process Calculations", which is concerned with the application of law of conservation of mass and energy, law of combining proportions, etc. to industrial processes, does not undergo drastic changes in its basic content in such a period. However, the scope and areas of application of the subject get widened with the introduction of new technologies into the realm of chemical engineering.

In this revised edition, we have briefly reviewed the new avenues and opportunities open to future chemical engineers and in order to equip them to meet this challenge, we have tried to extend the principles of stoichiometry and process calculations to new and emerging areas such as bioprocessing, energy and environment systems, etc. Material balance calculations for processes with and without chemical reactions involving bypass, recycle and purge operations have now been discussed separately in a new chapter. Considering the importance of exercise problems in a book dealing with process calculations, exercise problems at the end of all chapters are reorganized according to their difficulty level. Section on Objective Type Questions with answers is augmented by including questions on process calculations from the last 10 years' Graduate Aptitude Test in Engineering (GATE) questions.

The first edition of the book received appreciable recognition and acceptability from the academic community. We are extremely thankful to our former colleagues and students for their suggestions in improving the quality of the contents of the present edition. Earnest attempt is made to bring out this edition free of errors. However, we will be indebted to those who could point out errors that escaped our notice.

K.V. NARAYANAN
B. LAKSHMIKUTTY

Preface to First Edition

Chemical engineers are concerned with the transformation of raw materials into useful products on a commercial scale. Chemicai process industries achieve this transformation through chemical conversions and physical operations. As the major areas of their expertise lie in the development, design, construction, operation and management of process industries, chemical engineers have to answer questions such as: What is the raw material requirement for the specified production capacity or what yield of products results from a given supply of raw materials? If the unconverted materials are to be sent back to the processing unit, what should be the ratio of the recycled stream to the fresh feed admitted? If heat is to be supplied to the process, how many kilograms of steam would be necessary to maintain the temperature at the desired levels? How many kilojoules of heat are to be removed from the processing unit to prevent temperature from shooting up? The answers to these problems concerning the entire plant or individual units can be provided by performing process calculations involving material and energy balances.

Process calculations are also necessary to provide a variety of information such as the composition of flue gases that result from the combustion of fuels in the plant, the amount of liquid solvent to be used in an absorption column for removing undesired components from a gas stream, the rate at which air is to be sent through a dryer to reduce the moisture content of solid products before being packed for marketing, the temperature and humidity of such air, the rate of circulation of cooling water in the condenser and steam supply to the reboiler of a fractionating column, the fraction of the air to be subjected for dehumidification or the fraction to be bypassed if air of controlled humidity and temperature is desired for a certain operation in the plant and so on. The subject of "Stoichiometry and Process Calculations" covers these and many other problems related to chemical process industries.

This book is intended as a textbook for undergraduate courses in chemical engineering and related disciplines such as biotechnology, polymer technology, petrochemical engineering, electrochemical engineering, environmental engineering, safety engineering and industrial chemistry. The subject is taught in the early semesters of the undergraduate programme as "Stoichiometry", "Process Calculations" or "Chemical Process Principles". As the students at this stage are not yet exposed to different unit operations and unit processes, they usually find the subject difficult due to their inability to comprehend the physical situations being analyzed in the problems. An attempt is made in this book to introduce to the students the fundamentals of chemical engineering and the various unit operations and unit processes before problems on these topics are dealt with. The materials in this book are organized in such a way as to help the students to develop a systematic approach to making stoichiometric calculations.

The word "stoichiometry" in chemical engineering is not just the application of law of combining proportions to chemical reactions; it has got a wider meaning and encompasses all calculations involving material and energy balances related to process industries. A number of physical and chemical properties are required to make these calculations. These include density, heat capacity, latent heat, heat of reaction, calorific values, vapour pressure, humidity, etc. In the absence of experimental data on these properties, methods for their estimation should be known. In addition, it is essential to know the dependence of these properties on process parameters such as temperature and pressure. Other thermodynamic concepts such as the *P-V-T* behaviour of fluids, the phase behaviour of pure materials and solutions are also important to perform process calculations. The first eight chapters of the book deal with these concepts in detail and provide a solid foundation for developing skills necessary for solving complex chemical engineering process problems. Material and energy balances in processes with and without chemical reactions are covered in the last four chapters.

The basic philosophy of process calculations is to prepare the students to make analysis of processes through calculations and also to develop in them systematic problem-solving skills. Through these calculations, the students get equipped with fundamental information and skills that are repeatedly employed in subsequent courses as well as in professional life. With this purpose in mind, we have taken care to include a large number of exercise problems at the end of each chapter, and all of them are provided with answers. Objective type questions provided at the end of the book help the students prepare for competitive examinations such as GATE.

We are extremely thankful to our colleagues in the Department of Chemical Engineering at Government Engineering College, Thrissur for providing a helpful environment for writing this book. Suggestions and critical comments for further improvements are welcome.

K.V. NARAYANAN
B. LAKSHMIKUTTY

1 Introduction

Chemical engineering is concerned with the transformation of raw materials into useful products making use of chemical conversions and physical operations. Industries engaged in this transformation are known as *chemical process industries*. The reactions developed by the chemists in the laboratory are utilized by the chemical engineer for the commercial production of a wide variety of materials. A chemical engineer is interested more than anything else in the application of chemistry and the various branches of engineering sciences to the design, construction, operation, control and improvement of equipment for carrying out chemical processes on an industrial scale with the ultimate objective of making a profit out of the process. Therefore, we can treat chemical engineering as a branch of engineering which deals with the application of mathematics, physics and chemistry to the manufacturing processes of modern process industries. The official definition of chemical engineering put forward by the American Institute of Chemical Engineers is sufficiently broad to make the position clear regarding where a chemical engineer stands among other engineers and pure chemists: Chemical engineering is *the application of the principles of physical sciences together with the principles of economics and human relations to fields that pertain directly to processes and process equipment in which matter is treated to effect a change of state, energy content or composition.*

1.1 CHEMISTRY VERSUS CHEMICAL ENGINEERING

Both chemists and chemical engineers can be treated as members of the chemical profession since both should possess sound training in the principles of chemistry to be successful in their respective areas. However, there is a wrong tendency of equating a chemical engineer with a chemist forgetting the broad scope of problems encountered by the former. For a chemical engineer his or her interest in chemistry is from the point of view of its industrial applications. The bench scale techniques of the chemists as practised in the laboratory may be effective in making original and exploratory investigations. However, these techniques are unsuitable for large-scale production. It falls on the chemical engineer to convert the results of the laboratory scale investigation to commercial production. For effecting the transformation of the raw materials to the final desired products, a large number of physical changes are also involved. And processes involving only physical operations and which do not involve any chemical reactions such as distillation are also not uncommon in chemical engineering practice. A typical chemist has little knowledge of the theory and practice of such operations.

1.2 CHEMICAL PROCESS INDUSTRY

The chemical process industry is one whose principal products are manufactured by the processes based upon chemical and physical principles, now being recognized as chemical engineering principles. A chemical engineer is skilled in the development, design, construction and operation of process industries. He/she finds employment primarily in research and development, design and construction, manufacturing operations, management and administration, waste treatment and disposal, technical services and sales related to process industries. The chemical process industry generally employs "chemical changes" for the production of useful materials. We refer to chemical changes here in a broad sense as they include not only chemical reactions, but also many physiochemical changes such as the separation and purification of the components of a mixture. Purely mechanical changes are usually not considered part of a chemical process, unless they are essential to later chemical changes. For example, the manufacture of plastic polyethylene using ethylene produced from petroleum or natural gas involves a chemical process. On the other hand, the moulding and fabrication of the resulting plastic resin into final shapes for consumer products would not be considered part of a chemical process. The huge metallurgical industry is usually distinguished from the chemical process industry because of the special nature of the processes and the vast quantity of products, but the processing of metals might also be considered to be merely a segment of the chemical process industry. It follows that chemical engineering principles find applications in any industry which, though not producing chemicals, makes use of chemical change. Examples which suggest themselves are iron and steel production, the carbonization of coal, atomic power generation, ceramics and the pulp and paper industries. Although a chemical engineer working in such an industry might never be asked to carry out a complete plant design, his/her knowledge would be invaluable particularly in the development of new ideas and procedures. Table 1.1 lists the important chemical process industries and their end products.

Table 1.1 Major chemical process industries

Industry	*Products*	*End uses*
Inorganic chemicals	Sulphuric acid	Fertilizers, chemicals, petroleum refining, paint pigments, metal processing, explosives
	Nitric acid	Explosives, fertilizers
	Sodium hydroxide	Chemicals, rayon and film processing industries, petroleum refining, lye, cleansers, soap, metal processing

(*Contd.*)

Table 1.1 Major chemical process industries (Contd.)

Industry	*Products*	*End uses*
Organic chemicals	Acetic anhydride	Rayons, resins, plastics
	Ethylene glycol	Antifreeze, cellophane, dynamite, synthetic fibres
	Formaldehyde	Plastics
	Methanol	Formaldehyde manufacture, antifreeze, solvent
Petroleum and petrochemicals	Gasoline	Motor fuel
	Kerosene	Jet fuel, domestic fuel
	Oils	Lubricating, heating
	Ammonia	Fertilizers, chemicals
	Ethyl alcohol	Acetaldehyde, solvent, other chemical manufacture
	Alkyl aryl sulphonate	Detergents
	Styrene	Synthetic rubber and plastics
Pulp and paper	Paper, cardboard, fibre board, etc.	Books, records, news paper, boxes, building materials
Pigments and paint	Zinc oxide, titanium oxide, carbon black, lead chromate,	Pigments for paints, ink, plastic, rubber, ceramic, linoleum
	Linseed oil	Drying oil
	Phenolic, alkyd resins	Basic lacquers, varnishes and enamel paints
Rubber	Natural rubber, synthetic rubber (GR-S, Neoprene, Butyl)	Automobile tyres, moulded goods and sheeting, footwears, electrical insulation, etc.
Plastics	Phenol formaldehyde, polystyrene, polymethyl-methacrylate, PVC, polyethylene, polyesters	Various uses in all areas of everyday life
Synthetic fibres	Rayon, nylon, polyesters, acrylics	Cloth and clothing
Minerals	Glass, ceramic, cement	Windows, containers, bricks, pipes, concrete for construction of buildings, highways
Cleansing agents	Soaps, synthetic detergents, wetting agents	Household and industrial cleansing
Biochemicals	Pharmaceuticals and drugs	Health and medicinal applications
	Fermentation products: Penicillin, ethyl alcohol	Medicinal use, solvent, beverage
	Food products	Human sustenance
Metals	Steel, copper, aluminium, zirconium	Building material, machinery, etc.
	Uranium	Nuclear fuel

1.3 GREATEST ACHIEVEMENTS IN CHEMICAL ENGINEERING

Products of the chemical process industries are used in all areas of everyday life. The ability to bring once scarce materials to all members of society through industrial creativity is a defining characteristic of chemical engineering. The American Institute of Chemical Engineers has compiled a list of 10 greatest contributions of chemical engineering to improve the standard of living of modern human beings. They are as follows:

1. The contribution of chemical engineering towards isolation of isotopes by splitting atoms revolutionized such varied areas as biology, medicine, power generation, metallurgy and archaeology.
2. Mass production of plastics was made a viable economic reality by chemical engineers in the twentieth century that had had its positive impact on all aspects of modern life.
3. The concept of unit operations was extended to analyze the functioning of human body that have helped improve clinical care, diagnostic and therapeutic devices and development of artificial organs.
4. Chemical engineers could improve the yield of antibiotics through mutation and special brewing techniques. Low priced, high-volume drugs and pharmaceuticals owe their existence to the work of chemical engineers.
5. Introduction of synthetic fibres such as nylon, polyester, etc. had a profound impact on many areas of life and it also reduced the strain on natural sources of cotton and wool.
6. The liquefaction and subsequent separation of air into pure nitrogen and oxygen was another major contribution of chemical engineering. Nitrogen can be used to recover petroleum, freeze food, produce semiconductors or prevent unwanted reactions while oxygen is used to make steel, smelt copper, weld metals together and for artificial respiration.
7. Chemical engineers' contribution in the areas of waste treatment and pollution abatement helps keep the environment clean. Catalytic converters, reformulated gasoline and smoke stack scrubbers and new recycling technologies are notable achievements in this area.
8. A tremendous increase in food production was made possible by the manufacture of chemical fertilizers. Chemical engineers today play a very vital role in food processing also.
9. One of the greatest contributions of chemical engineering is in the area of petroleum processing and petrochemicals, which is now regarded as an enabling technology without which modern life would cease to function. Chemical engineers have developed processes such as catalytic cracking to break down complex organic molecules found in crude oils to simpler species which serve as building blocks for the production of many useful products such as gasoline, lubricating oils, plastics, synthetic rubber, and synthetic fibres.
10. Chemical engineers played a prominent role in developing today's synthetic rubber industry. Modern society is heavily dependent on synthetic rubber for products such as tires, gaskets, hoses, conveyor belts and shoes.

Among other products of importance are insecticides, food supplements and disinfectants that are essential for raising food plants and animals. Many building materials have been chemically processed, for example, metals, concrete, roofing materials, paints, and plastics. Clothing utilizes many synthetic fibres and dyes. Written communication uses paper and ink; and electronic communication requires many chemically processed insulators and conductors. Soaps, detergents, insecticides and disinfectants have a vital role in maintaining nation's health in addition to drugs and pharmaceuticals. There are many chemicals that never reach the consumer in their original form but are sold within the industry for further processing or use in the production of other chemicals for common use. It is often said that the chemical industry is its own best customer.

1.4 FUTURE AVENUES AND OPPORTUNITIES

Over the decades, chemical engineers were making great contributions to improve the quality of our life through their activities in diverse fields such as atomic science, polymers, paper, dyes, drugs, plastics, fertilizers, foods, petrochemicals, etc. What does the future hold for the next generation of chemical engineers? What are the challenges and opportunities that the new generation of chemical engineers have to face in the coming years?

The role of chemical engineers in the fields of biotechnology, pharmaceuticals, nanotechnology, molecular engineering, medical devices, and biomaterials is likely to expand dramatically in the coming years. They are expected to make significant contribution in developing improved polymer processing and devices relevant to bio-medical engineering. The physical and biological treatment of hazardous wastes is an area in which chemical engineers have a significant role to play. Development of separation techniques and technologies for solving separation problems in the chemical, environmental, food, pharmaceutical, and biotechnological industries offers further avenues for chemical engineers. With the broadening of their interest area to encompass genetic engineering, nanotechnology and molecular engineering, computer simulation and computational methods now extensively used in modeling studies are going to help the chemical engineers in a big way in the analysis of such processes. Large-scale centralized production that had been the hallmark of chemical processes in the past is going to be supplemented by small-scale distributed production, which will allow localized production of chemicals and energy.

The most dramatic change in chemical engineering is the result of the increasing importance of biology and biochemistry and the evolution of biotechnology as a real commercial activity. A switch from traditional sources of energy and raw materials to biological resources will be the most dominant change. Energy intensive reactors and separators can be replaced with a single vessel containing rationally engineered microorganisms capable of performing multiple catalytic steps. Chemical engineers are now focusing on the use of alternative feedstocks such as biomass and societal waste to produce chemicals and energy. Their aim is to use less carbon in the manufacturing processes by making them more carbon efficient and this lower carbon process is sometimes termed as "Green Chemistry" or sustainable processing. Great possibilities of advances in catalytic conversion of biomass and CO_2 into practical forms of energy and useful chemicals will require new advances in catalytic chemistry and processes most likely based on nanocomposite materials.

Nano-scale phenomena have always been a part of the profession because chemical engineers work with molecules. Advances in nanotechnology have opened the door to new possibilities, especially in materials. Some of these opportunities are products for biomedicine. Advanced materials, such as bifunctional materials, conducting polymers, biomaterial (tissue engineering), continue development for specialized applications. Microelectronicmechanical system (MEMS) devices are being commercialized. A successful approach for the chemical engineers will be in the integration of nano and biological platforms for commercial production of devices for specific applications.

Other exciting frontier areas of importance in chemical engineering include molecular and nano-scale engineering, molecular simulation, surface modification, protein separation processes and supercritical fluid extraction. Molecular engineering is an important part of pharmaceutical research and material science. Molecular engineering deals with any means of manufacturing molecules or creating new manufacturing materials using them. The field can be seen as a precision form of chemical engineering that includes protein engineering.

1.5 HISTORY OF CHEMICAL ENGINEERING

Chemical engineering was practised long before it was recognized as a distinct engineering profession. It can be said that even the ancient Greeks and Romans practised it when they were making soap or wine or were treating ores. However the evolution of chemical engineering as a distinct profession occurred concurrently with expansion in the chemical industry during the later half of the nineteenth century. The political changes that took place in France and Germany during the mid nineteenth century had its impact on the industrial processes that were in practice at that time. The chemical engineering profession emerged from the need to revise these processes with an emphasis on safer and more efficient methods. Despite these developments, education in these areas was not formalized. Students obtained some superficial knowledge about these processes in chemistry courses. In those days, the predominant chemical industries were built up in Germany whose achievements in the field of dyestuffs, synthetic nitrates, and coal hydrogenation were the work of chemists collaborating with mechanical and civil engineers. As a result, the process vessels of German chemical plants of that time and even of a later period, were larger size replicas of laboratory apparatus used in research. The "chemical engineer" of those years was either a mechanical engineer who had gained some knowledge of chemical process equipment with a lifetime of experience but little education, or an applied chemist with knowledge of large scale industrial chemical reactions. "From their experience in chemical plants, both chemists with an instinct for engineering and engineers with a taste for chemistry grew into chemical engineers without realizing it and without being willing to admit it" (Sir Harold Hartley, 1958).

One of the earliest attempts to organize the principles of chemical processing and to clarify the professional area of chemical engineering was made in England by George E. Davis (1850–1906). He was an industrial inspector by profession and the daily inspection rounds he used to take through various chemical plants of that time brought him close to the intricacies of chemical processes and the innovations in the operation of those plants. He was convinced of the necessity for a new branch of engineering that was equally comfortable with both applied chemistry and traditional engineering. In 1880 he acted upon these ideas and proposed formation of a "Society of Chemical Engineers" which proved unsuccessful. In 1887 he gave a series of 12 lectures, which were later published in the *Chemical Trade Journal.* The lecture was organized around individual chemical operations as practised by the British chemical industry. Though the material was quite empirical in nature, it convinced others that the time for chemical engineering had arrived.

In 1888, one year after Davis's lectures at Manchester, Professor Lewis M. Norton (1855–93) of the Chemistry Department of MIT started teaching a course in chemical engineering, which was the first four-year bachelor program in chemical engineering. (However, as a department, chemical engineering did not become independent until 1920!) Soon other colleges, such as University of Pennsylvania and Tulane University followed MIT's lead starting their own four-year programs. When Norton died in 1893, Professor Frank H. Thorpe (1864–1932) took responsibility of Norton's course at MIT and published in 1898 the first textbook in chemical engineering entitled *Outlines of Industrial Chemistry.*

In 1901 in England, Davis proceeded with the publication of his *Handbook of Chemical Engineering.* Davis was responsible for adopting the idea of "unit operations" in the second edition of his book (1904) although the term was coined by Arthur D. Little at MIT much later in 1915.

In the early years of the twentieth century, chemical engineering began to gain professional acceptance. To survive, chemical engineers had to claim industrial territory by defining themselves and demonstrating their uniqueness and worth. With this goal in mind, the American Institute of Chemical Engineers (AIChE) was formed in June 1908. Just weeks after the formation of AIChE, the American chemical society organized a "Division of Industrial Chemistry and Chemical Engineering" and authorized the publication of the *Journal of Industrial and Engineering Chemistry.*

1.6 CHEMICAL ENGINEERING IN INDIA

In India Dr. H.L. Roy introduced chemical engineering in the curriculum of the then Bengal Technical Institute, the nucleus of the present Jadavpur University as early as 1921 when chemical engineering was still at its infancy in the West. It started with a four-year diploma in chemical engineering. Full-fledged degree courses and master's programmes were introduced in the mid-forties and early fifties respectively. The University Department of Chemical Technology of the Bombay University (UDCT), which came into being in 1934 had a full-fledged chemical engineering section offering a two-year post BSc course in Textile Chemistry and Chemical Engineering. The chemical engineering section at the University Department of Chemical Technology has modernized the earlier courses and in 1951, the four-year post inter-sciences course offering B. Chem. Engg. was introduced. Andhra University College of Engineering established in 1933, offered graduate and postgraduate courses in Sugar Technology and Chemical Technology, and in 1958 the B.Tech degree course in chemical engineering with different electives was offered. The Indian Institute of Science founded in 1909 started offering courses in chemical engineering from the Department of General Chemistry. In 1951, the chemical engineering section got a separate status as the Department of Chemical Technology and Chemical Engineering. Harcourt Butler Technological Institute, Kanpur set up in 1921 though offered training in oil and paints, leather and sugar technology with chemical engineering input, a formal course in chemical engineering was started only in 1954. In the fifties, Indian Institutes of Technology (IITs) were established in different parts of the country all offering degree courses in chemical engineering. The Indian Institute of Chemical Engineers came into being in 1947.

1.7 UNIT OPERATIONS AND UNIT PROCESSES

The manufacturing processes of chemical and allied industries can usually be classified into coordinate series of unit physical operations and unit chemical processes—steps that are similar in principle if not identical in applications. The unit operations involve the addition or removal of some form of energy in the contacting, transport, and the conditioning of materials by physical means with or without any chemical changes taking place. The unit processes on the other hand, result in chemical changes—in composition, character, and properties—of materials and are most often affected or controlled by temperature and pressure changes, catalysis, intimacy of mixing, and other physical phenomena.

Chemical reactions that have common chemical characteristics underlying their industrial applications are grouped together for study as a unit process. This grouping of similar reactions into a unit process simplifies their study, since all the reactions in a group have similar requirements of raw materials, reaction conditions of concentration, temperature and pressure, and processing equipment. The process engineer applies the same basic principles to all the reactions. Also, the reactions require similar equipment, utilities, and technology. The principal unit processes are: combustion, oxidation, nitration, halogenation, sulphonation, ammonolysis, reduction, hydrogenation, esterification, hydrolysis, alkylation, polymerization, fermentation, etc. This classification of chemical conversions into unit processes is not as strong as the concept of unit operations for dealing with physical operations. Many exceptions and disparities exist among processes falling within a particular classification.

Table 1.2 Different types of unit operations and unit processes

Unit operations		*Unit processes*	
Fluid flow	Size reduction	Oxidation	Polymerization
Heat transfer	Sedimentation	Reduction	
Mass transfer	Filtration	Nitration	Pyrolysis
Drying	Mixing	Sulphonation	Hydrolysis
Distillation	Evaporation	Electrochemical reactions	
Crystallization	Sublimation	Industrial microbiological processes	
Extraction	Centrifugation	Halogenation	
Adsorption	Materials handling		
Gas absorption	Ion exchange		

The concept of unit operations distinguishes chemical engineering from all other professions and gives the chemical engineering education programmes a common focus. Practising chemical engineers come across a large number of chemical and physical operations for transforming matter from inexpensive raw materials to highly desired products. Many similarities exist in the manner in which the feed materials are converted to end products in different process industries. When we break the diverse processes employed by different industries into a series of separate and distinct steps called *unit operations*, we can find that many similarities exist between these operations employed by different industries for entirely different purposes. These seemingly different operations are common to all types of process industries. Identifying the operations such as filtration, drying, distillation, crystallization, grinding, sedimentation, heat exchange, evaporation, extraction, extrusion, etc. as unit operations became a convenient method for organizing chemical engineering knowledge and provided a central theme for chemical engineering education to rally around. The knowledge gained concerning a unit operation governing one set of materials can easily be applied to others. Whether one is using distillation for the manufacture of alcohol or for production of gasoline from petroleum, the underlying principle remains the same.

The unit operations concept became the basic theme in the chemical engineering profession ever since George Davis's lectures on the topic. However, it was Arthur D. Little in 1915 first recognized the potential of using this concept for distinguishing chemical engineering from other professions. While mechanical engineers focused on machinery, industrial chemists concerned themselves with products, and applied chemists studied individual reactions, no one before chemical engineers had concentrated upon the underlying process common to all chemical products, reactions and machinery. The uniqueness and worth of chemical engineers is now evident to all concerned with large-scale chemical manufacture. The important unit operations are discussed in the following paragraphs.

1.7.1 Fluid Flow Operations

These operations include the transportation of fluid from one point to the other. Since transportation, storage and handling of fluids are more convenient than those of solids, the fluid flow operations play a very significant role in process industries. A chemical engineer has to deal with movement of fluid through pipes, pumps and all kinds of process vessels. Sometimes reactant streams have to be passed through a bed of solid catalyst, sometimes a bed of solid will be fluidized to effect better conversion. The quantitative relationship between the rate of flow and pressure difference, the power requirement of the flow system, measurements of rate of flow, etc. are therefore very important to his/her profession than any other branch of engineering.

1.7.2 Heat Transfer Operations

As pointed out earlier all unit operations involve the transfer of energy into and out of the system. Also all chemical reactions are accompanied by characteristic and unavoidable heat effects. Heat transfer during boiling of liquids and condensation of vapours are frequently encountered by a chemical engineer. The transfer of heat at the desired rate is thus a major operation for carrying out all operations and reactions efficiently as is clear from the daily requirements of two important plant utilities—the cooling water and steam—by any process industry. Above all heat recovery will be a major concern from the point of view of conservation of energy and the environment. The major heat transfer equipment the chemical engineer has to work with includes a wide variety of tubular heat exchangers, plate heat exchangers, boilers and condensers. The transfer and control of heat in process plants, design and operation of heat transfer equipment employed by chemical process industries are therefore important areas of chemical engineering activity.

1.7.3 Evaporation

Evaporation is used in industries for concentrating aqueous solutions by vaporizing the solvent water and removing it as vapour. Typical industrial applications of evaporation include concentration of cane-sugar juice in a sugar factory, concentration of ammonium sulphate in a fertilizer unit, and concentration of spent soap *lye* to produce glycerine in a soap industry. Depending upon the properties of materials being handled, there are a number of different types of evaporators and different modes of feeding a multiple effect system consisting of a number of evaporators arranged in series. Short-tube vertical evaporators, long-tube evaporators, climbing film and falling film evaporators, etc. are some industrially important evaporators in use. Evaporators and evaporator accessories like steam-jet ejectors, condensers, steam trap, etc. require careful monitoring by chemical engineers.

1.7.4 Mass Transfer Operations

Mass transfer operations are employed by process industries for separation of mixtures into their component parts. There is no chemical process industry that does not require a preliminary purification of raw materials or final separation of products from by-products or other undesired contaminants. Mass transfer operations are thus very important in process industries. When a chemical reaction is implemented on a commercial scale, the investment on mass transfer equipment generally exceeds the capital investment associated with the reactions as such. The important mass transfer operations are reviewed in the following paragraphs.

Distillation: Distillation is used to separate liquid mixtures into component parts by boiling. The difference in the volatilities of the constituents is the property that is exploited to effect separation. The industrially important distillation method known as *fractionation* or *fractional distillation* has got very wide application in chemical and petroleum industries. The products obtained on distillation are commonly referred to as *distillate* (or *top product* which is rich in more volatile components) and *residue* (or *bottom product* which is rich in less volatile components).

Absorption: In absorption, the soluble constituents of a gas mixture are separated by absorbing in a suitable liquid solvent. The reverse process—the removal of certain constituents of a liquid mixture by contacting with a gas phase—is known as *desorption* or *stripping*. Ammonia is absorbed from a mixture of ammonia and air by contacting the gas with water in equipment known as *absorption columns*. Benzene vapours present in coke-oven gases can be absorbed in hydrocarbon oils and hydrogen sulphide can be absorbed from gas mixtures using ethanolamine solutions.

Liquid–liquid extraction: The process of separation of the components of a liquid mixture by treating with an immiscible liquid solvent in which the constituents are differently soluble is known as *liquid–liquid extraction*. Aqueous acetic acid solution is mixed with isopropyl ether solvent in order to extract the acid into the ether phase. Extraction results in two immiscible phases, the solvent rich phase called the *extract* and the original solution from which the solute is extracted known as the *raffinate*. The mutually insoluble extract and raffinate phases are then separated from one another by settling and gravity separation.

Leaching: Leaching is the separation of the components of a solid mixture by selectively dissolving the soluble components in the solid mixture in a liquid solvent. The recovery of minerals from naturally occurring ores, oils from cakes, tannin from wood barks, sugar from sugar beets, etc. are some of the industrial applications of leaching.

Adsorption: Components of a gas or liquid mixture can be adsorbed on the surface of a solid adsorbent. The adsorption of organic vapours on activated charcoal, decolourization of cane-sugar solution by adsorbing the colouring matter on activated carbon, drying of gases by adsorbing water on silica gel, etc. are examples to cite a few. The adsorbed constituent can be removed from the solid and thereby separation can be completed and the adsorbent regenerated for further use.

Humidification: Humidification and dehumidification operations are used by process industries for preparing air of desired temperature and humidity, water cooling, gas drying and other such purposes. A gas phase (usually air) is contacted with pure liquid (usually water). The transfer of vapour from liquid to gas occurs in humidification operations and the reverse process occurs in dehumidification. Cooling towers, spray chambers, humidifiers, etc. are pieces of equipment used for these operations.

Drying: Drying is usually one of the last operations in a process industry. Drying is an integral part of the process in industries such as paper industries, where as drying is done in other processes for reducing the cost of transportation of the product, to give some useful properties to the product like the free-flowing nature of salt, to prepare the product in a form that is suitable for handling and use. In drying a wet solid or a slurry is contacted with dry gas (usually air or flue gas) so that water is vaporized from the solid and is carried away by the gas. Depending upon the characteristics of the solid being dried, several types of dryers are in common use. Tray drier (for pasty materials and lumpy solids), rotary drier (for granular and free flowing solids), freeze driers (for foodstuffs and pharmaceuticals) and spray driers (for slurries and pastes) are typical driers in use.

Crystallization: Crystallization is the process in which solid particles are formed from liquid solution by evaporating and/or cooling of a saturated solution. The process is important since a variety of materials is marketed in crystalline form and also as a method of purification. Tank crystallizers, agitated batch crystallizers, Swenson–Walker crystallizer, vacuum crystallizers and Krystal crystallisers are typical industrial equipment used for crystallization.

Ion exchange: In ion exchange operations, the solute from a solution is retained on the solid by reaction with the solid ion-exchange resins. Ions in solution can be removed by this process as in purification of water.

In addition to the above separation methods in which the phases are in direct contact, separation can be achieved by transferring materials through membranes separating the two phases. Dialysis, electrodialysis, ultrafiltration, etc. are some membrane separation techniques.

1.7.5 Mechanical Separations

In a process plant the chemical engineer may have to deal with separation of solids from solids, solids from gases, solids from liquids and liquid drops from gases. Mechanical separations are based on physical differences among the particles such as size, shape or density. Important operations falling under this category are screening, filtration, sedimentation, centrifugation, etc.

Screening: Screening is a method of separating solid particles from a mixture of solids based on their size difference alone. Screens are available in variety of mesh sizes and depending upon the size of the feed handled by the screens, screening devices are classified as grizzlies, trommels, shaking screens and vibrating screens.

Filtration: In filtration, suspended solid particles in a fluid, either a gas or a liquid, is removed by passing through a filtering medium such as canvas cloth that retains the particles as a separate phase or cake and allows the passage of clear filtrate. Filter presses, leaf filters, rotary drum filters, etc. are the filters used for separating solids from liquids. Centrifugal filters are another class of filters in which the filtering medium is attached to a rotating basket and the centrifugal action forces the liquid through it.

Sedimentation: In settling and sedimentation, the particles are separated from the fluid by gravitational force acting on the particles. The particles can be solid particles or liquid drops. The separation of a dilute slurry or suspension by gravity settling into a clear fluid and a slurry of higher solids content is called *sedimentation.* Removal of solids from liquid sewage wastes, settling of crystals from mother liquor, separation of solid food particles from a liquid food, etc. are some important applications of settling for separation. Gravity settling tanks, Spitzkasten classifier and continuous thickener are pieces of equipment coming under this group. Settling of solid particles aided by centrifugal forces can be used to separate particles that will not settle easily in gravity settlers. Centrifugal separation is employed in many food industries such as breweries, vegetable oil processing, fruit juice processing, etc.

1.7.6 Size Reduction

Solids generally occur in sizes that are too large for conveniently handling them for either separation of constituents or for chemical processing. The reduction in size is achieved by means of crushing and grinding. Grinding operations are very prominent in ore processing and cement industries. Food processing industries also employ size reduction operations extensively as in grinding wheat, corn and rye to flour, in rolling, pressing and grinding of soybeans to produce oil and flour. Jaw crushers, gyratory crushers, roll crushers, hammer mills and ball mills are typical size reduction equipment used by processing industries.

1.7.7 Mixing

Mixing of solids and pastes and blending of liquids are among the fundamental operations in all chemical processes. Purpose of these operations are to produce product of uniform properties. Kneaders, mixing rolls, pugmills, ribbon blenders, screw mixers and tumbler mixers etc. are some typical vessels used for mixing pastes and dry powders.

1.8 STOICHIOMETRY AND PROCESS CALCULATIONS

There are certain fundamental principles underlying all unit operations and unit processes. These can be broadly categorized into four groups. They are the material and energy balances and the laws of equilibrium and the rate processes. As pointed out earlier, the unit operations deal mainly with the transfer and change of energy and the transfer and change of materials by physical or in some instances by physiochemical means. The study of material and energy balances in these operations is very vital for better assimilation of the course materials that constitute the undergraduate chemical engineering curriculum. The main thrust in the course offered under different names such as *Process Calculations*, *Stoichiometry*, or *Chemical Process Principles* is the study of material and energy balances involved in unit operations and unit processes. In addition, there are certain principles and techniques that the student should master at an early stage itself that will enable him/her a lot in the evolution into a good chemical engineer and later in his/her professional life. These include the behaviour of gases and gas mixtures, both real and ideal, estimation of their properties, phase behaviour of pure liquids and solutions, vapour pressure and how it is influenced by changes in temperature and pressure, humidity and saturation, application of psychrometric charts, steam tables, enthalpy-composition diagrams, etc. to cite a few. A course in Process Calculations or Stoichiometry covers all the above topics. It is clear that we use the word "stoichiometry" with a wider meaning: stoichiometry for a chemical engineer is not just the application of the laws of combining proportions of elements or compounds involved in chemical reactions.

The basic philosophy of process calculations is to assist the student in the analysis of processes through calculations and develop in him/her systematic problem-solving skills. Through these calculations the student gets equipped with fundamental information and skills that are repeatedly employed in subsequent courses as well as in professional life. Although the theory underlying the solution of these problems is well defined and unquestioned, the solution cannot be achieved by application of just some theoretical formulae or some semiempirical equations. Rather, using these principles for the solution of stoichiometric problems is an art, and like every art, its mastering requires practice.

2 Units and Dimensions

A given physical quantity can be measured using a variety of units. For example, the length of an object may be given in metres or in feet. The mass of an object may be given in kilograms or in pounds. In engineering calculations, it is frequently necessary to convert quantities measured in one system of units into another. In the absence of rigorous theoretical models, engineers may have to use many empirical equations developed based on the experience and experimental observations. These equations contain quantities that are measured in the specified units. Familiarity with the various systems of units and conversion of quantities from one system to another is, therefore, of prime importance in process calculations.

2.1 UNITS AND DIMENSIONS

A physical quantity is always expressed as the product of a number and a unit. For example, $P = 10^5$ N/m^2 or $L = 10$ ft. The numbers by themselves have no meaning in the language of measurements unless they are accompanied by units. For example, if you say the mass of an object is 50, the object may weigh 50 kg or 50 lb or even 50 g. To avoid confusion and to eliminate errors in calculations, the unit should be explicitly stated. *Unit* is any measure or amount used as a standard for measurement. By *dimension* we mean the measurable extent of a physical quantity. In the above example mass (represented by the symbol *M*) is a dimension and kilogram (kg) or pound (lb) is a unit. Units provide standards to measure the quantities called dimensions. Each unit is associated with a dimension which is unique. Or, a unit refers to one and only one dimension. The kilogram is a unit used to measure the dimension mass. It cannot be used to measure the dimension time or temperature. Thus the dimension of the unit metre is length (*L*). Primary quantities such as length, mass, time and temperature are commonly used as the basis of measurement. These are called *primary dimensions* and are represented by the symbols *L*, *M*, *T* and *K* respectively. The set of primary dimensions should satisfy the requirement that all quantities that need to be measured can be assigned one or a combination of primary dimensions. The *derived unit* is a combination of primary units and the *derived quantity* is a combination of two or more primary dimensions and their units are a combination of primary units. Thus the mass of an object has a dimension M and its unit is kilogram whereas the density of a substance has a dimension ML^{-3} derived from primary dimensions and its unit is kg/m^3 which is a derived unit. Quantities like heat capacity, viscosity, specific volume, thermal conductivity, etc. are derived quantities.

Though each unit is associated with a unique dimension, the unit used to describe a given dimension need not be unique. Thus kilogram and pound are just two of the different units available for expressing the mass, newtons per square metre (N/m^2) and pounds per square inch (psi) are just two of the different units available for expressing the quantity of pressure. This difference is due to the different system of units used for measurements.

2.2 SYSTEM OF UNITS

There are four systems of units in common use. They are the metric system (MKS system), the English system (FPS system), the CGS system, and the latest the Système International (SI system). A system of units adopts a set of primary units each of which corresponds to a primary dimension. All other units and dimensions allowed within a particular system are derived units and derived dimensions.

The FPS system developed in England uses the unit foot (ft) for the dimension of length, pound (lb) for the dimension of mass, second (s) for the dimension of time, and degree Fahrenheit (°F) for the dimension of temperature. In the MKS system (the metric system) developed in France in 1791, the unit for length is metre (m), the unit for mass is kilogram (kg), the unit for time is second (s) and the unit for temperature is degree Celsius (°C). India adopted the MKS system in 1957. The CGS system employs centimetre (cm), gram (g) and second (s) for the units of length, mass and time respectively and uses the same standards for the base units as the MKS system. In practice, it is difficult to work with MKS or CGS system alone, and so both were used in engineering practice depending upon the convenience.

The different systems of units not only have different sets of base units, they use different standards to physically represent these base units. Till May 20, 2019, the kelvin, the unit of thermodynamic temperature in SI, was taken as the fraction 1/273.16 of the thermodynamic temperature of the triple point of water. Similarly, a standard alloy block of platinum and iridium maintained at the International Bureau of Weights and Measures at Sevres, Paris was taken as the base unit of mass, the kilogram. The following table gives the primary quantities and their units and dimensions adopted in different systems of units.

Table 2.1 Primary dimensions and units in different system of units

Quantity	*Units*			*Symbol*			*Dimension*
	FPS	MKS	CGS	FPS	MKS	CGS	
Length	foot	metre	centimetre	ft	m	cm	L
Mass	pound	kilogram	gram	lb	kg	g	M
Time	second	second	second	s	s	s	T
Temperature	Fahrenheit	Celsius	Celsius	°F	°C	°C	K

2.2.1 SI Units

The International System of Units (abbreviated SI for Système International) was adopted at the Eleventh Conference Générale des Poids et Measures (CGPM) in 1960 with *length*, *mass*, *time, temperature, electric current and luminous intensity* as the six basic quantities. The SI system is a modification of the original metric system and it uses the metre for *length*, the kilogram for *mass*, the second for *time*, and the kelvin (K) for *temperature*. The primary quantity mass is long felt to be unsuitable for use in chemistry, where the number of molecules constituting the system is much more im portant than the mass. Taking this into consideration, the Fourteenth Conference Générale des Poids et Measures in 1971 expanded the list of primary dimensions by including the *amount of substance* as a primary quantity with unit *mole*. On May 20, 2019, four of them—the kilogram, the kelvin, the ampere and the mole—were redefined in terms of natural physical constants by setting exact numerical values for the Planck constant (h), the elementary charge (e), the Boltzmann constant (k) and the Avogadro constant (N_A). The other three—the second, the metre and the candela—are already based on universal constants. Table 2.2 gives the list of the seven primary dimensions and their units used in the SI system.

Table 2.2 SI base units

Quantity	*Unit*	*Symbol*	*Standard / Definition*	*Dimension*
Length	metre	m	The metre, symbol m, is the SI unit of length. It is defined by taking the fixed numerical value of the speed of light in vacuum c to be 299 792 458 when expressed in the unit m s^{-1}, where the second is defined in terms of the caesium frequency $\Delta\nu_{Cs}$.	L
Mass	kilogram	kg	The kilogram, symbol kg, is the SI unit of mass. It is defined by taking the fixed numerical value of the Planck constant h to be 6.626 070 15 × 10^{-34} when expressed in the unit Js, which is equal to kg m^2 s^{-1}, where the metre and the second are defined in terms of c and $\Delta\nu_{Cs}$.	M
Time	second	s	The second, symbol s, is the SI unit of time. It is defined by taking the fixed numerical value of the caesium frequency $\Delta\nu_{Cs}$, the unperturbed ground state hyperfine transition frequency of the caesium 133 atom, to be 9 192 631 770 when expressed in the unit Hz, which is equal to s^{-1}.	T
Electric current	ampere	A	The ampere, symbol A, is the SI unit of electric current. It is defined by taking the fixed numerical value of the elementary charge e to be 1.602 176 634 × 10^{-19} when expressed in the unit of C, which is equal to A s, where the second is defined in terms of $\Delta\nu_{Cs}$.	I
Thermodynamic temperature	kelvin	K	The kelvin, symbol K, is the SI unit of thermodynamic temperature. It is defined by taking the fixed numerical value of the Boltzmann constant k to be 1.380 649 × 10^{-23} when expressed in the unit J K^{-1}, which is equal to kg $m^2s^{-2}K^{-1}$ where kilogram, metre and second are defined in terms of h, c and $\Delta\nu_{Cs}$.	K
Amount of substance	mole	mol	The mole, symbol mol, is the SI unit of amount of substance. One mole contains exactly 6.022 140 76 × 10^{23} elementary entities. This number is the fixed	n

(Contd.)

Table 2.2 SI base units (Contd.)

			numerical value of the Avogadro constant N_A, when expressed in the unit mol^{-1} and is called the Avogadro number. The amount of substance of a system is a measure of the number of specified elementary entities. An elementary entity may be an atom, a molecule, an ion, an electron, any other particle or specified group of particles.	
Luminous intensity	candela	cd	The candela, symbol cd, is the SI unit of luminous intensity in a given direction. It is defined by taking the fixed numerical value of the luminous efficacy of monochromatic radiation of frequency 540 × 10^{12} Hz, K_{cd} to be 683 when expressed in the unit lmW^{-1}, which is equal to cd sr W^{-1},or cd sr kg^{-1} m^{-2} s^3, where the kilogram, metre and second are defined in terms of h, c and Δv_{Cs}	I_v

2.2.2 Derived Units

The derived units can be represented by a combination of two or more primary dimensions and their units are a combination of the primary units. For example, *velocity* can be defined as the time rate of change of distance. The dimension of velocity (u) can be written as length divided by time (LT^{-1}). The quantity *acceleration* (a) can be defined as the time rate of change of velocity. The dimension of acceleration could be written as u/T (velocity divided by time), or written in terms of the accepted fundamental dimensions, LT^{-2} and its unit in SI is ms^{-2}. Table 2.3 gives SI derived units having special names and symbols. Table 2.4 gives some derived quantities in different systems of units that are commonly used in chemical engineering calculations.

Table 2.3 SI derived units having special names

Derived quantity	*Name and symbol*	*In terms of derived* SI *units*	*In terms of* SI *base units*
Plane angle	radian (rad)		m m^{-1} = 1
Solid angle	steradian (sr)		m^2 m^{-2} = 1
Frequency	hertz (Hz)		s^{-1}
Force	newton (N)		m kg s^{-2}
Pressure, stress	pascal (Pa)	N m^{-2}	m^{-1} kg s^{-2}
Energy, work, quantity of heat	joule (J)	N m	m^2 kg s^{-2}
Power, radiant flux	watt (W)	J s^{-1}	m^2 kg s^{-3}
Electric charge, quantity of electricity	coulomb (C)		S A
Electric potential difference, electromotive force	volt (V)	W A^{-1}	m^2 kg s^{-3} A^{-1}
Capacitance	farad (F)	C V^{-1}	m^{-2} kg^{-1} s^4 A^2
Electric resistance	ohm (Ω)	V A^{-1}	m^2 kg s^{-3} A^{-2}
Electric conductance	siemens (S)	A V^{-1}	m^{-2} kg^{-1} s^3 A^2
Magnetic flux	weber (Wb)	V s	m^2 kg s^{-2} A^{-1}
Magnetic flux density	tests (T)	Wb m^{-2}	kg s^{-2} A^{-1}

Table 2.4 Derived units in SI

Quantity	Units (symbolic abbreviation in brackets) SI	MKS/CGS	FPS	Dimension (Based on SI)
Area	square metres (m^2)	square centimetre (cm^2)	square feet (ft^2)	L^2
Volume	cubic metres (m^3)	cubic centimetre (cm^3), litre (L)	cubic feet (ft^3)	L^3
Density	kilograms per cubic metre (kg m^{-3})	grams per cubic centimetre (g cm^{-3})	pounds per cubic foot (lb ft^{-3})	ML^{-3}
Mass flow rate	kilogram per second (kg s^{-1})	gram per second g s^{-1}	pounds per hour (lb h^{-1})	MT^{-1}
Molar flow rate	moles per second (mol s^{-1})	moles per second (mol s^{-1})	pound moles per hour (lb-mol h^{-1})	nT^{-1}
Volumetric flow rate	cubic metres per second (m^3 s^{-1})	cubic centimetre per second (cm^3 s^{-1}), litre per second (L s^{-1})	cubic feet per hour (ft^3 h^{-1})	L^3T^{-1}
Mass flux	moles per square metre per second (mol s^{-1} m^{-2})	moles per square metre per second (mol s^{-1} m^{-2})	pound moles per hour per square foot (lb-mol h^{-1} ft^{-2})	$nT^{-1}L^{-2}$
Force	newton (N)	kilogram-force (kgf)	pound force (lbf)	MLT^{-2}
Pressure	newton per square metre or pascal (N m^{-2} or Pa)	kilogram force per square centimetre (kgf cm^{-2})	pound force per square foot (lbf ft^{-2})	$ML^{-1}T^{-2}$
Energy	joule (J)	metre kilogram force (m kgf)	foot pound force (ft lbf)	ML^2T^{-2}
Power	watts or joules per second (W or J s^{-1})	horse power (HP)	foot pound force per second (ft lbf s^{-1})	ML^2T^{-3}
Heat, Enthalpy	joule (J)	kilocalories (kcal)	British Thermal Unit (Btu)	ML^2T^{-2}
Heat capacity	joules per kilogram per degree kelvin (J kg^{-1} K^{-1})	kilocalories per kilogram per degree celsius, (kcal kg^{-1} °C^{-1})	Btu per pound per degree Fahrenheit (Btu lb^{-1} °F^{-1})	$L^2T^{-2}K^{-1}$
Molar heat capacity	joules per mole per degree kelvin (J mol^{-1} K^{-1})	kilocalories per mole per degree celsius, (kcal mol^{-1} °C^{-1})	Btu per pound mole per degree Fahrenheit (Btu lb-mol^{-1} °F^{-1})	$ML^2T^{-2}K^{-1}n^{-1}$
Heat flux	joules per square metre per second (J s^{-1} m^{-2}, W m^{-2})	kilocalories per hour per square metre (kcal h^{-1} m^{-2})	Btu per hour per square foot (Btu h^{-1} ft^{-2})	MT^{-3}

2.2.3 SI Prefixes

Table 2.5 lists the SI prefixes that are used to form decimal multiples and submultiples of SI units. They allow very large or very small numerical values to be avoided. A prefix attaches directly to the name of a unit, and a prefix symbol attaches directly to the symbol for a unit. For example, one kilometre, symbol 1 km, is equal to one thousand meters, symbol 1000 m or 10^3 m. When prefixes are attached to SI units, the units so formed are called *multiples* and *submultiples* of SI units in order to distinguish them from the coherent system of SI units.

Table 2.5 SI prefixes

Factor	*Prefix*	*Symbol*	*Factor*	*Prefix*	*Symbol*
10^{24}	yotta	Y	10^{-1}	deci	d
10^{21}	zeta	Z	10^{-2}	centi	c
10^{18}	exa	E	10^{-3}	milli	m
10^{15}	peta	P	10^{-6}	micro	μ
10^{12}	tera	T	10^{-9}	nano	n
10^{9}	giga	G	10^{-12}	pico	p
10^{6}	mega	M	10^{-15}	femto	f
10^{3}	kilo	k	10^{-18}	atto	a
10^{2}	hecto	h	10^{-21}	zepto	z
10^{1}	deka	da	10^{-24}	yocto	y

2.2.4 General Guidelines on the Use of SI Units

The general guidelines for using SI units are given in Table 2.6.

Table 2.6 General guidelines on the use of SI units

Guideline	*Correct usage (examples)*	*Incorrect usage or usage not recommended (examples)*
Unit symbols are printed in lower-case letters. *Exception:* The symbol or the first letter of the symbol is an upper-case letter when the name of the unit is derived from the name of a person. Also, the recommended symbol for the litre is L.	m (metre) Pa (pascal) s (second) kW (kilowatts) V (volt)	M (Metre) pa (Pascal) S (Second) kw (Kilowatts) v (Volt)
Unit symbols are unaltered in the plural.	Length is 50 cm.	Length is 50 cms.
Unit symbols are not followed by a period. Unit symbols will be followed by a period at the end of a sentence.	Mass should be 25 kg or more. Mass should be 25 kg.	Mass is 25 kg. or more.
Symbols for units formed from other units by multiplication are indicated by means of either a centred dot or a space.	N·m or N m $m{\cdot}s^{-1}$ or m s^{-1}	As ms^{-1} means reciprocal milliseconds, it is incorrect to write ms^{-1} for indicating metre per second.
Symbols for units formed from other units by division are indicated by means of a solidus (oblique stroke, /), a horizontal line, or negative exponents. However, to avoid ambiguity, the solidus must not be repeated on the same line without using parentheses.	m/s, $\frac{m}{s}$, or $m{\cdot}s^{-1}$ m/s^2 or $m{\cdot}s^{-2}$ $m{\cdot}kg/(s^3{\cdot}A)$ or $m{\cdot}kg{\cdot}s^{-3}{\cdot}A^{-1}$	m/s/s $m{\cdot}kg/s^3/A$
Unit symbols and unit names are not used together.	W/m^2 or watts per square metre	watts/m^2 or W/square metre

(Contd.)

Table 2.6 General guidelines on the use of SI units (Contd.)

Guideline	*Correct usage (examples)*	*Incorrect usage or usage not recommended (examples)*
Because acceptable units generally have internationally recognized symbols and names, it is not permissible to use abbreviations for their unit symbols or names, such as sec (for either s or second), mins, (for either min or minutes), hrs (for either h or hours), lit (for either L or litre), amps (for either A or amperes), AMU (for either u or unified atomic mass unit), or mps (for either m/s or metre per second).	s or second mm^2 or square millimetre cm^3 or cubic centimetre min (minutes) h (hours) L or litre m/s or metre per second	sec or Sec sq. mm cc or ml mins hrs lit mps
Prefix symbols are printed in roman (upright) type regardless of the type used in the surrounding text, and are attached to unit symbols without a space between the prefix symbol and the unit symbol. This last rule also applies to prefixes attached to unit names.	mL (millilitre) pm (picometre) GΩ (gigaohm) THz (terahertz)	m L (milli litre) p m (pico metre) G Ω (giga ohm)
The prefix symbols Y (yotta), Z (zetta), E (exa), P (peta), T (tera), G (giga), and M (mega) are printed in upper-case letters while all other prefix symbols are printed in lower-case letters. Prefixes are normally printed in lower-case letters.	THz GΩ	tHz gΩ
The grouping formed by a prefix symbol attached to a unit symbol constitutes a new inseparable symbol (forming a multiple or submultiple of the unit concerned) which can be raised to a positive or negative power and which can be combined with other unit symbols to form compound unit symbols. Prefixes are also inseparable from the unit names to which they are attached.	millimetre (mm) 5 cm^3 = 5.0 $(cm)^3$ millimetre, micropascal, meganewton, are single words	milli metre (m m)
Compound prefix symbols, that is, prefix symbols formed by the juxtaposition of two or more prefix symbols, are not permitted. This rule also applies to compound prefixes.	nm (nanometer)	mμm (millimicrometer)
In a derived unit formed by division, the use of a prefix symbol (or a prefix) in both the numerator *and* the denominator may cause confusion. In a derived unit formed by multiplication, the use of more than one prefix symbol (or more than one prefix) may also cause confusion.	10 MV/m (preferred because it contains only one prefix symbol and it is in the numerator) 10 kV·s (preferred)	10 kV/mm (is acceptable but not preferred) 10 MV·ms (not preferred)
For historical reasons, the name *kilogram* for the SI base unit of mass contains the name *kilo*, the SI prefix for 10^3. Thus, because compound prefixes are unacceptable, symbols for decimal multiples and submultiples of the unit of mass are formed by attaching SI prefix symbols to g, the unit symbol for gram, and the names of such multiples and submultiples are formed by attaching SI prefixes to the name *gram*. *Example*: 1 mg *but not* 1 microkg.	10^{-6} kg = 1 mg (1 milligram)	10^{-6} kg = 1 μkg (1 microkilogram)

(Contd.)

Table 2.6 General guidelines on the use of SI units (Contd.)

Guideline	*Correct usage (examples)*	*Incorrect usage or usage not recommended (examples)*
When writing numbers, digits should be separated into groups of three, counting from the decimal marker towards the left and right, by the use of a thin, fixed space. Comma should not be used for the purpose. However, this practice is not usually followed for numbers having only four digits on either side of the decimal marker except when uniformity in a table is desired.	83 254 346 56 235.175 73 0.562 345 5 8765 or 8 765	83,254,522 56,235.175 73 0.5623455 8,765
The SI unit of volume is the cubic metre (m^3) and may be used to express the volume of any substance, whether solid, liquid, or gas. The litre (L) is a special name for the cubic decimetre (dm^3) but the CGPM recommends that the litre not be used to give the results of high accuracy measurements of volumes.		
Celsius temperature (t) is defined in terms of thermodynamic temperature (T) by the equation $t = T - T_0$, where $T_0 = 273.15$ K by definition. This implies that the numerical value of a given temperature interval or temperature difference whose value is expressed in the unit degree celsius (°C) is equal to the numerical value of the same interval or difference when its value is expressed in the unit kelvin (K); or $\{\Delta t\}_{°C} = \{\Delta T\}_K$. Thus temperature intervals or temperature differences may be expressed in either the degree celsius or the kelvin using the same numerical value. *Example*: The difference in temperature between the freezing point of gallium and the triple point of water is $\Delta t = 29.7546$°C $= \Delta T = 29.7546$ K.		

2.3 CONVERSION OF UNITS

Most of the physical or chemical data available in literature are expressed equivalently with a variety of units. Operating manuals of equipment and instruments using different systems of units still abound in the process plants. It is imperative that chemical engineering students or practising chemical engineers be well versed with the conversion of units from one system to another. Conversion factors which relate the units in one system of units to those in another system can be made use of for this purpose. Table 2.7 provides a condensed list of conversion factors of certain units frequently encountered in chemical engineering calculations. To convert the units to the desired system of units, the original expressions are multiplied or divided by the conversion factor. For example, the length measured in inches is multiplied by 2.54×10^{-2} to express it in metres. Here, 2.54×10^{-2} or to be accurate 2.54×10^{-2} m/in is the conversion factor for inch to metre. Thus a length measuring, say, 10 inches is equivalently written as

$$10 \text{ in} = 10 \text{ in} \times \left(\frac{2.54 \times 10^{-2} \text{ m}}{\text{in}}\right) = 0.254 \text{ m}$$

Here, the length in metres is multiplied by the conversion factor 2.54×10^{-2} m/in so that the resulting expression has the desired units of metre. The choice of whether the original expression is to be multiplied or divided by the conversion factor is to be made so as to cancel all the units in the original expression leaving only the desired units. Thus the mass expressed in kilogram is divided by $4.535\ 924 \times 10^{-1}$ kg/lb to obtain mass in pounds. Thus

$$50 \text{ kg} = 50 \text{ kg} \left(\frac{\text{lb}}{4.535924 \times 10^{-1} \text{ kg}}\right) = 110.23 \text{ lb}$$

2.3.1 Some Important Derived Quantities and Their Conversion

Force: According to Newton's second law of motion, the force acting on a body is directly proportional to the time rate of change of momentum. For a body of constant mass, Newton's law reduces to

$$F = cma \tag{2.1}$$

where F is the force, m is the mass of the body, a is the acceleration and c is a proportionality

Table 2.7 Conversion factors

To convert from	*To*	*Multiply by*
atmosphere, standard (atm)	pascal (Pa)	$1.013\ 25 \times 10^5$
bar (bar)	pascal (Pa)	1.0×10^5
British thermal unit (Btu)	joule (J)	$1.055\ 056 \times 10^3$
British thermal unit per pound degree Fahrenheit (Btu lb^{-1} $°F^{-1}$)	joule per kilogram kelvin (J kg^{-1} K^{-1})	4.1868×10^3
calorie (cal)	joule (J)	4.1868
centimetre (cm)	metre (m)	1.0×10^{-2}
centipoises (cP)	pascal second (Pa s)	1.0×10^{-3}
centistokes (cSt)	metre square per second (m^2 s^{-1})	1.0×10^{-6}
cubic foot (ft^3)	cubic metre (m^3)	$2.831\ 685 \times 10^{-2}$
cubic foot per minute (ft^3 min^{-1})	cubic metre per second (m^3 s^{-1})	$4.719\ 474 \times 10^{-4}$
dyne (dyn)	newton (N)	1.0×10^{-5}
erg (erg)	joule (J)	1.0×10^{-7}
foot (ft)	metre (m)	3.048×10^{-1}
foot per hour (ft h^{-1})	metre per second (m s^{-1})	$8.466\ 667 \times 10^{-5}$
gram per cubic centimetre (g cm^{-3})	kilogram per cubic metre (kg m^{-3})	1.0×10^3
horsepower (550 ft lbf s^{-1}), (hp)	watt (W)	$7.456\ 999 \times 10^2$
horsepower (metric)	watt (W)	$7.354\ 988 \times 10^2$
inch (in)	metre (m)	2.54×10^{-2}
kilocalorie (kcal)	joule (J)	4.1868×10^3
kilogram-force (kgf)	newton (N)	9.806 65
kilogram-force per square centimetre (kgf cm^{-2})	pascal (Pa)	$9.806\ 65 \times 10^4$
kilowatt hour (kW h)	joule (J)	3.6×10^6
mile (based on U.S. survey foot), (mi)	metre (m)	$1.609\ 347 \times 10^3$
mile per hour (mi h^{-1})	metre per second (m s^{-1})	4.4704×10^{-1}
poise (P)	pascal second (Pa s)	1.0×10^{-1}
pound (avoirdupois) (lb)	kilogram (kg)	$4.535\ 924 \times 10^{-1}$
poundal	newton (N)	$1.382\ 550 \times 10^{-1}$
pound-force (lbf)	newton (N)	4.448 222
pound-force per square inch, psi, (lbf in^{-2})	pascal (Pa)	$6.894\ 757 \times 10^3$
pound per cubic foot (lb ft^{-3})	kilogram per cubic metre (kg m^{-3})	$1.601\ 846 \times 10^1$
square foot (ft^2)	square metre (m^2)	$9.290\ 304 \times 10^{-2}$
stokes (St)	metre square per second (m^2 s^{-1})	1.0×10^{-4}

(*Contd.*)

Table 2.7 Conversion factors (Contd.)

To convert from	*To*	*Multiply by*
ton, metric (t)	kilogram (kg)	1.0×10^3
torr (Torr)	pascal (Pa)	$1.333\ 224 \times 10^2$
watt second (W s)	joule (J)	1.0
degree Celsius (temperature) (°C)	kelvin (K)	$T/\text{K} = t/°\text{C} + 273.15$
degree Fahrenheit (temperature) (°F)	degree Celsius (°C)	$t/°\text{C} = (t/°\text{F} - 32)/1.8$
degree Fahrenheit (temperature) (°F)	kelvin (K)	$T/\text{K} = (t/°\text{F} + 459.67)/1.8$
degree Rankine (°R)	kelvin (K)	$T/\text{K} = (T/°\text{R})/1.8$

constant. In the SI system, the constant c is unity and we have

$$F = ma \tag{2.2}$$

When a body of mass 1 kg is accelerated by 1 m/s², the force acting on the body is 1 kg m/s², which is designated as 1 *newton* or (1 N).

The force unit *poundal* in FPS system is similar to *newton*. It is the force that gives an acceleration of 1 ft/s² when applied to a body having a mass of one pound.

$$1 \text{ poundal} = 1 \text{ ft lb/s}^2$$

In engineering practice, two technical units of force, kilogram force (abbreviated as kgf) in MKS system and the pound force (abbreviated as lbf) in FPS system were in force for quite a long time. These are defined by letting c in Eq. (2.1) as equal to $1/g_c$, where g_c is called the Newton's law conversion factor. That is

$$F = \frac{ma}{g_c} \tag{2.3}$$

The value of g_c is 9.806 65 kg m (kgf)$^{-1}$ s^{-2} in MKS system and 32.174 lb ft (lbf)$^{-1}$ s^{-2} in FPS system. The numerical value of g_c corresponds to the acceleration due to gravity at mean sea level. Although g (acceleration due to gravity) varies from locality to locality, g_c does not. The dimension of g is LT^{-2} whereas the dimension of g_c is $MLF^{-1}T^{-2}$ where F stands for the technical dimension of force.

The force acting on a body under gravitational acceleration is called the *weight* of the body. Thus weight W is given as

$$W = \frac{mg}{g_c} \tag{2.4}$$

Although g is a local variable, in ordinary calculations g/g_c is taken as 1 kgf/kg or 1 lbf/lb, so that the weight of the body in technical force units is numerically equal to the mass of the body.

Pressure: The pressure is defined as the normal component of the force per unit area exerted by the fluid on a real or imaginary boundary. The unit of pressure in the SI system is *newton per square meter* (N/m²), also called the *pascal* (Pa). A multiple of *pascal*, called the *bar* is also used as a unit of pressure.

$$1 \text{ bar} = 10^5 \text{ Pa} = 10^5 \text{ N/m}^2$$

The pressure exerted by the atmosphere is called the *atmospheric pressure* and it varies with location and elevation on the earth's surface. One standard atmospheric pressure abbreviated as *atm* is used in all systems of units as an empirical unit of pressure. It is the average pressure exerted by the earth's atmosphere at sea level. Although actual atmospheric pressure varies from place to place, its value at mean sea level is taken to be $1.013\,25 \times 10^5$ N/m^2 (or 1.013 25 bar).

In engineering practice the pressure is frequently reported as gauge pressure. The absolute pressure is related to the gauge pressure by the following relation:

$$\text{absolute pressure} = \text{gauge pressure} + \text{atmospheric pressure}$$

Similarly, pressures below atmospheric (subatmospheric) are reported as vacuum such that

$$\text{absolute pressure} = \text{atmospheric pressure} - \text{vacuum}$$

Pressure can also be expressed in terms of the height of the column of mercury, which it will support at a temperature of 273.15 K in a standard gravitational field. At standard atmospheric pressure this height is 0.76 m (760 mm or 760 torr) with density of mercury taken as 13.5951×10^3 kg/m^3.

$$1 \text{ standard atmosphere (atm)} = 1.013\,25 \text{ bar} = 1.013\,25 \times 10^5 \text{ Pa}$$
$$= 1.013\,25 \times 10^5 \text{ N/m}^2 = 760 \text{ mm Hg}$$

Other important pressure units are the kgf/cm^2 in MKS system and lbf/in^2 (commonly written as *psi*) in FPS system.

Work: Energy is expended in the form of work when a force acts through a distance. Thus,

$$dW = F\,dZ \tag{2.5}$$

where W is the work done, F is the force acting and Z is the displacement. The unit of work in the SI system is N m (newton-metre) or J (joule).

Energy: Energy is a quantity that can be stored within the system and can be exchanged between the system and the surroundings. The exchange of energy occurs either as heat or as work. Heat and work cannot be stored within the system: they can be treated as energy in transit. Energy stored within the system due to its position above some arbitrary reference plane is referred to as its *potential energy* (PE). If mass m is at an elevation z above the ground, the potential energy of the mass is

$$PE = mgz \tag{2.6}$$

where g is the acceleration due to gravity. The energy possessed by the body by virtue of its motion is called its kinetic energy (KE). If a body of mass m is moving at a velocity u, the kinetic energy of the body is

$$\text{KE} = \frac{1}{2} mu^2 \tag{2.7}$$

The unit of energy in SI system is the joule (J).

$$1 \text{ J} = 1 \text{ N m} = 1 \text{ kg m}^2/\text{s}^2$$

Heat: Heat is that quantity which is transferred between bodies due to the difference in temperatures existing between them. Earlier, this quantity was thought of as a substance called *caloric*. Heat is now recognized as a form of energy that cannot be stored as such within the system. And like work, it is expressed in J (joule). Two other units used for heat are the *calorie* and the *Btu* (the British thermal unit).

$$1 \text{ calorie} = 4.1868 \text{ J}; \; 1 \text{ Btu} = 1055.06 \text{ J}$$

Power: The power is defined as the time rate of doing work. Its unit in the SI system is J/s, commonly designated as W (*watts*). In engineering calculations, power is generally expressed as *horsepower* (hp).

$$1 \text{ metric hp} = 75 \text{ m kgf s}^{-1} = 735.5 \text{ W}$$

$$1 \text{ British hp} = 550 \text{ ft lbf s}^{-1} = 745.7 \text{ W}$$

EXAMPLE 2.1 The flow rate of water through a pipe is reported as 15 cubic feet per minute. Taking density of water as 1 g/cm^3, calculate the mass flow rate in kg/s.

Solution Volumetric flow rate = 15 ft^3/min

Using the relationship 1ft = 0.3048 m and 1 min = 60 s, the volumetric flow rate can now be expressed in m^3/s.

$$\text{Volumetric flow rate} = \frac{15 \text{ ft}^3 \left(\dfrac{0.3048^3 \text{ m}^3}{\text{ft}^3}\right)}{\text{min}\left(\dfrac{60 \text{ s}}{\text{min}}\right)} = 7.079 \times 10^{-3} \text{ m}^3/\text{s}$$

$$\text{Density of water is } 1 \text{ g/cm}^3 = \frac{1 \text{ g}\left(\dfrac{10^{-3} \text{ kg}}{\text{g}}\right)}{\text{cm}^3\left(\dfrac{10^{-6} \text{ m}^3}{\text{cm}^3}\right)} = 1000 \text{ kg/m}^3$$

$$\text{Mass flow rate} = \text{Volumetric flow rate} \times \text{density of water}$$

$$= 7.079 \times 10^{-3} \text{ m}^3/\text{s} \times 1000 \text{ kg/m}^3 = 7.079 \text{ kg/s}$$

EXAMPLE 2.2 Determine the conversion factor for poundal to newton

Solution 1 poundal is 1 ft lb/s^2 $= 1\,(\text{ft})\left(\dfrac{0.3048 \text{ m}}{\text{ft}}\right)(\text{lb})\left(\dfrac{0.4536 \text{ kg}}{\text{lb}}\right)\Big/ \text{s}^2$

$$= 0.1383 \; \frac{\text{kgm}}{\text{s}^2} = 0.1383 \text{ N}$$

The conversion factor is 0.1383 N/poundal.

EXAMPLE 2.3 Express kgf/cm^2 and lbf/in^2 as N/m^2.

Solution 1 kgf = 9.806 65 N

$$1\frac{\text{kgf}}{\text{cm}^2} = 1\frac{(\text{kgf})\left(\dfrac{9.806\,65 \text{ N}}{\text{kgf}}\right)}{(\text{cm}^2)\left(\dfrac{10^{-4} \text{ m}^2}{\text{cm}^2}\right)} = 0.981 \times 10^5 \text{ N/m}^2$$

The Newton's law conversion factor relates lbf to poundal. 1 lbf = 32.174 $\frac{\text{lb ft}}{\text{s}^2}$. Using this and other conversion factors like 0.453 59 24 kg/lb, 0.3048 m/ft and 0.0254 m/in, we get the following.

$$1\,\frac{\text{lbf}}{\text{in}^2} = 1\,\frac{(\text{lbf})\left(\dfrac{32.174\text{ lb ft}}{\text{lbf s}^2}\right)\left(\dfrac{0.453\,59\,24\text{ kg}}{\text{lb}} \times \dfrac{0.3048\text{ m}}{\text{ft}}\right)}{(\text{in}^2)\left[\dfrac{(0.0254)^2\text{ m}^2}{\text{in}^2}\right]}$$

$$= 6894.75\,\frac{\text{kg m}}{\text{s}^2\text{ m}^2} = 6894.75\text{ N/m}^2$$

(Table 2.7 gives $6.894\,757 \times 10^3$ as the value for the conversion factor for lbf/in^2 to pascal.)

EXAMPLE 2.4 In the SI system, thermal conductivity has the unit W/(m K). The thermal conductivity of a solid material can be calculated as $k = \frac{xQ}{A\Delta T}$, where Q is the rate of heat transfer, x is the thickness of solid, A is the area of heat transfer and ΔT is the temperature difference across the solid. The following values were obtained experimentally: Q = 10000 kJ/h, A = 1 m^2, x = 100 mm and ΔT = 800 K.

(a) Calculate the thermal conductivity of the solid in W/(m K).
(b) Express the thermal conductivity in kcal/(h m °C).
(c) If thermal conductivity of a second material is 0.15 Btu/(h ft °F), which one will make a better thermal insulator?

Solution

(a) The rate of heat transfer, $Q = 10000\text{ kJ/h} = \dfrac{10000\text{ kJ}\left(\dfrac{1000\text{ J}}{\text{kJ}}\right)}{\text{h}\left(\dfrac{3600\text{ s}}{\text{h}}\right)} = 2777.78\text{ J/s}$

We have:

$$x = 100\text{ mm} = 0.1\text{ m},\ A = 1\text{ m}^2,\ \Delta T = 800\text{ K}$$

Substituting these values in the given equation, we get

$$k = \frac{xQ}{A\Delta T} = \frac{0.1 \times 2777.78}{1 \times 800}$$

$$= 0.347\,\frac{\text{J}}{\text{s m K}} = 0.347\,\frac{\text{W}}{\text{m K}}$$

(b)
$$0.347\,\frac{\text{W}}{\text{m K}} = 0.347\,\frac{\text{J}}{\text{s m K}}$$

$$= 0.347\,\frac{\text{J}\left(\dfrac{\text{cal}}{4.1868\text{ J}}\right)}{\text{s}\left(\dfrac{\text{h}}{3600\text{ s}}\right) \times \text{m} \times \text{K} \times \dfrac{°\text{C}}{\text{K}}} = 298.557\,\frac{\text{cal}}{\text{h m °C}}$$

(c) The thermal conductivity of the material in the FPS system is

$$0.347\ \frac{\text{W}}{\text{m K}} = 0.347\ \frac{\text{J}}{\text{s m K}}$$

$$= 0.347\ \frac{\text{J}\left(\dfrac{\text{Btu}}{1.055\,056\times10^3\ \text{J}}\right)}{\text{s}\left(\dfrac{\text{h}}{3600\ \text{s}}\right)\times \text{m}\times\dfrac{\text{ft}}{0.3048\ \text{m}}\times \text{K}\times\dfrac{1.8\,°\text{F}}{\text{K}}} = 0.2\ \frac{\text{Btu}}{\text{h ft °F}}$$

The second material will make a good thermal insulator as its thermal conductivity is 0.15 Btu/(h ft °F) which is less than the thermal conductivity of the first material calculated above (0.2 Btu/h ft °F).

EXAMPLE 2.5 The weight of an object is 300 N at a location where acceleration due to gravity is 9.81 m/s^2.

(a) Determine the mass of the object in kilograms.
(b) Express the mass in the FPS system.

Solution (a) Force is equal to the product of mass and acceleration, i.e.

$$F = ma$$

$$300 = m \times 9.81$$

Therefore,
$$m = \frac{300}{9.81} = 30.58\ \text{kg}$$

(b) The conversion factor for pound to kilogram is 4.535 924 × 10^{-1} kg/lb. Therefore the mass of the object in the FPS system is

$$30.58\ (\text{kg})\left(\frac{1\ \text{lb}}{4.535\,924\times10^{-1}\ \text{kg}}\right) = 67.42\ \text{lb}$$

EXAMPLE 2.6 The potential energy of a body at a height of 15 m is 2 kJ. If the body is moving at a velocity of 50 m/s what is its kinetic energy? (Take g = 9.8067 m/s^2)

Solution Potential energy (PE) = mgz
Here z = 15 m, PE = 2000 J.
Therefore,
$$2000 = m \times 9.8067 \times 15$$
Thus the mass of the body m = 13.596 kg.

$$\text{Kinetic energy, KE} = \frac{1}{2}mu^2$$

$$= \frac{1}{2}(13.596)(50)^2 = 16\,995\ \text{J} = 16.995\ \text{kJ}$$

EXAMPLE 2.7 Nitrogen gas is confined in a cylinder and the pressure of the gas is maintained by a weight placed on the piston. The mass of the piston and the weight together is 100 lb.

The acceleration due to gravity is 9.81 m/s^2 and the atmospheric pressure is one standard atmosphere. Assuming the piston is frictionless, determine:

(a) The force exerted by the atmosphere, the piston and the weight on the gas in N if the piston diameter is 4 in.
(b) The pressure of the gas in bar and psi.
(c) If the gas is allowed to expand pushing up the piston and the weight by 400 mm, what is the work done by the gas in kJ?
(d) What is the change in the potential energy of the piston and the weight after the expansion in part (c)?

Solution (a) One standard atmosphere = $1.013\,25 \times 10^5$ N/m^2

The area of the piston is

$$\frac{\pi}{4} \times D^2 = \frac{\pi}{4}(4 \times 2.54 \times 10^{-2})^2 = 8.1073 \times 10^{-3} \text{ m}^2$$

Force exerted by the atmosphere = pressure × area

$$= 1.013\,25 \times 10^5 \times (8.1073 \times 10^{-3}) = 821.47 \text{ N}$$

Force exerted by the piston and weight = $m \times g$

$$= (100 \times 0.4536) \times 9.81 = 444.98 \text{ N}$$

Total force acting on the gas is

$$821.47 + 444.98 = 1266.45 \text{ N}$$

(b) $$\text{Pressure} = \text{force/area} = \frac{1266.45}{(8.1073 \times 10^{-3})}$$

$$= 1.562 \times 10^5 \text{ N/m}^2 \; (= 1.562 \text{ bar})$$

From Table 2.7 the conversion factor for converting psi to pascal (Pa) is $6.894\,757 \times 10^3 \dfrac{\text{Pa}}{\text{psi}}$. Therefore, pressure in psi is

$$\frac{1.562 \times 10^5}{6.894\,757 \times 10^{-3}} = 22.65 \text{ psi}$$

(b) Work done = force × displacement = $1266.45 \times (400 \times 10^{-3}) = 506.58 \text{ J} = 0.507 \text{ kJ}$

(c) Change in the potential energy,

$$\Delta(\text{PE}) = mg\Delta z = 100 \times 0.4536 \times 9.81 \times (400 \times 10^{-3}) = 177.99 \text{ J}$$

2.3.2 Conversion of Empirical Equations

Empirical equations are available for estimating thermodynamic properties, transport coefficients, frictional losses and pressure drop for flow through different geometries, etc. These equations are correlations of experimental results by empirical means without regard to dimensional homogeneity and they provide the desired results only if we stick to the units specified for the quantities involved. For example, the rate of heat loss from a cylindrical pipe to the ambient air is correlated by the equation

$$q = 7.38 \times \frac{\Delta T^{1.25}}{D_P^{0.25}}$$

where q is the rate of heat loss in W/m², ΔT is the temperature difference in K, and D_P is the diameter of the pipe in mm. The dimension of q on the right-hand side is not the same as the dimensions on the left-hand side. Such equations are also known as *dimensional equations*. These give incorrect results when used for calculations in case we fail to use quantities in the specified units. If, for example, we want to use the FPS units in the calculation of heat loss using the equation given above, we have to convert the equation into the equivalent FPS form. In the following paragraphs, we illustrate through some examples the conversion of dimensional equations into their equivalent forms in other systems of units.

EXAMPLE 2.8 For the absorption of sulphur dioxide in water in towers packed with wood grids, the following empirical equation is proposed for evaluating mass transfer coefficient:

$$k_G = \frac{6.7 \times 10^{-4}}{d_S^{0.4}\, d_G^{0.2}} \left[\frac{G(d_S + d_T)}{d_S}\right]^{0.8}$$

where k_G is the mass transfer coefficient in lb-mol/(h ft² atm), G is the superficial mass velocity in lb/(ft² h), d_S represents clearance between grids in feet, d_G represents the height of grids in feet, and d_T is the thickness of the grid slats in feet. What is the corresponding equation if k_G is given in kmol/(m² h atm), G is in kg/(m² h) and the lengths are given in m?

Solution Suppose that the mass velocity is G' in kg/(m² h), and d'_S, d'_G *and* d'_T are the clearance between grids, height of grids and thickness of grid, respectively, all measured in m. Then

$$G = G' \frac{\text{kg}}{\text{m}^2\text{h}} \left(\frac{1\ \text{lb}}{0.4536\ \text{kg}}\right)\left(\frac{0.3048^2\ \text{m}^2}{1\ \text{ft}^2}\right) = 0.2048G' \frac{\text{lb}}{\text{ft}^2\text{h}}$$

$$d = d'(\text{m})\left(\frac{1\ \text{ft}}{0.3048\ \text{m}}\right) = 3.2808d'\ \text{ft}$$

Since G and d are in the units specified for the given equation, we can substitute these and get the following result.

$$k_G = \frac{6.7 \times 10^{-4}}{(3.2808d'_S)^{0.4}\,(3.2808d'_G)^{0.2}} \left[\frac{0.2048G' \times 3.2808\,(d'_S + d'_T)}{3.2808d'_S}\right]^{0.8}$$

$$k_G = \frac{9.2374 \times 10^{-5}}{d_S'^{\,0.4}\, d_G'^{\,0.2}} \left[\frac{G'(d'_S + d'_T)}{d'_S}\right]^{0.8}$$

where k_G is given in lb-mol/(ft² h atm). Since

$$1\frac{\text{lb-mol}}{\text{h ft}^2\text{ atm}} = 1\left(\frac{\text{lb-mol}}{\text{ft}^2\text{ h atm}}\right)\times\left(\frac{0.4536\text{ kmol}}{\text{lb mol}}\right)\times\left(\frac{\text{ft}^2}{0.3048^2\text{ m}^2}\right) = 4.8825\frac{\text{kmol}}{\text{m}^2\text{ h atm}},$$

the right-hand-side of the above result is to be multiplied by 4.8825 so that the mass transfer coefficient is expressed in kmol/(m^2·h·atm). Therefore, the desired equation is

$$k'_G = \frac{4.5102\times10^{-4}}{d_S'^{0.4}\,d_G'^{0.2}}\left[\frac{G'(d'_S+d'_T)}{d'_S}\right]^{0.8}$$

where k'_G is in kmol/(m^2 h atm), G' is in kg/(m^2 h), and the clearances and thickness are in metres.

Alternative

An alternative to the above technique is to first determine the units for the constant appearing in the original equation, in the present case it is 6.7×10^{-4} and convert the units into the desired system by multiplying with appropriate conversion factors. The constant gets a new numerical value after this manipulation. For example, let the constant be 6.7×10^{-4} (lb-mol)^a (lb)^b(ft)^c(atm)^d(h)^e. Now equate the exponents of the units on both sides and we get

$$\frac{\text{lb-mol}}{\text{ft}^2\text{ h atm}} = (\text{lb mol})^a\,(\text{lb})^b\,(\text{ft})^c\,(\text{atm})^d\,(\text{h})^e\left(\frac{1}{\text{ft}^{0.6}}\right)\left(\frac{\text{lb}}{\text{ft}^2\text{ h}}\right)^{0.8}$$

or

$$(\text{lb-mol})^1\,(\text{ft})^{-2}\,(\text{h})^{-1}\,(\text{atm})^{-1} = (\text{lb-mol})^a\,(\text{lb})^{b+0.8}\,(\text{ft})^{c-2.2}\,(\text{atm})^d\,(\text{h})^{e-0.8}$$

Thus we get, $a = 1$, $b = -0.8$, $c = 0.2$, $d = -1$, $e = -0.2$.

And the constant is 6.7×10^{-4} (lb-mol)^1(lb)$^{-0.8}$(ft)$^{0.2}$(h)$^{-0.2}$(atm)$^{-1}$.

$$6.7\times10^{-4}\,(\text{lb mol})(\text{lb})^{-0.8}(\text{ft})^{0.2}(\text{h})^{-0.2}(\text{atm})^{-1}$$

$$= 6.7\times10^{-4}\left(\text{lb mol}\times\frac{0.4536\text{ kmol}}{\text{lb mol}}\right)$$

$$\times\left(\text{lb}\times\frac{0.4536\text{ kg}}{\text{lb}}\right)^{-0.8}\times\left(\text{ft}\times\frac{0.3048\text{ m}}{\text{ft}}\right)^{0.2}\times(\text{h})^{-0.2}\times(\text{atm})^{-1}$$

$$= 4.51\times10^{-4}\,(\text{kmol})(\text{kg})^{-0.8}\,(\text{m})^{0.2}\,(\text{h})^{-0.2}\,(\text{atm})^{-1}$$

We can use the given equation with the constant 6.7×10^{-4} replaced by 4.51×10^{-4} where k_G is given in kmol/(m^2 h atm), G is in kg/(m^2 h) and the distances are given in m.

EXAMPLE 2.9 The effective heat capacity of a mixture of gases is given by

$$C_P = 7.13 + 0.577\times10^{-3}\,t + 0.0248\times10^{-6}\,t^2$$

where C_P is in Btu/(lb-mol °F) and t is in °F.

(a) What are the units of the constants in the equation?

(b) Change the equation into the form in which C_P is given in kJ/(kmol K) and temperature is in K.

Solution (a) Since the unit of C_P on the left-hand-side is Btu/(lb-mol °F), each term on the right-hand side of the equation has the same units. These terms are:

First term = 7.13 Btu/(lb-mol °F)
Second term = $0.577 \times 10^{-3}\, t$ Btu/(lb-mol °F)
Third term = $0.0248 \times 10^{-6}\, t^2$ Btu/(lb-mol °F)

The constant 7.13 has the same unit as heat capacity, i.e. $\dfrac{\text{Btu}}{\text{lb-mol °F}}$.

Now, the second term $0.577 \times 10^{-3} t \dfrac{\text{Btu}}{\text{lb-mol °F}}$ can be written as

$$0.577 \times 10^{-3} \frac{\text{Btu}}{\text{lb-mol (°F)}^2} \times t(\text{°F})$$

which means that the units of the constant 0.577×10^{-3} is $\dfrac{\text{Btu}}{\text{lb-mol (°F)}^2}$.

Similarly, the third term $0.0248 \times 10^{-6} t^2 \dfrac{\text{Btu}}{\text{lb-mol °F}}$ can be written as

$$0.0248 \times 10^{-6} \frac{\text{Btu}}{\text{lb-mol (°F)}^3} \times t^2(\text{°F})^2$$

Therefore, the units of the constant 0.0248×10^{-6} is $\dfrac{\text{Btu}}{\text{lb-mol (°F)}^3}$.

The units of the constant are respectively

$$\frac{\text{Btu}}{\text{lb-mol °F}}, \quad \frac{\text{Btu}}{\text{lb-mol (°F)}^2} \quad \text{and} \quad \frac{\text{Btu}}{\text{lb-mol (°F)}^3}.$$

(b) Suppose that the temperature of the gas is T K, then the equivalent Fahrenheit temperature is

$$t = (T - 273.15) \times 1.8 + 32.0 = 1.8T - 459.67$$

Substituting this in the equation for C_P, we get the heat capacity in Btu/(lb-mol °F) when T is in K.

$$C_P = 7.13 + 0.577 \times 10^{-3}\, (1.8T - 459.67) + 0.0248 \times 10^{-6}\, (1.8T - 459.67)^2$$
$$= 6.87 + 9.976 \times 10^{-4} T + 0.0804 \times 10^{-6} T^2$$

The heat capacity given by the above equation is in Btu/(lb-mol °F). Since

$$1 \frac{\text{Btu}}{\text{lb-mol °F}} = 1 \frac{\text{Btu} \times \dfrac{1.055\,056 \times 10^3 \text{ J}}{\text{Btu}}}{\text{lb-mol} \times \dfrac{0.453\,5924 \times 10^3 \text{ mol}}{1 \text{ lb-mol}}\, \text{°F} \times \dfrac{\text{K}}{1.8\text{°F}}}$$

$$= 4.1868 \frac{\text{J}}{\text{mol K}} = 4.1868 \frac{\text{kJ}}{\text{kmol K}}$$

the above equation is to be multiplied by 4.1868 to obtain the heat capacity in kJ/kmol K when the temperature is in K. The resulting equation is

$$C_P = 28.763 + 4.763 \times 10^{-3}T + 0.3366 \times 10^{-6}T^2$$

This equation evaluates heat capacity in kJ/kmol K when temperature T is in K.

2.4 DIMENSIONLESS EQUATIONS

Equations that are derived mathematically from basic principles are dimensionally homogeneous. All terms in such equations have the same dimensions and the constants are dimensionless. For example, the distance S covered in time t units by a body moving with an initial velocity u and an acceleration a is given as

$$S = ut + \frac{1}{2}at^2$$

In this equation all terms on both sides of the equation have the dimension of length and it is immaterial whether we are using FPS, MKS or SI system of units. In using dimensionally homogeneous equations, no conversion factors are required for using any system of units so long as we are using consistent units for the primary dimensions of length, mass, time and temperature. In contrast, *an empirical equation is not necessarily dimensionless and therefore must always be used with the same units in which it was developed.*

2.5 DIMENSIONAL ANALYSIS

The dimensional analysis serves as a very useful tool in chemical engineering practice, especially in heat flow, fluid flow and mass-transfer processes where even though the variables influencing the system are known, an exact theoretical equation cannot be derived in terms of these variables. The dimensional analysis stands intermediate between a fully empirical analysis and a systematic mathematical development starting from basic principles. The dimensional analysis is based on the principle of *dimensional similitude* by which it is possible to express all natural interrelationships in terms of functions and coefficients that are universal in form and numerical value, completely independent of the arbitrary system of units employed in making the measurements themselves, provided the system is inherently consistent. It means that if a theoretical equation exists among the variables affecting a physical process, that equation must be dimensionally homogeneous. Therefore, it is possible to group the variables of such an equation into a smaller number of dimensionless groups.

By dimensional analysis, the total number of variables affecting the system is reduced and the equation can be equivalently written in terms of a lesser number of dimensionless groups. Assume that a process is influenced by a large number of physical factors and we want to study the effect of each factor on the rate of the process. Experimental investigations to study the effect of each separate factor are to be carried out by systematically varying that factor while keeping all others constant. It means that a laborious series of experiments is necessary to obtain the exact nature of the functional relationship existing between the variables. It is difficult to establish a relationship of this sort even by conducting such a large number of experiments. Consider for instance, the experimental investigation of the variation of the heat transfer coefficient with velocity of a fluid flowing through a pipe by conducting a series of experiments with a pipe of diameter, say 25 mm. This is done by varying the velocity, say from 1 m/s to 30 m/s in steps, keeping all other factors constant. The results obtained can be utilized to predict the heat transfer coefficient for a given velocity. However, these results as such, cannot be used to predict the heat transfer coefficient in a pipe of a different diameter, say 50 mm for which another set of experiments will have to be done. It has been shown by the dimensional analysis that the variables, diameter, velocity, density and viscosity must appear as a dimensionless group, $Du\rho/\mu$, known as the Reynolds number. If the experimental data are reported in terms of this group, the evaluation of the heat transfer coefficient, which will appear in another dimensionless group for a fluid flowing through any diameter pipe is possible as long as the Reynolds number of the system falls within the range covered in the experiment. In deriving a working equation from the experimental results, it is necessary to derive it only in terms of the values of the dimensionless group, not in terms of each of the variables separately.

To conduct dimensional analysis on a given problem, the first step is to identify the variables affecting the system. These must be determined by the judgment of the investigator, exercised in the light of such knowledge of the mechanism as is available. For example, consider again the heat transfer from the walls of a pipe to the fluid flowing in turbulent flow through the inside of the pipe. It is known that the heat transfer process is affected by the following variables: the pipe diameter (D), the fluid velocity (u), the viscosity of the fluid (μ), the density of the fluid (ρ), the heat capacity of the fluid (C_P), the thermal conductivity of the fluid (k), and the heat transfer coefficient (h). The equation describing the heat transfer process can be written as

$$f\,(h,\ D,\ u,\ \mu,\ \rho,\ C_P, k) = 0$$

There are different techniques for conducting dimensional analysis. The method due to the Buckingham-π theorem is based on the fact that the functional relationship between n factors required to describe a system can be equivalently written in terms of $n–m$ dimensionless groups, where m is the number of primary dimensions required to represent these variables. In the heat transfer process described above there are seven variables and four primary dimensions are necessary to represent them. Thus $n = 7$, $m = 4$. Thus we can write the relationship for heat transfer as a function of three dimensionless groups, π_1, π_2 and π_3. Therefore

$$f\,'(\pi_1,\ \pi_2,\ \pi_3) = 0$$

The complete analysis, which is beyond the scope of this book, will show that these groups are the Nusselt number (hD/k), the Prandtl number ($C_P\mu/k$) and the Reynolds number ($Du\rho/\mu$).

The results of a dimensional analysis are, however, useless unless experiments are done to supplement the results. This is because the dimensional analysis does not give a numerical equation, and experiments are required to complete the solution. A dimensional analysis cannot be performed unless enough is known about the physics of the situation to decide what factors are important in the problem and what physical laws would be involved in a mathematical solution if such a solution were possible.

2.5.1 Dimensionless Groups in Chemical Engineering Practice

Some important dimensionless groups in chemical engineering practice, their basic definitions and the variables involved are given as follows.

1. **Biot number:** It is the ratio of conductive (internal) resistance to heat transfer to the convective (external) resistance to heat transfer. It is used in the analysis of unsteady-state conduction problems.
$$Bi = \frac{h(V/S)}{k}$$
where V is the volume, S is the surface area and k is the thermal conductivity of the solid object and h is the heat transfer coefficient of the surrounding fluid.
2. **Drag coefficient:** It is the ratio of the average drag per unit projected area to the product of the density of the fluid and velocity head.
$$C_D = \frac{F_D/A_P}{\rho(u^2/2)}$$
where F_D is the drag force exerted on the solid whose projected area in the direction of flow is A_P, ρ is the density of the fluid and u its velocity.
3. **Euler number:** It is the ratio of pressure force to inertial forces acting on a fluid element.
$$Eu = \frac{P}{\rho u^2}$$
where P is the pressure, ρ is the density of the fluid and u its velocity.
4. **Fanning friction factor:** It is the ratio of shear stress to the product of density and velocity head.
$$f = \frac{\tau_w}{\rho(u^2/2)}$$
where τ_w is the shear stress at the pipe wall, ρ is the density of the fluid and u its velocity.
5. **Fourier number:** It is used in unsteady-state heat conduction.
$$F_o = \frac{\alpha t}{s^2}$$
where α is thermal diffusivity of the solid, t is time and s is half the thickness of a slab or radius of a solid cylinder or sphere.
6. **Froude number:** It is the ratio of inertial force to gravity force.
$$Fr = \frac{u^2}{Lg}$$
where u is the velocity, g is the acceleration due to gravity, and L is the characteristic linear dimension for the flow system.

7. **Graetz number:** It is used for treating heat transfer to fluids in laminar flow.

$$Gz = \frac{mC_P}{kL}$$

where m is the mass flow rate, C_p and k are respectively, the heat capacity and thermal conductivity of the fluid, and L is the characteristics linear dimension for the flow system.

8. **Grashof number:** Used to explain natural convection heat transfer.

$$Gr = \frac{L^3 \rho^2 g \beta \, \Delta T}{\mu^2}$$

where L is the characteristic linear dimension, ρ is the density, g is the acceleration due to gravity, β is the coefficient of volume expansion of the fluid, ΔT is the temperature difference between the wall and bulk fluid, and μ is the viscosity of the fluid.

9. **Lewis number:** The ratio of thermal diffusivity to mass diffusivity.

$$Le = \frac{k}{C_P \rho D_{AB}}$$

where the ratio $\frac{k}{C_P \rho}$ is the thermal diffusivity, sometimes abbreviated as α, D_{AB} is the mass diffusivity of a mixture of components A and B, k is the thermal conductivity, C_P is the heat capacity and ρ is the density.

10. **Mach number:** It is the ratio of speed of the fluid to speed of sound in the fluid under conditions of flow.

$$Ma = \frac{u}{c}$$

where u is the speed of the fluid and c is the speed of sound in the fluid.

11. **Nusselt number:** It is the ratio of temperature gradient in the fluid in contact with a solid wall to a reference temperature gradient applicable to the entire fluid and represents a dimensionless heat transfer coefficient.

$$Nu = \frac{hL}{k}$$

where h is the heat transfer coefficient, L is the characteristic linear dimension, and k is the thermal conductivity.

12. **Peclet number:** It is the product of Reynolds number and Prandtl number.

$$Pe = \frac{LuC_P \rho}{k}$$

L is the characteristic linear dimension for the flow system, C_p, k and ρ are respectively, the heat capacity, thermal conductivity and density of the fluid and u is the velocity.

13. **Power number:** It is the ratio of the drag force to inertial force and is proportional to the drag coefficient.

$$P_o = \frac{P}{n^3 D^5 \rho}$$

where P, is the pressure, n is the rotational speed of impeller of diameter D and ρ is the density.

14. **Prandtl number:** It is the ratio of momentum diffusivity to thermal diffusivity.

$$Pr = \frac{C_P \mu}{k}$$

where C_P is the heat capacity of the fluid, μ is the viscosity of the fluid, and k is the thermal conductivity of the fluid.

15. **Reynolds number:** It is the ratio of inertial force to viscous forces acting on a fluid element.

$$Re = \frac{Lu\rho}{\mu}$$

where L is the characteristic linear dimension for the flow system, u is the velocity, ρ is the density and μ is the viscosity of the fluid.

16. **Schmidt number:** It is the ratio of momentum diffusivity to mass diffusivity.

$$Sc = \frac{\mu}{\rho D_{AB}}$$

where D_{AB} is the mass diffusivity of a species A in a mixture of A and B.

17. **Sherwood number:** It is represents a dimensionless mass transfer coefficient and is the ratio of concentration gradient in the fluid in contact with a solid wall to a reference concentration gradient applicable to the entire fluid.

$$Sh = \frac{k_C L}{D_{AB}}$$

where k_C is the mass transfer coefficient, L is the characteristic linear dimension for the flow system, and D_{AB} is the mass diffusivity.

18. **Stanton number:** The Stanton number for heat transfer (St_H) is the ratio of the Nusselt number to the product of the Reynolds number and the Prandtl number while the Stanton number for mass transfer (St_D) is the ratio of the Sherwood number to the product of the Reynolds number and the Schmidt number.

$$St_H = \frac{h}{C_P G}$$

$$St_D = \frac{k_C}{u}$$

where G is the mass velocity (ρu), h is the heat transfer coefficient, C_P is the heat capacity, k_C is the mass transfer coefficient and u is the velocity.

EXERCISES

Conversion of units

2.1 The viscosity of water is one centipoise. Express the viscosity of water in the following units?

(a) Pa s
(b) lb/(ft h)

2.2 Convert 1 bar into equivalent kgf/cm^2 and psi.

2.3 The thermal conductivity of an insulating brick is 0.15 Btu/(ft h °F). Express the thermal conductivity in SI units.

2.4 The heat transfer coefficient of oil flowing through a pipe is 300 W/(m^2 K). Determine the value of heat transfer coefficient expressed in

(a) kcal/(h m^2 °C)
(b) Btu/(h ft^2 °F)

2.5 Thermal conductivity of pure iron is 39 Btu/(ft h °F) and that of steel containing 1% C is 39 kcal/(m h °C). Which one is a good conductor of heat, iron or steel?

2.6 The mass velocity of a gas through a duct is 1000 kg/m^2 h. Express the velocity into lb/ft^2 s.

2.7 Mass flow rate is the product of volumetric flow rate and density. If the volumetric flow rate of a liquid of specific gravity 0.78 is 100 ft^3/min, find the flow rate in kg/s.

2.8 A man circling the earth in a spaceship weighed 300 N at a location where the local gravitational acceleration was 4.5 m/s^2. Calculate the mass of the man and his weight on the earth where the gravitational acceleration is 9.81 m/s^2.

2.9 A man whose weight is 600 N takes 2 min for climbing up a staircase. What is the power developed in him, if the staircase is made up of 20 stairs, each 0.18 m in height?

2.10 The potential energy of a body of mass 10.0 kg is 1.5 kJ. What is the height of the body from the ground? If a body of mass 10 kg is moving at a velocity of 50 m/s, what is its kinetic energy?

2.11 Nitrogen gas is confined in a cylinder and the pressure of the gas is maintained by a weight placed on the piston. The mass of the piston and the weight together is 50 kg. The acceleration due to gravity is 9.81 m/s^2 and the atmospheric pressure is 1.013 25 bar. Assume frictionless piston. Determine:

(a) The force exerted by the atmosphere, the piston and the weight on the gas if the piston is 100 mm in diameter.
(b) The pressure of the gas.
(c) If the gas is allowed to expand pushing up the piston and the weight by 400 mm, what is the work done by the gas in kJ?
(d) What is the change in the potential energy of the piston and the weight after the expansion in part (c)?

2.12 The volume of a microbial culture is observed to increase according to the relation V (cm^3) = e^t where t is in seconds. Find the equivalent expression where V is in inch^3 and t is in hours.

Conversion of empirical equations

2.13 Stefan–Boltzmann law states that the emissive power is proportional to the fourth power of absolute temperature. Stated mathematically,

$$E_b = \sigma T^4$$

where E_b is the emissive power in Btu/(ft^2 h), σ is the Stefan-Boltzmann constant, and T is the temperature in °R. What is the value of σ in W/(m^2 K^4), if its value in FPS units is 0.171×10^{-8} Btu/(ft^2 h °R^4)?

2.14 The following empirical equation was proposed to estimate the rate of heat loss from the walls of a cylindrical pipe to the ambient air:

$$q = 0.5\frac{\Delta T^{1.25}}{D^{0.25}}$$

where q is the rate of heat loss in Btu/ft^2 h, ΔT is the temperature difference in °F and D is the diameter in inches.

(a) Determine the units of the constant.

(b) Express the equation in SI units.

2.15 The van der Waals equation of state is given as

$$\left(P + \frac{a}{V^2}\right)(V - b) = RT$$

where a and b are known as the van der Waals constants. For CO_2 the constants are found to be a = 0.365 N m^4/mol^2 and $b = 4.28 \times 10^{-5}$ m^3/mol. If the pressure is given in atm and the molar volume in L/mol, what are the values of the constants?

2.16 Rate of a gas phase reaction ($A \Leftrightarrow B$) taking place at 298 K and 1 atm is given by

$$-\frac{dC_A}{dt} = k_1 p_A^{2.2} - k_2 p_B^{0.5}$$

with k_1= 17.3 and k_2 = 0.2, where C_A is the concentration in kmol/m^3, t is the time in minutes, and p_A, p_B are partial pressures in atm.

(a) Find the units of k_1 and k_2.

(b) If the rate is to be written in the form of

$$-\frac{dp_A}{dt} = k_1' C_A^{2.2} - k_2' C_B^{0.5}$$

what are the values of k_1' and k_2' $\left(\textit{Note}: \text{Use the relation, } C_i = \frac{p_i}{RT}\right)$

2.17 The pressure drop for a fluid flowing through a packed bed of solids is given by the relation:

$$\Delta P = 3.61\left[\frac{\mu^{0.15}\ H\rho^{0.85}v^{1.85}}{D_P^{1.15}}\right]$$

where ΔP is pressure drop in Pa, μ is fluid viscosity in poise, H is the bed depth in metres, ρ is fluid density in kg/m^3, v is fluid velocity in m/s and D_P is packing diameter in metres. Modify the relation such that the pressure drop is still in Pa, but other terms are in FPS units.

2.18 The Antoine equation predicts the effect of temperature on vapour pressure by the relation:

$$\ln P^s = A - \frac{B}{T - C}$$

where A, B, and C are Antoine constants. For water, the values of the constants when the vapour pressure is in kPa and temperature is in K are found to be A = 16.26205, B = 3799.887 and C = 46.854. What are these values when

(a) Vapour pressure is in mm Hg and temperature in °C.
(b) Vapour pressure is in psia and temperature in °F.

2.19 The heat capacity of sulphuric acid in Btu/(lb-mol °R) is given by

$$C_P = 23.06 + 2.071 \times 10^{-2}T$$

where T is in °R. What is the equivalent expression if heat capacity is given in (a) kJ/(kmol K) and T is in K (b) kcal/(kmol °C) and temperature in °C.

2.20 The following equation relates the heat capacity of CO gas with temperature.

$$C_P = 26.586 + 7.582 \times 10^{-3}T - 1.12 \times 10^{-6}T^2$$

where C_P is expressed in kJ/kmol K and T in K. Obtain equations which will express

(a) The heat capacity in cal/mol K, temperature in K.
(b) The heat capacity in cal/mol °C, temperature in °C.
(c) The heat capacity in J/kg K, temperature in K.
(d) The heat capacity in Btu/lb-mol °F, temperature in °F.

2.21 The heat capacity of carbon dioxide gas is given by

$$C_P = 0.1978 + 1.059 \times 10^{-4}t - 2.395 \times 10^{-8}t^2$$

where C_P is in Btu/(lb°F) and t is in °F. Change the equation into the form in which C_P is given in kJ/(kmol K) and temperature is in K.

2.22 For fluids in turbulent motion through tubes, the heat transfer coefficient is given as

$$h = a\left(\frac{C_P G^{0.8}}{D^{0.2}}\right)$$

The numerical value of the constant a = 10.1, when h is measured in Btu/(ft^2 h °F), C_P is the specific heat of the fluid given in Btu/(lb °F), G is the mass velocity in lb/(ft^2 s) and the diameter is in ft. Determine the value of a (a) when h is measured in W/(m^2 K), C_P is the specific heat of the fluid given in kJ/(kg K), G is the mass velocity in kg/(m^2 s) and (b) when h is measured in kcal/(m^2 h °C), C_P is the specific heat of the fluid given in kcal/(kg °C), G is the mass velocity in kg/(m^2 s) and the diameter in m.

2.23 The diffusivity of a gas pair A-B is given by

$$D_{AB} = \frac{1.8583 \times 10^{-3} T^{3/2} \left(\dfrac{1}{M_A} + \dfrac{1}{M_B} \right)^{1/2}}{P\sigma^2 \Omega}$$

where D_{AB} is in cm^2/s, T is the absolute temperature in K, M_A and M_B are molecular weights of species A and B, respectively, P is the pressure in atm, σ is the Lennard–Jones parameter in angstrom units and Ω is a dimensionless parameter. If the equation is written as

$$D_{AB} = \frac{C T^{3/2} \left(\dfrac{1}{M_A} + \dfrac{1}{M_B} \right)^{1/2}}{P\sigma^2 \Omega}$$

where D_{AB} is in m^2/s, T is in K, P is the pressure in Pa, σ is the Lennard–Jones parameter in nanometres, what is the value of C?

3 Fundamental Concepts of Stoichiometry

The word "stoichiometry" has its origin in the Greek words *stoicheion* (elementary constituent) and *metrein* (to measure). The conventional meaning of the word limits its application to calculations involving the quantities of materials involved in chemical reactions. But in chemical engineering the word stoichiometry is used in a broader sense; it is not just the application of the law of combining proportions of chemical reactions. It includes calculations involving the behaviour of gases and gas mixtures, the phase behaviour of liquids and solutions, the material and energy balance of unit operations and unit processes, etc. The present chapter introduces the reader to some basic concepts governing process calculations.

3.1 THE MOLE CONCEPT

A chemical element is made up of very minute, discrete and indivisible particles called *atoms*. An atom is the smallest particle that can take part in a chemical change. It is extremely small that the absolute mass of an atom cannot be determined easily. It is customary to indicate the mass of an atom in relation to the mass of an arbitrarily chosen reference substance. The *atomic weight* of an element is the mass of an atom based on a scale that assigns a mass of 12 for the carbon isotope ^{12}C. On this scale, the atomic weight of oxygen is 16 and that of hydrogen is 1.008. That is, the mass of an atom of oxygen is approximately 16 times the mass of an atom of hydrogen. Atomic weight is a dimensionless number. The atomic weights of elements are given in Table A.1 (Appendix A).

A molecule of an element or a compound is made up of a number of atoms of the constituent elements. The *molecular weight* is the sum of the atomic weights of the elements constituting the molecule. Thus the molecular weight of a monatomic element is its atomic weight, while that of a diatomic element such as hydrogen is twice its atomic weight. For hydrogen, the molecular weight is $2 \times 1.008 = 2.016$. A molecule of water is made up of 2 atoms of hydrogen and 1 atom of oxygen. The molecular weight of water is $2 \times 1.008 + 16 = 18.016$. Like atomic weight, molecular weight is also a dimensionless number.

A *gram-atom* of an element is the mass in grams that is numerically equal to the relative atomic weight of the element. Thus the mass of one gram-atom of carbon is 12 g and the mass of one gram-atom of oxygen is 16 g. A *gram-molecule* of a compound is the mass in grams of the compound numerically equal to its molecular weight. Thus one gram-molecule of water is 18.016 g of water. Since an atom of oxygen is 16/12 times heavier than an atom of carbon, the number of carbon atoms present in one gram-atom (12 g) of carbon and the number of oxygen atoms present in one gram-atom (16 g) of oxygen are the same. This is true for all elements, and in general, we can say that a gram-atom of any element contains the same number of atoms. This number is equal to $6.022\ 140\ 76 \times 10^{23}$, and is known as *Avogadro number*. Similarly, a gram-molecule of all compounds contains the same number of molecules and this number is $6.022\ 140\ 76 \times 10^{23}$.

The number of gram-atoms of an element or the number of gram-molecules of a compound can be determined from the mass of the element or the compound in grams.

$$\text{The number of gram-atoms of the element} = \frac{\text{the number of grams of the element}}{\text{atomic weight of the element}}$$

$$\text{The number of gram-molecules of the compound} = \frac{\text{the number of grams of the compound}}{\text{molecular weight of the compound}}$$

The term gram-molecule is generally abbreviated into *gram mole* and is referred simply as *mole* (symbol *mol* in SI units). In process calculations, for convenience, *gram mole* or *mole* is used for referring gram-atom of an element as well as gram-molecule of a compound. Thus one mole of aluminium represents 26.9815 g of aluminium and one mole of water means 18.016 g of water. A mole of O_2 signifies 32 g of oxygen, while a mole of O would mean 16 g of oxygen. In short, we can designate a mole as the amount of the substance that contains as many elementary entities as there are atoms in 12 g of carbon. The number of moles of a substance can be determined by the following formula.

$$\text{mol} = \frac{\text{mass in grams}}{\text{molecular weight}}$$

In engineering calculations a related quantity *kilogram mole* abbreviated as kilomole (symbol *kmol*) is also commonly used.

$$\text{kmol} = \frac{\text{mass in kilograms}}{\text{molecular weight}} = 10^3 \text{ mol}$$

In the FPS system, the quantity pound mole (abbreviated as *lb-mol*) is used to indicate the quantity of substance that contains as many elementary entities (atoms or molecules) as there are atoms in 12 pound of carbon. Thus the number of molecules contained by one pound mole of any compound is the same for all substances. This number will not be equal to 6.023×10^{23} which is the number of molecules contained by a mole of the substance.

$$\text{lb-mol} = \frac{\text{mass in lb}}{\text{molecular weight}} = 453.6 \text{ mol}$$

EXAMPLE 3.1 1000 pounds per minute of a gas (average molecular weight = 30.24) is being sent to an absorption column. What is the molar flow rate of the gas in kmol/h?

Solution 1000 lb = 1000 (lb) × 0.4536 $\left(\frac{\text{kg}}{\text{lb}}\right)$ = 453.6 kg

$$\text{Mass flow rate} = 453.6 \text{ kg/min} = 453.6 \times 60 = 27216 \text{ kg/h} = \frac{27216}{30.24} = 900 \text{ kmol/h}$$

EXAMPLE 3.2 How many molecules are present in 691 g K_2CO_3?

Solution The atomic weights of the elements are:
potassium = 39.1, carbon = 12.0, oxygen = 16.0
The molecular weight of K_2CO_3 is

$$39.1 \times 2 + 12.0 + 16 \times 3 = 138.2$$

$$691 \text{ g } K_2CO_3 = \frac{691}{138.2} = 5 \text{ mol } K_2CO_3$$

One mole of a substance contains 6.022 140 76 × 10^{23} molecules. Therefore, 5 mol K_2CO_3 contains 5 × 6.022 40 76 × 10^{23} = 30.110 703 8 × 10^{23} molecules.

EXAMPLE 3.3

(a) How many moles of sodium sulphate will contain 100 kg of sodium?
(b) How many kilograms of sodium sulphate will contain 100 kg of sodium?

Solution

(a) Since the atomic weight of sodium is 23,

$$100 \text{ kg sodium} = \frac{100 \times 10^3}{23} = 4.348 \times 10^3 \text{ g-atoms of sodium.}$$

The molecular formula of sodium sulphate is Na_2SO_4. One mole Na_2SO_4 contains 2 g-atom sodium. Therefore, the number of moles of Na_2SO_4 that contains 4.348 × 10^3 g-atom sodium is

$$\frac{4.348 \times 10^3}{2} = 2.174 \times 10^3 \text{ mol} = 2.174 \text{ kmol}$$

(b) 2.174 kmol sodium sulphate contains 100 kg sodium. The molecular weight of sodium sulphate is

$$23 \times 2 + 32.06 + 16 \times 4 = 142.06$$

Therefore, 2.174 kmol sodium sulphate is

$$2.174 \times 142.06 = 308.84 \text{ kg sodium sulphate.}$$

3.2 CHEMICAL CALCULATIONS AND USE OF MOLAL QUANTITIES

The conventional application of stoichiometry is concerned with the mass relations involved in chemical reactions between the reactants on the one hand and the products on the other. The mole concept is found to be very much helpful in dealing with calculations involving chemical reactions compared to those using ordinary mass units. Since a mole of a compound reacts with a simple multiple of moles of other reactants, it is easier to use molal quantities such as mol, kmol or lb-mol in mass balance calculations of processes in which chemical reactions are involved. This is particularly true when the desired change is brought about by several reactions in series, each with varying degrees of completion. An alternative to the use of molal quantities in stoichiometric calculations is the use of the combining weight ratios of the reactants and products. But this method is cumbersome and is error prone as shown in Example 3.4.

EXAMPLE 3.4 Sulphur trioxide gas is obtained by the combustion of iron pyrites (FeS_2) according to the following reaction:

$$4FeS_2 + 15O_2 \rightarrow 2Fe_2O_3 + 8SO_3$$

How many kilograms of pyrites are burned to obtain 100 kg of sulphur trioxide? How many kilograms of oxygen are consumed in the production of 50 kg of SO_3?

Solution Molecular weight of FeS_2 = 55.85 + 2 × 32.06 = 119.97
Molecular weight of Fe_2O_3 = 2 × 55.85 + 3 × 16 = 159.7
Molecular weight of SO_3 = 32.06 + 3 × 16 = 80.06

Using molal quantities

$$100 \text{ kg of } SO_3 = \frac{100}{80.06} = 1.2491 \text{ kmol.}$$

Since 4 kmol FeS_2 gives 8 kmol SO_3, pyrites needed to produce 100 kg of SO_3 (i.e., 1.2491 kmol) is

$$\frac{4}{8} \times 1.2491 = 0.6246 \text{ kmol} = 0.6246 \times 119.97 = 74.93 \text{ kg}$$

$$50 \text{ kg of } SO_3 = \frac{50}{80.06} = 0.6246 \text{ kmol}$$

Since 15/8 kmol oxygen is required for the production of 1 kmol SO_3, oxygen consumed in the production of 50 kg (0.6246 kmol) SO_3 is

$$\frac{15}{8} \times 0.6246 = 1.1711 \text{ kmol} = 1.1711 \times 32 = 37.48 \text{ kg}$$

Using the method of combining weight ratios

The relative weights of the reactants and products are calculated by multiplying the molecular weight of the constituents by the respective stoichiometric coefficients. These are written along with the chemical formula of the constituents taking part in the reaction:

$4FeS_2$	+	$15O_2$	→	$2Fe_2O_3$	+	$8SO_3$
(4 × 119.97)		(15 × 32)		(2 × 159.7)		(8 × 80.06)
479.88		480		319.4		640.48

479.88 kg of FeS_2 gives 640.48 kg of SO_3. Therefore, the amount of pyrites needed to produce 100 kg of SO_3 is

$$\frac{479.88}{640.48} \times 100 = 74.93 \text{ kg}$$

480 kg oxygen is needed to produce 640.88 kg SO_3. Therefore, oxygen consumed for the production of 50 kg SO_3 is

$$\frac{480}{640.48} \times 50 = 37.47 \text{ kg}$$

3.2.1 Molar Volume of Gaseous Substances

The volume occupied by one mole of a gaseous substance at a given set of temperature and pressure is called the *molar volume* of the gas under those conditions. The molar volume of an ideal gaseous substance is found to be a constant under standard conditions of temperature and pressure (STP). The standard temperature is 273.15 K and standard pressure is 1 atm (101.325 kPa). The concept of ideal gases is discussed in detail later. It has been verified experimentally that one mole of an ideal gas under standard conditions of temperature and pressure occupies a volume of 22.4143×10^{-3} m^3. Accordingly, the molar volume at STP (also known as *normal molar volume*) is 22.4143×10^{-3} m^3/mol or 22.4143 m^3/kmol at STP. The molar volumes of real gases at STP will be different from this value depending upon the extent to which the gases depart from ideal behaviour. The constancy of normal molar volume is another concept which is useful in the calculations involving gaseous substances.

EXAMPLE 3.5 Determine the volume of oxygen obtained under standard conditions, by the decomposition of 100 kg of potassium chlorate.

Solution Consider the reaction

$$2KClO_3 \rightarrow 2KCl + 3O_2$$

The molecular weight of potassium chlorate is $(39.10 + 35.46 + 3 \times 16) = 122.56$. 100 kg of potassium chlorate is equivalent to 0.816 kmol. When 2 mol of potassium chlorate is decomposed, 3 mol oxygen is produced. So 0.816 kmol potassium chlorate will liberate 1.224 kmol oxygen. Since one kmol oxygen under STP occupies a volume of 22.4143 m^3, the volume of oxygen produced at STP is $1.224 \times 22.4143 = 27.43$ m^3.

EXAMPLE 3.6 Iron reacts with steam according to the following reaction:

$$3Fe + 4H_2O \rightarrow Fe_3O_4 + 4H_2$$

(a) How many kilograms of iron and steam are required to produce 100 kg of hydrogen?
(b) What volume will the hydrogen occupy at standard conditions?

Solution

(a) 100 kg of hydrogen $= \dfrac{100}{2.016} = 49.60$ kmol

Since three moles of iron reacts giving 4 moles of hydrogen, the moles of iron required to produce 49.6 kmol (i.e. 100 kg) hydrogen is

$$\frac{3}{4} \times 49.6 = 37.2 \text{ kmol}$$

37.2 kmol iron is equivalent to $37.2 \times 55.85 = 2077.62$ kg.

As one mole steam reacts giving one mole hydrogen, the total steam required is 49.6 kmol = 893.6 kg.

(b) Since the normal molar volume of an ideal gas is 22.4143 m^3/kmol, the volume occupied by 100 kg (49.6 kmol) hydrogen at STP is

$$49.6 \times 22.4143 = 1111.7 \text{ m}^3.$$

3.2.2 Equivalent Weight

The *equivalent weight of an element* is the number of parts by weight of the element which can combine with or displace 1.008 parts by weight of hydrogen or the equivalent weight of any other element. Equivalent weight is a dimensionless number like atomic weight and molecular weight. When equivalent weight is expressed in grams, it is known as the gram equivalent weight. The equivalent weight of an element may be defined as the ratio of the atomic weight of the element to its valency. For example, consider the reaction

$$H_2 + \frac{1}{2}O_2 \rightarrow H_2O$$

2.016 parts by weight of hydrogen combines with 16 parts by weight of oxygen to form 18.016 parts by weight of water. Therefore 1.008 parts by weight of hydrogen combines with 8 parts by weight of oxygen. Hence the equivalent weight of oxygen is 8.

An element with more than one valency will have more than one equivalent weight. The element iron (atomic weight 55.85) has a valency 2 in ferrous compounds and 3 in ferric compounds. Thus iron has equivalent weights 27.925 and 18.617 respectively, in ferrous and ferric compounds.

The *equivalent weight of an acid* is the number of parts by weight of the acid which contains 1.008 parts by weight of replaceable hydrogen. The number of replaceable hydrogen atoms in a molecule of acid is known as the *basicity* of the acid. Thus the equivalent weight of an acid is the ratio of the molecular weight of the acid to its basicity. Since 36.46 parts by weight of HCl contains 1.008 parts by weight H^+, the equivalent weight of HCl is 36.46. Similarly the equivalent weight of H_2SO_4 is half its molecular weight, that is 49.038, since one molecular weight of H_2SO_4 contains 2 atomic weights of replaceable hydrogen.

The *equivalent weight of a base* is the ratio of its molecular weight to its acidity. The acidity of the base is the number of –OH groups in one molecule of the base. Thus the equivalent weight of a base is the number of parts by weight of the base that is required to react completely with one equivalent weight of an acid. One molecule of NaOH contains one –OH group. Therefore, the equivalent weight of NaOH is 40 which is the same as its molecular weight.

The *equivalent weight of a salt* is the number of parts by weight of the salt that is required to react with one equivalent weight of an acid or base or any other substance. For example, in the reaction

$$Na_2CO_3 + 2HCl \rightarrow 2NaCl + CO_2 + H_2O$$

106 parts by weight of Na_2CO_3 reacts with 2 × 36.46 parts by weight of HCl or two equivalents of HCl. Therefore, the equivalent weight of Na_2CO_3 is (molecular weight of Na_2CO_3)/2 = 53.

The *equivalent weight of an oxidiser* is the number of parts by weight of the substance which would give 8 parts by weight of oxygen for oxidation purposes, or it is the ratio of the molecular weight to the number of electrons gained per molecule of the oxidiser. The *equivalent weight of a reducer* is the number of parts by weight of the reducer which requires 8 parts by weight of oxygen for complete oxidation. Or it is the ratio of the molecular weight to the total number of electrons lost per molecule of the reducer. Potassium permanganate is a good oxidiser in acid medium

$$2KMnO_4 + 3H_2SO_4 \rightarrow K_2SO_4 + 2MnSO_4 + 3H_2O + 5O$$

The molecular weight of $KMnO_4$ = 158.04. Thus, 316.08 parts by weight of $KMnO_4$ liberates 80 parts by weight of oxygen. The equivalent weight of $KMnO_4$ is

$$316.08 \times \frac{8}{80} = 31.61$$

Now consider the oxidation of ferrous sulphate:

$$2FeSO_4 + H_2SO_4 + O \rightarrow Fe_2(SO_4)_3 + H_2O$$

2×151.91 parts by weight of $FeSO_4$ requires 16 parts by weight oxygen for complete oxidation. Therefore, the equivalent weight which is the number of parts by weight of the sulphate combining with 8 parts by weight of oxygen is 151.91.

EXAMPLE 3.7 Find the equivalent weight of $CaCO_3$

Solution

$$CaCO_3 + 2HCl \rightarrow CaCl_2 + H_2O + CO_2$$

One mole of $CaCO_3$ reacts with 2 equivalents of hydrochloric acid. The weight of $CaCO_3$ reacting with one gram-equivalent of acid is

$$\frac{\text{Molecular weight of } CaCO_3}{2} = \frac{100.08}{2} = 50.04 \text{ g}$$

Therefore, the equivalent weight of $CaCO_3$ is 50.04.

3.3 DENSITY AND SPECIFIC GRAVITY

The mass per unit volume of a substance is designated as the *density* of the substance. Since the volume of a given mass of a substance varies with temperature, density is a function of the temperature. The density of water at 277.2 K is 1000 kg/m^3 and that at 313.2 K is 992.25 kg/m^3. The density of mercury at 313.2 K is 13 496.9 kg/m^3. The densities of liquids and solids are considered to be independent of pressure for most calculations. However, the density of a gas depends strongly on temperature as well as on pressure. The term *specific gravity* refers to the relative density. It is a dimensionless ratio. For liquids and solids, it is defined as

$$\text{Specific gravity} = \text{sp gr } \frac{t_2}{t_1} = \frac{\text{density of the substance at temperature } t_2}{\text{density of water at temperature } t_1} \tag{3.1}$$

Thus, sp gr $\frac{288.8 \text{ K}}{288.8 \text{ K}}$ is the ratio of the density of the substance at 288.8 K (60°F) to the density of water at the same temperature. Since the density of water at 277.2 K is 1 g/cm^3 in CGS units, the specific gravity of a substance with reference to water at 277.2 K is numerically equal to its density expressed in CGS units. But, densities in SI units and in FPS units are not numerically equal to the specific gravities. This is because, the density of water in these units at 277.2 K is 1000 kg/m^3 and 62.43 lb/ft^3, respectively. For gases, the reference substance used for defining specific gravity is usually dry air at the same conditions as the gas.

Bulk density: For porous solid, the bulk volume of the solid includes the volumes of the pores. Thus by dividing the mass of the solid in air by the volume of the solid what we get is the bulk density or the apparent density of the solid. The true density of the solid depends on the porosity of the solid. The porosity or the void fraction is defined as the fraction of the total volume that is void. Let ε be the porosity and ρ_T and ρ_B be the true density and bulk density of the solid, respectively. Then

$$\varepsilon = 1 - \frac{\rho_B}{\rho_T} \tag{3.2}$$

3.4 SPECIFIC GRAVITY SCALES

For liquid solutions, the specific gravity may vary appreciably with the concentration of the solute. Therefore, many chemical process industries use specific gravity as a measure of the concentration of their products. Industries dealing with sulphuric acid, hydrochloric acid, caustic soda, etc. report the concentration in terms of specific gravity. These industries had been using certain special specific gravity scales for this purpose. Some of the important specific gravity scales in use are discussed as follows. Many of these scales are now being abandoned in favour of the SI units of density.

API scale: Many important physical properties of petroleum products, like the boiling point, molecular weight, viscosity, etc. depend strongly on their composition which in turn can be conveniently related to their specific gravities. The petroleum industries had been using the API (American Petroleum Institute) scale for this purpose. On this scale, one degree API is defined as

$$°\text{API} = \frac{141.5}{\text{sp gr} \dfrac{288.8 \text{ K}}{288.8 \text{ K}}} - 131.5 \tag{3.3}$$

For example, if the specific gravity of a liquid at 288.8 K with reference to water at the same temperature is 0.934 $\left(\text{sp gr} \frac{288.8 \text{ K}}{288.8 \text{ K}} = 0.934\right)$, it corresponds to 20 °API on this scale.

Baumé scale: The mathematical definition of °Bé (degree Baumé) for liquids lighter than water differs from that of liquids heavier than water. For liquids lighter than water,

$$°\text{Bé} = \frac{140}{\text{sp gr}\,\dfrac{288.8\text{ K}}{288.8\text{ K}}} - 130 \tag{3.4}$$

A liquid of specific gravity 0.933 $\left(\text{sp gr}\,\dfrac{288.8\text{ K}}{288.8\text{ K}} = 0.933\right)$ is 20 °Bé on this scale. For liquids heavier than water, the definition of Baumé degree is

$$°\text{Bé} = 145 - \frac{145}{\text{sp gr}\,\dfrac{288.8\text{ K}}{288.8\text{ K}}} \tag{3.5}$$

20 °Bé on this scale refers to a liquid of specific gravity 1.16. This scale is used to represent concentration of solutions heavier than water, like sulphuric acid, hydrochloric acid and caustic soda. 20 °Bé sulphuric acid represents a concentration of 22.25 % (weight) of H_2SO_4 in water.

50 °Bé sulphuric acid $\left(\text{sp gr}\,\dfrac{288.8\text{ K}}{288.8\text{ K}} = 1.5263\right)$ represents a concentration of 62.18 % (weight) H_2SO_4.

Twaddell scale: This scale is used for liquids heavier than water. One degree Twaddell (°Tw) is defined as

$$°\text{Tw} = 200\left(\text{sp gr}\,\frac{288.8\text{ K}}{288.8\text{ K}} - 1.00\right) \tag{3.6}$$

A specific gravity of 1.1 according to this scale would correspond to 20 °Tw. A 40% (weight) aqueous caustic soda solution has a specific gravity of 1.445 which is equivalent to 89 °Tw .

Brix scale: Degree Brix (°Brix) is used to indicate the strength of sugar solution. This scale is defined in such a way that the density of 1.0 % (weight) aqueous sugar solution is equivalent to one degree Brix.

$$°\text{Brix} = \frac{400}{\text{sp gr}\,\dfrac{288.8\text{ K}}{288.8\text{ K}}} - 400 \tag{3.7}$$

EXAMPLE 3.8 A body weighs 1.0 kg in air, 0.90 kg in water and 0.82 kg in a liquid. What is the specific gravity of the liquid?

Solution The loss in weight of a body in a liquid is equal to the weight of liquid displaced by the body. Since the weight of liquid displaced is equal to the product of the volume of liquid displaced and density of the liquid, the volume of liquid displaced is

$$\frac{\text{loss in weight}}{\text{density of the liquid}}$$

$$\text{Loss in weight in water} = 1.0 - 0.9 = 0.1\text{ kg}$$

$$\text{Loss in weight in the given liquid} = 1.0 - 0.82 = 0.18\text{ kg}$$

Since the volume displaced by the solid is the same in water as well as in the liquid, we have

$$\frac{0.1}{\rho_w} = \frac{0.18}{\rho_l}$$

where ρ_w and ρ_l are, respectively, the density of water and the liquid.

The specific gravity of the liquid is the ratio of its density to the density of water. Here this ratio is

$$\frac{\rho_l}{\rho_w} = \frac{0.18}{0.1} = 1.8$$

EXAMPLE 3.9 10 kg of liquid A of specific gravity 1.17 is mixed with 5 kg of liquid B of specific gravity 0.83. Assuming that there is no volume change on mixing, what is the specific gravity of the mixture? (Take density of water = 1000 kg/m^3.)

Solution Density of liquid A = 1.17 × 1000 = 1170 kg/m^3; density of liquid B = 0.83 × 1000 = 830 kg/m^3

Volume of 10 kg liquid A is

$$\frac{10}{1170} = 8.547 \times 10^{-3} \text{ m}^3$$

Volume of 5 kg liquid B is

$$\frac{5}{830} = 6.024 \times 10^{-3} \text{ m}^3$$

Assuming that there is no volume change on mixing, the volume of mixture is 14.571 × 10^{-3} m^3.

$$\text{Density of the mixture} = \frac{\text{mass of the mixture}}{\text{volume}} = \frac{(10+5)}{14.571 \times 10^{-3}} = 1029 \text{ kg/m}^3$$

EXAMPLE 3.10 What is the specific gravity on the Baumé scale for a 100°Tw solution?

Solution 200 (sp gr – 1.0) = 100

or

$$\text{sp gr} = \frac{100}{200} + 1.0 = 1.50$$

On the Baumé scale, this corresponds to

$$145 - \frac{145}{1.50} = 48.3 \text{ Bé}$$

EXAMPLE 3.11 250 cubic meters of 30 °API gas oil is blended with 1000 cubic meters of 15 °API fuel oil. What is the density of the resultant mixture in kg/m^3? The density of water at 288.5 K = 999 kg/m^3. Assume no volume change on mixing.

Solution The API gravity scale is defined by

$$°\text{API} = \frac{141.5}{\text{sp gr}\dfrac{288.8 \text{ K}}{288.8 \text{ K}}} - 131.5$$

For 30 °API gas oil, the specific gravity is

$$\frac{141.5}{131.5+30} = 0.8762$$

The density of gas oil is

$$0.8762 \times 999 = 875.3 \text{ kg/m}^3$$

Therefore, the mass of gas oil blended is

$$250 \times 875.3 = 2.1883 \times 10^5 \text{ kg}$$

Similarly, the specific gravity of fuel oil is

$$\frac{141.5}{131.5+15} = 0.9659$$

The density of fuel oil is

$$0.9659 \times 999 = 964.9 \text{ kg}$$

Therefore, the mass of fuel oil blended is

$$1000 \times 964.9 = 9.649 \times 10^5 \text{ kg}$$

The density of the mixture after blending is

$$\frac{(2.1883 \times 10^5 + 9.649 \times 10^5)}{(250+1000)} = 947 \text{ kg/m}^3$$

3.5 COMPOSITION OF SOLIDS, LIQUIDS AND GASES

3.5.1 Mass Fraction (Weight Fraction)

Compositions of solid and liquid mixtures are usually expressed as mass fraction or weight fraction. The mass fraction of a component in a mixture is the ratio of the mass of the component to the total mass of the mixture. Consider a binary mixture made up of components A and B. Let the mass of A and the mass of B in the mixture be m_A and m_B, respectively.

Then,

$$w_A, \text{ mass fraction of } A = \frac{m_A}{m_A + m_B} \tag{3.8}$$

In a solution containing several components (multi-component solutions),

$$w_A, \text{ mass fraction of } A = \frac{m_A}{\sum_{i=1}^{n} m_i} \tag{3.9}$$

where m_i is the mass of component i in the solution.

3.5.2 Mass Percent (Weight Percent)

The weight (mass) percent of a component in a mixture is the number of parts by weight (mass) of the constituent in 100 parts by weight (mass) of the mixture.

$$\text{weight percent of } A = \frac{m_A}{m_A + m_B} \times 100 \tag{3.10}$$

Thus, the weight per cent is the weight fraction multiplied by 100.

3.5.3 Mass Ratio

Another term used for expressing the concentration in binary mixtures is the mass ratio or weight ratio. In a mixture made up of constituents *A* and *B*, the mass ratio of *A* is the mass of *A* per unit mass of *B*. Thus,

$$\text{mass ratio of } A = \frac{m_A}{m_B} \tag{3.11}$$

We see that the mass ratio of *A* in a binary mixture of *A* and *B* is the mass fraction of *A* on *A*-free basis. Concentration expressed as mass ratio is particularly convenient in calculations involved in such operations where only certain of the constituents undergo a change in mass while the quantities of the others remain unchanged. For example, in drying operation, a wet solid is admitted to the dryer where the moisture content of the solid gets reduced where as the dry weight, i.e. the weight of the moisture-free solid remains unchanged. When ammonia is removed from an ammonia-air mixture by absorbing in water in an absorber, the quantities of ammonia-free air entering and leaving the absorber are the same. In these operations it is more convenient to carry out the mass balance calculations using mass ratios than the mass fractions. For example, let the mass ratio of moisture in a wet solid entering the drier is 0.25 and that in the solid leaving the drier is 0.05, and we are asked to determine the fraction of the original water that is removed in the drying operation. Since 0.25 kg is the moisture associated with 1 kg dry solid entering the drier and the solid leaving the drier contains only 0.05 kg moisture per kg dry solid, the water removed in the drying operation is $\frac{0.25 - 0.05}{0.25} \times 100 = 80\%$ of the water present in the feed. When we say that a wet material is being dried from an initial moisture content of 0.2 mass fraction to a final value of 0.0476, though we refer to the same problem, the calculation will not be as easy as we have illustrated with the mass ratio values.

3.5.4 Dry Basis and Wet Basis

In the conventional method of expressing the concentration—the wet basis—the percentage of water in the solid is the mass of moisture contained in 100 parts by weight of the wet material. The water content of a wet material can be expressed on a moisture-free basis or dry basis also. Percent moisture on a dry basis is the mass ratio of moisture to moisture-free solid expressed as percentage. Thus, when the moisture content of the solid is 25% on a dry basis, there are 25 parts by weight of water associated with 100 parts by weight of dry solid. Or, 125 parts by weight of the wet solid contains 25 parts by weight of water. Therefore, the percent water on wet basis is 20.

EXAMPLE 3.12 250 kg wet ammonium sulphate containing 50 kg moisture is sent to a dryer in order to remove 90% of the moisture in the feed. Calculate for the entrance and exit to the dryer, the following:

(a) The weight fraction of water
(b) The weight ratio of water
(c) The weight percentage of moisture on a wet basis
(d) The weight percentage of moisture on a dry basis.

Solution *Basis:* 250 kg wet charge.

Entrance to the dryer

Dry ammonium sulphate = 250 – 50 = 200 kg

Water in the charge = 50 kg

(a) The weight fraction of water is

$$\frac{\text{mass of water}}{\text{total mass}} = \frac{50}{250} = 0.20$$

(b) The weight ratio of water is

$$\frac{\text{mass of water}}{\text{mass of dry solid}} = \frac{50}{200} = 0.25$$

The weight percent of moisture on wet basis is

$$\frac{\text{mass of water}}{\text{total mass}} \times 100 = \frac{50}{250} \times 100 = 20.00\%$$

(*c*) The weight percent of moisture on dry basis is

$$\frac{\text{mass of water}}{\text{mass of dry solid}} \times 100 = \frac{50}{200} \times 100 = 25.00\%$$

Exit of the dryer

Mass of dry ammonium sulphate = 200 kg.

Since 90% of water in the charge is removed, the mass of water present in the solid = 0.01 × 50 = 5 kg.

(a) The weight fraction of water is

$$\frac{\text{mass of water}}{\text{total mass}} = \frac{5}{205} = 0.0244$$

(b) The weight ratio of water is

$$\frac{\text{mass of water}}{\text{mass of dry solid}} = \frac{5}{200} = 0.025$$

(c) The weight percent of moisture on a wet basis is

$$\frac{\text{mass of water}}{\text{total mass}} \times 100 = \frac{5}{205} \times 100 = 2.44\%$$

(d) The weight percent of moisture on a dry basis is

$$\frac{\text{mass of water}}{\text{mass of dry solid}} \times 100 = \frac{5}{200} \times 100 = 2.50\%$$

EXAMPLE 3.13 A wet stock of ammonium sulphate containing 25% water on dry basis is sent to a drier. The material leaving the dryer contains 2.5% moisture on dry basis. Determine the percentage of water removed in the drying operation.

Solution *Basis:* 100 kg of dry solids in the feed to the dryer

Water in the feed is 25 kg and that in the dried product is 2.5 kg. Therefore, the amount of water removed = 25 – 2.5 = 22.5 kg.

The percentage of water removed in the dryer is

$$\frac{22.5}{25} \times 100 = 90\%$$

EXAMPLE 3.14 A wet stock of ammonium sulphate containing 20% water[1] is sent to a drier. The material leaving the dryer contains 2.44% moisture. Determine how many kg of water is removed per kg of wet material charged. Also find the percent of original water in the feed that is removed by drying.

Solution *Basis:* 1 kg of wet ammonium sulphate

Weight of dry solid = 0.80 kg, weight of water = 0.20 kg

The material leaving the drier contains 0.80 kg dry solid which constitutes 100 – 2.44 = 97.56% of the total material leaving the dryer. Let x be the weight of total material. Then

$$x \times 0.9756 = 0.8 \quad \Rightarrow \quad x = 0.82 \text{ kg}$$

1 kg of material enters the dryer and 0.82 kg leaves. The amount of water removed is 1.00 – 0.82 = 0.18 kg.

The percent of water in the feed that is removed by drying is

$$\frac{0.18}{0.20} \times 100 = 90\%$$

3.5.5 Mole Fraction and Mole Percent

If a mixture is made up of distinct chemical entities, the composition can conveniently be expressed on a mole basis. The number of moles of a constituent is determined by dividing the mass by its molecular weight. The *mole fraction* is the ratio of the moles of the constituent to the total moles in the mixture. Consider a solution containing n_A moles of A and n_B moles of B. Then

$$x_A, \text{ mole fraction of } A = \frac{n_A}{n_A + n_B} \tag{3.12}$$

Let M_A and M_B be the molecular weights of A and B, respectively.

$$x_A, \text{ mole fraction of } A = \frac{m_A/M_A}{m_A/M_A + m_B/M_B} \tag{3.13}$$

[1] Unless otherwise specified, the percentage composition of solids and liquids may be treated as percentages on a wet basis.

where m_A and m_B are the masses of the components A and B, respectively. Suppose that the solution is made up of n constituents and n_i is the number of moles of component i in the solution. Then

$$x_A\text{, mole fraction of } A = \frac{n_A}{\sum_{i=1}^{n} n_i} \tag{3.14}$$

The mole ratio of component A in a binary mixture of A and B is the ratio of moles of A to moles of B in the mixture. The *mole percent* is the number of moles of a component present in 100 moles of the total mixture. It is equal to the mole fraction multiplied by 100. Composition of a gas mixture is generally expressed as mole percent or mole fraction.

3.5.6 Volume Fraction and Volume Percent

The *volume fraction* of a component in a solution is the ratio of its pure-component volume to the total volume of solution both at the same temperature and pressure. By pure-component volume, we mean the volume occupied by the component in its pure state. Thus if V_A is the pure-component volume of A and V is the volume of the solution, then

$$\text{volume fraction} = \frac{V_A}{V}$$

The volume fraction multiplied by 100 gives the volume percent of the component in solution. The volume percent is usually used to represent the composition of gases. In rare cases, the concentrations of liquids also are expressed on a volumetric basis, a typical example being the alcohol-water solution, the alcohol content being represented on a volumetric basis. It will be shown in later chapters that the volume occupied by an ideal gas at a given temperature T and pressure P in the pure state is the same as the volume occupied by the gas in mixtures with other ideal gases, provided the temperature and pressure are the same. This property of ideal gases can be attributed to the absence of intermolecular forces in such gases. Thus the total volume of the gas mixture V, is the sum of the pure-component volumes and the volumetric percentages of the constituent gases which add up exactly to 100. This is roughly true for real gases also, because at low pressures such as the atmospheric pressure at which the gas analysis is usually made, all real gases behave almost like ideal gases. Thus the volume percent and volume fraction are valid means for expressing the concentration of the gas mixtures. Composition expressed on a volumetric basis does not change with change in temperature, because both the pure-component volume and the total volume of the gas mixture vary with temperature to the same extent. Furthermore, as will be shown later in Chapter 4, *the volume percent and mole percent are identical for gas mixtures*. Unless otherwise indicated, the percentage composition of a gas mixture means the composition by volume, which also means the composition by mole as well.

A change in volume usually accompanies the formation of liquid solutions from its constituent pure liquids. The volume that a unit mass of a liquid in a solution contributes to the total solution volume is not the same as the volume occupied by the liquid in its pure state at the same temperature and pressure as the solution. The volume percentage by our definition is the ratio of the volume of the pure liquid to the volume of the solution expressed as a percentage.

The volume percentage as defined earlier will not add up to exactly 100 in the case of liquids because of the volume change accompanying the mixing. Therefore, the compositions of the liquid solutions are generally expressed on a mass basis rather than on a volume basis. Furthermore, the effect of temperature on the volumes of pure components vary from liquids to liquids. Therefore, the volumes of the pure components and the solution vary differently with changes in temperature. The volumetric composition of liquids is therefore a function of temperature. Hence, it is imperative that when we represent, say, for example, the ethyl alcohol content in an aqueous solution on a volumetric basis, the temperature be specified.

EXAMPLE 3.15 The solubility of sodium chloride in water at 290 K is 35.8 kg/100 kg of water. Express the solubility as the following:

(a) Mass fraction and mass percent of NaCl
(b) Mole fraction and mole percent of NaCl
(c) kmol NaCl per 1000 kg of water

Solution *Basis*: 100 kg of water and the salt dissolved in it in a saturated solution at 290 K.

(a) Mass of NaCl = 35.8 kg

$$\text{Mass of solution} = \text{mass of water} + \text{mass of NaCl} = 100 + 35.8 = 135.8 \text{ kg}$$

$$\text{Mass fraction of NaCl} = \frac{35.8}{135.8} = 0.2636$$

$$\text{Mass percent of NaCl} = \text{mass fraction} \times 100 = 0.2636 \times 100 = 26.36\%$$

(b)

$$\text{kmol of NaCl} = \frac{\text{mass of NaCl in kg}}{\text{molecular weight}} = \frac{35.8}{58.45} = 0.6125 \text{ kmol}$$

$$\text{kmol water} = \frac{\text{mass of water in kilograms}}{\text{molecular weight}} = \frac{100}{18.016} = 5.5494 \text{ kmol}$$

$$\text{Mole fraction} = \frac{\text{moles of NaCl}}{\text{moles of solution}} = \frac{0.6125}{(5.5506 + 0.6125)} = 0.0994$$

$$\text{Mole percent} = \text{mole fraction} \times 100 = 9.94\%$$

(c) 100 kg of water dissolves 0.6125 kmol NaCl. Therefore, kmol NaCl per 1000 kg of water is 6.125 kmol.

EXAMPLE 3.16 The absolute humidity of air is 0.015 kg water vapour/kg dry air. Assuming the average molecular weight of air to be 29, calculate the following:

(a) The mole percent of water vapour in the air
(b) The molal absolute humidity, which is same as the mole ratio of water vapour to dry air

Solution When the humidity is 0.015 kg water vapour/kg dry air, air contains

$$\frac{0.015}{18.016} = 8.326 \times 10^{-4} \text{ kmol water vapour in association with } \frac{1}{29} = 0.0345 \text{ kmol dry air}$$

(a) Mole per cent of water vapour is

$$\frac{8.326 \times 10^{-4}}{(8.326 \times 10^{-4} + .0345)} \times 100 = 2.36\%$$

(b) Molal absolute humidity is

$$\frac{8.326 \times 10^{-4}}{(0.0345)} = 0.0241 \text{ kmol water/kmol dry air}$$

3.6 OTHER EXPRESSIONS FOR CONCENTRATIONS

A solution is a homogeneous medium—gas, liquid or solid—made up of two or more components. In process calculations, the term *solution* is generally used to refer to liquids and gases made up of solutes and a solvent. The component present in larger proportions in the solution is the *solvent* and that present in smaller proportions is called the *solute*. In addition to the different methods of expressing compositions as discussed above, the concentration of solute in a solution can be expressed as parts per million, molarity, molality and normality.

Parts per million (ppm): It is used to represent extremely small amounts of solutes present in solutions. It is the ratio of the number of parts by weight of the solute to 10^6 parts by weight of solution. For very dilute aqueous solutions, the density of the solution being the same as density of pure water,

$$1 \text{ ppm} = \frac{1 \text{ g}}{10^6 \text{ g solution}} = \frac{1 \text{ g}}{10^6 \text{ mL solution}} = \frac{1 \text{ mg}}{10^3 \text{ mL solution}} = \text{mg/L}$$

ppm is the measure used to represent concentration of impurities present in trace quantities, such as hardness of water, BOD (biochemical oxygen demand) of effluents, etc.

Molarity: It is defined as the number of moles of the solute per litre (10^{-3} m^3) of solution.

Molality: It is the number of moles of the solute present per 1 kg of solvent.

Normality: It is the ratio of the number of gram equivalents of the solute per litre of solution. The product of normality and equivalent weight is equal to the weight in grams per litre of solution.

EXAMPLE 3.17 An aqueous solution of K_2CO_3 contains 50% salt and the specific gravity of the solution is 1.53. Determine the following:

(a) The mole percent of the salt in the solution
(b) The volume percent of water assuming density of water is 1000 kg/m^3 and there is no volume change on mixing
(c) The molality of the solution
(d) The molarity of the solution
(e) The normality of the solution

Solution *Basis:* 100 g of solution

Molecular weight of K_2CO_3 = 138.20

(a)
$$\text{Mass of } K_2CO_3 = 50 \text{ g} = \frac{50}{138.20} = 0.3618 \text{ mol}$$

$$\text{Mass of water} = 50 \text{ g} = \frac{50}{18.016} = 2.7753 \text{ mol}$$

$$\text{Mole percent of salt} = \frac{0.3618}{0.3618 + 2.7753} \times 100 = 11.53\%$$

(b)
$$\text{Volume of the solution} = \frac{\text{mass}}{\text{density}}$$

$$= \frac{100}{1.53} = 65.36 \text{ mL}$$

$$\text{Volume of water in the solution} = \frac{\text{mass}}{\text{density}}$$

$$= \frac{50}{1000 \times 10^{-3}} = 50 \text{ mL}$$

$$\text{Volume percent of water} = \frac{\text{volume of water}}{\text{solution volume}} \times 100$$

$$= \frac{50}{65.36} \times 100 = 76.50\%.$$

(c) 50×10^{-3} kg water contains 0.3618 mol K_2CO_3.

$$\text{Molality} = \frac{\text{mol solute}}{\text{kg solvent}}$$

$$= \frac{0.3618}{50 \times 10^{-3}} = 7.236 \text{ mol/kg}$$

(d) 65.36 mL solution contains 0.3618 mol K_2CO_3.

$$\text{Molarity} = \frac{\text{mol solute}}{\text{litre solution}}$$

$$= \frac{0.3618}{65.36 \times 10^{-3}} = 5.536 \text{ mol/L}$$

(e) Equivalent weight of K_2CO_3 = molecular weight/2 = 138.20/2 = 69.10

$$\text{Number of gram equivalents of } K_2CO_3 = \frac{\text{mass in grams}}{\text{equivalent weight}}$$

$$= \frac{50}{69.10} = 0.7236 \text{ gram equivalents}$$

$$\text{Normality} = \frac{\text{gram equivalents}}{\text{litre solution}}$$

$$= \frac{0.7236}{65.36 \times 10^{-3}}$$

$$= 11.07 \text{ gram equivalents/L or } 11.07 \text{ N}$$

EXAMPLE 3.18 Pure water and alcohol are mixed to get a 60% (weight) alcohol solution. The densities (kg/m^3) of water, alcohol and the solution may be taken to be 998, 798 and 895 respectively at 293 K. Calculate the following:

(a) The volume percent of ethanol in the solution at 293 K
(b) The molarity
(c) The molality

Solution *Basis:* 100 kg of 60% (weight) alcohol–water solution

(a) Volume of 100 kg of solution is

$$\frac{\text{mass}}{\text{density}} = \frac{100}{895} = 0.1117 \text{ m}^3$$

The volume of pure alcohol mixed to get the above solution is

$$60 \text{ kg} = \frac{60}{798} = 0.0752 \text{ m}^3$$

Volume percent of ethanol in the solution is

$$\frac{0.0752}{0.1117} \times 100 = 67.3\%$$

(*Note:* The volume of the solution resulting on mixing two pure liquids need not be equal to the sum of the volumes of the liquids. Therefore, the volume percent of the individual constituents will not add to give 100.)

(b)

$$\text{Molarity of the solution} = \frac{\text{moles of ethanol}}{\text{volume of solution in litres}}$$

$$\text{Moles of ethanol in 100 kg solution} = \frac{60}{46.048} = 1.303 \text{ kmol} = 1303 \text{ mol}$$

Volume of solution is 0.1117 m^3 = 111.7 L. Therefore,

$$\text{molarity} = \frac{1303}{111.7} = 11.67 \text{ mol/L}$$

(c) $$\text{Molality of the solution} = \frac{\text{moles of alcohol}}{\text{kilograms of water}}$$

Mass of water in 100 kg solution is 40 kg. Hence,

$$\text{molality} = \frac{1303}{40} = 32.575 \text{ mol/(kg of water)}$$

3.7 CHEMICAL REACTIONS AND PROCESS CALCULATIONS

The conversion of raw materials into useful products in most cases is brought about by means of chemical reactions. For solving material balance and energy balance problems involving chemical reactions, it is essential that a balanced chemical equation be written for the reaction under consideration. The chemical equation provides us with a variety of qualitative and quantitative information essential for solving material balance problems. For example, consider the determination of the quantity of iron pyrites (FeS_2) that is to be fed to the burner to produce say, 100 kg of sulphur dioxide by oxidation with air. The relevant chemical reaction is represented by the equation

$$4FeS_2 + 11O_2 \rightarrow 2Fe_2O_3 + 8SO_2$$

The coefficients of the chemical formulae of the different chemical species in the equation are called the *stoichiometric coefficients*. The ratio of the stoichiometric coefficients of any two substances involved in the reaction is the stoichiometric ratio of the substances. The stoichiometric ratio helps us to calculate the number of moles of one substance involved in a chemical reaction with a given number of moles of another substance. According to the equation, 4 mol of FeS_2 combines with 11 mol of oxygen to produce 2 mol of Fe_2O_3 and 8 mol of SO_2. The molecular weights of the reactants and products of the reaction can be calculated from the atomic weights of the constituent elements. Thus

$$\text{Molecular weight of } FeS_2 = 55.847 + 2 \times 32.065 = 119.977$$
$$\text{Molecular weight of } Fe_2O_3 = 2 \times 55.847 + 3 \times 16.00 = 159.694$$
$$\text{Molecular weight of } SO_2 = 32.064 + 2 \times 16.00 = 64.065$$

Accordingly, 479.9 kg (= 4 × 119.977) FeS_2 combines with 352 kg (= 11 × 32) oxygen producing 319.388 kg (= 2 × 159.694) Fe_2O_3 and 512.512 kg (= 8 × 64.065) SO_2. When the ratio of the pyrites charged to the oxygen supplied is 4:11 on a mole basis or 479.9: 352 on a weight basis, we say that the reactants are supplied in stoichiometric proportions.

3.7.1 Excess and Limiting Reactant

In industrial processes one would rarely come across the use of exact stoichiometric quantities for chemical conversion. Frequently, one of the reactants is supplied in excess to ensure completion of the reaction. Excess reactants may be necessary in certain cases to prevent the occurrence of undesired side reactions or to ensure the complete use of costly reactants. When the reactants are not in stoichiometric quantities, the reactant present in excess of the stoichiometric quantities is termed the *excess reactant* and the material that is not present in excess of that required to react with any of the other materials or the material that is present in least stoichiometric proportions is termed the *limiting reactant*. The limiting reactant is present in the smallest stoichiometric amount and if the reaction is allowed to go to completion, it would disappear first from the reaction mixture. It is the quantity of the limiting reactant present that determines the amount of the product obtained using the given feed mixture.

The synthesis of ammonia is represented by the equation

$$N_2 + 3H_2 \rightarrow 2NH_3$$

If the ammonia converter is charged with a mixture of nitrogen and hydrogen containing 20% nitrogen by volume, which one is the limiting reactant? If the reactants were in stoichiometric quantities, the ratio of moles of hydrogen to moles of nitrogen = 3/1 = 3.0. In the feed mixture, there is 20 moles of nitrogen for 80 moles of hydrogen. The ratio of moles of hydrogen to moles of nitrogen = 80/20 = 4.0. It means that for every mole of nitrogen, 4 moles of hydrogen is supplied which is one mole in excess of the stoichiometric requirement. Therefore, nitrogen is the limiting reactant and hydrogen is present in excess.

If more than two reactants are present one of the reactants may be chosen as the reference and the ratio of the moles of each reactant to the moles of the reference substance may be calculated. These are then compared with the respective stoichiometric quantities for finding out the limiting reactant. The excess of a reactant over that theoretically required to react with the limiting reactant is usually expressed as *percent excess*. It is defined as

$$\text{percent excess} = \frac{\text{excess amount}}{\text{theoretically required amount}} \times 100$$

The quantity of the desired substance formed by the reaction is determined from the quantity of the limiting reactant. Its consumption in a reaction will limit the extent to which the reaction can proceed. Frequently the limiting reactant may enter into many side reactions leading to the formation of undesired by-products. It may also form intermediates which are not completely decomposed and hence may be present in the reactor effluents.

Suppose that 500 kg of pyrites is oxidized in a supply of 400 kg of oxygen. The stoichiometric requirement of oxygen to combine with 500 kg of pyrites is (352/479.9) × 500 = 366.7 kg. Since oxygen supplied is greater than this amount, oxygen is the excess reactant and pyrites is the limiting reactant. The percent excess oxygen supplied for burning pyrites can be calculated as follows:

Pyrites charged	= 500 kg = 500/119.975	= 4.17 kmol
Oxygen theoretically required	= (11/4) × 4.17	= 11.47 kmol
Oxygen supplied	= 400 kg = 400/32	= 12.5 kmol
Excess of oxygen	= 12.5–11.47	= 1.03 kmol
Percent excess	= (1.03/11.47) × 100	= 9.0%

Oxygen theoretically required for the above reaction is 11.47 kmol. The term theoretical requirement in certain situations needs some clarification. Even if only part of the limiting reactant actually undergoes reaction, the theoretical requirement of other substances are calculated based on the complete conversion of the limiting reactant. Consider the combustion of one kilogram-atom of carbon. Assume that only 90% of the carbon charged is undergoing combustion and of the carbon converted, 80% is burned to CO_2 and the rest forms CO. The theoretical requirement of oxygen is to be calculated based on the complete conversion of carbon to CO_2.

Thus the theoretical requirement of oxygen is one kilomole. In some cases, as for example, in the combustion of a fuel containing oxygen, the theoretical requirement is based on *net oxygen demand*. That is, when the theoretically required quantity of oxygen by a given amount of the fuel is calculated, it is necessary to subtract the amount of oxygen already present in the fuel from the quantity of oxygen that is required by all the combustible components of the fuel. Suppose that one hundred kilogram coal containing 66.0% carbon, 12.0% hydrogen, 15.0% oxygen and 7.0% ash by weight is burned in air. How many moles of air are theoretically required by 100 kg coal? Assume a basis of 100 kg coal charged. The number of moles of carbon present in coal = 66/12 = 5.5 kmol. Since $C + O_2 \rightarrow CO_2$, one mol of carbon requires one of mol oxygen for complete combustion. Therefore, the amount of oxygen required by carbon = 5.5 kmol.

The number of moles of hydrogen present in coal = 12/2.016 = 5.95 kmol. Since $H_2 + 1/2O_2 \rightarrow H_2O$, one mol of hydrogen requires 0.5 mol of oxygen for complete combustion. Therefore, oxygen required by hydrogen in the fuel is 2.98 kmol. The total oxygen required is

$$5.5 + 2.98 = 8.48 \text{ kmol}$$

Oxygen present in the fuel = 15/32 = 0.47 kmol. Therefore, net oxygen demand = 8.48 –0.47 = 8.01 kmol. Since this oxygen should come from air which contains only 21% oxygen, the number of moles of air required theoretically for the combustion of 100 kg coal is 8.01 × (100/21) = 38.14 kmol.

3.7.2 Conversion

Chemical reactions never go to completion. Reactions occur within a given system till equilibrium between various constituents—reactants and products—is established. The extent of the reaction corresponding to this equilibrium condition is measured by the *equilibrium conversion* for the reaction. Reactions can never be carried out to a conversion beyond this equilibrium value. The actual conversion will be less than the equilibrium value due to such reasons as insufficient time allowed for the reaction. The *conversion* is the fraction of the feed or fraction of certain constituents of the feed that is converted into products.

$$\text{Percent conversion} = \frac{\text{moles of the feed reacted}}{\text{moles of the feed charged}} \times 100$$

Closely related to the percent conversion is the *degree of completion* of a reaction. Degree of completion is, in fact, the conversion of the limiting reactant and may be defined as the ratio of the amount of the limiting reactant that has undergone chemical transformation to the amount that was available for the reaction. In the oxidation of iron pyrites we have considered earlier, the charge consisted of 4.17 mol of pyrites. If the percent conversion of iron pyrites was 90%, only (90/100) × 4.17 = 3.75 mol of pyrites would undergo reaction. If the reactant takes part in more than one reaction, the percent conversion should consider the quantities involved in all the steps.

Frequently the limiting reactant may enter into many side reactions leading to the formation of undesired by-products. It may also form intermediates which are not completely decomposed and hence may be present in the reactor effluents.

3.7.3 Yield and Selectivity

Frequently, the reactants may enter into many side reactions, leading to the formation of undesired by-products. These may also form intermediates which are not completely decomposed, and hence, may be present in the reactor effluents. The terms *yield* and *selectivity* are used to indicate the degree to which the reactants are converted into desired products.

Yield is the ratio of the actual moles of the desired product to the moles which would have been resulted if the reactant was converted entirely to form the desired products. In other words,

$$\text{Percent yield} = \frac{\text{moles of the reactant converted to the desired product}}{\text{total moles of the reactant converted}} \times 100 \qquad (3.15)$$

The yield defined by Eq. (3.15) is also called the *reactor yield* to distinguish it from the overall yield or plant yield, which is defined as

$$\text{Plant yield, \%} = \frac{\text{moles of the reactant converted to the desired product}}{\text{total moles of the reactant fed to the process}} \times 100 \qquad (3.16)$$

Plant yield is a measure of the overall performance of the plant and includes all chemical and physical losses.

Selectivity is another term used to indicate the relative prominence of side reactions over the desired reactions. It is defined as

$$\text{Selectivity} = \frac{\text{moles of the reactant converted to the desired product}}{\text{moles of the reactant converted to undesired products}}$$

Consider that 60% of the reactant fed is converted in a reaction and 50% of that converted gives the desired product *A* and the rest undesired by-product *B*. Then, of the 100 moles of the reactant fed, 60 moles are converted of which 30 moles form desired product *A* and 30 moles form by-product *B*. Then

$$\text{Yield} = \frac{30}{60} \times 100 = 50\%, \text{ Plant yield} = \frac{30}{100} \times 100 = 30\% \text{ and selectivity} = \frac{30}{30} = 1.0$$

The difference between conversion and yield must be clear before attempting to solve any problem involving chemical reaction. It is clear that conversion refers to the reactants and yield to the products. Consider the production of methanol by the reaction of carbon monoxide and hydrogen. The reaction of CO with H_2 may also result in the formation of formaldehyde.

$$CO + 2H_2 \rightarrow CH_3OH$$

$$CO + H_2 \rightarrow HCHO$$

Starting with a mixture of 1 mol CO and 3 mol H_2, let 0.18 mol formaldehyde and 0.72 mol of methanol be present in the product mixture. 0.90 mol CO (the limiting reactant) has reacted or the conversion of CO is 90%. However, only 72% of the CO charged has been converted to the desired product (methanol). The yield of methanol is the ratio of the moles of CO converted to methanol to the moles of CO converted in any way. That is, the yield is (0.72/0.9) × 100 = 80%. However, the plant yield would be only (0.72/1.0) × 100 = 72%.

Consider another reaction. In the nitration of benzene to nitrobenzene, it is found that some dinitrobenzene is also formed as a by-product.

$$C_6H_6 + HNO_3 \rightarrow C_6H_5NO_2 + H_2O$$

$$C_6H_6 + 2HNO_3 \rightarrow C_6H_4N_2O_4 + 2H_2O$$

The yield in this process would be based on the nitrobenzene produced.

$$\text{Percent yield} = \frac{\text{moles of benzene converted to nitrobenzene}}{\text{total moles of benzene converted}} \times 100 \tag{3.17}$$

The conversion of benzene is given by

$$\text{Percent conversion} = \frac{\text{moles of benzene consumed in the reaction}}{\text{total moles of benzene charged}} \times 100 \tag{3.18}$$

3.7.4 Extent of Reaction

The *extent of reaction,* also called *reaction coordinate* measures the progress of a reaction and is defined as the degree to which a reaction has advanced. Let us denote the extent of reaction by ε. The extent of reaction and the number of moles taking part in the reaction are related as

$$dn_i = v_i \, d\varepsilon \tag{3.19}$$

The change in the extent of reaction $d\varepsilon$ is the same for each component whereas the changes in the number of moles are different for different species taking part in the reaction. In Eq. (3.19) v_i is the stoichiometric number of species i. The stoichiometric numbers are positive for products, negative for reactants and zero for inert species. Consider the reaction

$$2A + 3B \rightarrow L + 2M$$

Here $v_L = 1$, $v_M = 2$, $v_A = -2$ and $v_B = -3$. Let n_{i0} be the number of moles of the species initially present in the system and n_i the number of moles present after the reaction. Then $n_i = n_{i0} + \Delta n_i$ where Δn_i is the change in the number of moles of i due to the reaction.

$$\Delta n_i = v_i \varepsilon \tag{3.20}$$

Therefore,

$$n_i = n_{i0} + v_i \varepsilon \tag{3.21}$$

Let us examine how the extent of reaction is related to the conversion of reactants. Let the fraction of the limiting reactant converted be z_i. In terms of conversion, the number of moles of the reactant unconverted = $n_{i0} - z_i n_{i0}$ whereas in terms of extent of reaction this is $n_{i0} + v_i \varepsilon$. Comparing these two results we see that the conversion and the extent of reaction are related as

$$z = -\frac{v_i \varepsilon}{n_{i0}} \tag{3.22}$$

Since v_i is negative for reactants, the above equation may be written as

$$z = \frac{|v_i|\varepsilon}{n_{i0}} \tag{3.23}$$

EXAMPLE 3.19 Carbon monoxide combines with chlorine in the presence of a suitable catalyst to form phosgene according to the following reaction.

$$CO(g) + Cl_2(g) \rightarrow COCl_2(g)$$

After reaction, the products contained 12 moles of phosgene, 3 moles of chlorine and 8 moles of carbon monoxide. Assuming that the original reactant mixture is free of phosgene, calculate the following:

(a) The percent excess reactant used
(b) The percent conversion of the limiting reactant
(c) The moles of the total product per mole of the reactant mixture fed to the reactor

Solution The reaction product consisting of 12 moles of phosgene, 3 moles of chlorine, and 8 moles of carbon monoxide is considered. 12 moles of phosgene is produced by the reaction between 12 moles of CO and 12 moles of Cl_2. Therefore, the original reactant mixture contained 12 + 3 = 15 moles of Cl_2 and 12 + 8 moles = 20 moles of CO. Since the stoichiometric ratio between CO and Cl_2 is 1, and the reactant mixture contained CO and Cl_2 in the ratio 20:15, chlorine is the limiting reactant and CO is present in excess.

(a) $$\text{Percent excess of CO} = \frac{\text{excess amount}}{\text{theoretically required amount}} \times 100$$

$$= \frac{20-15}{15} \times 100 = 33.33\%$$

(b) Chlorine is the limiting reactant.

$$\text{Percent conversion} = \frac{\text{moles of } Cl_2 \text{ reacted}}{\text{moles of } Cl_2 \text{ charged}} \times 100 = \frac{12}{15} \times 100 = 80\%$$

(c) Total products = 12 + 3 + 8 = 23 mol
Total reactants = 20 + 15 = 35 mol

Moles of total products per mole of total reactants are

$$\frac{23}{35} = 0.657$$

EXAMPLE 3.20 A gas mixture containing 2 moles of nitrogen, 7 moles of hydrogen and 1 mole of ammonia initially, is undergoing the following reaction.

$$N_2 + 3H_2 \rightarrow 2NH_3$$

Derive expressions for the mole fractions of various components in the reaction mixture in terms of the extent of reaction.

Solution The total number of moles initially present is $n_0 = \Sigma\, n_{i0} = 2 + 7 + 1 = 10$ and the sum of the stoichiometric numbers is $v = \Sigma\, v_i = 2 - 1 - 3 = -2$. The number of moles of various components after the reaction is related to the extent of reaction by Eq. (3.23) and the ratio of the number of moles to the total moles gives the mole fraction y of the component.

$$y_{N_2} = \frac{n_{N_2,0} + \nu_{N_2}\varepsilon}{n_0 + \varepsilon\nu} = \frac{2-\varepsilon}{10-2\varepsilon}$$

$$y_{H_2} = \frac{n_{H_2,0} + \nu_{H_2}\varepsilon}{n_0 + \varepsilon\nu} = \frac{7-3\varepsilon}{10-2\varepsilon}$$

$$y_{NH_3} = \frac{n_{NH_3,0} + \nu_{NH_3}\varepsilon}{n_0 + \varepsilon\nu} = \frac{1+2\varepsilon}{10-2\varepsilon}$$

EXAMPLE 3.21 In the vapour-phase hydration of ethylene to ethanol, diethyl ether is obtained as a by-product.

$$C_2H_4 + H_2O \rightarrow C_2H_5OH$$

$$2C_2H_4 + H_2O \rightarrow (C_2H_5)_2O$$

A feed mixture consisting of 60% ethylene, 3% inerts and 37% water is sent to the reactor. The products analyzed 53.89% ethylene, 14.37% ethanol, 1.80% ether, 26.35% water, and 3.59% inerts. Calculate the conversion of ethylene, yield of ethanol and ether based on ethylene.

Solution *Basis:* 100 mol of feed mixture

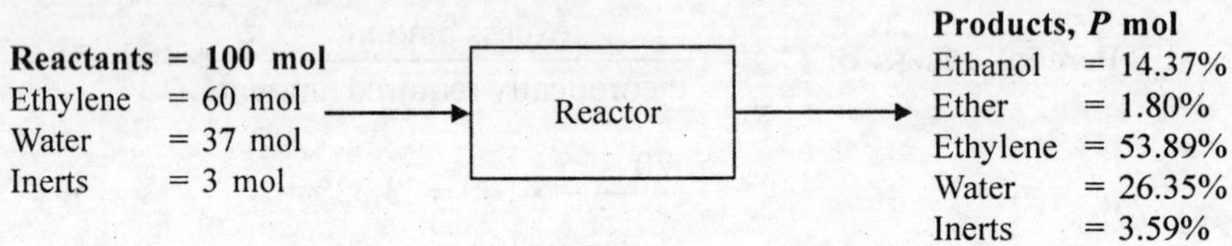

Figure. 3.1 Chemical process in Example 3.21.

Since the amount of inert materials does not change during reaction, inerts may be treated as the key component (for a detailed discussion on key components see Chapter 9) for making material balances. Let P mol of products are formed. Then inert balance gives

$$3 = 0.0359 \times P$$

Therefore, P = 83.57 mol. The product contains

$$\text{Ethylene} = \frac{53.89}{100} \times 83.57 = 45.04 \text{ mol}$$

$$\text{Ethanol} = \frac{14.37}{100} \times 83.57 = 12.01 \text{ mol}$$

$$\text{Ether} = \frac{1.8}{100} \times 83.57 = 1.5 \text{ mol}$$

$$\text{Water} = \frac{23.35}{100} \times 83.57 = 22.02 \text{ mol}$$

$$\text{Inerts} = 3 \text{ mol}$$

$$\text{Conversion of ethylene} = \frac{\text{moles of ethylene reacted}}{\text{moles of ethylene charged}} \times 100$$

$$= \frac{60 - 45.04}{60} \times 100 = 24.93\%$$

$$\text{Yield of ethanol} = \frac{\text{moles of ethylene converted to ethanol}}{\text{total moles of ethylene converted}} \times 100$$

$$= \frac{12.01}{60 - 45.04} \times 100 = 80.0\%$$

(The number of moles of ethylene converted to ethanol is equal to the number of moles of ethanol in the products because one mole of ethanol requires one mole of ethylene.)

$$\text{Yield of ether} = \frac{\text{moles of ethylene converted to ether}}{\text{total moles of ethylene converted}} \times 100$$

$$= \frac{1.5 \times 2}{60 - 45.04} \times 100 = 20.0\%$$

(The number of moles of ethylene converted to ether is equal to twice the number of moles of ether in the products as two moles of ethylene is needed for 1 mole of ether.)

EXERCISES

The mole concept

3.1 How many grams of NH_4Cl are there in 5 mol?

3.2 Convert 750 g $CuSO_4.5H_2O$ into moles. Find equivalent mol of $CuSO_4$ in the crystals.

3.3 How many kilogram of CS_2 will contain 3.5 kg-atom carbon?

3.4 How many grams of carbon are present in 264 g of CO_2?

3.5 The molecular formula of an organic compound is $C_{10}H_7Br$. Find the weight percentage of carbon, hydrogen and bromine in the solid.

3.6 Find the equivalents of 3 kmol of $FeCl_3$.

3.7 What is the equivalent weight of $Al_2(SO_4)_3$?

3.8 How many equivalents are there in 500 g $KMnO_4$?

3.9 Calculate the equivalent weight of H_3PO_4 in the reaction

$$Ca(OH)_2 + H_3PO_4 \rightarrow CaHPO_4 + 2H_2O$$

3.10 A certain organic compound is found to contain 81.5% C, 4.9% H, and 13.6% N by weight. If the molecular weight of the compound is 103, determine the molecular formula of the compound.

3.11 Aluminium chloride is made by the chlorination of molten aluminium metal in a furnace:

$$2Al(l) + 3Cl_2(g) \rightarrow 2AlCl_3(g)$$

(a) How many kilograms of $AlCl_3$ can be made from 100 kg of chlorine?
(b) How many grams of Al will react with 50 g of chlorine?

3.12 Sodium hydroxide is made by the electrolysis of brine solution. The overall reaction may be written as:

$$2NaCl + 2H_2O \rightarrow 2NaOH + H_2 + Cl_2$$

(a) How much NaOH can be made from 1000 kg NaCl?
(b) How much water is consumed in the production of 500 kg Cl_2?

3.13 Sulphur trioxide gas is obtained by the combustion of pyrites (FeS_2) according to the following reaction:

$$4FeS_2 + 15O_2 \rightarrow 2Fe_2O_3 + 8SO_3$$

The reaction is accompanied by the following side reaction:

$$4FeS_2 + 11O_2 \rightarrow 2Fe_2O_3 + 8SO_2$$

Assume that 80% (weight) of the pyrites charged reacts to give sulphur trioxide and 20% reacts giving sulphur dioxide.
(a) How many kilograms of pyrites charged will give 100 kg of SO_3?
(b) How many kilograms of oxygen will be consumed in the reaction?

3.14 Barium chloride reacts with sodium sulphate to precipitate barium sulphate:

$$BaCl_2 + Na_2SO_4 \rightarrow 2NaCl + BaSO_4$$

(a) How many grams of barium chloride are needed to react with 100 g of sodium sulphate?
(b) For precipitating 50 g of barium sulphate, how many grams of the reactants are consumed?
(c) How many grams of sodium chloride would be obtained when 50 g of barium sulphate is precipitated?

3.15 Chromite ore analyzed 30.4% Cr_2O_3. Determine the theoretical amount of lead chromate ($PbCrO_4$) that can be obtained from 1000 kg of the ore.

3.16 The alloy brass contains lead as an impurity in the form of lead sulphate ($PbSO_4$). By dissolving brass in nitric acid, lead sulphate is precipitated. A sample of brass weighing 5.0 g is dissolved in nitric acid and 0.03 g of precipitate is formed. Calculate the percentage of lead present in the brass sample.

3.17 How many kilograms of CO_2 are obtained by the decomposition of 100 kg of limestone containing 94.5% $CaCO_3$, 4.2% $MgCO_3$ and 1.3% inert materials? What is the volume of CO_2 obtained at STP?

3.18 Sulphur dioxide is obtained by the following reaction

$$Cu + 2H_2SO_4 \rightarrow CuSO_4 + 2H_2O + SO_2$$

(a) When 50 kg Cu dissolves in sulphuric acid what volume of sulphur dioxide is produced at standard conditions?

(b) How many kilograms of 94% sulphuric acid will be required for the above reaction?

3.19 Crude calcium carbide, CaC_2, is made in an electric furnace by the following reaction.

$$CaO + 3C \rightarrow CaC_2 + CO$$

The product contains 85% CaC_2 and 15% unreacted CaO.

(a) How much CaO is to be added to the furnace charge for each 1000 kg CaC_2?
(b) How much of CaO is to be added to the furnace charge for each 1000 kg of crude product?

Specific gravity and specific gravity scales

3.20 A 1-kg lead ball of density 11.34×10^3 kg/m^3 is immersed in water. The density of water is 1000 kg/m^3. Calculate the buoyant force on the body.

3.21 A body weighs 1.0 kg in air, 0.90 kg in water and 0.85 kg in a liquid. What is the specific gravity of the liquid?

3.22 10 kg of liquid *A* of specific gravity 1.2 is mixed with 3 kg of liquid *B* of specific gravity 0.8. Assuming that there is no volume change on mixing, what is the specific gravity of the mixture?

3.23 An alloy contains metal *A* and *B* in the ratio 5:3. If metal *A* has a specific gravity 10 and metal *B* has a specific gravity 5 in the pure state, what would be the specific gravity of the alloy?

3.24 An aqueous solution of a valuable chemical (molecular weight = 180) leaves the reactor at a rate of 60×10^{-3} m^3/h. The solution concentration is 40% (weight) and its specific gravity is 1.05. Determine (a) the concentration of the solution in kg/m^3 and (b) the flow rate in kmol/h.

3.25 A certain solution has a specific gravity of 1.15 at 288.8 K referred to water at 288.8 K. Express the specific gravity as °Bé and °Tw.

3.26 What is the specific gravity on Baumé scale for a 90 °Tw solution?

3.27 The specific gravity of a hydrocarbon oil is 0.88 at 288.8 K. What are the corresponding values in the Baumé and API scales?

3.28 The bulk density of a solid is 1.125 g/mL and the true density is 1.5 g/mL. What is the porosity of the solid?

3.29 500 cubic meters of 30 °API gas oil is blended with 2000 cubic meters of 15 °API fuel oil. What is the density of the resultant mixture in kg/m^3? The density of water at 288.5 K = 0.999 g/ml. Assume no volume change on mixing.

3.30 100 litres each of gasoline (55°API), kerosene (40°API), gas oil (31°API), and isopentane (96°API) are mixed. The density of water at 288.5 K = 0.999 g/mL.

(a) Determine the density of the mixture in kg/m^3.
(b) What is the specific gravity in °API?
(c) Express the composition of the mixture in weight percent.

3.31 The specific gravity $\frac{288.5\ \text{K}}{288.5\ \text{K}}$ of an ammonia–water solution is 0.9180. What would be the specific gravity $\frac{288.5\ \text{K}}{300\ \text{K}}$ if the density of water at 288.5 K and 300 K are respectively, 0.998 g/mL and 0.989 g/mL?

3.32 An analysis of sea water showed 2.8% NaCl, 0.5% $MgCl_2$ and 0.0085% NaBr by weight. The average specific gravity of the water is 1.03. What mass of magnesium, sodium and chlorine can be obtained from 100 m^3 of sea water?

Composition of solids, liquids and gases

3.33 What is the weight percentage of CaO in $Ca(OH)_2$?

3.34 Determine the weight percentage of the constituent elements of potassium sulphate?

3.35 What is the percentage of water in $Al_2(SO_4)_3{\cdot}17H_2O$?

3.36 Compare the percentages of iron in ferrous chloride and ferric chloride?

3.37 An aqueous solution contains 40% by weight $NaNO_3$. Determine the composition in mole percent.

3.38 How many kg of Glauber's salt ($Na_2SO_4{\cdot}10H_2O$) will be obtained from 250 kg Na_2SO_4?

3.39 A sample of urea (NH_2CONH_2) contains 42.0% nitrogen by weight. What is the percent purity of the sample?

3.40 Determine the mass fraction and mole fraction of chlorine in the substance $Ca(ClO)_2$.

3.41 The strength of phosphoric acid is usually represented as weight percent of P_2O_5. A sample of phosphoric acid analyzed 40% P_2O_5. What is the percent by weight of H_3PO_4 in the sample?

3.42 A blast furnace treats 10^6 kg per day hematite ore which contains 50% pure ferric oxide. Determine the weight of pig iron produced per day. Pig iron contains 94% iron.

3.43 A liquid mixture contains three components A (MW = 72), B (MW = 58) and C (MW = 56) in which A and B are present in the mole ratio 1.5:1 and the weight percent of B is 25%. A sample of the mixture is found to contain 10 kg of C. Calculate the total number of moles of the mixture.

3.44 A portland cement sample contained 20% SiO_2 by weight derived from two silicate compounds, $SiO_2{\cdot}2CaO$ and $SiO_2{\cdot}3CaO$ that are present in the cement in the mole ratio 3:4. Determine the percent by weight of each silicate compound in the cement.

3.45 An ethanol–water mixture forms an azeotrope* at 89.43 mole percent ethanol at 101.3 kPa and 351.4 K. What is the composition of the azeotrope in weight percent?

3.46 A 20% (weight) aqueous solution of monoethanolamine (MEA, $NH_2CH_2CH_2OH$) is used as a solvent for absorbing CO_2 from a gas stream. The solution leaving contains 0.25 mol CO_2 per mol MEA. Determine

(a) The mole percent of CO_2 in the solution leaving the absorber.

(b) The mass percent of CO_2 in the solution.

---

*Azeotropes are constant boiling mixtures that boil without a change in composition till the entire liquid is vaporized.

3.47 A water soaked cloth is dried from 45% to 9% moisture on dry basis. Find the weight of water removed from 2000 kg of dry fabric.

3.48 A solution of sodium chloride is saturated in water at 289 K. Calculate the weight of salt in kg that can be dissolved in 100 kg of this solution if it is heated to a temperature of 343 K. The solubility of sodium chloride at 289 K = 6.14 kmol/1000 kg water. The solubility at 343 K = 6.39 kmol/1000 kg of water.

3.49 The solubility of benzoic acid (C_6H_5COOH) is found out to be 66 parts in 100 parts by weight of ether ($C_2H_5OC_2H_5$). Find the mole fraction of benzoic acid in the saturated solution with ether.

3.50 The solubility of benzoic acid (C_6H_5COOH) in ether ($C_2H_5OC_2H_5$) is found to be 28.59% (by mole). What is the solubility in weight percent? What is the weight ratio of acid to ether in the saturated solution?

3.51 Hydrogen chloride is made by the action of sulphuric acid on sodium chloride. Hydrogen chloride being readily soluble in water forms hydrochloric acid. Calculate the following:

(a) The weight in grams of HCl formed by the action of excess sulphuric acid on 1 kg of salt which is 99.5% pure
(b) The volume of hydrochloric acid solution (specific gravity 1.2) containing 40% by weight HCl that can be produced
(c) The weight in kilograms of sodium sulphate obtained

3.52 An excess of $NaNO_3$ is treated with 25 kg sulphuric acid solution which contains 93.2% by weight of pure H_2SO_4. Calculate the following:

(a) The number of kilomoles of pure nitric acid obtained
(b) The mass of nitric acid containing 70% by weight HNO_3 obtained
(c) The number of kilograms of Na_2SO_4 produced

3.53 A liquid mixture contains three components *A* (MW = 72), *B* (MW = 58) and *C* (MW = 56) in which *A* and *B* are present in the mole ratio 1.5:1 and the weight percent of *B* is 25%. The specific gravities of the pure liquids are 0.67, 0.60 and 0.58 respectively, for *A*, *B* and *C* and there is no volume change on mixing. Calculate the following:

(a) The analysis of the mixture in mole percent
(b) The molecular weight of the mixture
(c) The volume percent of *C* on a *B*-free basis
(d) The specific gravity of the mixture

3.54 An alcohol–water solution contains 20% (volume) ethanol at 300 K. The densities of ethanol and water at 300 K are 0.798 g/mL and 0.998 g/mL respectively. What is the weight percent of alcohol?

3.55 Calculate the concentration in mol/L of pure methanol at 298 K if the density of methanol at 298 K is 0.9842 g/mL.

3.56 A company has a contract to buy NaCl of 98 percent purity for ₹ 300 per 1000 kg salt delivered. Its last shipment of 1000 kg was only of 90% purity. How much they should pay for the shipment?

3.57 A compound is found to contain 62.4% Ca and 37.6% C.
(a) How many gram atoms of Ca and C are present in 100 g of the compound?
(b) Suggest an empirical formula for the compound.

3.58 It is desired to prepare a 40% solution of NaCl in water at 300 K.
(a) How many kg of anhydrous sodium chloride should be added to 0.05 cubic metres of pure water having a density of 0.998 g/mL at 300 K?
(b) If the salt contains 10% water, how many kg of salt is required?

3.59 Absolute humidity of air is 0.02 kg water vapour/kg dry air. Assuming the average molecular weight of air to be 29, calculate the following:
(a) The mole percent of water vapour in the air
(b) The molal absolute humidity, which is same as the mole ratio of water vapour to dry air.

3.60 Assuming that dry air contains 21% oxygen and 79% nitrogen, calculate the following:
(a) The composition in weight percent
(b) The average molecular weight of dry air

3.61 By electrolysing brine, a mixture of gases is obtained at the cathode having the following composition by weight: chlorine 67%, bromine 28% and oxygen 5%. Calculate composition of gases by volume.

Other expressions for concentrations

3.62 Determine the weight percent of NaOH in an aqueous solution of molality 2.

3.63 Calculate the molality of a solution of 93% H_2SO_4 (W/V). The density of the solution is 1840 kg/m^3.

3.64 A 6.9 molar solution of KOH in water contains 30% by weight of KOH. Calculate the density of the solution.

3.65 The concentration of SO_2 in the flue gases from a boiler is found to be 0.2 kg/m^3 at STP. Determine the concentration of SO_2 in parts per million by volume at STP. Assume that the gases are perfect.

3.66 A benzene solution of an organic compound *A* analyses 10% of *A*. The molality of the solution is reported to be 0.62. Calculate the following:
(a) The molecular weight of the compound
(b) The mole fraction of the compound in the solution

3.67 An aqueous solution of NaCl contains 20% NaCl. The density of the solution is 1.16 g/mL. 500 ml water of density 1 g/mL is added to 1 litre of the solution. What will be the molality and molarity of the resulting solution?

3.68 A solution of $ZnBr_2$ in water contains 130 g salt per 100 mL solution at 293 K. The specific gravity of the solution is 2.00. Calculate the following:
(a) The concentration of $ZnBr_2$ in mole percent
(b) The concentration of $ZnBr_2$ in weight percent
(c) The molarity
(d) The molality

3.69 The molality of an aqueous solution of LiCl in water is 10. The density of the solution is 1.16 g/mL at 350 K. Determine the following:
(a) The weight percent of LiCl in the solution
(b) The molarity of the solution at 350 K
(c) The normality of the solution at 350 K
(d) The composition of the solution in mole percent

3.70 The molarity of an aqueous solution of $MgCl_2$ at 300 K is 4.0. The specific gravity of the solution is 1.3 at 300 K. Determine the following:
(a) The concentration of $MgCl_2$ in weight fraction
(b) The concentration of $MgCl_2$ in mole fraction
(c) The molality of the solution
(d) The normality of the solution at 300 K

3.71 Pure water and alcohol are mixed to get a 50% alcohol solution. The density (g/mL) of water, alcohol and the solution may be taken to be 0.998, 0.780 and 0.914, respectively at 293 K. Calculate the following:
(a) The volume percent of ethanol in the solution at 293 K
(b) The molarity
(c) The molality

3.72 A solution of potassium chloride in water contains 384 g KCl per litre of the solution at 300 K. The specific gravity of the solution is 1.6. Determine the following:
(a) The concentration in weight percent
(b) The mole fraction of KCl
(c) The molarity of the solution
(d) The molality of the solution

3.73 Silver nitrate reacts with metallic Zn depositing silver according to the reaction

$$Zn + 2AgNO_3 \rightarrow 2Ag + Zn(NO_3)_2$$

With 0.05 kg metallic Zn is added to 10^{-3} m^3 of silver nitrate solution, it was found that after all silver in the solution is deposited in metallic form some Zn metal is left unreacted. The total weight of the unreacted Zn and deposited silver was found to be 0.07 kg. Determine the following:
(a) The mass of silver deposited
(b) The molarity of the silver nitrate solution

Chemical reactions and process calculations

3.74 1 kg nitrogen is mixed with 3.5 m^3 of hydrogen at 300 K and 101.3 kPa and sent to the ammonia converter. The product leaving the converter analyzed 13.7% ammonia, 70.32% hydrogen and 15.98% nitrogen.
(a) Identify the limiting reactant.
(b) What is the percent excess of the excess reactant?
(c) What is the percent conversion of the limiting reactant?

3.75 In the chlorination of ethylene to dichloroethane, the conversion of ethylene is 99.0%. If 94 mol of dichloroethane is produced per 100 mol of ethylene fed, calculate the overall yield and the reactor yield based on ethylene.

$$C_2H_4 + Cl_2 \rightarrow C_2H_4Cl_2$$

3.76 In the manufacture of methanol by the reaction of carbon monoxide and hydrogen, some formaldehyde is also formed as a by-product.

$$CO + 2H_2 \rightarrow CH_3OH$$

$$CO + H_2 \rightarrow HCHO$$

A mixture consisting of CO and H_2 is allowed to react and the product analyzed 2.92% CO, 19.71% methanol, 6.57% formaldehyde and 70.80% hydrogen. Calculate the following:

(a) The percent conversion of limiting reactant
(b) The percent excess of any reactant
(c) Percent yield of methanol

3.77 Water vapour decomposes according to the following reaction

$$H_2O \rightarrow H_2 + \tfrac{1}{2}O_2$$

What is the mole fraction of oxygen in the reaction mixture in terms of the extent of reaction if the system contained n_0 moles of water vapour initially?

3.78 The following reaction occurs in a mixture consisting of 2 mol methane, 1 mol water, 1 mol carbon monoxide and 4 mol hydrogen initially.

$$CH_4 + H_2O \rightarrow CO + 3H_2$$

What is the mole fraction of hydrogen in the reaction mixture in terms of the extent of reaction.

3.79 A system consisting of 2 mol methane and 3 mol water is undergoing the following reaction:

$$CH_4 + H_2O \rightarrow CO + 3H_2$$

$$CH_4 + 2H_2O \rightarrow CO_2 + 4H_2$$

Derive expressions for the mole fraction of hydrogen in terms of the extent of reactions.

3.80 The following gas-phase reactions occur in a mixture initially containing 3 mol ethylene and 2 mol oxygen.

$$C_2H_4 + \tfrac{1}{2}O_2 \rightarrow (CH_2)_2O$$

$$C_2H_4 + 3O_2 \rightarrow 2CO_2 + 2H_2O$$

Derive an expression for the mole fraction of ethylene in terms of the extent of reactions.

3.81 In the vapour-phase hydration of ethylene to ethanol, diethyl ether is obtained as a by-product.

$$C_2H_4 + H_2O \rightarrow C_2H_5OH$$

$$2C_2H_4 + H_2O \rightarrow (C_2H_5)_2O$$

A feed mixture consisting of 55% ethylene, 5% inerts and 40% water is sent to the reactor. The products analyzed 52.26% ethylene, 5.49% ethanol, 0.16% ether, 36.81%

water, and 5.28% inerts. Calculate the conversion of ethylene, yield of ethanol and ether based on ethylene.

3.82 Elemental phosphorous is produced from phosphate rock in an electric furnace by the following reaction:

$$2Ca_3(PO_4)_2 + 10C + 6SiO_2 \rightarrow P_4 + 6CaSiO_3 + 10CO$$

The furnace is fed with 1000 kg phosphate. Carbon charged is 25% in excess and silica charged is 50% in excess. The reaction goes to 95% completion. The unconverted reactants along with the calcium silicate formed constitute the slag. Calculate the following:

(a) The mass of carbon and silica charged (in kilograms)
(b) The amount of phosphorous obtained (in kilograms)
(c) The mass of slag produced (in kilograms)

3.83 Iron pyrites is burned in 50% excess air. The following reaction occurs:

$$4FeS_2 + 11O_2 \rightarrow 2Fe_2O_3 + 8SO_2$$

For 100 kg of iron pyrites charged, calculate the following:

(a) The amount of air supplied (in kilograms)
(b) The composition of exit gases if the percent conversion of iron pyrites is 80%

3.84 Ammonia reacts with sulphuric acid giving ammonium sulphate:

$$2NH_3 + H_2SO_4 \rightarrow (NH_4)_2SO_4$$

(a) 20 m^3 of ammonia at 1.2 bar and 300 K reacts with 40 kg of sulphuric acid. Which is the excess reactant and what is the percent excess?
(b) How much ammonium sulphate is obtained?

3.85 Sulphur dioxide reacts with oxygen producing sulphur trioxide:

$$SO_2 + \tfrac{1}{2}O_2 \rightarrow SO_3$$

In order to ensure complete reaction, twice as much oxygen is supplied than that required theoretically. However, only 60% conversion is obtained. The pressure was 500 kPa and temperature 800 K. 100 kg of SO_2 is charged to the converter. Determine the following:

(a) The volume of pure oxygen supplied at 1.5 bar and 300 K
(b) The volume of sulphur trioxide produced
(c) The volume of gases leaving the converter
(d) The composition of gases leaving the converter
(e) The average molecular weight of the gas leaving the converter

3.86 Nitrogen dioxide shows a tendency to associate and form nitrogen tetroxide.

$$2NO_2 \rightarrow N_2O_4$$

One cubic metre nitrogen dioxide at 100 kPa and 300 K is taken in a closed rigid container and allowed to attain equilibrium at constant temperature and volume. The pressure inside the container has fallen to 85 kPa at equilibrium.

(a) What is the degree of association?
(b) What is the partial pressure of N_2O_4 in the final mixture?

3.87 Ammonium chloride in the vapour phase dissociates into ammonia and hydrogen chloride according to

$$NH_4Cl \rightarrow NH_3 + HCl$$

10.7 g of ammonium chloride is taken in a container. When dissociation is complete and equilibrium is attained the pressure, the volume and temperature of the gas mixture were measured to be 1.2 bar, 7.764×10^{-3} m^3 and 400 K, respectively. Determine the following:
(a) The fraction of ammonium chloride dissociated
(b) The partial pressure of HCl in the products

3.88 A gaseous mixture consisting of 50% hydrogen and 50% acetaldehyde (C_2H_4O) is initially contained in a rigid vessel at a total pressure of 1.0 bar. Methanol is formed according to

$$C_2H_4O + H_2 \rightarrow C_2H_6O$$

After a time it was found that the total pressure in the vessel has fallen to 0.9 bar while the temperature was the same as that of the initial mixture. Assuming that the products are still in the vapour phase, calculate the degree of completion of the reaction.

3.89 Ammonia is made by the reaction between hydrogen and nitrogen according to the following reaction:

$$N_2 + 3H_2 \rightarrow 2NH_3$$

(a) For complete conversion of 100 cubic metres of nitrogen at 20 bar and 350 K, what volume of hydrogen at the same conditions of temperature and pressure is theoretically required?
(b) If hydrogen is available at 5 bar and 290 K, what is the volume required which is stoichiometrically equivalent to 100 m^3 of nitrogen at 20 bar and 350 K?
(c) If the reaction is carried out at 50 bar and 600 K, what volumes of nitrogen and hydrogen at these conditions are theoretically required for producing 1000 kg ammonia and what will be the volume of ammonia produced at the reactor conditions?

3.90 Carbon dioxide dissociates into carbon monoxide and oxygen at 1 bar and 3500 K.

$$CO_2 \rightarrow CO + \tfrac{1}{2}O_2$$

25 L of CO_2 at 1 bar and 300 K is heated to 3500 K at constant pressure. If all gases behave ideally, determine the following:
(a) The final volume of the gas if no dissociation is occurred
(b) The fraction of CO_2 is dissociated if the final volume is found to be 0.35 m^3.

4 Ideal Gases and Gas Mixtures

Operations in which gases and gas mixtures are to be handled are quite numerous in chemical process industries. Chemical conversions involving gases are quite common, the synthesis of ammonia from a mixture of nitrogen and hydrogen, combustion of gaseous fuels in air being two typical examples. Unit operations such as absorption, distillation, drying, adsorption, gas cooling, humidification, etc., transportation and storage of gases are some of the important areas where gases undergo a change of state. The calculations involving the quantities of the material and energy required in such processes as well as the effect of operating parameters on such processes are frequently encountered by chemical engineers.

The gaseous state is the most simple form of matter among the three states—solids, liquids and gas. The general characteristics of a gas are the following:

- Gases do not have a definite shape or volume. These are determined by the shape and volume of the container in which the gas is taken.
- Compared to solids and liquids, gases have very low densities. Consequently, the molecules in a gas are far removed from one another.
- Even a small increase in pressure will compress the gas to a very great extent. In contrast, the effect of pressure on solids and liquids is negligible.

4.1 IDEAL GAS

On a molecular level, an ideal gas may be treated as the one for which (a) the size of the molecules is very small compared to the distance between them so that the volume of the molecules is negligible in comparison with the total volume of the gas and (b) the intermolecular forces are negligibly small. The concept of ideal behaviour coupled with a mechanism to estimate the extent of deviation from ideality is very useful in predicting the behaviour of real systems which are either very complex or are not mathematically amenable. Though no gas is ideal in the true sense of the term, experiments have shown that at low pressures and fairly high temperatures, all ordinary gases behave ideally. A pressure-volume-temperature relationship for a real fluid is not easily explained mathematically. However, for an ideal gas such a relationship is easily established as described by a series of laws such as Boyle's law, Charles' law and Avogadro's hypothesis.

Boyle's law: Boyle in 1662 proposed that for a fixed quantity of a gas the volume is inversely proportional to pressure at constant temperature.

$$PV = \text{constant} \qquad \text{(at fixed } T \text{ and } n) \tag{4.1}$$

where P = pressure, V = volume, T = temperature and n is the number of moles of the gas.

Charles' law: The dependence of the volume of a gas on temperature was reported by Charles in 1787. A quantitative relationship between volume and temperature was proposed by Gay–Lussac in 1802. The laws of Gay–Lussac and Charles state that the volume of a fixed quantity of gas at constant pressure varies directly with temperature.

$$\frac{V}{T} = \text{constant} \qquad \text{(at fixed } P \text{ and } n) \tag{4.2}$$

Avogadro's principle: According to Avogadro's hypothesis equimolal quantities of all gases at the same temperature and pressure occupy the same volume.

4.1.1 Ideal Gas Equation

The ideal gas equation is obtained by combining Boyle's law and Charle's law. Assume that one mole of a gas is initially at temperature T_0, and pressure P_0 and let the volume occupied by the gas under these conditions be V_0. Now let the gas be brought to the final state where the pressure, volume and temperature are, respectively, P, V and T. Let this change be accomplished in two steps. First, an isothermal process in which the temperature is kept constant at T_0, and pressure is changed to P. In the second step, the gas pressure is kept constant at P and the temperature is changed from T_0 to T.

In the first step the volume of the gas changes from V_0 to V' at constant temperature T_0 and Boyle's law is applicable.

$$P_0V_0 = PV' = \text{constant} \tag{4.3}$$

Rearranging this equation, we get

$$V' = \frac{P_0V_0}{P} \tag{4.4}$$

In the second step, the volume changes from V' to V at constant pressure P and therefore, Charles' law is applicable.

$$\frac{V'}{T_0} = \frac{V}{T} = \text{constant} \tag{4.5}$$

Substituting Eqs. (4.4) into Eq. (4.5) and rearranging, we get

$$\frac{PV}{T} = \frac{P_0V_0}{T_0} \tag{4.6}$$

The state of the gas as given by its pressure P, volume V and temperature T may be thought of as attained through the two successive steps, one isothermal and the other isobaric, from an arbitrarily chosen reference state at which pressure is P_0 and temperature is T_0. Assigning standard values for the reference conditions P_0 and T_0, the volume occupied by any gas at the reference state V_0 would be constant according to Avogadro's principle, so that we can take the quantity P_0V_0/T_0 on the right hand side of Eq. (4.6) to be a universal constant which is denoted by R.

$$\frac{PV}{T} = \frac{P_0V_0}{T_0} = R \tag{4.7}$$

or

$$PV = RT \tag{4.8}$$

R is known as the *gas constant* and its magnitude depends on the units of standard pressure P_0, standard volume V_0 and standard temperature T_0. As pointed out in Chapter 2, experiments have shown that the volume occupied by one mole of an ideal gas at standard temperature T_0 = 273.15 K and standard pressure P_0 = 1.01325 × 10^5 N/m^2 is 22.4143 × 10^{-3} m^3/mol. In the FPS system, the molar volume at standard conditions (P_0 = 14.7 psia, T_0 = 492°R) is 359 ft^3/(lb-mol). In SI units the ideal gas constant is equal to 8.314 J/(mol K) or 8.314 kJ/(kmol K). The values of R in other units are given in Table 4.1. Equation (4.8) gives the volume of one mole of an ideal gas at pressure P and temperature T. For n moles of the gas, the total volume V is n times the molar volume, and Eq. (4.8) can be generalized as

$$PV = nRT \tag{4.9}$$

Table 4.1 Ideal gas constant in various units

P	T	V	n	R	Units of R
N/m^2 (Pa)	K	m^3	mol	8.314	J/(mol K)
10^3 N/m^2 (kPa)	K	m^3	kmol	8.314	kJ/(kmol K)
atm	K	L	mol	8.206 × 10^{-2}	L atm/(mol K)
atm	K	m^3	mol	8.206 × 10^{-5}	m^3 atm/(mol K)
atm	K	m^3	kmol	8.206 × 10^{-2}	m^3 atm/(kmol K)
atm	K	cm^3	mol	82.06	cm^3 atm/(mol K)
atm	°R	ft^3	lb mol	0.73	ft^3 atm/(lb-mol °R)
psia (pounds per square inch)	°R	ft^3	lb mol	10.73	ft^3 psia/(lb-mol °R)

Gases such as hydrogen, helium, nitrogen and oxygen at room temperatures follow the perfect gas law very closely and they can be treated as ideal gases for engineering calculations. However, the only state at which the real gases follow ideal behaviour is at zero pressure. For engineering purposes all gases at pressures in the neighbourhood of atmospheric pressure are treated as ideal.

EXAMPLE 4.1 (a) Find the value of the gas constant R in $\frac{\text{m}^3 \text{ mmHg}}{\text{mol K}}$ (b) In the MKS system energy is usually expressed as calories or kilocalories. What will be the value of R when it is expressed in $\frac{\text{cal}}{\text{mol K}}$?

Solution

(a) For one mole of ideal gas at the standard conditions, $R = \dfrac{P_0 V_0}{T_0}$. The standard conditions are P_0 = 760 mm Hg, T_0 = 273.15 K and V_0 = 22.4143 × 10^{-3} m^3/mol. Substituting these values in the above equation, we get

$$R = \frac{760 \times 22.4143 \times 10^{-3}}{273.15} = 6.2364 \times 10^{-2} \frac{m^3 \text{ mm Hg}}{\text{mol K}}$$

(b) When P_0 = 1.01325 × 10^5 N/m^2, T_0 = 273.15 K and V_0 = 22.4143 × 10^{-3} m^3/mol, we get

$$R = \frac{1.01325 \times 10^5 (22.4143 \times 10^{-3})}{273.15} = 8.314 \frac{\text{N.m}}{\text{mol K}} = 8.314 \frac{\text{J}}{\text{mol K}}$$

Since 1 cal (thermo-chemical calorie[1]) = 4.184 J, $R = \dfrac{8.314}{4.184} = 1.987 \dfrac{\text{cal}}{\text{mol K}}$

4.1.2 Calculations for Ideal Gas

In the application of ideal gas equations, two types of problems are generally encountered. In the first type, the pressure (P_1), temperature (T_1) and volume (V_1) at an initial state will be known. The gas is undergoing a change of state and the conditions in the final state are pressure (P_2), temperature (T_2) and volume (V_2). You are required to calculate any one of the three quantities in the final state P_2, T_2 or V_2 for the same mass of the gas. Using Eq. (4.9), we can write for both states,

$$P_1 V_1 = nRT_1$$
$$P_2 V_2 = nRT_2$$

Dividing the first equation by the second and rearranging the result, we get

$$\frac{P_1 V_1}{T_1} = \frac{P_2 V_2}{T_2} \tag{4.10}$$

Equation (4.10) may be used for the solution of the problems of this type.

In the second class of problems, the mass of the gas is also involved. It would be required to find the pressure, volume or temperature of a given mass (that is, given number of moles) of a gas. Alternatively, for given conditions of pressure, volume and temperature, the mass is to be determined. Equation (4.9) can be used to solve such problems. However, in using Eq. (4.9), care should be taken to use the appropriate value of R depending upon the units of the four variables P, V, T and n. Otherwise, the units of P, V, T and n have to be converted into appropriate units so that they are consistent with the units of R chosen for calculations. This difficulty can be circumvented by using the following relation

[1] In older tables of thermodynamic properties of substances, the calorie known as the *thermo-chemical calorie* is equivalent to 4.184 J, whereas in the *International Steam Table* it is equivalent to 4.1868 J.

$$\frac{PV}{T} = \frac{P_0 V_0}{T_0} \tag{4.11}$$

and calculating the volume V_0 occupied by the gas at standard conditions P_0 and T_0

$$V_0 = V \times \frac{P}{P_0} \times \frac{T_0}{T} \tag{4.12}$$

In Eq. (4.12), V_0 is the volume occupied by a given quantity of an ideal gas in the standard state. Any units of pressure, volume, or temperature may be used in this equation, provided the units in both the given state and the standard state are the same. In the second step, the volume V_0 so calculated is converted to moles by dividing it by the normal molal volume. In the usual standard state (the one corresponding to P_0 = 1 atm and T_0 = 273.15 K), the normal molal volume is 22.4143 × 10^{-3} m^3/mol or 359 ft^3/(lb-mol).

EXAMPLE 4.2 Assuming air to behave as an ideal gas, calculate the molar volume of air at 350 K and 1 bar.

Solution One mole of a gas at STP (standard conditions of temperature and pressure) occupies 22.4143 × 10^{-3} m^3 (That is P_0 = 1.01325 bar, T_0 = 273.15 K, and V_0 = 22.4143 × 10^{-3} m^3). Now we can use Eq. (4.11) to calculate the volume (V) at 1 bar (P) and 350 K (T).

$$V = V_0 \times \frac{P_0}{P} \times \frac{T}{T_0}$$

$$= 22.4143 \times 10^{-3} \left(\frac{1.01325}{1}\right) \frac{350}{273.15} = 2.91 \times 10^{-2} \text{ m}^3$$

Thus the molar volume of air at 350 K and 1 bar = 2.91 × 10^{-2} m^3/mol.

[Alternatively, Eq. (4.9) can be used with temperature T = 350 K and pressure P = 1 bar = 10^5 N/m^2. Accordingly, one mole of air at the stated conditions will have a volume V, where

$$V = \frac{RT}{P}$$

$$= \frac{8.314 \times 350}{1 \times 10^5} = 2.91 \times 10^{-2} \text{ m}^3$$

EXAMPLE 4.3 A 150 L oxygen cylinder contains gas at 300 K and 10 bar. What is the mass of oxygen in the cylinder?

Solution Here P =10 bar, T = 300 K and V = 150 L

The volume at STP is found out using Eq. (4.12):

$$V_0 = V \times \frac{P}{P_0} \times \frac{T_0}{T}$$

$$= 150 \left(\frac{10}{1.01325}\right)\left(\frac{273.15}{300}\right) = 1347.9 \text{ L}$$

As the normal molal volume is 22.4143 L/mol, the number of moles of oxygen = 60.14 mol. Thus, the mass of oxygen in the cylinder = 60.14 × 32 = 1924.36 g = 1.9244 kg.

EXAMPLE 4.4 An automobile tyre is inflated to a pressure of 195 kPa at 273 K. If the pressure inside the tyre is not to exceed 250 kPa, what is the maximum temperature to which the tyre may be heated?

Solution Using Eq. (4.10), we can write

$$T_2 = T_1 \left(\frac{V_2}{V_1}\right)\left(\frac{P_2}{P_1}\right)$$

It is assumed that the volume of air inside the tyre remains constant, so that $V_2/V_1 = 1$. Therefore,

$$T_2 = 273 \times 1 \times \frac{250}{195} = 350 \text{ K}$$

EXAMPLE 4.5 Carbon dioxide is contained in a 250 L cylinder at a temperature of 300 K. The gas from the cylinder is allowed to enter an evacuated chamber of capacity 750 L by opening a valve. The flow of gas into the chamber stops when the pressures inside the chamber and the cylinder equal 100 kPa. The temperature of the gas is uniform throughout and it is equal to 310 K. What was the original pressure inside the cylinder?

Solution After CO_2 is admitted into the chamber, the total volume of the gas (V_2) = 1000 L (T_2 = 310 K, P_2 = 100 kPa). Initial volume (V_1) = 250 L (T_1 = 300 K, P_1 = ?). Eq. (4.10) can be rearranged to give

$$P_1 = P_2 \left(\frac{V_2}{V_1}\right)\left(\frac{T_1}{T_2}\right)$$

Substituting the values, we get

$$P_1 = 100 \left(\frac{1000}{250}\right)\left(\frac{300}{310}\right) = 387.1 \text{ kPa}$$

4.2 MIXTURES OF IDEAL GASES

Consider a mixture of different gases *A*, *B*, *C*, etc. at temperature *T* and pressure *P* and let the total volume of the mixture be *V*. Let n_A, n_B, n_C, etc. be the number of moles of the different components. If all the component gases are ideal, then the behaviour of the individual species is not influenced in the presence of other gases in the mixture. This is because of the absence of intermolecular forces in an ideal gas. The pressure exerted by n_A moles of *A* in the mixture will be the same as the pressure exerted when n_A moles of *A* alone occupy a volume *V* at temperature *T*. This is true for the other constituents *B*, *C*, etc. as well. The pressure exerted by a constituent in a mixture of gases when it alone occupies the total volume at a given temperature is called the *partial pressure* of that constituent in the mixture. The partial pressure of constituent i is denoted by p_i. In the case of ideal gases this is also equal to the pressure exerted by the constituent i in the mixture.

Pure-component volume is defined as the volume occupied by a component if it alone were present at the total pressure and temperature of the mixture. The volume V_i occupied by the pure component i at T and P, the temperature and pressure of the mixture, is the pure-component volume of i in the mixture. This is in fact a fictitious volume as the constituents cannot have individual volumes in the mixture as they get distributed throughout the container. However, it can be viewed as the contribution of individual constituents to the total volume of the mixture, in which case it is also known as the *partial volume*. Even this is strictly true only for ideal gases. The partial volume of a component in the mixture is the increase in the volume of the mixture resulting by the addition of a given number of moles of the constituent at the temperature and pressure of the mixture. For ideal gases, this increase in volume will be the same as the pure-component volume because the intermolecular forces are negligible. For real gases, however, there may be volume change on mixing and the partial volume and pure-component volume will be different.

4.2.1 Dalton's Law

Dalton proposed the law of additive pressures in 1805 which states that the *total pressure of a mixture of gases is equal to the sum of the partial pressures of the individual components.* Stated mathematically,

$$P = p_A + p_B + p_C + \ldots = \sum_i p_i$$

Dalton's law is strictly true only for ideal gases. For real gases, the partial pressure of a component as defined above and the pressure exerted by it in the mixture may be different because of the differences arising due to the change in intermolecular forces in the new environment in the mixture as compared to those in the pure state.

If the constituent gases are ideal, the mixture also will be ideal so that

$$PV = nRT \tag{4.13}$$

where

$$n = n_A + n_B + n_C + \ldots = \sum_i n_i$$

n is the total number of moles in the mixture. Also, by the definition of partial pressures,

$$\begin{aligned} p_A V &= n_A RT \\ p_B V &= n_B RT \\ p_C V &= n_C RT \end{aligned} \tag{4.14}$$

Adding the terms on both sides of Eq. (4.14), we get

$$(p_A + p_B + p_C + \ldots)V = (n_A + n_B + n_C + \ldots)RT \tag{4.15}$$

or

$$\left(\sum_i p_i\right)V = \left(\sum_i n_i\right)RT = nRT \tag{4.16}$$

Comparing Eq. (4.13) with Eq. (4.16), we see that

$$P = \sum_i p_i$$

thus verifying the applicability of Dalton's law for ideal gases.

Another useful result that follows from Dalton's law is that the *partial pressure of a component gas in an ideal gas mixture is the product of its mole fraction and the total pressure.* Divide Eq. (4.14) by Eq. (4.13).

$$\frac{p_i}{P} = \frac{n_i}{n} = y_i \tag{4.17}$$

where $i = A, B, C, \ldots$ and y_i is the mole fraction of i in the mixture. Equation (4.17) can be written as

$$p_i = y_i P \tag{4.18}$$

4.2.2 Amagat's Law of Additive Volumes

Amagat's law states that the total volume of a mixture of gases at a given temperature and pressure is equal to the sum of the pure component volumes of the constituents at the same temperature and pressure. This law is strictly true only for ideal gases. It can be stated mathematically as

$$V = V_A + V_B + V_C + \ldots = \sum_i V_i \tag{4.19}$$

Using the definition of pure component volume and assuming the gases are ideal, we can write

$$\begin{aligned} PV_A &= n_A RT \\ PV_B &= n_B RT \\ PV_C &= n_C RT \\ &\ldots\ldots \end{aligned} \tag{4.20}$$

Adding the terms on both sides of Eq. (4.20), we get

$$P(V_A + V_B + V_C + \ldots) = (n_A + n_B + n_C + \ldots)RT \tag{4.21}$$

That is

$$P\left(\sum_i V_i\right) = \left(\sum_i n_i\right) RT = nRT \tag{4.22}$$

Since Eq. (4.13) is applicable for the mixture as a whole, we see that

$$V = \sum_i V_i$$

which shows that Amagat's law is applicable for ideal gases. Also, by dividing Eq. (4.20) by Eq. (4.13), we get

$$\frac{V_i}{V} = \frac{n_i}{n} = y_i \qquad \text{(for } i = A, B, C, \ldots) \tag{4.23}$$

where y_i is the mole fraction of gas i in the mixture. Equation (4.23) means that the volume fraction and mole fraction in an ideal gas mixture are identical, the result we have used in Chapter 3. Combining Eqs. (4.17) and (4.23) the following general result of great applicability is deduced:

$$\text{mole fraction} = \text{volume fraction} = \text{pressure fraction}$$

4.2.3 Average Molecular Weight

Through the use of average molecular weight, a complex gas mixture can be treated as though it is a pure gas. If m is the mass of a mixture of gases and M_{av} its average molecular weight, then the number of moles of the mixture n is given by

$$m = n\, M_{av} \tag{4.24}$$

We can calculate the average molecular weight of a mixture of gases knowing its molal composition. Assume a suitable basis, say 100 moles of the mixture, and evaluate the number of moles of each constituent in the mixture. The mass of each constituent in the mixture is obtained by multiplying the number of moles by the respective molecular weight.

$$m_i = n_i M_i \tag{4.25}$$

where m_i is the mass and n_i is the number of moles of constituent i in the mixture and M_i its molecular weight. The total mass of the mixture is the sum of the individual masses.

$$m = \Sigma\, m_i = \Sigma\, n_i M_i \tag{4.26}$$

The average molecular weight is calculated as

$$M_{av} = \frac{m}{n} \tag{4.27}$$

4.2.4 Density and Specific Gravity

The density of a mixture of gases is the mass per unit volume of the mixture at the specified conditions of temperature and pressure. It is expressed in kg/m^3, g/L, lb/ft^3, etc. From the molal composition of the mixture, the mass of n moles of the mixture can be evaluated by the method outlined above. The volume of n moles of the gas mixture at the specified temperature and pressure is determined by using Eq. (4.9) or Eq. (4.11). The density is now evaluated as the ratio of mass to volume.

Let M_{av} be the average molecular weight of the gas mixture. Then, the mass of n moles of the gas is $m = nM_{av}$. Volume is obtained from the ideal gas equation $V = \dfrac{nRT}{P}$. Therefore,

$$\text{density} = \frac{m}{V} = \frac{PM_{av}}{RT} \tag{4.28}$$

As volume of gas depends strongly on temperature and pressure, the density is very much dependent on the temperature and pressure of the gas. When the standard temperature and pressure are used in the computation, the value so calculated is the density of the gas at STP.

The specific gravity of a gas is the ratio of the density of the gas to the density of dry air at the same temperature and pressure. Using Eq. (4.28) we can show that the specific gravity of a gas is the ratio of the average molecular weight of the gas to that of dry air.

EXAMPLE 4.6 An analysis of the vent gases from the chlorinator in a plant for making chlorinated rubber showed 70% by volume HCl, 20% by volume Cl_2 and the rest CCl_4. Determine the following:

(a) The percent composition by weight
(b) The average molecular weight of the gas
(c) The density at standard conditions (in kg/m^3)

Solution For gas mixtures, the volume percent and mole percent are the same. Therefore, the gas contains 70% (mole) HCl, 20% (mole) Cl_2 and 100 – 70 – 20 = 10% (mole) CCl_4.

Basis: 100 kmol gas mixture

(a) The calculations are presented as follows:

Component	*No. of* kmol	*Molecular weight*	*Mass in* kg	*Weight percent*
HCl	70	36.453	2551.71	46.33
Cl_2	20	70.906	1418.12	25.75
CCl_4	10	153.812	1538.12	27.92
Total			5507.95	100.00

Mass in kg is obtained by multiplying the number of kmol of each constituent by its molecular weight. Weight percent is calculated by dividing the mass in kg of each component by the total mass and by multiplying by 100. For example, for chlorine, the mass is 1418.12 kg and the mass percent is $\frac{1418.12}{5507.95} \times 100 = 25.75\%$.

(b) Average molecular weight of the gas is obtained by dividing the total mass by the number of moles. Here, 100 kmol has a mass = 5507.95 kg. Therefore,

$$\text{average molecular weight} = \frac{5507.95}{100} = 55.08$$

(c) 100 kmol of an ideal gas occupies a volume of 100 × 22.414 m^3 at STP. We have already found out that the mass of 100 kmol is 5507.95 kg. Since density is mass/volume,

$$\text{density at standard conditions} = \frac{5507.95}{100 \times 22.414} = 2.4574 \text{ kg/m}^3$$

EXAMPLE 4.7 Natural gas is piped from the well at 300 K and 400 kPa. The gas is found to contain 93.0% methane, 4.5% ethane and the rest nitrogen. Calculate the following:

(a) The partial pressure of nitrogen
(b) The pure-component volume of ethane in 10 m^3 of the gas
(c) The density at standard conditions in kg/m^3
(d) The density of the gas as piped in kg/m^3

(e) The average molecular weight of the gas
(f) The composition in weight percent

Solution

(a) Mole percent of nitrogen = 100 – 93.0 – 4.5 = 2.5. Therefore,

$$\text{mole fraction of nitrogen} = 2.5/100 = 0.025$$

Using Eq. (4.18),

$$\text{partial pressure of nitrogen} = \text{mole fraction of nitrogen} \times \text{total pressure}$$
$$= 0.025 \times 400 = 10 \text{ kPa}$$

(b) Using Eq. (4.23),

$$\text{pure-component volume of ethane} = \text{mole fraction of ethane} \times \text{total volume}$$
$$= (4.5/100) \times 10 = 0.45 \text{ m}^3$$

(c) *Basis:* 100 mol natural gas

The calculation are presented as follows:

Component	*Molecular weight*	*No. of moles*	*Weight* (g)	*Weight fraction*
Methane	16.032	93.0	16.032 × 93.0 = 1490.98	1490.98/1696.24 = 0.8790
Ethane	30.048	4.5	30.048 × 4.5 = 135.22	135.22/1696.24 = 0.0797
Nitrogen	28.014	2.5	28.014 × 2.5 = 70.04	(1 – 0.8790 – 0.0797) = 0.0413
Total		100.0	1696.24	1.00

$$\text{density} = \frac{\text{mass}}{\text{volume}}$$

The mass of 100 moles = 1696.24 g = 1.696 kg.
The volume of 100 moles at STP is

$$100(22.414 \times 10^{-3}) = 2.2414 \text{ m}^3$$

Therefore,

$$\text{density} = \frac{1.696}{2.2414} = 0.7567 \text{ kg/m}^3$$

(d) The volume at 300 K and 400 kPa is

$$V_0 \times \frac{P_0}{P} \times \frac{T}{T_0} = 2.2414\left(\frac{101.325}{400}\right)\left(\frac{300}{273.15}\right) = 0.6236 \text{ m}^3$$

Therefore, the density at 300 K and 400 kPa is

$$\frac{1.696}{0.6236} = 2.7197 \text{ kg/m}^3$$

(e) The mass of 100 moles of the gas = 1696 g. Therefore,

$$\text{average molecular weight} = 1696.0/100 = 16.96$$

(f) Composition in weight percent is obtained by multiplying the weight fraction by 100. Thus

Methane = 87.90%, Ethane = 7.97%, and Nitrogen = 4.13%

4.2.5 Calculations Involving Change in the Mass

Material and energy balance calculations usually deal with physical operations in which a gas stream undergoes change in mass. This change in mass results due to condensation of certain constituents in the mixture, due to absorption of certain soluble constituents in a liquid solvent, due to evaporation of liquids into a gas stream, or due to the adsorption of certain constituents on a solid adsorbent surface. Separation techniques such as humidification, dehumidification, gas absorption, drying and adsorption are some examples of such operations. Chapter 9 gives brief explanation of the principle behind these operations. In all these operations it can be seen that even when the total mass of the mixture changes, the mass of certain components in the gas remains unchanged. For example, in the removal of water vapour present in an inert gas stream by adsorbing the vapour on silica gel, the amount of inert gas in the stream entering the adsorber as well as in the stream leaving the adsorber will be the same. Only the amount of water vapour is reduced due to adsorption. In such operations where the amount of one of the components in the mixture remains unchanged, it will be convenient to base the calculations on such components. Such materials are known as *tie elements* or *key components*. If the temperature and pressure of the gases entering and leaving the unit are the same, the change in the volume of the gas directly gives the change in the mass of the non-tie elements. If the conditions at the entrance and the exit are different, appropriate corrections are necessary. There are several methods for solving the problems involving the changes in the mass of the gas mixture. The choice of a suitable method depends largely on the type of data available. Example 4.8 illustrates the different techniques used for solving such problems.

EXAMPLE 4.8 100 m^3/h of an ammonia-air mixture containing 20% ammonia by volume is admitted to an absorption column at 120 kPa and 300 K in order to recover ammonia by absorbing in water. 90 percent of ammonia in the entering gas is absorbed and the gas leaves the column at 100 kPa and 280 K. What is the volume of gas leaving in one hour?

Solution *Basis:* 1 hour operation

Calculation on a mole basis: The number of moles of the gas entering is determined using the ideal gas equation. Since volume percent and mole percent are the same, moles of ammonia and air in the gas entering the column and the moles of ammonia absorbed can be calculated. The moles of air entering will be equal to the moles of air leaving as air is not absorbed. The sum of the moles of air leaving and the moles of ammonia that is left unabsorbed gives the total moles of the gas leaving. This is converted to volume at the desired temperature and pressure using the ideal gas equation.

Gas entering = 100 m^3 at 120 kPa and 300 K.

Using Eq. (4.12), the volume at standard conditions is determined which is divided by 22.414 to give the number of moles of the gas entering.

$$\text{Moles of gas entering} = \frac{V}{22.4143} \times \frac{P}{P_0} \times \frac{T_0}{T}$$

$$= \frac{100}{22.4143} \times \frac{120}{101.325} \times \frac{273.15}{300} = 4.8108 \text{ kmol}$$

Moles of air entering = moles of air leaving = 0.80 × 4.8108 = 3.8486 kmol

Moles of ammonia entering = 0.20 × 4.8108 = 0.9622 kmol

Moles of ammonia absorbed = 0.90 × 0.9622 = 0.8659 kmol

Moles of ammonia leaving = 0.9622 – 0.8659 = 0.0963 kmol

Total gas leaving = 3.8486 + 0.0963 = 3.9449 kmol

Volume of gas leaving at standard conditions = 3.9449 × 22.4143 = 88.422 m^3

Using Eq. (4.11), the volume of gas leaving at 100 kPa and 280 K is

$$V_0 \times \frac{P_0}{P} \times \frac{T}{T_0} = 88.422 \times \frac{101.325}{100} \times \frac{280}{273.15} = 91.84 \text{ m}^3$$

Calculation on a pure-component volume basis: It is assumed that the absorption of ammonia (that results in the reduction in the number of moles of ammonia in the gas) occurs at the temperature and pressure of the entering gas. As there is no change in the number of moles of air, the pure component volume of air remains the same as that before absorption. The total volume of the gas decreases by the pure-component volume of ammonia absorbed, which is determined directly from the given data. The sum of the pure component volume of ammonia remaining unabsorbed and the pure-component volume of air makes up the total volume of the gas after absorption at the initial temperature and pressure. This is then converted to the volume at the final temperature and pressure.

Gas entering (120 kPa, 300 K):

Pure-component volume of air = 80 m^3

Pure-component volume of ammonia = 20 m^3

Gas after absorption (120 kPa, 300 K):

Pure-component volume of air = 80 m^3

Pure component volume of ammonia absorbed = 0.90 × 20 = 18 m^3

Pure-component volume of ammonia remaining = 20 – 18 = 2 m^3

Total volume after absorption (120 kPa, 300 K) = 80 + 2 = 82 m^3

Equation (4.10) gives

$$V_2 = V_1 \times \frac{P_1}{P_2} \times \frac{T_2}{T_1}$$

Therefore, the volume of gas leaving (100 kPa, 280 K) is

$$82 \times \frac{120}{100} \times \frac{280}{300} = 91.84 \text{ m}^3$$

Calculation on a partial pressure basis: Assume that the absorption is taking place at the temperature of the gas entering without any change in the total volume. Therefore the reduction in the number of moles due to absorption will lead to the reduction in the partial pressure of ammonia in direct proportion to the decrease in its number of moles. However, the partial pressure of the tie element—the air—remains the same as that before absorption. At this pressure, the volume occupied by air is equal to the total volume of the gas. The total volume of the gas leaving at the specified temperature and pressure can now be computed using Eq. (4.10). Alternatively, the total pressure after absorption can be found out as the sum of the partial pressures of the air and the partial pressure of ammonia that is not absorbed. This can be used to calculate the total volume of the gas leaving the absorber.

Since the partial pressure is given by the product of mole fraction (or, volume fraction) and the total pressure, we have for the gas entering the absorber,

$$\text{Partial pressure of air} = 0.80 \times 120 = 96 \text{ kPa}$$

$$\text{Partial pressure of ammonia} = 0.20 \times 120 = 24 \text{ kPa}$$

$$\text{Partial pressure of ammonia after absorption} = 24 \times 0.10 = 2.4 \text{ kPa}$$

$$\text{Total pressure after absorption} = 96 + 2.4 = 98.4 \text{ kPa}$$

We have 100 m^3 of the gas at 98.4 kPa and 300 K after absorption. Therefore using Eq. (4.10), the volume of gas leaving the absorber at 100 kPa and 280 K

$$V_1 \times \frac{P_1}{P_2} \times \frac{T_2}{T_1} = 100 \times \frac{98.4}{100} \times \frac{280}{300} = 91.84 \text{ m}^3$$

EXAMPLE 4.9 Air is to be dehumidified by condensing the water vapour present in it by cooling at constant pressure. 100 m^3 of air at 100 kPa and 300 K contains water vapour which exerts a partial pressure of 4 kPa. Keeping the pressure constant, this air is cooled to 275 K and the condensed water is removed. The partial pressure of water in the air after cooling is found to be 1.8 kPa. Calculate the following:

(a) The volume of air after dehumidification in m^3

(b) The mass of water removed in kg

Solution *Basis:* 100 m^3 air entering

(a) In the partial pressure method, the change in the number of moles is taken into account by a corresponding change in the pressure:
Partial pressure of dry air entering = 100 – 4 = 96 kPa
Partial pressure of dry air leaving = 100 – 1.8 = 98.2 kPa
Dry air at 96 kPa and 300 K occupies a volume 100 m^3. Therefore, the volume occupied by the dry air at 98.2 kPa and 275 K is

$$V_1 \times \frac{P_1}{P_2} \times \frac{T_2}{T_1} = 100 \times \frac{96}{98.2} \times \frac{275}{300} = 89.61 \text{ m}^3$$

This will be the volume occupied by the air-water vapour mixture after condensation.

(b) The volume of water vapour entering at its partial pressure is the same as the volume of total gas entering. Therefore,

$$\text{moles of water vapour entering} = \frac{\text{volume of water vapour at STP}}{\text{normal molal volume}}$$

$$\frac{V_0}{22.414} = \frac{V}{22.414} \times \frac{P}{P_0} \times \frac{T_0}{T}$$

$$= \frac{100}{22.414} \times \frac{4}{101.325} \times \frac{273.15}{300} = 0.1604 \text{ kmol}$$

Similarly, the volume of water vapour leaving at its partial pressure is the volume of total mixture leaving. Therefore, the moles of water vapour leaving is

$$\frac{V_0}{22.414} = \frac{V}{22.414} \times \frac{P}{P_0} \times \frac{T_0}{T}$$

$$= \frac{89.61}{22.414} \times \frac{1.8}{101.325} \times \frac{273.15}{275} = 0.0705 \text{ kmol}$$

Mass of water vapour removed = moles × molecular weight of water

$$= (0.1604 - 0.0705)18.02 = 1.619 \text{ kg}$$

EXAMPLE 4.10 The feed to an absorption column consists of 20% H_2S and the balance inerts. Only H_2S is removed from the gas by absorbing in an alkaline solution. The gas enters the absorber at 600 kPa and 310 K and leaves at 500 kPa and 290 K containing 2% H_2S. If H_2S is removed at a rate of 100 kg/h, calculate the following:

(a) The volume of gas (in m^3) entering per hour
(b) The volume of gas (in m^3) leaving per hour
(c) Percentage recovery of H_2S

Solution *Basis:* 100 m^3 gas entering

The volume of gas entering at STP is

$$V_0 = V \times \frac{P}{P_0} \times \frac{T_0}{T} = 100 \times \frac{600}{101.325} \times \frac{273.15}{310} = 521.76 \text{ m}^3$$

Moles of gas entering is

$$\frac{V_0}{22.414} = \frac{521.76}{22.414} = 23.28 \text{ kmol}$$

Mole fraction of H_2S in the gas = volume fraction = 0.20

Moles of H_2S entering = 0.20 × 23.28 = 4.656 kmol

Moles of inerts entering = 23.28 – 4.656 = 18.624 kmol

Moles of inerts leaving will be the same as those entering. But inerts constitute 98% of the gas leaving. Therefore,

$$\text{Moles of total gas leaving} = \frac{18.624}{0.98} = 19.00 \text{ kmol}$$

Moles of H_2S in the gas leaving = 0.02 × 19.00 = 0.38 kmol

Mass of H_2S absorbed = moles absorbed × molecular weight

$$= (4.656 - 0.38)34.08 = 145.726 \text{ kg}$$

(a) 145.726 kg of H_2S is absorbed per 100 m^3 of gas entering. For the removal of 100 kg/h of H_2S, gas entering per hour is

$$\frac{100}{145.726} \times 100 = 68.622 \text{ m}^3/\text{h}$$

(b) 19 kmol gas leaves per 100 m^3 of gas entering, moles of gas leaving per 68.622 m^3/h of gas entering is

$$\frac{19}{100} \times 68.622 = 13.038 \text{ kmol/h}$$

The volume of gas leaving at STP is

$$13.038 \times 22.414 = 292.23 \text{ m}^3/\text{h}$$

The volume of gas leaving at 500 kPa and 290 K is

$$V_0 \times \frac{P_0}{P} \times \frac{T}{T_0} = 292.05 \times \frac{101.325}{500} \times \frac{290}{273.15} = 62.87 \text{ m}^3/\text{h}$$

(c) For 4.656 kmol H_2S in the gas entering, 0.38 kmol is present in the gas leaving. Therefore, the percent recovery of H_2S is

$$\frac{(4.656 - 0.38)}{4.656} \times 100 = 91.84\%$$

4.3 REACTIONS INVOLVING GASES

Gases take part in chemical reactions as reactants, products or as both. The law of combining proportions indicates that the number moles of the substances taking part in a chemical reaction bears a simple ratio of whole numbers. For example, the ratio of moles of nitrogen to hydrogen reacting to form ammonia is 1:3. If both nitrogen and hydrogen are at the same conditions of temperature and pressure, the ratio of their volumes will also be 1:3. This is because of Avogadro's law which means that equal volumes of all gases contain the same number of molecules. Thus, 1 m^3 of nitrogen reacts with 3 m^3 of hydrogen forming 2 m^3 of ammonia if the temperature and pressure are constant. If the temperature and pressure of the reactants and products are not the same, the volumes may be corrected using the ideal gas equation.

The use of molal quantities is recommended for performing these calculations. The advantage is that for calculations carried out in molal quantities, the complications arising due to the change in volumes accompanying the changes in temperature and pressure can be avoided. Once the results are obtained in molal quantities, they may be converted to the volumetric basis, using the normal molal volume constant and the ideal gas law. The choice of a suitable basis for the calculation depends on the type of data available. A unit weight or unit mole (a basis of 100 moles is preferred for gas mixtures) of a substance may be chosen as the basis depending upon whether the given data for the substance is in weight units or volume units. The following examples will illustrate the calculations involving ideal gases in chemical reactions.

EXAMPLE 4.11 Ammonia is made by the reaction between hydrogen and nitrogen according to the following reaction:

$$N_2 + 3H_2 \rightarrow 2NH_3$$

(a) For complete conversion of 100 cubic metres of nitrogen at 20 bar and 350 K, what volume of hydrogen at the same conditions of temperature and pressure is theoretically required?

(b) What is the volume of hydrogen at 5 bar and 290 K required which is stoichiometrically equivalent to 100 m^3 of nitrogen at 20 bar and 350 K?

(c) If the reaction is carried out at 50 bar and 600 K, what volumes of nitrogen and hydrogen at these conditions are theoretically required for producing 1000 kg of ammonia and what will be the volume of ammonia produced at the reactor conditions?

Solution

(a) According to Avogadro's principle, equal volumes of all gases under similar conditions of temperature and pressure contain the same number of moles. Since one mole nitrogen requires three moles hydrogen, 100 m^3 of nitrogen at 20 bar and 350 K requires 300 m^3 hydrogen at the same temperature and pressure.

(b) The volume of hydrogen that is stoichiometrically equivalent to 100 m^3 of nitrogen at 20 bar and 350 K is 300 m^3 at 20 bar and 350 K. Therefore, at 5 bar and 290 K the equivalent volume is

$$V_1 \times \frac{P_1}{P_2} \times \frac{T_2}{T_1} = 300 \times \frac{20}{5} \times \frac{290}{350} = 994.29 \text{ m}^3$$

(c) *Basis:* 1000 kg ammonia

Moles of ammonia = 1000/17.03 = 58.72 kmol

2 moles of ammonia requires 1 mole of nitrogen and 3 moles of hydrogen. Therefore, 58.72 kmol requires 29.36 kmol nitrogen and 88.08 kmol hydrogen.

29.36 kmol nitrogen will have a volume 29.36 × 22.414 m^3 at STP. At 50 bar and 600 K, the volume of nitrogen is

$$V_0 \times \frac{P_0}{P} \times \frac{T}{T_0} = 29.36 \times 22.414\left(\frac{1.01325}{50}\right)\left(\frac{600}{273.15}\right) = 29.29 \text{ m}^3$$

$$\text{Volume of hydrogen} = 29.29 \times 3 = 87.88 \text{ m}^3$$

$$\text{Volume of ammonia} = 29.29 \times 2 = 58.59 \text{ m}^3$$

EXAMPLE 4.12 Producer gas has the following composition by volume: Carbon monoxide = 25.0%, carbon dioxide = 4.0%, oxygen = 3.0% and nitrogen = 68.0%.

(a) Determine the volume of the gas at 1 bar and 290 K per kg carbon.

(b) 100 m^3 of the gas at 1 bar and 290 K is to be burned with air 20% in excess of that is theoretically required for complete combustion. What volume of air is required at 1 bar and 290 K? (Assume that air is a mixture of 21.0% oxygen and 79.0% nitrogen.)

(c) For the conditions in part (b), what would be the composition of the gas after complete combustion?

Solution *Basis:* 100 kmol producer gas

(a) The gas contains 25.0 kmol CO_2 and 4.0 kmol CO. Moles of carbon present = 25.0 + 4.0 = 29.0 kmol.

Therefore, the mass of carbon per 100 kmol gas is

$$29.0 \times 12 = 348.0 \text{ kg}$$

Moles of producer gas that contains 1 kg of carbon is

$$\frac{100}{348} = 0.2874 \text{ kmol}$$

The volume of 0.2874 kmol gas at STP is

$$0.2874 \times 22.414 = 6.44 \text{ m}^3$$

The volume of gas at 1 bar and 290 K is

$$V_0 \times \frac{P_0}{P} \times \frac{T}{T_0} = 6.44\left(\frac{1.01355}{1}\right)\left(\frac{290}{273.15}\right) = 6.93 \text{ m}^3$$

(b)
$$CO + \frac{1}{2}O_2 \rightarrow CO_2$$

100 kmol gas contains 25.0 kmol CO which requires 12.5 kmol oxygen for complete combustion. 3.0 kmol oxygen is already present in the gas mixture. Therefore, the theoretical requirement of oxygen is 12.5 – 3.0 = 9.5 kmol.

Oxygen supplied is 20% in excess of the theoretical requirement. Hence,

$$\text{oxygen supplied} = 1.2 \times 9.5 = 11.4 \text{ kmol}$$

Since air contains 21.0% oxygen, air supplied is

$$11.4 \times \frac{100}{21} = 54.286 \text{ kmol}$$

100 kmol gas requires 54.286 kmol air. Since air and the fuel are at the same conditions of temperature and pressure, 100 m^3 of the fuel gas will require 54.286 m^3 of air.

(c) After complete combustion, the products contain:

Carbon dioxide = carbon dioxide present in the fuel
+ carbon dioxide produced by combustion of CO
= 4.0 + 25.0 = 29.0 kmol

Oxygen = oxygen supplied + oxygen present in the fuel – oxygen consumed
= 11.4 + 3.0 – 12.5 = 1.9 kmol

Nitrogen = nitrogen present in the fuel + nitrogen in the air supplied
= 68.0 + (54.286 – 11.4) = 110.886 kmol

Total products = 29.0 + 1.9 + 110.886 = 141.786 kmol

Percent composition (mole):

$$\text{Carbon dioxide} = \frac{29.0}{141.786} \times 100 = 20.45\%$$

$$\text{Oxygen} = \frac{1.9}{141.786} \times 100 = 1.34\%$$

$$\text{Nitrogen} = \frac{110.886}{141.786} \times 100 = 78.21\%$$

Percent composition by volume and percent composition by mole are identical.

EXAMPLE 4.13 Chlorine is produced by the oxidation of hydrogen chloride gas with air

$$4HCl + O_2 \rightarrow 2Cl_2 + 2H_2O$$

The reaction is carried out at 1.2×10^5 N/m^2 and 400 K. 50 percent excess air is used and the reaction is only 80 percent complete. Calculate the following:

(a) The volume of air admitted per 100 m^3 of HCl if both air and HCl enter the reactor at 1.0 bar and 290 K
(b) The volume of gases leaving the reactor per 100 m^3 of HCl entering
(c) Kilograms of chlorine produced per 100 m^3 of HCl entering
(d) The percent composition by volume of the exit gas on a dry basis

Solution *Basis:* 1 mol HCl entering the reactor

(a) Oxygen required = 0.25 mol

$$\text{Oxygen supplied} = 1.5 \times 0.25 = 0.375 \text{ mol}$$

$$\text{Air supplied} = \frac{100}{21} \times 0.375 = 1.786 \text{ mol}$$

1 mol HCl requires 1.786 mol air. Since air and HCl enter at same temperature and pressure, the volumes of HCl and air bear the same ratio 1 : 1.786. Therefore, for 100 m^3 of HCl, the air admitted is

$$\frac{1.786}{1} \times 100 = 178.6 \text{ m}^3$$

(b) Only 80% of the HCl is converted. HCl converted is 0.8 mol. Oxygen consumed = 0.8/4 = 0.2 mol, chlorine produced = 0.8/2 = 0.4 mol, and water produced = 0.4 mol. Therefore, the gases leaving the reactor are as follows:

Compound	*No. of moles*
HCl	1 – 0.8 = 0.20
Oxygen	0.375 – 0.20 = 0.175
Chlorine	0.40
Water	0.40
Nitrogen	1.786 – 0.375 = 1.411
Total	2.586

Moles of gas leaving per mol HCl entering is 2.586 mol.

For 100 m^3 of HCl entering at 1 bar and 290 K, the volume of gas leaving will be 2.586 × 100 = 258.6 m^3 at 1 bar and 290 K. Therefore, volume of gas leaving at 1.2 bar and 400 K is

$$V_1 \times \frac{P_1}{P_2} \times \frac{T_2}{T_1} = 258.6 \times \frac{1}{1.2} \times \frac{400}{290} = 297.24 \text{ m}^3$$

(c) Per mol HCl entering, chlorine produced is 0.4 mol. That is, per 100 m^3 of HCl entering, chlorine produced will be 40 m^3 at 1 bar and 290 kPa.

This is equivalent to

$$40 \times \frac{1}{1.01325} \times \frac{273.15}{290} = 37.18 \text{ m}^3 \text{ at STP}$$

or
$$37.18/22.414 = 1.66 \text{ kmol}$$

Therefore, the mass of chlorine produced = 1.66 × 70.90 = 117.62 kg

(d) The percent composition on a dry basis does take into account the water present in the mixture. On a water-free basis, the composition is evaluated as follows:

Compound	*No. of moles*	*Mole percent*
HCl	1 – 0.8 = 0.20	(0.2/2.186) × 100 = 9.15
Oxygen	0.375 – 0.20 = 0.175	(0.175/2.186) × 100 = 8.00
Chlorine	0.40	(0.40/2.186) × 100 = 18.30
Nitrogen	1.786 – 0.375 = 1.411	(1.411/2.186) × 100 = 64.55
Total	2.186	100.00

EXAMPLE 4.14 Carbon dioxide dissociates into carbon monoxide and oxygen at 1 bar and 3500 K.

$$CO_2 \rightarrow CO + \frac{1}{2}O_2$$

25 L of CO_2 at 1 bar and 300 K is heated to 3500 K at constant pressure. If all gases behave ideally, determine the following:

(a) The final volume of the gas if no dissociation had occurred.

(b) If the final volume is found to be 0.35 m^3 what fraction of CO_2 is dissociated?

Solution

(a) If no dissociation had occurred, the final volume is given by the ideal gas equation

$$V_1 \times \frac{P_1}{P_2} \times \frac{T_2}{T_1} = 25 \times 10^{-3} \left(\frac{1}{1}\right)\left(\frac{3500}{300}\right) = 0.292 \text{ m}^3$$

(b) Moles of CO_2 initially taken is

$$\frac{V_0}{22.414} = \frac{V}{22.414} \times \frac{P}{P_0} \times \frac{T_0}{T}$$

$$= \frac{25}{22.414} \times \frac{1}{1.01325} \times \frac{273.15}{300} = 1.00 \text{ mol}$$

Let z be fraction of CO_2 dissociated. Then after dissociation the mixture contains:
$CO_2 = 1.00 - z$ mol, $CO = z$ mol, $O_2 = z$ mol
Therefore, the total moles after decomposition = $(1.00 + 0.5z)$ mol. (A)
Since the gas mixture occupies 0.35 m^3 at 1 bar and 3500 K, the number of moles present is

$$\frac{V_0}{22.414} = \frac{V}{22.414} \times \frac{P}{P_0} \times \frac{T_0}{T}$$

$$= \frac{350}{22.414} \times \frac{1}{1.01325} \times \frac{273}{3500} = 1.20 \text{ mol} \qquad \text{(B)}$$

Comparing (A) and (B), we get

$$(1.0 + 0.5z) = 1.2$$

Solving this, we get $z = 0.40$, which means 40% of CO_2 is converted.

EXERCISES

Ideal gas

4.1 How many kilograms of nitrogen will occupy a volume of 1000 m^3 at a pressure of 1.2 bar and a temperature of 400 K?

4.2 A gas occupies a volume of 100 m^3 at 375 K and 1.25×10^5 N/m^2. What volume will the gas occupy at 275 K and 1 bar?

4.3 20 kg of chlorine is to be stored in a container at 0.9 bar and 295 K. What should be the volume of the container?

4.4 (a) How many kilograms of liquid water will be obtained by the complete condensation of 5 m^3 water vapour at 200 kPa and 1000 K?
(b) The steam tables give the specific volume of water vapour to be 2.937 m^3/kg. Using this data calculate the mass of water vapour obtained and compare with the results in part (a).

4.5 Steam tables give the specific volume of superheated steam at 1200 kPa and 925 K to be 0.3534 m^3/kg. How does it compare with the value obtained by the ideal gas equation?

4.6 It is desired to compress 7.00 kg of nitrogen gas to a volume of 0.5 m^3 at 305 K. Calculate the pressure required.

4.7 A fire-extinguisher tank can hold 1 m^3 of gas. The tank contains 10 kg of CO_2 at 300 K. What will be the pressure of the gas in N/m^2?

4.8 A 20 L glass vessel contains 28 g nitrogen. If the vessel cannot withstand a pressure exceeding 120 kPa, up to what temperature the gas in the vessel can be heated?

4.9 Calculate the density at standard temperature and pressure of (a) methane (b) sulphur dioxide.

4.10 A 50 m diameter spherical balloon contains air at 300 K and 1 bar. How many kilograms of air is present in the balloon? Take the molecular weight of air to be 29.

4.11 Find the mass of methane in a 100 m^3 gas holder at 293 K if the pressure on the holder is 20 kPa gauge. The barometric pressure is 100 kPa.

4.12 How many kilograms of hydrogen can be stored at 325 K and 100 bar in a compressed gas cylinder that normally holds 0.5 kg hydrogen at 300 K and 130 bar?

4.13 A tank contains oxygen gas at 1.40×10^7 N/m^2. On withdrawing 140 L of oxygen measured at 1.01×10^5 N/m^2, the pressure in the tank is observed to fall to 1.30×10^5 N/m^2. Assume that there is no change in the temperature of the gas. What is the volume of the tank in m^3?

4.14 A cooking gas cylinder can withstand a pressure of 15.0 atm. The pressure inside the cylinder is 12 atm at 300 K. During a sudden fire in the building the temperature starts rising. At what temperature will the cylinder explode?

4.15 Calculate the volume occupied at 400 K and 2 atm pressure by the gas evolved from 2 ml of solid CO_2. The density of solid CO_2 = 1500 kg/m^3.

Mixtures of ideal gases

4.16 A vessel of volume 0.02 m^3 contains a mixture of hydrogen and helium at 320.15 K and 4.15×10^5 N/m^2 pressure. The mass of the mixture is 10 g. Calculate the individual masses of hydrogen and helium in the mixture.

4.17 The flow rate of a gas mixture consisting of 60% ethane, 25% hydrogen and 15% carbon dioxide is found to be 200 m^3/h at 300 K and 1.2 bar.
(a) Determine the composition of the gas in weight percent
(b) Express the composition in mol/L.
(c) What is the flow rate in kg/h?

4.18 2.0 cubic metres of oxygen at 200 kPa and 280 K is mixed with 10 kg of nitrogen at 95 kPa and 300 K and the resulting mixture is brought to 110 kPa and 290 K.
(a) What is the partial pressure of oxygen in the final mixture?
(b) What is the final volume of the mixture?

4.19 In an air-naphthalene ($C_{10}H_8$) mixture at 70 kPa and 350 K, the partial pressure of naphthalene is found to be 7 kPa. What volume of this mixture will contain 1 kg of naphthalene?

4.20 15 kg each of nitrogen and hydrogen are mixed together at 300 kPa and 298 K. Determine the following:
(a) The partial pressure of nitrogen
(b) The pure-component volume of nitrogen
(c) The specific volume of the mixture (m^3/kg)

4.21 0.66 g of a gas containing 81.8% C and 18.2% H by weight measures 0.369 L at 100 kPa and 295 K. What is the formula of the hydrocarbon?

4.22 N_2O_4 decomposes to NO_2 at high temperatures. 20 g of N_2O_4 when heated to 373 K at 96 kPa, is found to occupy a volume of 0.0125 m^3. Assuming the applicability of ideal gas law, calculate the percent dissociation of N_2O_4 to NO_2.

4.23 In the manufacture of formaldehyde by the Monsanto process, air, methyl alcohol, and steam are used in the ratio 4:2:1 by weight at 800 kPa and 380 K. Calculate the partial pressure of the gases in N/m^2.

4.24 What is the average molecular weight of a flue gas having the following composition by volume?
CO_2: 13.0%, CO: 3.0%, O_2: 4.0% and N_2: 80.0%

4.25 Assuming that air contains 78.1% nitrogen, 21.0% oxygen and 0.9% argon, calculate the average molecular weight and density of air at 298 K and 1.01325×10^5 N/m^2.

4.26 A natural gas has the following composition by volume: CO_2: 0.8%, N_2: 3.2%, and CH_4: 96.0%. Calculate
(a) The composition in weight percent
(b) The average molecular weight
(c) The density at standard conditions in kg/m^3

4.27 The average molecular weight of a gas mixture containing oxygen and other gases is 40. A student evaluated the average molecular weight as 36.8. He got the wrong result because he used an incorrect value of 16 as the molecular weight of oxygen in the calculation. What is the mole percentage of oxygen in the mixture?

4.28 A flue gas has the following percent composition by volume:
CO_2 = 14.00, SO_2 = 0.50, CO = 2.00, O_2 = 2.5 and N_2 = 81.00.
Determine
(a) The average molecular weight of the gas
(b) The composition of gas in weight percent
(c) The density of the gas at 320 K and 1.5 bar
(d) The specific gravity at 320 K and 1.5 bar

4.29 A gas mixture contains 10.0% CO_2, 15.0% Cl_2, 12.5% N_2 and the rest hydrogen. Calculate the following:
(a) The average molecular weight of the gas
(b) The gas composition in weight percent

4.30 200 kg of a gas mixture containing 20% SO_2, 20% NO_2 and 60% N_2 is contained in a gas cylinder. Determine the average molecular weight of the mixture
(a) If the analysis is on a mass basis
(b) If the analysis is on a volume basis

4.31 A sample withdrawn from a gas mixture at 120 kPa and 300 K gave the following results when separated into its constituents: C_2H_6 17.93 g, H_2 0.50 g, CO_2 6.57 g. Determine the following:
(a) The average molecular weight
(b) The density of the gas mixture

4.32 In a gas mixture consisting of hydrogen, nitrogen and carbon dioxide, the partial pressures are 25 kPa for hydrogen, 35 kPa for nitrogen and 140 kPa for CO_2. For 50 m^3 of the gas mixture at 400 K, determine

(a) The number of moles of the mixture
(b) The number of moles and mole fraction of hydrogen
(c) The mass and mass fraction of hydrogen
(d) The pure-component volume of hydrogen
(e) The average molecular weight of the mixture

4.33 The average molecular weight of a mixture of oxygen and sulphur dioxide is found to be 44.8. For 5 kg of this mixture at 298 K and 200 kPa calculate

(a) The partial pressure of oxygen
(b) The volume of the mixture
(c) The density at the standard conditions

4.34 An ammonia converter is charged with a 1:3 mixture of nitrogen and hydrogen at 1000 bar and 500 K. Calculate the following:

(a) The partial pressures of nitrogen and hydrogen
(b) The average molecular weight
(c) The density of the mixture

4.35 A blue gas has the following composition by volume percent: CO: 34.8, H_2: 42.0, CH_4: 0.4, CO_2: 5.5, O_2: 0.2, and N_2: 17.1. Calculate

(a) The composition in weight percent
(b) The average molecular weight
(c) The density at standard conditions in kg/m^3

4.36 A gas mixture has the following composition by volume: ethylene: 30.6%, benzene: 24.5%, oxygen: 1.3%, methane: 15.5%, ethane: 25%, and nitrogen: 3.1%. Calculate the following:

(a) The average molecular weight
(b) The composition by weight
(c) The density of the mixture at STP in kg/m^3

4.37 Air is dried from a partial pressure of 7 kPa of water vapour to a partial pressure of 1.5 kPa. The temperature of the entering air is 475 K. The pressure remains constant at 1 bar at inlet and outlet. How much water in kilograms is removed per cubic metre of entering gas?

4.38 An ammonia-air mixture containing 10% ammonia is sent to an absorber in which 90% of ammonia is absorbed.

(a) What is the mole fraction of ammonia in the gas leaving?
(b) Express the composition of the exit gas from the absorber in weight percent.
(c) What is the average molecular weight of the gas entering and leaving the column? (Average molecular weight of air is 29.)

Calculations involving change in the mass of the gas

4.39 Flue gas analysing 80.0% nitrogen, 7.0% oxygen and 13.0% carbon dioxide is sent to a dryer at 500 K and 95 kPa. In the dryer it picks up moisture from the wet material and leaves at 90 kPa and 360 K with the following composition: N_2 = 50%, O_2 = 4.5%, CO_2 = 8.0% and H_2O = 37.5%. Per 100 m^3 of gas entering the dryer, determine the following:

(a) The volume of gas leaving the dryer
(b) The weight of water evaporated

4.40 Hydrochloric acid is obtained by absorbing HCl gas in water. A gas mixture analyzing 30% HCl and 70% air enters the absorber at 95 kPa and 320 K and leaves at 90 kPa and 300 K. 95% of HCl present in the entering gas is absorbed. For 100 m^3 of gas entering the absorber, determine the following:

(a) The volume of gas leaving in m^3
(b) The mass of HCl absorbed in kg
(c) The percent composition by volume of the gas leaving

4.41 Hydrochloric acid is obtained by absorbing HCl gas in water. A gas mixture consisting of HCl and air enters the absorber at 95 kPa and 320 K with a partial pressure of HCl of 28.5 kPa and leaves at 90 kPa and 300 K with a partial pressure of HCl of 2 kPa. For 100 m^3 of gas entering the absorber determine the following:

(a) The volume of gas leaving in m^3
(b) The mass of HCl absorbed in kg

4.42 Air is to be dehumidified by cooling and condensation of water vapour present in it. 100 m^3 of air at 100 kPa and 305 K contains water vapour which exerts a partial pressure of 3 kPa. Keeping the pressure constant, this air is cooled to 285 K and the condensed water is removed. The partial pressure of water in the air after cooling is found to be 1.5 kPa. Calculate the following:

(a) The volume of air after dehumidification in m^3
(b) The mass of water removed in kg

4.43 The feed to an absorption column consists of 30% H_2S and 70% inerts. Only H_2S is removed from the gas by absorbing in an alkaline solution. The gas enters the absorber at 700 kPa and 350 K and leaves at 600 kPa and 300 K containing 5% H_2S. If H_2S is removed at a rate of 100 kg/h, calculate the following:

(a) Cubic meters of gas entering per hour
(b) Cubic meters of gas leaving per hour
(c) Percentage recovery of H_2S

4.44 A flue gas containing 18% CO_2 enters an absorber at 120 kPa and 400 K. Only CO_2 is absorbed and the gas leaves the unit at 95 kPa and 310 K containing 2% CO_2. Assuming that 100 m^3 of gas is admitted to the unit, calculate

(a) The volume of gas leaving in m^3
(b) The weight of CO_2 absorbed in kg

4.45 100 m^3 of a gas mixture in which benzene and nitrogen are present in the weight ratio 1 : 3 is sent to an absorption column to recover benzene. The gas enters the column at 120 kPa and 350 K. The gas leaving the absorber measures 95 m^3 at 101.3 kPa and 300 K. Determine the following:

(a) The average molecular weight of the gas entering
(b) The average molecular weight of the gas leaving
(c) Weight in kilograms of benzene absorbed
(d) The percent recovery of benzene

4.46 50 cubic metres per minute of a gas containing 5% oxygen is flowing through a pipe. It is desired to increase the oxygen content in the stream to 10% by (a) the addition of pure oxygen and (b) the addition of air containing 21% oxygen and 79% nitrogen. Determine the volumetric rate of addition (m^3/min) in both cases.

4.47 100 m^3/h of an ammonia-air mixture containing 20% ammonia by volume is admitted to an absorption column at 120 kPa and 300 K in order to recover ammonia by absorbing it in water. The gas leaves the column at 100 kPa and 280 K with a partial pressure of ammonia of 2 kPa.

(a) What is the volume of gas leaving in one hour?
(b) How many kilograms of ammonia are recovered in one hour?
(c) What is the percent recovery?

4.48 100 m^3 per hour of a mixture of toluene and air at 120 kPa and 375 K is passed through a cooler where some of the toluene is condensed. Toluene exerts a partial pressure of 50 kPa in the entering stream. The gases leaving the cooler measured 70 m^3 at 100 kPa and 325 K. Assuming an average molecular weight of 29 for air, determine the following:

(a) The average molecular weight of gas leaving the cooler
(b) The amount of toluene (C_7H_8) condensed (in kilograms).

4.49 Two tanks are initially sealed off from one another by means of a valve. Tank I initially contains 1 m^3 of air at 7 bar and 340 K. Tank II initially contains a mixture of oxygen and nitrogen containing 95% nitrogen at 14 bar and 360 K. The valve is then opened allowing the contents of the tank to mix. After complete mixing had been done, the gas was found to contain 85% nitrogen. Calculate the volume of Tank II.

Reactions involving gases

4.50 Ammonia reacts with sulphuric acid giving ammonium sulphate

$$2NH_3 + H_2SO_4 \rightarrow (NH_4)_2SO_4$$

(a) 20 m^3 of ammonia at 1.2 bar and 300 K reacts with 40 kg of sulphuric acid. Which is the excess reactant and what is the percent excess?
(b) How many kilograms of ammonium sulphate will be obtained?

4.51 Sulphur dioxide reacts with oxygen producing sulphur trioxide.

$$SO_2 + \tfrac{1}{2}O_2 \rightarrow SO_3$$

In order to ensure complete reaction, twice as much oxygen is supplied than that required theoretically. However, only 60% conversion is obtained. The pressure was 500 kPa and temperature 800 K. 100 kg of SO_2 is charged to the converter. Determine the following:

(a) The volume of pure oxygen supplied at 1.5 bar and 300 K
(b) The volume of sulphur trioxide produced
(c) The volume of gases leaving the converter
(d) The composition of gases leaving the converter
(e) The average molecular weight of the gas leaving the converter

4.52 Nitrogen dioxide shows a tendency to associate and form nitrogen tetroxide.

$$2NO_2 \rightarrow N_2O_4$$

One cubic metre nitrogen dioxide at 100 kPa and 300 K is taken in a closed rigid container and allowed to attain equilibrium at constant temperature and volume. The pressure inside the container has fallen to 85 kPa at equilibrium.

(a) What is the degree of association?
(b) What is the partial pressure of N_2O_4 in the final mixture?

4.53 Ammonium chloride in the vapour phase dissociates into ammonia and hydrogen chloride according to

$$NH_4Cl \rightarrow NH_3 + HCl$$

10.7 g ammonium chloride is taken in a container. When dissociation is complete and equilibrium is attained the pressure, volume and temperature of the gas mixture were measured to be 1.2 bar, 7.764×10^{-3} m^3 and 400 K, respectively. Determine the following:

(a) The fraction of ammonium chloride dissociated
(b) The partial pressure of HCl in the products

4.54 One kilomole of methane is completely burnt to form CO_2 and water vapour using 10% excess air than that required for complete combustion. The resulting gas mixture is treated to remove all the water content.

(a) What will be the volume of dry gas leaving the burner at 110.3 kPa and 275 K?
(b) What mass of water is removed from the products of combustion?

4.55 Pure sulphur is burned with excess air to give sulphur trioxide. Due to incomplete oxidation some sulphur dioxide is also formed. The analysis of the burner gases showed 0.8% SO_3, 7.8% SO_2, 12.2% O_2 and 79.2% N_2. Determine the following:

(a) The percent excess air supplied
(b) The volume of burner gases at 1.2 bar and 600 K per kg sulphur burned
(c) The volume of air supplied at 1.013 bar and 288 K per kg sulphur burned

4.56 An organic ester of formula $C_{19}H_{36}O_2$ is to be hydrogenated at a rate of 100 kg/h to yield $C_{19}H_{38}O_2$. The hydrogen required for the plant which runs continuously, is available as 50 L cylinders in which the gas is contained at 70 bar and 300 K. How many cylinders the company should order per week?

4.57 Formaldehyde is produced by the gas phase oxidation of methanol with air over a catalyst

$$CH_3OH + \frac{1}{2}O_2 \rightarrow HCHO + H_2O$$

100 m^3 of methanol vapour at 1.013 × 10^5 N/m^2 and 550 K is to be treated. If 10% excess air is supplied and the reaction is only 80% complete, calculate

(a) The composition of the product gas
(b) The volume of product gases at 1.5 × 10^5 N/m^2 and 800 K

4.58 A gaseous mixture consisting of 50% hydrogen and 50% acetaldehyde (C_2H_4O) is initially contained in a rigid vessel at a total pressure of 1.0 bar. Methanol is formed according to

$$C_2H_4O + H_2 \rightarrow C_2H_6O$$

After a time it was found that the total pressure in the vessel has fallen to 0.9 bar while the temperature was the same as that of the initial mixture. Assuming that the products are still in the vapour phase, calculate the degree of completion of the reaction.

4.59 A sample of a gas mixture containing NO and NO_2 is contained in a standard cell of volume 10 L at 170 kPa and 300 K. If the mixture weighed 28.087 g, what percent of the gas mixture is NO?

4.60 A mixture of ethane (C_2H_6) and ethylene (C_2H_4) occupies 4.0 × 10^{-2} m^3 at 1atm and 400 K. The mixture reacts completely with 0.130 kg of O_2 to produce CO_2 and H_2O. Assuming ideal gas behaviour, calculate the mole fractions of C_2H_4 and C_2H_6 in the mixture.

4.61 Pure methane is completely burned with air. The gas leaving the burner which contain no oxygen is passed through a cooler where some of the water is removed by condensation. The mole fraction of nitrogen in the gas leaving the cooler was 0.8335. Calculate the following:

(a) The analysis of the gas leaving the cooler
(b) Water condensed in kg per kmol of methane burned
(c) The partial pressure of water in the gas leaving the cooler at 325 K and 1.5 bar.

4.62 In the nitric acid manufacture, 100 m^3/h of ammonia at 290 K and 1 bar is mixed with air and passed over a catalyst in a converter and the following reaction takes place:

$$4NH_3 + 5O_2 \rightarrow 6H_2O + 4NO$$

The gases from the converter are passed into an oxidizing tower where the oxidation completed according to the following reaction:

$$2NO + O_2 \rightarrow 2NO_2$$

The gases from the oxidizing tower are passed into a cooling tower where NO_2 is absorbed in 50 kg/h of water and forms nitric acid according to the reaction

$$3NO_2 + H_2O \rightarrow 2HNO_3 + NO$$

The overall reaction for the process is given by

$$NH_3 + 2O_2 \rightarrow HNO_3 + H_2O$$

Air supplied is 20% in excess of that required for complete oxidation of the ammonia to nitric acid and water. The degree of completion of the reaction in the converter is 85% and 90% of the nitric oxide entering the oxidising tower is converted to nitrogen peroxide. Assume that the cooling tower exit gases contain no water vapour. Calculate the following:

(a) The amount of air to be used in m^3/h at 1 bar and 290 K
(b) The amount of gases leaving the converter in m^3/h at 0.95 bar and 1000 K
(c) The amount in kg/h and concentration in weight % of nitric acid produced

5 Properties of Real Gases

The significance of process calculations in which the change in quantity and state of the gases are involved was pointed out in Chapter 4, while discussing the behaviour of ideal gases. Since an ideal gas model provides a very simple equation of state, those calculations were quite simple. But in practice, very few gases can be approximated as ideal gases. The assumption of ideal gas behaviour in other cases leads to erroneous results depending upon the extent of deviation of the gases from ideal behaviour. Instead of ideal gas equation, realistic models—equations of state—which predict the actual behaviour of gases should be used in dealing with real gases.

5.1 REAL GASES

At very low pressures and at very high temperatures, the molal volume of a gas becomes so large that the molecules exist far removed from one another and the intermolecular forces become negligibly small. These are the requirements for a gas to behave as an ideal gas. Thus at very low pressures and at high temperatures real gases tend to exhibit ideal behaviour. However, under ordinary pressures the molecules possess appreciable volume in comparison with the volume of the gas as a whole. Also, the intermolecular forces existing in the gas will not be negligible as in an ideal gas. Therefore, ideal gas equations will be inadequate to represent the behaviour of real gases. For example, the volumes calculated by the ideal gas law is too large, sometimes even five times the actual volume. The following methods are generally used for calculations involving real gases:

1. Experimental *P-V-T* data
2. Equation of state
3. Generalized compressibility charts.

Before we consider these various options, a brief discussion on the critical conditions of a pure substance is quite in order, as the critical state is important in the development of the equations of state.

5.2 CRITICAL PROPERTIES

The thermodynamic state of a pure fluid such as water can be represented as a function of pressure and volume as shown in Figure 5.1. The figure shows the variation in molar volume with pressure at various constant temperatures. Assume that the initial state of the system is represented by point *p*. The change in volume of water with pressure at a constant temperature T_1 is along the isotherm *p-q-r-s*. In the region from *p* to *q*, the change in volume of the liquid with change in pressure is negligibly small as the liquid is incompressible. At point *q*, the liquid begins to vaporize and the volume of the system increases sharply. The pressure remains constant till the entire liquid is vaporized. This pressure is known as the *saturation pressure*. At point *r* the liquid is completely transformed into vapour. At point *q* the system is in the saturated liquid state and at point *r* it is in the saturated vapour state. Between points *q* and *r* the system consists of saturated liquid and saturated vapour in equilibrium. The volumes corresponding to point *q* and *r* are, respectively, the specific volumes of saturated liquid and saturated vapour. In the region between points *q* and *r* (the two-phase region), the volume of the two-phase mixture depends upon the *quality* of the mixture, which is the fraction of the mixture that is liquid. On further reduction of pressure from point *r*, the vapour becomes superheated and the volume increases along the isotherm *r-s*.

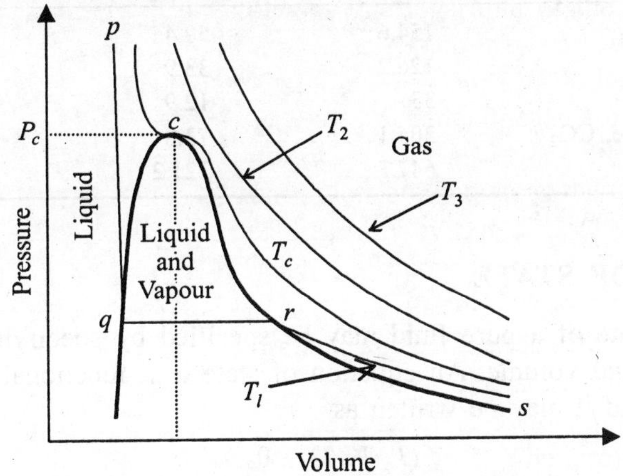

Figure 5.1 *P-V-T* diagram of a pure fluid.

If the liquid were at a temperature greater than T_1, the saturation pressure at which the two phases coexist in equilibrium would be above that corresponding to T_1 and the length of the horizontal section joining the saturated liquid and vapour phases would be smaller. The locus of the saturated phases is represented by the dome-shaped curve *q-c-r*. The area under this dome represents the two-phase region, the area to the left of the curve *q-c* the liquid region and the area to the right of the curve *c-r* the vapour region.

At temperature T_c known as the *critical temperature* the length of the two-phase region is so short that the PV isotherm passes through a point of inflection at point C, which is called the *critical point*. The pressure at the critical point is P_c, the *critical pressure*. At the critical point, the liquid and vapour phases have identical properties and are therefore indistinguishable. A liquid at a temperature above its critical temperature cannot be vaporized by reducing the pressure at constant temperature. The isotherms at T_2 and T_3 in the figure are above the critical temperature. The two-phase region can be approached from the liquid state by increasing the temperature at constant pressure as well, provided the pressure is below the critical pressure P_c. The critical properties of water are T_c = 647.3 K, P_c = 221.2 bar and $V_c = 57.1 \times 10^{-6}$ m^3/mol.

The critical temperature and critical pressure represent the highest temperature and pressure respectively, at which a pure material can exist in vapour–liquid equilibrium. If the temperature is less than the critical temperature, the substance to the right of the saturated vapour line is called a *vapour* and if the temperature is above the critical temperature the substance is called a *gas*. A vapour can be condensed to a liquid by compression at constant temperature or by cooling at constant pressure whereas a gas can be condensed only by the reduction of temperature at constant pressure. It can not be condensed by compression at constant temperature. If the temperature and the pressure of the substance are greater than T_c and P_c respectively, it is in the fluid region and it can neither be condensed by any of the above methods nor can it be vaporized by reduction in pressure as is possible with liquids. The substance in this state is neither a gas nor a liquid. Table 5.1 gives the critical constants of a few substances.

Table 5.1 Critical properties of some typical materials

Substance	T_c (K)	P_c (bar)	$V_c \times 10^6$ (m^3/mol)
Oxygen, O_2	154.6	50.4	73.4
Nitrogen, N_2	126.2	33.9	89.8
Hydrogen, H_2	33	12.9	64.3
Carbon dioxide, CO_2	304.1	73.8	93.9
Water, H_2O	647.3	221.2	57.1

5.3 EQUATIONS OF STATE

The thermodynamic state of a pure fluid may be specified by specifying properties such as pressure, temperature and volume. An equation of state is a functional relationship between these three variables and it may be written as

$$f(P, V, T) = 0 \tag{5.1}$$

The simplest equation of state is the perfect gas law given by Eq. (4.9), i.e. $PV = nRT$. But it is applicable only to ideal gases and is inadequate to explain the behaviour of real gases. For real gases to behave ideally, the molecular interactions should be negligible. At low molar volumes or high pressures, molecules come very close to each other and molecular interactions cannot be neglected. The perfect gas law may be viewed as an approximation describing the behaviour of real gases at ordinary pressures or as describing the limiting behaviour of real gases at low pressures. Many equations of state have been proposed to explain the actual behaviour of gases. An equation of state, in general, should satisfy certain limiting conditions. For example, all equations reduce to the ideal gas equation at low pressures. That is,

$$PV = RT \quad \text{as} \quad P \to 0$$

Also the *P-V* curve at constant temperature should exhibit a point of inflection at the critical point. Stated mathematically, this means

$$\left(\frac{\partial P}{\partial V}\right)_{T_C} = 0 \tag{5.2}$$

$$\left(\frac{\partial^2 P}{\partial V^2}\right)_{T_C} = 0 \tag{5.3}$$

In the following sections we describe some important equations of state.

5.3.1 The van der Waals Equation

The perfect gas equation fails to explain the *P-V-T* behaviour of real gases. The van der Waals equation is a powerful equation of state that represents the behaviour of actual gases reasonably accurately. The perfect gas equation was developed on the premise that the molecular interactive forces are negligible and the molecules of the gas occupy a negligibly small volume in comparison to the total volume of the gas. The van der Waals equation takes into account these two non-ideal features of real gases by incorporating certain correction factors in the pressure and volume terms of the ideal gas equation.

The pressure exerted by the molecules of a gas will be less than the ideal gas pressure because of the attractive forces between molecules. This attractive force is proportional to $1/V^2$ where V is the molar volume of the gas. The actual pressure is thus proposed to be less than the ideal gas pressure by a/V^2, a fact verified to be true by the kinetic theory. In using the ideal gas equation, the pressure term should be replaced by $p + a/V^2$. In an ideal gas, a molecule is treated as point mass occupying no volume on its own. But in reality the molecules have a finite volume, though it is only a small fraction of the total volume of the gas. The actual volume of a gas is thus greater than the volume occupied by the gas if it were ideal. The volume term in the ideal gas equation is to be replaced by $V - b$, where b indicates the excluded volume per mole of the gas. Incorporating these corrections in the ideal gas equation, van der Waals proposed the following equation for the *P-V-T* behaviour of real gases:

$$\left(P + \frac{a}{V^2}\right)(V - b) = RT \tag{5.4}$$

where a and b are called van der Waals constants. The van der Waals equation is cubic in volume and at temperatures below the critical temperature, there are three real roots. When P is the saturation pressure, the smallest and largest roots correspond to the molar volumes of saturated liquid and saturated vapour respectively. The intermediate root has no physical significance.

The constants in the van der Waals equation can be evaluated utilising the limiting conditions which are obeyed by any equation of state. Equation (5.4) can be written as

$$P = \frac{RT}{V-b} - \frac{a}{V^2} \tag{5.5}$$

The limiting conditions, Eqs. (5.2) and (5.3) can be applied by finding the first and second derivatives of P with respect to V at constant T. These are as follows:

$$\left(\frac{\partial P}{\partial V}\right)_{T_c} = 0 = \frac{2a}{V_c^{\,3}} - \frac{RT_c}{(V_c - b)^2}$$

$$\left(\frac{\partial^2 P}{\partial V^2}\right)_{T_c} = 0 = \frac{2RT_c}{(V_c - b)^3} - \frac{6a}{V_c^{\,4}}$$

Solving these two equations simultaneously, we get

$$V_c = 3b \tag{5.6}$$

$$T_c = \frac{8a}{27Rb} \tag{5.7}$$

$$P_c = \frac{RT_c}{2b} - \frac{a}{9b^2} \tag{5.8}$$

The last two of the above identities can be solved to get the van der Waals constants as

$$a = \frac{27R^2T_c^{\,2}}{64P_c} \tag{5.9}$$

$$b = \frac{RT_c}{8P_c} \tag{5.10}$$

where T_c and P_c are, respectively the critical temperature and pressure of the fluid.

EXAMPLE 5.1 Calculate the pressure developed by one kmol gaseous ammonia contained in a vessel of 0.6 m^3 capacity at a constant temperature of 473 K by the following methods:

(a) Using the ideal gas equation

(b) Using the van der Waals equation given that a = 0.4233 N m^4/mol^2; b = 3.73 × 10^{-5} m^3/mol

Solution

(a) Use the ideal gas equation, $P = \dfrac{RT}{V}$, where V is the molar volume of the gas. Here, V = 0.6 × 10^{-3} m^3/mol. Therefore, the pressure given by the ideal gas equation is

$$P = \frac{8.314 \times 473}{0.6 \times 10^{-3}} = 6.554 \times 10^6 \text{ N/m}^2 = 65.54 \text{ bar}$$

(b) The van der Waals equation may be rearranged as

$$P = \frac{RT}{V-b} - \frac{a}{V^2}$$

Substituting of the values gives

$$P = \frac{8.314\,(473)}{(0.6\times10^{-3})-(3.73\times10^{-5})} - \frac{0.4233}{(0.6\times10^{-3})^2} = 5.8128 \times 10^6 \text{ N/m}^2 = 58.13 \text{ bar}$$

EXAMPLE 5.2 Estimate the molar volume of CO_2 at 500 K and 100 bar using the (a) ideal gas equation and (b) the van der Waals equation. The van der Waals constants are 0.364 N m^4/mol^2 and 4.267 × 10^{-5} m^3/mol.

Solution

(a) Using the ideal gas equation,

$$V = \frac{RT}{P} = \frac{8.314\,(500)}{(1\times10^7)} = 0.4157\times10^{-3}\ \text{m}^3$$

(b) Volume can be determined using the van der Waals equation by trial and error. Here, a = 0.364 N m^4/(mol)^2 and b = 4.267 × 10^{-5} m^3/mol. The van der Waals equation gives

$$\left(P + \frac{a}{V^2}\right)(V - b) = RT$$

Substituting the values, we get

$$\left(100\times10^5 + \frac{0.364}{V^2}\right)(V - 4.267\times10^{-5}) = 8.314\times500$$

This can be simplified as

$$V^3 - 4.5837 \times 10^{-4}\, V^2 + 3.64 \times 10^{-8} V - 1.5582 \times 10^{-12} = 0$$

Solving this, the molar volume of CO_2 in the present case is found to be

$$V = 3.717 \times 10^{-4} \text{ m}^3\text{/mol.}$$

5.3.2 Other Equations of State

A number of equations of state have been proposed by various researchers such as the Redlich–Kwong equation, the Soave's modification for the Redlich–Kwong equation, the Peng–Robinson equation, etc. The Benedict–Webb–Rubin equation and the virial equations are prominent among them. A review of these equations of state can be found in books on thermodynamics.

Redlich–Kwong equation: This is a two-parameter equation of state widely used in engineering calculations.

$$P = \frac{RT}{V - b} - \frac{a}{T^{0.5}V(V + b)} \tag{5.11}$$

The constants a and b are related to the critical properties as given below:

$$a = \frac{0.4278R^2T_c^{2.5}}{P_c}; \qquad b = \frac{0.0867RT_c}{P_c} \tag{5.12}$$

Redlich–Kwong–Soave equation: Soave's modification of the Redlich–Kwong equation is as follows:

$$P = \frac{RT}{V-b} - \frac{a'(T)}{V(V+b)} \tag{5.13}$$

where $a'(T) = a\alpha$ where α depends on the temperature. The constant b is given by Eq. (5.12), whereas the constant a is evaluated using Eq. (5.12) with the exponent of T_c changed to 2.

Peng–Robinson equation: It is written as

$$P = \frac{RT}{V-b} - \frac{a\alpha}{V(V+b)+b(V-b)} \tag{5.14}$$

where a and b are constants, and α depends on reduced temperature T.

Benedict–Webb–Rubin equation: This equation contains a more number of parameters compared to the equations given above and hence is more complex. Being more accurate than the cubic equations of state discussed above, it is widely used for estimating the properties of light hydrocarbons and their mixtures in petroleum and natural gas industries.

$$P = \frac{RT}{V} + \frac{B_0RT - A_0 - C_0/T^2}{V^2} + \frac{bRT - a}{V^3} + \frac{a\alpha}{V^6} + \frac{c}{V^3T^2}\left(1+\frac{\gamma}{V^2}\right)\exp\frac{-\gamma}{V^2} \tag{5.15}$$

where A_0, B_0, C_0, a, b, c, α and γ are constants.

Virial equations: Virial equations express the compressibility factor Z (= PV/RT) as functions of pressure or volume. The compressibility factor is the ratio of the volume of a real gas (V) to the volume if the gas behaved ideally at the stated temperature and pressure (RT/P). Virial equations are available in the following two forms:

$$Z = \frac{PV}{RT} = 1 + B'P + C'P^2 + D'P^3 + \cdots \tag{5.16}$$

$$Z = \frac{PV}{RT} = 1 + \frac{B}{V} + \frac{C}{V^2} + \frac{D}{V^3} + \cdots \tag{5.17}$$

Equations (5.16) and (5.17) are known as virial equations and the constants B, C, D and B', C', D' are known as virial coefficients. These equations find greatest application in predicting P-V-T behaviour at low to moderate pressures.

EXAMPLE 5.3 Assume that gaseous ammonia follows the Redlich–Kwong equation of state. Calculate the pressure developed by one mole of NH_3 contained in a vessel of volume 0.6×10^{-3} m^3 at 473 K given that the critical pressure and temperature are 112.8 bar and 405.5 K.

Solution The constants a and b in the Redlich–Kwong equation are evaluated using Eq. (5.12), i.e.

$$a = \frac{0.4278R^2T_c^{2.5}}{P_c} = \frac{(0.4278)(8.314)^2(405.5)^{2.5}}{112.8 \times 10^5} = 8.68\,\frac{\text{N m}^4\ \text{K}^{0.5}}{\text{mol}^2}$$

$$b = \frac{0.0867RT_c}{P_c} = \frac{(0.0867)(8.314)(405.5)}{112.8 \times 10^5} = 2.59 \times 10^{-5} \text{ (m}^3\text{/mol)}$$

Equation (5.11) gives the pressure developed by the gas.

$$P = \frac{RT}{V - b} - \frac{a}{T^{0.5}V(V + b)}$$

$$= \frac{(8.314)(473)}{(0.6 \times 10^{-3}) - (2.59 \times 10^{-5})}$$

$$- \frac{8.68}{(473^{0.5})(0.6 \times 10^{-3})(0.6 \times 10^{-3} + 2.59 \times 10^{-5})}$$

$$= 5.787 \times 10^6 \text{ N/m}^2 = 57.87 \text{ bar}$$

EXAMPLE 5.4 Calculate the molar volume of gaseous ammonia at 373 K and 10 bar using the Redlich–Kwong equation given that the critical pressure and temperature are 112.8 bar and 405.5 K.

Solution The constants in the Redlich–Kwong equation were evaluated in Example 5.3. Now the molar volume is given by Eq. (5.11).

$$P = \frac{RT}{V - b} - \frac{a}{T^{0.5}V(V + b)}$$

$$10^6 = \frac{(8.314)(373)}{V - 2.59 \times 10^{-5}} - \frac{8.68}{373^{0.5}V(V + 2.59 \times 10^{-5})}$$

Solving this equation, we get $V = 3.0 \times 10^{-3}$ m^3/mol.

EXAMPLE 5.5 Using the virial equation of state, calculate the molar volume of methanol vapour at 500 K and 10 bar. The virial coefficients are, $B = -2.19 \times 10^{-4}$ m^3/mol; $C = -1.73 \times 10^{-8}$ m^6/mol^2.

Solution The virial equation truncated to three terms is

$$Z = \frac{PV}{RT} = 1 + \frac{B}{V} + \frac{C}{V^2}$$

which can be rearranged as

$$V = \frac{RT}{P}\left(1 + \frac{B}{V} + \frac{C}{V^2}\right)$$

$$= 4.157 \times 10^{-3}\left(1 - \frac{2.19 \times 10^{-4}}{V} - \frac{1.73 \times 10^{-8}}{V^2}\right) \tag{A}$$

Assume a value for V and calculate the right-hand side of this equation. The assumed value will be equal to the molar volume if the right-hand side so evaluated and the volume assumed are equal. Otherwise repeat the calculation with a new value for V. The volume calculated using the ideal gas equation can be taken as the initial guess value for V. The ideal gas volume is

$$V = RT/P = 4.157 \times 10^{-3} \text{ m}^3$$

The results of these calculations are:

Assumed V (m^3):	4.157×10^{-3}	3.934×10^{-3}	3.921×10^{-3}	3.920×10^{-3}
RHS (m^3) :	3.934×10^{-3}	3.921×10^{-3}	3.920×10^{-3}	3.920×10^{-3}

Since the assumed and calculated values are equal at $V = 3.92 \times 10^{-3}$ m^3, it may be accepted as the molar volume of methanol.

5.4 COMPRESSIBILITY CHARTS

5.4.1 Compressibility Factor

The compressibility factor measures the deviation of a gas from ideal behaviour. It is defined as $Z = PV/RT$. It can be interpreted as the ratio of the actual volume of a gas (V) to the volume that the gas will occupy if it were an ideal gas (RT/P) at the given temperature and pressure. Compressibility charts are plots of the compressibility factor against pressure with temperature as parameter as shown in Figure 5.2. For a perfect gas, the compressibility factor is unity and the curve of Z versus P is a straight horizontal line at Z = 1.0. The shape of the curve for a real gas depends on the nature of the gas. However, since a real gas behaves ideally as pressure is reduced, the compressibility factor of a real gas approaches unity as the pressure is reduced to zero. This is clear from Figure 5.2. At other conditions, the compressibility factor may be less than or more than unity depending on the temperature and pressure of the gas.

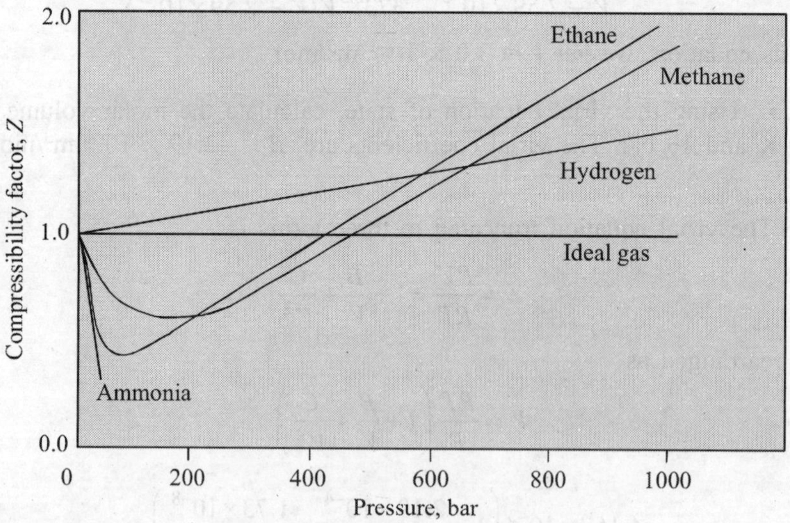

Figure 5.2 Compressibility factor.

5.4.2 Principle of Corresponding States

The principle of corresponding states can be stated as that *all gases when compared at the same reduced temperature and reduced pressure, have approximately the same compressibility factor and all deviate from the ideal behaviour to the same extent.* The reduced variables are obtained by dividing the actual variables by the corresponding critical constants. Thus the reduced pressure $P_r = P/P_c$, the reduced temperature $T_r = T/T_c$, and reduced volume $V_r = V/V_c$.

According to the principle of corresponding states, the compressibility factor of any substance is a function of the reduced temperature and reduced pressure alone. For example, consider the compressibility factors of methane with the critical properties $T_c = 191$ K, $P_c = 46$ bar and nitrogen with the critical properties $T_c = 126$ K, $P_c = 34$ bar. Methane at 382 K and 69 bar and nitrogen at 252 K and 51 bar have identical reduced properties $T_r = 2$, $P_r = 1.5$. Then according to the law of corresponding states, nitrogen at 252 K and 51 bar and methane at 382 K and 69 bar should have the same compressibility factors. Thus the principle of corresponding states enables one to coordinate the properties of a range of gases to a single diagram such as the generalised compressibility chart.

5.4.3 Generalized Compressibility Charts

These charts enable engineering calculations to be made with considerable ease and also permit the development of thermodynamic functions of gases for which no experimental data are available. The generalised compressibility chart (Figure 5.3) is constructed based on the principle of corresponding states. Based on experimental *P-V-T* data on 30 gases, Nelson and Obert* prepared these charts which give the compressibility factors with an average deviation of less than 5% from the experimental results.

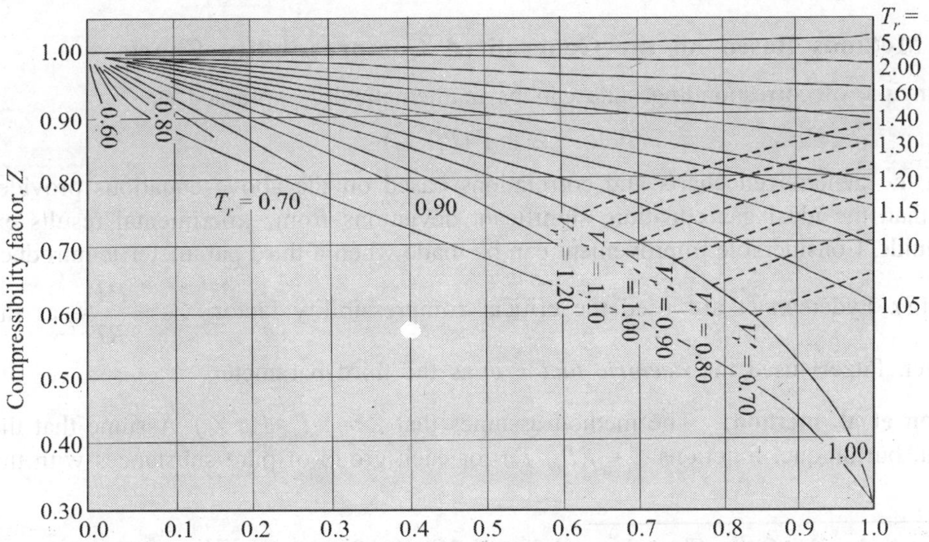

Figure 5.3(a) Generalized compressibility chart (lower pressures).

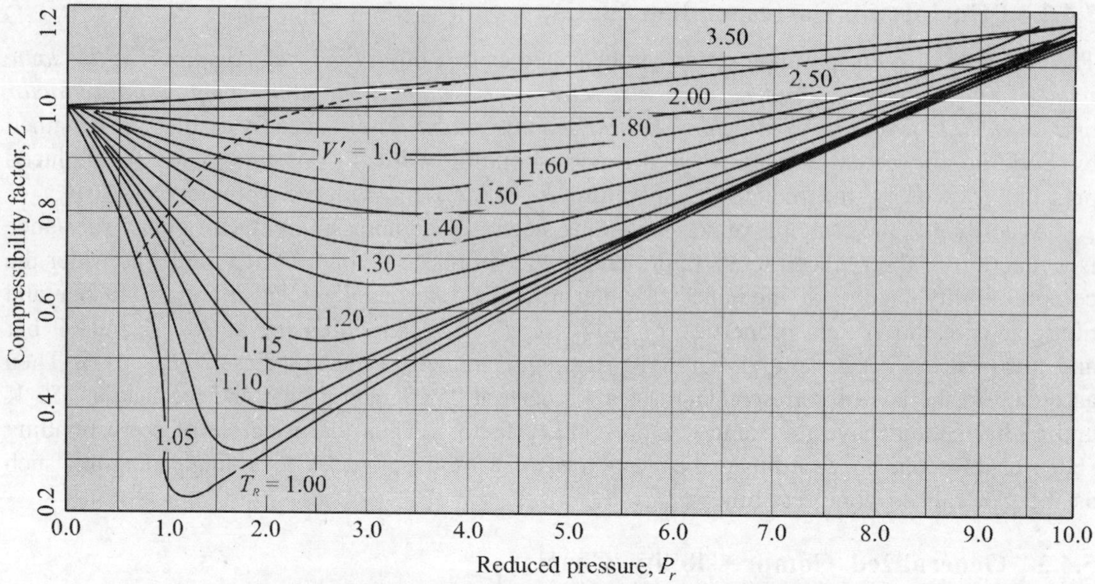

Figure 5.3(b) Generalized compressibility chart.

On these charts, the compressibility factor Z is plotted as a function of reduced temperature and reduced pressure. The generalized compressibility charts provide one of the best means of expressing the deviation from ideal behaviour. When precise equations of state are not available to represent the *P-V-T* behaviour, these charts provide a convenient method for the evaluation of compressibility factor.

5.4.4 Methods Based on the Generalized Compressibility Charts

The principle of corresponding state can be mathematically expressed as

$$Z = f\,(P_r,\ T_r) \tag{5.18}$$

Though the generalized charts and correlations based on the above equations provide better results than the ideal gas equation, significant deviations from experimental results exist for many fluids. Considerable improvement can be made when a third parameter is introduced into Eq. (5.18). Lyderson et al.[†] used the critical compressibility factor, $Z_c = \dfrac{P_c V_c}{RT_c}$ as the third parameter. Pitzer[‡] used the *acentric factor* ω as the third parameter.

Lydersen et al. method: The method assumes that $Z = f(T_r, P_r, Z_c)$. Assume that there are different, but unique, functions $Z = f(P_r, T_r)$ for each group of pure substances with the same

---

*Nelson, L.E. and E.F. Obert, *Chem. Eng.*, 61(7), 203–208 (1954).

[†]See Hougen, O.A., K.M. Watson and R.A. Ragatz, *Chemical Process Principles,* vol. 2, Asia Publishing House (1963).

[‡]Pitzer, K.S., *J. Am. Chem. Soc.*, 77, 3427 (1955).

Z_c. Then, for each Z_c we have a different set of compressibility charts such as the one shown in Figure 5.3. All fluids with the same Z_c values then follow the Z-T_r-P_r behaviour shown on charts drawn for that particular Z_c. The generalized charts were drawn for $Z_c = 0.27$ as most materials fell in the range of 0.26 to 0.28 and correction tables were developed to correct for values of Z_c different from 0.27. The generalized charts for compressibility is given in Figure 5.4.

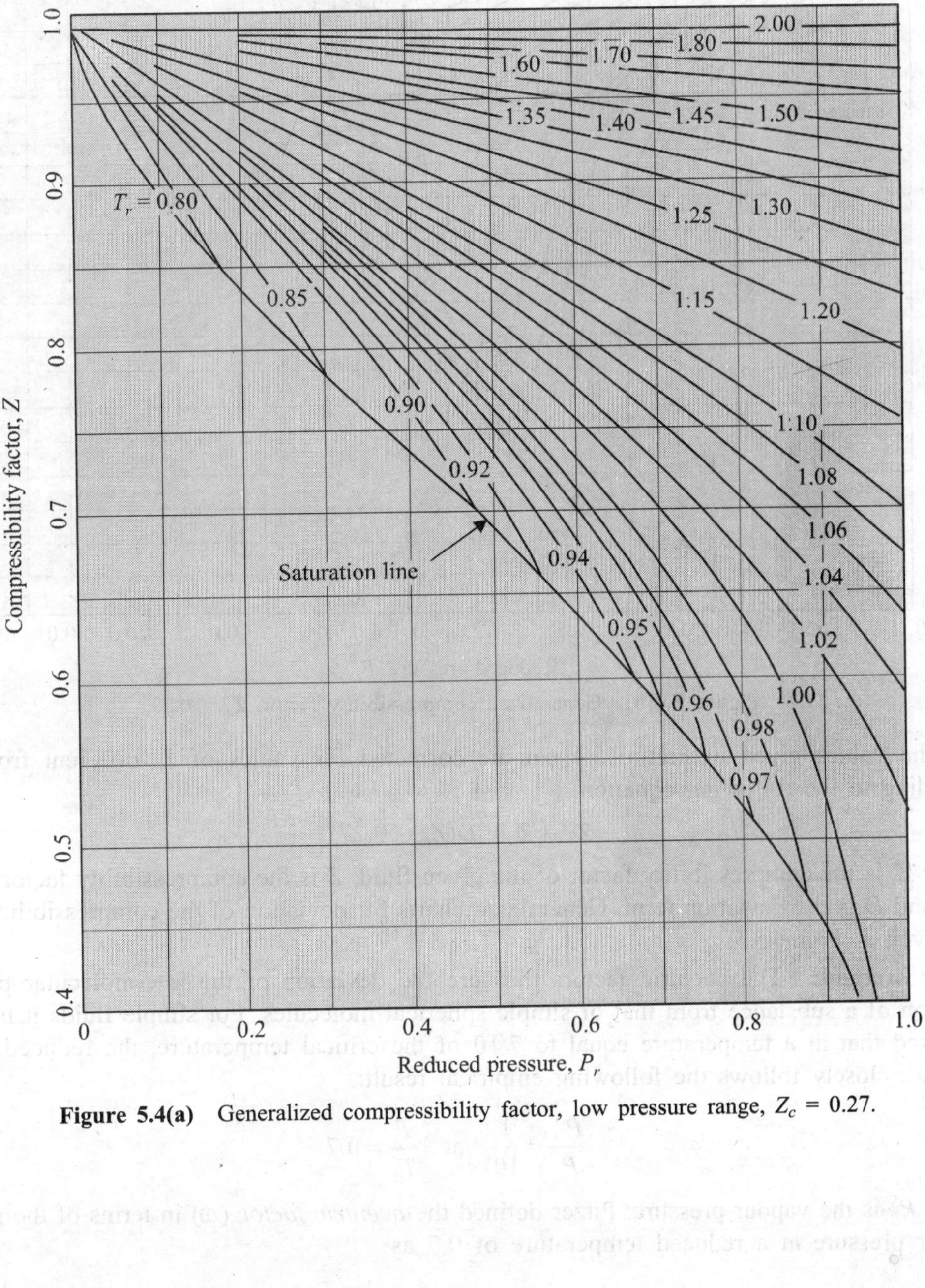

Figure 5.4(a) Generalized compressibility factor, low pressure range, $Z_c = 0.27$.

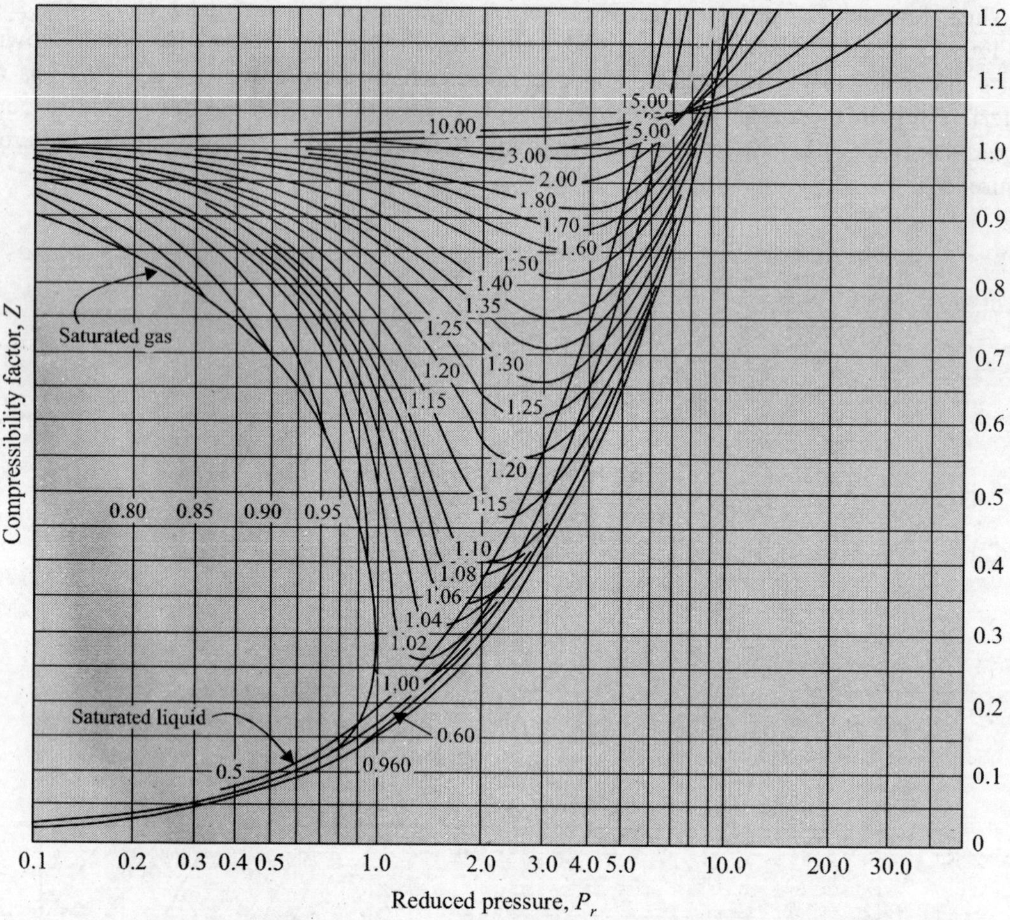

Figure 5.4(b) Generalized compressibility factor, Z_c = 0.27.

The values given in Figure 5.4 can be corrected for values of Z_c different from 0.27 according to the following equation.

$$Z' = Z + D(Z_c - 0.27) \tag{5.19}$$

where Z' is the compressibility factor of the given fluid, Z is the compressibility factor at Z_c = 0.27 and D is the deviation term. Generalized charts for deviation of the compressibility factor are given in Figue 5.5.

Pitzer method: The acentric factors measure the deviation of the intermolecular potential function of a substance from that of simple spherical molecules. For simple fluids it has been observed that at a temperature equal to 7/10 of the critical temperature, the reduced vapour pressure closely follows the following empirical result:

$$\frac{P^S}{P_c} = \frac{1}{10} \quad \text{at} \quad \frac{T}{T_c} = 0.7$$

where P^S is the vapour pressure. Pitzer defined the *acentric factor* (ω) in terms of the reduced vapour pressure at a reduced temperature of 0.7 as

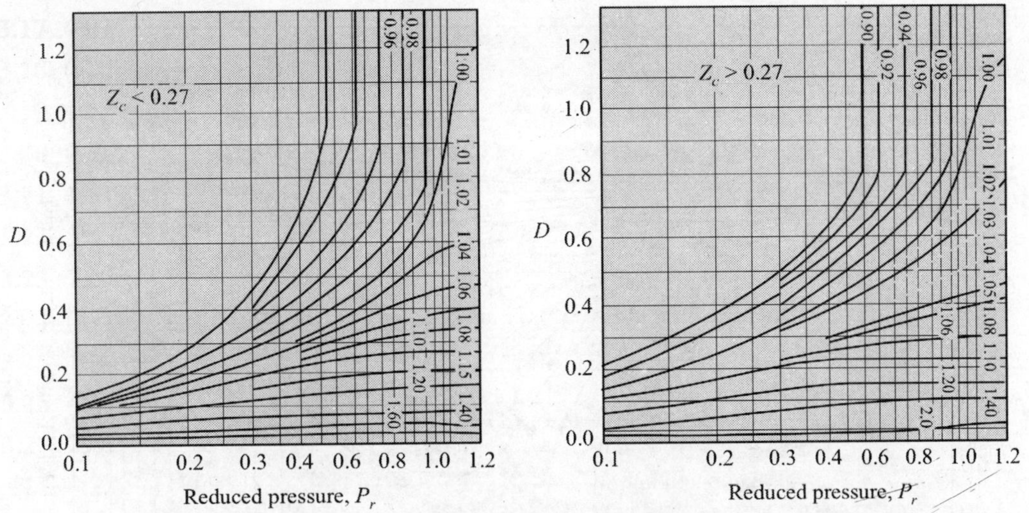

Figure 5.5 Generalized deviation D for compressibility factor at Z_c = 0.27.

$$\omega = -1.00 - \log\left(\frac{P^S}{P_c}\right)_{T_r = 0.7} \tag{5.20}$$

For simple fluids the acentric factor = 0; for more complex fluids the acentric factor > 0. The acentric factors are listed in standard references. [See, for example, Prausnitz, J.M., *Molecular Thermodynamics of Fluid Phase Equilibria*, Prentice-Hall Inc. (1986).] Table 5.2 gives the acentric factors of some common substances.

Table 5.2 Acentric factors

Molecule	ω	*Molecule*	ω
CH_4	0.008	$n\text{-}C_4H_{10}$	0.193
O_2	0.021	C_6H_6	0.212
N_2	0.040	CO_2	0.225
CO	0.049	NH_3	0.250
C_2H_4	0.085	$n\text{-}C_6H_{14}$	0.296
C_2H_6	0.098	H_2O	0.344

Pitzer proposed the following functional form for the compressibility factor.

$$Z = f(T_r, P_r, \omega) \tag{5.21}$$

It was suggested that instead of preparing separate Z-T_r-P_r charts for different values of ω, a linear relation could be used

$$Z = Z^{(0)}(T_r, P_r) + \omega Z^{(1)}(T_r, P_r) \tag{5.22}$$

Here, the function $Z^{(0)}$ would apply to spherical molecules, and the $Z^{(1)}$ term is a deviation function. Pitzer presented tabular and graphical values of the function $Z^{(0)}$ and the $Z^{(1)}$. Figures 5.6 and 5.7 give these plots incorporating the Lee–Kessler* modifications.

*Lee, B.L. and M.G. Kessler, *AIChE J.*, 21, 510 (1975).

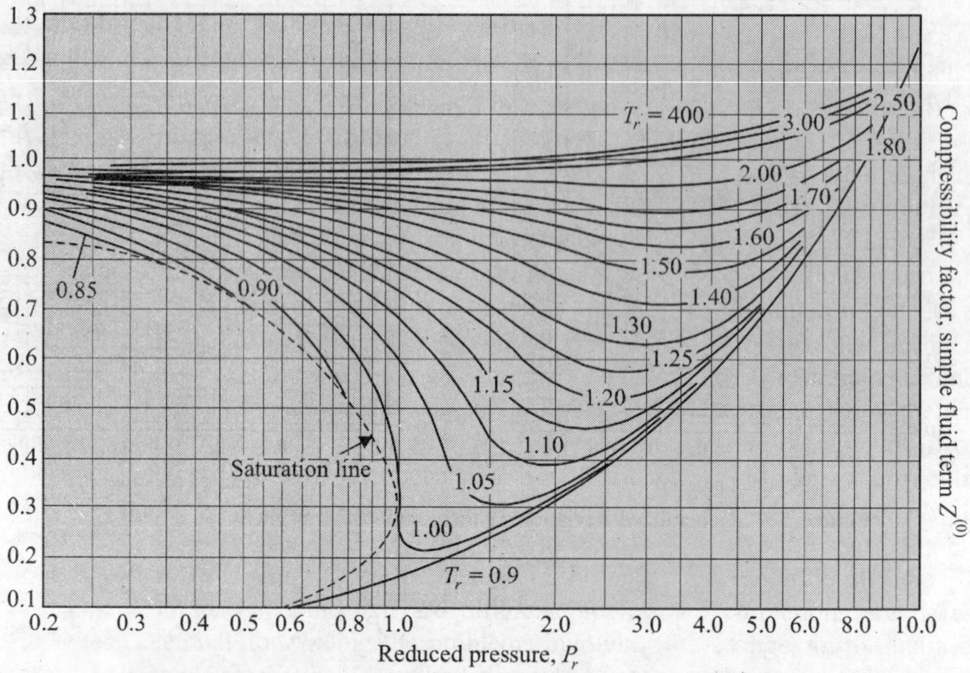

Figure 5.6 Generalized compressibility factor for simple fluid.

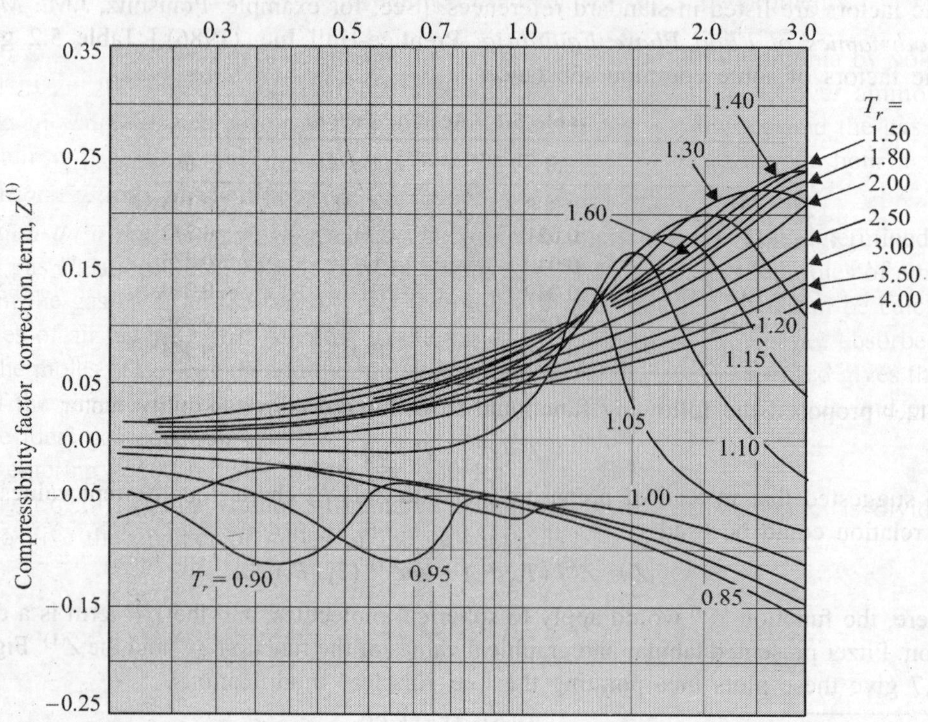

Figure 5.7 Generalized compressibility factors correction term.

EXAMPLE 5.6 Using the Lydersen method, determine the molar volume of n-butane at 510 K and 26.6 bar. The critical pressure and temperature of *n*-butane are 38 bar and 425.2 K respectively, and the critical compressibility factor is 0.274.

Solution The reduced properties of *n*-butane are:

$$P_r = \frac{26.6}{38} = 0.70$$

$$T_r = \frac{510}{425.2} = 1.20$$

From Figures 5.4 and 5.5

$$Z = 0.865$$

$$D = 0.15$$

Using Eq. (5.19),

$$Z' = 0.865 + 0.15(0.274 - 0.27) = 0.8656$$

The molar volume is

$$V = \frac{RTZ}{P} = \frac{(8.314)(510)(0.8656)}{26.6 \times 10^5}$$

$$= 1.3798 \times 10^{-3} \text{ m}^3\text{/mol}$$

EXAMPLE 5.7 Using the Pitzer correlation, determine the molar volume of *n*-butane at 510 K and 26.6 bar. The critical pressure and temperature of n-butane are 38 bar and 425.2 K, respectively and the acentric factor is 0.193.

Solution The reduced properties of *n*-butane are:

$$P_r = \frac{26.6}{38} = 0.70$$

$$T_r = \frac{510}{425.2} = 1.20$$

From Figures 5.6 and 5.7

$$Z^{(0)} = 0.855$$

$$Z^{(1)} = 0.042$$

Using Eq. (5.22)

$$Z = 0.855 + 0.193(0.042) = 0.8631$$

The molar volume is

$$V = \frac{RTZ}{P} = \frac{(8.314)(510)(0.8631)}{26.6 \times 10^5}$$

$$= 1.3798 \times 10^{-3} \text{ m}^3\text{/mol}$$

5.5 MIXTURE OF REAL GASES

We have already seen several methods for the evaluation of the *P-V-T* relationship of pure gases. For mixtures of ideal gases, the properties can be determined by adding together the properties of the individual components. This simple technique fails in the case of mixtures of real gases. However, it is possible to estimate the *P-V-T* properties of real gaseous mixtures from the properties of pure constituents. Three such methods are discussed in the following sections.

5.5.1 Equations of State

An equation of state may be used to evaluate the properties of the mixture provided the constants appearing in the equations are obtained by combining the constants of the pure components by an appropriate method. We will discuss only the van der Waals equation of state in this section. For example, the constants a and b in the van der Waals equation are evaluated as

$$a = \left(\Sigma\, y_i a_i^{1/2}\right)^2 \tag{5.23}$$

$$b = \Sigma\, y_i b_i \tag{5.24}$$

where a_i and b_i are the van der Waals constants for the component i in the mixture and y_i is its mole fraction in the mixture. The constants a and b, calculated as above, may be treated as the van der Waals constants for the mixture, which may be used in Eq. (5.4) for evaluation of *P-V-T* properties of the mixture. Alternatively, the pseudo-critical temperature and pressure (T_c' and P_c') discussed later in this chapter may be used for the evaluation of the van der Waals constants of the mixture as given below.

$$a = \frac{27R^2(T_c')^2}{64P_c'}$$

$$b = \frac{RT_c'}{8P_c'}$$

The constants are substituted into the van der Waals equation and the pressure, volume or temperature of the gas mixture is determined.

5.5.2 Mean Compressibility Factor

A mean compressibility factor Z_m may be defined such that the *P-V-T* relation for the mixture is written as

$$PV = Z_m RT \tag{5.25}$$

Z_m is a function of the pressure, temperature and composition of the mixture. It may be evaluated as

$$Z_m = \Sigma\, y_i Z_i \tag{5.26}$$

Since the compressibility factor of component i, Z_i is a function of the reduced pressure, the question may naturally arise: what pressure should be used for determining Z_i? If volumes are additive at constant temperature and pressure (i.e., Amagat's law is applicable), it is recommended that Z_i be evaluated at the temperature and the total pressure of the mixture. When Dalton's law is applicable, Z_i is evaluated at the temperature and the partial pressure of the component i in the mixture. Thus

$$p_{r,i} = \frac{P}{P_{c,i}} \quad \text{(Amagat's law)} \tag{5.27}$$

$$p_{r,i} = \frac{p_i}{P_{c,i}} = \frac{y_i P}{P_{c,i}} \quad \text{(Dalton's law)} \tag{5.28}$$

5.5.3 Pseudo-critical Properties

We define an effective critical parameter for a gas mixture as the mole fraction weighted average of the critical properties of the constituent gases. Since it is different from the true critical property of the mixture, it is called pseudo-critical property. Thus the pseudo-critical temperature (T_c') of the mixture is

$$T_c' = \Sigma\, y_i T_{c,i} \tag{5.29}$$

The pseudo-critical pressure $\left(P_c' = \Sigma\, P_{c,i}\right)$ is

$$P_c' = \Sigma\, y_i P_{c,i} \tag{5.30}$$

The pseudo-reduced temperature (T_r') and pressure (P_r') are now evaluated:

$$T_r' = \frac{T}{T_c'} \tag{5.31}$$

$$P_r' = \frac{P}{P_c'} \tag{5.32}$$

In the Kay's method for the evaluation of *P-V-T* behaviour of mixtures of real gases, these pseudo-reduced parameters are used to find the compressibility factor of the gas mixture with the help of a generalized compressibility chart.

EXAMPLE 5.8 A gas mixture contains 40.0% nitrogen and 60.0% ethane at 6000 kPa and 325 K. The van der Waals constants: nitrogen (a = 0.1365 N m^4/mol^2 and b = 3.86 × 10^{-5} m^3/mol), ethane (a = 0.557 N m^4/mol^2 and b = 6.51 × 10^{-5} m^3/mol). The critical constants: nitrogen (P_c = 3394 kPa, T_c = 126.1 K), ethane (P_c = 4880 kPa, T_c = 305.4 K). Calculate the molar volume of the mixture using the following:

(a) The ideal gas equation
(b) The van der Waals equation
(c) The mean compressibility factor based on (i) Amagat's law and (ii) Dalton's law.
(d) The Kay's method

Solution (a) One mole of an ideal gas will occupy volume

$$V = \frac{RT}{P} = \frac{8.314 \times 325}{6000 \times 10^3} = 4.503 \times 10^{-4}\ \text{m}^3/\text{mol}$$

(b) The van der Waals constants for the mixture are evaluated using Eqs. (5.23) and (5.24). The mole fractions of nitrogen and ethane are $y_{N_2} = 0.4$ and $y_{ethane} = 0.6$.

$$a = \left(\Sigma\, y_i a_i^{1/2}\right)^2 = [0.4(0.1365)^{1/2} + 0.6(0.557)^{1/2}]^2 = 0.3547 \text{ N m}^4/\text{mol}^2$$

$$b = \Sigma\, y_i b_i = 0.4(3.86 \times 10^{-5}) + 0.6(6.51 \times 10^{-5}) = 5.45 \times 10^{-5} \text{ m}^3/\text{mol}$$

Substituting these values in the van der Waals equation, we get

$$\left(6000 \times 10^3 + \frac{0.3547}{V^2}\right)(V - 5.45 \times 10^{-5}) = 8.314 \times 325$$

This may be rearranged as

$$V^3 - 5.0484 \times 10^{-4} V^2 + 5.9117 \times 10^{-8}\, V - 3.2217 \times 10^{-12} = 0$$

On solving this equation, we get

$$V = 3.680 \times 10^{-4} \text{ m}^3/\text{mol}$$

(c) (i) Assume that the mixture follows Amagat's law. Then the compressibility factors of the components are evaluated as

$$p_{r,i} = \frac{P}{P_{c,i}}$$

For nitrogen,

$$p_{r,i} = \frac{6000}{3394} = 1.768; \quad T_{r,i} = \frac{325}{126.1} = 2.5773$$

For ethane,

$$p_{r,i} = \frac{6000}{4880} = 1.230; \quad T_{r,i} = \frac{325}{305.4} = 1.0642$$

Using the compressibility charts, we get $Z_{N_2} = 1.0$ and $Z_{ethane} = 0.42$.

Combining these,

$$Z_m = 0.40 \times 1.00 + 0.60 \times 0.42 = 0.652$$

Now,

$$V = \frac{Z_m RT}{P} = \frac{0.652 \times 8.314 \times 325}{6000 \times 10^3} = 2.94 \times 10^{-4} \text{ m}^3/\text{mol}$$

(ii) Alternatively, if Dalton's law is used, i.e.

$$p_{r,i} = \frac{p_i}{P_{c,i}} = \frac{y_i P}{P_{c,i}}$$

For nitrogen,

$$p_{r,i} = \frac{0.4 \times 6000}{3394} = 0.707; \quad T_{r,i} = \frac{325}{126.1} = 2.5773$$

For ethane,

$$p_{r,i} = \frac{0.6 \times 6000}{4880} = 0.738; \quad T_{r,i} = \frac{325}{305.4} = 1.0642$$

Using the compressibility charts, we get $Z_{N_2} = 1.0$ and $Z_{ethane} = 0.76$.

Combining these,

$$Z_m = 0.40 \times 1.0 + 0.60 \times 0.76 = 0.856$$

Therefore, $$V = \frac{Z_m RT}{P} = \frac{0.856 \times 8.314 \times 325}{6000 \times 10^3} = 3.855 \times 10^{-4} \text{ m}^3/\text{mol}$$

(d) The pseudo-critical properties are:

$$T_c' = \Sigma\, y_i T_{c,i} = 0.4 \times 126.1 + 0.6 \times 305.4 = 233.68 \text{ K}$$

$$P_c' = \Sigma\, y_i P_{c,i} = 0.4 \times 3394 + 0.6 \times 4880 = 4285.6 \text{ kPa}$$

The pseudo-reduced properties are calculated as:

$$T_r' = \frac{T}{T_c'} = \frac{325}{233.68} = 1.391$$

$$P_r' = \frac{P}{P_c'} = \frac{6000}{4285.6} = 1.4$$

For these reduced parameters, $Z = 0.83$ and the molar volume is

$$V = \frac{Z_m RT}{P} = \frac{0.83 \times 8.314 \times 325}{6000 \times 10^3} = 3.738 \times 10^{-4} \text{ m}^3/\text{mol}$$

EXAMPLE 5.9 The molar volume of a gas mixture analyzing 40% nitrogen and 60% ethane at 325 K is 4.5 × 10^{-4} m^3/mol. Determine the pressure of the gas by

(a) The ideal gas equation
(b) The van der Waals equation

The van der Waals constants: nitrogen (a = 0.1365 N m^4/mol^2 and b = 3.86 × 10^{-5} m^3/mol), ethane (a = 0.557 N m^4/mol^2 and b = 6.51 × 10^{-5} m^3/mol).

Solution

(a) $$P = \frac{RT}{V}$$

$$= \frac{8.314 \times 325}{4.5 \times 10^{-4}} = 6.005 \times 10^6 \text{ N/m}^2 \ (= 6005 \text{ kPa})$$

(b) The van der Waals constants for the mixture are evaluated using Eqs. (5.23) and (5.24). The mole fractions of nitrogen and ethane are $y_{N_2} = 0.4$ and $y_{ethane} = 0.6$.

$$a = \left(\Sigma\, y_i a_i^{1/2}\right)^2 = [0.4(0.1365)^{1/2} + 0.6(0.557)^{1/2}]^2 = 0.3547 \text{ N m}^4/\text{mol}^2$$

$$b = \Sigma\, y_i b_i = 0.4(3.86 \times 10^{-5}) + 0.6(6.51 \times 10^{-5}) = 5.45 \times 10^{-5} \text{ m}^3/\text{mol}$$

Substituting these values in the van der Waals equation, we get,

$$P = \frac{RT}{V-b} - \frac{a}{V^2} = \frac{8.314 \times 325}{4.5 \times 10^{-4} - 5.45 \times 10^{-5}} - \frac{0.3547}{(4.5 \times 10^{-4})^2}$$

$$= 5.08 \times 10^6 \text{ N/m}^2 \; (= 5080 \text{ kPa})$$

EXERCISES

Equations of state

5.1 A pressure gauge connected to a cylinder containing 175 kg propane indicates 50 bar absolute pressure. Using van der Waals equation, determine the temperature of the gas in the cylinder. The capacity of the cylinder is 1 m^3. The van der Waals constants are a = 0.9252 N m^4/mol^2 and b = 9.05 × 10^{-5} m^3/mol.

5.2 One mole of a gas obeying van der Waals equation is contained in a cylinder of volume 0.05 m^3 at 100 kPa. If the van der Waals constants are a = 0.135 N m^4/mol^2 and b = 3.22 × 10^{-5} m^3/mol, calculate the temperature of the gas.

5.3 One kmol CO_2 occupies a volume of 0.381 m^3 at 313 K. Compare the pressures given by the
(a) Ideal gas equation
(b) van der Waals equation

Take the van der Waals constants to be a = 0.365 N m^4/mol^2 and b = 4.28 × 10^{-5} m^3/mol.

5.4 Use the Redlich–Kwong equation to calculate the pressure of 0.5 kg gaseous ammonia contained in a vessel of 0.03 m^3 at a constant temperature of 338 K. The critical temperature and pressure are 405.5 K and 112.8 bar respectively.

5.5 Calculate the pressure exerted by one mole gaseous ammonia contained in a vessel of volume 0.03 m^3 at a temperature of 340 K using the Redlich–Kwong equation of state. The critical constants of ammonia are 405.5 K and 112.8 bar. Compare the result with the value obtained using the ideal gas equation.

5.6 Ethylene at 500 bar and 375 K is contained in a cylinder of volume 0.03 m^3. The critical properties of ethylene are 282.4 K and 50.4 bar. How many kilograms of ethylene are there in the cylinder? Use the van der Waals equation.

5.7 A fire extinguisher of capacity 0.04 m^3 contains CO_2 at 101.3 kPa and 300 K. It is to be charged with CO_2 to a pressure of 20 bar at the constant temperature of 300 K. Determine the kilograms of gas to be admitted assuming the van der Waals equation. The critical constants of CO_2 are 304.2 K and 73.8 bar.

5.8 Calculate the molar volume and the compressibility factor of methanol vapour at 450 K and 10 bar using

(a) The truncated virial equation, given the following virial coefficients:

$$B = -2.19 \times 10^{-4}\ \text{m}^3/\text{mol} \text{ and } C = -1.73 \times 10^{-8}\ \text{m}^6/\text{mol}^2$$

(b) The Redlich–Kwong equation, given the critical constants of methanol as 512.6 K and 81.0 bar

5.9 Calculate the volume of vessel that should be provided to store 1 kmol of methane at 325 K and 600 bar using the ideal gas equation and the van der Waals equation. The critical constants of methane are 190.6 K and 46.0 bar.

5.10 Calculate the molar volume of a gas at 9 MPa and 373 K using the van der Waals equation given that the constants a and b of the gas are respectively, 0.231 N m^4/mol^2 and 4.49×10^{-5} m^3/mol.

5.11 Calculate the volume occupied by one mole of oxygen at 300 K and 100 bar using

(a) The ideal gas law

(b) The van der Waals equation.

Take a = 0.1378 N m^4/mol^2, and $b = 3.18 \times 10^{-5}$ m^3/mol

5.12 Calculate the molar volume of ammonia at 373 K and 10 bar using (a) the van der Waals equation and (b) the Redlich–Kwong equation given that the critical temperature is 405.5 K and the critical pressure is 112.8 bar.

5.13 Determine the van der Waals constants and the molar volume of ethane at the critical point given that the critical temperature and pressure are respectively 305.2 K and 49.4 bar.

5.14 Using the Redlich–Kwong equation calculate the molal volumes of saturated liquid and saturated vapour of methyl chloride at 333 K. The saturation pressure of methyl chloride at 333 is 13.76 bar. The critical temperature and pressure are respectively 416.3 K and 66.8 bar.

5.15 Using the virial equation calculate the molar volume and compressibility factor of isopropanol vapour at 473 K and 10 bar. The virial coefficients are:

$$B = -3.88 \times 10^{-4}\ \text{m}^3/\text{mol} \quad \text{and} \quad C = -2.6 \times 10^{-8}\ \text{m}^6/\text{mol}^2.$$

5.16 Determine the molar volume of gaseous methane at 300 K and 600 bar by the following methods:

(a) Using the ideal gas equation

(b) Using the van der Waals equation given that a = 0.2285 N m^4/mol^2; $b = 4.27 \times 10^{-5}$ m^3/mol

(c) Using the Redlich–Kwong equation given that T_C = 191.1 K and P_C = 46.4 bar.

Generalized compressibility charts

5.17 Using the Nelson–Obert chart calculate the pressure developed by 4.0 kg oxygen contained in a vessel of volume 0.03 m^3 at 250 K. The critical constants of oxygen are 154.6 K and 50.5 bar.

5.18 A 6.0 m^3 tank contains 25 kmol of a gas at 275 bar and 882 K. Determine the critical temperature of the gas using a generalized method given that the critical pressure is 50 bar.

5.19 Using the Nelson–Obert chart, estimate the volume occupied by 1 kg water vapour at 17 bar and 500 K; given the critical pressure and temperature of water as 221.2 bar and 647.3 K respectively. Compare the result with the volume calculated using the ideal gas equation.

5.20 Estimate the molar volume of water at 973.2 K and 200 bar using

(a) The Lydersen method

(b) The Pitzer method

The critical pressure and temperature of water are 221.2 bar and 647.3 K respectively, and the critical compressibility factor and the acentric factor are, 0.234 and 0.3448 respectively.

5.21 Calculate the molar volume and the compressibility factor of isopropyl alcohol at 475 K and 10 bar using

(a) The van der Waals equation

(b) The Lydersen method

(c) The Pitzer correlation

The critical pressure and temperature of isopropyl alcohol are 47.64 bar and 508.3 K. The critical compressibility factor and the acentric factor are 0.248 and 0.6689 respectively.

5.22 Calculate the molar volume of isobutene at 428 K and 86.2 bar using

(a) The van der Waals equation

(b) The Lydersen method

(c) The Pitzer correlation

The critical pressure and temperature of isobutene are 36.48 bar and 408.1 K. The critical compressibility factor and the acentric factor are 0.2627 and 0.282 respectively.

Mixture of real gases

5.23 A gas mixture containing 20% ethanol and 80% carbon dioxide has a molar volume equal to 2.0×10^{-4} m^3/mol at 500 K. The critical parameters are: ethanol (P_c = 6383 kPa, T_c = 516.3 K) and carbon dioxide (P_c = 7382 kPa, T_c = 304.1 K). What is the pressure of the gas? Use pseudo-critical method.

5.24 The molar volume of a gas mixture is experimentally found to be 1.4×10^{-4} m^3/mol at 121.6×10^5 N/m^2 and 298 K. The analysis of the mixture showed that it contained 57% ethylene, 40% argon and 3% helium. By how much does the molar volume calculated using the pseudo-critical properties deviate from the experimental result? The critical properties are: ethylene (P_c = 5032 kPa, T_c = 282.4 K), argon (P_c = 4898 kPa, T_c = 150.9 K) and helium (P_c = 228 kPa, T_c = 5.2 K).

5.25 20 kg of a gas mixture analyzing 10% carbon dioxide, 40% methane and 50% ethylene is to be contained in a cylinder at a maximum pressure of 165 bar at 355 K. What should be the volume of the cylinder? Use Kay's method. The critical properties are: carbon dioxide (P_c = 7382 kPa, T_c = 132.9 K), methane (P_c = 4604 kPa, T_c = 190.6 K) and ethylene (P_c = 5032 kPa, T_c = 282.4 K).

5.26 A gas mixture analyzed 20% methane, 30% ethylene and 50% nitrogen. The molar volume of the mixture at 100 bar is found to be 3×10^{-4} m^3/mol. The critical parameters are: methane (P_c = 4604 kPa, T_c = 190.6 K), ethylene (P_c = 5032 kPa, T_c = 282.4 K) and nitrogen (P_c = 3394 kPa, T_c = 126.1 K). Using the pseudo-critical method, determine the temperature of the gas.

5.27 A gas mixture analyzes 20% methane, 30% ethylene and the rest nitrogen. Determine the molar volume of the gas at 100 bar and 375 K using

(a) The ideal gas equation
(b) The van der Waals equation. The constants a and b are: methane (a = 0.225 N m^4/mol^2, b = 3.159×10^{-5} m^3/mol), ethylene (a = 0.448 N m^4/mol^2, b = 5.72×10^{-5} m^3/mol) and nitrogen (a = 0.1347 N m^4/mol^2, b = 3.86×10^{-5} m^3/mol).
(c) The mean compressibility factor, given the critical constants as: methane (P_c = 4604 kPa, T_c = 190.6 K), ethylene (P_c = 5032 kPa, T_c = 282.4 K) and nitrogen (P_c = 3394 kPa, T_c = 126.1 K)
(d) The pseudo-critical properties.

5.28 A gas mixture consists of 70% nitrogen, 20% oxygen, 9% ammonia and 1% water. Determine the density of the gas at 800 kPa and 925 K

(a) Using the van der Waals equation with average values of the constants. The van der Walls constants are: nitrogen (a = 0.1347 N m^4/mol^2, b = 3.86×10^{-5} m^3/mol), oxygen (a = 0.1360 N m^4/mol^2, b = 3.19×10^{-5} m^3/mol), ammonia (a = 0.419 N m^4/mol^2, b = 3.73×10^{-5} m^3/mol) and water vapour (a = 0.548 N m^4/mol^2, b = 3.06×10^{-5} m^3/mol).
(b) Using the van der Waals equation with the constants evaluated using the pseudo-critical properties. The following are critical properties: nitrogen (P_c = 3394 kPa, T_c = 126.1 K), oxygen (P_c = 5043 kPa, T_c = 154.6 K), ammonia (P_c = 1128 kPa, T_c = 405.7 K) and water (P_c = 22090 kPa, T_c = 647.3 K).

6 Vapour Pressure

All solids and liquids exert definite vapour pressures which are characteristic of the substances. Vapour pressure is considered to be one of the important thermodynamic properties of the substance. For pure substance, it is defined as the pressure exerted by its vapours in thermodynamic equilibrium with its condensed phase (solid or liquid) in a closed system at a given temperature. Vapour pressure indicates the tendency of the particles to escape into vapour phase from the liquid or solid phase. A liquid with a high vapour pressure is referred to as highly volatile. For solutions, vapour pressure determines the volatility of the components in a solution, which in turn decides the relative volatility of the system. The relative volatility of a solution is a measure of the easiness of separating liquid mixtures into their components by distillation. Distillation is the most widely used technique in chemical engineering for separation and purification of solutions. Besides, vapour pressure data is useful for the estimation of many physical properties required in the design of chemical equipment. Thus, a good knowledge about the methods for the estimation of vapour pressure in the absence of experimental data as well as the methods to predict the effect of temperature on vapour pressure, etc. is very essential for chemical engineers.

6.1 VAPOUR PRESSURE AND BOILING POINT

6.1.1 Vapour Pressure

Vapour pressure of a substance depends only on temperature. For solids, the vapour pressure may be immeasurably low under normal circumstances, whereas liquids exert reasonably high vapour pressures. Consider a pure liquid taken in a closed container at a given temperature. As the potential energy due to molecular interaction far exceeds the kinetic energy of translation, the molecules are not free to escape into the vapour phase and the substance exists as liquid. When the temperature is increased, the molecules attain translational kinetic energy that is sufficient to overcome the intermolecular attractive forces and the molecules escape into the vapour phase. This process is known as *vaporization*. The pressure exerted by the vapour increases with the formation of more and more vapour. This also means that with the formation of more and more vapour there will be a corresponding increase in the number of molecules colliding with the surface of liquid. Some of these molecules get attracted by the molecules in the liquid and returns to the liquid phase. This process is just the reverse of vaporization and is known as *condensation*. Eventually, a dynamic equilibrium is established at which the number of molecules of vapour that condense to form the liquid equals the number of molecules of liquid that vaporize at constant temperature. At this state, as the rate of vaporization is equal to the rate of condensation, the pressure attained by the vapour remains unchanged. This pressure is the vapour pressure of the substance at the given temperature.

The magnitude of the vapour pressure is not affected by the quantity of liquid and vapour as long as a free liquid surface is present in the system. It is a characteristic physical property of the substance, which is determined by the potential energy of molecular interaction that is to be overcome for vaporization to occur. This molecular interaction depends on the size and nature of the molecules. Among chemically similar materials, the size of the molecules and the attractive force between molecules increase with increase in the molecular weight. Consequently, substances with higher molecular weights have lower vapour pressures at any specified temperature.

6.1.2 Boiling Point

Consider a liquid being heated in an open vessel. As temperature increases the vapour pressure also increases, and at a particular temperature, it matches the external pressure. Vaporization now takes place throughout the bulk of the liquid and the vapour so produced can displace the surroundings. We treat the vaporization under this condition as *boiling* of the liquid and the temperature at which this occurs as the *boiling point* of the liquid at the given external pressure. The boiling point of a liquid when the external pressure is 101.3 kPa (atmospheric pressure) is designated as the normal boiling point (NBP) of the liquid. For water, the normal boiling point is 373.2 K, as at this temperature, the vapour pressure of water is 101.3 kPa.

6.2 PHASE BEHAVIOUR OF PURE SUBSTANCES

Gases are substances that are normally in the gaseous state at room temperature and atmospheric pressure. If the temperature of a gas is below its critical temperature, it is called vapour. The vapour can be condensed to a liquid by compression at constant temperature or by cooling at constant pressure. A gas whose temperature is above the critical temperature cannot be condensed by increasing the pressure at constant temperature.

When a vapour is compressed, the condensation starts when the pressure becomes equal to the saturation pressure or the vapour pressure. The vapour at the point of condensation is called a *saturated vapour* and the liquid condensate is called a *saturated liquid*. The saturation pressure at a given temperature is a characteristic property of the substance and will not change till the entire vapour is condensed. For water, the saturation pressure at 373.2 K is 1.013 bar. Alternatively, the vapour pressure of water is 1.013 bar at 373.2 K. If the pressure of the vapour is below its vapour pressure, it is called a *superheated vapour*. Alternatively, the temperature of a superheated vapour is above the saturation temperature at the specified pressure. For example, the water vapour at, say, 0.75 bar and 373.2 K is a superheated vapour, because the vapour pressure at

373.2 K is 1.013 bar or because the saturation temperature at 0.75 bar is 365 K. When water vapour at 0.75 bar and 373.2 K is compressed at constant temperature, condensation begins when the pressure reaches 1.013 bar. The liquid and vapour phases coexist in equilibrium at this condition. The amount of liquid formed increases at the expense of the vapour and eventually the entire vapour is condensed.

For a pure fluid there is a unique set of saturation temperature and saturation pressure values at which the two phases exist in equilibrium. At constant temperature, further increase in pressure is possible only when the vapour is completely condensed. The substance is now a subcooled liquid—a liquid whose temperature is below the saturation temperature at a given pressure or one whose pressure is above the vapour pressure at the specified temperature.

The ideas presented above can be illustrated clearly with the help of a *P-T* diagram, as shown in Figure 6.1.

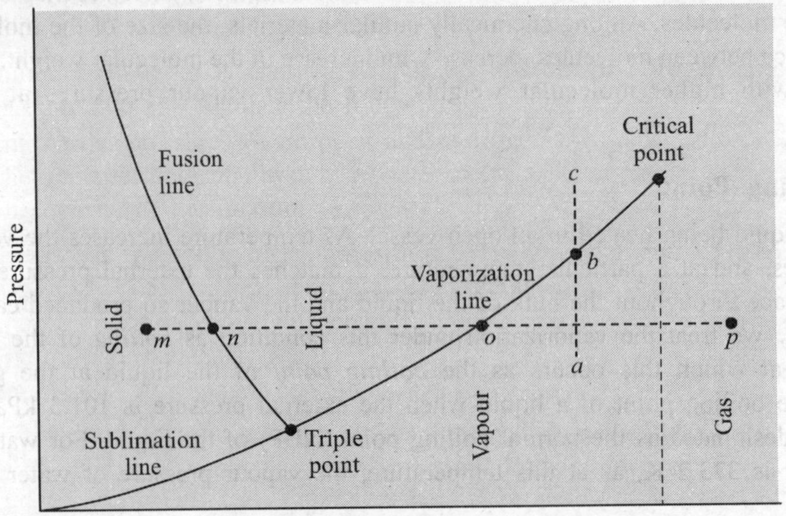

Figure 6.1 Pressure–temperature diagram of a pure substance.

The compression of a superheated vapour initially at point *a* is represented by a vertical line like the dotted line *a-b-c* in the figure. At point *b*, the liquid and vapour are in equilibrium. The pressure and temperature at this point represent the saturation conditions. On complete condensation, the material enters the subcooled liquid region represented by the line *b-c*. The vaporization of a liquid is possible by the reverse process, i.e. by reducing the pressure over the liquid at constant temperature. This is represented by the line *c-b-a*.

Figure 6.1 includes the solid region also. A horizontal line such as the line *m-n-o-p* in the figure represents the phase changes occurring by heating at constant pressure. Consider unit mass of solid at pressure and temperature corresponding to point *m* in the figure. When it is heated at constant pressure, the temperature increases and at point *n* the solid starts to melt. The temperature corresponding to point *n* is the fusion temperature. The fusion line represents the state of the system at which the solid is in equilibrium with the liquid. If transfer of heat is continued the temperature remains constant till the entire solid is liquefied. The fusion temperature is a function only of the pressure. On further heating, the substance enters the liquid region, and when it attains the temperature corresponding to point *o* lying on the vaporization curve, the liquid begins to vaporize. This temperature is the saturation temperature at the specified pressure and it remains constant till the vaporization is completed. The saturation temperature is also known as the boiling point of the liquid at the given pressure. Any point on the vapour liquid-equilibrium curve indicates the saturation temperature and the pressure at which a vapour and liquid coexist in equilibrium. As seen from Figure 6.1, the vapour-liquid equilibrium curve extends only up to the critical point. When all liquid has been vaporized, temperature increases along the line *o-p* in the vapour region. Heating beyond the critical temperature T_c takes the substance into the gas region.

The fusion curve and the vaporization curve meet at the triple point where the three phases coexist in equilibrium. For water at the triple point, $T = 273.21$ K and $P = 6.1 \times 10^{-3}$ bar. If the pressure of the system is below the triple point pressure, the heating of the solid results in sublimation, i.e., the transformation of solid directly into vapour. The sublimation curve in Figure 6.1 represents the state of the system in which the vapour is in equilibrium with solid. It may be noted that the fusion curve extends up to infinity whereas the vaporization curve terminates at the critical point.

Quality of vapour and use of steam tables: A pure fluid in vapour–liquid equilibrium exists as a mixture of saturated liquid and saturated vapour, the amounts of which depend on the extent of vaporization of the liquid. Quality of vapour measures the fraction of the mixture that is vapour. For water and water vapour in equilibrium, the weight fraction of the mixture that is dry steam is called the *quality* or *dryness fraction* of steam.

Steam tables provide a tabular representation of the thermodynamic properties of water in the saturated liquid state or in the saturated or superheated vapour state. A detailed discussion on the features of steam tables is provided in Chapter 12. Steam tables can be used to read the vapour pressure of water at any given temperature or the saturation temperature of water at a given pressure.

Refer to steam tables (thermodynamic properties of saturated steam) given in Appendix A3. Consider the data given in the first four columns of the table. They are the saturation pressure, saturation temperature and specific volumes of saturated liquid and saturated vapour, respectively, of water.

Pressure (kPa)	*Temperature* (K)	*Specific volume* (×10³ m³/kg)	
		Saturated liquid	*Saturated vapour*
40.00	349.02	1.027	3993
50.00	354.48	1.030	3240

At 40 kPa pressure, the saturation temperature of water is 349.02 K and the saturated liquid at this condition has a specific volume of 1.027×10^{-3} m³/kg and saturated vapour has a specific volume of 3.993 m³/kg. Alternatively, the steam table can be used to determine the vapour pressure of pure water at a given temperature. For example, at 354.48 K, the vapour pressure of water is 50 kPa. The steam tables which give temperature in the first column and other properties in the subsequent columns facilitate determination of vapour pressure as a function of temperature easily.

Suppose that at 40 kPa, a mixture of saturated liquid and saturated vapour exist in equilibrium. The specific volume of the mixture depends on the quality y or the fraction of the mixture that is dry steam.

$$V = (1 - y)V_L + yV_G = V_L + y(V_G - V_L) \tag{6.1}$$

where V_L is the specific volume of saturated liquid, V_G is the specific volume of saturated vapour and y is the quality of the steam. For a two-phase mixture of water at 50 kPa consisting of 40% dry saturated steam and 60% saturated liquid, $y = 0.4$ and the specific volume is

$$V = V_L + y(V_G - V_L) = 1.030 \times 10^{-3} + 0.4(3.240 - 1.030 \times 10^{-3}) = 1.297 \text{ m}^3/\text{kg}$$

EXAMPLE 6.1 The specific volume of wet steam at 500 kPa is 0.2813 m³/kg. Using the data from steam tables, determine the quality of steam.

Solution The following data are obtained from steam tables:

The specific volume of saturated liquid at 500 kPa = 1.093×10^{-3} m³/kg and the specific volume of saturated vapour = 0.3747 m³/kg. Let y be the quality of steam. Then y is the fraction that is vapour and $1 - y$ is the fraction of liquid.

$$(1 - y)(1.093 \times 10^{-3}) + y(0.3747) = 0.2813$$

Solving for y, we get $y = 0.75$. Therefore, the steam is 75% vapour and 25% liquid.

6.3 VAPOUR PRESSURE AND TEMPERATURE

The vapour pressure of a liquid increases with increase in temperature. The variation is as shown in Figure 6.1. The following paragraphs deal with some of the different correlations that have been proposed to represent the dependence of vapour pressure on temperature.

6.3.1 The Clapeyron Equation

The Clapeyron equation predicts the dependence of vapour pressure on temperature. It is given by

$$\frac{dP^S}{dT} = \frac{\lambda}{T\Delta V} \tag{6.2}$$

where P^S is the vapour pressure of the liquid at temperature T, λ is the latent heat of vaporization and ΔV is the volume change accompanying vaporization.

6.3.2 The Clausius–Clapeyron Equation

Equation (6.2) can be modified to yield the Clausius–Clapeyron equation applicable for vapour-liquid equilibrium by using the following simplifications:

1. The latent heat of vaporization is constant and independent of temperature.
2. The molar volume of liquid is negligible compared to that of vapour.
3. The vapour behaves as an ideal gas.

If the temperature is not too near the critical point, the volume of the liquid is small in comparison with the volume of vapour. The volume change accompanying vaporization $\Delta V = V_G - V_L$, is therefore, approximately equal to V_G, the molar volume of vapour. Equation (6.2) now becomes

$$\frac{dP^S}{dT} = \frac{\lambda}{TV_G} \tag{6.3}$$

The vapour pressure in regions well below the critical point is relatively so small that the vapour can be assumed to behave as an ideal gas. The molar volume V_G can now be replaced by RT/P^S so that Eq. (6.3) becomes

$$\frac{dP^S}{dT} = P^S \frac{\lambda}{RT^2}$$

or

$$\frac{d \ln P^S}{dT} = \frac{\lambda}{RT^2} \tag{6.4}$$

This result is known as the *Clausius–Clapeyron equation.* Assuming that the heat of vaporization remains constant in the temperature range of T_1 to T_2, Eq. (6.4) can be integrated to give the following equation.

$$\ln \frac{P_2^S}{P_1^S} = \frac{\lambda}{R}\left(\frac{1}{T_1} - \frac{1}{T_2}\right) \tag{6.5}$$

P_2^S and P_1^S are the saturation pressures at temperatures T_2 and T_1 respectively.

The assumption of constant latent heat of vaporization involved in the derivation of Clausius–Clapeyron can be expected to be true only for moderate temperature ranges. Besides, at high vapour pressures the vapour deviates considerably from ideal behaviour. Therefore, the Clausius–Clapeyron equation is only an approximation and experimental data should be relied upon if they are available. However, for rough estimates, the equation is found to be quite satisfactory in many cases.

EXAMPLE 6.2 Calculate the vapour pressure of water at 363 K if the vapour pressure at 373 K is 101.3 kPa. The mean heat of vaporization in this temperature range is 2275 kJ/kg.

Solution Refer to Eq. (6.5). $T_1 = 363$ K; $T_2 = 373$ K; $P_2^S = 101.3$ kPa

$\lambda = 2275 \times 18$ kJ/kmol; $R = 8.314$ kJ/kmol K; $P_1^S = ?$

Substituting the values in Eq. (6.5), we get

$$\ln \frac{101.3}{P_1^S} = \frac{2275 \times 18}{8.314}\left(\frac{1}{363} - \frac{1}{373}\right) = 0.3638$$

$$P_1^S = 70.41 \text{ kPa}$$

EXAMPLE 6.3 The vapour pressure of acetone at 273 K is 8.52 kPa and that at 353 K is 194.9 kPa. Dry air initially at 101.3 kPa and 300 K is allowed to get saturated with the vapours of acetone at constant temperature and volume. Determine

(a) The final pressure of the mixture
(b) The mole percent of acetone in the final mixture

Assume that the Clausius–Clapeyron equation is applicable to acetone.

Solution Using the Clausius–Clapeyron equation, i.e. Eq. (6.5), we get

$$\ln \frac{8.52}{194.9} = \frac{\lambda}{R}\left(\frac{1}{353} - \frac{1}{273}\right)$$

Therefore, λ/R = 3770.52, and it is assumed constant in the given interval.

The vapour pressure at 300 K is evaluated as

$$\ln \frac{P^S}{194.9} = \frac{\lambda}{R}\left(\frac{1}{353} - \frac{1}{300}\right)$$

Substituting λ/R = 3770.52 in the above equation, we get P^s = 29.53 kPa at 300 K.

(a) When air is saturated with the acetone vapours at 300 K, the partial pressure exerted by acetone will be equal to its vapour pressure at 300 K. The partial pressure of air is 101.3 kPa at constant volume. Therefore, total pressure = 101.3 + 29.53 = 130.83 kPa.

(b) Mole percent = $\dfrac{\text{moles of acetone}}{\text{total moles}} \times 100 = \dfrac{\text{partial pressure}}{\text{total pressure}} \times 100$

$$= \frac{29.53}{130.83} \times 100 = 22.6\%$$

6.2 The Antoine Equation

The Clausius–Clapeyron equation, i.e. Eq. (6.4) can be put in the following form

$$\frac{dP^S}{P^S} = \frac{\lambda}{R}\left(\frac{dT}{T^2}\right) = -\frac{\lambda}{R}\, d\left(\frac{1}{T}\right) \tag{6.6}$$

Integrating this equation, we get

$$\ln P^S = A - \frac{B}{T} \tag{6.7}$$

where A and B are constants. The Antoine equation is an empirical correlation for vapour pressure and resembles Eq. (6.7). It is written as

$$\ln P^S = A - \frac{B}{T - C} \tag{6.8}$$

where A, B, and C are known as Antoine constants. Equation (6.8) reduces to Eq. (6.7) when $C = 0$. The Antoine constants for a few systems are listed in the Table 6.1 with P^S in kPa and T in K. An exhaustive list is available in Reid, R.C, J.M. Prausnitz and B.E. Pauling, *The Properties of Gases and Liquids*, 4th ed., McGraw-Hill (1988).

Table 6.1 The Antoine constants

Name	*A*	*B*	*C*	*Range of temperature* (K)
Water	16.26205	3799.887	46.854	284–441
Acetone	14.39155	2795.817	43.198	241–350
Ammonia	14.9331	2940.46	56.34	179–261
Benzene	13.8858	2788.51	52.36	280–377
Chloroform	13.9582	2696.79	46.16	26–370
Ethyl alcohol	16.5092	3578.91	50.50	270–369
n-Heptane	13.8587	2911.32	56.51	270–400
n-Hexane	13.8216	2697.55	48.78	245–370
Methyl alcohol	16.5725	3626.55	34.29	257–364
Toluene	13.9987	3096.52	53.67	280–410
n-Pentane	13.8183	2477.07	39.94	220–330
Acrylonitrile	13.9103	2782.21	51.15	255–385
Nitromethane	14.2043	2972.64	64.15	
Ethyl benzene	14.0045	3279.47	59.95	
Chlorobenzene	13.9926	3295.12	55.60	335–405
Acetonitrile	14.2724	2945.47	49.15	
Ethylene glycol	18.23507	6022.18	28.25	364–494
Glycerol	15.22417	4487.04	140.2	440–600
1,4-Dioxane	14.1177	2966.88	62.15	275–410
Methyl isopropyl ketone	12.16287	1993.12	103.2	271–406
Benzoic acid	15.14837	4190.70	125.2	405–560

It is to be remembered that the use of the Antoine equation outside the recommended range of temperature leads to incorrect results. The Antoine equation provides an excellent correlating equation for vapour pressures lying in the range of 1 to 200 kPa (Reid et al.).

More complex multi-parameter empirical correlations are available in literature for accurate representation of vapour pressure of pure substances. The Riedel equation is one such model, i.e.

$$\ln P^S = A - \frac{B}{T} + C \ln T + DT^6$$

where *A*, *B*, *C* and *D* are constants.

Lee and Kesler (*AIChE J.*, 21, 510, 1975) have proposed the following equation for predicting the vapour pressure of pure hydrocarbons knowing the critical properties and the acentric factor. The equation gives the vapour pressure at the reduced temperature T_r as

$$\ln P_r^S = (\ln P_r)^{(0)} + \omega(\ln P_r)^{(1)}$$

where $P_r^S = \dfrac{P^S}{P_c}$ and $T_r = \dfrac{T}{T_c}$ and ω is the acentric factor.

With P and P_c in kPa, the following equations may be used to evaluate $(\ln P_r)^{(0)}$ and $(\ln P_r)^{(1)}$.

$$(\ln P_r)^{(0)} = 5.927\,14 - \frac{6.096\,48}{T_r} - 1.288\,62 \ln T_r + 0.169\,347 T_r^6$$

$$(\ln P_r)^{(1)} = 15.2518 - \frac{15.6875}{T_r} - 13.4721 \ln T_r + 0.435\,77 T_r^6$$

This equation predicts the vapour pressure of pure hydrocarbons within 3 to 4 percent if $0.30 < T_r < 1.0$.

EXAMPLE 6.4 The Antoine constants for *n*-heptane are A = 13.8587, B = 2911.32, and C = 56.56. P^s is in kPa and T is in K. Calculate

(a) The vapour pressure of *n*-heptane at 325 K
(b) The normal boiling point of *n*-heptane

Solution (a) The Antoine equation is

$$\ln P^S = A - \frac{B}{T - C}$$

For *n*-heptane at 325 K, the vapour pressure is

$$\ln P^S = 13.8587 - \frac{2911.32}{325 - 56.56} = 3.0134$$

Therefore, P^S = 20.36 kPa.

(b) The normal boiling point is the temperature at which the vapour pressure is 101.3 kPa. Substitute P^S = 101.3 kPa in the Antoine equation.

$$\ln 101.3 = 13.8587 - \frac{2911.32}{T - 56.56}$$

Solving this equation, T = 371.62 K.

6.4 VAPOUR PRESSURE PLOTS

Vapour pressure, as we have seen, is a strong function of temperature. Generally, a large number of experimental data would be necessary to establish a functional relationship between vapour pressure and temperature graphically. Frequently in chemical engineering calculations, it becomes necessary to interpolate or extrapolate vapour pressure data from a very few experimental observations, these being easily carried out if a straight line results when the vapour pressure is plotted against temperature. The Clapeyron equation, i.e. Eq. (6.2) provides a relationship which indicates that a plot of the logarithm of the vapour pressure against the reciprocal of temperature is a straight line. It is observed that this relationship predicts the dependence of vapour pressure on temperature accurately when the pressure is low. But, there is a considerable deviation from the linear behaviour at higher pressures. The simplifications that were used in arriving at this relationship are no longer valid if the pressure is high. At high pressures, the molar volume of the liquid is comparable with that of the vapour and cannot be neglected. Also, the vapour deviates from ideal behaviour and the latent heat of vaporization is not constant. The reference substance plots for vapour pressures are useful in such cases, as these provide the vapour pressure as linear functions of temperature even at higher pressures. Two such plots are in common use: The equal-pressure reference-substance plots and the equal-temperature reference-substance plots.

6.4.1 Equal-Temperature Reference-Substance Plots

In equal-temperature reference-substance plots, the logarithm of the vapour pressure of a substance is plotted against the logarithm of the vapour pressure of a reference substance, both at the same temperature. Since the vapour pressure versus temperature data for water are readily available, water is usually chosen as the reference substance.

The Clapeyron equation (Eq. 6.6) may be written as

$$d \ln P^S = -\frac{\lambda}{R} d\left(\frac{1}{T}\right) \tag{6.9}$$

where P^S is the vapour pressure of the substance. The vapour pressure of the reference substance (P_R^S) at the same temperature may be written as

$$d \ln P_R^S = -\frac{\lambda_R}{R} d\left(\frac{1}{T}\right) \tag{6.10}$$

Dividing Eq. (6.9) by Eq. (6.10), we get

$$\frac{d \ln P^S}{d \ln P_R^S} = \frac{(\lambda/R)d(1/T)}{(\lambda_R/R)d(1/T)} \tag{6.11}$$

Equation (6.11) leads to

$$\ln P^S = \frac{\lambda}{\lambda_R} \ln P_R^S + \text{constant} \tag{6.12}$$

If the ratio λ/λ_R is constant, this equation predicts a linear behaviour when the logarithm of vapour pressure of a substance is plotted against the logarithm of the vapour pressure of the reference substance. Though the latent heat of vaporisation may vary with temperature, it is a reasonably good approximation to assume that λ and λ_R both vary to the same extent that their ratio remains constant. Thus, Eq. (6.12) provides a better means for predicting the vapour pressure variation with temperature than Eq. (6.1)

The equal-temperature reference-substance plots are known as the *Cox chart* because these are prepared based on the method suggested originally by E.R. Cox (*Ind. Eng. Chem.*, 15, 592, 1923). Figure 6.2 gives the Cox chart in which the vapour pressure of different substances are given in mm Hg. and temperature is given in °C. In these plots, the logarithm of the vapour pressure of the substance is plotted as the ordinate against the logarithm of the vapour pressure of the reference substance at the same temperature as the abscissa. For convenience, the abscissa scale is converted to read temperature instead of vapour pressure. This non-linear temperature axis is created as follows. Refer Figure 6.3. The ordinate reads the vapour pressure on the logarithmic scale. A straight line with positive slope is drawn, which represents the vapour pressure of the reference substance, say water. Take the vapour pressure of water at a desired temperature on the *y*-axis, and move horizontally to the straight line representing the vapour pressure. For example, the vapour pressure of water at 75°C is 289.2 mm Hg (38.55 kPa). From 289.2 on the *y*-axis, move horizontally to point *C* on the straight line *AB*. From point *C* move vertically downwards to point *D* on the *x*-axis. Point *D* is marked to read 75°C (348 K). The temperature 50°C (323 K) can be marked by taking the vapour pressure 92.6 mm Hg (12.34 kPa) on the vertical axis, moving horizontally to the line *AB* and then

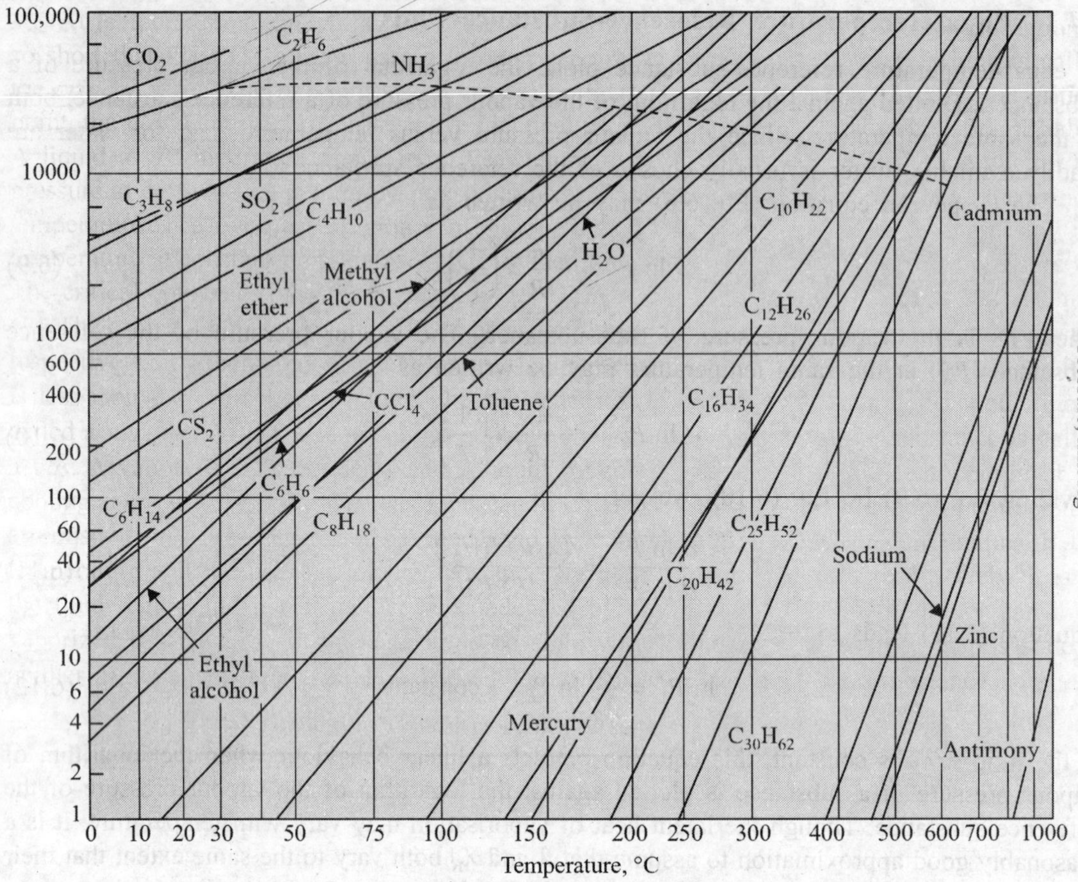

Figure 6.2 The Cox chart.

vertically to the x-axis to locate 50°C (323 K). Similarly, temperatures 10°C, 25°C, 100°C, 150°C, etc., can be marked off by taking the vapour pressures of water at the respective temperatures. Figure 6.3 illustrates the construction of the Cox chart.

The vapour pressures of other substances plotted on this special coordinate scales will yield straight lines over a considerable range of temperature. These lines can be drawn from the vapour pressure of a substance at just two temperatures. Another feature of the Cox chart is that the lines of vapour pressures of the members of a homologous series of compounds or members of a group of closely related compounds converge at a point, called the *infinite point* which is characteristic of that group. This feature makes it possible to establish the complete vapour pressure–temperature relationship of a member of the group with only one experimental value and the infinite point.

It has been shown that the temperature scale on the Cox chart with water as the reference substance is nearly equivalent to the function $(T+C)^{-1}$ where C is approximately equal to –43 K. Thus the Cox chart resembles the Antoine vapour pressure equation:

$$\ln P^S = A - \frac{B}{T-43} \tag{6.13}$$

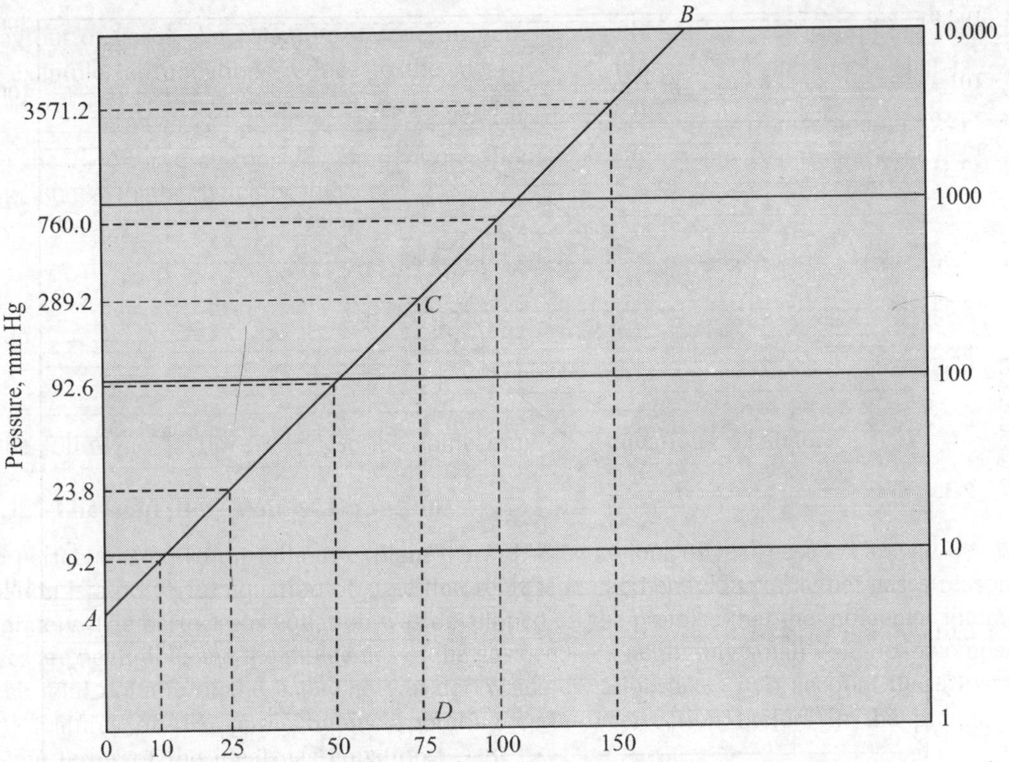

Figure 6.3 Construction of Cox chart.

EXAMPLE 6.5 Construct the Cox chart for the vapour pressure of acetone given that the vapour pressures are 8.52 kPa and 194.9 kPa at 273 K and 353 K respectively. The vapour pressure of water is obtained from steam tables as given below:

T (K)	273	293	313	323	333	353	373
P^S (kPa)	0.61	2.33	7.37	12.34	19.90	47.35	101.3

Compare the vapour pressure at 323 K given by the Cox chart with the experimental value of 73.94 kPa.

Solution Pressures are marked on the logarithmic scale on the vertical axis. A straight line *AB* is drawn with a convenient slope as shown in Figure 6.4. The temperature scale is established by the procedure explained earlier. For example, consider the second data point. At 293 K, the vapour pressure is 2.33 kPa. From 2.33 kPa on the vertical axis, move horizontally to line *AB* and make point *C*. From point *C*, move vertically downwards and mark T = 293 K, on the horizontal axis. Thus all temperatures can be marked. Points *D* and *E* are located such that they give the vapour pressures of acetone at 273 K and 353 K respectively. Line *DE* is drawn which can be used for interpolating and extrapolating the vapour pressure data of acetone at any temperature. Read the vapour pressure of acetone at 323 K from the graph. Vapour pressure of acetone at 323 K = 74.5 kPa.

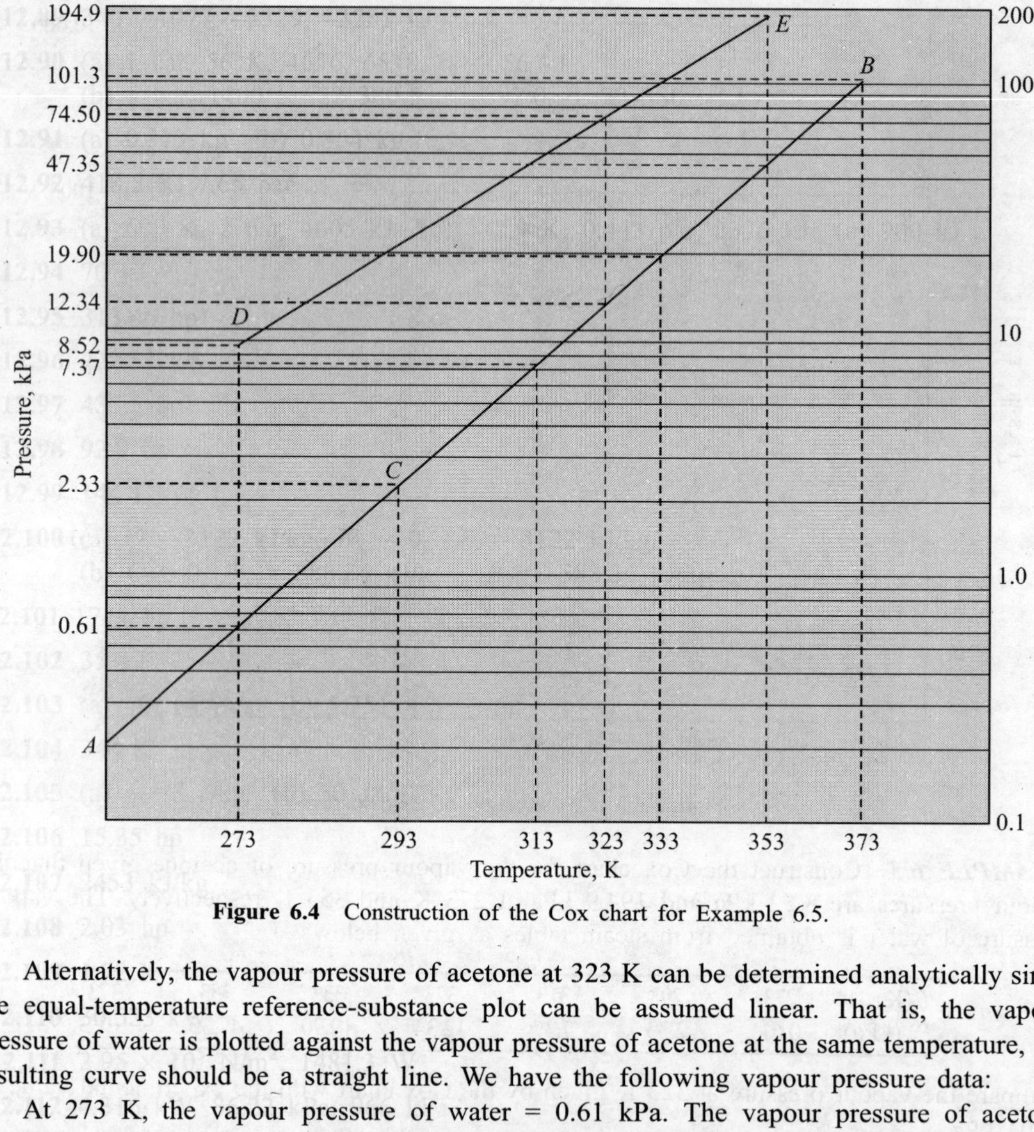

Figure 6.4 Construction of the Cox chart for Example 6.5.

Alternatively, the vapour pressure of acetone at 323 K can be determined analytically since the equal-temperature reference-substance plot can be assumed linear. That is, the vapour pressure of water is plotted against the vapour pressure of acetone at the same temperature, the resulting curve should be a straight line. We have the following vapour pressure data:

At 273 K, the vapour pressure of water = 0.61 kPa. The vapour pressure of acetone = 8.52 kPa. At 353 K, the vapour pressure of water = 47.35 kPa. The vapour pressure of acetone = 194.9 kPa. The vapour pressure of water at 323 K is 12.34 kPa.

The vapour pressure of acetone at 323 K (P^S) is now obtained by linear interpolation, knowing that the vapour pressure of water at this temperature is 12.34 kPa.

$$\frac{\log 194.9 - \log 8.52}{\log 47.35 - \log 0.61} = \frac{\log P^S - \log 8.52}{\log 12.34 - \log 0.61}$$

$$\frac{2.2898 - 0.9304}{1.6753 + 0.2147} = \frac{\log P^S - 0.9304}{1.0913 + 0.2147}$$

$\log P^S = 1.8698$. Therefore, the vapour pressure of acetone at 323 K, $P^S = 74.09$ kPa.

The vapour pressure given by the equal-temperature reference-substance method is greater than the experimental value of 73.94 kPa by 0.15 kPa. Therefore,

$$\text{percent deviation} = \frac{0.15}{73.94} \times 100 = 0.20\%$$

6.4.2 Equal-Pressure Reference-Substance Plots

In equal-pressure reference-substance plots, the temperature at which a substance has a given vapour pressure is plotted against the temperature at which the reference substance has the same vapour pressure. These are analogous to the Cox chart and the underlying equations can be derived in the same way as Eq. (6.11) was derived.

It is observed that if boiling point of a substance is plotted against the boiling point of a reference substance at the same pressure, a straight line results. This is found to be true for moderate ranges of pressures, though there may be variation over wide ranges of pressures. The Dühring lines are examples of equal-pressure reference-substance plots.

In Figure 6.5 the Dühring lines of various concentrations of sodium hydroxide are plotted with water as the reference substance. The temperature of water is taken as ordinate against the

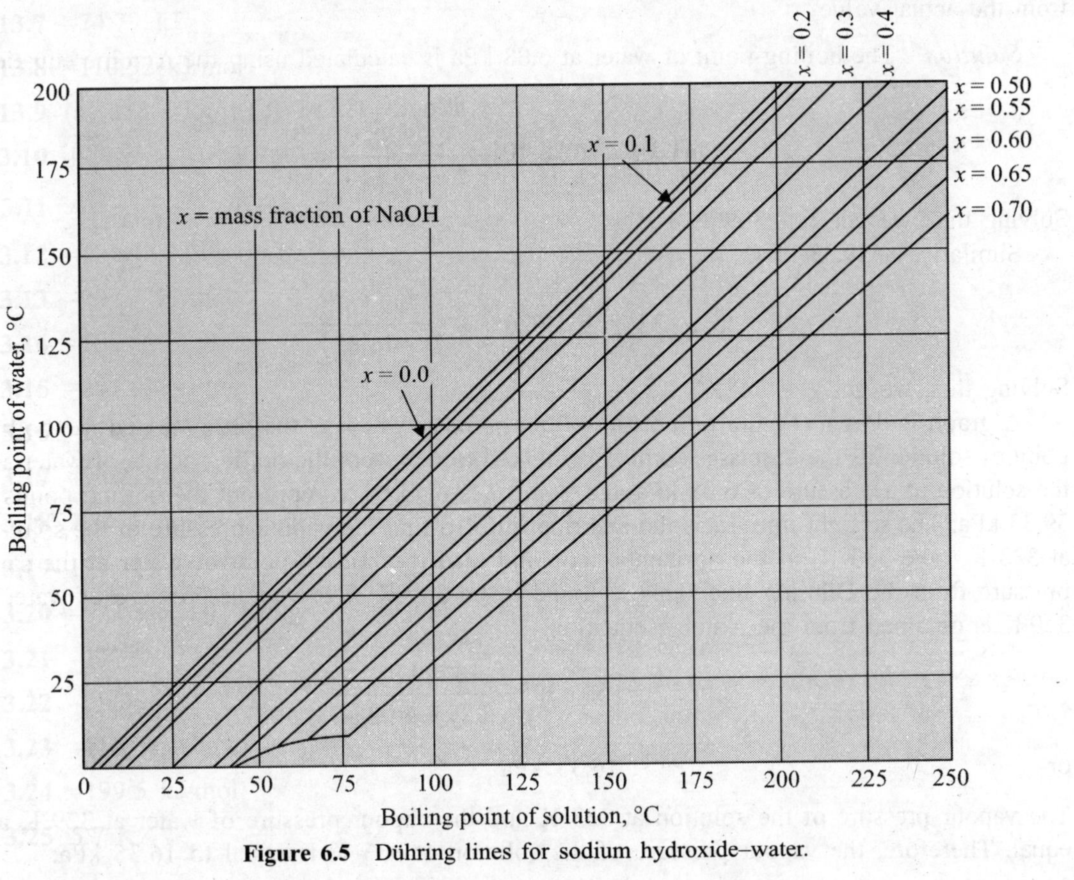

Figure 6.5 Dühring lines for sodium hydroxide–water.

temperature of solution having the same vapour pressure as the abscissa. These charts can be used to find the boiling point of a solution of known composition at a given pressure or the vapour pressure of the solution at a given temperature. For example, let the boiling point of a 30% solution of sodium hydroxide in water at a pressure of say, 80 kPa is to be measured. At 80 kPa, the boiling point of water is 93.3°C (366.5 K). From 93.3°C on the vertical axis move horizontally to the Dühring line of 30% NaOH and then vertically to the horizontal axis to get the boiling point of 30% solution at 80 kPa as 106.8°C (380 K).

EXAMPLE 6.6 The vapour pressure over a 25% solution of sulphuric acid in water is 6.08 kPa at 313 K and 39.33 kPa at 353 K. Assuming that the Dühring line is linear over this range, predict the vapour pressure over the solution at 333 K. The vapour pressure of water is given by the Antoine equation

$$\ln P^S = 16.26205 - \frac{3799.887}{T - 46.854}$$

where pressure is in kPa and temperature in K. If the experimental value of the vapour pressure of a 25% solution of sulphuric acid in water at 333 K is 16.53 kPa, what is the percent deviation of the vapour pressure predicted by the equal-pressure reference-substance method from the actual value?

Solution The boiling point of water at 6.08 kPa is calculated using the Antoine equation, i.e.

$$\ln 6.08 = 16.26205 - \frac{3799.887}{T - 46.854}$$

Solving this, we get T = 309.69 K.

Similarly, at 39.33 kPa,

$$\ln 39.33 = 16.26205 - \frac{3799.887}{T - 46.854}$$

Solving this, we get T = 348.67 K.

A graph is drawn (Figure 6.6) with boiling point of water as the ordinate and the boiling point of solution as the abscissa. Point A is marked to represent the boiling points of water and the solution at a pressure of 6.08 kPa and point B is marked to represent the boiling points at 39.33 kPa. The straight line AB is the Dühring line. To find the vapour pressure of the solution at 333 K, take 333 K on the horizontal axis and read the boiling point of water at the same pressure from the Dühring line. This is found to be 329 K. The vapour pressure of water at 329 K is obtained from the Antoine equation.

$$\ln P^S = 16.26205 - \frac{3799.887}{329 - 46.854}$$

or

$$P^S = 16.35 \text{ kPa}$$

The vapour pressure of the solution at 333 K and the vapour pressure of water at 329 K are equal. Therefore, the vapour pressure of the solution at 333 K is equal to 16.35 kPa.

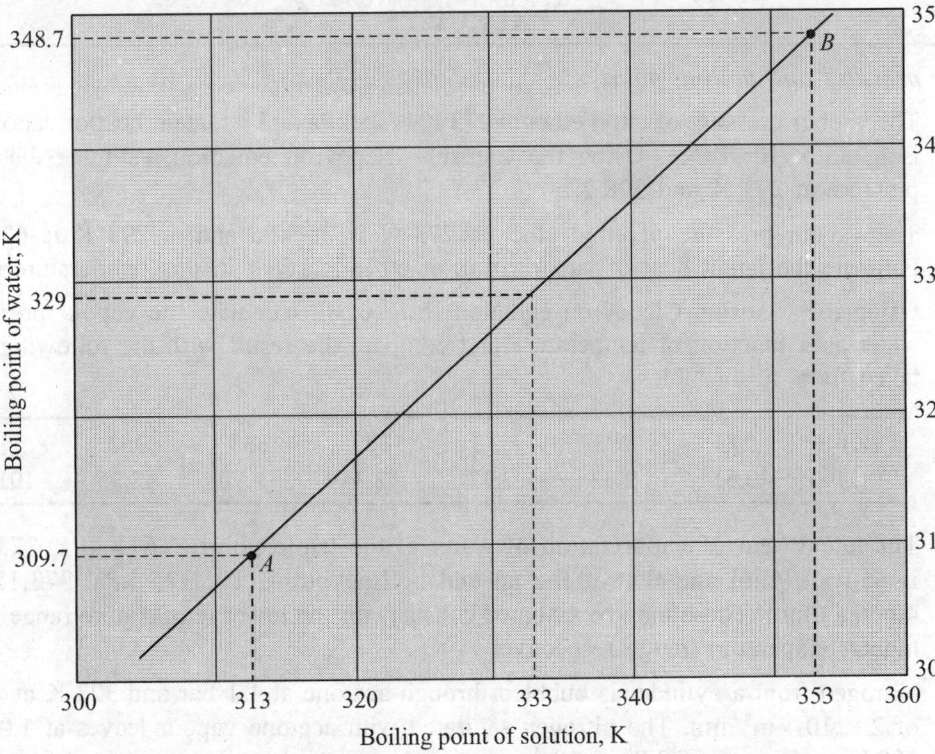

Figure 6.6 Equal pressure reference plot for sulphuric acid (Example 6.6).

Alternatively, assuming the Dühring line to be a straight line, the boiling point of water (T_R) can be calculated as

$$\frac{348.67 - 309.69}{353 - 313} = \frac{T_R - 309.69}{333 - 313}$$

which gives $T_R = 329.18$ K. Water at 329.18 K and 25% sulphuric acid at 333 K have the same vapour pressure. The vapour pressure of water at 329.18 K is determined using the Antoine equation.

$$\ln P^S = 16.26205 - \frac{3799.887}{329.18 - 46.854}$$

Therefore, P^S = 16.49 kPa. Thus, the vapour pressure of the solution at 333 K = 16.49 kPa. This is less than the experimental value of 16.53 kPa by 0.04 kPa. Therefore, the percent deviation is as follows:

$$\% \text{ deviation } = \frac{0.04}{16.53} \times 100 = 0.24\%$$

EXERCISES

Vapour pressure and boiling point

6.1 The vapour pressure of ethyl ether at 273 K is 25 kPa and its latent heat of vaporization is 4.185×10^2 kJ/kg. Using the Clausius–Clapeyron equation, estimate the vapour pressure at 293 K and 308 K.

6.2 The vapour pressure of ethyl ether at 273 K is 25 kPa and at 293 K is 63.5 kPa. Estimate the latent heat of vaporization of ether in kJ/kg in this temperature range.

6.3 Using the Clausius–Clapeyron equation [Eq. (6.5)], calculate the vapour pressure of water as a function of temperature and compare the result with the following values taken from steam tables.

T (K)	273	293	313	323	333	353	373
P^S (kPa)	0.61	2.33	7.37	12.34	19.90	47.35	101.3

The latent heat of vaporization of water at the triple point (0.611 kPa, 273.16 K) is 45.05 kJ/mol and that at the normal boiling point (101.325 kPa, 373.15 K) is 40.66 kJ/mol. These may be assumed constant for the lower temperature range and the higher temperature range respectively.

6.4 Nitrogen from a cylinder is bubbled through acetone at 1.1 bar and 323 K at the rate of 2×10^{-4} m^3/min. The nitrogen saturated with acetone vapour leaves at 1.013 bar, 308 K at the rate of 3.83×10^{-4} m^3/min. What is the vapour pressure of acetone at 308 K?

6.5 Mercury has a density of 13.69×10^3 kg/m^3 in the liquid state and 14.193×10^3 kg/m^3 in the solid state both measured at the melting point of 234.33 K at 1 bar. If the heat of fusion of mercury is 9.7876 kJ/kg, what is the melting point of mercury at 10 bar?

6.6 The vapour pressure of chloroform is given by the Antoine equation

$$\ln P^S = 13.9582 - \frac{2696.79}{T - 46.16}$$

where pressure is in kPa and temperature in K. Determine (a) the boiling point of chloroform at 50 kPa and (b) the vapour pressure at 300 K.

6.7 The Antoine constants of acrylonitrile are A = 13.9103, B = 2782.21, and C = 51.15, when the vapour pressure is in kPa and temperature in K. Calculate the vapour pressure at 300 K.

6.8 How many kilograms of acetone must have vaporized before an atmospheric chamber of volume 10 cubic metres gets saturated with the vapours of acetone at 313 K? The Antoine constants (pressure in kPa and temperature in K) for acetone are A = 14.5463, B = 2940.46 and C = 49.15.

6.9 The vapour pressure of water (kPa) is represented by the equation

$$\log P^S = -\frac{A}{T_r} + B - e^{-20(T_r - b)^2}$$

where $A = 3.142$ and $b = 0.163$ and temperature is in K. If the vapour pressure of water at 303.15 K is 4.241 kPa and the critical temperature of water is 647.3 K, what is the vapour pressure of water at 350 K? Get the vapour pressure of water from steam tables and calculate the percent deviation.

6.10 Air at 300 K and 1.0 bar containing 0.02 moles water per mole dry air is compressed and stored at 100 bar and 300 K for use in an experimental wind tunnel. How much water will condense in the high-pressure storage vessel per 100 cubic metres of air compressed? The vapour pressure of water at 300 K = 3.564 kPa.

6.11 A stream of gas at 302 K and 100 kPa, 50% saturated with water vapour is passed through a drying tower where 90% of the water vapour is removed. For 100 cubic metres of gas admitted through the tower, how many kilograms of water is removed? The vapour pressure of water at 302 K is 4.0 kPa.

Reference-substance plots

6.12 Prepare the Cox chart for *n*-heptane if the vapour pressures are 12.28 kPa at 313 K and 105.60 kPa at 373 K. Estimate the vapour pressure of *n*-heptane at 333 K using the chart. If the experimental value is 27.91 kPa, what is the percent deviation in the estimate? The following data for vapour pressure of water is taken from steam tables:

T (K)	273	293	313	323	333	353	373
P^S (kPa)	0.61	2.33	7.37	12.34	19.90	47.35	101.3

6.13 Construct the Cox chart for predicting the vapour pressure of ethanol and using the chart estimate the normal boiling point of ethanol and the vapour pressure at 323 K. By how much % these values depart from the actual values? The vapour pressure of ethanol at 293 K and 363 K are respectively, 5.76 kPa and 157.03 kPa. The vapour pressure of water is as follows:

T (K)	273	293	313	323	333	353	373
P^S (kPa)	0.61	2.33	7.37	12.34	19.90	47.35	101.3

The experimental value of the normal boiling point is 351.7 and the vapour pressure at 323 K is 29.2 kPa.

6.14 The following data refer to the vapour pressure of sulphur dioxide:

T (K)	279.5	328.7
P^S (kPa)	202.7	1013

(a) Fit the data into an equation of the form $\ln P^S = A - \dfrac{B}{T - 43}$ where P^S is in kPa and T in K.

(b) What is the vapour pressure of sulphur dioxide at 373 K?

6.15 The boiling points of methanol, ethanol and *n*-propanol at 8 kPa are 285.25 K, 299.15 K and 316.65 K respectively. Construct the Cox chart for methanol and ethanol given that their normal boiling points are 337.85 K and 351.7 K respectively. Using this chart predict the normal boiling point of *n*-propanol. How does your result compare with the experimental value of 370.95 K? The vapour pressure of water is given below:

T (K)	273	293	313	323	333	353	373
P^S (kPa)	0.61	2.33	7.37	12.34	19.90	47.35	101.3

6.16 The following table gives the vapour pressures of benzene and water. Plot the reference temperature vapour pressure curve with the logarithm of vapour pressure of benzene on the *y*-axis and the logarithm of vapour pressure of water on the *x*-axis.

T (K)	288.6	299.3	315.4	333.8	353.3	377.0	415.7	452
P_B^S (kPa)	8.0	13.3	26.7	53.3	101.3	202.7	506.6	1013.0
P_W^S (kPa)	1.8	3.4	8.3	26.1	49.0	116.3	382.5	980.0

(a) Determine the vapour pressure of benzene at 373 K.

(b) Determine the latent heat of vaporization of benzene at 298 K given that the latent heat of vaporization of water at 298 K is 2443 kJ/kg. Compare your result with the experimental value of 434 kJ/kg.

6.17 Construct the Dühring line for ethylene glycol if it is given that its normal boiling point is 470.45 K and the boiling point at a pressure of 8 kPa is 402.65 K. The vapour pressure of water is given below: $\ln P^S = 16.26205 - \dfrac{3799.887}{T - 46.854}$.

(a) By how much does the vapour pressure of ethylene glycol at 432 K deviate from the experimental value of 26.66 kPa?

(b) What is the boiling point of ethylene glycol at 2.67 kPa?

6.18 Construct the Dühring line for ethanol if it is known that the vapour pressure of ethanol at 299.1 K is 8 kPa and that at 351.55 K is 101.3 kPa. The vapour pressure of water is given by the Antoine equation

$$\ln P^S = 16.26205 - \frac{3799.887}{T - 46.854}$$

where pressure is in kPa and temperature is in K. Compare the vapour pressure of ethanol at 321.6 K given by the Dühring line with the experimental value 26.66 kPa.

7 Solutions and Phase Behaviour

Chemical engineering unit operations are concerned with transport, storage, concentration and separation of solutions containing two or more components. The components constituting the solution may exist as a miscible liquid solution or as two immiscible or partially miscible liquid phases. The components in the solution may be volatile or certain of the constituents may be non-volatile. The vapour pressure over the solutions is a very important consideration in the design and operation of process equipment. In this chapter, we are concerned with the evaluation of vapour pressure over solutions, calculation of the composition of the vapour in equilibrium with the given liquid, the evaluation of the bubble-point and dew-point temperatures of binary and multi-component mixtures.

7.1 IDEAL SOLUTIONS AND RAOULT'S LAW

7.1.1 Ideal Solutions

The vapour pressure of a constituent of a solution at any temperature will be different from the vapour pressure it exerts in the pure state. Consider a pure liquid *A* in equilibrium with its vapour. The number of molecules of *A* per unit area of the liquid surface is more than the number of molecules when *A* is in solution with another liquid, say, *B*. This is the reason why the vapour pressure over a solution is less than that of the pure liquid. Also, the forces that hold the molecules together in the liquid and thus prevent their escape into the vapour phase are not the same as the forces that are present in the pure liquid state. If the molecular forces of interaction between the like molecules (*A-A* or *B-B* interaction) were of the same order of magnitude as those between unlike molecules (the *A-B* interaction), the forces that a molecule of *A* experiences in the solution with *B* will be the same as it experiences in the pure state. If this is the case, the difference between the vapour pressure of a constituent in the solution and that in the pure state will be entirely due to the decrease in the number of molecules per unit area of the liquid surface, which in turn depends on the concentration of the particular component in the solution. Therefore, it is natural to expect the vapour pressure exerted by the constituents in the solution formed by chemically similar substances with approximately equal intermolecular forces to be directly proportional to the concentration in the solution. Such solutions are known as *ideal solutions*. A solution conforming to the ideal behaviour has the following general characteristics, all interrelated.

1. The components are chemically similar. The average intermolecular forces of attraction and repulsion in the pure state and in the solution are of approximately the same order of magnitude.
2. There is no volume change on mixing or the volume of the solution varies linearly with composition.
3. There is neither absorption nor evolution of heat when an ideal solution is formed by mixing the constituents; that is, there is no enthalpy change and no temperature change on mixing.
4. The components in an ideal solution obey Raoult's law.

7.1.2 Raoult's Law

Raoult's law states that the vapour pressure of a component in a solution (or the equilibrium partial pressure of the component) is directly proportional to its concentration. That is,

$$p_i = x_i P_i^S \tag{7.1}$$

where p_i is the partial pressure of component i and x_i is its mole fraction in the liquid. P_i^S is the vapour pressure of pure i at the same temperature as that of the solution.

Raoult's law is obeyed by ideal solutions over the entire range of compositions. However, it should be understood that there exists no ideal solution in the strict sense of the word; but actual mixtures approach ideality as a limiting behaviour. *Ideality* requires that the molecules of the constituents are similar in size, structure and chemical nature; only optical isomers of organic compounds meet these requirements. Thus a mixture of *ortho*-, *meta*-, and *para*-xylene conforms very closely to the ideal solution behaviour. Practically, adjacent or nearly adjacent members of the homologous series of organic compounds can be expected to form ideal solutions. Thus, in engineering calculations, mixtures of benzene and toluene, *n*-octane and *n*-hexane, ethyl alcohol and propyl alcohol, acetone and acetonitrile, paraffin hydrocarbons in paraffin oils, etc. can be treated as ideal solutions. Raoult's law applies as fair approximation to mixtures of hydrocarbons that show a reasonable similarity in molecular structure such as the ones encountered in petroleum industry. In most other cases Raoult's law applies only over a limited concentration range.

7.1.3 Raoult's Law and Vapour–Liquid Equilibrium Calculation

Consider a liquid mixture in equilibrium with its vapour. The vapour composition will be different from that of the liquid with which it is in equilibrium. The determination of the compositions of the vapour and liquid in equilibrium at specified conditions of temperature and pressure is important in many chemical engineering calculations, particularly in computations involved in separation operations. Also important is the establishment of conditions of temperature and pressure at which the vapour and liquid of known composition exist in equilibrium. These calculations are relatively simple for solutions obeying Raoult's law.

Consider a binary mixture of components A and B. Assume that Raoult's law is applicable to the solution over the entire composition range. When the solution is in equilibrium with its vapour, we can write for the equilibrium partial pressure of the constituents, the following:

$$p_A = x_A P_A^S \tag{7.2}$$

$$p_B = x_B P_B^S \tag{7.3}$$

where p_A and p_B are the partial pressures of components A and B respectively, and x_A and x_B are their respective mole fractions in the liquid. We know that for a binary solution,

$$x_A + x_B = 1 \qquad \text{or} \qquad x_B = (1 - x_A) \tag{7.4}$$

By Dalton's law, the total pressure of the vapour (P) is equal to the sum of the partial pressures of the constituents. Therefore,

$$P = p_A + p_B$$

That is,

$$P = x_A P_A^S + x_B P_B^S = x_A P_A^S + (1 - x_A) P_B^S \tag{7.5}$$

Using Eq. (7.5), the total pressure over a solution of known composition can be calculated. The mole fraction of a component in a gas phase, by Dalton's law, is the ratio of partial pressure to the total pressure. Let y_A and y_B be the mole fractions of components A and B respectively, in the vapour phase. By Dalton's law,

$$y_A = \frac{p_A}{P} \tag{7.6}$$

Equation (7.6) can be rewritten as given below using Equations (7.2) and (7.5).

$$y_A = \frac{x_A P_A^S}{x_A P_A^S + x_B P_B^S} \tag{7.7}$$

Similarly, for component B,

$$y_B = \frac{x_B P_B^S}{x_A P_A^S + x_B P_B^S} \tag{7.8}$$

It may be noted that $y_A + y_B = 1$.

7.1.4 *P-x-y* Diagram

Equations (7.2) and (7.3) indicate that when the partial pressures over the solution at constant temperature are plotted against the mole fraction x_A, we would get two straight lines shown as dotted lines in Figure 7.1.

We can write Eq. (7.5) as given below which indicates that the total pressure versus x_A also will be a straight line.

$$P = x_A P_A^S + (1 - x_A) P_B^S = (P_A^S - P_B^S) x_A + P_B^S \tag{7.9}$$

Equation (7.9) means that the total pressure over an ideal solution also varies linearly with liquid composition. Figure 7.1 also shows the total pressure versus y_A at constant temperature. The *P-x-y* diagram of Figure 7.1 is one method of representing the vapour–liquid equilibrium data at constant temperature.

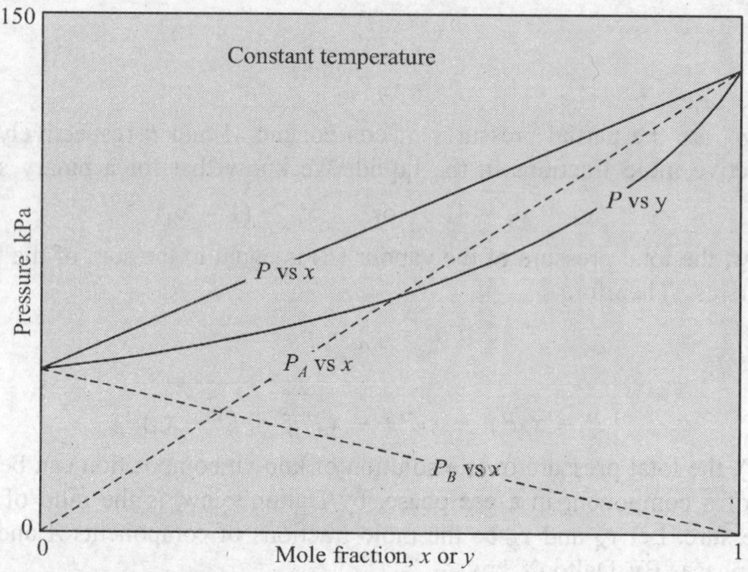

Figure 7.1 The *P-x-y* diagram of ideal solutions.

7.1.5 Boiling-point Diagram (*T-x-y* Diagram)

The *P-x-y* diagram gives the vapour–liquid equilibrium relation at constant temperature. Another method of expressing equilibrium is the *T-x-y* diagram, also known as the *boiling point diagram*. This refers to constant pressure conditions and is important in distillation calculations. Consider Eq. (7.5). The vapour pressures of the pure components depend on the temperature and therefore the equilibrium temperature at constant pressure is determined by trial and error so that at equilibrium Eq. (7.5) is satisfied for a given liquid concentration x_A. Equilibrium vapour composition y_A is then calculated using Eq. (7.6). This is repeated for different x_A values ranging from 0 to 1. When $x_A = 0$, the equilibrium temperature is the boiling point of pure *B*, and when $x_A = 1.0$, the equilibrium temperature is the boiling point of pure *A*. The plot of *T* versus x_A and *T* versus y_A are known as the *bubble-point curve* and the *dew-point curve* of the mixture at the given pressure *P*. These two curves together constitute the boiling point diagram (Figure 7.2).

The calculations for establishing the boiling point diagram can be simplified if we can eliminate the trial-and-error route for calculation of equilibrium temperature. This is achieved by assuming a temperature and using the vapour pressure values at this temperature in Eq. (7.5) to calculate the liquid composition, and Eq. (7.6) to calculate the vapour composition. For this purpose, Eq. (7.5) and Eq. (7.6) may be rearranged as

$$x_A = \frac{P - P_B^S}{P_A^S - P_B^S} \tag{7.10}$$

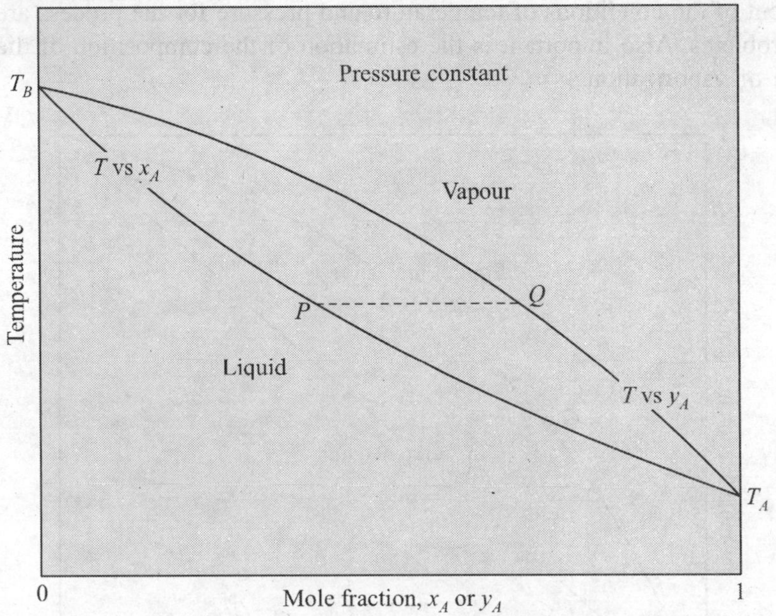

Figure 7.2 The boiling point diagram.

$$y_A = \frac{x_A P_A^S}{P} \tag{7.11}$$

It may be noted that the temperature assumed should lie between the boiling points of pure components at the given total pressure.

7.1.6 Equilibrium Diagram (*y-x* Diagram)

A horizontal line such as the line *PQ* in Figure 7.2 connects the composition of liquid and vapour in equilibrium at a specified temperature. Such a line connecting the composition of the equilibrium phases is called a *tie line*. Several such tie lines can be drawn, each giving a set of *x* and *y* values representing the composition of the liquid and vapour phases in equilibrium. On a plot with vapour composition (*y*) as the ordinate against the liquid composition *x* as the abscissa, these equilibrium compositions are represented as points such as point *M* in Figure 7.3.

A smooth curve drawn through such points gives the *equilibrium distribution diagram* or simply, the *equilibrium curve*. The equilibrium diagram also refers to a specified constant total pressure. The farther the equilibrium curve from the diagonal, the richer the vapour in component *A*. Separation of the solution into its components by distillation will be difficult if the equilibrium curve is very close to the diagonal.

7.1.7 Flash Vaporization

In flash vaporization, also known as *equilibrium distillation*, a liquid mixture is partially vaporized producing liquid and vapour products that are in equilibrium. The determination of the degree or extent of vaporization for producing a distillate (and a residue) of specified composition and the establishment of the conditions of temperature and pressure for the process are typical flash vaporization problems. Also important is the estimation of the composition of the products for a given degree of vaporization.

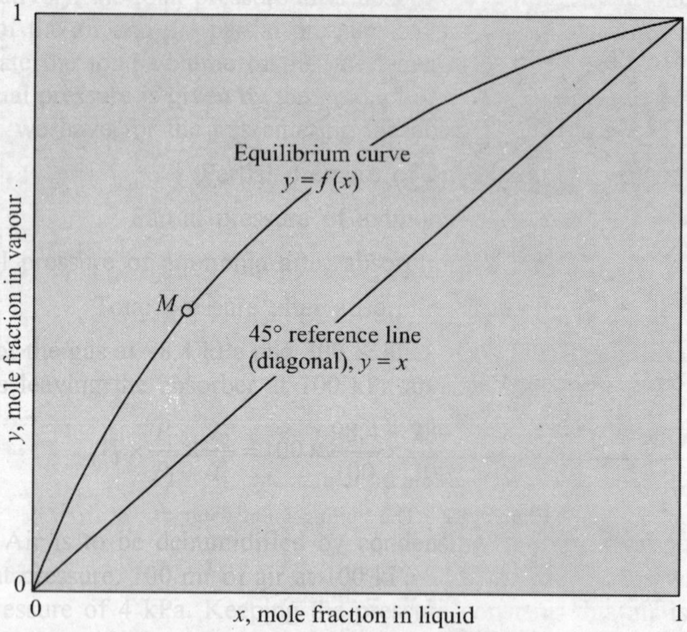

Figure 7.3 The equilibrium diagram.

Consider the flash vaporization process which is schematically represented in Figure 7.4. The liquid is heated by passing through a heat exchanger and the pressure is then reduced by passing the mixture through a valve. The sudden reduction in pressure results in partial vaporization of the liquid. The product liquid and vapour streams can be considered to be in equilibrium with each other owing to the intimate contact which prevails during the flashing operations. The products are then introduced into a vapour–liquid separating vessel and separated as distillate and residue.

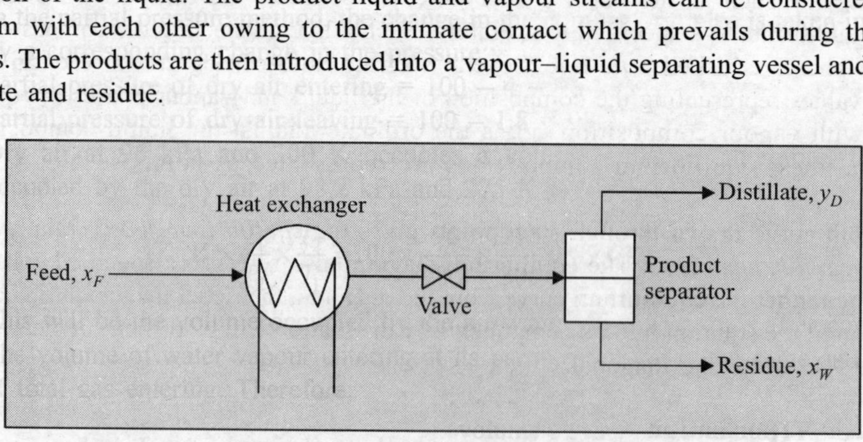

Figure 7.4 Flash distillation process.

If f is the fraction of the feed that is vaporized in a flash vaporization process, and if x_W and y_D are the mole fraction of component A in the liquid and vapour products respectively, then moles of A in the feed = moles of A in the vapour product + moles of A in the liquid product

$$x_F = fy_D + (1 - f)x_W$$

where x_F is the mole fraction of A in the feed liquid. This equation may be rearranged as

$$f = \frac{x_F - x_W}{y_D - x_W} \tag{7.12}$$

Equation (7.12) gives the fraction of the feed that is vaporized. Here x_W and y_D are equilibrium compositions, which are interrelated through Raoult's law if the solution is ideal.

EXAMPLE 7.1 At 360 K the vapour pressures of *n*-heptane and toluene are respectively 71.2 kPa and 48.9 kPa. Determine the composition of the liquid and vapour in equilibrium at 360 K and 65 kPa, if it is given that *n*-heptane and toluene form ideal solutions.

Solution Refer to Eq. (7.9)

$$P_A^S = 71.2; \quad P_B^S = 48.9, \quad P = 65.0$$

$$65 = 48.9 + (71.2 - 48.9)x$$

where x is the mole fraction of heptane in the liquid. On solving this, $x = 0.722$.

Using Eq. (7.6),

$$y = \frac{0.722 \times 71.2}{65} = 0.791$$

The liquid and the vapour at the given conditions contain respectively 72.2% (mol) and 79.1% (mol) heptane.

EXAMPLE 7.2 The vapour pressures of two pure liquids A and B are respectively 100 kPa and 60 kPa at 320 K. The concentration of A in the vapour in equilibrium with a solution of A and B is found to be 50% (mol). Determine the following:

(a) The composition of the liquid
(b) The total pressure of the vapour

Solution Partial pressures of the components are obtained by using Eqs. (7.2) and (7.3), i.e.

$$p_A = x_A P_A^S = 100x_A$$

$$p_B = x_B P_B^S = 60x_B$$

Equation (7.5) gives the total pressure:

$$P = x_A P_A^S + (1 - x_A)P_B^S = 100x_A + (1 - x_A)60 = 60 + 40x_A \tag{A}$$

(a) By Dalton's law

$$y_A = \frac{p_A}{P}$$

Substitution of the values gives

$$0.5 = \frac{100x_A}{60 + 40x_A}$$

Solving the above equation, we get $x_A = 0.375$. The liquid contains 37.5% (mole) *A*.

(b) Using Eq. (A), the total pressure is

$$60 + 40 \times 0.375 = 75 \text{ kPa}$$

EXAMPLE 7.3 A liquid mixture containing 25% *A*, 30% *B* and the rest *C* is in equilibrium with the vapour which contains 50% *B*. All percentages are on a mole basis. The equilibrium pressure and temperature are 200 kPa and 350 K. At 350 K the vapour pressure of *C* is 50 kPa. What is the percentage of *A* in the vapour?

Solution $x_A = 0.25$, $x_B = 0.30$, $x_C = (1.0 - 0.25 - 0.30) = 0.45$

$P_C^S = 50$ kPa. Using Raoult's law, the partial pressure of *C* in the vapour is

$$p_C = x_C P_C^S = 0.45 \times 50 = 22.5 \text{ kPa}$$

$$y_C = \frac{p_C}{P} = \frac{22.5}{200} = 0.1125$$

It is given that $y_B = 0.50$. Therefore, $y_A = 1 - 0.1125 - 0.50 = 0.3875$. The vapour contains 38.75% *A*.

EXAMPLE 7.4 A liquid mixture containing 65 mol % benzene and 35 mol % toluene is subjected to flash vaporization at 363 K and 101.3 kPa. The vapour pressure of benzene at this temperature is 136.09 kPa and the vapour pressure of toluene is 54.21 kPa. Flash vaporization is essentially an equilibrium stage operation. Calculate the mole percent of the feed that is vaporized.

Solution The concentration of liquid and vapour phases that exist in equilibrium at 363 K and 101.3 kPa are calculated using Eqs. (7.10) and (7.11):

$$x_A = \frac{P - P_B^S}{P_A^S - P_B^S} = \frac{101.3 - 54.21}{136.09 - 54.21} = 0.575$$

$$y_A = \frac{x_A P_A^S}{P} = \frac{0.575 \times 136.09}{101.3} = 0.773$$

These are the compositions of the liquid and vapour products that result on flash vaporization. Therefore, $x_W = 0.575$ and $y_D = 0.773$. Substituting these in Eq. (7.12), we can calculate the fraction of the feed that is vaporized.

$$f = \frac{x_F - x_W}{y_D - x_W} = \frac{0.65 - 0.575}{0.773 - 0.575} = 0.379$$

That is, 37.9% of the feed is vaporized during flash vaporization.

EXAMPLE 7.5 Mixtures of *n*-heptane (*A*) and *n*-octane (*B*) are expected to behave ideally. The total pressure over the system is 101.3 kPa. Using the vapour pressure data given below, (a) Construct the boiling point diagram and, (b) construct the equilibrium diagram.

T (K)	371.4	378	383	388	393	398.6
P_A^S (kPa)	101.3	125.3	140.0	160.0	179.9	205.3
P_B^S (kPa)	44.4	55.6	64.5	74.8	86.6	101.3

Solution

Sample calculation. Consider the second set of data:

$$T = 378 \text{ K}; \ P_A^S = 125.3 \text{ kPa}; \ P_B^S = 55.6 \text{ kPa}.$$

Using Eq. (7.9),

$$101.3 = 55.6 + x_A (125.3 - 55.6)$$

Therefore, $x_A = 0.656$. Using Eq. (7.6), we see

$$y_A = 0.656 \times 125.3/101.3 = 0.811$$

These calculations are repeated for other temperatures. The results are tabulated below:

T (K)	371.4	378	383	388	393	398.6
x_A	1.000	0.656	0.487	0.312	0.157	0
y_A	1.000	0.811	0.674	0.492	0.279	0

(a) A plot of T versus x and y gives the boiling point diagram [Figure 7.5(a)].

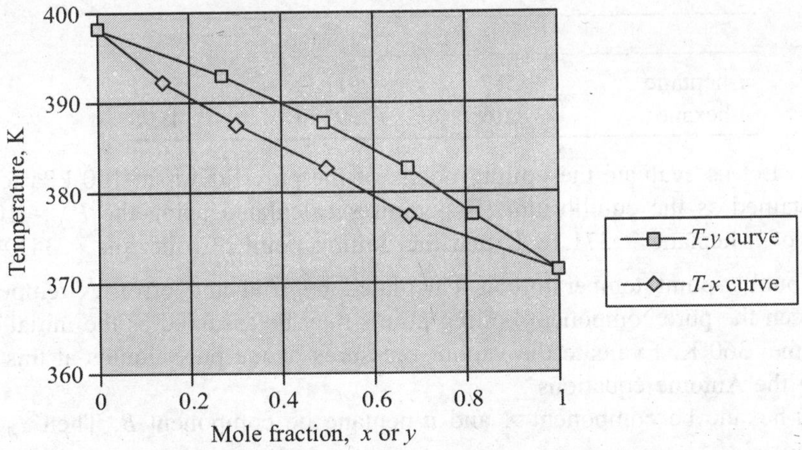

Figure 7.5(a) Boiling point diagram for Example 7.5.

(b) A plot of y against x gives the equilibrium diagram [Figure 7.5(b)].

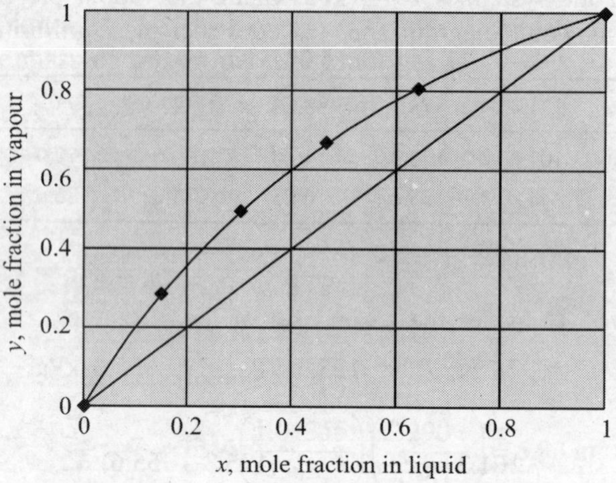

Figure 7.5(b) Equilibrium curve for Example 7.5.

EXAMPLE 7.6 A mixture of *n*-hexane and *n*-heptane containing 25% *n*-hexane is vaporized at a pressure of 100 kPa. Determine the following:

(a) The bubble point temperature
(b) The composition of the vapour

The vapour pressures are given by the Antoine equation

$$\ln P^S = A - \frac{B}{T - C}$$

where P is in kPa and T in K. The Antoine constants are given as follows:

	A	*B*	*C*
n-heptane	13.8587	2911.32	56.51
n-hexane	13.8216	2697.55	48.78

Solution Let us evaluate the boiling points of the pure liquids at 100 kPa total pressure. These are obtained as the equilibrium temperatures calculated using the P^S = 100 kPa. The boiling point of *n*-heptane = 371.13 K and the boiling point of *n*-hexane = 341.47 K.

(a) The bubble-point temperature is calculated by trial and error. A temperature lying between the pure component boiling points may be assumed as the initial guess value. Assume, 360 K. Evaluate the vapour pressures of the pure liquids at this temperature using the Antoine equations.

Let *n*-hexane be component A and *n*-heptane be component B. Then x_A = 0.25.

$$T = 360 \text{ K}$$

$$\ln P_A^S = 13.8216 - \frac{2697.55}{360 - 48.78} \Rightarrow P_A^S = 173.11 \text{ kPa}$$

$$\ln P_B^S = 13.8587 - \frac{2911.32}{360 - 56.51} \Rightarrow P_B^S = 71.23 \text{ kPa}$$

$$P = x_A P_A^S + (1 - x_A) P_B^S = 0.25 \times 173.11 + 0.75 \times 71.23 = 96.7 \text{ kPa}$$

Since the total pressure is less than 100 kPa, equilibrium temperature is not equal to 360 K.

Assume another value for T, say, 365 K.

$$T = 365 \text{ K}$$

$$\ln P_A^S = 13.8216 - \frac{2697.55}{365 - 48.78} \Rightarrow P_A^S = 198.54 \text{ kPa}$$

$$\ln P_B^S = 13.8587 - \frac{2911.32}{365 - 56.51} \Rightarrow P_B^S = 83.21 \text{ kPa}$$

$$P = x_A P_A^S + (1 - x_A) P_B^S = 0.25 \times 198.54 + 0.75 \times 83.21 = 112.04 \text{ kPa}$$

As the pressure is greater than 100 kPa, the temperature lies between 360 K and 365 K. By a few more trials we can see that at

$$T = 361.125 \text{ K}$$

$$\ln P_A^S = 13.8216 - \frac{2697.55}{361.125 - 48.78} \Rightarrow P_A^S = 178.60 \text{ kPa}$$

$$\ln P_B^S = 13.8587 - \frac{2911.32}{361.125 - 56.51} \Rightarrow P_B^S = 73.80 \text{ kPa}$$

$$P = x_A P_A^S + (1 - x_A) P_B^S = 0.25 \times 178.60 + 0.75 \times 73.80 = 100 \text{ kPa}$$

Therefore, the bubble point temperature = 361.125 K

(b) The partial pressure of *n*-Hexane in the vapour at the bubble point is

$$p_A = x_A P_A^S = 0.25 \times 178.60 = 44.65 \text{ kPa}$$

Since $y_A P = p_A = 44.65$ kPa, the mole fraction of hexane in the vapour is

$$y_A = \frac{p_A}{P} = \frac{44.65}{100} = 0.4465$$

The vapour contains 44.65% hexane and 55.35% heptane.

EXAMPLE 7.7 The vapour pressures of acetone (A) and acetonitrile (B) can be evaluated by the following Antoine equation.

$$\ln P_A^S = 14.39155 - \frac{2795.817}{T - 43.198}$$

$$\ln P_B^S = 14.2724 - \frac{2945.47}{T - 49.15}$$

where T is in K and P is in kPa. Assume that acetone and acetonitrile form ideal solutions. For a vapour phase consisting of 40% acetone and 60% acetonitrile, calculate the following:

(a) The dew point temperature and concentration of the first drop of liquid at 65 kPa
(b) The dew point pressure and concentration of the first drop of liquid formed at 327 K

Solution

(a) Equation (7.10) gives

$$x_A = \frac{P - P_B^S}{P_A^S - P_B^S}$$

Consider Eq. (7.11)

$$y_A = \frac{x_A P_A^S}{P}$$

When Eq. (7.10) is substituted in Eq. (7.11) this, we get

$$y_A = \frac{P_A^S}{P} \times \frac{P - P_B^S}{P_A^S - P_B^S}$$

Here $y_A = 0.4$ and $P = 65$. Therefore,

$$\frac{P_A^S}{65} \times \frac{65 - P_B^S}{P_A^S - P_B^S} = 0.4$$

Assume a temperature and calculate the vapour pressures using the Antoine equation. Substitute the vapour pressure values in the above equation. See that the LHS is equal to 0.4. This is repeated till the LHS of the above equation becomes equal to 0.4.

At $T = 333.55$ K, $P_A^S = 117.03$ kPa and $P_B^S = 50.18$ kPa

$$\frac{P_A^S}{P} \times \frac{P - P_B^S}{P_A^S - P_B^S} = 0.3991$$

As this is very close to the required value of 0.4, $T = 334.15$ K. Using Eq. (7.11),

$$x_A = \frac{y_A P}{P_A^S} = \frac{0.4 \times 65}{117.03} = 0.222$$

(b) At 327 K, we have $P_A^S = 93.71$ kPa and $P_B^S = 39.31$ kPa. Here $y = 0.4$

Equation (7.7) relates y to x, i.e.

$$y_A = \frac{x_A P_A^S}{x_A P_A^S + x_B P_B^S}$$

This can be rearranged to the following form.

$$x_A = \frac{y_A P_B^S}{P_A^S - y_A (P_A^S - P_B^S)}$$

Substituting the given values, we get $x_A = 0.2185$. From Eq. (7.11), dew-point pressure is

$$\frac{x_A P_A^S}{y_A} = \frac{0.2185 \times 93.71}{0.4} = 51.2 \text{ kPa}$$

7.2 NON-IDEAL SOLUTIONS

7.2.1 Deviation from Ideal Behaviour

Components in an ideal solution obey Raoult's law over the full concentration range. For solutions to be ideal, the components must be chemically similar and the intermolecular forces must be of the same order of magnitude whether they are in the pure state or in the solution. Only a very few systems approach ideal behaviour in the true sense of the term, although many can be treated as ideal solutions for engineering calculations. The non-ideal behaviour of liquid mixtures arises due to the dissimilarity among molecules. The dissimilarities may be due to the difference in the molecular structure or due to the difference in the molecular weight. For such solutions, the plot of partial pressure against liquid composition will not be a straight line as is the case with ideal solutions. The partial pressure curves will be above or below the straight line representing ideal solution behaviour. If the partial pressure is greater than that given by Raoult's law, the solution is said to exhibit a *positive deviation from ideality*, and if it is less than the partial pressure given by Raoult's law, the solution is said to exhibit a *negative deviation from ideality*. The total pressure versus liquid composition also will be above or below the Raoult's law line depending upon whether the solution exhibits a positive or negative deviation from ideality. The general nature of the vapour pressure curves showing positive and negative deviations are shown in Figure 7.6.

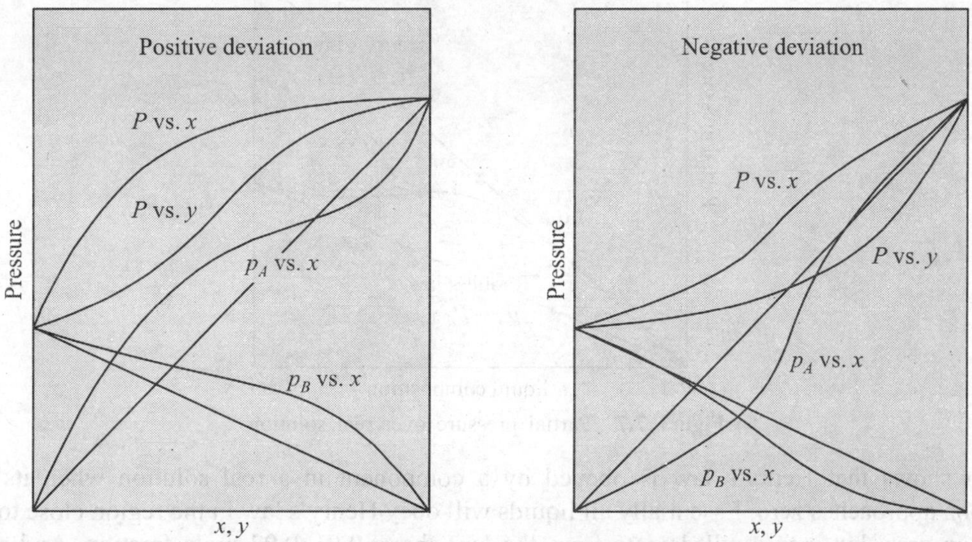

Figure 7.6 *P-x-y* curves for non-ideal solutions.

These diagrams refer to constant temperature conditions. The uppermost curves give the total vapour pressure as a function of liquid composition (*P-x* curve). The corresponding curves as a function of the vapour composition (*P-y* curve) lie below the *P-x* curves as the vapour is rich in the more volatile component. An important characteristic of real solutions is evident from these curves. As the mole fraction of the components approach unity, the partial pressure becomes asymptotic with the straight line that is applicable to ideal solutions. That is, all real solutions exhibit ideal behaviour as the mole fraction of the components tends to unity or mathematically,

$$p_A = x_A P_A^S \quad \text{as} \quad x_A \to 1 \tag{7.13}$$

In dilute solutions, since the solvent is present in larger proportions, we can say that the solvent in a dilute solution obeys Raoult's law.

7.2.2 Henry's Law

Henry's law describes another form of ideal behaviour shown by all real solutions. Henry's law states that the partial pressure of a component over a liquid solution is directly proportional to the concentration of the component in the liquid. Mathematically,

$$p_i = x_i H_i \tag{7.14}$$

the proportionality constant H_i is known as the Henry's law constant. Like vapour pressure, the constant depends on temperature. When it is numerically equal to the vapour pressure of the component, Henry's law reduces to Raoult's law. Figure 7.7 gives the partial pressure over a real solution.

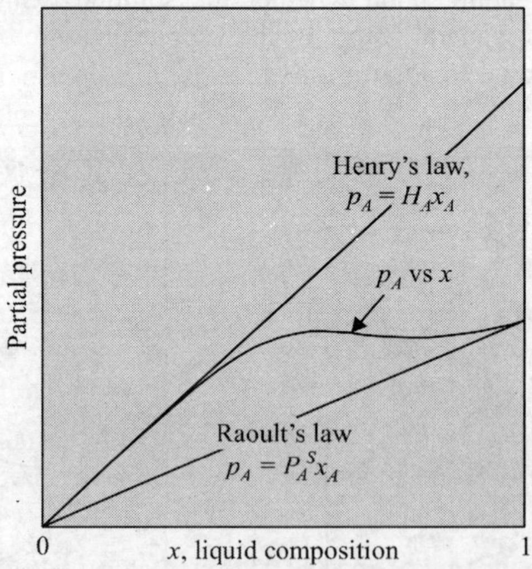

Figure 7.7 Partial pressure over real solution.

It shows that Henry's law is obeyed by a component in a real solution when its mole fraction approaches zero. Essentially all liquids will obey Henry's law in the region close to mole fraction zero, but many will deviate from the law above 0.01–0.02 mole fraction. And almost all liquids deviate above 0.1 mole fraction. But in some exceptional cases, Henry's law is found to be obeyed quite well up to $x_i = 0.5$. In a dilute solution as the solute is present in very small proportions it obeys Henry's law. Combining it with our earlier observation, we can generalize that *in dilute solution, the solute obeys Henry's law and the solvent obeys Raoult's law.*

7.2.3 Henry's Law and Gas Solubility

The solubility of a gas in a liquid can be predicted using the Henry's law constant of the gas. Since the solubility of the gas in a liquid is usually very low, the mole fraction of a gas in a saturated liquid solution is very small. The solute gas obeys Henry's law and therefore its partial pressure will be directly proportional to its mole fraction, the proportionality constant being the Henry's law constant. It means that the mole fraction or the solubility of the gas in the liquid is proportional to the partial pressure of the gas over the liquid as given by

$$x_i = \frac{p_i}{H_i} \tag{7.15}$$

EXAMPLE 7.8 An aqueous solution of acetaldehyde contains 2% acetaldehyde by weight. The partial pressure of acetaldehyde over the solution is found to be 41.4 kPa at 367 K. What will be the partial pressure over a 0.1 molal solution at the same temperature?

Solution The mole fraction of acetaldehyde (MW = 44.032) in the 1% aqueous solution

$$= \frac{2/44.032}{(2/44.032)+(98/18.016)} = 8.28 \times 10^{-3}$$

The Henry's law constant for acetaldehyde H_A is given by Eq. (7.14)

$$41.4 = 8.28 \times 10^{-3}\ H_A$$

or

$$H_A = \frac{41.4}{8.28 \times 10^{-3}} = 5.00 \times 10^{3} \text{ kPa}$$

The mole fraction of acetaldehyde in 0.1 molal solution is

$$\frac{0.1}{1000/18.016 + 0.1} = 1.8 \times 10^{-3}$$

Partial pressure over 0.1 molal solution is computed by Eq. (7.14):

$$p_A = H_A x_A = 5.00 \times 10^{-3} \times (1.8 \times 10^{-3}) = 9 \text{ kPa}$$

7.3 VAPOUR–LIQUID EQUILIBRIUM CALCULATONS

7.3.1 Vaporization Equilibrium Constants

For high-pressure vapour–liquid equilibrium calculations, it is convenient to express the phase equilibrium relations in terms of vaporization equilibrium constants or K factors. It is defined as the ratio of mole fraction in the vapour phase y to that in the liquid x or $K = (y/x)$. The vaporization equilibrium constant is a complex function of temperature, pressure, and the vapour phase and liquid phase compositions. If Raoult's law is applicable to the liquid solution and Dalton's law is obeyed by the vapour phase, the vaporization constant becomes

$$K_i = \frac{y}{x} = \frac{p_i / P}{p_i / P_i^S} = \frac{P_i^S}{P} \tag{7.16}$$

Thus, K factors of ideal solutions depend only on temperature and pressure. Equation (7.16) can be applied to components below their critical temperatures and low pressures even for non-ideal solutions. However, at high pressures and high temperatures, Eq. (7.16) will give incorrect results. Many correlations are available for estimating K values. DePriester [*Chem. Eng. Progr., Symposium Ser.*, 7, 49 (1953)] nomographs provide such correlations for many hydrocarbons. Calculations of the temperature and pressure at the bubble point and dew point of a multicomponent mixture as well as flash vaporization calculations are easily done using the vaporization equilibrium constants as detailed below.

7.3.2 Bubble-Point Temperature and Pressure

The temperatures at which a given mixture can exist as liquid and vapour phases in equilibrium at a specified pressure lie between the *bubble point* and *dew point* of the mixture. At the bubble point, the mixture begins to vaporize and the liquid phase has the same composition as the original mixture. The vapour in equilibrium with this liquid will have composition

$$y_i = K_i x_i$$

Since the sum of the mole fractions is unity, we have

$$\Sigma\, y_i = \Sigma\, K_i x_i = 1 \tag{7.17}$$

If the vaporization constants can be expressed as a function of temperature, Eq. (7.17) can be solved for the equilibrium temperature which in this case is the bubble point of the mixture. If K values are obtained from charts or nomographs, a temperature is assumed and the K values at this temperature are substituted into Eq. (7.17). The bubble point is determined by trial such that Eq. (7.17) is satisfied. For ideal solutions, the bubble point is determined by solving the following equation:

$$\Sigma\, P_i^S x_i = P \tag{7.18}$$

which is a generalization of Eq. (7.5) for multicomponent solutions. To find the bubble-point pressure, a similar procedure as above is adopted, but by assuming various values of pressure until $\Sigma\, K_i x_i = 1$.

7.3.3 Dew-Point Temperature and Pressure

The temperature at which the first drop of condensate is formed when a vapour is cooled at constant pressure is the *dew point* of the vapour. The vapour in equilibrium with the liquid at the dew point has the same composition as the original mixture. In order to find the dew-point temperature, a temperature is assumed arbitrarily and the K_i is determined for this temperature. Then,

$$x_i = \frac{y_i}{K_i}$$

At the dew point,

$$\Sigma x_i = \Sigma \frac{y_i}{K_i} = 1.00 \quad (7.19)$$

If the sum on the left-hand side of the equation is not equal to unity, repeat the calculation by assuming another temperature till this equation is satisfied. The trial-and-error solution is not needed if the vaporization constants are expressed as an explicit function of temperature. For ideal solutions, Eq. (7.19) can be put in the following form:

$$\Sigma x_i = P \Sigma \frac{y_i}{P_i^S} = 1.00 \quad (7.20)$$

Given the vapour pressure as a function of temperature through an equation such as the Antoine equation, Eq. (7.20) can be directly solved for the dew-point temperature.

7.3.4 Flash Vaporization Calculations

The general flash vaporization problem can be stated thus: Given a mixture of known overall composition z_i at a temperature T and a pressure P, what is the fraction that is vapour (V) and what are the compositions of the liquid and vapour phases in equilibrium?

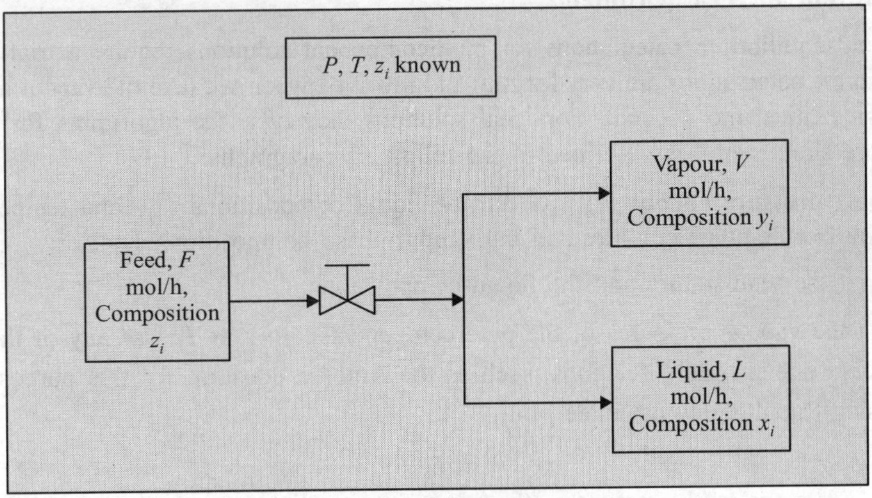

Figure 7.8 Flash vaporization calculation.

The overall material balance for the system is

$$F = V + L$$

where F is the total number of moles of the initial mixture. The component i balance for the system is

$$Fz_i = Vy_i + Lx_i \quad (7.21)$$

Since $y_i = K_i x_i$, it can be eliminated from Eq. (7.20) to get the following:

$$x_i = \frac{F z_i}{K_i V + L} = \frac{F}{V}\left(\frac{z_i}{K_i + L/V}\right) \tag{7.22}$$

Since $\Sigma x_i = 1$, we have

$$\sum \frac{z_i}{K_i + L/V} = \frac{V}{F} \tag{7.23}$$

The above equation can be used for the calculation of T, P or fractional vaporization in an iterative procedure. Alternatively x_i can be eliminated from Eq. (7.21) which leads to

$$y_i = \frac{F z_i}{V + L/K_i} = \frac{F}{V}\left(\frac{z_i}{1 + L/(VK_i)}\right) \tag{7.24}$$

As $\Sigma y_i = 1$

$$\sum \frac{z_i}{1 + L/(VK_i)} = \frac{V}{F} \tag{7.25}$$

Equation (7.25) also can be utilized in an iterative procedure to estimate T, P or the fraction of the initial mixture that is vaporized.

7.3.5 Solution of VLE Problems

Vapour-liquid equilibrium calculations for multicomponent solutions require a trial-and-error solution and the calculations are very lengthy and involved when we take the vapour and liquid phase non-idealities into account. For ideal solutions, however, the algorithms for the VLE problems are simpler and is described in the following paragraphs.

Bubble-point pressure (Problem): Given the liquid compositions $\{x_i\}$ and temperature T, calculate the bubble-point pressure and the vapour-phase compositions $\{y_i\}$.

1. Input the temperature and the liquid compositions.
2. Get the vapour pressures of the pure components $\{P_i^S\}$ at T. Use any of the vapour pressure-temperature relations, such as the Antoine equation for this purpose.
3. Using Raoult's law calculate p_i.

$$p_i = x_i P_i^S$$

4. Calculate the total pressure as the sum of the partial pressures.

$$P = \Sigma p_i = \Sigma x_i P_i^S$$

5. The vapour-phase compositions are calculated using Dalton's law:

$$y_i = \frac{p_i}{P} = \frac{x_i P_i^S}{\Sigma x_i P_i^S}$$

Bubble-point temperature (Problem): Given the liquid compositions $\{x_i\}$ and pressure P, calculate the bubble-point temperature and the vapour-phase compositions $\{y_i\}$.

1. Input the pressure and liquid compositions.
2. Evaluate the boiling points of the pure species $\{T_i^S\}$ at the given pressure.
3. Start with an initial guess value for the bubble point temperature. The mole fraction weighted average of the boiling points of the pure components may be chosen as the initial guess value.

$$T = \Sigma\, x_i T_i^S$$

4. Get the vapour pressure at temperature T and calculate $x_i K_i$, which for an ideal solution is $x_i \dfrac{P_i^S}{P}$. Since $y_i = x_i K_i$,

$$\Sigma\, x_i \frac{P_i^S}{P} = 1.0 \quad \text{or} \quad \frac{P}{\Sigma\, x_i P_i^S} = 1.0 \tag{7.26}$$

5. If the above equality is satisfied, go to step 7. Otherwise change the temperature assumed for the bubble point. To ensure faster convergence to the desired value, first evaluate P_k^S using the relation

$$\frac{P}{\Sigma\, x_i \alpha_{ik}} = P_k^S$$

which is obtained by multiplying both sides of Eq. (7.26) with P_k^S, the vapour pressure of the least volatile component and by letting $\alpha_{ik} = \dfrac{P_i^S}{P_k^S}$, the relative volatility of component i with reference to component k.

6. Using the P_k^S calculated in step 5, evaluate the saturation temperature using the Antoine equation or any other vapour pressure-temperature relation. Let this be the new guess value for bubble-point temperature and go to step 4.
7. Calculate the vapour phase compositions using

$$y_i = K_i x_i = x_i \frac{P_i^S}{P}$$

Dew-point pressure (Problem): Given the vapour compositions $\{y_i\}$ and temperature T, calculate the dew-point pressure and the liquid-phase compositions $\{x_i\}$.

1. Input the temperature and the vapour compositions.
2. Get the vapour pressures of the pure components $\{P_i^S\}$ at T. Use any of the vapour pressure–temperature relations, such as the Antoine equation for this purpose.

3. Calculate the total pressure using Eq. (7.20):

$$P = \frac{1}{\Sigma \frac{y_i}{P_i^S}}$$

4. The liquid-phase compositions are calculated using the following relation:

$$x_i = \frac{y_i}{K_i} = \frac{y_i P}{P_i^S}$$

Dew-point temperature (Problem): Given the vapour-phase compositions $\{y_i\}$ and pressure P, calculate the dew-point temperature and the liquid-phase compositions $\{x_i\}$.

1. Input the pressure and liquid compositions.
2. Evaluate the boiling points of the pure species $\{T_i^S\}$ at the given pressure.
3. Start with an initial guess value for the dew-point temperature. The mole fraction weighted average of the boiling points of the pure components is chosen as the initial guess value.

$$T = \Sigma\, y_i T_i^S$$

4. Get the vapour pressure at temperature T and calculate $P\Sigma \frac{y_i}{P_i^S}$.
5. If $P\Sigma \frac{y_i}{P_i^S} = 1$ go to step 7. Otherwise change the temperature assumed for the dew point. To ensure faster convergence to the desired value, first evaluate P_k^S using the relation

$$P\Sigma \frac{y_i}{\alpha_{ik}} = P_k^S$$

which is obtained by multiplying both sides of Eq. (7.20) with P_k^S, the vapour pressure of the least volatile component and by letting $\alpha_{ik} = \frac{P_i^S}{P_k^S}$, the relative volatility of component i with reference to component k.

6. Using the P_k^S calculated in step 5, evaluate the saturation temperature using the Antoine equation or any other vapour pressure–temperature relations. Let this be the new guess value for dew-point temperature and go to step 4.
7. Calculate the liquid phase compositions using

$$x_i = \frac{y_i}{K_i} = y_i \frac{P}{P_i^S}$$

Flash vaporization (Problem 1): Determine the fraction of the feed vaporized (V/F) from a mixture of composition $\{z_i\}$, given temperature T and pressure P of flash vaporization.

1. Input the composition of the feed and the temperature and pressure.
2. Get the vapour pressures of the pure liquids at the given temperature.
3. Check whether vapour and liquid phases exist in equilibrium under the given condition. This is done by evaluating the bubble-point and the dew-point pressures at the given temperature. The bubble-point pressure is

$$P_b = \Sigma\, x_i P_i^S$$

and the dew-point pressure is $P_d = \dfrac{1}{\Sigma \dfrac{z_i}{P_i^S}}$. If the given pressure lies between P_b and P_d flash vaporization is possible and proceed with step 4. Otherwise stop.

4. Assume fractional vaporization V and calculate

$$\Sigma \frac{z_i \dfrac{P_i^S}{P}}{1 + V\left(\dfrac{P_i^S}{P} - 1\right)}$$

The above expression results on rearranging the left-hand-side of Eq. (7.24). If the assumed fractional vaporization is the correct value, the expression evaluates to unity. Otherwise change V and repeat the step 4 calculation.

5. The vapour phase and liquid phase compositions are evaluated using the following relations:

$$y_i = \frac{z_i \dfrac{P_i^S}{P}}{1 + V\left(\dfrac{P_i^S}{P} - 1.0\right)}$$

$$x_i = y_i \frac{P}{P_i^S}$$

Problem 2: Determine the temperature and composition of the liquid and vapour products on flash vaporization of a mixture of composition $\{z_i\}$, given the fraction of the feed vaporized (V/F) and the pressure of vaporization.

1. Input the composition of the feed, the pressure and the fraction of the feed vaporized.
2. Evaluate the boiling points of pure liquids $\{T_i^S\}$ at the given total pressure P. Start with an initial guess value T for the equilibrium temperature, where

$$T = \Sigma\, z_i T_i^S$$

3. Get the vapour pressures of the pure liquids at the given temperature and calculate

$$\sum \frac{z_i \dfrac{P_i^S}{P}}{1+V\left(\dfrac{P_i^S}{P}-1\right)}$$

4. If the assumed temperature is the equilibrium temperature, the above expression evaluates to unity. Otherwise change temperature and repeat the step 4 calculation.
5. The vapour phase and liquid phase compositions are evaluated using the following relations:

$$y_i = \frac{z_i \dfrac{P_i^S}{P}}{1+V\left(\dfrac{P_i^S}{P}-1.0\right)}$$

$$x_i = y_i \frac{P}{P_i^S}$$

EXAMPLE 7.9 A solution contains 60.0% *n*-pentane, 25.0% *n*-hexane and 15.0% *n*-heptane. The Antoine constants for the constituents are given below:

Name	*A*	*B*	*C*
n-pentane	13.8183	2477.07	39.94
n-hexane	13.8216	2697.55	48.78
n-heptane	13.8587	2911.32	56.51

Assuming that Raoult's law is applicable, determine the following:

(a) Bubble-point temperature at 400 kPa
(b) The composition of the vapour formed at the bubble point in part (a)
(c) Bubble-point pressure at 300 K

Solution

(a) As Raoult's law is applicable, $K_i = y_i/x_i = P_i^S/P$ and $y_1 = \dfrac{x_i P_i^S}{P}$

$$\sum y_i = \sum K_i x_i = \sum \frac{x_i P_i^S}{P} = 1$$

Assume a temperature, evaluate the vapour pressure at this temperature and calculate $\sum y_i$. If the temperature assumed is the bubble point of the mixture, $\sum y_i = 1$. If this equality is not satisfied, another temperature is assumed and the calculations are repeated till the above equality will be satisfied.

Let us assume that $T = 369$ K. The vapour pressures are evaluated using the Antoine equation.

$$P_A^S = 539.47 \text{ kPa}, \qquad P_B^S = 220.86 \text{ kPa}, \qquad P_C^S = 93.90 \text{ kPa}$$

Vapour-phase compositions are calculated and $\sum y_i$ evaluated.

$$\sum y_i = \sum \frac{x_i P_i^S}{P}$$

$$= \frac{(0.6 \times 539.47 + 0.25 \times 220.86 + 0.15 \times 93.90)}{400} = 0.9825$$

Since $\sum y_i \neq 1.0$, let us assume that $T = 370$ K.

$$P_A^S = 551.9 \text{ kPa}, \qquad P_B^S = 226.73 \text{ kPa}, \qquad P_C^S = 96.73 \text{ kPa}$$

$$\sum y_i = \sum \frac{x_i P_i^S}{P}$$

$$= \frac{(0.6 \times 551.91 + 0.25 \times 226.73 + 0.15 \times 96.73)}{400} = 1.0058$$

It is clear that the bubble-point temperature lies between 369 and 370 K. By a few more trials, we can find that at 369.75 K, $\Sigma y_i = 1.0$. At 369.75 K,

$$P_A^S = 548.78 \text{ kPa}, \qquad P_B^S = 225.25 \text{ kPa}, \qquad P_C^S = 96.01 \text{ kPa}$$

$$\sum y_i = \sum \frac{x_i P_i^S}{P}$$

$$= \frac{(0.6 \times 548.78 + 0.25 \times 225.25 + 0.15 \times 96.01)}{400} = 1.0000$$

Therefore, the bubble-point temperature of the mixture is 369.75 K.

(b) The vapour formed at the bubble-point in part (a) has the composition

$$y_A = \frac{x_A P_A^S}{P} = \frac{(0.6 \times 548.78)}{400} = 0.8232$$

$$y_B = \frac{x_B P_B^S}{P} = \frac{(0.25 \times 225.25)}{400} = 0.1408$$

$$y_C = \frac{x_C P_C^S}{P} = \frac{(0.15 \times 96.01)}{400} = 0.0360$$

The vapour formed at the bubble point contains 82.32% *n*-pentane, 14.08% *n*-hexane and 3.60% *n*-heptane.

(c) The bubble-point pressure is directly found out. At the given temperature, calculate the vapour pressures of the components and evaluate the partial pressure using Raoult's law. The sum of the partial pressures is equal to the bubble-point pressure.

At 300 K, the vapour pressures are:

$$P_A^S = 73.21 \text{ kPa}, \quad P_B^S = 21.84 \text{ kPa}, \quad P_C^S = 6.70 \text{ kPa}$$

Substituting the values, we get

$$p_i = x_i P_i^S$$

$$p_A = 43.93 \text{ kPa}, \quad p_B = 5.46 \text{ kPa}, \quad p_C = 1.005 \text{ kPa}$$

$$P = \sum p_i = 43.93 + 5.46 + 1.005 = 50.4 \text{ kPa}$$

Therefore, the bubble point pressure is 50.4 kPa.

EXAMPLE 7.10 A vapour contains 60.0% *n*-pentane, 25.0% *n*-hexane and 15.0 % *n*-heptane. The Antoine constants for the constituents are given below:

Name	*A*	*B*	*C*
n-pentane	13.8183	2477.07	39.94
n-hexane	13.8216	2697.55	48.78
n-heptane	13.8587	2911.32	56.51

Assuming that Raoult's law is applicable, determine the following:

(a) Dew-point temperature at 400 kPa
(b) Dew-point pressure at 300 K

Solution

(a) Dew-point temperature is the temperature at which the first drop of condensate is formed when the vapour is cooled at constant pressure. At the dew point the vapour will have the composition of the original mixture and the liquid will have the composition in equilibrium with the original mixture. The temperature is to be obtained by trial. The initial guess value for the dew point may be taken to be the mole fraction weighted average of the pure component boiling points at the specified pressure. The pure component boiling points are evaluated using the Antoine equation with P = 400 kPa. These are as follows:

$$T_A = 356.42 \text{ K}, \quad T_B = 393.29 \text{ K}, \quad T_C = 426.57 \text{ K}$$

The mole fraction weighted average value is

$$T = (0.6 \times 356.42 + 0.25 \times 393.29 + 0.15 \times 426.57) = 376.16 \text{ K}$$

Take this to be the initial guess value for the dew point. At this temperature, the vapour pressures are:

$$P_A^S = 633.27 \text{ kPa}, \quad P_B^S = 265.55 \text{ kPa}, \quad P_C^S = 115.68 \text{ kPa}$$

Since

$$x_i = y_i \frac{P}{P_i^S}$$

$$\sum x_i = 400\left(\frac{0.6}{633.27} + \frac{0.25}{265.55} + \frac{0.15}{115.68}\right) = 1.274$$

Since the sum of the mole fractions is not equal to unity, the temperature is not the correct dew point. Assume another temperature and repeat the calculations. With a few trials the dew-point temperature can be found to be 385.94 K.

At 385.94 K, the vapour pressures are

$$P_A^S = 779.8 \text{ kPa}, \qquad P_B^S = 337.24 \text{ kPa}, \qquad P_C^S = 151.60 \text{ kPa}$$

$$\sum x_i = 400\left(\frac{0.6}{779.88} + \frac{0.25}{336.24} + \frac{0.15}{151.60}\right) = 1.000$$

Therefore, the dew-point temperature is 385.94 K.

(b) The dew-point pressure is calculated using the equation

$$P = \frac{1}{\sum \frac{y_i}{P_i^S}}$$

where y_i and P_i^S are the mole fraction and vapour pressure of component i.

At 300 K, the vapour pressures are:

$$P_A^S = 73.21 \text{ kPa}, \qquad P_B^S = 21.84 \text{ kPa}, \qquad P_C^S = 6.70 \text{ kPa}$$

Therefore, the dew-point pressure is

$$P = \frac{1}{\left(\frac{0.6}{73.21} + \frac{0.25}{21.84} + \frac{0.15}{6.70}\right)} = \frac{1}{0.0420} = 23.79 \text{ kPa}$$

EXAMPLE 7.11 A mixture contains 45 mol % methanol (A), 30 mol % ethanol (B) and the rest n-propanol (C). Liquid solution may be assumed to be ideal and the perfect gas law is valid for the vapour phase. Calculate the following at a total pressure of 101.3 kPa:

(a) The bubble-point and the vapour composition
(b) The dew-point and the liquid composition.

The vapour pressures of the pure liquids are given as follows:

Temperature (K)	333	343	353	363
P_A^S, kPa	81.97	133.29	186.61	266.58
P_B^S, kPa	49.32	73.31	106.63	166.61
P_C^S, kPa	39.32	62.65	93.30	133.29

Solution The vapour pressures of the components are plotted against temperature so that the interpolation of vapour pressure can be done easily.

(a) If the vapour phase can be treated as an ideal gas and liquid phase, an ideal solution, the *K*-values can be written as $K_i = y_i/x_i = P_i^S/P$. Equation (7.17) can be written as

$$\sum y_i = \sum K_i x_i = \sum \frac{x_i P_i^S}{P} = 1$$

Now temperatures are assumed till the above equality is satisfied. It is seen that at 344 K,

$$\sum (x_i P_i^S/P) = 0.98$$

and at 345 K,

$$\sum \frac{x_i P_i^S}{P} = 1.015$$

Therefore, the bubble-point lies between 344 and 345 K. By interpolation, the bubble-point is obtained as 344.6 K. At this temperature the vapour pressures are obtained from the *P* vs *T* plots. P_A^S = 137.3 kPa, P_B^S = 76.20 kPa and P_C^S = 65.40 kPa.

Component	x_i	P_i^S	$K_i = P_i^S/P$	$y_i = K_i x_i$
Methanol	0.45	137.30	1.355	0.610
Ethanol	0.30	76.20	0.752	0.226
Propanol	0.25	65.40	0.646	0.162
			$\sum K_i x_i$	0.998

The equilibrium vapour contains 61% methanol, 22.6% ethanol and 16.2% propanol.

(b) Equation (7.20) for the present case becomes

$$\sum x_i = \sum \frac{y_i}{K_i} = \frac{y_i P}{P_i^S} = 1.00$$

The dew-point temperature is to be determined by trial such that the above relation is satisfied. By trial it can be seen that at 347.5 K P_A^S = 153.28 kPa, P_B^S = 85.25 kPa and P_C^S = 73.31 kPa.

Component	y_i	P_i^S	$K_i = P_i^S/P$	$x_i = y_i/K_i$
Methanol	0.45	153.28	1.5131	0.2974
Ethanol	0.30	85.25	0.8416	0.3565
Propanol	0.25	73.31	0.7237	0.3454
			$\sum y_i/K_i$	0.9993

The values in the last column are the liquid composition at the dew point. Thus the liquid contains 29.7% methanol, 35.7% ethanol and 34.5% propanol.

EXAMPLE 7.12 A hydrocarbon mixture contains 25 mol % propane, 40 mol % *n*-butane and 35 mol % *n*-pentane at 1447.14 kPa. Assume ideal solution behaviour and calculate the following:

(a) The bubble-point temperature and composition of the vapour
(b) The dew-point temperature and the composition of the liquid
(c) The temperature and the composition of the liquid and vapour in equilibrium when 45 mol % of the initial mixture is vaporized. (The values of K_i can be obtained from Figure 13.6 of *Chemical Engineer's Handbook,* 5th ed.)

Solution

(a) Assume temperature say, 355.4 K and the K_i values are found out from the nomograph (Figure 13.6b in *Chemical Engineer's Handbook*). The products of K_i and x_i are calculated and their sum $\sum x_i K_i$ is found out. The results for two temperatures 355.4 K and 366.5 K are shown below.

		$T = 355.4$ K		$T = 366.5$ K	
Component	x_i	K_i	$K_i x_i$	K_i	$K_i x_i$
Propane	0.25	2.000	0.500	2.30	0.575
n-butane	0.40	0.780	0.312	0.90	0.360
n-pentane	0.35	0.330	0.116	0.40	0.140
			$\sum K_i x_i$ 0.928		$\sum K_i x_i$ 1.075

The bubble-point temperature lies between 355.4 K and 366.5 K. By interpolation the temperature is found out to be 361 K. The calculations are carried out at this temperature and the results are as follows:

Component	x_i	K_i	$K_i x_i$
Propane	0.25	2.12	0.530
n-butane	0.40	0.85	0.340
n-pentane	0.35	0.37	0.130
			$\sum K_i x_i$ 1.000

Since $\sum x_i K_i$ is approximately 1.00, the bubble-point temperature is 361 K. The values in the last column are the mole fraction of various components in the vapour. At the bubble-point the vapour contains 53% propane, 34% butane and 13% pentane.

(b) At the dew-point temperature $\sum y_i / K_i = 1$. At 377.6 K this value is 1.1598, and at 388.8 K it is 0.9677.

		$T = 377.6$ K		$T = 388.8$ K	
Component	y_i	K_i	y_i / K_i	K_i	y_i / K_i
Propane	0.25	2.6	0.0962	2.9	0.0862
n-butane	0.40	1.1	0.3636	1.3	0.3077
n-pentane	0.35	0.5	0.7000	0.61	0.5738
			$\sum y_i/K_i$ 1.1598		$\sum y_i/K_i$ 0.9677

By interpolation the dew-point temperature is found to be 387 K. The calculations for this temperature is given below.

Component	y_i	K_i	y_i/K_i
Propane	0.25	2.85	0.0877
n-butane	0.40	1.25	0.3200
n-pentane	0.35	0.59	0.5932
		$\sum y_i/K_i$	1.0009

The last column in the above table is the liquid compositions. The equilibrium liquid at the dew-point contains 8.77% propane, 32.0% butane and 59.32% pentane.

(c) In the following calculations temperature is assumed so as to satisfy Eq. (7.25). For a basis of 100 mol of the initial mixture, F = 100 mol, V = 45 mol and L = 55 mol. Equation (7.25) becomes

$$\sum \frac{z_i}{1+(L/VK_i)} = 0.45$$

		T = 366.5 K		T = 377.6 K	
Component	z_i	K_i	$z_i/[1+(L/VK_i)]$	K_i	$z_i/[1+(L/VK_i)]$
Propane	0.25	2.30	0.1632	2.6	0.1701
n-butane	0.40	0.90	0.1697	1.1	0.1895
n-pentane	0.35	0.40	0.0863	0.5	0.1016
		$\sum z_i/[1+(L/VK_i)]$	0.4192	$\sum z_i/[1+(L/VK_i)]$	0.4612

From the calculations given above, we see that the equilibrium temperature is between 366.5 K and 377.6 K. By interpolation, T = 374.6 K.

		T = 374.6 K	
Component	z_i	K_i	$z_i/[1+(L/VK_i)]$
Propane	0.25	2.50	0.1679
n-butane	0.40	1.08	0.1876
n-pentane	0.35	0.48	0.0987
		$\sum z_i/[1+(L/VK_i)]$	0.4542

Comparing Eqs. (7.23) and (7.24) we can see that

$$y_i = \frac{z_i/[1+(L/VK_i)]}{\sum z_i/[1+(L/VK_i)]}$$

These are calculated using the values in the last column. Corresponding x_i values are found out using the material balance [Eq. (7.21)].

$$Fz_i = Vy_i + Lx_i$$

The results of the calculation are given as follows:

Component	y_i	x_i
Propane	0.3697	0.1521
n-butane	0.4130	0.3894
n-pentane	0.2173	0.4586

7.4 VAPOUR PRESSURE OF IMMISCIBLE SYSTEMS

When a mixture of two immiscible liquids is in equilibrium with its vapour, each component in the solution contributes its full vapour pressure towards its partial pressure in the vapour phase. In a mixture of two immiscible liquids, the liquid surface will be made up of two pure liquid surfaces and vaporization occurs as if it is occurring from a pure liquid. The vapour pressure exerted is independent of the amount of the surface, but depends only on the nature and temperature of the liquid. Therefore, both components are free to exert their full vapour pressure as though they alone are present in the liquid phase.

We know that a pure liquid boils when its vapour pressure becomes equal to the surrounding pressure. Since both components contribute their full vapour pressures, the total pressure exerted by the vapour over a mixture of two immiscible liquids becomes equal to the surrounding pressure at a lower temperature than the boiling points of either of its constituents. Therefore, the boiling point of a mixture of immiscible liquids will be lower than the boiling points of its components. This property is made use of in *steam distillation* of high-boiling and heat-sensitive materials to free off non-volatile impurities. Assume that the impure material is mixed with water with which it is immiscible and heated. When the liquid attains the temperature at which the sum of the vapour pressures is equal to the existing pressure, the liquid boils. This temperature is definitely less than the boiling point of water irrespective of the material. Thus, any high-boiling material which is immiscible with water can be distilled below 373 K at atmospheric pressure.

The temperature at which an immiscible liquid mixture boils is known as the *three-phase temperature* as two liquid phases and one vapour phase coexist in equilibrium at this condition. Consider two immiscible liquids A and B in equilibrium with its vapour. At the three-phase temperature, we have

$$P = P_A^S + P_B^S \tag{7.26}$$

By the Gibbs phase rule, the system is univariant (the number of components, $C = 2$, the number of phases, $\pi = 3$ and therefore the number of degrees of freedom, $F = C - \pi + 2 = 1$), which means that by fixing any one phase rule variable, the system is uniquely determined. Thus at a given total pressure P, the three-phase temperature as well as the vapour composition are uniquely determined and cannot be altered at will. The equilibrium temperature can be determined by plotting the vapour pressures P_A^S, P_B^S and $P_A^S + P_B^S$ against temperature as the abscissa. To find the three-phase temperature at a given total pressure P, move horizontally from P on the vertical axis to the $P_A^S + P_B^S$ curve and read the temperature on the x-axis. Alternatively, P_A^S versus temperature and $(P - P_B^S)$ versus temperature can be drawn as shown in Figure 7.9 (Hausbrandt Chart).

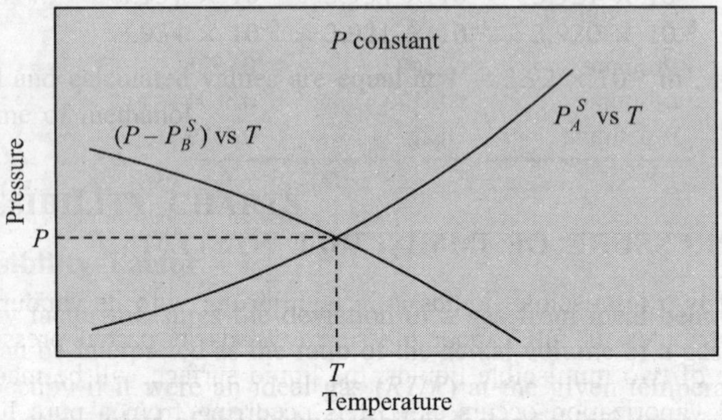

Figure 7.9 Determination of three-phase temperature.

The temperature corresponding to the point of intersection of these curves in Figure 7.9 gives the three-phase equilibrium temperature (T^*). This is because at the point of intersection,

$$P_A^S = P - P_B^S$$

or

$$P = P_A^S + P_B^S$$

The mole ratio of A to B in the vapour is equal to the ratio of the respective vapour pressures at the three-phase temperature. Let n_A and n_B be the number of moles of A and B in the vapour. Then

$$\frac{n_A}{n_B} = \frac{P_A^S}{P_B^S} \tag{7.27}$$

$$y_A = \frac{P_A^S}{P_A^S + P_B^S} \tag{7.28}$$

The phase diagram for an immiscible system is shown in Figure 7.10. When a mixture of two immiscible liquids having an overall composition represented by point M is heated, at temperature T^*, the sum of the vapour pressures equals the surrounding pressure and the vapour of composition represented by point E is formed. On further addition of heat, the temperature remains constant at T^* and more vapour of the same composition as given by point E is formed. This continues till one of the components disappears from the liquid, and the system becomes a two-phase mixture either L_A-V or L_B-V depending upon the initial composition.

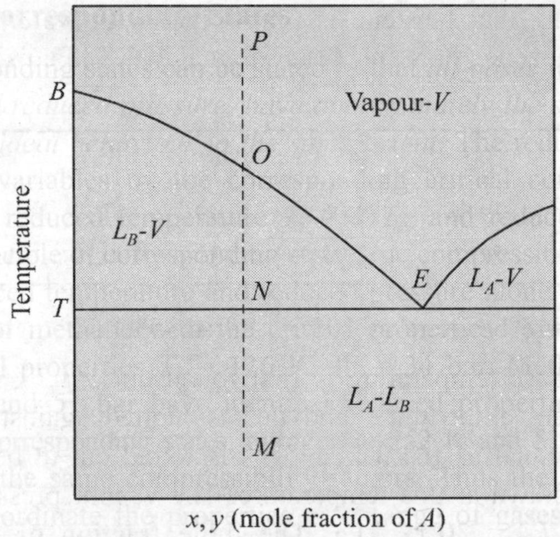

Figure 7.10 Temperature composition diagram of immiscible system.

EXAMPLE 7.13 A high boiling organic liquid is purified from non-volatile impurities by allowing it to mix with steam directly at a total pressure of 93.30 kPa. The vapour pressure data are as follows:

Temperature (K)	353	373
Vapour Pressure of water (kPa)	47.98	101.3
Vapour pressure of liquid (kPa)	2.67	5.33

Assume that water and the organic liquid are immiscible and the impurities do not affect the vaporization characteristics. The vapour pressures vary linearly with temperature. Calculate the following under three-phase equilibrium:

(a) The equilibrium temperature
(b) The composition of the resulting vapour

Solution

(a) At 353 K, the sum of the vapour pressures is 50.65 kPa, and at 373 K it is 106.63 kPa. Assuming that the vapour pressures vary linearly, the temperature at which the sum of vapour pressures is 93.3 kPa is obtained by interpolation.

$$T^* = 353 + \frac{373 - 353}{106.63 - 50.65}(93.3 - 50.65) = 368.2 \text{ K}$$

(b) At 368.2 K the vapour pressure of water is 88.50 kPa and that of the liquid is 4.80 kPa. Since at three-phase equilibrium, the partial pressure is equal to the vapour pressure, the ratio of mole fractions of the components will be same as the ratio of vapour pressures. Let y be the mole fraction of water in the vapour. Then

$$y = \frac{88.5}{93.30} = 0.9486$$

The vapour contains 94.86 mol % water vapour.

EXAMPLE 7.14 Assuming that benzene is immiscible with water, prepare a temperature-composition diagram for benzene (1)–water (2) system at 101.3 kPa using the following vapour pressure data:

T (K)	323	333	343	348	353	363	373
P_2^S (kPa)	12.40	19.86	31.06	37.99	47.32	70.11	101.3
P_1^S (kPa)	35.85	51.85	72.91	85.31	100.50	135.42	179.14

The boiling point of pure benzene at 101.3 kPa is 353.1 K.

Solution The three-phase temperature is first found out. At T^*, $P = P_1^S + P_2^S.P_1^S + P_2^S$ is calculated for each given temperature, and this is plotted against temperature. T^* is the temperature at which P is equal to 101.3 kPa. This is found out to be 342 K. The horizontal line CD in Figure 7.11 is drawn at this temperature. The vapour pressures at this temperature are P_1^S = 71.18 kPa and P_2^S = 30.12 kPa. The mole fraction of benzene in the vapour represented by point E in Fig. 7.11 is 71.18/101.3 = 0.70.

The dew-point curve BE is plotted by choosing a temperature lying between 373 K (boiling point of water) and 342 K (the three-phase temperature). For example, take T = 353 K. The partial pressure of water at the dew point is equal to the vapour pressure. For the dew-point temperature of 353 K the partial pressure of water is

$$47.32 = (1 - y)P$$

where y is the mole fraction of benzene in the vapour. y = 0.5329.

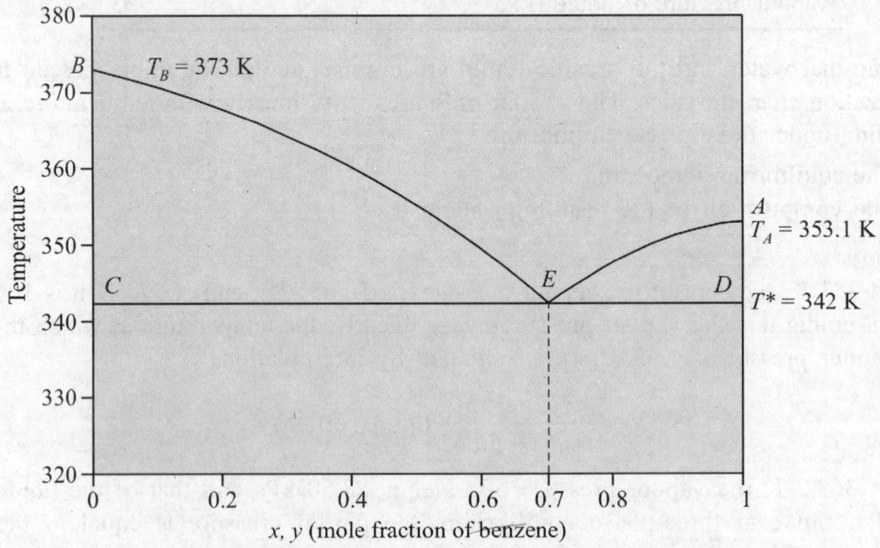

Figure 7.11 Temperature-composition diagram for Example 7.12.

This calculation is repeated for various temperature and the entire curve BE is drawn. For getting the curve AE the procedure is the same. Here temperatures are assumed between 342 K and 353.1 K, the latter being the boiling point of pure benzene. On the curve *AE*, the partial pressure of benzene in the vapour phase equals its vapour pressure. For example, for a dew-point temperature of 348 K,

$$85.31 = P \times y \quad \text{or} \quad y = 0.8422$$

The following table gives the results of a few such calculations:

T (K)	342	348	353	353.1	363	373
y (curve *AE*)	0.70	0.84	0.99	1.00		
y (curve *BE*)	0.70		0.53		0.31	0

7.5 VAPOUR PRESSURE OF SOLUTIONS OF NON-VOLATILE SOLUTES

The presence of non-volatile solutes in the solution lowers the vapour pressure of the volatile components and raises the boiling point of the solvent. The difference between the boiling point of the solution and the boiling point of the pure solvent is known as the *boiling point elevation.* Unless the solute and the solvent are chemically similar, the equilibrium vapour pressure cannot be determined as product of mole fraction and the vapour pressure of the pure solvent. When Raoult's law is used, the vapour pressure of the solvent is independent of the nature of the solute. This is not true for ionizing solutes. For ionizing solutes the effective mole fraction in the liquid will be greater than that of the non-ionizing type. It is found that an aqueous 0.1 M solution of sodium chloride is about twice as effective as a 0.1 M solution of sucrose in raising the boiling point. Assumption of ideal solution behaviour will therefore lead to erroneous results when estimating the vapour pressure over a solution containing non-volatile solutes. The vapour pressure and the boiling point of such solutions are usually estimated using reference substance plots. The Dühring lines described earlier in Chapter 6 provide a convenient method for predicting the vapour pressure and the boiling point elevation. If the vapour pressures of the solution are known at two temperatures, the temperatures at which the pure solvent also has the same vapour pressures can be determined from standard tables of experimental values or through correlations. When the temperature of pure solvent is plotted against the temperature of the solution having the same vapour pressure as the pure solvent, a straight line results. Several such lines can be drawn with the concentration of the solute as the parameter. Though it cannot be generalized, for sodium hydroxide–water solutions these lines are found to be approximately parallel, indicating that the difference between the boiling point of the solution and the boiling point of pure water, or the boiling point elevation, is independent of temperature or pressure.

7.5.1 Relative Vapour Pressure

Frequently the vapour pressure over a solution containing non-volatile solutes is expressed as

$$p = kP^S$$

where p is the effective vapour pressure of the solution, P^S is the vapour pressure of pure solvent and k is the relative vapour pressure of the solution which depends on concentration, temperature or pressure. For ideal solutions, the factor k is equal to the mole fraction of the solvent and is independent of temperature or pressure. For non-ideal solutions k will be different from mole fraction. Assuming that the factor k is independent of temperature or pressure, the value of k can be determined from a single measurement of the vapour pressure or the boiling point of the solution.

EXAMPLE 7.15 The normal boiling point of a 5-molal solution of sodium chloride in water is found to be 379.2 K. What will be the boiling point of the solution at 70 kPa? The vapour pressure of water is given by

$$\ln P_W^S = 16.26205 = \frac{3799.887}{T - 46.854}$$

where P is in kPa and T is in K. Assume that the relative vapour pressure is independent of temperature.

Solution The relative vapour pressure of a 5-molal solution can be computed as

$$k = \frac{p}{P_W^S} \tag{7.30}$$

where p is the vapour pressure of the solution at 379.2 K and P_W^S is the vapour pressure of water at 379.2 K. The latter value is determined using the Antoine equation:

$$\ln P_W^S = 16.26205 = \frac{3799.887}{379.2 - 46.854} \Rightarrow P_W^S = 125.03 \text{ kPa}$$

Substituting this in Eq. (7.30), we can compute the relative vapour pressure.

$$k = \frac{101.3}{125.03} = 0.8102$$

When the solution boils at 70 kPa, the vapour pressure of the solution, p = 70 kPa. Since the relative vapour pressure is assumed constant, the vapour pressure P_W^S at the solution boiling point is from Eq. (7.30)

$$P_W^S = \frac{p}{k} = \frac{70}{0.8102} = 86.4 \text{ kPa}$$

The temperature at which the vapour pressure of water is 86.4 kPa is calculated using the Antoine equation.

$$\ln 86.4 = 16.26205 - \frac{3799.887}{T - 46.854} \Rightarrow T = 368.8 \text{ K}$$

Therefore, the boiling point of the solution at 70 kPa = 368.8 K.

EXERCISES

Ideal solutions and Raoult's law

7.1 *n*-Heptane and toluene form an ideal solution. At 373 K their vapour pressures are 106 kPa and 74 kPa respectively. Determine the composition of the liquid and the vapour in equilibrium at 373 K and 101.3 kPa.

7.2 The vapour pressures of benzene and toluene are given by the Antoine equation

$$\ln P^S = A - \frac{B}{T - C}$$

where P is in kPa and T in K. The Antoine constants are

	A	B	C
Benzene	13.8858	2788.51	52.36
Toluene	13.9987	3096.52	53.67

Determine the partial pressures and the weight composition of the vapour in equilibrium with a liquid mixture consisting of equal weights of the two components at 300 K.

7.3 At 300 K, the vapour pressures of two pure liquids A and B are respectively 80 kPa and 50 kPa. The concentration of A in the vapour in equilibrium with a solution of A and B is found to be 35% (mol). Determine

(a) The composition of the liquid
(b) The total pressure of the vapour

7.4 Air is cooled to 80 K at 101.3 kPa. Calculate the composition of the liquid and the vapour phases at this condition assuming that the mixture behaves ideally. The vapour pressure of nitrogen at 80 K = 135.74 kPa and the vapour pressure of oxygen at 80 K = 30.04 kPa.

7.5 A mixture of A and B conforms closely to Raoult's law. The pure component vapour pressures P_A^S and P_B^S in kPa at temperature T in K are given by

$$\ln P_A^S = 14.27 - \frac{2945}{T - 49}$$

$$\ln P_B^S = 14.20 - \frac{2973}{T - 64}$$

If the bubble point of a certain mixture of A and B is 349 K at a total pressure of 80 kPa, find the composition of the first vapour that forms.

7.6 The vapour pressures of substances A and B can be evaluated by the Antoine equations

$$\ln P_A^S = 14.5463 - \frac{2940.46}{T - 35.93}$$

$$\ln P_B^S = 14.2724 - \frac{2945.47}{T - 49.15}$$

where T is in K and P is in kPa. Assuming that the solutions formed by these are ideal, calculate the compositions of the vapour and the liquid in equilibrium at 327 K and 65 kPa.

7.7 A mixture of *n*-hexane and *n*-heptane containing 10% *n*-hexane is vaporized at a pressure of 100 kPa. Determine the following:

(a) The bubble point temperature
(b) The composition of the vapour

The vapour pressures are given by the Antoine equation

$$\ln P^S = A - \frac{B}{T - C}$$

where P is in kPa and T in K. The Antoine constants are

Name	*A*	*B*	*C*
n-heptane	13.8587	2911.32	56.51
n-hexane	13.8216	2697.55	48.78

7.8 The vapour pressures of benzene and toluene are given below.

T (K)	353.1	358	363	368	373	378	383	383.6
P_A^S (kPa)	101.3	116.9	135.4	155.7	179.1	204.2	233.0	240.0
P_B^S (kPa)	39.6	46.0	54.0	63.3	74.2	86.0	99.0	101.3

Calculate the equilibrium data for the system at 101.3 kPa.

7.9 At 303 K the vapour pressures of benzene (A) and toluene (B) are 15.75 kPa and 4.89 kPa respectively. Determine the partial pressures and the weight composition of the vapour in equilibrium with a liquid mixture consisting of equal weights of the two components.

7.10 The vapour pressures of benzene (A) and chlorobenzene (B) are given below:

T (K)	352.8	363.2	373.2	383.2	393.2	405.3
P_A^S (kPa)	101.3	135.1	178.7	232.5	298.0	395.3
P_B^S (kPa)	18.2	27.7	39.06	53.7	72.3	101.3

Assume that benzene and chlorobenzene form ideal solution.

(a) Prepare the *P-x-y* diagram for the system at 373.2 K.
(b) Using the *P-x-y* diagram of part (a), determine the bubble point pressure and dew-point pressure of an equimolar mixture of benzene and chlorobenzene at 373.2 K.

7.11 An equimolar mixture of benzene and toluene is contained in a cylinder at 65 kPa. Determine the temperature range in which the mixture exists as two phases. The vapour pressure of benzene and toluene are given by the Antoine equation with the following constants. Pressure is in kPa and temperature in K.

Name	*A*	*B*	*C*
Benzene	13.8858	2788.51	52.36
Toluene	13.9987	3096.52	53.67

7.12 A 40% (mol) solution of methanol in water boils at 348.5 K. The vapour analyzed 72.9% methanol. How does this compare with the vapour composition given by Raoult's law? The vapour pressures (kPa) are given by the Antoine equation with following constants:

Name	*A*	*B*	*C*
Water	16.26205	3799.887	46.854
Methyl alcohol	16.5725	3626.55	34.29

7.13 An equimolar solution of benzene and toluene is totally evaporated at a constant temperature of 363 K. At this temperature, the vapour pressures of benzene and toluene are 135.4 kPa and 54 kPa respectively. What are the pressures at the beginning and at the end of the vaporization process?

7.14 A liquid mixture containing 15% *A*, 25% *B* and the rest *C* is in equilibrium with the vapour which contains 40% *B*. All the percentages are on a mole basis. The equilibrium pressure and temperature are 200 kPa and 350 K. At 350 K the vapour pressure of *C* is 20 kPa. What is the percentage of *A* in the vapour?

7.15 A liquid mixture contains 10% *A*, 15% *B* and the rest *C*, all compositions being in mole percent. At equilibrium the pressure was 230 kPa and temperature was 340 K, and the vapour was found to contain 34% *B*. The vapour pressure of *C* at the equilibrium temperature is 3.5 kPa. Assuming ideal behaviour for the gas and the liquid, calculate the vapour composition.

7.16 Three liquids *A* (MW = 40), *B* (MW = 60) and *C* (MW = 80) form an ideal solution. At 370 K, the vapour pressures are found to be 50 kPa, 25 kPa and 12.5 kPa respectively for *A*, *B* and *C*. A vapour analyzing 40% *A*, 40% *B* and 20% *C* on a volume basis is in equilibrium with the liquid at 370 K. Determine the following:

(a) The equilibrium pressure
(b) The composition of the liquid in weight percent

7.17 Liquids *A* and *B* form ideal solutions while liquid *C* is immiscible with both *A* and *B*. A mixture of *A*, *B* and *C* is found to boil at 240 K under a pressure of 100 kPa. What is the composition of the liquid on a *C*-free basis? The vapour pressures at 240 K are 15 kPa, 3 kPa and 95 kPa respectively for *A*, *B* and *C*.

7.18 Three liquids A, B and C form an ideal solution. The dew point of a vapour mixture consisting of A, B and C at pressure of 15 kPa is found to be 323 K. The vapour pressures of pure liquids A, B and C are correlated by the Antoine equations. With vapour pressure in kPa and temperature in K, the Antoine constants are as follows:

Liquid	*A*	*B*	*C*
A	13.9	2800	52.4
B	14.0	3100	53.7
C	16.3	3800	46.9

Determine the following:

(a) The composition of the vapour assuming that the vapour contains equimolar quantities of A and B.

(b) The composition of the first drop of liquid formed at 323 K and 15 kPa

7.19 The vapour pressures of substances A and B can be evaluated by the Antoine equations

$$\ln P_A^S = 14.5463 - \frac{2940.46}{T - 35.93}$$

$$\ln P_B^S = 14.2724 - \frac{2945.47}{T - 49.15}$$

where T is in K and P is in kPa. Assuming that the solutions formed by these are ideal, calculate the following:

(a) The bubble-point temperature of a liquid containing 40% (mole) A at 65 kPa and the composition of the vapour in equilibrium with this liquid.

(b) The bubble-point pressure for a liquid containing 40% (mole) A and the composition of vapour in equilibrium at 327 K.

7.20 The vapour pressures of substances A and B can be evaluated by the Antoine equations

$$\ln P_A^S = 14.5463 - \frac{2940.46}{T - 35.93}$$

$$\ln P_B^S = 14.2724 - \frac{2945.47}{T - 49.15}$$

where T is in K and P is in kPa. Assuming that the solutions formed by these are ideal, calculate the fraction of the system that is liquid and the composition of the liquid and the vapour in equilibrium at 327 K and 65 kPa when the overall composition of the system is 70 mole percent A.

7.21 A solution of methanol in water containing 0.158 mole fraction methanol boils at 357.3 K and 101.3 kPa and produces a vapour containing 0.553 mole fraction methanol.

Is Raoult's law valid? If not determine the percent deviation from it for each component. The vapour pressure of methanol may be estimated using the Antoine equation

$$\ln P^S = 16.5725 - \frac{3626.55}{T - 34.29}$$

7.22 Two substances A and B are known to form ideal liquid solutions. A vapour mixture containing 50 mol % A and 50 mol % B is at 311 K and 101.3 kPa. This mixture is compressed isothermally until condensation occurs. At what pressure does condensation occur and what is the composition of the liquid that forms? The vapour pressures of A and B are 142 kPa and 122 kPa respectively.

7.23 A liquefied fuel gas consists of 60% n-butane, 35% propane and 5% ethane. All compositions are on mole basis. The vapour pressures at 303 K are: ethane = 4.72×10^3 kPa, propane = 1.09×10^3 kPa and n-butane = 3.45×10^2 kPa. Calculate the following:

(a) The vapour pressure of the liquid at 303 K
(b) The composition of the vapour at 303 K
(c) Repeat parts (a) and (b) if all ethane is removed from the liquid

7.24 Hexane (A) and heptane (B) form an ideal solution. The vapour pressures of pure hexane and heptane are:

T (K)	342	343	348	353	358	363	368	372.4
P_A^S (kPa)	101.3	104	122	141.3	163.3	187.3	210.2	235.3
P_B^S (kPa)	39.3	40.3	46.4	56.8	66.4	78.4	90	101.3

(a) Plot the boiling point diagram for the system at 101.3 kPa.
(b) Plot the equilibrium diagram at 101.3 kPa.

7.25 The binary system, acetone (A)-acetonitrile (B) may be treated as an ideal solution. Using the vapour pressure data given below prepare the following:

(a) A plot of P versus x and P versus y at 323 K
(b) Boiling point diagram

T (K)	311.45	315	319	323	327	331	335.33
P_A^S (kPa)	53.32	61.09	70.91	81.97	94.36	108.2	124.95
P_B^S (kPa)	21.25	24.61	28.90	33.79	39.35	45.62	53.32

7.26 The vapour pressures of benzene (A) and chlorobenzene (B) are given below:

T (K)	352.8	363.2	373.2	383.2	393.2	405.3
P_A^S (kPa)	101.3	135.1	178.7	232.5	298.0	395.3
P_B^S (kPa)	18.2	27.7	39.06	53.7	72.3	101.3

Assume that benzene and chlorobenzene form an ideal solution.

(a) What are the normal boiling points of benzene and chlorobenzene?
(b) Prepare the *T-x-y* diagram for the system at 101.3 kPa.
(c) Using the *T-x-y* diagram of part (b) determine the bubble point and dew-point of an equimolar mixture of benzene and chlorobenzene at 101.3 kPa.

7.27 Assuming Raoult's law to be valid for the system benzene (*A*)-ethyl benzene (*B*) and the vapour pressures are given by the Antoine equations:

$$\ln P_A^S = 13.8858 - \frac{2788.51}{T - 52.41}$$

$$\ln P_B^S = 14.0045 - \frac{3279.47}{T - 60.00}$$

where *P* is in kPa and *T* is in K.

(a) Construct the *P-x-y* diagram at 370 K.
(b) Construct the *T-x-y* diagram at 100 kPa.

7.28 A liquid mixture containing 65 mol % benzene and 35 mol % toluene is subjected to flash vaporization at 363 K and 101.3 kPa. The vapour pressure of benzene at this temperature is 136.09 kPa and the vapour pressure of toluene is 54.21 kPa. Flash vaporization is essentially an equilibrium stage operation. Calculate the following:

(a) The exit vapour composition
(b) The exit liquid composition
(c) The mole percent of the feed that is vaporized

7.29 A solution containing 25% benzene (MW = 78.048), 35% toluene (MW = 92.064) and 40% xylene (MW = 106.08) is in equilibrium with its vapour at 373 K. All percentages are on a weight basis. Determine the following:

(a) The total pressure
(b) The composition of the vapour
(c) The average molecular weight of the liquid and vapour

The vapour pressures at 373 K are: benzene = 178.7 kPa, toluene = 74.7 kPa and xylene = 28 kPa.

7.30 For the system *n*-pentane (*A*)-*n*-heptane (*B*) the vapour pressures are given by the Antoine equation

$$\ln P = A - \frac{B}{T - C}$$

where *P* is in kPa and *T* is in K. The constants are as follows:

System	*A*	*B*	*C*
n-pentane	13.8183	2477.07	40.00
n-heptane	13.8587	2911.32	56.56

Assuming that the solution formed is ideal, calculate the following:

(a) The composition of the liquid and the vapour in equilibrium at 130 kPa and 335 K
(b) The composition of the vapour in equilibrium with a liquid containing 30% (mol) pentane and the equilibrium temperature at P = 100 kPa
(c) The total pressure and the vapour composition in equilibrium with a liquid of composition x = 0.40 at T = 335 K

7.31 The vapour pressures of ethanol and water are given by the Antoine equation. The Antoine constants are:

Name	*A*	*B*	*C*
Water	16.26205	3799.887	46.854
Ethyl alcohol	16.5092	3578.91	50.50

Compare the equilibrium data calculated using Raoult's law with the experimental values given below:

T (K)	373.2	368.7	359.9	357.3	355.5	353.9	352.9	351.94	351.61	351.35	351.61
x	0	0.019	0.097	0.166	0.261	0.397	0.52	0.676	0.747	0.894	1
y	0	0.17	0.438	0.509	0.558	0.612	0.66	0.739	0.782	0.894	1

An ethanol–water solution containing 25% by mole ethanol at 313 K is being heated in a closed container at a constant pressure of 101.3 kPa. Calculate the following using the experimental values:

(a) The temperature at which vaporization begins
(b) The composition of the first bubble of vapour formed
(c) The composition of the residual liquid when 25 mole % is vaporized

Non-ideal solutions

7.32 An aqueous solution of acetaldehyde contains 1% acetaldehyde by weight. The partial pressure of acetaldehyde over the solution is found to be 20.6 kPa at 367 K. What will be the partial pressure over a 0.1 molal solution at the same temperature?

7.33 The Henry's law constant for oxygen in water at 298 K is 4.4 × 10^4 bar. Estimate the solubility of oxygen in water at 298 K for a partial pressure of oxygen of 0.25 bar.

7.34 How many cubic metres of oxygen at 100 kPa can be dissolved in 10 kg water at 293 K if the Henry's law constant for oxygen in water at 293 K is 4.4 × 10^9 N/m^2.

7.35 At 293 K, the solubility of oxygen in water is found to be 1 g in 100 g water for a partial pressure of 0.25 bar for oxygen. Estimate the Henry's law constant of oxygen in water at 293 K.

7.36 The partial pressure of acetone (A) and chloroform (B) were measured at 298 K and are reported below:

x_B	0	0.2	0.4	0.6	0.8	1.0
$\bar{p}_A$ (bar)	0.457	0.355	0.243	0.134	0.049	0
$\bar{p}_B$ (bar)	0	0.046	0.108	0.187	0.288	0.386

Determine the Henry's law constants.

7.37 Calculate the concentration of nitrogen in water exposed to air at 298 K and 1 bar if the Henry's law constant for nitrogen in water is 8.68×10^4 bar at this temperature. Express the result in moles of nitrogen per kg water (*Hint*: Air is 79 percent nitrogen by volume).

7.38 The partial pressure of methyl chloride in a mixture varies with its mole fraction at 298 K as detailed below:

X	0.0005	0.0009	0.0019	0.0024
$\bar{p}$ (bar)	0.27	0.48	0.99	1.24

Estimate the Henry's law constant of methyl chloride.

Vapour–liquid equilibrium calculations

7.39 A solution contains 50.0% *n*-pentane, 30.0% *n*-hexane and 20.0% *n*-heptane. The Antoine constants for the constituents are given below:

Name	*A*	*B*	*C*
n-pentane	13.8183	2477.07	39.94
n-hexane	13.8216	2697.55	48.78
n-heptane	13.8587	2911.32	56.51

Assuming that Raoult's law is applicable, determine the following:

(a) The dew-point pressure of the mixture at 295 K

(b) The dew-point temperature at 400 kPa

(c) The bubble-point temperature at 400 kPa

(d) The bubble-point pressure at 295 K

7.40 A solution of hydrocarbons contains *n*-propane 5.0%, *n*-butane 30.0%, *n*-pentane 40.0% and *n*-hexane 25.0%. Compute the bubble-point and the dew-point at 350 kPa. The *K*-values can be taken from the DePriester nomographs.

7.41 A mixture of hydrocarbon vapours containing 21.8% ethane, 66.5% propane, 10.73% isobutane and 0.97% *n*-butane is cooled at constant temperature of 311 K. Using the following *K* values determine the dew-point pressure and the liquid composition.

Component	$P = 1310$ kPa	$P = 1380$ kPa	$P = 1450$ kPa
Ethane	3.22	3.07	2.92
Propane	1.005	0.973	0.92
Isobutane	0.45	0.43	0.41
n-butane	0.315	0.305	0.295

7.42 A solution has the following composition in mol %: ethane 0.25%, propane 25.00%, isobutane 18.5%, *n*-butane 56.0%, isopentane 0.25%. For a pressure of 10 bars, calculate the following:

(a) The bubble-point
(b) The dew-point
(c) The composition of the liquid and vapour when 40% of the mixture is vaporized

7.43 1000 moles of a mixture of the following composition is contained in a vessel at 2200 kPa and 355 K.

Component	*Mole percent*	K *value*
Methane	16.0	8.8
Ethane	23.0	1.82
Propane	36.0	0.71
Isobutane	11.0	0.35
n-butane	14.0	0.27

How many moles of liquid and vapour will be there at these conditions? Also calculate the liquid and vapour compositions.

7.44 A stream of gas in a natural gasoline plant has the following composition by volume: ethane 10%, propane 14%, isobutane 19%, *n*-butane 54% and isopentane 3%.

(a) Calculate the pressure necessary to condense this gas completely at 311 K.
(b) For a condenser operating at the pressure in part (a), calculate the temperature at which condensation starts and the temperature at which 50 mol % of the vapour gets condensed. Also calculate the composition of the first liquid to condense and the composition of the liquid and vapour phases at 50% condensation.

7.45 Determine the composition of the vapour in equilibrium with the liquid and the pressure of the system at 313 K for a liquid mixture of 5 mol % methane, 10 mol % ethane, 30 mol % propane, 25 mol % iso-butane, and 30 mol % *n*-butane. Determine the pressure and the composition of the liquid in equilibrium with a vapour mixture of the above composition.

7.46 Calculate the pressure at which condensation starts and the pressure at which condensation is complete when a vapour mixture of the following composition is subjected to condensation at a temperature of 300 K: 20 mol % ethylene, 20 mol % ethane, 40 mol % propane and 20 mol % *n*-butane.

7.47 A vapour mixture containing 15% ethane, 20% propane, 60% isobutane and the rest *n*-butane is subjected to partial condensation so that 75% of the vapour is condensed. If the condenser temperature is 300 K, determine the pressure.

7.48 An equimolar mixture of propane (1) and *n*-butane (2) is partially condensed so that 50 mol % of the mixture is in the liquid state at 311 K. Using the DePriester nomograph, determine the following:

(a) The pressure
(b) The vapour and liquid compositions
(c) The pressure at which condensation begins at the constant temperature of 311 K

7.49 The following table gives the vapour pressures of alcohols in kPa at different temperature (K).

T (K)	*Vapour pressure* (kPa)			
	Methanol	*Ethanol*	*n-Propanol*	*n-Butanol*
323	55.3	29.3	11.9	4.5
333	83.9	46.9	19.9	7.9
338	102.2	58.4	25.3	10.4
343	123.9	72.3	32.1	13.3
348	149.2	88.7	40.3	17.5
353	178.5	108.3	50.1	22.0
358	212.4	131.2	62.0	27.5
363	251.2	158.0	76.1	30.1
373	346.4	227.4	112.4	51.7

For an alcohol mixture of composition 30% methanol, 20% ethanol, 15% *n*-propanol and 35% *n*-butanol, calculate the following for a total pressure of 101.3 kPa:

(a) The bubble-point temperature and vapour composition
(b) The dew-point temperature and liquid composition

Vapour pressure of immiscible systems

7.50 Ethyl ether and water are charged into a still in the weight ratio 3:1 and allowed to boil. Ether and water are immiscible. The temperature of the contents is maintained at 303 K at which the vapour pressures of the pure liquids are 4.2 kPa for water and 80 kPa for ether. Calculate the following:

(a) The pressure inside the still
(b) The composition of the residual liquid when half the original mixture is vaporized

7.51 A vapour mixture contains 85% (mol) benzene and 15% (mol) water at 101 kPa and 373 K. Assume that liquid benzene and water are immiscible. The vapour pressures are given by Antoine equations, the constants of which are:

	A	*B*	*C*
Benzene	13.8858	2788.51	52.36
Water	16.26205	3799.887	46.854

(a) On cooling this vapour at constant pressure, at what temperature does condensation begin and what is the composition of the condensate?
(b) If cooling is continued, at what temperature do both liquids condense together?

7.52 A liquid of molecular weight 150 is insoluble in water and is to be purified by steam distillation at a total pressure of 101.3 kPa. The vapour pressure of the liquid at the temperature of distillation is 3.5 kPa. How many kilograms of steam is required theoretically to distil one kilogram of the liquid?

7.53 A high-boiling organic liquid (MW = 185) is being steam-distilled. The vapour temperature is measured to be 374.2 K.

(a) What is the pressure in the still?
(b) If the vapour analyzed 87% water by weight, what is the vaporization efficiency? (The vaporization efficiency may be defined as the ratio of the minimum steam required for steam distillation to the actual steam consumption.)

The vapour pressure of water at 374.2 K is 105 kPa and that of the organic liquid is 2.3 kPa.

7.54 It is proposed to purify benzene from small amounts of non-volatile impurities by subjecting it to distillation with saturated steam at 99.3 kPa. Calculate the temperature at which distillation will proceed and the weight of steam accompanying 1 kg benzene. The vapour pressure data is given below:

T (K)	323	333	343	348	353	363	373
P_2^S (kPa)	12.40	19.86	31.06	37.99	47.32	70.11	101.3
P_1^S (kPa)	35.85	51.85	72.91	85.31	100.50	135.42	179.14

7.55 Benzene (1) is to be purified from non-volatile impurities by subjecting it to steam distillation at a pressure of 100 kPa. It may be assumed that benzene is immiscible with water (2). The following vapour pressure data are available:

T (K)	323	333	343	348	353	363	373
P_1^S (kPa)	35.85	51.85	72.91	85.31	100.50	135.42	179.14
P_2^S (kPa)	12.40	19.86	31.06	37.99	47.32	70.11	101.3

Determine the following:

(a) The temperature of distillation
(b) The mole fraction of benzene in the vapour
(c) The mass of steam required to distil 1 kg benzene

7.56 Ethyl aniline is purified from dissolved non-volatile impurities by steam distillation at 101.3 kPa pressure. The vapour pressure of aniline and water are as follows:

T (K)	353.8	369.2	372.3	386.4
Aniline (kPa)	1.33	2.67	3.04	5.33
Water (kPa)	48.5	87.7	98.3	163.3

Calculate the following:

(a) The boiling points of the mixture
(b) The composition of distillate

7.57 A mixture containing chlorobenzene and water is being distilled. The liquids may be assumed immiscible. The distillation is carried out under such conditions that two liquid layers are present in the still. The vapour pressure data as follows:

Chlorobenzene: $$\ln P^S = 13.9926 - \frac{3295.12}{T - 55.60}$$

Water: $$\ln P^S = 16.26205 - \frac{3799.887}{T - 46.854}$$

Determine the temperature of distillation and the composition of the vapour if

(a) The pressure is 4 kPa.
(b) The pressure is 40 kPa.

7.58 It is desired to separate $CHCl_3$ from a non-volatile insoluble impurity by steam distillation at a total pressure of 175 kPa. Saturated steam at 200 kPa is introduced directly into the liquid. Calculate:

(a) The boiling point of the mixture
(b) The composition of the distillate
(c) The composition of the distillate if 20% more steam than the theoretical amount distils over

The vapour pressure data are as follows:

T (K)	299.1	315.9	334.5	357.05	393.2
Chloroform (kPa)	26.7	53.3	101.3	202.7	506.6
Water (kPa)	3.3	8.5	21.1	55.3	198.2

7.59 A gas mixture contained 2×10^5 kg of hexane and 5×10^3 kg of water vapour at 373 K and 200 kPa. The mixture is cooled at constant pressure till water begins to condense. Assume that hexane and water are immiscible in the liquid phase.

(a) What is the initial partial pressure of hexane in the gas?
(b) What is the temperature at which condensation begins?

The vapour pressure of *n*-hexane is given by the Antoine equation with the following constants when pressure is in kPa and temperature in K:

$$A = 13.8216 \qquad B = 2697.55 \qquad \text{and} \qquad C = 48.78$$

7.60 A vapour mixture contains 85 mol% CS_2 and 15 mol% H_2O at 101.3 kPa and 373 K. The liquids CS_2 and H_2O are immiscible. Vapour pressure data for the pure fluids are given as follows:

T (K)	283.2	293.2	303.2	313.2	323.2	333.2
CS_2 (kPa)	26.34	39.69	57.95	82.35	114.26	155.15
Water (kPa)	1.227	2.337	4.241	7.375	12.34	19.92

The vapour is cooled at 101.3 kPa till condensation begins. Determine the following:

(a) The temperature at which condensation begins
(b) The composition of the first condensate
(c) The temperature at which both liquids condense together

7.61 Dimethylaniline is distilled with steam at 90 kPa to free it from non-volatile impurities. Assuming it to be completely immiscible with water, determine the following:

(a) The distillation temperature
(b) The composition of the vapour produced

The vapour pressure data are as follows:

T (K)	343.2	374.8	399
P^S (aniline), kPa	1.333	5.332	13.329
P^S (water), kPa	31.19	106.91	285.86

The plot of $\ln P^S$ versus $1/T$ may be assumed linear.

7.62 A stream contains 30 mol % toluene, 40 mol % ethyl benzene and 30 mol % water. Assuming that mixtures of ethylbenzene and toluene obey Raoult's law and they are completely immiscible in water, calculate the following for a total pressure of 101.3 kPa:

(a) The bubble-point temperature and the composition of the vapour
(b) The dew-point temperature and the composition of the liquid.

The vapour pressure data are given below:

T (K)	352.6	361.0	366.5	383.2	388.8
P^S (water), kPa	46.29	64.33	79.42		
P^S (toluene), kPa	38.49	50.65	60.78	99.27	116.50
P^S (ethyl benzene), kPa	16.41	22.79	27.35	47.61	56.73

7.63 *n*-heptane (1) and water (2) are essentially immiscible as liquids. A vapour mixture containing 65 mol % water at 373 K and 101.3 kPa is cooled slowly at constant pressure until condensation is complete. Construct a plot for the process showing temperature versus equilibrium mole fraction of heptane in the residual vapour. For *n*-heptane,

$$\ln P_1^S = 13.87770 - \frac{2918.738}{T - 56.404}$$

The vapour pressure of water is

$$\ln P^s = 16.26205 - \frac{3799.887}{T - 46.854}$$

where P is in kPa and T in K.

7.64 Toluene (1) and water (2) are essentially immiscible in the liquid state. Determine the dew-point temperature and the composition of the first drops of liquid formed when the vapour mixtures of these species containing

(a) 23 mol % toluene

(b) 77 mol % toluene at 101.3 kPa.

At 101.3 kPa, what are the bubble-point temperature and the composition of the last drop of vapour in each case? The vapour pressure of toluene is

$$\ln P_1^S = 14.00976 - \frac{3103.010}{T - 53.413}$$

The vapour pressure of water is

$$\ln P^S = 16.26205 - \frac{3799.887}{T - 46.854}$$

where P is in kPa and T is in K.

7.65 Components 1 and 2 are insoluble in the liquid phase. Estimate the dew-point temperature and the compositions of the first drops of liquid formed when vapour mixtures of components 1 and 2 containing

(a) 75 mol % component 1 and

(b) 25 mol % component 1

are cooled at a constant pressure of 101.33 kPa. The vapour pressures of the pure components in kPa are given against temperature in K in the following table.

T	358	363	368	373	378	383	388	393	398.6
P_1^S	23.53	33.56	39.58	46.59	54.62	63.97	74.32	86.34	101.33
P_2^S	57.82	70.13	84.53	101.33	120.79	143.28	169.02	198.51	239.54

Vapour pressure of solutions of non-volatile solutes

7.66 An aqueous solution containing 20% by weight dissolved non-volatile solute (MW = 120) is sent to a flash distillation chamber maintained at 5 kPa and 330 K. The effective vapour pressure of the solution (p) is given as

$$p = x_W P_W^S$$

where x_W and P_W^S are the mole fraction of water in the solution and the vapour pressure of pure water respectively. The vapour pressure of water at 330 K is 17.147 kPa. Per 100 kg of the solution, determine the following:

(a) The amount of pure water obtained as vapour
(b) The concentration of the solute in the liquid leaving the chamber

7.67 A liquid mixture consisting of 15% *A*, 70% *B* and 15% *C* is flashed at 250 K and 10 kPa. All compositions are on a mole basis. The vapour pressure of component *C* is negligible in comparison with those of *A* and *B*. The vaporization equilibrium constants of *A* and *B* may be taken as 5.25 and 0.65 respectively. Determine the following:

(a) The moles of liquid and vapour at equilibrium
(b) The mole fractions in the liquid and vapour

7.68 The normal boiling point of a 5-molal solution of sodium chloride in water is found to be 379.2 K. The vapour pressure of water at this temperature is 124.8 kPa. Assuming that the relative vapour pressure is independent of temperature, determine the vapour pressure of the solution at 300 K if the vapour pressure of water at 300 K is 3.5 kPa.

7.69 The normal boiling point of a 5-molal solution of sodium chloride in water is found to be 379.2 K. What will be the boiling point of the solution at 80 kPa? The vapour pressure of water is given by

$$\ln P_W^S = 16.26205 - \frac{3799.887}{T - 46.854}$$

where *P* is in kPa and *T* is in K. Assume that the relative vapour pressure is independent of temperature.

7.70 The normal boiling point of a 10 molal solution of sodium nitrate in water is 382 K. Determine the following:

(a) The boiling point elevation of the solution
(b) The boiling point elevation at 50 kPa

Assume that the vapour pressure of water is given by the Antoine equation

$$\ln P_W^S = 16.26205 - \frac{3799.887}{T - 46.854}$$

where *P* is in kPa and *T* is in K, and the relative vapour pressure is independent of temperature.

8 Humidity and Humidity Chart

Air conditioning, water cooling, humidification, drying operations, etc. are processes in which one has to deal with vapours that will be present in association with a non-condensable gas or a mixture of non-condensable gases. In good number of such situations, the vapour is water vapour and non-condensable gas is air. *Air conditioning* is the process of maintaining the water content in the air at some desired level. Maintaining the humidity and temperature of air at some specified levels is important not only for houses and buildings, but also for certain plant areas in process industries. This is achieved by either condensation of water vapour from air or by evaporation of liquid water into an unsaturated air stream. Cooling of process water is achieved in cooling towers by evaporation of water into air stream. Air conditioning and water cooling operations fall under the general category of operations designated in chemical engineering parlance as *humidification-dehumidification operations*. Another important unit operation which also makes use of an air stream or a stream of hot flue gases picking up moisture by evaporation is the drying operation. In drying, a wet material will be exposed to a relatively dry hot air or flue gas. The material gets dried by giving off its moisture to the unsaturated gas stream with which it is contacted. The air-water vapour system is not the only system of importance involving mixtures of condensable vapours and non-condensable gases. Frequently, vapours of organic compounds such as benzene, carbon tetrachloride, acetone, etc. present in air or process gases are to be recovered as part of a solvent recovery operation or are to be removed from the gas stream as part of a purification step. A clear understanding of the concepts of humidity and saturation, degree of saturation and the conditions under which the gas stream remains saturated are therefore very essential to a chemical engineer.

8.1 SATURATION

Consider a liquid A (say, water) in contact with a non-condensable gas B (say, air) in a closed container at constant temperature and pressure. Initially the air is dry so that the partial pressure of water vapour in the air is zero. When the dry air comes in contact with liquid water, it picks up moisture from the liquid and as a result the partial pressure of water in the gas increases. Finally when equilibrium between the liquid and the gas is established, the partial pressure of water vapour in the gas stream remains at a constant value. Even if we provide an infinitely long period for contact between the liquid and the gas, the equilibrium partial pressure exerted by the vapours of the liquid in the gas will not change so long as the temperature is not altered. This is the state at which the gas is *saturated* with the vapours of the liquid. The partial pressure exerted by the vapour under saturation is equal to the *vapour pressure* of the liquid at the given temperature. Since the total pressure P is constant, we have under saturation, the partial pressure of vapour, $p_A = P_A^S$, where P_A^S is the vapour pressure of component A, and the partial pressure of the gas, $p_B = P - P_A^S$. If the gas is *unsaturated*, the partial pressure of the vapour is less than the vapour pressure of the liquid at the temperature of the gas.

EXAMPLE 8.1 The vapour pressure of acetone at 280 K is 13.25 kPa. For a mixture of nitrogen gas saturated with the vapours of acetone at 280 K and 105 kPa, calculate the following:

(a) The mole percent of acetone in the mixture
(b) The percent composition by weight
(c) The amount of vapour in kilograms per m^3 of the mixture.

Solution Since the nitrogen gas is saturated with acetone vapours, the partial pressure of acetone in the gas at 280 K, p_A = the vapour pressure of acetone at 280 K, P_A^S = 13.25 kPa.

(a) The mole fraction of acetone is

$$\frac{\text{partial pressure}}{\text{total pressure}} = \frac{p_A}{P} = \frac{P_A^S}{P} = \frac{13.25}{105} = 0.1262$$

The mole percent = mole fraction × 100 = 12.62%

(b) The molecular weight of acetone = 58.048 and the molecular weight of nitrogen = 28.0. One mole of the gas mixture contains 0.1262 mole of acetone and 0.8738 mole of nitrogen.
Or one mole of the gas mixture contains 0.1262 × 58.048 = 7.3257 g of acetone and 0.8738 × 28 = 24.4664 g of nitrogen. Therefore, the composition in weight percent is:

$$\text{acetone} = \frac{7.3257}{(7.3257 + 24.4664)} \times 100 = 23.04\%$$

$$\text{nitrogen} = \frac{24.4664}{(7.3257 + 24.4664)} \times 100 = 76.96\%$$

(c) Volume of one kmol gas at 280 K and 105 kPa is

$$V_0 \times \frac{P_0}{P} \times \frac{T}{T_0} = 22.4 \times \frac{101.3}{105} \times \frac{280}{273.15} = 22.17 \text{ m}^3$$

The mass of vapour present in one kmol of gas is 7.3257 kg.

$$\text{Concentration of the vapour} = \frac{7.3257}{22.17} = 0.3304 \text{ kg/m}^3$$

EXAMPLE 8.2 It is desired to prepare a 10% benzene vapour–air mixture by saturating dry air with benzene from a container of the liquid at 101.3 kPa. What temperature should be used to achieve the desired composition? The vapour pressure of benzene is given by the Antoine equation

$$\ln P^S = 13.8858 - \frac{2788.51}{T - 52.36}$$

Solution The partial pressure of benzene in a 10% benzene vapour–air mixture at 101.3 kPa is 101.3 × 0.1 = 10.13 kPa. For saturated air, the partial pressure is equal to the vapour pressure. Using the Antoine equation, the temperature of benzene at which the vapour pressure is 10.13 kPa is determined.

$$\ln 10.13 = 13.8858 - \frac{2788.51}{T - 52.36} \Rightarrow T = 293.4 \text{ K}$$

EXAMPLE 8.3 Dry air is blown through acetone at 285 K and a constant pressure of 101.3 kPa. If it is desired that 5 kg of acetone be evaporated what is the minimum amount of dry air required in kilograms? The vapour pressure of acetone at 285 K is 16.82 kPa.

Solution The minimum quantity of air that is required to contain 5 kg of acetone corresponds to the condition when air is saturated with 5 kg of acetone vapours. Under the saturated condition, the ratio of moles of acetone to moles of dry air is

$$\frac{P_A^S}{P - P_A^S} = \frac{16.82}{101.3 - 16.82} = 0.1991$$

The ratio of mass of acetone to mass of dry air is

$$0.1991 \times \frac{58.048}{29} = 0.3985$$

where 58.048 is the molecular weight of acetone and 29 is the molecular weight of dry air. One kilogram of dry air under saturated state contains 0.3985 kg of acetone vapour. The amount of dry air required to evaporate 5 kg of acetone is

$$\frac{5}{0.3985} = 12.55 \text{ kg}$$

8.2 HUMIDITY, PERCENT HUMIDITY AND DEW POINT

8.2.1 Humidity

The term *humidity* is used to indicate the concentration of the vapour in a gas–vapour mixture. For the air-water vapour system, humidity is defined as the amount of water vapour present per unit quantity of dry air and can be expressed on a mole basis or a mass basis. The *molal humidity* is defined as the number of moles of water vapour present per one mole of dry air.

Let there be n_A moles of water vapour and n_B moles of dry air in a given sample of air–water vapour mixture. Then,

$$\text{molal humidity, } Y = \frac{\text{moles of water vapour}}{\text{moles of dry air}} = \frac{n_A}{n_B} \tag{8.1}$$

Assuming that the air–water vapour mixture behaves as an ideal gas, the following equations can be used to determine the number of moles of the constituents.

$$\begin{aligned} n_A &= \frac{p_A V}{RT} \\ n_B &= \frac{p_B V}{RT} \end{aligned} \tag{8.2}$$

where V is the volume of the mixture and T is the temperature. Combining Eqs. (8.1) and (8.2), we get

$$\text{molal humidity, } Y = \frac{n_A}{n_B} = \frac{p_A}{p_B} = \frac{p_A}{P - p_A} \tag{8.3}$$

Equation (8.3) means that the molal humidity is equal to the ratio of the partial pressure of water vapour to the partial pressure of dry air. Though the definitions given here and in the following sections refer to the air–water vapour system, these are equally applicable for all vapour–gas mixtures.

If the gas is saturated with the vapours, the molal humidity means the *molal humidity at saturation* denoted as Y_s. It is equal to the moles of water vapour per mole of dry air under saturation. Since at saturation, the partial pressure of the vapour is equal to the vapour pressure, the molal saturation humidity can be written as

$$Y_s = \frac{P_A^S}{P - P_A^S} \tag{8.4}$$

where P_A^S is the vapour pressure of water.

The *absolute humidity* is defined as the mass of vapour present per unit mass of vapour-free gas. For the air–water vapour system, it is the kilograms of water present per one kg of dry air. If m_A and m_B are the weights of the vapour and vapour-free gas, then

$$\text{absolute humidity, } Y' = \frac{m_A}{m_B} \tag{8.5}$$

Absolute humidity is related to the molal humidity as:

$$Y' = \frac{m_A}{m_B} = \frac{n_A}{n_B} \times \frac{M_A}{M_B} = Y \times \frac{M_A}{M_B} \tag{8.6}$$

where M_A and M_B are the molecular weights of A and B respectively. Combining with Eq. (8.3) and noting that for air the average molecular weight is approximately 29 and the molecular weight of water is 18, the absolute humidity of the air–water vapour system can be related to the partial pressure as

$$Y' = \frac{p_A}{P - p_A} \times \frac{18}{29} \tag{8.7}$$

The *absolute saturation humidity* (Y'_s) denotes the kilograms of water vapour that one kg of dry air can hold at saturated conditions. Analogous to Eq. (8.4), we can write

$$Y'_s = \frac{P_A^S}{P - P_A^S} \times \frac{M_A}{M_B} \tag{8.8}$$

At the boiling point of the liquid, the vapour pressure is equal to the total pressure so that the denominator in Eq. (8.8) reduces to zero. It means that as the boiling point of the liquid is approached, the saturation humidity tends to infinity. We can see from Eqs. (8.4) and (8.8) that at a given temperature, the molal humidity at saturation is independent of the nature of the gas and depends only on the pressure whereas the absolute saturation humidity depends on the characteristics of the gas as well.

EXAMPLE 8.4 A mixture of acetone vapour and nitrogen gas at 101.3 kPa and 295 K contains acetone vapour to the extent that it exerts a partial pressure of 15 kPa. The vapour pressure of acetone at 295 K is 26.36 kPa. Determine the following:

(a) The mole fraction of acetone in the mixture
(b) The weight fraction of acetone in the mixture
(c) The molal humidity
(d) The absolute humidity
(e) The molal saturation humidity
(f) The absolute saturation humidity
(g) The mass of acetone in 100 m³ of the mixture.

Solution

(a) The partial pressure of acetone in the gas is 15 kPa.

$$\text{Mole fraction of acetone} = \frac{\text{partial pressure}}{\text{total pressure}} = \frac{p}{P} = \frac{15}{101.3} = 0.1481$$

(b) One mole of the mixture contains 0.1481 mole acetone and the rest (0.8519 mole) nitrogen. The molecular weight of acetone = 58.048 and the molecular weight of nitrogen = 28.

Therefore, the weight fraction of acetone in the mixture is

$$\frac{0.1481 \times 58.048}{0.1481 \times 58.048 + 0.8519 \times 28} = 0.2649$$

(c) Molal humidity is the ratio of moles of acetone to moles of dry nitrogen, i.e.

$$Y = \frac{0.1481}{0.8519} = 0.1738$$

(*Note:* Molal humidity may be calculated as the ratio of partial pressure of acetone to the partial pressure of nitrogen, i.e.

$$Y = \frac{15}{101.3 - 15} = 0.1738 \frac{\text{moles of acetone}}{\text{moles of nitrogen}})$$

(d) Absolute humidity is the ratio of kilograms of acetone to kilograms of nitrogen. It is obtained by multiplying the molal humidity by the ratio of the molecular weights

$$Y' = 0.1738 \times \frac{58.048}{28} = 0.3603 \frac{\text{kilograms of acetone}}{\text{kilograms of nitrogen}}$$

(e) The vapour pressure of acetone at 295 K is, $P^S = 26.36$ kPa

Saturation humidity, $Y_s = \dfrac{P^S}{P - P^S}$

$$= \frac{26.36}{101.3 - 26.36} = 0.3517 \frac{\text{moles of acetone}}{\text{moles of nitrogen}}$$

(f) Absolute saturation humidity, $Y_s' = 0.3517 \times \dfrac{58.048}{28} = 0.7292 \dfrac{\text{kilograms of acetone}}{\text{kilograms of nitrogen}}$

(g) The number of moles of the mixture in 100 m^3 is

$$\frac{V}{22.414} \times \frac{P}{P_0} \times \frac{T_0}{T} = \frac{100}{22.414} \times \frac{101.3}{101.3} \times \frac{273.15}{295} = 4.13 \text{ kmol}$$

Since the mole fraction of acetone is 0.1481, the number of moles of acetone in the mixture is

$$0.1481 \times 4.13 = 0.6118 \text{ kmol}$$

The mass of acetone in 100 m^3 of the total gas is

$$0.6118 \times 58.048 = 35.5 \text{ kg}$$

8.2.2 Relative Humidity and Percent Humidity

The degree of unsaturation of a gas–vapour mixture can be expressed in two ways: the first is the percent relative saturation and the second is the percent saturation. The *percent relative saturation* or *percent relative humidity* is the ratio of the actual partial pressure exerted by the vapour to its vapour pressure at the same temperature, which is expressed as a percentage. Thus,

$$\text{relative saturation (relative humidity)} = \frac{p_A}{P_A^S} \times 100\% \tag{8.9}$$

On the other hand, the percent saturation or percent humidity is the ratio of moles of vapour present in the gas per mole of dry gas to the moles of vapour present per mole of a vapour-free gas if it is saturated at the temperature of the system, expressed as a percentage. It is same as the ratio of kilograms of vapour present per kilogram of dry gas at the given temperature to kilograms of vapour that would be associated with one kilogram of dry gas if the gas mixture were saturated with the vapours at the same temperature.

$$\text{Percent saturation (percent humidity)} = \frac{(n_A/n_B)}{(n_A/n_B)_{\text{saturation}}} \times 100\% \tag{8.10}$$

The numerator in Eq. (8.10) is the molal humidity and the denominator is the molal humidity at saturation. Therefore, the percent humidity can be defined as the ratio of the existing humidity of the mixture to the humidity if the gas mixture is saturated with the vapours.

$$\text{Percent saturation} = \frac{Y}{Y_s} \times 100\% = \frac{Y'}{Y'_s} \times 100\% \tag{8.11}$$

A percent saturation of 50% means that the water content per one kilogram of dry air is only half of the water that one kilogram dry air will contain if the air is saturated with water vapour at the same temperature. On the other hand, the relative saturation of 50 percent means that the partial pressure of water in the air–water vapour mixture is only half the vapour pressure of water at the temperature of the system. The relationship between percent saturation (PS) and percent relative saturation (RS) will be clear by substituting Eqs. (8.3) and (8.4) into Eq. (8.11).

$$\text{PS} = \frac{p_A}{P_A^S} \times \frac{P - P_A^S}{P - p_A} \times 100\% \tag{8.12}$$

Using Eq. (8.9), the above equation can be modified as

$$\text{PS} = \text{RS} \times \frac{P - P_A^S}{P - p_A} \tag{8.13}$$

Since the ratio $(P - P_A^S)/(P - p_A)$ in Eq. (8.13) is always less than 1, PS < RS.

EXAMPLE 8.5 Calculate the percent saturation and the relative saturation of the vapour–gas mixture for the conditions given in Example 8.4.

Solution The percent relative humidity is the ratio of the partial pressure of acetone to the vapour pressure expressed as a percentage, i.e.

$$\text{RS} = \frac{p_A}{P_A^S} \times 100\% = \frac{15}{26.36} \times 100 = 56.9\%$$

The percent humidity is given by Eq. (8.11)

$$\text{PS} = \frac{Y}{Y_s} \times 100\% = \frac{0.1738}{0.3517} \times 100 = 49.42\%$$

EXAMPLE 8.6 Moist air contains 0.0109 kg water vapour per cubic metre of the mixture at 300 K and 101.3 kPa. Calculate the following:

(a) The partial pressure of water vapour
(b) The relative saturation
(c) The absolute humidity of the air
(d) The percent saturation

(e) The temperature to which the mixture be heated so that its percent saturation becomes 10%.

The vapour pressure of water (in kPa) is approximated by the Antoine equation as

$$\ln P^S = 16.26205 - \frac{3799.887}{T - 46.854}$$

Solution The number of moles of the mixture in one cubic metre at 300 K and 101.3 kPa is

$$\frac{V}{22.414} \times \frac{P}{P_0} \times \frac{T_0}{T} = \frac{1}{22.414} \times \frac{101.3}{101.3} \times \frac{273.15}{300} = 0.0406 \text{ kmol}$$

$$0.0109 \text{ kg of water} = 0.0109/18.016 = 6.05 \times 10^{-4} \text{ kmol}$$

(a) The mole fraction of water vapour is

$$\frac{n_W}{n} = \frac{6.05 \times 10^{-4}}{0.0406} = 0.0149$$

The partial pressure of water vapour is

$$\text{mole fraction} \times \text{total pressure} = 0.0149 \times 101.3 = 1.51 \text{ kPa}$$

(b) The vapour pressure at 300 K is calculated using the Antoine equation:

$$\ln P^S = 16.26205 - \frac{3799.887}{300 - 46.854} \Rightarrow P^S = 3.5 \text{ kPa}$$

Relative saturation is the ratio of partial pressure to vapour pressure, i.e.

$$\text{RS} = \frac{1.51}{3.5} \times 100 = 43.14\%$$

(c) Absolute humidity can be evaluated by Eq. (8.7):

$$Y' = \frac{p_A}{P - p_A} \times \frac{18}{29} = \frac{1.51}{101.3 - 1.51} \times \frac{18}{29}$$

$$= 9.39 \times 10^{-3} \text{ kg water/kg dry air}$$

(d) Saturation absolute humidity is

$$Y_s' = \frac{P_A^S}{P - P_A^S} \times \frac{18}{29} = 0.0222 \text{ kg water/kg dry air}$$

$$\text{Percent saturation (PS)} = \frac{Y'}{Y_S'} \times 100 = 42.27\%$$

(e) When the mixture is heated at constant pressure, its humidity (Y') will not change. However, saturation humidity at the new temperature is different from the earlier value. Since the percent saturation at the new temperature is given to be 10%, we have

$$\text{PS} = \frac{Y'}{Y_S'} \times 100 = 10.0\%$$

$$Y'_s = \frac{Y'}{0.1} = \frac{9.39 \times 10^{-3}}{0.1} = 0.0939 \text{ kg water/kg dry air}$$

But, by Eq. (8.8),

$$Y'_s = \frac{P_A^S}{101.3 - P_A^S} \times \frac{18}{29} = 0.0939 \text{ kg water/kg dry air}$$

Therefore, P_A^S = 13.31 kPa. The temperature corresponding to this vapour pressure is calculated using the Antoine equation. This temperature is 324.8 K.

EXAMPLE 8.7 A fuel gas saturated with water vapour at 300 K and 100 kPa has a heating value of 25000 kJ/m³ of the total gas. What will be the heating value in kJ/m³ of the gas at 295 K and 105 kPa if its relative saturation is only 50%? The vapour pressure of water at 300 K and 295 K are respectively 3.5 kPa and 2.6 kPa.

Solution *Basis*: One cubic metre of the saturated fuel gas–water vapour mixture at 300 K and 100 kPa.

The number of moles in the mixture is

$$\frac{V}{22.414} \times \frac{P}{P_0} \times \frac{T_0}{T} = \frac{1}{22.414} \times \frac{100}{101.3} \times \frac{273.15}{300} = 0.0401 \text{ kmol}$$

Since the partial pressure of water vapour in the saturated gas is equal to the vapour pressure of water vapour,

the number of moles of the fuel = total moles × mole fraction

$$= 0.0401 \times \frac{100 - 3.5}{100} = 0.0387 \text{ kmol}$$

The heat liberated by combustion of one cubic metre of the saturated gas at 300 K and 100 kPa is the heat liberated by the combustion of 0.0387 kmol of the fuel gas. Therefore, heating value

of the fuel = $\frac{25\,000}{0.0387} = 646043$ kJ/kmol.

Now consider as a basis of one cubic metre of the gas at 295 K and 105 kPa. The total number of moles present in the gas is

$$\frac{V}{22.414} \times \frac{P}{P_0} \times \frac{T_0}{T} = \frac{1}{22.414} \times \frac{105}{101.3} \times \frac{273.15}{295} = 0.0428 \text{ kmol}$$

The partial pressure of water vapour in the gas is

RS × vapour pressure = 0.5 × 2.6 = 1.3 kPa.

The number of moles of the fuel in the gas is

$$0.0428 \times \frac{105 - 1.3}{105} = 0.0423 \text{ kmol}$$

Since the heating value is 646 043 kJ/kmol, 0.0423 kmol on combustion gives out 0.0423 × 646 043 = 27 328 kJ. That is, the heat liberated by the combustion of one cubic metre of the gas at 295 K and 105 kPa = 27 328 kJ. Therefore, the heating value of the fuel at 105 kPa and 295 K = 27 328 kJ/m^3.

8.2.3 Dew Point

Let us consider a process in which a given air–water system at certain temperature and pressure is cooled at constant pressure out of contact with water. For illustration, consider a mixture of air and water vapour with a relative humidity of 30% at 314 K and atmospheric pressure. The vapour pressure of water at 314 K is 7.78 kPa. Therefore, the partial pressure exerted by water in the air is 2.33 kPa and the absolute humidity is 0.0146 kg water vapour per kg dry air. Suppose that the air is cooled at constant pressure out of contact with liquid water. On cooling, the humidity or the partial pressure of water vapour will not change initially, but its relative saturation will increase. This is because the vapour pressure decreases with decrease in temperature and the relative saturation is measured as the ratio of partial pressure to the vapour pressure. When the temperature of air drops to, say, 300 K at which the vapour pressure of water is 3.56 kPa, the relative saturation of air becomes 65%. On further cooling, a temperature will be reached at which the vapour pressure of water is equal to the existing partial pressure of water vapour in the air and the air becomes saturated with the water vapour present in it. This temperature is known as the *dew point* of the air. From the tables of vapour pressure data of water, we can see that at 293 K the vapour pressure of water is 2.33 kPa, same as the existing partial pressure of water vapour in the air. Thus, the dew point of the air is 293 K. If air is cooled below this temperature, water vapour will condense. Suppose that the temperature is brought down to 288 K. The maximum partial pressure the water vapour can exert at 288 K is 1.7 kPa, the vapour pressure of water at 288 K. The humidity corresponding to this partial pressure can be calculated to be 0.0106 kg water per kg dry air. Since the initial humidity was 0.0146 kg water per kg dry air, cooling the air by 5 K below the dew point results in condensation of 0.004 kg water vapour per every kg of dry air in the given sample of humid air.

Thus, we see that dew point is the lowest temperature to which a gas–vapour mixture can be cooled without condensation at constant pressure and out of contact with the liquid. Or it is the temperature at which the vapour begins to condense when cooled at constant pressure. At the dew point the existing vapour content is sufficient to saturate the gas. Stated differently, the dew point is that temperature at which the vapour pressure of the liquid is equal to the partial pressure of vapour in the given vapour–gas mixture. Knowing the dew point, the humidity of the air is readily evaluated using the vapour pressure data. This is possible because the saturation humidity at the dew point and the humidity of the given air–water vapour system are the same. However, to establish the state of the given mixture uniquely, one more specification such as the temperature, the relative saturation or the percent saturation also is necessary.

Dew point of air can be determined experimentally by cooling a well-polished metal surface in the presence of air and measuring the temperature at which water vapour begins to condense and fog appears on the metallic surface. As the dew point is a measure of the humidity, this is one of the methods for measuring atmospheric humidity.

EXAMPLE 8.8 A mixture of nitrogen and benzene has a dew point of 300 K. The temperature and the pressure of the mixture are 335 K and 150 kPa respectively. The vapour pressure of benzene (kPa) is

$$\ln P^S = 13.8858 - \frac{2788.51}{T - 52.36}$$

Calculate the following:

(a) Percent saturation of the mixture
(b) Kilogram of benzene per kilogram of nitrogen
(c) Kilogram of benzene per m³ of mixture
(d) Dew point of the mixture if the total pressure is 100 kPa
(e) Pressure required to condense 60% of the benzene from the mixture at 300 K.

Solution

(a) Using the Antoine equation, the vapour pressure of benzene at 300 K = 13.811 kPa and at 335 K the vapour pressure is 55.693 kPa. Since the vapour pressure at the dew point is equal to the partial pressure of the vapour, the partial pressure of benzene at 335 K and 150 kPa = 13.811 kPa. Using the partial pressure and the vapour pressure, the humidity and saturation humidity at 335 K and 150 kPa can be calculated using Eqs. (8.3) and (8.4):

$$Y = \frac{p_A}{P - p_A} = \frac{13.811}{150 - 13.811} = 0.1014 \text{ kmol benzene/kmol nitrogen}$$

$$Y_s = \frac{P_A^S}{P - P_A^S} = \frac{55.693}{150 - 55.693} = 0.5906 \text{ kmol benzene/kmol nitrogen}$$

$$\text{Percent saturation} = \frac{Y}{Y_S} \times 100 = 17.17\%$$

(b) The quantity of benzene per kilogram of nitrogen is

$$Y \times \frac{M_A}{M_B} = 0.1014 \times \frac{78.048}{28} = 0.2826 \frac{\text{kg benzene}}{\text{kg nitrogen}}$$

(M_A = molecular weight of benzene, M_B = molecular weight of nitrogen)

(c) *Basis*: One cubic metre of the mixture

The number of moles of the mixture is

$$\frac{V}{22.414} \times \frac{P}{P_0} \times \frac{T_0}{T} = \frac{1}{22.414} \times \frac{150}{101.3} \times \frac{273.15}{335} = 0.0539 \text{ kmol}$$

Since the molal humidity is 0.104 kmol benzene per kmol nitrogen, the mole fraction of benzene is

$$y = \frac{Y}{1 + Y} = \frac{0.1014}{1.1014} = 0.09206$$

Therefore, the number of moles of benzene in 1 m³ (i.e., 0.0539 kmol) of the mixture = 0.0539 × 0.09206 = 4.9623 × 10^{-3} kmol. Concentration of benzene in kg/m³ = 4.9623 × 10^{-3} × 78.048 = 0.3873 kg/m³.

(d) Since the humidity remains the same, the partial pressure of benzene at 100 kPa = mole fraction × total pressure = 0.0926 × 100 = 9.26 kPa. Using the Antoine equation, the temperature at which the vapour pressure of benzene is 9.26 kPa, is found out. This temperature is the dew point at 100 kPa. The dew point = 291.51 K.

(e) The mixture initially contains 0.1014 kmol benzene/kmol nitrogen. When 60% of this is condensed the humidity becomes 0.4 × 0.1014 = 0.04056 kmol benzene per kmol nitrogen. At 300 K, if this value is the saturation humidity at a pressure of P kPa,

$$Y = \frac{p_A}{P - p_A} = \frac{13.811}{P - 13.811} = 0.04056 \text{ kmol benzene/kmol nitrogen}$$

Therefore, P = 354.3 kPa.

EXAMPLE 8.9 One cubic metre of a gas at STP is to be dried from a dew point of 300 K to a dew point of 285 K. (a) How much water must be removed? (b) What will be the volume of the gas at STP after drying? The vapour pressure of water is 1.4 kPa at 285 K and 3.56 kPa at 300 K.

Solution *Basis:* 1 m^3 of gas at STP. Moles of gas $= \dfrac{1}{22.414} = 0.0446$ kmol

(a) For air at a dew point of 300 K, the partial pressure of water in the air is equal to the vapour pressure of water at 300 K. Since the ratio of partial pressure of water to the partial pressure of air is the molal humidity, the molal humidity of the given air

$$Y_1 = \frac{p_A}{P - p_A} = \frac{3.56}{101.3 - 3.56} = 0.0364 \text{ kmol water per kmol dry air}$$

For air at a dew point of 285 K, the molal humidity

$$Y_2 = \frac{p_A}{P - p_A} = \frac{1.4}{101.3 - 1.4} = 0.0140 \text{ kmol water per kmol dry air}$$

Therefore, the moles of water removed per kmol of dry air on drying is

$$Y_1 - Y_2 = 0.0364 - 0.0140 = 0.0224 \text{ kmol}$$

The number of kmoles of dry air present in 1 m^3 of the sample is

$$0.0446 \times \frac{1}{1.0364} = 0.0430 \text{ kmol}$$

The amount of water removed is

$$0.0430 \times 0.0224 \times 18.016 = 0.0174 \text{ kg}$$

(b) After drying, the mixture contains 0.0430 kmol dry air and $0.0430 \times \dfrac{0.0140}{1} = 6.02 \times 10^{-4}$ kmol water vapour. Therefore, the volume at STP is

$$0.0430 \times 1.0140 \times 22.414 = 0.9773 \text{ m}^3$$

8.3 HUMIDITY CHART

Humidity charts or psychrometric charts simplify the calculations involved in operations such as humidification, drying, condensation of vapours from non-condensable gases and evaporation of liquids into gas streams, etc. The humidity charts usually refer to air–water vapour systems although charts for any vapour–gas combination can be constructed. One limitation of these charts is that they are applicable only for the total pressure for which they are constructed. For other pressures, separate charts are to be prepared or appropriate correction terms are to be used. The humidity chart for an air–water system at atmospheric pressure is shown in Figure 8.1. The ordinate represents the absolute humidity Y' (kg water vapour per kg dry air) and the abscissa represents the dry-bulb temperature. The temperature of a gas–vapour mixture that is measured by an ordinary thermometer immersed in the mixture is called the *dry-bulb temperature* (DBT). The absolute saturation humidity at any given temperature can be read from the 100 percent saturation curve: Take the temperature on the horizontal axis and move vertically up to the 100 percent curve and then horizontally to the humidity axis. To find the humidity of air of 50 percent saturation at a given temperature, the same procedure is adopted but using the 50 percent saturation curve instead of the 100 percent curve. Note that in some psychrometric charts, the percent relative saturation curves are plotted instead of percent saturation curves.

The humidity charts for an air–water system can be constructed using the vapour pressure versus temperature data for water. Refer to Figure 8.2.

To draw the 100 percent saturation humidity curve (*BCD*), assume a temperature, find the vapour pressure at that temperature, and calculate the saturation humidity using Eq. (8.8). For example, for temperature corresponding to point *A*, the vertical height *AC* gives the saturation humidity. By repeating the calculation at different temperatures below the boiling point of water at the given pressure, other points on the saturation curve can be marked on the diagram. A smooth curve drawn through these points establishes the 100 percent saturation curve. Saturation humidity at the boiling point is infinity and therefore the curve rises asymptotically to infinity near the boiling point. Other percent saturation curves are drawn by dividing the vertical distance between this curve and the temperature axis equally into as many divisions as required. For example, the distance *AC* can be divided into 4 equal parts, then the curves passing through these parts can be marked as 25%, 50% and 75% saturation lines as shown in Figure 8.2.

Specifying the dry-bulb temperature and the percent humidity uniquely determines the state of the air–water vapour system such as the one represented by point *P* where the dry-bulb temperature is T_B and the percent saturation is 25%. The dew point of the air at point *P* is obtained by moving horizontally to point *Q* on the 100 percent curve and by reading the corresponding temperature on the horizontal axis. Alternatively if the dew point of the air, say T_D is known, the humidity will be the same as that at *Q*. Then the state of the system can be anywhere on the horizontal line passing through *Q*. Though by specifying the dew point the humidity is known, to define the state completely one more property need be specified in addition to the dew point.

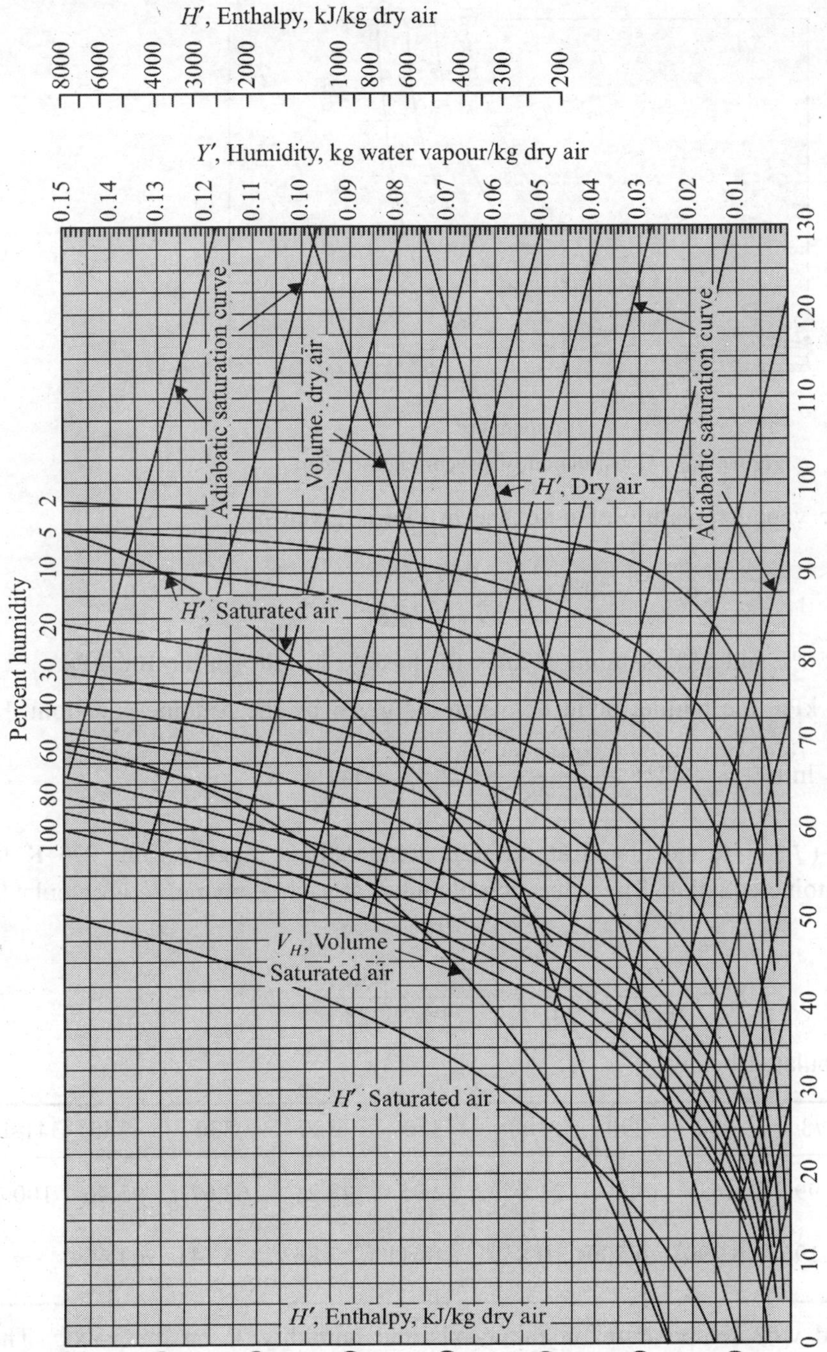

Figure 8.1 Psychrometric chart for air-water vapour system at 1 std atmosphere.

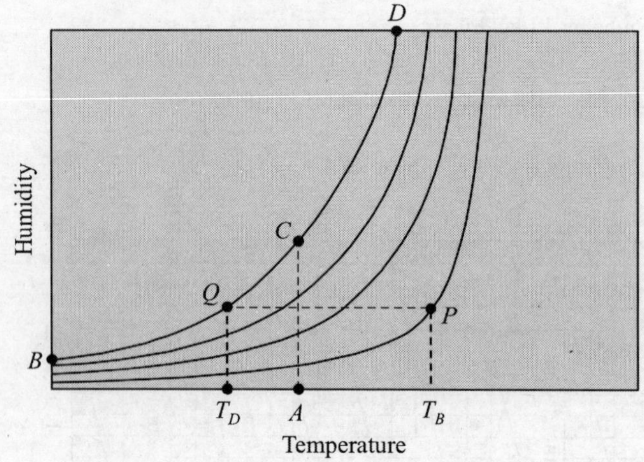

Figure 8.2 Construction of psychrometric chart.

EXAMPLE 8.10 The vapour pressure of *n*-hexane in kPa is given by

$$\ln P^S = 13.8216 - \frac{2697.55}{T - 48.78}$$

Plot the 100%, 75%, 50% and 25% saturation lines for hexane in nitrogen at 100 kPa.

Solution At 100 kPa, the boiling point of hexane is given by the Antoine equation:

$$\ln 100 = 13.8216 - \frac{2697.55}{T - 48.78} \Rightarrow T = 341.47 \text{ K}$$

The vapour pressures (P_A^S) are calculated at different temperatures ranging from 273 K to 341.47 K using the Antoine equation. The saturation humidity at each temperature is calculated using Eq. (8.8):

$$Y_s' = \frac{P_A^S}{P - P_A^S} \times \frac{M_A}{M_B} = \frac{P_A^S}{P - P_A^S} \times \frac{86.11}{28}$$

The results are tabulated below:

T (K)	273	280	290	300	310	320	330	340	341.47
P_A^S (kPa)	5.99	8.63	13.99	21.84	32.95	48.21	68.67	95.46	100.0
$Y_s' = \dfrac{\text{kg hexane}}{\text{kg nitrogen}}$	0.196	0.290	0.500	0.859	1.511	2.863	6.741	64.664	∞

A graph is plotted with temperature on the *x*-axis and humidity Y' on the *y*-axis. The saturation humidity values at each temperature is marked and a smooth curve is drawn through these points. This is the 100% saturation curve. The ordinate at each temperature is now divided into 4 equal parts, representing 25%, 50% and 75% saturation humidity values. Joining the 25% saturation humidity values by a smooth curve, the 25% saturation curve is established.

Similarly, the 50% and 75% curves are also drawn. The psychrometric chart so created is shown in Figure 8.3.

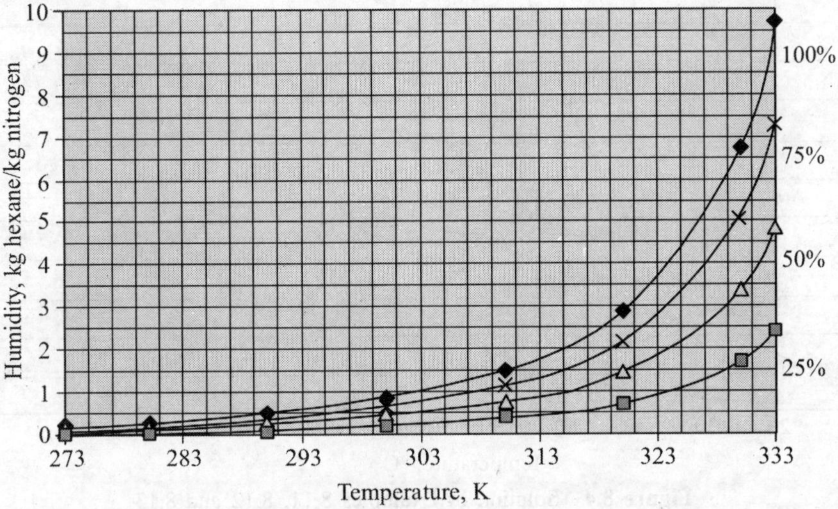

Figure 8.3 Psychrometric chart for nitrogen-hexane system at 100 kPa (Example 8.10).

EXAMPLE 8.11 An air–water vapour sample at 101.3 kPa has a dry-bulb temperature of 328 K and is 10% saturated with water vapour. Using the psychrometric chart determine the following:

(a) The absolute humidity, kg water vapour per kg dry air
(b) The partial pressure of water vapour
(c) The absolute saturation humidity at 328 K
(d) The vapour pressure of water at 328 K
(e) The percent relative saturation
(f) The dew point of the system

Solution

(a) Refer to Figure 8.4. Point *P* in the psychrometric chart corresponds to 328 K and 10% saturation. The *y*-coordinate of this point is read from the chart. This is the humidity at this condition. This is equal to 0.012 kg water vapour/kg dry air. $Y' = 0.012$ kg water per kg dry air.

(b) Partial pressure is calculated using Eq. (8.7):

$$Y' = \frac{p_A}{P - p_A} \times \frac{18}{29} = 0.012$$

$0.0193(P - p_A) = p_A$ where P is 101.3 kPa. Solving this, we get $p_A = 1.921$ kPa.

(c) The saturation humidity at 328 K is obtained directly from the psychrometric chart by reading the *y*-coordinate of point *Q* where point *Q* is the point on the 100% saturation curve corresponding to temperature 328 K. This is found to be 0.115 kg water per kg dry air. $Y_s' = 0.115$ kg water per kg dry air.

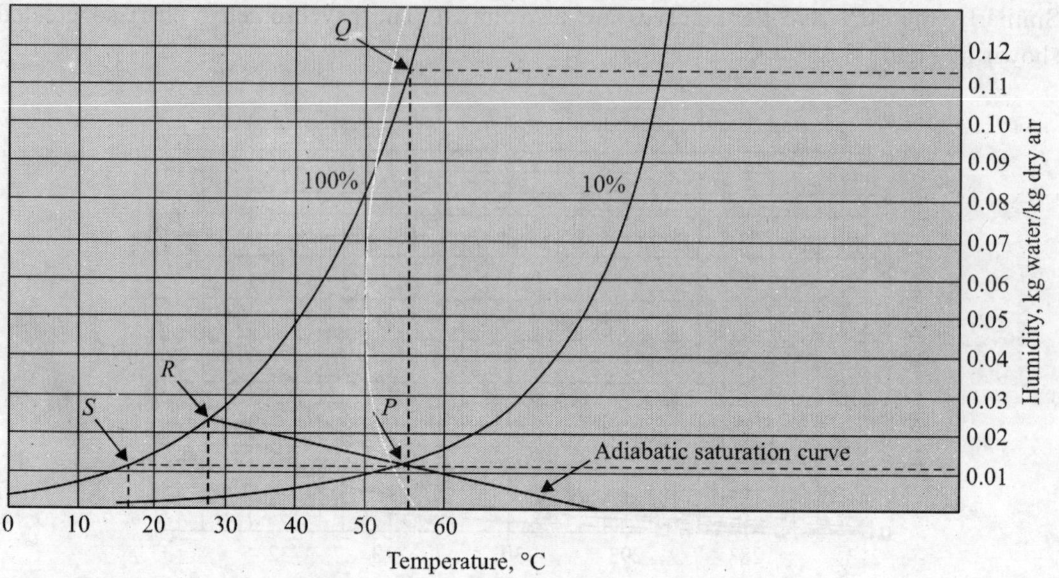

Figure 8.4 Solution of Examples 8.11, 8.12 and 8.13.

(d) Equation (8.8) can be used to calculate vapour pressure:

$$Y_s' = \frac{P_A^S}{P - P_A^S} \times \frac{M_A}{M_B} = \frac{P_A^S}{P - P_A^S} \times \frac{18}{29}$$

$$0.115 = \frac{P_A^S}{101.3 - P_A^S} \times 0.6207$$

Solving this, we get $P_A^S = 15.8$ kPa.

(*The vapour pressure at* 328 K *is found to be* 15.7 kPa *from steam tables, which compares well with the value calculated above.*)

(e) The percent relative saturation is obtained using Eq. (8.9):

$$\text{RS} = \frac{p_A}{P_A^S} \times 100\% = \frac{1.921}{15.8} \times 100\% = 12.16\%$$

(f) The dew point is obtained by moving from point P horizontally to the saturation curve and noting the temperature coordinate of point S. The dew point is found to be 290 K. (*The vapour pressure at the dew point should be equal to the partial pressure of water in the air, which is obtained as* 1.921 *in part (b). Using steam tables, the vapour pressure at* 290 K *is found to be* 1.93 kPa.)

8.4 HUMID HEAT

The heat capacity of the gas–vapour system at constant pressure is designated as the *humid heat* (C_s). It is the amount of heat required to raise the temperature of one kg of dry gas and the accompanying vapour by one degree. If the specific heat of the vapour is C_A(kJ/kg K) and that of the dry gas is C_B(kJ/kg K), then the humid heat is

$$C_s = C_B + Y'C_A \tag{8.14}$$

For the air–water vapour system, $C_A = 1.884$ kJ/kg K and $C_B = 1.005$ kJ/kg K. Knowing the humid heat of the air–water system, the calculation of the heat requirement in operations such as drying and humidification can be done easily as

$$Q = G_s C_s \Delta T \tag{8.15}$$

In Eq. (8.15) Q is the quantity of heat required for heating G_s kg of dry air and the accompanying vapour by (ΔT) K.

EXAMPLE 8.12 Determine the humid heat of the sample of air in Example 8.11, and calculate the amount of heat to be supplied to heat 100 m^3 of the air to 373 K.

Solution The humid heat is calculated using Eq. (8.14): Noting that $C_A = 1.884$ kJ/kg K and $C_B = 1.005$ kJ/kg K,

$$C_s = 1.005 + 0.012 \times 1.884 = 1.0276 \text{ kJ/kg K}$$

$$\text{Number of moles in 100 m}^3 \text{ of the air} = \frac{PV}{RT} = \frac{101.3 \times 100}{8.314 \times 328} = 3.7147 \text{ kmol}$$

$$\text{Number of moles of dry air in 100 m}^3 = 3.7147 \times \frac{p}{P} = 3.7147 \times \frac{101.3 - 1.921}{101.3}$$

$$= 3.6443 \text{ kmol}$$

$$\text{Weight of dry air} = 3.6443 \times 29 = 105.68 \text{ kg}$$

$$\text{Heat to be supplied} = G_s C_s \Delta T = 105.68 \times 1.0276 \times (373 - 328) = 4887 \text{ kJ.}$$

8.5 WET-BULB TEMPERATURE

A wet-bulb thermometer such as the one shown schematically in Figure 8.5 can be used to measure the humidity of air. The temperature indicated by an ordinary thermometer is the dry-bulb temperature of the air. Temperature indicated by the wet-bulb thermometer is called the *wet-bulb temperature* (WBT). Wet-bulb temperature is the steady-state temperature attained by a small quantity of liquid evaporating into a large quantity of unsaturated gas stream. Let us see what happens when an unsaturated air stream is passed across an ordinary thermometer whose bulb is covered with a wet wick, the end of which is immersed in water as shown in Figure 8.5.

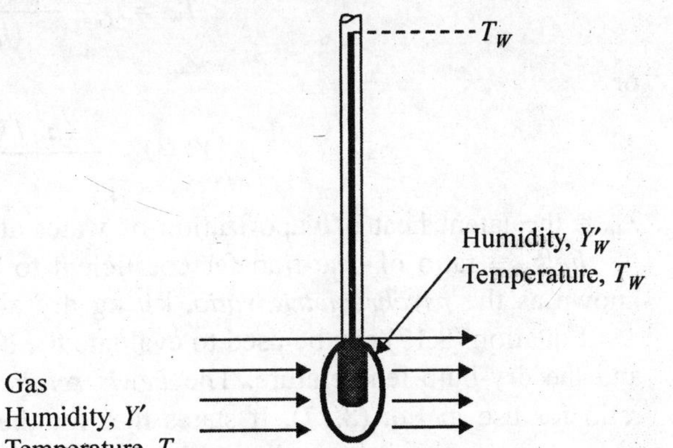

Figure 8.5 Wet-bulb thermometer.

The air film in contact with the bulb of the thermometer will be saturated with water vapour which exerts a partial pressure equal to the vapour pressure of water. The partial pressure of water vapour in the air passed across the thermometer is less than the vapour pressure unless the air is saturated. This results in vaporization of water and diffusion of the vapour into the air stream. Vaporization requires heat and the heat of vaporization is initially supplied by the water film surrounding the bulb and consequently the water cools down lowering the temperature indicated by the thermometer. When the temperature of the water film falls, heat transfer occurs from the bulk of the air to the thermometer bulb. It means that there are two opposing processes occurring in the system. One is the simultaneous transfer of heat and mass to the air stream from the thermometer and the other is the transfer of sensible heat from the air to the thermometer. Eventually a dynamic equilibrium will be established at which the rate of sensible heat transfer towards the thermometer and the rate at which the latent heat of vaporization is lost from the thermometer are equal. Under this condition, the thermometer indicates a constant temperature that remains unaltered with further passage of air. The temperature indicated under this steady-state condition is known as the *wet-bulb temperature of the air*.

The difference between the dry-bulb and wet-bulb temperature is known as the *wet-bulb depression* and is a measure of the humidity of the air stream. If the air were saturated, there would not be any vaporization and consequent lowering of temperature; the dry-bulb and wet-bulb temperature would be the same and wet-bulb depression would be zero. The higher the wet-bulb depression, the larger the degree of unsaturation of the air.

The wet-bulb temperature can be estimated by equating the rate of loss of latent heat of vaporization and the rate of transfer of sensible heat. Though the derivation is beyond the scope of this book, the mathematical equation for estimating the wet-bulb temperature is given below. Let T_G and Y' be the dry-bulb temperature and humidity of the air respectively. Let T_W be the wet-bulb temperature of the air and Y'_W be the saturation humidity at the wet-bulb temperature. Then

$$T_W = T_G - \frac{\lambda_W (Y'_W - Y')}{(h_G/k_Y)} \tag{8.16}$$

or

$$Y' = Y'_W - \frac{(h_G / k_Y)(T_G - T_W)}{\lambda_W} \tag{8.17}$$

λ_W = the latent heat of vaporization of water at the wet-bulb temperature, kJ/kg.

h_G/k_Y = ratio of heat-transfer coefficient to the mass transfer coefficient of air film, also known as the *psychrometric ratio*, kJ/(kg dry air K).

Equation (8.17) can be used to evaluate the humidity of air, given the wet-bulb temperature and the dry-bulb temperature. The *Lewis relation* can be used to evaluate the psychrometric ratio for use in Eq. (8.17). It states that for the air–water system the psychrometric ratio is approximately equal to the humid heat (C_s), the latter being easily evaluated using Eq. (8.14). For any given wet-bulb temperature (T_W), the air–water system can exist in different states each being represented by a point on the humidity chart with a specific dry-bulb temperature (T_G) and humidity (Y') combination. These points with coordinates T_G and Y' should satisfy Eq. (8.17) and all these points can be marked on the humidity chart and joined by a smooth curve. This curve is known as the *wet-bulb temperature line* or the *psychrometric line*.

In Figure 8.6, *CE* is a psychrometric line for the wet-bulb temperature T_C. Several such lines are drawn on the psychrometric chart. The psychrometric lines facilitate the determination of humidity of air from the knowledge of dry-bulb and wet-bulb temperatures. Strictly speaking, these lines are neither straight nor parallel, but by slight modification of the coordinates, they are rendered straight and parallel on the psychrometric charts. In this way the interpolation of data is rather simple.

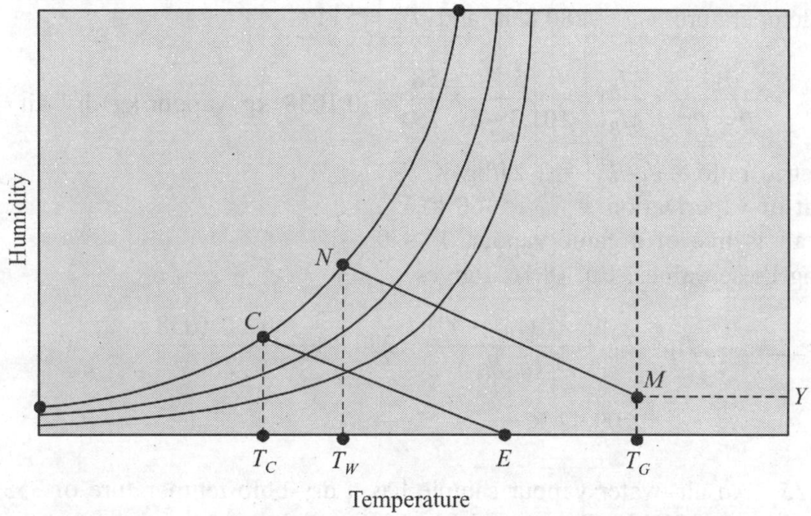

Figure 8.6 Determination of humidity using wet-bulb temperature.

To find the humidity and other properties of air, given the wet-bulb temperature and the dry-bulb temperature T_W and T_G, respectively, take the wet-bulb temperature on the temperature axis and move vertically to the 100 percent saturation curve. Locate point *N* on the saturation curve and move along the psychrometric line passing through *N* downwards to its intersection (point *M*) with the vertical line at the given dry-bulb temperature. This point (*M*) determines the state of the air–water system and the humidity and other properties corresponding to this point can be read from the chart directly.

EXAMPLE 8.13 For the air–water vapour system of the conditions in Example 8.11, find the wet-bulb temperature of air.

Solution Refer to Figure 8.4. Locate the psychrometric line passing through point *P* and move along this line to the hundred percent saturation curve. Let point *R* represents the intersection of the psychrometric line with the saturation curve. The temperature coordinate of point *R* gives the wet-bulb temperature. From the chart, the wet-bulb temperature is found to be 301 K.

EXAMPLE 8.14 Air at 101.3 kPa is blown across the bulb of a mercury thermometer. The bulb is covered with a wick. The wick is immersed in an organic liquid (MW = 58). The reading of the thermometer is 280.8 K. At this temperature the vapour pressure of the liquid is 5 kPa. Find the air temperature given that the psychrometric ratio is 2 kJ/kg K and the latent heat of vaporization of the liquid is 360 kJ/kg. Assume that the air which is blown is free from the organic vapour.

Solution Equation (8.16) is used to evaluate the temperature, i.e.

$$T_W = T_G - \frac{\lambda_W (Y_W' - Y')}{(h_G/k_Y)}$$

Wet-bulb temperature T_W = 280.8 K and P_W^S = 5 kPa

$$Y_W' = \frac{P_W^S}{P - P_W^S} \times \frac{M_A}{M_B} = \frac{5}{101.3 - 5} \times \frac{58}{29} = 0.1038 \text{ kg vapour/kg dry air}$$

Psychrometric ratio = h_G/k_Y = 2 kJ/kg K
Latent heat of vaporization = λ_W = 360 kJ/kg.
Since the air is free of organic vapour, $Y' = 0$.
Substituting these values, Eq. (8.16) gives

$$T_G = T_W + \frac{\lambda_W (Y_W' - Y')}{(h_G/k_Y)} = 280.8 + \frac{360 \times 0.1038}{2}$$

$$= 299.48 \text{ K}$$

EXAMPLE 8.15 An air–water vapour sample has a dry-bulb temperature of 353.2 K and a wet-bulb temperature of 308 K. Compute the humidity of the sample given that the latent heat of vaporization of water at 308 K is 2418.0 kJ/kg and the psychrometric ratio is 0.950 kJ/kg K. Compare the result with the value read directly from the psychrometric chart. The vapour pressure of water at 308 K is 5.62 kPa.

Solution From the psychrometric chart, the humidity is obtained directly. This is found to be 0.018 kg water per kg dry air.

Equation (8.17) can be used to calculate the humidity.

$$Y' = Y_W' - \frac{(h_G / k_Y)(T_G - T_W)}{\lambda_W}$$

Here, $$Y_W' = \frac{P_W^S}{P - P_W^S} \times \frac{M_A}{M_B} = \frac{5.62}{101.3 - 5.62} \times \frac{18}{29} = 0.0365 \text{ kg water per kg dry air}$$

$$Y' = 0.0365 - \frac{(0.950)(353.2 - 308)}{2418.0} = 0.0187 \text{ kg water per kg dry air}$$

This compares well with the humidity obtained from the psychrometric chart.

8.6 ADIABATIC SATURATION TEMPERATURE

As we have seen, the wet-bulb temperature is the steady-state temperature attained by a small quantity of liquid evaporating into a large quantity of unsaturated gas–vapour stream. The gas is passed in such large quantities that its temperature and humidity are not changed by evaporation of the liquid. Now consider the situation where the quantity of unsaturated gas mixture is small in comparison with the liquid with which it is in contact. The temperature as well as the humidity of the air changes during the process whereas the liquid temperature is maintained constant at the equilibrium value by recirculation. Figure 8.7 shows an adiabatic humidification chamber in which water is sprayed into the unsaturated air–water vapour mixture.

In a continuous operation, water attains the equilibrium temperature and this temperature is maintained constant by recirculation. The loss of water by evaporation is compensated by the addition of make-up water also at the equilibrium temperature. The chamber is assumed to be sufficiently large for equilibrium to be established between the gas stream and the liquid. The operation is adiabatic, that is, there is no addition to or removal of heat from the chamber.

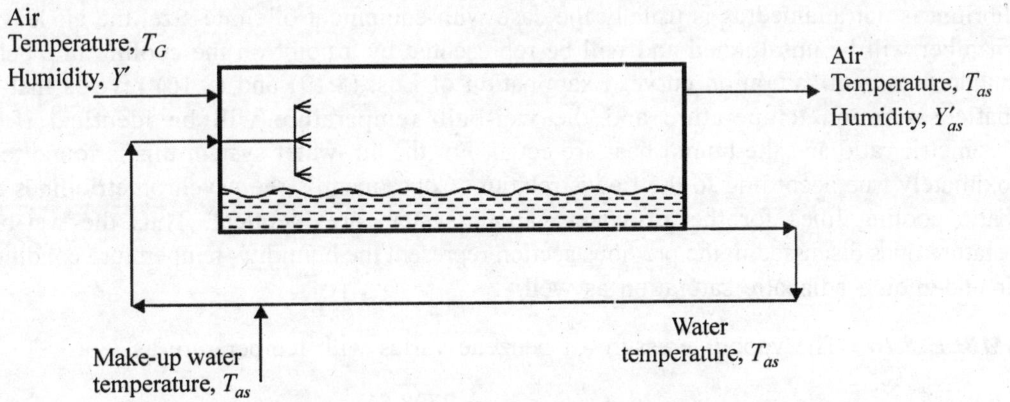

Figure 8.7 Adiabatic saturation chamber.

When an unsaturated air stream is brought into contact with water, the air gets humidified and cooled. Humidification of the gas is achieved by the evaporation of water. Evaporation requires heat of vaporization. Since the process is adiabatic and water temperature is maintained at a constant value, the latent heat of vaporization is to be met by the sensible heat removed from the air and as a result the air gets cooled. Humidification and cooling of the gas continue till the equilibrium between water and the gas is reached. Under equilibrium, the temperature of air and water will be equal and humidity of air will be the saturation humidity at the equilibrium temperature. The equilibrium temperature attained by a gas–vapour mixture when it is adiabatically humidified in contact with the liquid is known as the *adiabatic saturation temperature*, T_{as}.

An energy balance over the system can be written by equating the heat given out by the gas stream by cooling from its initial temperature T_G to the adiabatic saturation temperature T_{as} and the heat required for vaporization of water so that the humidity of the air stream increases from its initial value Y' to the final value Y'_{as}. The humidity is the saturation humidity at the adiabatic saturation temperature T_{as}. Energy balance gives for one kg dry air circulated

$$C_s(T_G - T_{as}) = (Y'_{as} - Y')\lambda_{as} \tag{8.18}$$

where λ_{as} is the latent heat of vaporization of water at the adiabatic saturation temperature and C_s is the humid heat of the air. Equation (8.18) can be rearranged as

$$T_{as} = T_G - \frac{\lambda_{as}\,(Y'_{as} - Y')}{C_s} \tag{8.19}$$

Equation (8.19) gives the adiabatic saturation temperature of a gas–vapour mixture at the dry-bulb temperature T_G and humidity Y'.

On the humidity chart, Eq. (8.18) gives a family of curves known as the *adiabatic saturation curves* or simply *adiabatic cooling curves*. The points on an adiabatic cooling line give the temperature–humidity history of an unsaturated air subjected to adiabatic humidification. When equilibrium is established, the temperature attained is the adiabatic saturation temperature and the humidity attained is the saturation humidity at T_{as}. The adiabatic saturation temperature and the saturation humidity attained on adiabatic humidification are given by the coordinates of the point of intersection of the adiabatic cooling line with the hundred percent saturation curve. If equilibrium is not attained as is usually the case with equipment of finite size, the air leaving the chamber will be unsaturated and will be represented by a point on the cooling line below the hundred percent saturation curve. Examination of Eqs. (8.19) and (8.16) reveals that the adiabatic saturation temperature and the wet-bulb temperature will be identical if the psychrometric ratio and the humid heat are equal. For the air–water system this is found to be approximately true according to the Lewis relation. Consequently, the psychrometric lines and adiabatic cooling lines for the air–water vapour system are identical. Thus the wet-bulb temperature lines discussed in the previous section represent the humidity–temperature conditions of air undergoing adiabatic saturation as well.

EXAMPLE 8.16 The vapour pressure of benzene varies with temperature as

$$\ln P^S = 13.8858 - \frac{2788.51}{T - 52.36}$$

The specific heat of benzene vapour may be taken as 1.2 kJ/kg K and that of dry air is 1.005 kJ/kg K. The latent heat of vaporization of benzene at 283 K is 435.4 kJ/kg. Calculate and plot

(a) The saturation curve for the benzene–air system at 101.3 kPa for the temperature range of 283 K to 313 K with temperature increments of 10 K

(b) The adiabatic cooling line for the adiabatic saturation temperature of 283 K

Solution

(a) The vapour pressures at different temperatures are calculated using the Antoine equation. The saturation humidity values are determine at each temperature using Eq. (8.8):

$$Y'_S = \frac{P_A^S}{P - P_A^S} \times \frac{M_A}{M_B} = \frac{P_A^S}{P - P_A^S} \times \frac{78.048}{29}$$

The results are tabulated as follows:

T(K)	283	293	303	313
P_A^S (kPa)	6.02	9.95	15.80	24.22
Y'_s (kg benzene/kg dry air)	0.17	0.293	0.497	0.846

A graph is plotted with Y' on the *y*-axis and temperature on the *x*-axis as shown in Figure 8.8.

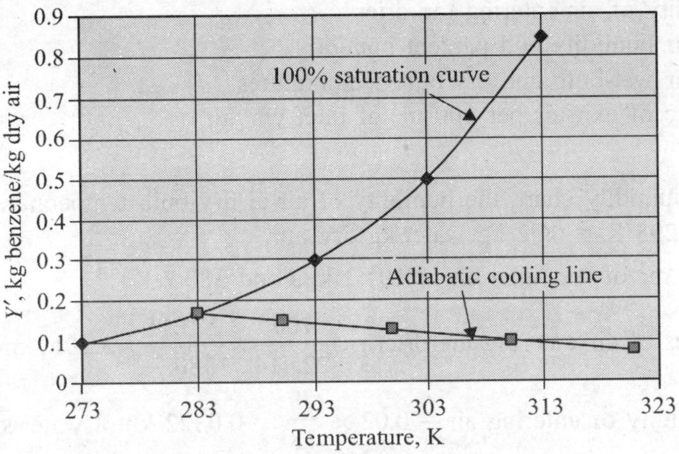

Figure 8.8 Solution to Example 8.16.

(b) For each temperature, an adiabatic cooling line can be drawn using Eq. (8.19).

$$T_{as} = T_G - \frac{\lambda_{as}(Y'_{as} - Y')}{C_s}$$

Here $C_s = C_B + Y'C_A = 1.005 + 1.2Y'$ and $\lambda_{as} = 435.4$ kJ/kg K. To draw the adiabatic cooling line, for 283 K, put T_{as} = 283 K and Y'_{as} 0.17 in Eq. (8.19) and find the sets of T_G and Y that satisfy the equation. To make the calculations simpler, assume Y'(Y' should be lower than Y'_{as}) and calculate T_G that satisfies the equation. For example, let us assume Y' = 0.15 kg benzene/kg dry air. Then

$$T_G = T_{as} + \frac{\lambda_{as}\,(Y'_{as} - Y')}{C_s} = 283 + \frac{435.4(0.17 - Y')}{1.005 + 1.2Y'}$$

$$= 283 + \frac{435.4(0.17 - 0.15)}{1.005 + 1.2 \times 0.15} = 290.35 \text{ K}$$

Now let Y' = 0.125, Y' = 0.10 etc. and repeat the calculations. The results are tabulated below:

T_G (K)	290.4	300	310.1	320.8
Y' (kg benzene/kg dry air)	0.15	0.125	0.10	0.075

The above points are marked and a smooth curve is drawn joining these points as shown in Figure 8.8, which gives the adiabatic cooling line for the adiabatic saturation temperature of 283 K. The curve passes through (T_{as}, Y'_{as}) on the 100% saturation curve. (These calculations can be repeated for T_{as} = 293 K, 303 K and 313 K.)

EXAMPLE 8.17 An adiabatic drier is used to dry a wet material. The drying air enters at 380.7 K and 101.3 kPa with a dew point of 298 K. Measurements show that 2.25 kg of water is evaporated per 100 m^3 of wet inlet air. Calculate using psychrometric chart

(a) The humidity of air entering the drier
(b) The exit air humidity and percent humidity
(c) The exit air wet-bulb and dry-bulb temperatures
(d) The volume of exit air per 100 m^3 of inlet wet air

Solution

(a) From the humidity chart, the humidity of air at dry-bulb temperature of 380.7 K and dew point 298 K = 0.02 kg water/kg dry air.

(b) *Basis:* 100 m^3 of wet inlet air at 101.3 kPa and 380.7 K

$$\text{Moles of air entering the drier} = \frac{100}{22.414} \times \frac{273.15}{380.7} = 3.2011 \text{ kmol}$$

$$\text{Molal humidity of entering air} = 0.02 \times \frac{29}{18} = 0.0322 \text{ kmol water/kmol dry air}$$

$$\text{Moles of dry air entering the drier} = 3.2011 \times \frac{1}{1.0322} = 3.1012 \text{ kmol}$$

$$\text{Weight of dry air in the entering air} = 3.2011 \times \frac{1}{1.0322} \times 29 = 89.934 \text{ kg}$$

Mass of water present in the air entering the drier = 89.934 × 0.02 = 1.7987 kg

Mass of water in the air leaving the drier = 1.7987 + 2.25 = 4.0487 kg

$$\text{Exit air humidity} = \frac{4.0487}{89.934} = 0.045 \text{ kg water/kg dry air}$$

In the psychrometric chart, move along the adiabatic cooling line passing through the point representing the condition of air entering the drier to the point where the humidity is 0.045 kg water per kg dry air. The percent humidity at this condition is read from the chart. It is equal to 55%.

(c) From the humidity chart, the exit air wet bulb temperature = 313.2 K, and the exit air dry bulb temperature = 322.2 K.

(d) The molal humidity of air leaving the drier is

$$0.045 \times \frac{29}{18} = 0.0725 \text{ kmol water per kmol dry air}$$

Since the number of kmoles of dry air entering the drier per 100 m^3 of air is equal to 3.1012 kmol, the number of kmoles of gas leaving the drier is

$$3.1012 \times \frac{1.0725}{1} = 3.3261 \text{ kmol}$$

The volume of exit air (101.3 kPa, 322.2 K) is

$$3.3261 \times 22.414 \times \frac{322.2}{273.2} = 87.92 \text{ m}^3$$

8.7 HUMID VOLUME AND ENTHALPY

8.7.1 Humid Volume

The volume of one kg of dry gas and the accompanying vapour is known as the *humid volume* of the vapour–gas mixture. Assuming the ideal gas law to be applicable to the gas–vapour mixture, for the gas at the dry-bulb temperature T_G, pressure P and having humidity Y', the humid volume V_H is calculated using

$$V = V_0 \times \frac{P_0}{P} \times \frac{T}{T_0}$$

or

$$V_H = \left(\frac{1}{M_B} + \frac{Y'}{M_A}\right) \times 22.414 \times \frac{101.325}{P} \times \frac{T_G}{273.15} \tag{8.20}$$

V_H is the humid volume in m^3/kg dry air. M_A and M_B are the molecular weights of the vapour and dry gas respectively. For the air–water vapour system, Eq. (8.20) gives

$$V_H = 8.314\left(\frac{1}{29} + \frac{Y'}{18}\right) \times \frac{T_G}{P} \tag{8.21}$$

where P is in kPa and V_H is in m^3/kg dry air. Given the humidity of air, the humid volume at any temperature can be calculated using this equation. Equation (8.21) gives the specific volume of dry air (V_D) if we put $Y' = 0$ and the humid volume of saturated air (V_s) if we put $Y' = Y'_s$, the saturation humidity. The former varies linearly with temperature and can be represented by a straight line on the humidity chart while the latter is shown as a curve that is slightly concave upwards. For any given temperature, the specific volume of dry air and humid volume of saturated air can be directly read from the chart, and the humid volume of air of known percent humidity calculated using a linear interpolation as

$$V_H = V_D + \frac{\text{PS}}{100}(V_s - V_D) \tag{8.22}$$

8.7.2 Enthalpy of Humid Air

Enthalpy of dry air and the accompanying water vapour is determined with reference to an arbitrarily chosen reference state. The reference temperature chosen for enthalpy calculation is 273.15 K at which water exists as saturated liquid. For the air–water vapour system with humidity Y' and temperature T_G, the enthalpy can be calculated as the heat required for bringing one kg of dry air and Y' kg of liquid water from the initial temperature T_0 to the final state at T_G at constant pressure. Thus the enthalpy of the air

$$H = C_B(T_G - T_0) + Y'[\lambda_0 + C_A(T_G - T_0)] \tag{8.23}$$

where λ_0 is the latent heat of vaporization of water at the reference temperature T_0. Using Eq. (8.14) for the humid heat, Eq. (8.23) can be modified as

$$H = C_s(T_G - T_0) + Y'\lambda_0 \tag{8.24}$$

Equation (8.24) reveals that the enthalpy can be kept constant in a process where humidity is increased and temperature is reduced, as is the case with adiabatic humidification operations.

Put $Y' = 0$ in Eq. (8.23) and we get the enthalpy of dry air (H_D) which is a linear function of temperature. This is plotted as a straight line on the psychrometric chart. When $Y' = Y'_s$, we get the enthalpy of saturated air (H_s) as a function of temperature. Saturated enthalpy curve also is shown on the psychrometric chart. For air of known humidity, the enthalpy can be calculated using Eq. (8.24) or can be obtained from the chart by interpolation as

$$H = H_D + \frac{\text{PS}}{100}(H_s - H_D) \tag{8.25}$$

EXAMPLE 8.18 Air at 101.3 kPa has a dry-bulb temperature of 303 K and a wet-bulb temperature of 288 K. It is desired to heat the air in a preheater and then spray water into it adiabatically so that final air will be saturated at 303 K. Using the psychrometric chart, determine

(a) The humidity of the initial air
(b) The percent humidity of the initial air
(c) The temperature to which the air is to be heated in the preheater
(d) The heat to be supplied per kg of dry air

Solution The process is represented schematically in Figure 8.9.

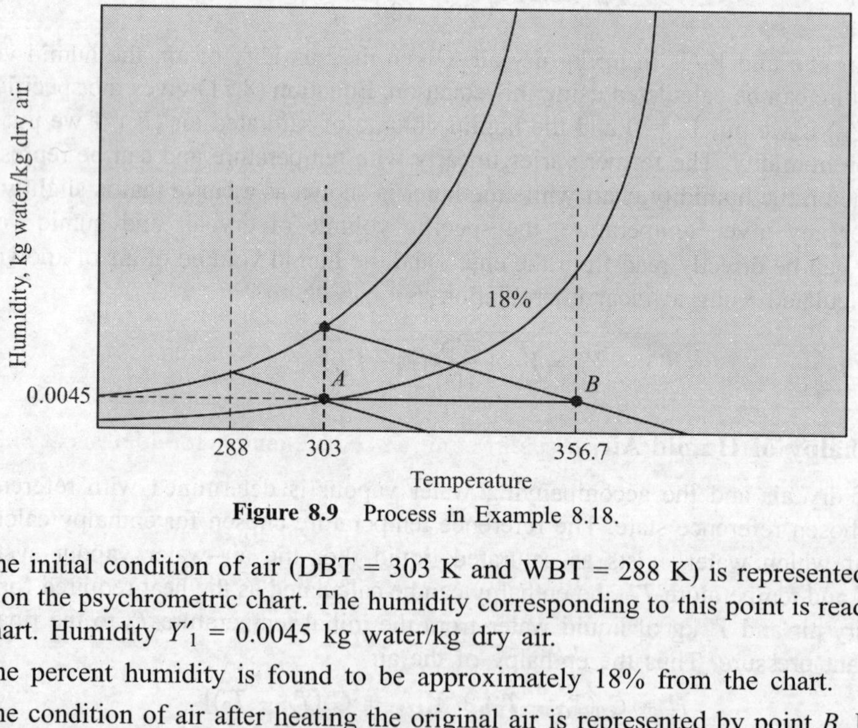

Figure 8.9 Process in Example 8.18.

(a) The initial condition of air (DBT = 303 K and WBT = 288 K) is represented by point *A* on the psychrometric chart. The humidity corresponding to this point is read from the chart. Humidity $Y'_1 = 0.0045$ kg water/kg dry air.

(b) The percent humidity is found to be approximately 18% from the chart.

(c) The condition of air after heating the original air is represented by point *B*. Point *B* is determined in such a way that the air at this condition has the adiabatic saturation temperature of 303 K (the temperature attained by the air when water is sprayed into it adiabatically) and the humidity is the same as the original air. The temperature corresponding to this point is 356.7 K.

(d) The humid heat of the original air is found by Eq. (8.14):

$$C_s = C_B + Y'C_A = 1.005 + 0.0045 \times 1.884 = 1.0135 \frac{\text{kJ}}{\text{kg K}}$$

The heat requirement in the preheater to heat the air from 303 K to 356.7 K per kilogram of dry air is obtained by Eq. (8.15):

$$Q = G_s C_s \Delta T = 1 \times 1.0135 \times (356.7 - 303) = 54.43 \text{ kJ}$$

EXAMPLE 8.19 The wet-bulb and dry-bulb temperature of air are 313 K and 333 K respectively. Determine the following using psychrometric chart:

(a) The absolute humidity
(b) The percent humidity
(c) The humid volume
(d) The enthalpy of wet air

Solution

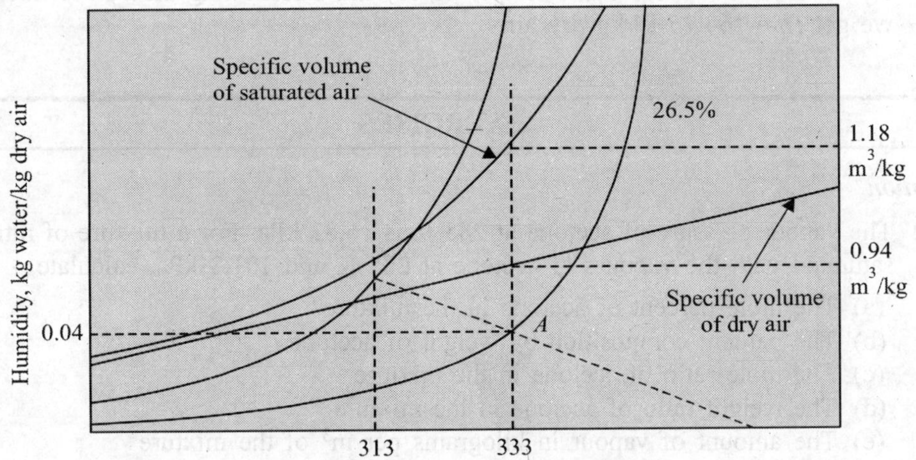

Figure 8.10 Solution to Example 8.19.

(a) Obtain point A on the humidity chart such that the psychrometric line at the WBT 313 K intersects the vertical at the DBT 333 K at point A. Refer Figure 8.10. The humidity at point A is read from the chart and is equal to 0.04 kg water per kg dry air.

(b) The percent humidity at point A is 26.5% from the chart.

(c) At 333 K the volume of saturated air = 1.18 m^3/kg dry air (V_S).
The volume of dry air at 333 K = 0.94 m^3/kg dry air (V_D).
By Eq. (8.22), humid volume is given by

$$V_H = V_D + \frac{\text{PS}}{100}(V_H - V_D)$$

$$= 0.94 + 0.265(1.18 - 0.94) = 1.0036 \text{ m}^3\text{/kg dry air}$$

[*Note*: Equation (8.21) can be used to calculate the humid volume, i.e.

$$V_H = 8.314\left(\frac{1}{29}+\frac{Y'}{18}\right)\times\frac{T_G}{P}$$

Here, Y' = 0.04 kg water per kg dry air, T_G = 333 K and P = 101.3 kPa. Substituting the values, V_H = 1.0032 m^3/kg dry air.]

(d) At 333 K the enthalpy of saturated air = 470 J/kg dry air (H_s). The enthalpy of dry air at 333 K = 60 kJ/kg dry air (H_D).
By Eq. (8.25), the enthalpy of the wet air is

$$H = H_D + \frac{\text{PS}}{100}(H_s - H_D)$$

$$= 60 + 0.265(470 - 60) = 168.65 \text{ kJ/kg dry air}$$

[*Note*: Enthalpy of humid air can be calculated by Eq. (8.24), i.e.

$$H = C_s(T_G - T_0) + Y'\lambda_0$$

C_s = 1.005+ 0.04 × 1.884 = 1.08036 kJ/kg K; T_0 = 273.15 K; λ_0 = 2502.3 kJ/kg; T_G = 333 K; Y' = 0.04 kg water per kg dry air. Substituting the values in Eq. (8.24), we get H = 164.75 kJ/kg dry air.]

EXERCISES

Saturation

8.1 The vapour pressure of acetone at 285 K is 16.82 kPa. For a mixture of nitrogen gas saturated with the vapours of acetone at 285 K and 101.3 kPa, calculate

(a) The mole percent of acetone in the mixture
(b) The percent composition by weight of acetone
(c) The mole ratio of acetone in the mixture
(d) The weight ratio of acetone in the mixture
(e) The amount of vapour in kilograms per m^3 of the mixture

8.2 Carbon dioxide gas at 300 K and 100 kPa is saturated with water vapour. The vapour pressure of water at 300 K is 3.56 kPa. Calculate

(a) The mole percent of water vapour in the mixture
(b) The percent composition by weight
(c) The mole ratio of water vapour in the mixture
(d) The weight ratio of water vapour in the mixture
(e) The amount of vapour in kilograms per m^3 of the mixture

8.3 A gas holder contains dry nitrogen at 300 K and 101.3 kPa. Water is injected into the vessel till the gas is saturated. If the temperature after saturation is 300 K and the vapour pressure of water at 300 K is 3.56 kPa, calculate

(a) The pressure inside the vessel
(b) The mole ratio of water to nitrogen in the saturated gas

8.4 A gas holder contains 1000 cubic metres of a gas saturated with water vapour at 300 K and 105 kPa. The vapour pressure of water at 300 K is 3.56 kPa. How many kilograms of water are present in the holder?

8.5 A certain exhaust gas consists of benzene vapour mixed with air. The partial pressure of benzene vapour is 300 Pa at a total pressure of 107 kPa and 298 K. Does this concentration exceed the ambient air quality stipulation of a maximum hydrocarbon concentration of 1.60×10^{-7} kg/m^3?

8.6 What is the minimum volume (cubic metres) of dry air at 293 K and 100 kPa necessary to evaporate 6 kg of ethyl alcohol if the total pressure remains constant at 100 kPa? Assume that the air is blown through the alcohol to remove it in such a way that the exit pressure of the air–alcohol mixture is 100 kPa. The vapour pressure of alcohol (kPa) is given as

$$\ln P^S = 16.5092 - \frac{3578.91}{T - 50.50}$$

8.7 Dry air at 295 K and 100 kPa is bubbled through benzene. If the saturated air leaves at 300 K and 100 kPa, how many kilograms of benzene are evaporated per 100 m^3 of entering air? The vapour pressure of benzene (kPa) is given by

$$\ln P^S = 13.8858 - \frac{2788.51}{T - 52.36}$$

8.8 10 kg $KClO_3$ is decomposed and the oxygen evolved is collected over water at 300 K and 100 kPa. Determine the weight of saturated oxygen obtained if the vapour pressure of water at 300 K is 3.5 kPa.

8.9 Dry air is saturated with hexane vapours at 253 K and 101.3 kPa. The vapour pressure of hexane at 253 K is 1.85 kPa. What is the percent excess air for combustion?

8.10 It is desired to prepare a saturated methanol vapour–air mixture containing 2% methanol by bringing dry air in contact with methanol kept in a container at 100 kPa. What temperature should be used to achieve the desired composition? The vapour pressure of methanol is given by the Antoine equation

$$\ln P^S = 16.5725 - \frac{3626.25}{T - 34.29}$$

8.11 Air at 303 K and 101.3 kPa is 60 percent saturated with water vapour. If it is desired that 90 percent of water is to be removed by compression at constant temperature, what should be the pressure to which it is compressed? The vapour pressure of water at 303 K is 4.24 kPa.

Humidity, percent humidity, dew point

8.12 A mixture containing benzene vapour and nitrogen in the weight ratio 0.1790:1 is found to have a percent humidity of 40% at 101.325 kPa. What is its relative humidity?

8.13 A mixture of benzene vapour and nitrogen gas at 105 kPa and 330 K contains benzene vapour to the extent that it exerts a partial pressure of 15 kPa. The vapour pressure of benzene is given by the Antoine equation

$$\ln P^S = 13.8858 - \frac{2788.51}{T - 52.36}$$

Determine the following:

(a) The mole fraction of benzene in the mixture
(b) The weight fraction of benzene in the mixture
(c) The molal humidity
(d) The absolute humidity
(e) The molal saturation humidity
(f) The absolute saturation humidity
(g) The mass of benzene in 100 m^3 of the mixture
(h) The percent humidity
(i) The percent relative humidity

8.14 A mixture of benzene vapour and air contains 10% benzene by volume at 300 K and 101.3 kPa. The vapour pressure of benzene at 300 K is 13.8 kPa. Calculate the following:

(a) The partial pressure of benzene in the mixture
(b) The weight fraction of benzene in the mixture
(c) The molal humidity
(d) The absolute humidity
(e) The percent saturation
(f) The percent relative saturation
(g) The mass of air in 100 m^3 of the mixture

8.15 The percent saturation of a mixture of acetone vapour and nitrogen at 105 kPa and 300 K is found to be 80%. The vapour pressure (kPa) of acetone is given by the Antoine equation with constants, $A = 14.39155$, $B = 2795.817$ and $C = 43.198$. Calculate the following:

(a) The molal humidity
(b) The absolute humidity
(c) The partial pressure of acetone
(d) The relative saturation
(e) The dew point

8.16 Moist air contains 0.025 kg water vapour per cubic metre of the mixture at 313 K and 101.3 kPa. Calculate the following:

(a) The partial pressure of water vapour
(b) The relative saturation
(c) The absolute humidity of the air
(d) The percent saturation
(e) The temperature to which the mixture be heated so that its percent saturation becomes 10%

The vapour pressure of water (in kPa) is approximated by the Antoine equation as

$$\ln P^S = 16.26205 - \frac{3799.887}{T - 46.854}$$

8.17 Air at 288 K and 100 kPa contains 1 kg of water vapour. Water is completely removed by adsorbing on silica gel. The dry air after adsorption measures 1000 m^3 at 108 kPa and 293 K. What was the relative humidity of air before adsorption? The vapour pressure of water at 288 K is 1.7 kPa.

8.18 The absolute humidity of a carbon dioxide–water vapour mixture at 310 K and 100 kPa is measured to be 0.022 kg water per kg dry carbon dioxide. Calculate the following:

(a) The molal humidity
(b) The percent relative saturation
(c) The percent saturation
(d) The temperature to which the gas is to be heated at constant pressure to reduce its percent saturation to 30 percent

The vapour pressure of water in kPa is approximated by the Antoine equation as

$$\ln P^S = 16.26205 - \frac{3799.887}{T - 46.854}$$

8.19 A fuel gas saturated with water vapour at 288 K and 101.3 kPa has a heating value of 20000 kJ/m^3 of the total gas. What will be the heating value of the gas in kJ/m^3 of the gas at 295 K and 100 kPa if its relative saturation is only 50%? The vapour pressure of water at 288 K and 295 K are respectively, 1.7 kPa and 2.6 kPa.

8.20 A mixture of acetone vapour and air contains 8.3% acetone by volume. Determine the dew point of the mixture if

(a) The temperature and pressure are 300 K and 100 kPa
(b) The temperature and pressure are 310 K and 90 kPa

The vapour pressure of acetone (kPa) is given by the Antoine equation with constants, $A = 14.39155$, $B = 2795.817$ and $C = 43.198$.

8.21 The dew point of air at an atmospheric pressure of 101.3 kPa is 300 K. The vapour pressure of water at 300 K is 3.56 kPa. Determine the mole fraction of water vapour in the air.

8.22 If the dew point of air at 1.013 bar is 278 K, what will be the dew point at 10 bar? The Antoine constants of water are $A = 16.26205$, $B = 3799.887$, $C = 46.854$.

8.23 Air at 298 K, 55% saturated with water vapour, is initially at 100 kPa. It is then compressed to 1000 kPa and cooled to a temperature so that 90% of the water vapour is condensed. Calculate the following:

(a) The final temperature of the air
(b) The volume of final air per cubic metre of original air

The vapour pressure of water (kPa) is given by the Antoine equation as

$$\ln P^S = 16.26205 - \frac{3799.887}{T - 46.854}$$

8.24 The dew point of a mixture of benzene vapour and dry air at 303 K and 100 kPa is found to be 288 K. The vapour pressure of benzene at 288 K is 7.8 kPa and at 303 K is 15.8 kPa. Determine the following:

(a) The absolute humidity and molal absolute humidity of the mixture
(b) The composition of the mixture in volume percent and weight percent
(c) The relative saturation and the percent saturation

8.25 Methane is completely burned with air. After a portion of the water in the outlet gases from the burner which contains no oxygen is removed by absorption, the gas is found to contain 85% nitrogen at a temperature of 328 K and a pressure of 140 kPa. If more water is to be removed by condensation, determine the following:

(a) The temperature to which this gas must be cooled at constant pressure
(b) The pressure to which the gas must be compressed at constant temperature

The vapour pressure of water in kPa is given by

$$\ln P^S = 16.26205 - \frac{3799.887}{T - 46.854}$$

8.26 A hydrocarbon fuel is burned with dry air. Determine the dew point of the flue gases at 533 K and 100 kPa if the dry flue gases analyzed 13.0% CO_2, 0.05% CO, 2.5% O_2 and 84.45% N_2. The vapour pressure of water is given by the Antoine equation with A = 16.26205, B = 3799.887 and C = 46.854 when pressure is in kPa and temperature in K.

8.27 Determine the mass of 10 m^3 of air at 300 K and 115 kPa if the relative humidity is 65% and the vapour pressure of water at 300 K is 3.56 kPa.

8.28 In a process in which benzene is used as a solvent, it is evaporated into dry nitrogen. The resulting mixture at a temperature of 300 K and a pressure of 101.3 kPa has a relative humidity of 60%. It is required to recover 80% of the benzene present by cooling to 283 K and compressing to a suitable pressure. The vapour pressure of benzene at 283 K is 6.02 kPa and that at 300 K is 13.81 kPa. To what pressure should the gas be compressed?

8.29 Hydrogen saturated with the vapours of water at 101.3 kPa and 305 K is passed through a cooler in which the temperature of the gas is reduced to 280 K in order to condense a part of the vapour. The gas leaving the cooler is heated to 293 K. Calculate

(a) The weight of water condensed per kg of moisture-free hydrogen
(b) The percent humidity of the gas in the final state

The vapour pressure of water (in kPa) may be estimated using the equation

$$\ln P^S = 16.26205 - \frac{3799.887}{T - 46.854}$$

8.30 Air at 303 K, relative humidity 40% is mixed adiabatically with air at 313 K, relative humidity 40% in the ratio of 1 kg of the former to 2 kg of the latter on a dry basis. Find the final temperature and humidity of air. The vapour pressure of water (in kPa) may be estimated using the equation

$$\ln P^S = 16.26205 - \frac{3799.887}{T - 46.854}$$

8.31 One cubic metre of a gas at STP with a dew point of 293 K is to be dried to a dew point of 278 K. How much water must be removed and what will be the volume of the gas at STP after drying? The vapour pressure of water is 0.872 kPa at 278 K and 2.337 kPa at 293 K.

8.32 Water is cooled by direct contact with ambient air in cooling towers. In a certain cooling operation, air enters the tower at 300 K with a partial pressure of water vapour of 670 Pa and leaves at 294 K with a partial pressure of water 2.2 kPa. The total pressure is 100 kPa. Calculate the following:

(a) The relative humidity of air–water vapour at the inlet and exit of the tower
(b) The percent composition by volume at the inlet and exit of the tower
(c) The percent humidity of air at the inlet and exit of the tower
(d) The mass of water vapour in kilograms at the inlet and exit per 100 m^3 of dry air entering the tower
(e) The mass of water vapour in kilograms at the inlet and exit per 100 m^3 of wet air entering the tower.

The vapour pressure (in kPa) of water is given by

$$\ln P^S = 16.26205 - \frac{3799.887}{T - 46.854}$$

8.33 A mixture of nitrogen and methanol has a dew point of 313 K. The temperature and pressure of the mixture are 353 K and 200 kPa respectively. The vapour pressure of methanol (kPa) is

$$\ln P^S = 16.5725 - \frac{3626.55}{T - 34.29}$$

Calculate the following:

(a) The percent saturation of the mixture
(b) Kilograms of alcohol per kilogram nitrogen
(c) Kilograms of alcohol per m^3 of mixture
(d) Dew point of the mixture if the total pressure were 100 kPa
(e) Pressure required to condense 60% of the alcohol from the mixture at 300 K

8.34 A mixture of acetone vapour and nitrogen gas at 300 K has a dew point of 283 K. It enters an evaporator where it picks up acetone vapour. The mixture leaving the evaporator is at 298 K and has a dew point of 293 K. The pressure is kept constant at 100 kPa. Determine the following:

(a) The absolute humidity of the mixture entering and leaving the evaporator
(b) The mass of acetone in kilograms evaporated per kilogram of nitrogen entering the evaporator
(c) The mass of acetone in kilograms evaporated per 100 m^3 of gas entering the evaporator
(d) The volume of gas leaving the evaporator per 100 m^3 of gas entering the unit

The vapour pressure of acetone at 283 K is 15.4 kPa and 24.5 kPa at 293 K.

8.35 A mixture of air and acetone vapour at 303 K and 100 kPa has a dew point of 298 K. 90 percent of acetone in the mixture is to be condensed and removed by cooling and compression. If the mixture is cooled to 278 K, what should be the final pressure? The vapour pressure of acetone (kPa) may be estimated using the Antoine equation

$$\ln P^S = 14.39155 - \frac{2795.817}{T - 43.198}$$

8.36 Air supplied to a drier is at 340 K and 101.3 kPa and has a dew point of 278 K. The air leaves the drier at 310 K and 100 kPa with a dew point of 298 K. At what rate (m^3 per hour measured at the drier inlet conditions) the air should be admitted to the drier if it is desired that 100 kg of water is removed per hour? The vapour pressure of water at 278 K is 0.87 kPa and at 298 K is 3.17 kPa.

8.37 Air at a rate of 25 m^3 per minute is supplied to a drier in which it removes moisture at a rate of 0.45 kg per minute. The air enters the drier at 333 K and 101.3 kPa with a percentage humidity of 10 and leaves at 310 K and 100 kPa. The vapour pressure of water at 310 K = 6.3 kPa and at 333 K = 19.9 kPa.

(a) What is the percent humidity of air leaving the drier?
(b) At what rate does air leave the drier in m^3 per minute?

8.38 The atmosphere in the afternoon during a humid period is at 313 K and is 70% saturated. During the night the temperature falls to 303 K. The pressure is 101.3 kPa. The vapour pressure of water at 303 K is 4.24 kPa and 7.38 kPa at 313 K. What percent of water in the afternoon air is deposited as dew during the night.

8.39 A room of volume 100 m^3 is to be maintained at 300 K and 60% relative humidity. For this fresh air from the outside saturated at 260 K is to be admitted to the room at such a rate that the air is completely renewed each hour. The pressure is 100 kPa. The vapour pressure of water is 3.56 kPa at 300 K and 0.199 kPa at 260 K. Calculate the weight of water evaporated in one hour.

8.40 A mixture of air and water vapour has a relative humidity of 80% at 293 K and 100 kPa. It is then cooled and compressed to 283 K and 250 kPa. The vapour pressure of water is 1.23 kPa at 283 K and 2.34 kPa at 293 K. Determine the following:

(a) Molal humidity of air at initial and final conditions
(b) Kilograms of water condensed per 100 m^3 of initial wet air
(c) Final volume of wet air per 100 m^3 of initial air.

8.41 A gas mixture at 300 K and 100 kPa contains acetone vapour such that the percentage saturation is 70%. The vapour pressure of acetone in kPa is given as

$$\ln P^S = 14.39155 - \frac{2795.817}{T - 43.198}$$

If it is desired that 40% of the acetone is to be removed by condensation, to what temperature the gas must be cooled at constant pressure?

8.42 A mixture of nitrogen gas and benzene vapour at 300 K and 100 kPa has a percent saturation of 70%. It is desired that 80% of the benzene is to be removed by cooling and compression. If the gas is cooled to 273 K, what should be the pressure to which it is compressed? The vapour pressure of benzene at 273 K is 3.5 kPa and 13.8 kPa at 300 K.

8.43 An acetone vapour–nitrogen gas mixture at 310 K and 101.3 kPa with a partial pressure of acetone of 1.5 kPa is passed over a solution containing acetone as part of a solvent-recovery operation. Acetone is to be recovered from the solution at a rate of 25 kg per hour by evaporating into the gas stream. If the gas stream leaves the unit at 287 K and 100 kPa with a percent saturation of 85%, calculate the volumetric rate at which the gas is admitted in m^3 per hour. The vapour pressure of acetone at 287 K is 8.86 kPa.

Wet-bulb temperature and adiabatic saturation temperature

8.44 For an air–water vapour mixture with a dry-bulb temperature of 335 K and a wet-bulb temperature of 308 K, determine the humidity of air at 101.3 kPa The latent heat of vaporization of water at 308 K is 2419 kJ/kg and the psychrometric ratio is 0.950 kJ/kg K. The vapour pressure of water at 308 K is 5.62 kPa.

8.45 An air-water vapour sample has a dry-bulb temperature of 333 K and a wet-bulb temperature of 313 K. Compute the humidity of the sample given that the latent heat of vaporization of water at 313 K is 2407 kJ/kg and the psychrometric ratio is 0.950 kJ/kg K. Compare the result with the value read directly from the psychrometric chart. The vapour pressure of water at 313 K is 7.37 kPa.

8.46 A mixture of air and benzene vapour containing 0.7 kg benzene per kg dry air is to be cooled from 343 K to 288 K in a tubular cooler condenser. The specific heat of dry air is 1.005 kJ/kg K and the specific heat of benzene vapour is 1.1 kJ/kg K. The vapour pressure of benzene in kPa is given as

$$\ln P^S = 13.8858 - \frac{2788.51}{T - 52.36}$$

The psychrometric ratio and the latent heat of vaporization of benzene in the range of interest may be taken to be constant at 1.7 kJ/kg K and 250 kJ/kg, respectively. The pressure is 101.3 kPa.

Determine the following:

(a) The wet-bulb temperature of the inlet stream

(b) The humidity of the exit stream

(c) The total amount of heat to be transferred per kg dry air

8.47 The vapour pressure of benzene varies with temperature as

$$\ln P^S = 13.8858 - \frac{2788.51}{T - 52.36}$$

The latent heat of vaporization of benzene at 293 K is 435.4 kJ/kg and the psychrometric ratio is 1.717 kJ/kg K. Calculate and plot the following:

(a) The saturation curve for the benzene–air system at 101.3 kPa for the temperature range 283 K to 313 K with temperature increments of 10 K
(b) The psychrometric line for the wet-bulb temperature of 293 K

8.48 A toluene–air mixture at 330 K and 101.3 kPa contains 0.05 kg toluene vapour per kg dry air. The Antoine equation for vapour pressure of toluene is

$$\ln P^S\,(\text{kPa}) = 13.9987 - \frac{3096.52}{T - 53.67}$$

The specific heats of toluene vapour and air are, respectively, 1.256 kJ/kg K and 1.005 kJ/kg K. Assume a constant value of 404 kJ/kg for the latent heat of vaporization of toluene.

(a) Estimate the adiabatic saturation temperature.
(b) If the psychrometric ratio for toluene–air system is 1.842 kJ/kg K, estimate the wet-bulb temperature.

8.49 The vapour pressure and the latent heat of vaporization of carbon tetrachloride as function of temperature is given below:

T (K)	278	300	322	333	344	350
P^S (kPa)	5.6	16.4	39.3	57.3	84.7	101.3
λ (kJ/kg)	215.7	206.3	197.5	193.1	188.6	186.4

The specific heat of dry air and that of CCl_4 vapour may be assumed constant. They are respectively, 1.005 kJ/kg K and 0.942 kJ/kg K. For a pressure of 101.3 kPa, calculate following:

(a) The adiabatic saturation temperature for dry air at 350 K
(b) The adiabatic saturation temperature for air at 350 K and having a humidity of 1.5 kg vapour per kg dry air.

8.50 The vapour pressure of chloroform (kPa) is given as

$$\ln P^S = 13.9582 - \frac{2696.79}{T - 46.16}$$

Plot the 100%, 50% and 25% saturation curves for chloroform in air at 101.3 kPa.

Humidity chart

8.51 An air–water vapour sample at 101.3 kPa has a dry-bulb temperature of 333 K and an absolute humidity of 0.01 kg water vapour per kg dry air. Using the psychrometric chart, determine the following:

(a) The percent saturation

(b) The partial pressure of water vapour
(c) The absolute saturation humidity at 333 K
(d) The vapour pressure of water at 333 K
(e) The percent relative saturation
(f) The dew point of the system
(g) The humid heat
(h) The wet-bulb temperature

8.52 The wet-bulb and dry-bulb temperature of air are 313 K and 333 K respectively. Determine the following using the psychrometric chart:

(a) Absolute humidity
(b) Molal humidity
(c) The percent humidity
(d) The humid volume
(e) The dew point
(f) The enthalpy of wet air

8.53 The dry-bulb temperature and the dew point of air are 343 K and 308 K respectively. Estimate the following using the psychrometric chart:

(a) Absolute humidity
(b) Molal humidity
(c) The percent humidity
(d) The humid volume
(e) The wet-bulb temperature
(f) The enthalpy of wet air

8.54 Moist air at 313 K and 50% saturation is heated to 362 K at a constant pressure of 101.3 kPa. Using the psychrometric chart, determine the heat added per cubic metre of initial air.

8.55 An air stream at 101.3 kPa and 333 K is 25% saturated with water vapour. Determine the humid volume and enthalpy of the air. If the air is adiabatically humidified to 70% saturation, what will be its enthalpy in the final state?

8.56 Air with a dry bulb temperature of 363 K and a wet-bulb temperature of 308 K is passed over a wet material in a drier. The rate of evaporation is found to be 0.015 kg water per kg dry air. Assuming the process to be adiabatic, calculate the following using the psychrometric chart:

(a) The absolute humidity, molal humidity and percent saturation of air entering the drier.
(b) The dry-bulb temperature, the wet-bulb temperature, the absolute humidity and the percent saturation of air leaving the drier.

8.57 The dry-bulb temperature and dew point of an air sample at 101.3 kPa are 328 K and 308 K respectively. Calculate the following using psychrometric chart:

(a) The absolute humidity, molal humidity and percent saturation
(b) The partial pressure of water vapour and relative saturation

(c) The wet bulb temperature
(d) The humid heat, humid volume and total enthalpy

8.58 Construct a psychrometric chart for an acetone–air system in the temperature range of 260 K to 330 K which includes the following details.

(a) 25%, 50%, 75% and 100% saturation curves
(b) Specific volume of dry air and humid volume of saturated gas
(c) Enthalpy of dry and saturated gas in kJ/kg dry air referred to liquid acetone and dry air at 260 K
(d) Adiabatic cooling line for an adiabatic saturation temperature of 300 K
(e) Psychrometric line for a wet-bulb temperature of 300 K

The vapour pressure (kPa) of acetone is given by the Antoine equation with constants, $A = 14.5463$, $B = 2940.46$ and $C = 49.19$. The specific heat of acetone vapour and air are 1.47 kJ/kg K and 1.005 kJ/kg K respectively. The latent heat of vaporization of acetone at 300 K is 544 kJ/kg and the psychrometric ratio for air-acetone system is 1.7 kJ/kg K.

Adiabatic saturation in spray chambers

8.59 Air at 313 K dry-bulb temperature and 295 K wet-bulb temperature is admitted to a spray chamber where water is sprayed into it adiabatically. If the exit air is at 300 K, determine, using the psychrometric chart, the quantity of water added per kg dry air.

8.60 It is desired to maintain the temperature of air entering a building at 308 K and humidity at 60%. This is achieved by humidifying air available at 305 K with 65% percent humidity to saturation adiabatically in a spray chamber and by heating the gas leaving the chamber to the desired final condition. Assume the pressure to be atmospheric. Using the psychrometric chart, determine the following:

(a) The temperature of air leaving the spray chamber
(b) The amount of water picked up by one kg dry air passing through the chamber.

8.61 Air at 303 K and with 75% saturation is to be prepared from a supply of warm summer air at 310.7 K and 35% humidity. This is achieved by spraying water to the inlet air and by heating the saturated air leaving the spray chamber to 303 K. Assume all the operations are at atmospheric pressure. Using the psychrometric chart, determine the following:

(a) The temperature of the air leaving the chamber
(b) The amount of water evaporated per 1000 m^3 of air entering the chamber

8.62 An air stream at 101.3 kPa and 333 K is 26.5% saturated with water vapour. It is sent to a spray chamber where it is adiabatically humidified. The temperature of the air leaving the chamber is 5 K above the adiabatic saturation temperature. Determine the following:

(a) The absolute humidity of the initial air
(b) The wet-bulb temperature of the air
(c) The dew point of the initial air

(d) The percent saturation of the air leaving the spray chamber
(e) The humidity and temperature of the air leaving the chamber
(f) The quantity of water picked up by 100 m^3 of wet air admitted to the chamber

8.63 Air at 101.3 kPa has a dry-bulb temperature of 303 K and a wet-bulb temperature of 293 K. It is desired to heat the air in a preheater and then spray water into it adiabatically so that final air will be saturated at 308 K. Using the psychrometric chart, determine the following:

(a) The humidity of the initial air
(b) The percent humidity of the initial air
(c) The temperature to which the air is to be heated in the preheater
(d) The heat to be supplied per kg of dry air.

8.64 Air is dehumidified by spraying into it water whose temperature is below the dew point of the air. Air with a dew point of 294 K and a percent humidity of 40% enters a dehumidifier and leaves at 287 K and with a dew point of 285 K. How much water is removed per kg dry air? How much heat is removed per kg dry air? Solve using psychrometric chart.

8.65 10000 kg/h of air at 326 K and 20% humidity is to be produced from air at 301 K and 20% humidity by first heating, then humidifying adiabatically to the desired humidity and finally reheating the air to 326 K. The air leaves the adiabatic humidification chamber with 95% saturation. Using the psychrometric chart, calculate the following:

(a) The temperature to which air is preheated
(b) The temperature of the air leaving the humidification chamber
(c) The heat required in the preheater
(d) The heat required in the preheater

8.66 Air with a dry-bulb temperature of 310 K and a wet-bulb temperature of 302.5 K is passed into an evaporator. The air leaves the evaporator at 325 K with a wet-bulb temperature of 320 K. The pressure is constant at 101.3 kPa. Calculate the following using psychrometric charts:

(a) The molal humidity, percent saturation and dew point of the initial air
(b) The molal humidity, percent saturation and the dew point of the final air
(c) The amount of water evaporated per 100 m^3 of wet air entering the evaporator

8.67 It is desired that the air entering a building be at 308 K and 30% saturated with water vapour. This is achieved by passing air at 298 K and 30% saturated, through a series of water sprays where it is cooled and saturated. The air emerging from the spray chamber is then heated to 308 K. The water and air leave the chamber at the same temperature. Using psychrometric chart, calculate the following:

(a) The temperature of water
(b) Humidity of air entering the building
(c) The quantity of water condensed or evaporated in the spray chamber per kg moisture free air.

Adiabatic saturation in drying operations

8.68 A wet material containing 70% moisture is to be dried at a rate of 500 kg/h to 5% moisture in a counter current drier, both moisture content being on a wet basis. The air enters the drier at 333 K with a partial pressure of water of 1.0 kPa and leaves at 313 K and 70% saturated. Calculate the air circulation rate in kg/h. The total pressure is 100 kPa and the vapour pressure of water at 313 K = 7.38 kPa.

8.69 A drier is to be designed for removing 100 kg/h of water from a wet material. The following conditions are expected for the air:

Inlet: temperature = 340 K, pressure = 101.3 kPa, dew point = 278 K

Exit: temperature = 310 K, pressure = 100.0 kPa, dew point = 298 K

Vapour pressure of water = 0.872 kPa (278 K), 3.166 kPa (298 K)

At what rate air must be admitted (m^3/h) at the inlet conditions?

8.70 Air at 333 K and pressure of 100 kPa, 10% saturated with water vapour is sent to a drier at a rate of 20,000 m^3/h where it picks up moisture from the solids at a rate of 25 kg/h. If the air leaves at 308 K and 98 kPa calculate the following:

(a) The percent saturation of the air leaving the drier
(b) The volume of air leaving the drier per hour

The vapour pressure of water is 5.62 kPa (308 K), 19.92 kPa (333 K).

8.71 A wet material is being dried by removing water at a rate of 100 kg in a drier. Water evaporates into an air stream entering the drier at 350 K and 101.3 kPa with a dew point of 278 K. The air leaves the drier at 308 K and 100 kPa with a relative humidity of 90%. The vapour pressure of water at 278 K and 308 K are respectively 0.87 and 5.62 kPa. Determine the following:

(a) The rate at which air is admitted, m^3/h
(b) The rate at which air leaves the drier, m^3/h

8.72 Air at 365 K, 101.32 kPa and a dew point of 283 K is supplied to a drier which operates adiabatically. Determine the following using psychrometric charts:

(a) The minimum temperature to which the air gets cooled in the drier
(b) On the basis of 10 m^3 of air entering the drier, the maximum quantity of water evaporated in the drier
(c) The volume of air leaving the drier for the conditions in part (b)

8.73 Air enters a drier at 388 K, 101.3 kPa and with a dew point of 286 K. The drier operates adiabatically. If the air leaves the drier saturated with water vapour, how many kilograms of water will be evaporated per 100 m^3 of air entering? Solve using psychrometric charts.

8.74 Air is supplied to a drier at a dry-bulb temperature of 305 K and a wet-bulb temperature of 298 K. It is heated to 363 K before admitting to the drier. The drier is operated adiabatically. Air leaves the drier fully saturated. Using the psychrometric chart, determine the following:

(a) The humidity, percent humidity and the dew point of the initial air
(b) The heat to be supplied in the heater per 100 m^3 of initial air
(c) Water evaporated per 100 m^3 of air entering the drier
(d) The temperature of air leaving the drier

8.75 An adiabatic drier is used to dry paper. The drying air enters at 371 K and 101.3 kPa with a dew point of 298 K. Measurements show that 2 kg of water is evaporated per 100 m^3 of wet inlet air. Calculate the following using psychrometric chart:

(a) The exit air wet-bulb and dry-bulb temperatures
(b) The exit air percent humidity
(c) The volume of exit air per 100 m^3 of inlet wet air

8.76 A rotary drier operating at 101.3 kPa dries 500 kg/h of wet grain at 295 K from 10% moisture to 1% moisture. Air-flowing counter current to the wet material enters at 380 K dry-bulb and 316 K wet-bulb temperature and leaves at 325 K dry-bulb temperature saturated with water vapour. The grains leave the drier at 316 K and its specific heat is 0.75 kJ/kg K. Using the psychrometric chart, determine the following:

(a) The humidity of the air entering and leaving the drier
(b) Water removed in kg/h
(c) The heat input to the drier

8.77 Air, with a dry-bulb temperature of 311 K and a wet-bulb temperature of 300 K, is scrubbed with water to remove dust. The water is maintained at 298 K. The air leaving the scrubber is in equilibrium with water. It is then heated to 366 K in an air pre-heater and admitted to an adiabatic rotary drier. The air leaves the drier at 322 K. The material to be dried enters and leaves the drier at 319 K. The material losses moisture at a rate of 0.05 kg of water per kg of product. The total product is 1000 kg/h. Using the psychrometric chart, determine the following:

(a) The humidity and percent humidity of the initial air, air after scrubbing, after preheating and air leaving the drier
(b) The total weight of dry air used per hour
(c) The total volume of air leaving the drier

8.78 Wet material is spread over trays stacked one above the other. Air containing 0.005 kg water/kg dry air is heated to 323 K and passed over the lower tray. It leaves the tray at 60% humidity and is then reheated to 323 K and passed over the upper tray after which the air leaves the drier. On the assumption that material on each tray has reached the wet-bulb temperature and that heat losses from the drier can be neglected, determine

(a) The temperature of the material on the trays
(b) The amount of water removed in kg/h if 1000 m^3/h of moist air leaves the drier
(c) The temperature to which the air is to be preheated, if the final condition of air is to be achieved in the first tray itself.

8.79 A laboratory drier produces 50 kg of dry solid containing 6 weight percent water from a wet feed material containing 20.5% water. The drier operates adiabatically, with the wet solid entering the drier at the wet-bulb temperature of the dryer air. A total of

500 m^3/h of hot air at 365.7 K and 101.3 kPa with a dew point of 303.2 K is supplied to the drier. Calculate the following:

(a) The temperature of the air leaving the drier
(b) The percent humidity of the air leaving the drier

8.80 Air at 101.3 kPa enters an adiabatic drier at 372 K with a dew point of 287 K and leaves at 80% humidity. Wet paper enters the drier with 25% moisture and leaves with 5% moisture. Determine the following:

(a) The temperature of air leaving the drier
(b) Water evaporated in kilograms per 100 cubic metres of air entering
(c) Kilograms of finished product per 100 cubic metres of air entering

8.81 It is required to evaporate 100 kg/h of water from a wet solid. The drying is carried out in an adiabatic drier. Air is available at 101.3 kPa and 295 K at 70% humidity. Its temperature is raised by passing through a preheater to such a value that that the air leaving the drier is at 311.7 K and 90% saturated. Calculate the following:

(a) The adiabatic saturation temperature of air leaving the drier
(b) The temperature of air leaving the preheater
(c) The rate of circulation of moist air (kg/h)
(d) The volume of moist air before preheating

9 Material Balance in Unit Operations

Material balance calculation is an essential part of the solution of many complex chemical engineering problems. Material balance assists in the planning and design of processes, in the economic evaluation of the proposed and existing processes, in process control, and in process optimization. In chemical process industries, it is possible to produce a given end product from different raw materials. For an appropriate choice of a process, it is very essential to compute the material requirement for these different routes. A properly listed material balance enables one to estimate the material requirement for an existing process or for a process which is being planned. For instance, in the extraction of oil from oil seeds, it predicts the amount of solvent required for treating a given quantity of seeds. It can evaluate the quantity of water to be evaporated from an aqueous NaOH solution to produce a concentrated solution of given strength. Such information can be used in the design of equipment, in the search for alternative processes or in the evaluation of the economics of the process. Material balance can assist in the simulation of processes based on which certain financial decisions can be made. For existing processes, if direct measurements were made of the weight and compositions of various streams entering and leaving a process, and of the change in material inventory in the system, no material balance calculations would be necessary. As this is not always feasible, the calculation of unknown quantities will always be necessary. Thus, material balance can be used in the hourly and daily operating decisions to be made for running the process efficiently and economically.

9.1 BASIC MATERIAL BALANCE PRINCIPLES

Material balance is an expression of law of conservation of mass. The law of conservation of mass states that the mass can neither be created nor destroyed, but may undergo phase changes, i.e. transformation or transfer from one phase to another and also chemical changes. In an industrial process, material balance provides an exact accounting of all the materials that enter and leave and of the changes in the inventory of the material in the course of a given interval of time of operation.

The material balance for processes with or without chemical reaction can be written in the following form:

$$\begin{Bmatrix}\text{material input}\\ \text{to the system}\end{Bmatrix} - \begin{Bmatrix}\text{material output}\\ \text{from the system}\end{Bmatrix} + \begin{Bmatrix}\text{material generation}\\ \text{within the system}\end{Bmatrix} - \begin{Bmatrix}\text{material consumption}\\ \text{within the system}\end{Bmatrix}$$

$$= \begin{Bmatrix}\text{accumulation of material}\\ \text{within the system}\end{Bmatrix} \tag{9.1}$$

Material balances can be written on mass basis or on mole basis.

9.1.1 Total and Component Balances

Total material balance refers to a balance written for the total mass involved in the process and the component balance establishes the quantitative relationship between the individual chemical compounds or atomic species involved in the process. When a total material balance is written on mass basis, the generation and consumption terms are zero, whether chemical reaction occurs or not. This is because, for all ordinary processes, the transformation of mass into energy can be neglected and the law of conservation of mass is universally valid. The general material balance equation reduces to

$$\text{Input} - \text{Output} = \text{Accumulation} \tag{9.2}$$

Equation (9.2) is valid when the component balance is written on mass basis for any compound whose quantity is not changed by chemical reaction. For substances undergoing chemical transformation, the component mass balance should take into account the generation and consumption terms in Eq. (9.1). For writing component balance for any chemical element, Eq. (9.2) is applicable whether or not it has participated in the chemical reaction.

If mole basis is used, Eq. (9.2) is valid for writing the total material balance for processes involving no chemical reaction. On the other hand, if chemical reaction occurs within the system, the material balance equation [Eq. (9.2)] is not applicable on molar units either for total balance or for component balances. The consumption and generation terms are to be taken into account, and Eq. (9.1) would be the proper choice. However, Eq. (9.2) is still valid if the balance is written for the atomic species. Applicability of material balance equations are summarised in Table 9.1.

Table 9.1 Choice of Material Balance Equation

Basis	*Material balance*	*Chemical reaction*	*Applicability of Eq.* (9.2)
Mass basis	Total material balance	Without chemical reaction	Eq. (9.2) is applicable
Mole basis	Total material balance	Without chemical reaction	Eq. (9.2) is applicable
Mass basis	Component balance	Without chemical reaction	Eq. (9.2) is applicable
Mole basis	Component balance	Without chemical reaction	Eq. (9.2) is applicable
Mass basis	Component balance for elements/atomic species	Without chemical reaction	Eq. (9.2) is applicable
Mole basis	Component balance for elements/atomic species	Without chemical reaction	Eq. (9.2) is applicable
Mass basis	Total material balance	With chemical reaction	Eq. (9.2) is applicable
Mole basis	Total material balance	With chemical reaction	Not applicable. Use Eq. (9.1)
Mass basis	Component balance	With chemical reaction	Not applicable. Use Eq. (9.1)
Mole basis	Component balance	With chemical reaction	Not applicable. Use Eq. (9.1)
Mass basis	Component balance for elements/atomic species	With chemical reaction	Eq. (9.2) is applicable
Mole basis	Component balance for elements/atomic species	With chemical reaction	Eq. (9.2) is applicable

With the chosen basis, write the overall material balance and the individual balances for the components. For problems involving chemical reactions, it becomes necessary to develop material balances based on chemical elements or radicals constituting the species that undergo chemical transformation. However, for compounds or substances that are not decomposed, altered or formed in the process, nothing is gained by writing material balance for the elements present. For these substances and for problems involving no chemical reactions, material balance can be established in terms of the chemical compounds or substances with distinct characteristics and fixed compositions like the ash present in coal and the inert gangue present in mineral ores. For problems involving no chemical reactions, the weight units such as the kilograms can be used for calculations, whereas for problems involving chemical reactions, molar units (mol or kmol) are preferred.

9.1.2 Steady-state and Unsteady-state Processes

For material balance purposes, industrial processes may be classified as *steady-state* or *unsteady-state* processes and as *batch* or *continuous processes*. A steady-state process is one in which there is no change in conditions (pressure, temperature, composition) or rates of flow with time at any given point in the system. All other processes are unsteady state processes. (See Figure 9.1.)

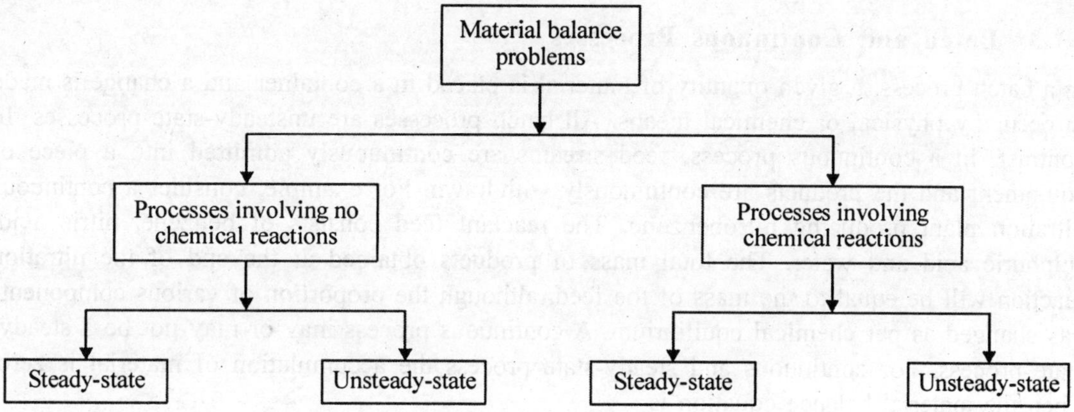

Figure 9.1 Classification of material balance problems.

One simple example of an unsteady-state process is differential distillation or simple distillation. Refer to Figure 9.2.

A liquid mixture consisting of two or more components is slowly vaporized in a closed container. The resulting vapour is condensed and stored in another tank. Because of the difference in volatilities, the vapour will have a composition different from the liquid from which it is produced. The liquid will become depleted in more volatile components as more and more vapour is produced. The amount of the liquid remaining in the container as well as its composition change with time. When the vapour is condensed and collected as distillate, the quantity and composition of the distillate product in the receiver change continuously with time.

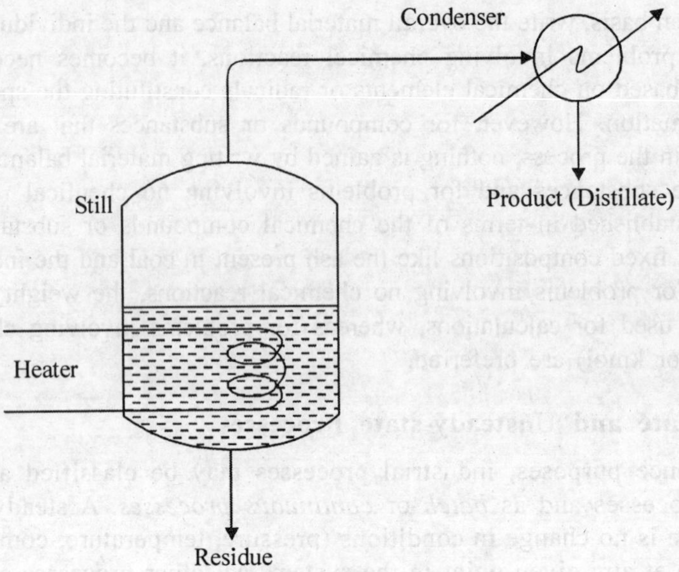

Figure 9.2 Simple distillation.

9.1.3 Batch and Continuous Processes

In a batch process, a given quantity of material is placed in a container and a change is made to occur by physical or chemical means. All batch processes are unsteady-state processes. In contrast, in a continuous process, feed streams are continuously admitted into a piece of equipment and the products are continuously withdrawn. For example, consider a continuous nitration plant producing nitrobenzene. The reactant feed consists of benzene, nitric acid, sulphuric acid and water. The total mass of products obtained at the end of the nitration reaction will be equal to the mass of the feed, although the proportion of various components has changed as per chemical equilibrium. A continuous process may or may not be a steady-state process. For continuous and steady-state process the accumulation of material is zero. Then the material balance equation is

$$\text{input} = \text{output} \tag{9.3}$$

9.1.4 Tie Element

The *tie element* or *key component* is the material which enters and leaves the system in a single stream or one which enters and leaves without any change. An element, compound or an unreactive mass like the ash present in coal can be treated as a key component. Proper choice of key components makes the calculation of the quantitative ratio of the two streams easier. Consider, for example, absorption of ammonia from a mixture of air and ammonia by contacting it with pure water as the solvent. Assume that 100 kmol of the gas mixture containing 10% ammonia is treated to reduce the ammonia content to, say, 1%. The quantity of the final gas leaving the absorber is easily obtained by taking ammonia-free air as the tie element. Since air is insoluble in water, all the air in the entering stream, that is 90 kmol, should be present in the exit gas as well. Since the air in the exit gas constitute 99.0% of the exit stream, the total exit gas would be 90/0.99 = 90.91 kmol.

Though the solution of material balance problems are very much simplified through the use of tie element concept, it is not always necessary that certain component be present as key component for solving material balance problems. Also, in complex processes more than one key component will be necessary for solution.

EXAMPLE 9.1 Coal contains 85% carbon and 15% ash. The cinder formed in the combustion of coal contains 80% ash and 20% carbon. Determine the weight of cinder formed by the combustion of 100 kg of coal and the percent of fuel value that is lost.

Solution *Basis*: 100 kg of coal (refer to Figure 9.3)

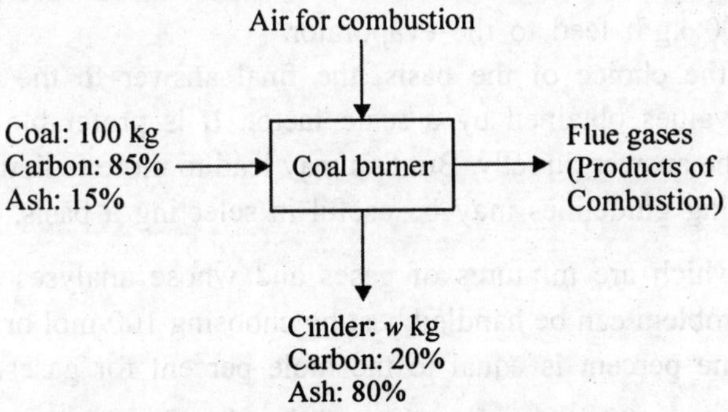

Figure 9.3 Combustion of coal (Example 9.1).

Ash enters the system in a single stream, that is, with the coal fired. It leaves the system in a single stream without any change, with the cinder formed by the combustion. Therefore, ash can be used as a key component for getting the quantity of cinder formed.

The weight of ash entering the system is

$$\frac{15}{100} \times 100 = 15 \text{ kg}$$

Let w be the weight of cinder formed.

The weight of ash leaving is

$$w \times \frac{80}{100} \text{ kg}$$

Taking an ash balance, we get

$$15 = w \times \left(\frac{80}{100}\right) \quad \Rightarrow \quad w = 18.75 \text{ kg}$$

Hence, 18.75 kg cinder is formed by the combustion of 100 kg coal.

Since 20% of cinder is carbon, weight of carbon that is lost unburned when 100 kg of coal is charged is equal to (20/100) × 18.75 = 3.75 kg. As the coal originally contained 85 kg of carbon, the percent fuel value that is lost in the cinder is

$$(3.75/85) \times 100 = 4.41\%$$

9.1.5 Basis for Calculation

In complex problems involving a number of different materials and quantities, the choice of a *basis* is critical for simplifying the solution. Usually, a certain quantity of one of the pertinent components is chosen as the basis. This may be the quantity of one of the streams entering or leaving the process, the quantity of one of the active components entering or leaving the process, or the quantity of one of the inert components entering or leaving the process. In Example 9.1 we have used 100 kg of coal burned as the basis for calculation which was an obvious choice for solving that problem. We could have used 100 kg of carbon in the coal as the basis and obtained the same result as above; but the solution will not be that easy as the one provided above. In many situations the statement of the problem makes the choice of the basis quite obvious. Consider, for example, 1000 kg/h of 40% NaOH solution being concentrated to 5% by evaporation. How much water is to be evaporated? The basis to be chosen is one hour operation or 1000 kg/h feed to the evaporator.

Depending upon the choice of the basis, the final answer to the problem may require multiplication of the values obtained by a scale factor. It is preferable to choose as basis a quantity which gives the answer directly. But this may lead to more tedious calculations in some situations. The following guidelines may be useful in selecting a basis.

(a) For streams which are mixtures of gases and whose analyses are given on volume percent, the problem can be handled best by choosing 100 mol or kmol as a basis. Note that the volume percent is equal to the mole percent for gases.

(b) For solid streams 100 kg or 1 mol (or 1 kmol) of a pure component solid can be chosen as a basis. For solid streams containing substances like ash, gangue, inert cake, etc. the weight basis should be used for computation. Liquid streams also can be handled by choosing 100 kg or 1 mol as a basis.

(c) If the problem demands answers as a time rate, as for example the production rate in kg/h or the flow rate in cubic metres per hour, it would be convenient to make the computation on the basis of the quantity of material flowing in unit time.

(d) In processes like drying and humidification operations, unit weight (or unit mole) of dry gas (air) and the accompanying vapour will serve as a good basis for material balance calculations.

EXAMPLE 9.2 Wood containing 40% moisture is dried to 5% moisture. What mass of water in kilograms is evaporated per kg of dry wood?

Solution *Basis*: 1 kg of dry wood in the feed to the drier (Figure 9.4)

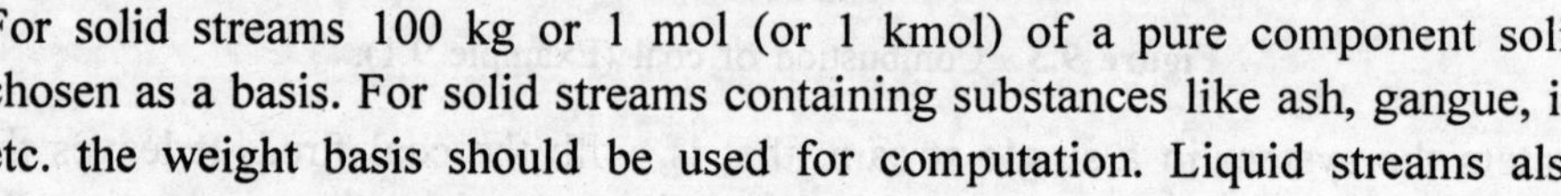

Figure 9.4 Drying of wood (Example 9.2).

The entering material contains 40/60 = 2/3 kg water per kg dry wood and the leaving material contains 5/95 kg water per kg dry wood. If w be the weight of water evaporated, then water balance about the drier yields

$$1\times\frac{40}{60}=w+\frac{5}{95} \quad \Rightarrow \quad w=\frac{40}{60}-\frac{5}{95}=0.614 \text{ kg}$$

The amount of water evaporated per kg dry wood is 0.614 kg.

If we were required to find the amount of water evaporated per kilogram of wet wood entering the drier, the basis of 1 kg of dry wood could still be chosen for solving the problem. Knowing that 1 kg of dry wood is associated with 100/60 kg of wet wood, the water evaporated per kg of wet material is

$$0.614\times\frac{60}{100}=0.368 \text{ kg}$$

EXAMPLE 9.3 The liquid effluent from a processing unit is discharged into a stream. The flow rate and BOD of the stream before the discharge point are respectively 6 m^3/s and 3×10^{-5} g/L. The measurements made immediately below the discharge point indicated a BOD of 5×10^{-3} g/L. If the plant discharges the effluents at a rate of 16×10^3 m^3/day, what is the BOD of the effluent from the plant? (Note: Biochemical Oxygen Demand, BOD, is a measure of the oxygen utilized by microorganisms during the oxidation of organic materials. BOD is a direct measure of oxygen requirement and is directly proportional to the amount of organic waste which has to be broken down.)

Solution *Basis*: One hour of operation (Figure 9.5)

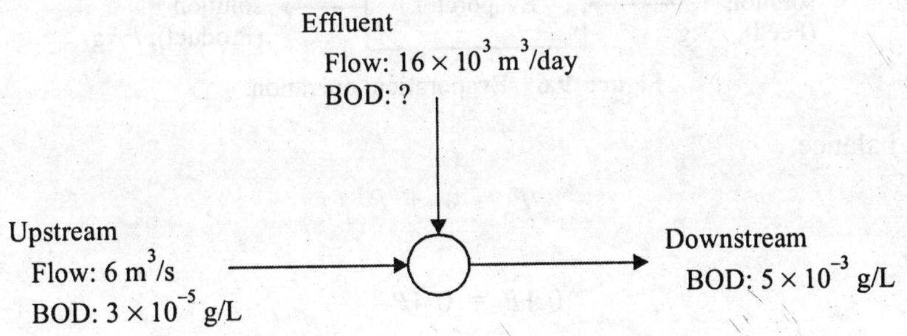

Figure 9.5 Effluent discharge (Example 9.3).

The rate of discharge of effluent is

$$\frac{16\times10^3}{24}=666.67 \text{ m}^3\text{/h} \quad (= 666.67 \times 10^3 \text{ L/h})$$

Let x be the BOD expressed in g/L. Consider the BOD balance at the point where the discharge from the plant joins the stream.

$$6 \times 10^3 \times 3600 \times 3 \times 10^{-5} + 666.67 \times 10^3 x = (21600 + 666.67) \times 10^3 \times 5 \times 10^{-3}$$

Solving this we get $x = 0.1660$. Hence, the BOD of the effluents = 0.1660 g/L.

9.1.6 Independent Material Balance Equations and Degrees of Freedom

The total material balance and the component balances provide a set of equations which can be solved to determine the unknown quantities. Before attempting the solution, check whether the number of unknowns and the number of equations formulated are equal. The number of unknown quantities cannot exceed the number of independent material balance equations. Otherwise, the problem is indeterminate. If the number of independent equations exceeds the number of unknown quantities, the choice of the equations to be used for solving the problem is left to one's judgement. However, some guidelines will make this choice easy. If all the analytical data used for formulating the balance equations are free from errors, it is immaterial on which data we base our calculations. If the data are of doubtful accuracy, the equations based on components that are present in larger proportions may be selected for solving the problem.

The recognition of the maximum number of truly independent equations is important. Any material balance equation that can be derived from other equations already written cannot be regarded as an additional independent equation. For example, consider evaporation of water from a 10% NaOH solution to concentrate the solution to say, 40% NaOH. Assuming F kg to be the amount of solution admitted to the evaporator, w kg be the amount of water evaporated and P kg be the amount of final concentrated solution obtained (Figure 9.6), the following material balance equations can be written.

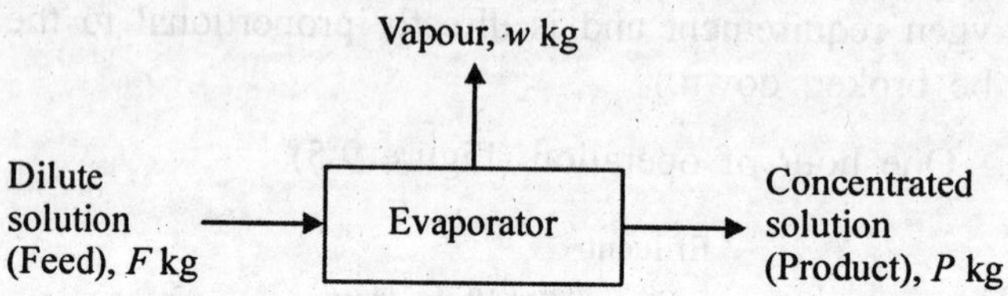

Figure 9.6 Evaporation operation.

Total material balance:

$$F = w + P \tag{9.4}$$

NaOH balance:

$$0.1F = 0.4P \tag{9.5}$$

Water balance:

$$0.9F = w + 0.6P \tag{9.6}$$

We can see that only two of the above three equations are independent, as one of them is obtained by algebraic manipulation and combination of the other two. For example, Eq. (9.6) is the resultant when Eq. (9.5) is subtracted from Eq. (9.4).

If two or more substances exist in a fixed ratio with respect to one another in each stream where they appear, only one independent material balance equation may be written with respect to these substances. Although a balance may be written for any one substance in question, it is best to combine the substances appearing in constant ratio into a single entity and establish a balance equation on this item. Before attempting a solution, we should be sure that the set of equations has a unique solution.

Certain points regarding the number of independent equations may be noted. In general, for processes involving no chemical reaction, the number of independent equations is equal to the number of chemical compounds present, and for processes involving chemical reactions, the number of independent equation is equal to the number of atomic species present. The difference between the number of unknown variables and the number of equations is known as the *number of degrees of freedom*. If the number of degrees of freedom is positive it means that the number of variables is more than the number of equations, and we should try to get additional equations or variable specifications to get a unique solution to the problem. If the number of degrees of freedom is negative, we have too many equations than there are variables. This may arise if we have overlooked to include some variables in the flow diagram. If the number of degrees of freedom is zero, it means that the material balance problem is properly specified, and we can proceed to solve the equations for evaluating the unknown variables.

EXAMPLE 9.4 In the azeotropic distillation of an ethanol–water solution, a feed mixture containing 95.6% alcohol is mixed with pure benzene and distilled. The benzene forms a ternary azeotrope with alcohol–water with a composition of 74.1% benzene, 7.4% water and 18.5% alcohol, which is distilled over as the overhead product. Absolute alcohol is obtained as the residue product. Determine the quantity of benzene required for producing 100 kg of absolute alcohol.

Solution Let us first try to solve this problem by the direct and most obvious method, i.e. by taking a basis of 100 kg of absolute alcohol product.

Basis: 100 kg of absolute alcohol

Let the feed be F kg of the mixture of 95.6% alcohol and pure benzene. The overhead product is D kg and the bottom is W kg. Here W = 100 kg. Let the weight fractions of alcohol, water and benzene in the feed be respectively $w_{F,a}$, $w_{F,w}$ and $w_{F,b}$.

We can now formulate the balance equations. The overall material balance is

$$F = D + W \quad \Rightarrow \quad F = D + 100 \tag{A}$$

Alcohol balance:

$$Fw_{F,a} = 0.185D + W \quad \Rightarrow \quad Fw_{F,a} = 0.185D + 100 \tag{B}$$

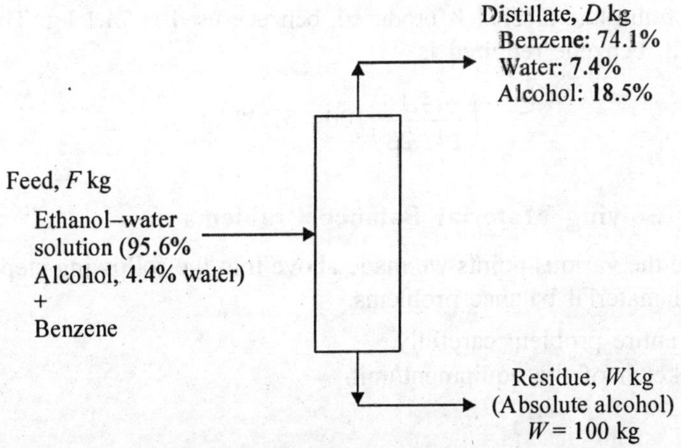

Figure 9.7 Distillation (Example 9.4).

Water balance:

$$Fw_{F,w} = 0.074D \tag{C}$$

Benzene balance:

$$Fw_{F,b} = 0.741D \tag{D}$$

Since the sum of the weight fractions is unity, we have

$$w_{F,a} + w_{F,w} + w_{F,b} = 1 \tag{E}$$

Examination of Eqs. (A) through (E) reveals that not all of them are independent. Combining together Eqs. (B) through (E) gives Eq. (A). Thus we have only four independent equations—Eqs. (B) through (E)—in five unknown quantities. Since the feed is formed from 95.6% alcohol, the weight fractions $w_{F,a}$ and $w_{F,w}$ are related as

$$\frac{w_{F,a}}{w_{F,w}} = \frac{95.6}{4.4} = 21.7273 \quad \Rightarrow \quad w_{F,a} = 21.7273 w_{F,w} \tag{F}$$

Since the weight fractions of alcohol and water in the feed are interrelated as given by Eq. (F), the number of independent variables to be determined reduces to four. These four variables can be determined by the solution of the four equations. Combining Eqs. (B), (C) and (F), we get $D = 70.28$ kg. Using this in Eq. (D), the amount of benzene in the feed $= Fw_{F,b} = 52.08$ kg.

The problem can be solved more easily by taking a basis of 100 kg of distillate.

Basis: 100 kg of overhead product

$D = 100$ kg. Choose water as the key component. Let x be the weight of 95.6% alcohol in the feed. Then, water balance gives

$$\frac{4.4}{100} x = 7.4 \quad \Rightarrow \quad x = 168.1818 \text{ kg}$$

The amount of absolute alcohol is obtained from alcohol balance.

$$168.1818 \times 0.956 = 18.5 + W$$

$$W = 168.1818 \times 0.956 - 18.5 = 142.28 \text{ kg}$$

When 142.28 kg of absolute alcohol is produced, benzene used is 74.1 kg. Therefore, for 100 kg of absolute alcohol, benzene required is

$$\left(\frac{74.1}{142.28}\right) 100 = 52.08 \text{ kg}$$

9.1.7 Steps for Solving Material Balance Problems

We can summarize the various points we made above into the following step-by-step procedure for the solution of material balance problems.

1. Read the entire problem carefully.
2. Draw a sketch of the equipment/unit.

3. Label the flow of each stream and associated composition with symbols.
4. List all the available data on composition and stream flows in the picture. Calculate additional compositions from the given data as necessary.
5. Establish the key component or tie element.
6. Select a suitable basis for calculation.
7. With the chosen basis, write the overall material balance and the individual balances for the components.
8. Before attempting to solve, check whether the number of unknowns and the number of equations formulated are equal. The number of unknown quantities cannot exceed the number of independent material balance equations. Otherwise, the problem is indeterminate. If the number of independent equations exceeds the number of unknown quantities, the choice of the equations to be used for solving the problem is left to your judgement.
9. Solve the equations.
10. Check the answers by substituting them, or at least some of them into the material balances.

EXAMPLE 9.5 1000 kg of mixed acid of composition 40% H_2SO_4, 45% HNO_3 and 15% H_2O is to be produced by strengthening waste acid of composition 30% H_2SO_4, 36% HNO_3 and 34% H_2O by weight. Concentrated sulphuric acid of strength 95% and concentrated nitric acid containing 80% acid are available for this purpose. How many kilograms of spent acid and concentrated acids are to be mixed together?

Solution Let W kg of waste acid, S kg of concentrated sulphuric acid and N kg of concentrated nitric acid are mixed together to produce M kg of mixed acid. The streams in quantities and compositions are indicated in Figure 9.8.

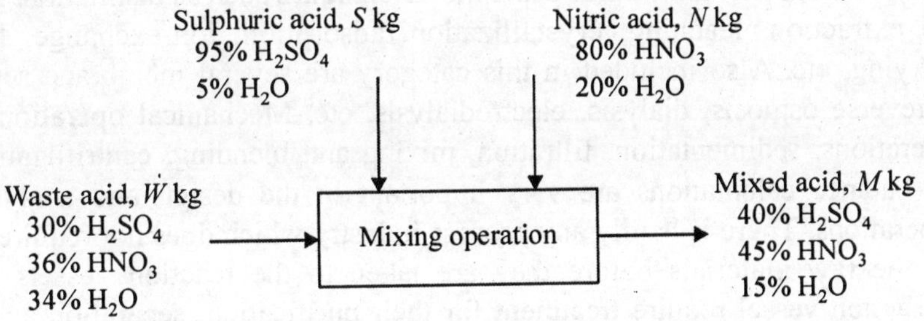

Figure 9.8 Fortification of waste acid (Example 9.5).

Basis: 1000 kg of mixed acid. M = 1000 kg.

We can write four material balance equations—one total material balance, three component balances, one each for H_2SO_4, HNO_3 and H_2O.

Total material balance:

$$W + S + N = 1000 \tag{A}$$

H_2SO_4 balance:

$$0.3W + 0.95S = 0.4 \times 1000 \tag{B}$$

HNO_3 balance:

$$0.36W + 0.8N = 0.45 \times 1000 \tag{C}$$

H_2O balance:

$$0.34W = 0.05S + 0.20N = 0.15 \times 1000 \tag{D}$$

There are only three independent material balance equations as Eq. (D) can be obtained by adding together Eqs. (B) and (C) and subtracting the sum from Eq. (A). The number of unknown quantities is three. They are W, S and N. Equations (A), (B) and (C) are solved simultaneously to determine the quantities W, S and N. We get $W = 70.22$ kg, $S = 398.88$ kg, $N = 530.90$ kg.

Quantities to be mixed are:

Waste acid = 70.22 kg

Concentrated H_2SO_4 = 398.88 kg

Concentrated HNO_3 = 530.90 kg

9.2 MATERIAL BALANCE WITHOUT CHEMICAL REACTIONS

In chemical process industries, we employ a large number of chemical and physical operations for transforming matter from inexpensive raw materials to highly desired products. The physical operations are generally treated in chemical engineering parlance as *unit operations*. These involve the addition or removal of some form of energy in the contacting, transport, and conditioning of materials mainly by physical means. Unit operations in chemical engineering practice can be broadly classified as fluid flow operations, heat transfer operations, mass transfer operations and mechanical operations involving particle handling and conditioning. The mass transfer operations include various separation techniques such as distillation, gas absorption, liquid–liquid extraction, leaching, crystallization, adsorption, ion-exchange, humidification operations, drying, etc. Also included in this category are several membrane separations such as osmosis, reverse osmosis, dialysis, electrodialysis, etc. Mechanical operations include size reduction operations, sedimentation, filtration, mixing and blending, centrifugation, etc.

Material balance calculations are very important in the design and analysis of various separation operations. There is hardly any process industry which does not require a preliminary treatment of the raw materials before they are taken to the reaction vessels. The products leaving the reaction vessel require treatment for their purification, separation and for recovery of by-products. Separation techniques are also employed to give definite properties to the products, to render the products in a form suitable for packaging and handling and to reduce the cost of transportation of the products. The material balance problems encountered in the separation operations are discussed in this chapter as these operations are purely physical in nature, and no chemical conversions are involved. Before we embark on the solution of the material balance problems, we will briefly discuss the salient features of some of the important separation methods which are quite in order for a better understanding of the material balance systems. The present discussion is limited to the systems that are important from a material balance point of view.

9.2.1 Evaporation

Evaporation is used in industries for concentrating aqueous solutions by vaporizing the solvent water and removing it as vapour. Typical industrial applications of evaporation include concentration of cane-sugar juice in a sugar factory, concentration of ammonium sulphate in a fertilizer unit, and concentration of spent soap *lye* to produce glycerine in a soap industry. Depending upon the mode of providing heat energy to the liquid being concentrated, evaporators fall into different types. Evaporators heated by direct fire, evaporators with heating medium in jackets and steam heated tubular evaporators are common in industries. Among the tubular evaporators, short-tube vertical evaporators, long-tube evaporators, climbing film and falling film evaporators, etc. are some industrially important ones in use. A schematic sketch of a standard short-tube evaporator is shown in Figure 9.9.

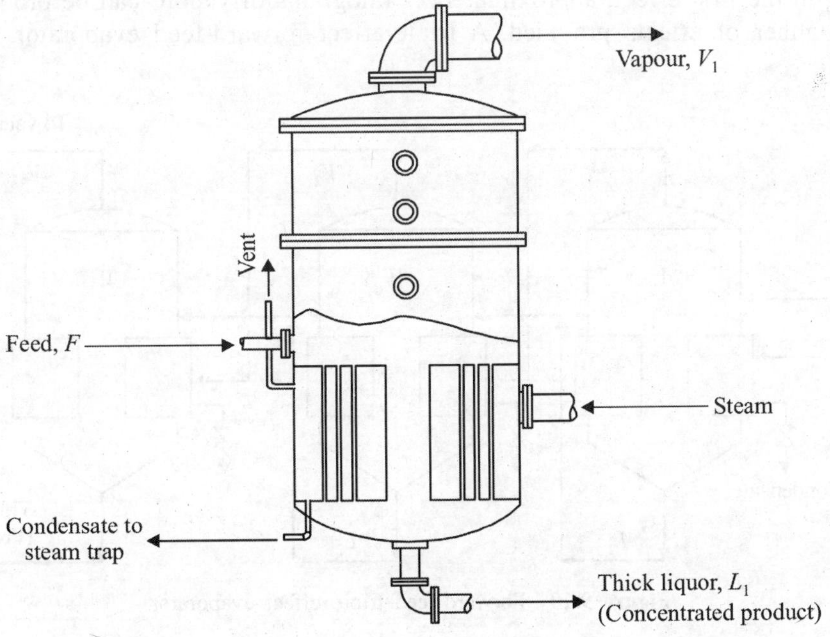

Figure 9.9 Schematic sketch of a short-tube vertical evaporator.

The feed solution is admitted as shown. The solution flows up through the inside of the tubes and is heated by steam condensing in the steam chest, outside the tubes. The heat liberated by the condensing steam is transferred through the tubes to the liquid. The vapour space of the evaporator is maintained at a lower pressure. When the temperature of the liquid reaches the boiling point of the solution at this reduced pressure, the liquid boils. The vapour generated is taken out through the vapour outlet which is condensed and discarded. The concentrated solution known as *thick liquor* is withdrawn as product from the bottom of the evaporator.

Let F be the feed rate in kg/h and x_F be the mass fraction of the solute in the feed. Let the thick liquor be withdrawn at a rate of L_1 kg/h and the concentration of the solute in the final product be x_1. Let V_1 be the quantity of vapour produced in kg/h. The overall material balance can be written as

$$F = V_1 + L_1$$

Since the vapour produced is free of solute, the solute balance gives

$$Fx_F = L_1x_1$$

Simultaneous solution of these two equations provide answer to the material balance problems on single-effect evaporators.

In a single-effect evaporator, with one kilogram of steam condensing in the steam chest, roughly only one kilogram of vapour is produced. The cost of steam consumption is therefore the major cost of operation of the evaporator unit. To reduce the steam consumption, several evaporators can be connected in series to form a multiple-effect evaporator system. The vapour produced in the first effect is used as the heating medium in the second effect and the vapour from the second is the heating medium for the third and so on. Thus by one kilogram of steam condensing in the first effect, approximately N kilograms of vapour can be produced, where N is the number of effects provided. A triple-effect forward-feed evaporator is shown in Figure 9.10.

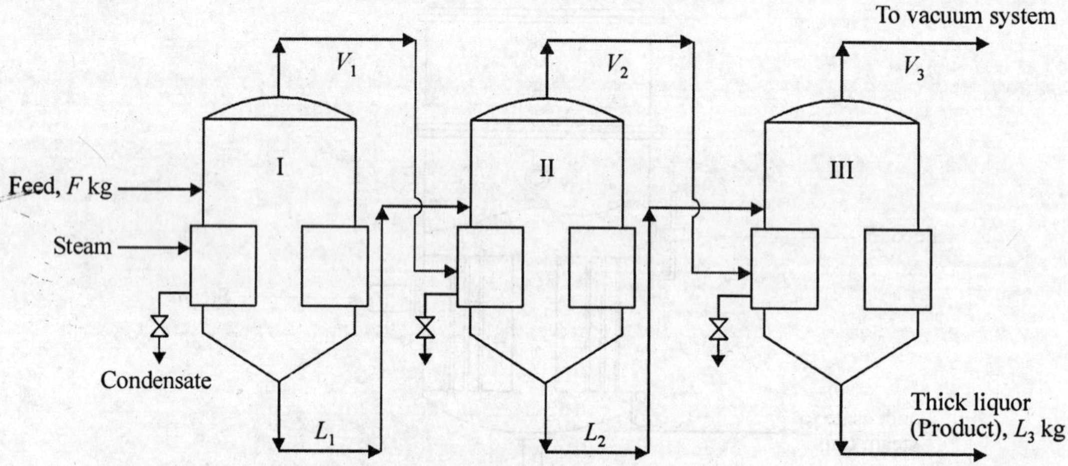

Figure 9.10 Forward-feed triple effect evaporator.

The overall material balance for a triple-effect evaporator is

$$F = V_1 + V_2 + V_3 + L_3$$

where V_1, V_2, and V_3 are the amounts of vapour produced in the first, second and third effect respectively, in kg/h, and L_3 is the product rate in kg/h. Let x_1, x_2, and x_3 be the concentration of the liquid leaving each effect. The solute balance for the combined system is

$$Fx_F = L_3x_3$$

The following solute balance equations can be written for each effect.

$$Fx_F = L_1x_1$$
$$L_1x_1 = L_2x_2$$
$$L_2x_2 = L_3x_3$$

The total material balance for individual effects are

$$F = V_1 + L_1$$
$$L_1 = V_2 + L_2$$
$$L_2 = V_3 + L_3$$

EXAMPLE 9.6 A triple effect evaporator (Figure 9.11) is used to concentrate 1000 kg of aqueous solution from a concentration of 20% solute to 80% solute. Assuming an equal amount of vaporization in each effect, calculate the composition and weight of the solution entering the second and third effects.

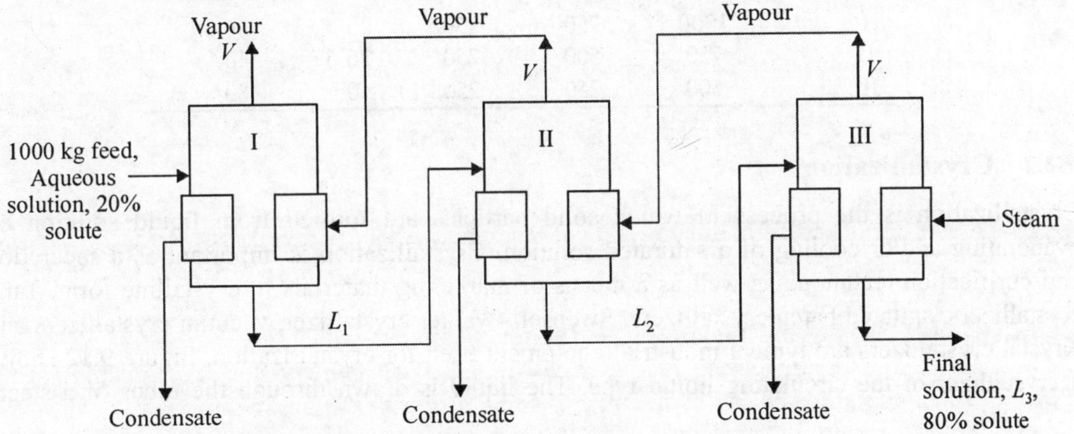

Figure 9.11 Triple effect evaporator for Example 9.6.

Solution *Basis:* 1000 kg feed to the first effect

Let L_1, L_2 and L_3 be the rate at which the concentrated solution is leaving each effect, x_1, x_2 and x_3 respectively be the concentrations of solute in these effects. Let V be the amount of vapour produced in each effect. Consider the combined system consisting of all the three effects. The total material balance yields

$$1000 = 3V + L_3 \tag{A}$$

The solute balance is

$$1000 \times 0.2 = L_3 x_3$$
$$200 = 0.8L_3 \tag{B}$$

Solving Eqs. (A) and (B), we get $L_3 = 250$ kg and $V = 250$ kg.
Now we can write the material balance equation for the first effect.
Total balance:

$$1000 = L_1 + V = L_1 + 250 \tag{C}$$

Solute balance:

$$1000 \times 0.2 = L_1 x_1 \tag{D}$$

From Eq. (C), $L_1 = 750$ kg. Substituting this in Eq. (D), we get $x_1 = 0.2667$.

For the second effect, the corresponding equations are

$$L_1 = L_2 + V \Rightarrow 750 = L_2 + 250 \quad \text{(E)}$$

$$L_1x_1 = L_2x_2 \Rightarrow 750 \times 0.2667 = L_2x_2 \quad \text{(F)}$$

From Eqs. (E) and (F), we get $L_2 = 500$ kg and $x_2 = 0.4$.

The results of material balance calculation for the triple-effect evaporator system are summarized as follows:

Effect	*Feed* (kg)	*Products* (kg)		*Liquid concentration* (%)	
		Liquid	*Vapour*	*Feed*	*product*
I	1000	750	250	20	26.67
II	750	500	250	26.67	40
III	500	250	250	40	80

9.2.2 Crystallization

Crystallization is the process in which solid particles are formed from liquid solution by evaporating and/or cooling of a saturated solution. Crystallization is important as a separation and purification technique as well as a means of marketing materials in crystalline form. Tank crystallizers, agitated batch crystallizers, Swenson–Walker crystallizer, vacuum crystallizers and Krystal crystallizers are typical industrial equipment used for crystallization. Figure 9.12 shows a crystallizer of the circulating liquid type. The liquid is drawn through the tubes of a steam

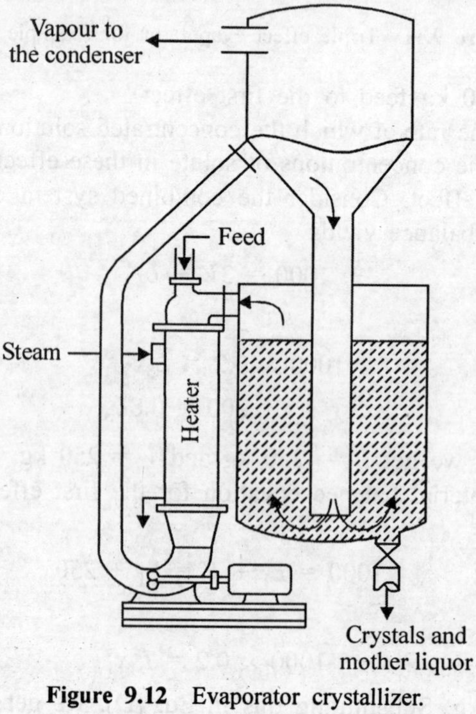

Figure 9.12 Evaporator crystallizer.

heated exchanger where it is heated to sufficiently high temperatures without evaporation. The heated liquid then flows into the vapour space where flash evaporation occurs leading to the supersaturation of the solution. The supersaturated liquid is made to flow down through the downcomer and then up through the bed of crystals which grows in size. The saturated liquid leaving the bed joins the fresh feed solution before entering the heater.

Crystallization occurs when the solution is concentrated and usually cooled until the solute concentration becomes greater than its solubility at that temperature. Equilibrium is said to be attained when the solution is saturated.

In general, in industrial crystallization processes, equilibrium is attained between the solution (mother liquor) and the crystals, and hence the mother liquor leaving the process is a saturated solution at the final temperature of crystallization. The yield of crystals can be calculated knowing the initial concentration of the solute, final temperature and the solubility at this temperature. The material balance calculations are straightforward when the solute crystals are anhydrous. Simple water and solute balances are sufficient. When the crystals are hydrated, some of the water in the solution is removed with the crystals as a hydrate.

Let W_1 kg of hydrated crystals are formed in which the weight fraction of solute is x_1 from F kg of solution containing x_F weight fraction of solute. Let x_2 be the solubility of the solute in weight fraction of solute, and W_2 and W_3 be the weights of mother liquor remaining after crystallization and the weight of water evaporated during crystallization operation (see Figure 9.13), respectively.

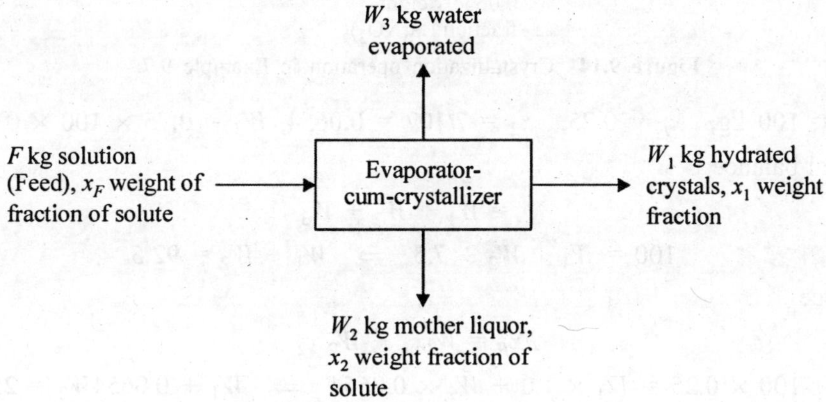

Figure 9.13 Crystallization operation.

The material balance for the crystallization operation yield the following equations:
Total balance:

$$F = W_1 + W_2 + W_3$$

Solute balance:

$$Fx_F = W_1x_1 + W_2x_2$$

Solvent balance:

$$F(1 - x_F) = W_1(1 - x_1) + W_2(1 - x_2) + W_3$$

EXAMPLE 9.7 A crystallizer is charged with 100 kg of a solution containing 25% $Ba(NO_3)_2$ in water. On cooling 10% of the original water present evaporates. Calculate the yield of crystals when the solution is cooled to 283 K. The solubility at 283 K is 7.0 kg $Ba(NO_3)_2$/100 kg total water.

Solution *Basis:* 100 kg of feed solution

Let W_1 be the weight of crystals, F the weight of feed, and x_F weight fraction of solute in the feed. Let x_2 be the solubility of the solute in weight fraction of solute, and W_2 and W_3 be the weights of mother liquor remaining after crystallisation and the weight of water evaporated during crystallization operation (see Figure 9.14).

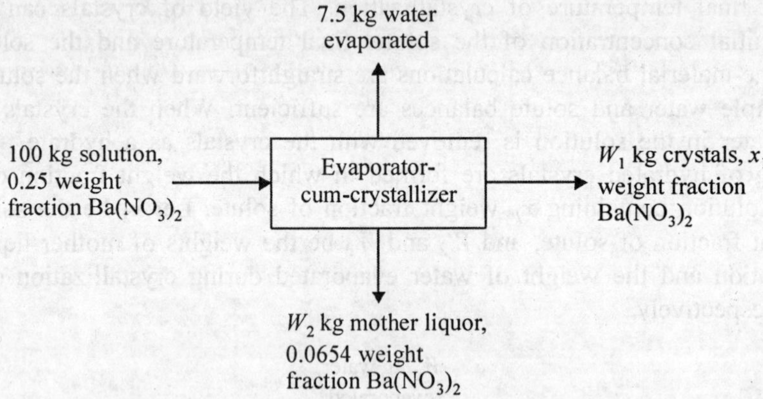

Figure 9.14 Crystallization operation in Example 9.7.

$$F = 100 \text{ kg}, \quad x_F = 0.25, \quad x_2 = 7/107 = 0.0654, \; W_3 = 0.75 \times 100 \times 0.1 = 7.5 \text{ kg}$$

Total material balance

$$F = W_1 + W_2 + W_3$$

That is,

$$100 = W_1 + W_2 + 7.5 \quad \Rightarrow \quad W_1 + W_2 = 92.5 \tag{A}$$

Solute balance

$$Fx_F = W_1x_1 + W_2\,x_2$$

$$100 \times 0.25 = W_1 \times 1.0 + W_2 \times 0.0654 \quad \Rightarrow \quad W_1 + 0.0654W_2 = 25 \tag{B}$$

Solving Eqs. (A) and (B) simultaneously, we get $W_1 = 20.28$ kg.

EXAMPLE 9.8 An aqueous solution of Na_2CO_3 contains 15% carbonate by weight. 80% of the carbonate is recovered as $Na_2CO_3{\cdot}10H_2O$ by evaporation of water and subsequent cooling to 278 K. The solubility of Na_2CO_3 at 278 K is 9.0% (weight). On the basis of 100 kg of solution treated, determine the following:

(a) The quantity of crystals formed
(b) The amount of water evaporated

Solution *Basis:* 100 kg solution treated

See Figure 9.13. Let W_1 be the weight of hydrated crystals, x_1 the weight fraction of Na_2CO_3 in the crystals, F the weight of feed and x_F the weight fraction of Na_2CO_3 in the feed. Let x_2 be the solubility of Na_2CO_3 in weight fraction Na_2CO_3, and W_2 and W_3 be the weights of mother liquor remaining after crystallization and the weight of water evaporated during crystallization operation.

Total material balance:

$$F = W_1 + W_2 + W_3 \tag{A}$$

Na_2CO_3 balance:

$$Fx_F = W_1x_1 + W_2x_2 \tag{B}$$

Here F = 100 kg, x_F = 0.15, the molecular weight of Na_2CO_3 = 106 and the molecular weight of $Na_2CO_3 \cdot 10H_2O$ = 286.

(a) The weight fraction of Na_2CO_3 in the crystals is

$$x_1 = \frac{106}{286} = 0.3706$$

The mass of Na_2CO_3 recovered in the crystals is

$$W_1x_1 = 0.8Fx_F = 0.8 \times 100 \times 0.15 = 12 \text{ kg}$$

Therefore, the quantity of crystals formed is

$$W_1 = \frac{12}{0.3706} = 32.28 \text{ kg}$$

(b) Substituting the values in Eq. (B), we get,

$$100 \times 0.15 = 12 + W_2x_2$$

The mass of Na_2CO_3 in the mother liquor is

$$W_2x_2 = 15 - 12 = 3 \text{ kg}$$

The weight fraction of Na_2CO_3 in the mother liquor = x_2 = 0.09 (from solubility)

The weight of mother liquor is

$$W_2 = \frac{W_2x_2}{x_2} = \frac{3}{x_2} = \frac{3}{0.09} = \frac{100}{3} \text{ kg}$$

From Eq. (A), we get

$$W_3 = F - (W_1 + W_2)$$
$$= 100 - (32.38 + 33.33) = 34.29 \text{ kg}$$

Hence, the amount of water evaporated = 34.29 kg

EXAMPLE 9.9 An aqueous solution containing 60% $Na_2S_2O_3$ and 1% soluble impurities is diluted with water and fed to a crystallizer where it is cooled in order to crystallize $Na_2S_2O_3 \cdot 5H_2O$. The crystals carry 0.05 kg of solution (excluding impurities) per kg of crystals. The free water present in the adhering solution is removed on drying the crystals. The final dried product contains not more than 0.1% impurity. The solubility of the pentahydrate is 1.5 kg $Na_2S_2O_3{\cdot}5H_2O$/kg free water. On the basis of 100 kg of 60% solution, calculate the following:

(a) The amount of water in kilograms added before cooling
(b) The amount of crystals formed
(c) The percentage recovery of the $Na_2S_2O_3$ in the dried hydrated crystals

Solution *Basis:* 100 kg 60% solution

Refer to Figure 9.15. Let W kg pure water is added to the original solution. W_1, W_2 and W_3 are the weights of $Na_2S_2O_3{\cdot}5H_2O$ crystallized, the mother liquor obtained and the solution carried away by the crystals, respectively. The impurities present in the feed get dissolved in the free water and is present in the products in association with W_2 kg mother liquor and W_3 kg adhering solution. When the crystals are dried, the impurities in the adhering solution gets deposited on the crystals, the concentration of which in the final dried crystals being limited to 0.1%.

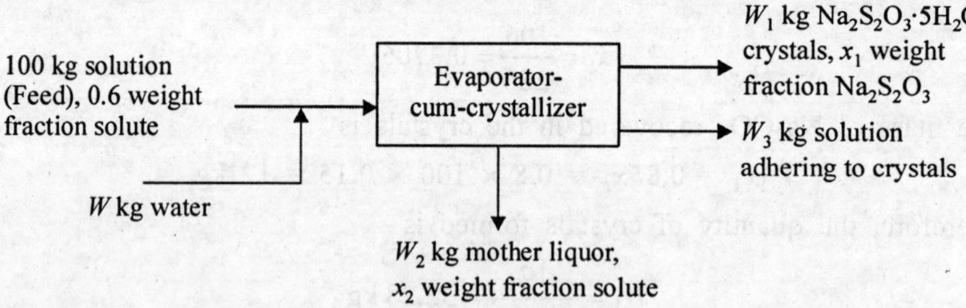

Figure 9.15 Crystallization operation in Example 9.9.

The molecular weight of $Na_2S_2O_3$ = 158 and the molecular weight of $Na_2S_2O_3{\cdot}5H_2O$ = 248. Since no water is lost by evaporation, all the free water present in the combined feed will be present in the mother liquor. 60 kg of anhydrous salt forms $60 \times \dfrac{248}{158} = 94.117$ kg of pentahydrate, the free water present in the system is $100 + W - 1 - 94.177 = W + 4.823$ kg. Therefore, the concentration of impurity in the mother liquor is $x_1 = \dfrac{1}{W + 4.823}$ kg of impurity/kg of free water. Now consider the material balance over the crystallizer.

Total (impurities-free) balance:

$$100 - 1 + W = W_1 + W_2 + W_3$$

That is,

$$W_1 + W_2 + W_3 - W = 99 \tag{A}$$

$Na_2S_2O_3$ balance:

$$60 = \left(W_1 + W_2 \times \frac{1.5}{2.5} + W_3 \times \frac{1.5}{2.5} \right) \times \frac{158}{248}$$

which simplifies as

$$W_1 + 0.6W_2 + 0.6W_3 = 94.177 \tag{B}$$

Since each kilogram of crystals carry 0.05 kg solution, $W_3 = 0.05W_1$. Substituting this in Eqs. (A) and (B), we get

$$1.05W_1 + W_2 - W = 99 \qquad \text{(C)}$$

$$1.03W_1 + 0.6W_2 = 94.177 \qquad \text{(D)}$$

Impurity in the adhering solution is

$$W_3 \times \frac{1}{2.5} \times \frac{1}{W + 4.823} \text{ kg}$$

Substituting $W_3 = 0.05W_1$, the amount of impurity in the adhering solution is

$$\frac{0.05W_1}{2.5(W + 4.823)} \text{ kg}$$

However, this should not be more than 0.1% of the final weight of crystals. The final weight of impurity-free crystals is

$$W_1 + W_3 \frac{1.5}{2.5} = W_1 + 0.05W_1 \frac{1.5}{2.5} = 1.03W_1$$

Therefore,

$$\frac{\text{weight of impurities in the crystals}}{\text{weight of pure crystals}} = 0.001$$

$$= \frac{0.05W_1/2.5(W + 4.823)}{1.03W_1} = \frac{0.0194}{W + 4.823}$$

Solving the above equation, we get $W = 14.577$ kg. Substituting this in Eq. (C), we get

$$1.05W_1 + W_2 = 113.577 \qquad \text{(E)}$$

Solving Eqs. (D) and (E), we get $W_1 = 65.08$ kg, $W_2 = 45.25$ kg and, $W_3 = 0.05W_1 = 45.25$ kg

(a) The amount of water added before cooling = 14.58 kg.

(b) The amount of $Na_2S_2O_3 \cdot 5H_2O$ crystals formed = 65.08 kg

(c) $Na_2S_2O_3$ in the adhering solution gets deposited on the crystals on drying. Therefore, total anhydrous $Na_2S_2O_3$ produced

$$W_1 \times \frac{158}{248} + W_3 \times \frac{1.5}{2.5} \times \frac{158}{248} = 42.701 \text{ g}$$

$$\text{The percentage recovery of } Na_2S_2O_3 = \frac{42.70}{60} \times 100 = 71.2\%$$

9.2.3 Leaching

Leaching is the separation of the components of a solid mixture by selectively dissolving the soluble components of the solid in a liquid solvent. Recovery of minerals from naturally occurring ores, oils from oilseeds, tannin from wood barks, vitamins and pharmaceutical products from plant leaves and roots, sugar from sugar beets, etc. are some of the industrial applications of leaching. A Bollman extractor (Figure 9.16) is a typical industrial unit used for leaching oil from seeds.

Solids are moved in perforated baskets attached to a chain conveyor. As the solids move downwards, solvent (half miscella) moves in parallel flow. The perforations on the basket permit passage of liquid from basket to basket. The strong solution of oil (full miscella) collects at the bottom and is removed. The leaching is completed when the solids move upward countercurrent to the fresh solvent which is admitted at the top as shown in the figure.

Consider a leaching operation such as the recovery of oil from oilseeds by treating with an organic solvent like hexane. Let C be the component that is being transferred from the solid phase to the liquid phase. The inert insoluble components in the feed is denoted as B. A is the solvent used for leaching. In the present example, A is hexane, B is insoluble materials in the feed, and C is oil. The streams entering and leaving a leaching unit are shown in Figure 9.17. After the solute has been transferred into the liquid, the mixture is allowed to settle. On settling, a clear solution known as *overflow* (R) containing the solute C dissolved in the solvent and an *underflow* which is a slurry of leached solids B suspended in the solution (E) that is not removed as overflow are obtained.

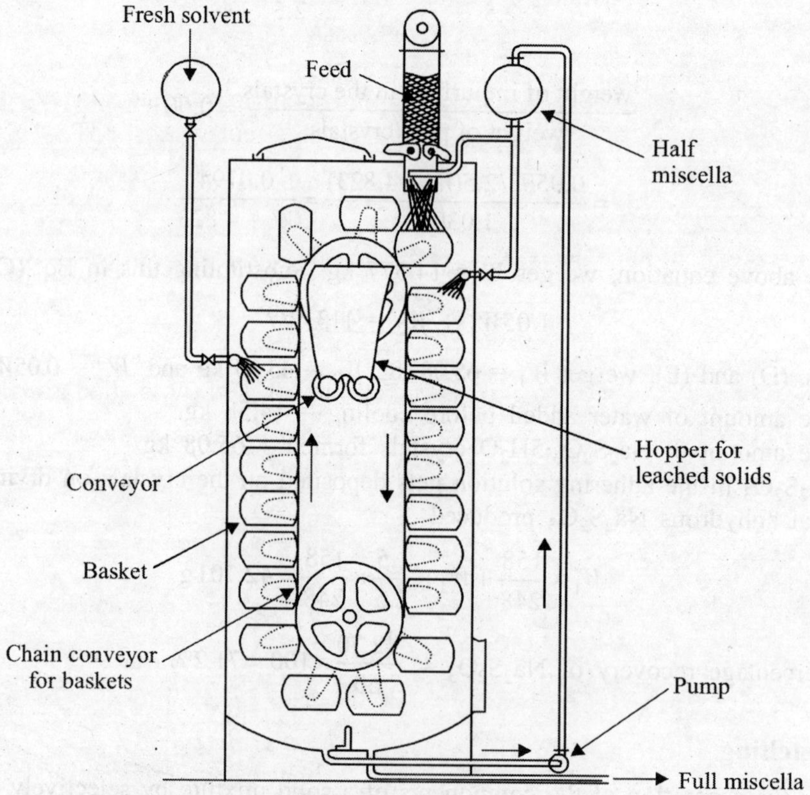

Figure 9.16 Bollman extractor.

Feed is a mixture of B kg inert solids and F kg solute. S is the mass of solvent added. The overflow is R kg solution consisting of A and C. Underflow is B kg inert solids suspended in E kg clear solution. Concentration of inert material in any stream can be represented as weight ratio of B on B-free basis and is denoted by N. Concentration of solute is represented as the weight ratio of solute on B-free basis. x represents solute concentration in overflow solution and y represents solute concentration in underflow solution. Thus, x and y are calculated as the ratio of C to $(A + C)$ in a mixture and N is calculated as the ratio of B to $(A + C)$ in the mixture. For the leaching system shown in Figure 9.17, we can write the inert balance as

$$B = N_F F = NE$$

where N_F and N are the inert concentration in the feed and underflow on an inert-free basis. The solution $(A + C)$ balance can be written as

$$F + S = E + R$$

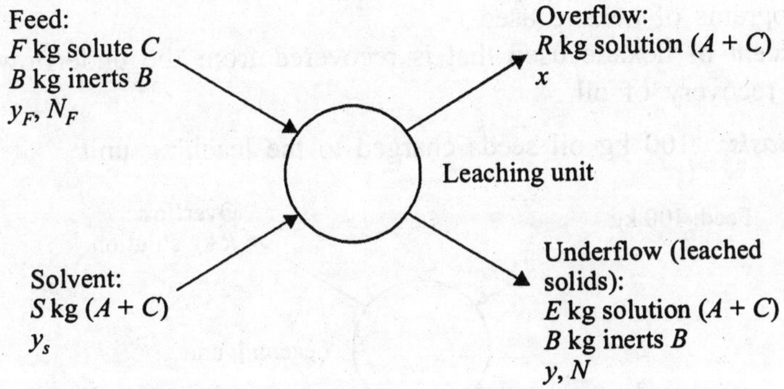

Figure 9.17 Streams in a leaching operation.

The component-C (solute) balance is

$$Fy_F + Sy_S = Ey + Rx$$

EXAMPLE 9.10 A tannery extracts certain wood barks which contains 40% tannin, 5% moisture, 23% soluble non-tannin materials and the rest insoluble lignin. The residue removed from the extraction tanks contain 50% water, 3% tannin and 1% soluble non-tannin materials. What percent of the original tannin remains unextracted?

Solution *Basis:* 100 kg of bark

The feed to the leaching unit contains 100 – 40 – 5 – 23 = 32 kg of lignin. Since lignin is insoluble in the solvent used for leaching, all this will be present in the residue. Therefore, lignin can be treated as the key component for material balance. Let the residue left after the leaching operation be W kg. The weight percent of lignin in the residue = 100 – 50 – 3 – 1 = 46%. Take a balance on the key component lignin:

$$32 = W \times 0.46$$

which gives

$$W = \frac{32}{0.46} = 69.57 \text{ kg}$$

Since the residue contains 3% unextracted tannin, the percentage of the tannin in the bark that is unextracted is

$$\frac{69.57 \times 0.03}{100 \times 0.40} \times 100 = 5.2\%.$$

EXAMPLE 9.11 Oilseeds containing 49% oils, 40% pulp, 3% mineral salts and the rest moisture are leached with hexane as the solvent. The underflow from the leaching operation contains 25% hexane, 2.5% salts, 15% oil and 7.5% moisture. The overflow contains 25% oil which is distilled to recover the entire hexane in pure form leaving behind the oil, water and salt. The underflow is subjected to steam distillation which recovers 95% hexane. For treating 100 kg seeds, calculate the following:

(a) The kilograms of hexane used
(b) The percent of hexane used that is recovered from the underflow
(c) Percent recovery of oil

Solution *Basis:* 100 kg oil seeds charged to the leaching unit

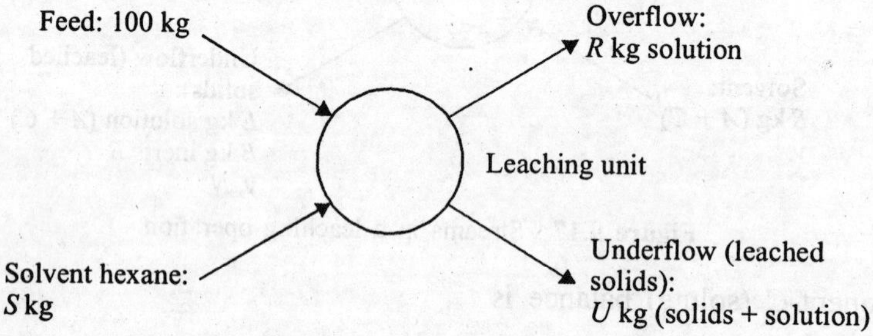

Figure 9.18 Streams in a leaching operation.

See Figure 9.18. Let S be the kilograms of hexane used for leaching. The underflow from the operation contains all the insoluble pulp present in the feed in addition to the unextracted oil, salt and moisture. Hexane dissolves some oil, salt and moisture. A portion of this solution is adhering with the insoluble solids in the underflow and let U be the combined weight of the underflow solids and solution adhering to it. The other portion forms the clear solution (R).

(a) Total balance:

$$100 + S = U + R \tag{A}$$

The feed contains 49 kg of oil, oil concentration in the underflow is 15% and that in the overflow is 25%. Hence, the oil balance:

$$49 = U \times 0.15 + R \times 0.25 \tag{B}$$

The feed contains 40 kg of pulp. The percentage of pulp in the underflow is 100 – 25 – 15 – 2.5 – 7.5 = 50%. Therefore, the inert balance:

$$40 = 0.5U \tag{C}$$

The insoluble pulp is the tie element in this problem as it is entering in a single stream and leaves as a single stream. The inert balance [Eq. (C)] gives

$$U = 80 \text{ kg}$$

Substituting $U = 80$ kg in Eq. (B), we get

$$R = 148 \text{ kg}$$

Substituting U and R in Eq. (A), we get $S = 128$ kg. Thus the amount of solvent used for extraction = 128 kg.

(b) The hexane in the underflow = $0.25U$ = 20 kg. The amount of hexane recovered by steam distillation is 0.95 × 20 = 19 kg. This constitutes only

$$\frac{19}{128} \times 100 = 14.84\%$$

(c) Oil recovered in the clear solution is

$$\frac{37}{49} \times 100 = 75.51\%$$

9.2.4 Adsorption

Components of a gas or liquid mixture can be adsorbed on the surface of a solid adsorbent. Two types of adsorption phenomena exist: *physical adsorption* occurring due to the intermolecular forces and *chemisorption* occurring due to the chemical interaction between the solid and the fluid. Physical adsorption is utilized as a separation method. The solid used for adsorption is known as *adsorbent*, and the fluid adsorbed on the surface is the *adsorbate*. The adsorption of organic vapours on activated charcoal, decolourization of cane-sugar solution by adsorbing the colouring matter on activated carbon, drying of gases by adsorbing water on silica gel, etc. are examples to cite a few. The adsorbed constituent can be removed from the solid and thereby separation can be completed and the adsorbent regenerated for further use. Fuller's earth, activated clays, activated charcoal, bone char, bauxite, alumina, silica gel, etc. are some important adsorbents. Adsorption of a fluid on the surface of a solid occurs till equilibrium is attained. The adsorption equilibrium is a relationship between the concentration of solute in the fluid and the quantity adsorbed per unit weight of adsorbent. This relationship for a constant temperature is known as *adsorption isotherm.*

9.2.5 Drying

Drying is usually one of the last operations in a process industry. Drying operation is unavoidable and an integral part of the process in industries such as paper industries, whereas it is done in other processes to reduce the cost of transportation of the product, to give some useful properties to the product like the free flowing nature of salt, and to prepare the product in a form that is suitable for handling and use. In drying, a wet solid or slurry is contacted with dry gas (usually air or flue gas) so that water is vaporized from the solid and is carried away by the gas. When the vapour pressure exerted by the moisture in the solid equals the partial pressure of water in the gas exposed to the solid, the solid is said to have attained its *equilibrium moisture content*, and drying stops. Depending upon the characteristics of the solid being dried, several types of driers are in common use. Tray driers (for pasty materials and lumpy solids), rotary driers (for granular and free flowing solids), freeze driers (for foodstuffs and pharmaceuticals) and spray driers (for slurries and pastes) are typical driers in use. The continuous driers are smaller in size in relation to the quantity being dried, and the dried product has a uniform moisture content. A rotary continuous drier is shown in Figure 9.19 and a general flow diagram is shown in Figure 9.20.

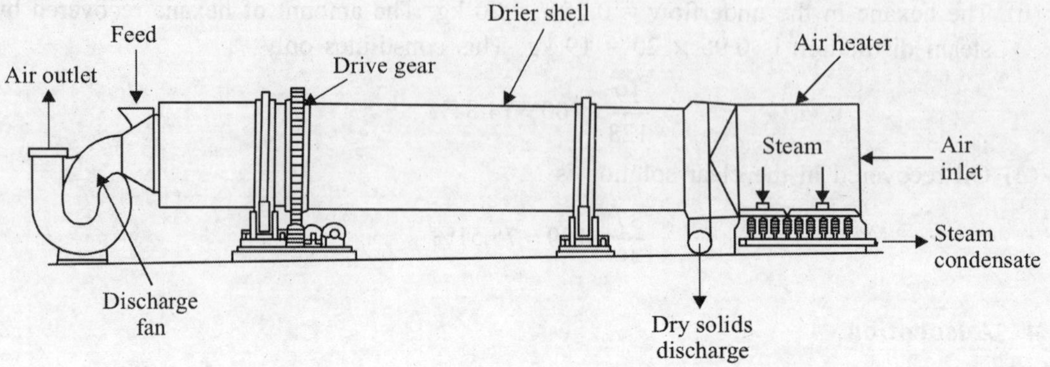

Figure 9.19 Counter-current direct heated rotary drier.

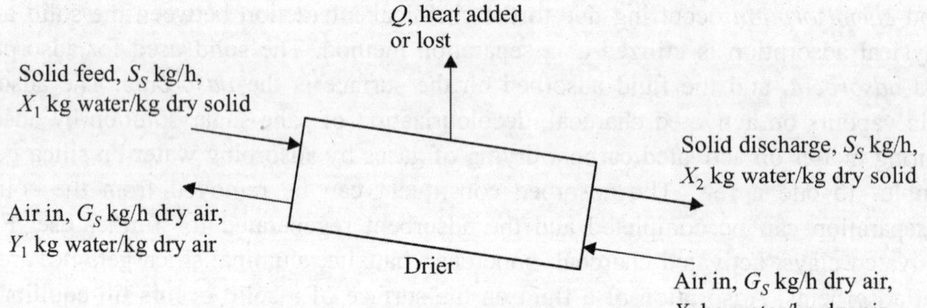

Figure 9.20 Continuous counter-current drying.

The drier is a long cylinder installed at a slight inclination with the floor to facilitate the free flow of the material to be dried that is introduced at one end. The cylinder is made to rotate at a very low speed. The drying gas, which may be hot air or dry flue gas is admitted at the other end, flows counter current to the solid, the gas being pulled through the drier by means of exhaust fans.

Let S_S kg/h be the mass flow rate of dry solid in the wet material admitted to the drier and let it be dried from an initial moisture content of X_1 to a final moisture content of X_2, the moisture content being expressed as kilograms of moisture per kilogram of dry solid. The gas flow rate is G_S kilograms of dry gas per hour and it receives the moisture evaporated from the solid. The humidity of the gas stream increases from Y_2 to Y_1 kg moisture per kilogram of dry gas. A moisture balance on the drier yields

$$S_S(X_1 - X_2) = G_S(Y_1 - Y_2)$$

EXAMPLE 9.12 A drier is fed with wet solid to reduce the moisture content from 80% to 15%. The product leaving the drier is admitted into an oven which further brings down the moisture to 2%. If the drier can handle 1000 kg of wet solid per day, calculate

(a) The weight of products leaving the drier and the oven per day
(b) The percentage of the original water that is removed in the drier and the oven

Solution *Basis:* 1000 kg of wet solid admitted to the drier.

Let V_1 kg and V_2 kg be the amount of water removed in the drier and the oven respectively. Let x_1 and x_2 be the weight fraction of moisture in the solid leaving the drier and the oven respectively. Let F be the amount of wet material fed to the drier and x_F be the moisture content in the feed. The total material balance for the drier is

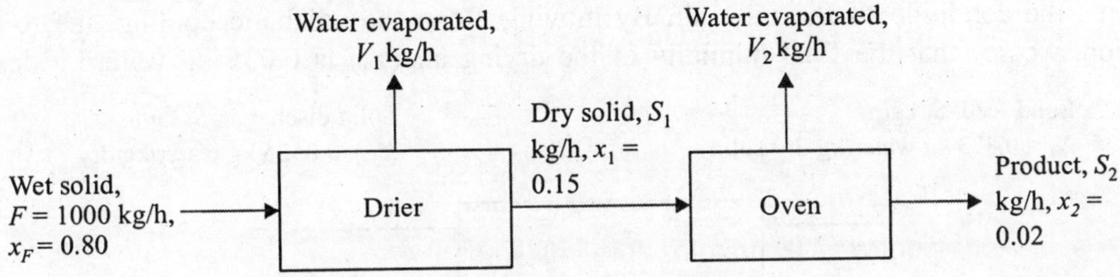

Figure 9.21 Drying operation in Example 9.12.

$$F = S_1 + V_1 \tag{A}$$

The corresponding moisture-free solid balance is

$$F(1 - x_F) = S_1(1 - x_1) \tag{B}$$

Similarly for the oven we can write

$$S_1 = S_2 + V_2 \tag{C}$$

$$S_1(1 - x_1) = S_2(1 - x_2) \tag{D}$$

Here, F = 1000 kg, x_F = 0.80, x_1 = 0.15 and x_2 = 0.02. Equations (B) and (D) are balances for the tie element—moisture free solid—and substituting the given values, they give S_1 = 235.29 kg and S_2 = 204.08 kg.

Substituting the value of S_1 in Eq. (A), we get V_1 = 764.71 kg.

Substituting the values of S_1 and S_2 in Eq. (C), we get V_2 = 31.21 kg.

(a) The weight of products leaving the drier = S_1 = 235.29 kg.
The weight of products leaving the oven = S_2 = 204.08 kg.

(b) The percentage of original water that is removed in the drier is

$$\frac{V_1}{Fx_F} \times 100 = \frac{764.71}{800} \times 100 = 95.6\%$$

The percentage of original water that is removed in the oven is

$$\frac{V_2}{Fx_F} \times 100 = \frac{31.21}{800} \times 100 = 3.90\%$$

EXAMPLE 9.13 Air at 101.3 kPa enters an adiabatic drier at 87.5°C with a dew point of 20°C and leaves at 70% humidity. Wet paper enters the drier with 25% moisture and leaves with 5% moisture. Determine the following:

(a) Water evaporated in kilograms per 100 cubic metres of air entering
(b) Finished product in kilograms per 100 cubic metres of air entering.

Solution The operation is schematically represented as shown in Figure 9.22.

The drying air enters with a dry-bulb temperature of 87.5°C and a dew point of 20°C. From the humidity charts for the air–water system at 101.3 kPa, we can see that these conditions correspond to a humidity of 0.015 kg water/kg dry air (Y_2). In an adiabatic drier, air absorbs moisture from the wet solid, and the humidity and temperature of the air change along the adiabatic cooling line passing through the point on the psychrometric chart that represents the condition of the inlet air. By moving along the adiabatic cooling line to 70% saturation, we see that the final humidity of the drying air (Y_1) is 0.035 kg water/kg dry air.

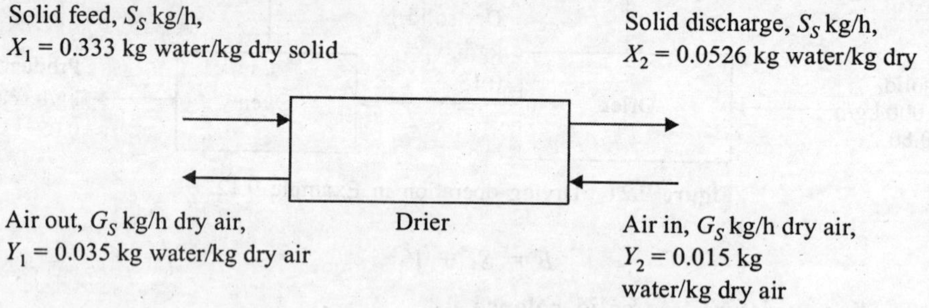

Figure 9.22 Drying in Example 9.13.

Since the amount of dry solid present in the product stream remains the same as that in the feed stream, and the amount of dry air present in the gas stream also remains constant, it is better to do the material balance on a dry basis. Let S_S kg/h be the mass flow rate of dry solid in the wet material admitted to the drier and it is dried from an initial moisture content of X_1 to a final moisture content of X_2 on a dry basis.

$$X_1 = \frac{25}{75} = 0.333 \text{ kg water/kg dry solid}$$

$$X_2 = \frac{5}{95} = 0.0526 \text{ kg water/kg dry solid}$$

Basis: 1 kg of dry air passed through the drier

The moisture balance for the drier is

$$S_S X_1 + Y_2 = S_S X_2 + Y_1$$

$$S_S (X_1 - X_2) = Y_1 - Y_2 \Rightarrow S_S = \frac{Y_1 - Y_2}{X_1 - X_2}$$

$$= \frac{0.035 - 0.015}{0.333 - 0.0526} = 0.0713 \text{ kg of dry solid}$$

That is, with 1 kg of dry air and accompanying water, the moisture associated with 0.0713 kg of dry solid can be reduced by the desired extent.

(a) 100 m^3 air enters the drier at 101.3 kPa and 87.5°C (360.65 K).
Moles of wet air entering the drier is

$$\frac{100}{22.414} \times \frac{101.3}{101.325} \times \frac{273.15}{360.65} = 3.3782 \text{ kmol}$$

Moisture content in the air admitted = 0.015 kg water/kg dry air, i.e.

$$\frac{0.015}{1} \times \frac{29}{18} = 0.0242 \text{ kmol water/kmol dry air}$$

Moles of dry air present in 100 m^3 of air admitted is

$$3.3782 \times \frac{1}{1 + 0.0242} = 3.2984 \text{ kmol}$$

Mass of dry air is 3.2984 × 29 = 95.65 kg

Since the moisture content of the air increases from 0.015 kg water/kg dry air to 0.035 kg water per kg dry air, the total moisture evaporated is

$$95.65(0.035 - 0.015) = 1.913 \text{ kg}$$

(b) Since 1 kg of dry air is needed for treating 0.0713 kg of dry solids, 95.65 kg of dry air can treat 0.0713 × 95.65 = 6.82 kg of dry solid. Since the product contains 0.0526 kg of moisture per kg of dry solid, the quantity of finished product is

$$6.82 \times \frac{1.0526}{1} = 7.18 \text{ kg}$$

9.2.6 Liquid–Liquid Extraction

The process of separation of the components of a liquid mixture by treating with an immiscible liquid solvent in which the constituents are differently soluble is known as *liquid–liquid extraction.* The liquid–liquid extraction operation involves bringing the feed liquid in intimate contact with an immiscible or partially miscible solvent, allowing the solute to get distributed between the phases and finally allowing the phases to separate from each other. The resulting solvent-rich phase is called the *extract* and the residual solution from which the solute is extracted is known as the *raffinate*. The mutually insoluble extract and raffinate phases are then separated from one another by settling and gravity separation. For example, aqueous acetic acid solution is mixed with isopropyl ether solvent in order to extract the acid into the ether phase. A single-stage extraction operation may be represented as shown in Figure 9.23.

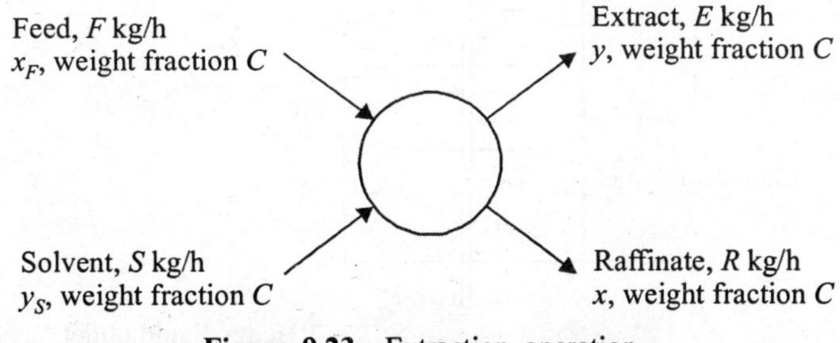

Figure 9.23 Extraction operation.

Consider a simple case in liquid–liquid extraction in which solute C to be extracted is present in the feed in association with solvent A. A solvent B is used for extraction of C from the feed. B is either insoluble in A or is partially miscible with A. When the solvent is contacted with the feed, the solute gets transferred to the solvent. The products leaving the extraction unit are an A-rich phase known as raffinate and a B-rich phase known as extract. The solute C will be present in the raffinate and extract phases. If A and B are partially miscible, traces of B will be present in the raffinate and traces of A will be present in the extract. In a most general extraction operation, therefore, we have to deal with ternary mixtures as raffinate and extract phases. Denoting the weight fraction of solute in the raffinate and extract phases by x and y respectively, that in the feed and solvent by x_F and y_S respectively, and the mass flow rates of feed, solvent, raffinate and extract phases by F, S, R and E respectively, the total material balance and component-C balance can be written as

$$F + S = E + R$$

$$Fx_F + Sy_S = Ey + Rx$$

Gravity settlers, mixer-settler cascades, sieve-tray columns, packed columns, spray towers, rotating-disk contactors, etc. are the common types of extraction equipment. Figure 9.24 shows a rotating-disk extractor schematically.

For good mass transfer rates, one liquid should be thoroughly dispersed in the other and the liquid dispersion is facilitated in rotating-disk contactors by mechanical means. The mechanical agitation is provided by high-speed rotating disks.

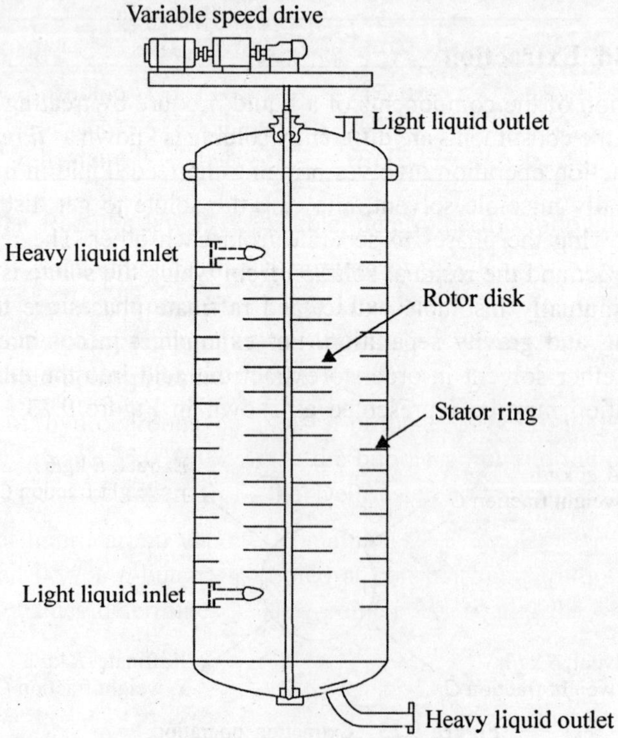

Figure 9.24 Rotating-disk contactor.

EXAMPLE 9.14 Isopropyl alcohol and water can be separated by extraction with ethylene tetrachloride (C_2Cl_4). 100 kg of a solution containing 30% (weight) isopropyl alcohol and the rest water is mixed with the solvent ethylene tetrachloride. After extraction, the raffinate phase analyzed 71% water, 28.1% isopropyl alcohol and 0.9% ethylene tetrachloride. The extract phase analyzed 94% ethylene tetrachloride, 5.2% isopropyl alcohol and the rest water. Calculate the following:

(a) The amount of solvent
(b) The quantities of raffinate and extract phases
(c) The percent extraction of isopropyl alcohol

Solution *Basis:* 100 kg of feed
The extraction operation is schematically represented in Figure 9.25.

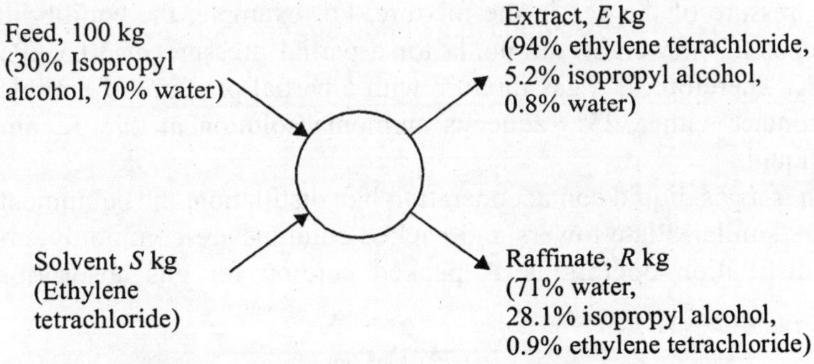

Figure 9.25 Extraction operation in Example 9.14.

Let R kg be the amount of raffinate (water-rich phase) and E kg be the amount of extract (solvent-rich phase). Let x_A, x_B and x_C be the weight fractions of water, ethylene tetrachloride and isopropyl alcohol, respectively, in the raffinate. Let y_A, y_B and y_C the corresponding values in the extract. Let S be the amount of solvent used.

Total balance:

$$R + E = 100 + S \tag{A}$$

Isopropyl alcohol balance:

$$0.281R + 0.052E = 30 \tag{B}$$

Ethylene tetrachloride balance:

$$0.009R + 0.94E = S \tag{C}$$

Simultaneous solution of the above equations gives $S = 45.10$, $E = 47.04$ and $R = 98.06$.

(a) The amount of solvent used = S = 45.10 kg.
(b) The amount of extract = 47.04 kg; the amount of raffinate = 98.06 kg.
(c) The amount of isopropyl alcohol extracted is

$$Ey_c = 47.04 \times 0.052 = 2.45 \text{ kg}$$

The percent extraction of isopropyl alcohol is

$$\frac{2.45}{30} \times 100 = 8.17\%$$

9.2.7 Absorption

In absorption, soluble constituents of a gas mixture are separated by absorbing in a suitable liquid solvent. The reverse process—the removal of certain constituents of a liquid mixture by contacting with a gas phase—is known as *desorption* or *stripping*. Ammonia is absorbed from a mixture of ammonia and air by contacting the gas with water in equipment known as *absorption columns*. Benzene vapours present in coke-oven gases can be absorbed in hydrocarbon oils and hydrogen sulphide can be absorbed from gas mixtures using ethanolamine solutions. The essential difference between absorption and distillation is that in absorption, the liquid phase used for effecting separation is a foreign substance introduced as a solvent whereas in distillation, the vapour and liquid phases are produced from the feed material by boiling.

The rate at which a gaseous constituent of a mixture is absorbed in a liquid depends upon the equilibrium solubility of the gas in the liquid at the given temperature, which in turn depends upon the partial pressure of the gas in the mixture. For example, the equilibrium solubility of ammonia in water is 25% (weight) ammonia for a partial pressure of 30.3 kPa at 293 K and 46.9 kPa at 303 K. Therefore, if a gas mixture with a partial pressure of ammonia greater than 30.3 kPa is in contact with a 25% aqueous ammonia solution at 293 K, ammonia will be absorbed in the liquid.

As absorption is a gas–liquid contact operation like distillation, the equipment for absorption and distillation are similar. Plate towers and packed columns are commonly employed for gas absorption and distillation operations. A packed column for gas absorption is shown in Figure 9.26.

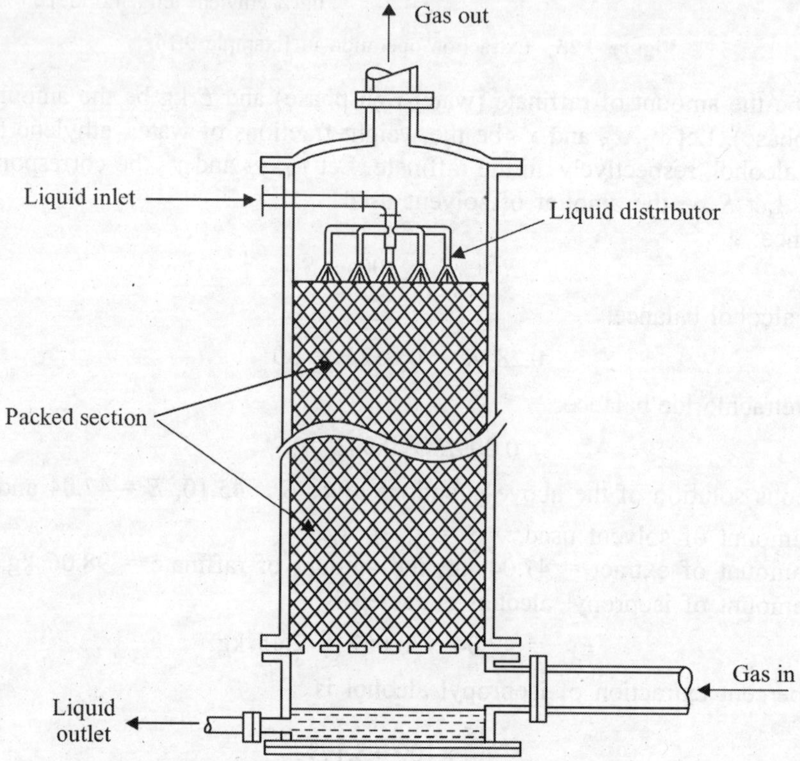

Figure 9.26 Packed bed absorption column.

Packed columns are vertical columns filled with some packing materials. Tower packings are devices of large surfaces, low weight per unit volume, and they provide a large free cross section for flow of fluids. Random packings such as Raschig rings, Lessing rings, Berl saddles, Intalox saddles, etc. and regular packings such as wood grids, stacked Raschig rings, etc. are used for this purpose. The liquid is distributed over and trickles down through the packing. Gas is admitted from the bottom and flows counter-current to the liquid. The gas is in contact with the liquid throughout the column unlike in a plate column where the gas is in contact with the liquid only intermittently. The mass transfer occurs and compositions of the gas and liquid change continuously in a packed column.

Consider absorption of solute A from a gas mixture into a liquid solvent. Let G and L represent the flow rate of the gas and liquid in kmol/h, and x and y the mole fraction of A in the liquid and gas, respectively. The flow rates as well as the concentrations vary throughout the column. The conditions at the bottom of the column are represented by suffix 1 and conditions at the top of the column are indicated by suffix 2. Then the total material balance for the column is

$$G_1 + L_2 = G_2 + L_1$$

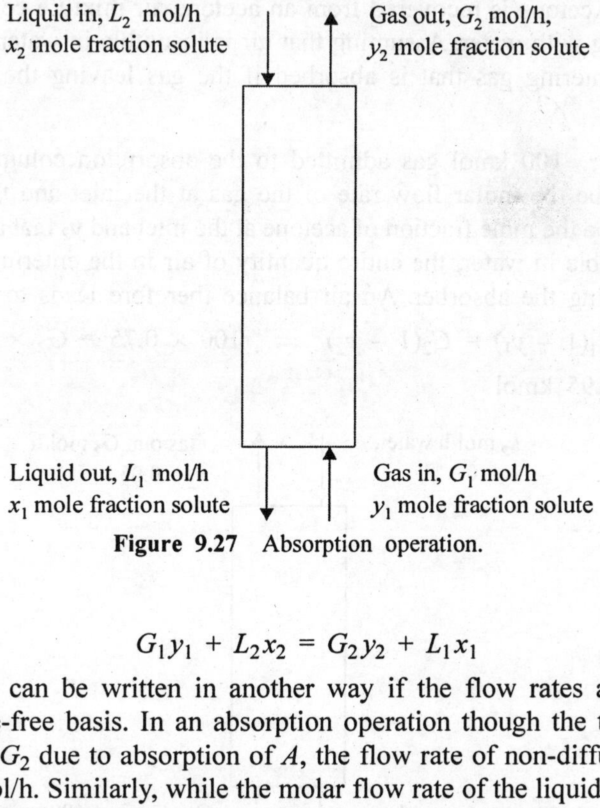

Figure 9.27 Absorption operation.

The solute balance is

$$G_1y_1 + L_2x_2 = G_2y_2 + L_1x_1 \tag{9.7}$$

The material balance can be written in another way if the flow rates and compositions are expressed on a solute-free basis. In an absorption operation though the total flow rate of gas decreases from G_1 to G_2 due to absorption of A, the flow rate of non-diffusing gas is unaltered and let this be G_S kmol/h. Similarly, while the molar flow rate of the liquid phase increases from L_2 to L_1 the flow rate of pure solvent is constant at, say, L_S kmol/h. Let X and Y represent the mole ratio of solute in the liquid and gas respectively. Then

$$X = \frac{\text{moles of } A \text{ in the liquid}}{\text{moles of solvent}} = \frac{x}{1-x}$$

$$Y = \frac{\text{moles of } A \text{ in the gas}}{\text{moles of solute free gas}} = \frac{y}{1-y}$$

The number of moles of solute present in a gas stream can be determined as the product of G_S and Y, and the number of moles of solute in a liquid stream can be obtained as the product of L_S and X. Thus,

$$G_1 y_1 = G_S Y_1 \quad \text{and} \quad G_2 y_2 = G_S Y_2$$

$$L_1 x_1 = L_S X_1 \quad \text{and} \quad L_2 x_2 = L_S X_2$$

where X_1 and X_2 are the mole ratio of solutes in the liquid at the bottom and top of the column, and Y_1 and Y_2 are the mole ratio of solute in the gas stream at the inlet and the exit of the gas. Substituting these values in Eq. (9.7) we get,

$$G_S (Y_1 - Y_2) = L_S (X_1 - X_2) \tag{9.8}$$

EXAMPLE 9.15 Acetone is recovered from an acetone-air mixture containing 25% (volume) acetone by scrubbing with water. Assuming that air is insoluble in water, determine the percent of acetone in the entering gas that is absorbed if the gas leaving the scrubber analyzes 5% acetone.

Solution *Basis:* 100 kmol gas admitted to the absorption column

Let G_1 and G_2 be the molar flow rate of the gas at the inlet and the exit of the absorber respectively, and y_1 be the mole fraction of acetone at the inlet and y_2 that at the exit (Figure 9.28). The air being insoluble in water, the entire quantity of air in the entering stream will appear in the gas stream leaving the absorber. An air balance therefore leads to

$$G_1(1 - y_1) = G_2(1 - y_2) \quad \Rightarrow \quad 100 \times 0.75 = G_2 \times 0.95$$

Therefore, $G_2 = 78.95$ kmol

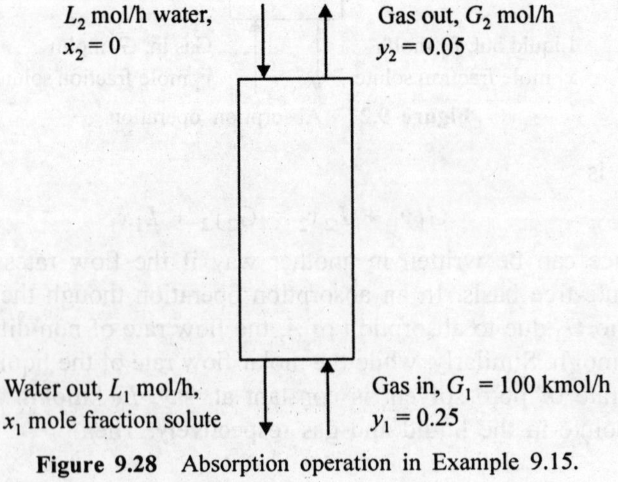

Figure 9.28 Absorption operation in Example 9.15.

The amount of acetone in the gas leaving is

$$G_2y_2 = 78.95 \times 0.05 = 3.9475 \text{ kmol}$$

The amount of acetone in the gas entering is

$$G_1y_1 = 100 \times 0.25 = 25 \text{ kmol}$$

Acetone absorbed is

$$\frac{25 - 3.9475}{25} \times 100 = 84.21\%$$

EXAMPLE 9.16 A gas mixture consisting of 65% N_2 and 35% SO_3 by volume is admitted to an absorption column at a rate of 4500 kg/h. It is contacted with a stream of 50% H_2SO_4 flowing counter-current to the gas stream at a rate of 5000 kg/h. The gases leave at 101.3 kPa. Water lost with the exit gases exerts a partial pressure of 25.0 kPa. If the concentrated acid leaving the bottom of the column contained 75.0% H_2SO_4, what percent of the entering SO_3 is absorbed and converted to acid?

Solution *Basis:* One hour operation

The absorption operation of this example is shown schematically in Figure 9.29.

The average molecular weight of gas entering the column = $0.65 \times 28 + 0.35 \times 80 = 46.2$.

The molar flow rate of gases entering the absorber is

$$\frac{4500}{46.2} = 97.40 \text{ kmol/h}$$

Therefore, the moles of nitrogen entering is

$$97.4 \times 0.65 = 63.31 \text{ kmol}$$

$$\text{Moles of } SO_3 \text{ entering} = 97.4 \times 0.35 = 34.09 \text{ kmol}$$

Let W kg be the mass of 75% H_2SO_4 produced and x be the moles of SO_3 that leaves with the exit gases unabsorbed.

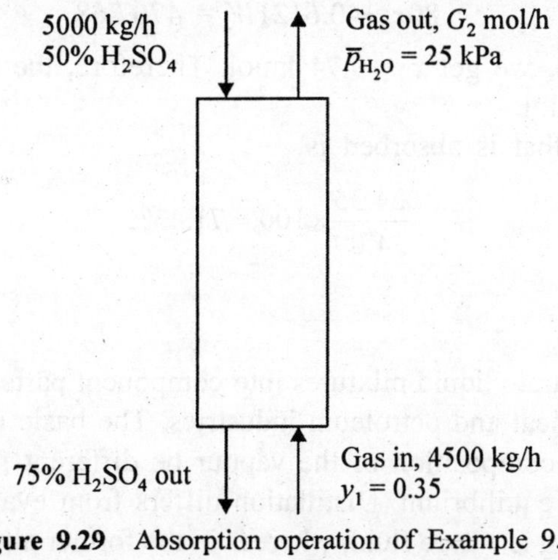

Figure 9.29 Absorption operation of Example 9.16.

Since no nitrogen is absorbed, the gases leaving will contain 63.31 kmol nitrogen, x kmol SO_3 and water vapour, which exerts a partial pressure of 25 kPa. Let y be the moles of water vapour present in the gases. Then

$$\frac{25}{101.3} = \frac{y}{63.31 + x + y}$$

which means that

$$y = 0.32765x + 20.744 \tag{A}$$

Total balance:

Kilograms of H_2SO_4 (50%) entering + kilograms of gas mixture entering
= kilograms of gas leaving + kilograms of con. H_2SO_4 (75%) leaving

That is,

$$5000 + 4500 = (63.31 \times 28 + x \times 80 + y \times 18.016) + W$$

$$\Rightarrow \quad W + 80x + 18.016y = 7727.32$$

Substituting Eq. (A) into the above the total material balance reduces to

$$84.9174x + W = 7352.68 \tag{B}$$

Since the molecular weight of H_2SO_4 is 98.016, 98.016 kg of sulphuric acid contains 80 kg of SO_3. Now consider an SO_3-balance for the absorber:

Kilograms of SO_3 entering with 50% H_2SO_4 + kilograms of SO_3 in the feed gas
= kilograms of SO_3 in the gas leaving + kilograms SO_3 in the 75% H_2SO_4 leaving

$$5000 \times 0.5 \times \frac{80}{98.016} + 34.09 \times 80 = 80x + 0.75W \times \frac{80}{98.016}$$

The above equation reduces to

$$80x + 0.6121W = 4767.68 \tag{C}$$

Solving Eqs. (B) and (C), we get $x = 9.74$ kmol. Therefore, the amount of SO_3 absorbed = 34.09 − 9.74 = 24.35 kmol.

The percent of SO_3 that is absorbed is

$$\frac{24.35}{34.09} \times 100 = 71.43\%$$

9.2.8 Distillation

Distillation is used to separate liquid mixtures into component parts by boiling and is one of the major operations in chemical and petroleum industries. The basic requirement for a separation by distillation is that the composition of the vapour be different from the composition of the liquid with which it is in equilibrium. Distillation differs from evaporation in that in the latter vapour produced on boiling is a pure fluid, whereas in the former all the components constituting the liquid will be present in the vapour but in different proportions. The vapour will be richer in the more volatile components compared to the liquid with which it is in equilibrium. The products obtained on distillation are commonly referred to as distillate (or top product), which is rich in more volatile components and residue (or bottom product), which is rich in less volatile components.

When a liquid mixture at a high temperature and pressure is subjected to a sudden reduction in pressure by passing through a valve, the liquid gets partially vaporized. The resulting liquid–vapour mixture is separated into the distillate and the bottom products as shown in Figure 9.30. The process is known as *flash distillation* or *flash vaporization*. The flash distillation is an ideal single-stage operation in which the distillate and bottom products have equilibrium compositions. The products will have an appreciable difference in their compositions only if the volatilities of the constituents are appreciably different. Therefore, it is generally impossible to obtain pure products by a single-stage flash vaporization.

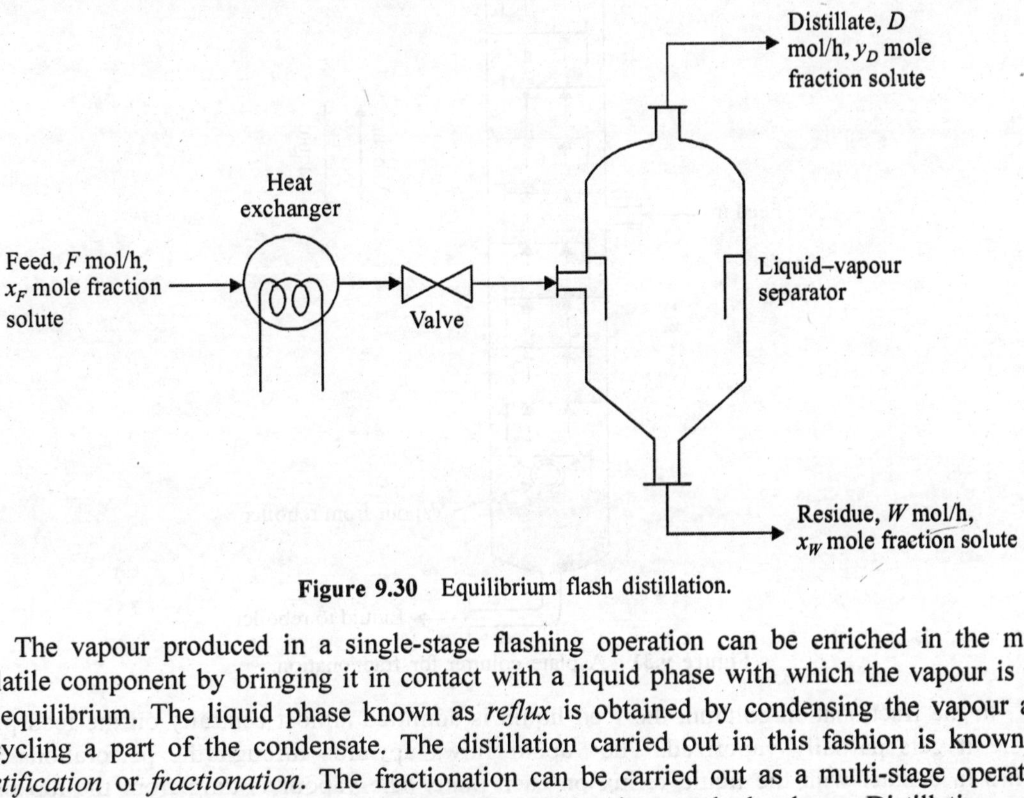

Figure 9.30 Equilibrium flash distillation.

The vapour produced in a single-stage flashing operation can be enriched in the more volatile component by bringing it in contact with a liquid phase with which the vapour is not in equilibrium. The liquid phase known as *reflux* is obtained by condensing the vapour and recycling a part of the condensate. The distillation carried out in this fashion is known as *rectification* or *fractionation*. The fractionation can be carried out as a multi-stage operation in a plate column or as a continuous contact operation in a packed column. Distillation using a plate column is schematically represented in Figure 9.31.

A fractionating column is a cylindrical vertical tower which is divided into a number of sections by means of plates or trays. Different designs for trays are available, the simplest one being a sieve tray. Sieve trays are flat plates with a large number of small perforations that permit upward flow of vapour. The liquid flows across the plate and run over an overflow weir through the downcomer in the plate to the plate below. The vapour issuing through the perforations provides turbulence in the pool of liquid present on the tray, thereby facilitating mass transfer on the plate. On each plate more volatile components are vaporized and transferred to the vapour and the less volatile components are condensed and transferred to the liquid. Thus, the vapour leaving the tray is richer in more volatile components than the vapour coming to that tray. The liquid leaving the tray is leaner in more volatile components compared to the liquid entering the tray, it is still capable of enriching the vapour coming to the lower plate.

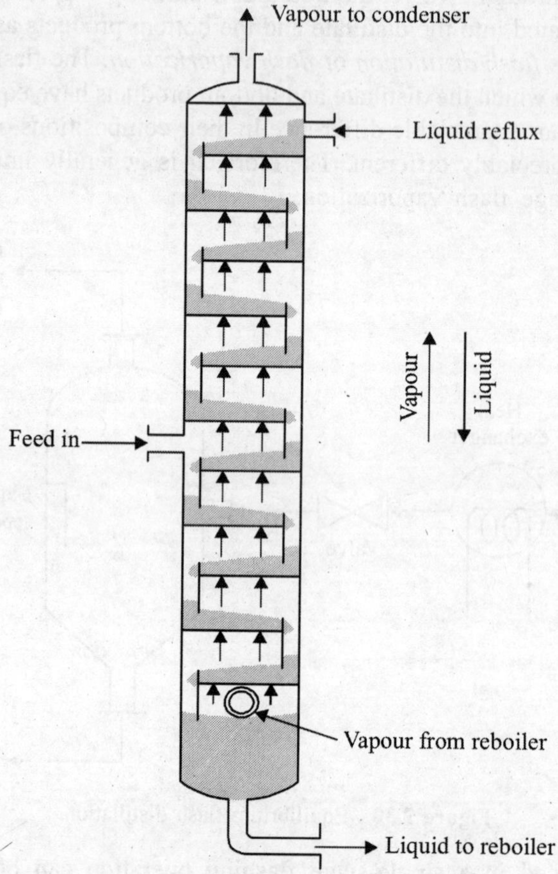

Figure 9.31 A plate column for fractionation.

In the fractionation column, the feed liquid is admitted almost centrally on the feed plate where it gets partially vaporized. The vapour flows upwards through the perforations and comes in contact with the liquid reflux on each plate. The vapour issuing from the top plate is taken to a condenser. The vapours are condensed by exchange of heat with a coolant, the most frequently used cooling medium being water. A part of the condensate is returned to the column as reflux and the other part is withdrawn as distillate or top product. The ratio of the moles of liquid recycled as reflux to the moles of distillate product is known as *reflux ratio*. The bottom of the column is equipped with a reboiler which receives the liquid flowing downwards from the feed plate and vaporizes it partially. The vapour so produced is sent back to the bottom plate. This vapour when comes in contact with the liquid flowing downwards strips away more volatile components from the liquid and as a result the bottom product can be made to be almost free of more volatile components. Steam is used as the heating medium in the reboiler. The liquid leaving the reboiler is collected as the residue or the bottom product.

Consider the distillation of a binary mixture of components A and B in which A is the more volatile component. Let F, D and W be the molar flow rates of the feed, distillate and residue respectively, and let x_F, x_D and x_W be their compositions expressed in terms of mole fractions of A. Considering the column as a whole, the total material balance and the component-A balance may be written as

$$F = D + W \tag{9.9}$$

$$Fx_F = Dx_D + Wx_W \tag{9.10}$$

Let R denotes the reflux ratio for the column, and let L be the reflux in kmol/h. Then, $L = RD$.

EXAMPLE 9.17 A continuous distillation column is used to regenerate solvent for use in a solvent extraction unit. The column treats 200 kmol/h of a feed containing 10% (mol) ethyl alcohol and the rest water. The overhead product is 89% (mol) alcohol and the bottom product is 0.3% (mol) alcohol. The overhead is sent to the extraction unit and the bottom is wasted. What is the daily requirement of make-up alcohol in the solvent extraction unit?

Solution *Basis:* 200 kmol/h feed

Let F, D and W be the flow rates of the feed, distillate and residue respectively, and x_F, x_D and x_W be the mole fraction of ethanol in the feed, distillate and residue respectively. F = 200 kmol/h, $x_F = 0.10$, $x_D = 0.89$ and $x_W = 0.003$.

The total material balance:

$$F = D + W$$

Alcohol balance:

$$Fx_F = Dx_D + Wx_W$$

Substituting the given values in these equations, we get

$$D + W = 200$$

$$0.89D + 0.003W = 200 \times 0.10 = 20$$

Solving the above equations, we get D = 21.87 kmol and W = 178.13 kmol.

$$\text{Alcohol wasted} = \text{alcohol present in the residue}$$
$$= Wx_W = 178.13 \times 0.003 = 0.5344 \text{ kmol/h}$$

Therefore, the make-up alcohol required per day = 0.5344 × 24 × 46 = 589.98 kg

EXAMPLE 9.18 An aqueous solution of methanol containing 20% (weight) methanol is to be separated into a distillate product containing 97% (weight) methanol and a bottom product containing 2% (weight) methanol. For treating 100 kg of feed with a reflux ratio of 3.5 on a weight basis, calculate the following:

(a) The amounts of distillate and bottom products
(b) The amount of vapour condensed in the condenser per kg of distillate
(c) The amount of vapour condensed in the condenser per kg of feed.

Note: The reflux ratio is defined as the ratio of the amount of condensate recycled to the amount of condensate withdrawn as distillate product.

Solution *Basis:* 100 kg feed

The distillation operation of this example is represented schematically in Figure 9.32.

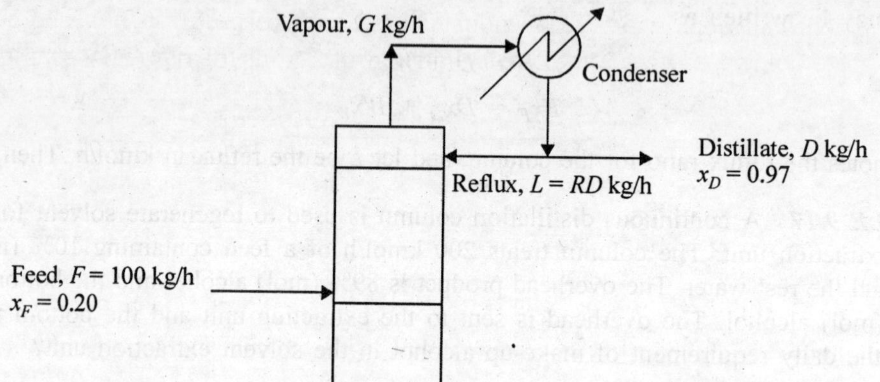

Figure 9.32 Distillation operation of Example 9.18.

(a) The total material balance:

$$F = D + W$$

Alcohol balance:

$$Fx_F = Dx_D + Wx_W$$

Substitute $x_F = 0.20$, $x_D = 0.97$ and $x_w = 0.02$ in Eq. (9.10) and eliminate W using Eq. (9.9), we get

$$0.20F = 0.97D + 0.02(F - D)$$

Put $F = 100$ and solve the above equation for D. We get $D = 18.95$ kg/h. Putting values of F and D in Eq. (9.9), we get $W = 81.05$ kg/h.

(b) Take a total material balance around the condenser.

$$G = L + D$$

Since the reflux ratio $R = 3.5$, the reflux $L = RD = 3.5D$, and the above equation gives, $G = 4.5D$. For 1 kg distillate product, the amount of vapour condensed, $G = 4.5$ kg.

(c) For one kg of feed, the distillate obtained is $D = 0.1895$ kg [from part (a)]. Therefore, the amount of vapour condensed per kg of feed is

$$4.5 \times 0.1895 = 0.853 \text{ kg}$$

EXERCISES

Basic material balance principles

9.1 Wet sewage sludge enters a continuous thickener at a rate of 100 kg per hour and dehydrated sludge leaves the thickener at a rate of 75 kg per hour. Determine the amount of water removed in the thickener in one hour, assuming steady-state operation.

9.2 Air is dehumidified at a constant pressure of 101.3 kPa. The partial pressure of water in the air admitted to the dehumidifier is 7 kPa and the temperature is 350 K. The partial pressure of water in the air leaving is 1.5 kPa. How much water (in kilograms) is removed from 100 cubic meters of air admitted?

9.3 Urea, phosphoric acid and potassium chloride are mixed together to obtain a mixed fertilizer having NPK content 10 : 26 : 26 as %N, $\%P_2O_5$ and $\%K_2O$ by weight, balance being the weight of filler materials. Calculate the quantities to be mixed to get 1000 kg of mixed fertilizer.

9.4 Formulate the independent material balance equations for determining the flow rate of benzene in the feed under the following circumstances. An aqueous acetic acid solution containing 80% acetic acid and the rest water is charged into a still along with pure benzene. 400 kg/h pure acetic acid is withdrawn as product from the still. The top product leaving the still consists of 11.0% acetic acid, 21.5% water and the rest benzene.

9.5 A soap plant produces raw soap containing 50% moisture. This is to be dried to 20% moisture before it is pressed into cakes for sale. How many 100-g soap pieces can be obtained from 1000 kg of original raw soap?

9.6 A weak acid containing 12.5% H_2SO_4 and the rest water is fortified by adding 500 kg of concentrated acid containing 80% H_2SO_4. Determine the amount of the solution obtained if it contains 18.5% H_2SO_4.

9.7 A cellulose solution contains 5.0% cellulose by weight in water. It is to be diluted to 4.0% using a 1.0% solution of cellulose in water. Determine the kilograms of 1.0% solution required to dilute 100 kg of the 5.0% solution.

9.8 One hundred kilograms of a dilute waste acid containing 30.0% sulphuric acid is to be fortified to 50.0% sulphuric acid using concentrated sulphuric acid of strength 96.0%. How many kilograms of concentrated sulphuric acid are required for this process?

9.9 Waste acid from a nitrating process contains 25% HNO_3, 55% H_2SO_4 and 20% H_2O by weight. This is to be concentrated to get fortified acid containing 27% HNO_3, 60% H_2SO_4 and 13% water. This is done by adding concentrated H_2SO_4 of strength 93% H_2SO_4 and concentrated HNO_3 of strength 90% HNO_3 in suitable quantities to the waste acid. If 1000 kg fortified acid is to be produced, calculate the kg of the various solutions mixed.

9.10 The liquid effluent from a processing plant having a BOD of 0.15 g/L is discharged at a rate of 4000 cubic metres per day into a stream flowing at a rate of 0.3 cubic metres per second and having an initial BOD of 5×10^{-3} g/L. Determine the BOD in the stream immediately below the discharge point?

9.11 It is decided to measure the flow rate of a pure air stream by injecting pure CO_2 at a rate of 10 mol/h into the flowing stream. The resultant mixture analyzed 8.6% CO_2 on a mole basis. What is the flow rate of air?

9.12 Pure oxygen at 120 kPa and 300K is injected into an ammonia pipeline at a rate of 0.02 m^3/s. The concentration of oxygen in the pipeline at a point far removed from the point of injection is found to be 10% (volume). What is the flow rate of ammonia through the pipe in kg/h?

9.13 Two tanks which are connected to each other is initially sealed off from one another by means of a valve. Tank I initially contains 1 m^3 of air at 600 kPa and 343.2 K. Tank II contains a mixture of oxygen and nitrogen containing 95% (mole) nitrogen at 1200 kPa and 363.2 K. The valve is now opened and the contents of the tanks are allowed to mix. After complete mixing, the tanks contained 85% (mole) nitrogen. Calculate the volume of tank II.

9.14 It is desired to produce a gas mixture analyzing 40% methane, 35% ethane and 25% propane by blending the following three gas mixtures in suitable proportions.

Constituent	Mixture I	Mixture II	Mixture III
CH_4	25	35	55
C_2H_6	35	20	40
C_3H_8	40	45	5

Determine the proportion in which the gases are to be mixed.

9.15 A process stream of given flow rate is obtained by mixing together two separate streams. In order to determine the ratio in which the streams are mixed, a soluble salt is added to one of the streams at a steady rate. The concentration of salt in the stream is determined to be 5% (W) and that in the combined stream is found to be 0.45 % (W). Determine the ratio in which the streams are mixed.

9.16 A laundry can purchase soap containing 30% by weight water at a rate of ₹14 per kg f.o.b. factory. The same manufacturer offers a soap containing 5% water by weight. If the freight rate is ₹1.40 per 10 kg, what is the maximum price that the laundry can pay the manufacturer for the soap containing 5% water?

9.17 Rubber latex containing 15% (weight) rubber solids is to be coagulated by treating with 66°Bé H_2SO_4. If the ratio of the weights of acid to latex solids is 1:50, determine the following:

(a) Cubic metres/hour of 66°Bé acid required for treating 1000 kg/h of latex
(b) How many kilograms of solution to be removed per hour for recovering the solids

9.18 A company has drawn a contract for the purchase of paper at a price of ₹10 per kg of paper containing 5% moisture by weight and to adjust the price if the moisture content varies, so that the price of bone dry paper is constant. Also if the moisture content exceeds 5%, the additional freight charge incurred due to the excess moisture will be deducted from the total cost. The freight charge is ₹2 per kg of paper. Calculate the following:

(a) The total amount to be paid for 10000 kg of paper containing 3% moisture
(b) The total amount to be paid for 8000 kg of paper containing 9% moisture.

9.19 A water sample taken from a stream contained 200 ppm Na_2SO_4. To measure the flow rate of the running stream, 5 kg Na_2SO_4 is added to the stream uniformly over a one-hour period. The analysis of water taken from the stream from a spot down stream from the point of addition of the salt showed 350 ppm Na_2SO_4. What is the rate of flow of water in the stream?

9.20 Ion-exchange process shown in Figure 9.33 is used for water purification.

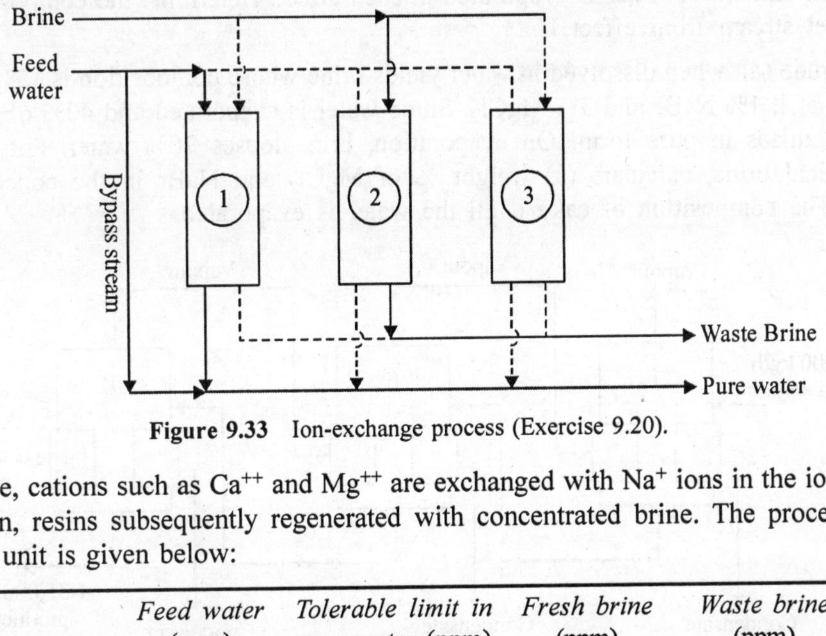

Figure 9.33 Ion-exchange process (Exercise 9.20).

Here, cations such as Ca^{++} and Mg^{++} are exchanged with Na^+ ions in the ion-exchange resin, resins subsequently regenerated with concentrated brine. The process data for the unit is given below:

	Feed water (ppm)	*Tolerable limit in pure water* (ppm)	*Fresh brine* (ppm)	*Waste brine* (ppm)
Ca^{++}	400	10	500	41500
Mg^{++}	250	15	300	25000
Na^+	60	800	$\sim 2 \times 10^5$	$\sim 2 \times 10^5$

At any given time when one bed purifies hard water, one is engaged in the regeneration of the used-up bed and the third one is kept idle as a standby unit. Determine the average brine consumption per 1000 kg of process water being treated.

Evaporation

9.21 In a process for producing caustic (NaOH), 4000 kg/h of a solution containing 10 wt% NaOH is evaporated in the first evaporator, giving a 20% NaOH solution. This is then fed into a second evaporator, which gives a product of 50% NaOH. Calculate the following:

(a) The amount of water removed from each evaporator
(b) The feed to the second evaporator, kg/h
(c) The amount of product, kg/h

9.22 An aqueous solution containing 15% NaOH and 0.5% NaCl is concentrated at a rate of 100 kg/min in an evaporator. The concentrated solution is then mixed with 2000 kg of aqueous NaOH solution in a mixer. At the end of one hour a sample is collected from the mixer and analyzed. The analysis shows 40% NaOH and 0.8571% NaCl. Calculate the following:

(a) The concentration of the aqueous solution in the mixer
(b) The composition of the concentrate from the evaporator
(c) The mass of water (in kilograms) evaporated in one hour

9.23 A 50% NaCl solution is to be concentrated in a triple-effect evaporator (Figure 9.45). Equal amount of water is evaporated in each effect. Determine the composition of the outlet stream from effect II.

9.24 A crude salt when dissolved in water yields brine whose composition is 15% by weight of NaCl, 1% NaBr and 3% $MgCl_2$. Some water is evaporated and 40% of salt (NaCl) crystallises in pure form. On evaporation, brine looses 70% water. For 100 kg of original brine, calculate (a) weight % of $MgCl_2$ and NaBr in the concentrate and (b) The composition of cake if all the water is evaporated.

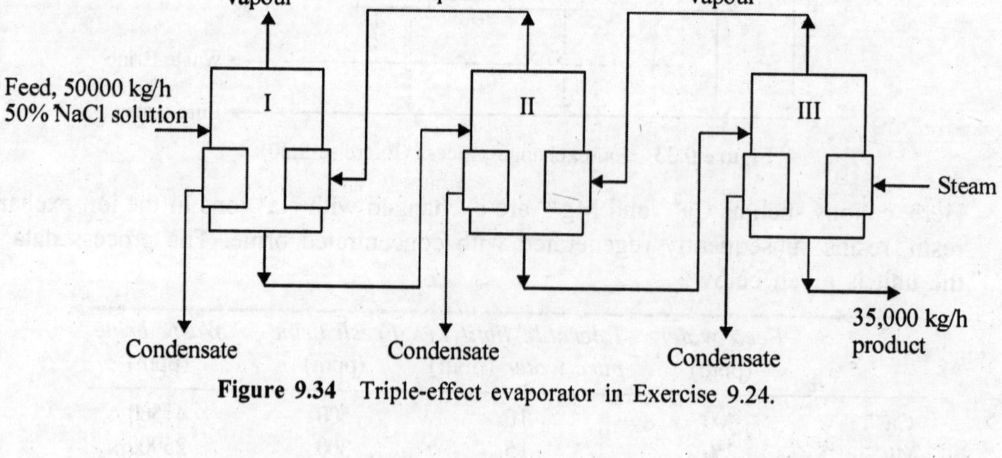

Figure 9.34 Triple-effect evaporator in Exercise 9.24.

9.25 An evaporator is fed continuously with 50000 kg/h of a solution containing 10% NaOH, 10% NaCl, and the rest water by weight. During evaporation, water is removed as vapour and salt NaCl precipitates as crystals and is removed by filtration. The concentrated liquor leaving the evaporator contains 50% NaOH, 2% NaCl, and the rest water. Determine

(a) the mass of water evaporated per hour;
(b) the mass of salt precipitated per hour;
(c) the mass of concentrated liquor produced per hour

Crystallization

9.26 0.05 kg of a slightly soluble salt is mixed with 0.1 kg of water. The undissolved salt is removed by filtration. The filter cake weighed 0.045 kg as obtained and 0.040 kg after drying. What is the solubility of the salt in water expressed in kg salt/100 kg water? What assumptions are inherent in the solution?

9.27 Fifty kilograms of dry sodium bicarbonate is to be crystallized and removed from 1000 kg of a saturated solution at 333 K. To what temperature the solution be cooled, if the solubility data is as follows?

Temperature (K)	333	323	313	303	293	283
Solubility $\left(\dfrac{\text{kg bicarbonate}}{\text{100 kg water}}\right)$	16.4	14.5	12.7	11.1	9.6	8.2

9.28 200 kg of a 15% and 100 kg of 5% solutions of sodium sulphate by weight are mixed in a crystallizer and crystallization takes place. If 50 kg $Na_2SO_4 \cdot 10H_2O$ crystals are formed, compute the composition of the magma.

9.29 One hundred kilograms of a mixture of $Na_2CO_3 \cdot 10H_2O$ and $Na_2SO_4 \cdot 10H_2O$ is heated to drive away the water of hydration. The anhydrous salt mixture weighed 39.335 kg. What is the mole ratio in which the two salts are present in the mixture?

9.30 A 10-kg mixture of $Ca(NO_3)_2 \cdot 4H_2O$ and $CuSO_4 \cdot 5H_2O$ is heated to drive off the water of hydration. The residue weighed 6.5591 kg. Determine the percent composition of the mixture of hydrated salts.

9.31 A saturated solution of calcium chloride in water is to be prepared by dissolving $CaCl_2 \cdot 6H_2O$ in 100 kg water at 293 K. If the solubility of calcium chloride in water at 293 K is 75 kg of anhydrous salt per 100 kg of water, what mass of hydrated crystals is required?

9.32 If 0.1 kg of Na_2SO_4 is dissolved in 0.2 kg of water and the resulting solution is cooled until 0.1 kg of $Na_2SO_4 \cdot 10H_2O$ crystallizes out. Determine

(a) The composition of the mother liquor
(b) The amount of anhydrous crystals obtained per 0.1 kg original solution

9.33 What will be the yield of hypo ($Na_2S_2O_3 \cdot 5H_2O$) if 100 kg of a 50% solution of $Na_2S_2O_3$ is cooled to 293 K. The solubility at 293 K is 70 parts $Na_2S_2O_3$ per 100 parts water.

9.34 After a crystallization process, a solution of $CaCl_2$ in water contains 62 kg of salt per 100 kg of water. Calculate the weight of solution necessary to dissolve 250 kg of $CaCl_2{\cdot}6H_2O$ at 298 K. The solubility at 298 K is 7.38 kmol $CaCl_2$ in 1000 kg of water.

9.35 A saturated solution of barium nitrate is to be prepared from 100 kg of the salt at 373 K.

(a) Determine the amount of water required.

(b) If this solution is cooled to 273 K, how much $Ba(NO_3)_2$ crystals will be obtained if the precipitated crystals carry 4 kg of water per 100 kg of dry crystals?

The solubility data of $Ba(NO_3)_2$ in water: at 273 K, 5.0 kg/100 kg water; at 373 K, 34 kg/100 kg water.

9.36 A saturated solution of sodium chloride is prepared at 373 K using 100 kg of salt.

(a) How much water is required?

(b) If the solution is cooled to 273 K, how much salt is precipitated out of the solution? Solubility of sodium chloride is 39.8 kg/100 kg water at 373 K and 35.7 kg/100 kg water at 273 K.

9.37 A batch of 1000 kg of KCl is dissolved in sufficient water to make a saturated solution at 363 K (solubility is 35 wt% KCl in water). The solution is cooled to 293 K, at which its solubility is 25.4 wt%.

(a) What is the weight of water required for solution and the weight of crystals of KCl obtained?

(b) What is the weight of crystals obtained if 5% of the original water evaporates on cooling?

9.38 A salt solution weighing 10000 kg with 30 wt% Na_2CO_3 is cooled to 293 K. The salt crystallizes as decahydrate.

(a) What will be the yield of $Na_2CO_3{\cdot}10H_2O$ crystals if the solubility is 21.5 kg of anhydrous Na_2CO_3/100 kg of total water if no water is evaporated?

(b) What will be the yield if 3% of total weight of the solution is lost by evaporation in cooling?

9.39 A hot solution of $Ba(NO_3)_2$ from an evaporator contains 30.6 kg of $Ba(NO_3)_2$ per 100 kg of water and goes to a crystallizer where the solution is cooled and $Ba(NO_3)_2$ crystallizes. On cooling 10% of the original water present evaporates. For a feed solution of 100 kg, calculate the following:

(a) The yield of crystals if the solution is cooled to 290 K, if the solubility is 8.6 kg $Ba(NO_3)_2$/100 kg total water at 290 K

(b) The yield if cooled to 283 K, if the solubility is 7.0 kg $Ba(NO_3)_2$/100 kg total water at 283 K

9.40 A solution of sodium sulphate in water is saturated at a temperature of 313 K. Calculate the weight of crystals and the percentage yield obtained when cooling 1000 kg of this solution to a temperature of 278 K. At 278 K decahydrate is the stable crystalline form. The solubility at 313 K is 32.6% and at 278 K is 5.75% Na_2SO_4.

9.41 A saturated solution containing 1500 kg of potassium chloride at 360 K is cooled in an open tank to 290 K. If the specific gravity of the solution is 1.2, the solubility of KCl per 100 parts of water is 53.55 at 360 K and 34.5 at 290 K, calculate

(a) The capacity of the tank required

(b) The weight of crystals obtained neglecting the loss of water by evaporation

9.42 Sodium carbonate is recovered as decahydrate from 1000 kg of a 5% solution of sodium carbonate in water. If it is desired that 95% of Na_2CO_3 is to be recovered as decahydrate by cooling to 278 K, determine the following:

(a) The mass of water evaporated

(b) The mass of crystals obtained

The solubility of Na_2CO_3 in water at 278 K is 9.0% (weight).

9.43 One thousand kilograms of a 30% solution of Na_2CO_3 in water is cooled slowly to 293 K. During cooling, a portion of water is evaporated and removed. 700 kg of $Na_2CO_3 \cdot 10H_2O$ crystals are formed in the process. What percent of total water in the feed is evaporated? The solubiltiy of anhydrous Na_2CO_3 at 293 K is 21.5 kg/100 kg of water.

9.44 An aqueous solution of Na_2CO_3 contains 12% carbonate and small amounts of soluble impurities. 80% of the carbonate is recovered as $Na_2CO_3 \cdot 10H_2O$ by evaporation of water and subsequent cooling to 278 K. The solubility of Na_2CO_3 at 278 K is 9.0% (weight). On the basis of 100 kg solution treated, determine the following:

(c) The amount of water evaporated

(d) The quantity of crystals formed

9.45 In the recovery of glycerine from spent soap lye, 10000 kg/h of lye containing 10.5% glycerine and 11.5% salt (NaCl) is concentrated by evaporation. The concentrated solution leaving the evaporator contained 80% glycerine and 5% salt. Entrainment loss is estimated to be 2% of the glycerine in the charge. Calculate the quantity of

(a) The water evaporated per hour

(b) The salt crystallised per hour

9.46 A solution containing 25% $MgSO_4$ and 75% water is cooled so that $MgSO_4 \cdot 7H_2O$ crystals are obtained on crystallization. During this process, 6.5% of the total water present in the feed is lost due to evaporation. The solubility of anhydrous $MgSO_4$ in water at this temperature is such that 35% of the anhydrous salt is present in the saturated solution. For treating 1000 kg of solution, calculate the following:

(a) The weight of $MgSO_4 \cdot 7H_2O$ crystals obtained

(b) The solubility of anhydrous salt in water (kg salt/100kg water)

9.47 The solubility of $NaNO_3$ in water is 1.76 kg/kg water at 373 K and 0.88 kg/kg water at 293 K. A 25% (weight) solution of $NaNO_3$ in water is concentrated to saturation by evaporation at 373 K. It is then cooled to 293 K. The crystals formed are separated. It is found that each kilogram of crystals so removed carries with it 0.1 kg of solution.

When the crystals are dried, $NaNO_3$ in the adhering solution gets deposited on the crystals. For 1000 kg of the original dilute solution, calculate the following:

(a) The amount of water evaporated for attaining saturation at 373 K
(b) The weight of dry crystals obtained

9.48 A dilute salt solution containing 6% salt is fed to an evaporator at a rate of 10000 kg/h for a trial run. Crystals formed along with the saturated solution adhering with them are sent to a centrifuge for the recovery of the solution. 60% of the adhering solution is recovered. It is found that the saturated solution that is withdrawn from the evaporator is 850 kg and that is recovered by centrifuging is 200 kg. The crystals are dried to drive off the remaining water. The dry crystals obtained weighed 360 kg. Determine the following:

(a) The solubility of the salt in kg salt/100 kg water
(b) The mass of salt crystallized in the evaporator
(c) The amount of water evaporated in the evaporator
(d) The amount of water removed in the drier

9.49 An aqueous solution containing 60% $Na_2S_2O_3$ and 1% soluble impurities is diluted with water and fed to a crystallizer where it is cooled to 283 K in order to crystallize $Na_2S_2O_3{\cdot}5H_2O$. The crystals carry 0.06 kg of solution (excluding impurities) per kg of crystals. The free water present in the adhering solution is removed on drying the crystals. The final dried product contains not more than 0.1% impurity. The solubility of the pentahydrate is 1.4 kg of $Na_2S_2O_3{\cdot}5H_2O$/kg free water. On the basis of 100 kg of 60% solution, calculate the following:

(a) The mass of water added before cooling
(b) The percentage recovery of the $Na_2S_2O_3$ in the dried hydrated crystals

Leaching

9.50 Oilseeds containing 40% oil and the rest inert insolubles are extracted with hexane to recover oil. Oil is dissolved in the solvent and is removed as a clear solution. The underflow sludge analyzed 10.53% oil and 26.32% hexane. Identify the key component. Determine the percent recovery of oil.

9.51 Tannin is extracted from certain wood bark which contains 6% moisture, 11.0% tannin, 8.0% soluble non-tannin materials and the rest insoluble lignin. After tannin is extracted, the solid residue analyzes 1% tannin and 0.25% soluble non-tannin on a dry basis. What is the percent extraction of tannin?

9.52 Black ash containing 45% Na_2CO_3, 5% water soluble materials and 50% inert insolubles is extracted with water to recover soda ash. The solid residue leaving the extraction unit contains 5% Na_2CO_3, 0.5% water soluble materials, 85% insolubles and the rest water. Calculate per 1000 kg of black ash treated

(b) The mass of residue produced
(a) The mass of sodium carbonate extracted

9.53 Oil is extracted from seeds by leaching with organic solvents. Soybean seeds containing 20% oil, 65% inert solids and 15% water are leached with hexane and after extraction the solid residue is removed from the solution of oil in hexane. The residue analyzed 1.0% oil, 88% inert cake and 11% water. What percent of oil in the seeds is recovered?

9.54 A counter-current extractor is employed to extract oil from a solid meal using ethyl ether as the solvent. The fresh meal is charged to the unit at a rate of 1000 kg/h and contains 25.0% (weight) oil. Pure solvent enters the bottom of the extractor. The overflow from the unit contains 60% (weight) oil. The underflow contains 0.25 kg of solution per kg of oil-free solids and the concentration of oil in the underflow is 12%. Calculate the following:

(a) The solvent requirement
(b) The percentage oil recovery

9.55 100 kg of an ore containing 70% solute and 30% inert solid is to be extracted by washing it twice with 50 kg of batches of fresh water by a simple multiple contact method. The underflow retains 0.6 kg of solution per kg of inert solid. Calculate the following:

(a) Concentration of solute in the final underflow
(b) Concentration of the solute in the combined extract
(c) Percentage recovery of the solute after two washings

Drying

9.56 Soap as produced contains 50% moisture on a wet basis. Before it can be pressed into cakes for sale, the moisture should be reduced to 20%. How many 100 g cakes can be pressed from 1000 kg of wet soap?

9.57 A wet paper pulp containing 70% water is dried in order to remove 60% of the water present. Determine the following:

(a) The mass of water removed per 100 kg of wet pulp
(b) The composition of the dried pulp

9.58 A wet granular material containing 80% water is dried in a rotary counter-current drier. The charge is admitted at one end and hot dry air is passed from the other end. In a typical operation, it is found that 100 kg of water is removed from the material giving a dry product containing 40% water. What is the weight of the wet material charged to the drier?

9.59 Wood containing 40% moisture is dried to 20% moisture, both moistures expressed on a wet basis. Determine the quantity of water evaporated per kilogram dry wood.

9.60 A batch of leather leaving a drier weighs 1000 kg and contains 5% moisture. During drying the leather loses 50% of its original weight. Determine the following:

(a) The moisture content of the leather entering the drier on a dry basis
(b) The amount of moisture removed per kg of bone dry leather
(c) Water removed as percent of the original water present

9.61 A drier is fed with wet solid to reduce the moisture content from 90% to 20%. The product leaving the drier is admitted to an oven which further brings down the moisture to 2%. If the drier can handle 1000 kg of wet solid per day, calculate the following:

(a) The weight of products leaving the drier and the oven per day
(b) The percentage of the original water that is removed in the drier and the oven

9.62 Air supplied to an adiabatic drier is at 101.3 kPa and 369.15 K with a dew point of 301.15 K. 2.0 kg of water is evaporated per 100 cubic metres of wet air entering the drier. Calculate the following:

(a) The exit air wet-bulb temperature and dry-bulb temperature
(b) The exit air percent humidity
(c) The volume of exit air per 100 cubic metres of inlet air

9.63 A laboratory drier produces 50 kg of dry solid containing 6% (weight) water from a wet-feed material containing 20.5% water. The drier operates adiabatically, with the wet solid entering the drier at the wet-bulb temperature of the drier air. A total of 500 m^3/h of hot air at 365.7 K and 101.3 kPa with a dew point of 303.2 K is supplied to the drier. Calculate the following:

(a) The temperature of the air leaving the drier
(b) The percent humidity of the air leaving the drier

9.64 Air at 101.3 kPa enters an adiabatic drier at 372 K with a dew point of 287 K and leaves at 80% humidity. Wet paper enters the drier with 25% moisture and leaves with 5% moisture. Determine the following:

(a) The temperature of air leaving the drier
(b) Water evaporated in kilograms per 100 cubic metres of air entering
(c) The mass of finished product (in kilograms) per 100 cubic metres of air entering.

9.65 Wet lumber (5% moisture) is dried to 1% moisture in a hot-air drier. Air fed to the drier contains 0.5% water. The moist air leaving the drier contains 2% (weight) water. How much air is required to dry 2000 kg/h of lumber?

9.66 1000 kg/h of precipitated chalk slurry containing 10% $CaCO_3$ is to be filtered and dried continuously. The filter cake is expected to carry 0.8 kg of water per kg of chalk. Hot air would enter the adiabatic drier at 350 K and 1 bar and 5% humidity. At the exit the air would be saturated. The dried chalk would contain only 5% water on a wet basis. Calculate on an hourly basis the following:

(a) The weight of filtrate
(b) The weight of the cake after dying
(c) The weight of air at the drier inlet

9.67 $CaCO_3$ slurry is to be dried. The drier is designed to remove 100 kg of moisture per hour. Air at 293 K and 40% relative humidity enters the drier, and leaves at 338 K and 65% relative humidity.

(a) What is the weight in kilograms of bone dry air required per hour? The atmospheric pressure is 103 kPa.

(b) If the humidity of the air entering the drier can be varied, what will be the minimum amount of dry air required? The constants for the Antoine equation for vapour pressure of water in kPa may be taken as A = 16.26205, B = 3799.887 and C = −46.854.

9.68 A solid material wet with toluene is dried in an adiabatic drier to recover toluene and to produce dry solids. The drier is fed with 100 m^3/h of an air-toluene mixture at 101.3 kPa and 333 K with a wet-bulb temperature of 305 K. The gases leave the drier at 310 K DBT with relative saturation of 95%. The vapour pressure of toluene is given by the Antoine equation

$$\ln P^S = 13.9987 - \frac{3096.52}{T - 53.67}$$

where pressure is in kPa and temperature is in K. The latent heat of vaporization is 404 kJ/kg and the psychrometric ratio for the toluene-air system is 1.88 kJ/kg K. Determine the following:

(a) The toluene evaporated in the drier, (kg/h)
(b) The volume of gases leaving the drier, (m^3/h)

Extraction

9.69 Two liquids A and B are only partially miscible. At a certain temperature, 41 kg of A and 59 kg of B are mixed well and the mixture is allowed to settle. The mixture separates into two immiscible phases, one rich in A and the other rich in B. The A-rich phase analyses 90% A and the B-rich phase analyzes 80% B. What are the weights of the A-rich and B-rich phases?

9.70 A mixture of phenol and water under certain conditions of temperature and composition forms two separate layers, one rich in phenol and the other rich in water. At 300 K, the composition of the upper and lower layers are 70% and 9% by weight of phenol respectively. If 40 g of phenol and 60 g of water are mixed and the layers are allowed to separate at 300 K, what will be the weights of the two layers?

9.71 Acetic acid is extracted from an aqueous solution containing 42.86% (weight) acetic acid using benzene as the solvent. The mutual solubility of benzene and water may be neglected. When equal weights of benzene and the aqueous solution are mixed and the phases separated, the aqueous phase analyzed 16.25% acetic acid. On the basis of 100 kg of the aqueous solution being treated, calculate the following:

(a) The masses of the aqueous phase and benzene phase
(b) The percent recovery of acetic acid into benzene

9.72 Toxic impurities A and B present in an oil is to be removed before it is subjected to hydrogenation. 1000 kg of oil is being charged. A portion of this oil containing 3% A and 2% B is sent through a purifier where it is treated with a mixture of two solvents S_1 and S_2. 100 kg of S_1 can dissolve a maximum of 15.0 kg of A and 2 kg of B whereas 100 kg of S_2 can dissolve a maximum of 3 kg of A and 12 kg of B. The solvent mixture containing dissolved impurities is then separated from the oil. The oil free of impurities is mixed with a portion of the original feed that has bypassed the purifier and sent to the hydrogenation unit. The feed to the hydrogenator should contain A and B not more than 0.1% each. Calculate the following:

(a) The amounts of solvents S_1 and S_2 required
(b) The amount of oil bypassed the purifier

9.73 A mixture containing 30% acetone and 70% chloroform is extracted with a mixed solvent containing acetic acid and water. The two immiscible phases—the raffinate and extract phases—that result after extraction had the following analysis:
Extract: acetone 7.5%, chloroform 6.06%, acetic acid 31.88% and water 54.56%
Raffinate: acetone 20.0%, chloroform 67.0% acetic acid 10.0% and water 3.0%
For the basis of 100 kg of the mixture extracted, determine the following:
(a) The composition of the mixed solvent on a weight basis
(b) The quantities of raffinate and extract phases
(c) The amount of mixed solvent used

Absorption

9.74 Ammonia is recovered from a gas mixture containing 25% (volume) CO_2 and 75% (volume) NH_3 by scrubbing with water. Assuming that CO_2 is insoluble in water, determine the percent of ammonia in the entering gas that is absorbed if the gas leaving the scrubber analyzes 35% NH_3.

9.75 Air at 540 K and 101.3 kPa is dried from a partial pressure of 7000 Pa of water vapour to a partial pressure of 1500 Pa of water vapour at constant total pressure. How much water in kilograms is removed per cubic metre of wet air entering?

9.76 A stream of gas at 302 K and 100 kPa, 50% saturated with water vapour is passed through a drying tower where 90% of the water vapour is removed. For 100 cubic metres of gas admitted through the tower, how many kilograms of water are removed? The vapour pressure of water at 302 K is 4.0 kPa.

9.77 One hundred kilograms per hour of 98% (weight) sulphuric acid is used for the absorption of SO_3 in the contact sulphuric acid plant. If 20% oleum is the product leaving the absorption tower, what mass of SO_3 is absorbed per hour?

9.78 A gas mixture at 300 K and 1 bar analyzing by volume 20% N_2 and 80% CH_4 is subjected to liquefaction at a rate of 1500 kg/h. It is found that only 30% (weight) of the entering gas is liquefied and the concentration of N_2 in the liquid is 60% by weight. The unliquefied gas leaves the unit at 273 K and 1 bar. Determine (a) the volume of unliquefied gas, m^3/h (b) the composition of the gas leaving expressed as volume %.

9.79 Hydrogen sulphide is absorbed from a gaseous mixture containing 26% H_2S and 74% inerts by a solution in a tower. The tower operates at 4 bar, and 330 K. The gases leave the tower with an H_2S content of 8%. Assuming that H_2S is alone removed and nothing is added as the gases pass through the tower and if the feed to the tower is 3000 m^3/h, calculate
(a) The amount of H_2S recovered from the gas
(b) The percentage recovery of H_2S

9.80 A gas mixture consisting of 60% N_2 and 40% SO_3 is admitted to an absorption column at a rate of 100 kmol/h. It is contacted with a stream of 50% H_2SO_4 flowing counter-current to the gas stream at a rate of 6000 kg/h. The gases leave at 101.3 kPa. The water lost with the exit gases exerts a partial pressure of 26.66 kPa. If the concentrated acid leaving the bottom of the column contains 74.73% H_2SO_4 what percent of the entering SO_3 will be absorbed and converted to acid?

9.81 In the preparation of cooking liquor for a sulphite pulp mill, an absorption column is used to absorb SO_2 in a weak liquor. The weak liquor enters the top of the column at a rate of 1.5 m^3/min with SO_2 concentration of 0.5% (weight) and leaves with SO_2 concentration of 1.0% (weight). The gas stream entering the bottom of the column and passing in the counter-current direction to the liquid stream contains 15% SO_2. When the gas leaves the top of the column, 80% SO_2 is absorbed. The pressure in the column is 1.5 bar and the temperature is 310 K. Assuming that the liquor has a specific gravity of 1.0, determine the following:

(a) The amount of SO_2 absorbed per minute
(b) The molar flow rate of gas entering the absorber
(c) The percentage of SO_2 in the gas leaving the column
(d) The volumetric flow rate of gas leaving the absorber

9.82 Figure 9.35 represents the flow sheet for the recovery of acetone from air. All compositions are on a weight basis. Make a material balance and determine the quantities of the following streams:

(a) Water added in the absorber
(b) Acetone-free air leaving the absorber
(c) Aqueous solution of acetone leaving the absorber
(d) Distillate product
(e) Bottom product

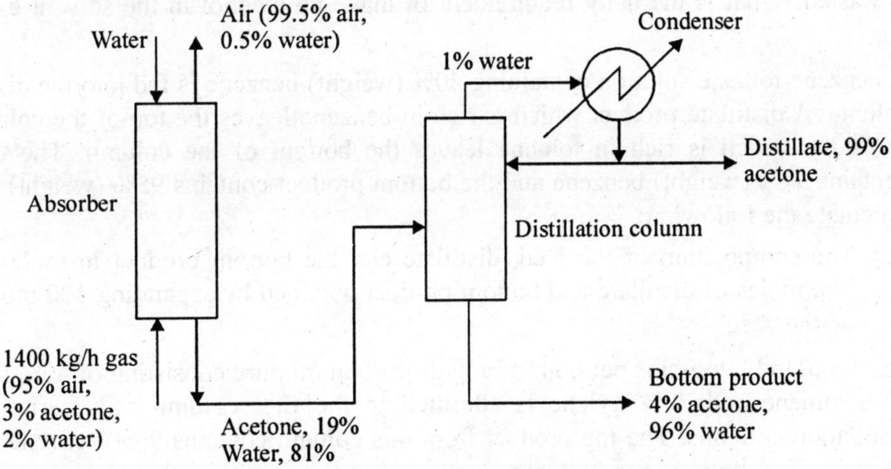

Figure 9.35 Recovery of acetone from air (Exercise 9.82).

Distillation

9.83 A mixture of benzene and toluene containing 10% by mole benzene is continuously distilled at a rate of 1000 kmol/h in a distillation column. 95% of the benzene in the feed is recovered as distillate product which contains 98% benzene and 2% toluene. Calculate the following:

(a) The moles of bottom product
(b) The composition of the bottom product

9.84 An aqueous solution of ethanol containing 10% by weight ethanol is continuously distilled at a rate of 1000 kg/h in a distillation column. Ten percent of the feed is recovered as distillate product which contains 60% ethanol and 40% water. Calculate the following:

(a) The weight of alcohol lost in the bottom product
(b) The composition of the bottom product

9.85 A distillation column is charged with aqueous solution of ethanol containing 35% ethanol by weight. The concentrated alcohol is withdrawn as the distillate containing 85% alcohol. The bottom product (residue) contains 5% ethanol. Determine the following:

(a) The mass of distillate per 100 kg of feed
(b) The ratio of the mass of the distillate to mass of the residue

9.86 A continuous distillation column is used to regenerate solvent for use in a solvent extraction unit. The column treats 100 kmol/h of a feed containing 15% (mol) ethyl alcohol and the rest water. The overhead product is 89.43% alcohol and the bottom product is 0.5% alcohol. The overhead is sent to the extraction unit and the bottom is wasted. What is the daily requirement of make-up alcohol in the solvent extraction unit?

9.87 A benzene-toluene solution containing 40% (weight) benzene is fed into the distillation column. A distillate product which is rich in benzene leaves the top of the column and a residue which is rich in toluene leaves the bottom of the column. The distillate contains 97% (weight) benzene and the bottom product contains 95% (weight) toluene. Calculate the following:

(a) The composition of the feed, distillate and the bottom product in mole percent
(b) The moles of distillate and bottom product obtained by separating 100 moles/hour of the feed

9.88 One thousand kilograms per hour of a hydrocarbon mixture consisting of 40% benzene, 40% toluene and 20% xylene is admitted to the first column of a series of two distillation columns. The top product from this column contains 99% benzene and 1% toluene. The bottom product enters the second column in the series where it is subjected to further purification. The distillate leaving the second column is 95% toluene and 5% benzene and the bottom product is 90% xylene and 10% toluene. Calculate (a) the quantity and (b) the composition of the bottom product from the first column. All concentrations are on a weight basis.

9.89 One hundred moles of a hydrocarbon mixture consisting of 20% ethane, 40% propane and 40% butane is admitted to the first column of a series of two distillation columns. The top product from this column contains 95% ethane, 4% propane and 1% butane. The bottom product enters the second column in the series where it is subjected to further purification. The distillate leaving the second column is 99% propane and 1% butane and the bottom product is 8.4% propane and 91.6% butane. Calculate (a) the quantity and composition of the bottom product from the first column and (b) the quantity of the distillate from the second column.

9.90 The feed to a distillation column is separated into net overhead product containing nothing with a boiling point higher than that of isobutane and bottoms containing nothing with a boiling point below that of propane. The composition of the feed is

Component	*mol%*
Ethylene	2.0
Ethane	3.0
Propylene	5.0
Propane	15.0
Isobutane	25.0
n-Butane	35.0
n-Pentane	15.0

The composition of isobutane in the overhead is 5.0 mol%, and the concentration of propane in the bottom is 0.8 mol%. Calculate the composition of the overhead and bottom streams per 100 moles of feed.

9.91 Refer to Figure 9.36. Feed enters the first column of the two-stage unit with a flow rate of 100 kg/s. The top product from this column is withdrawn at a rate of 40 kg/s.

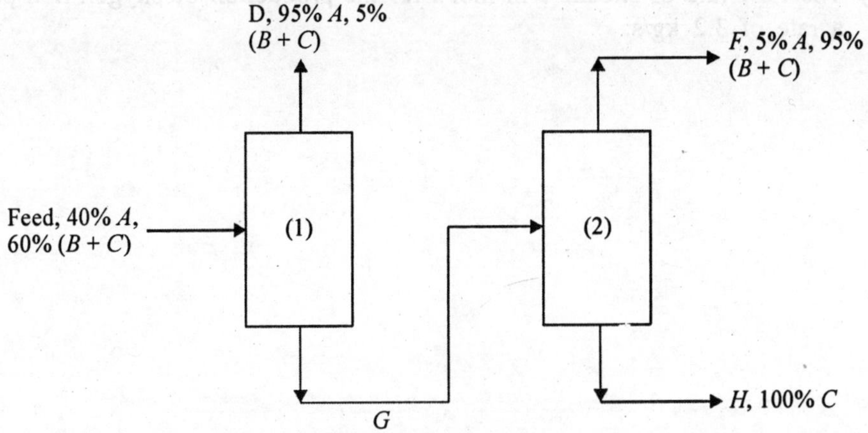

Figure 9.36 Operation in Exercise 9.91.

Streams *D* and *F* contain *B* and *C* in the ratio 10 : 1 by weight. All percentages are on a weight basis. Determine the following:

(a) Flow rates of streams *F*, *G* and *H*
(b) Percent of *B* and *C* in the feed, *D* and *F*

9.92 Oxygen is produced by liquefying air and distilling it at low temperatures. Liquefied air (21% O_2, 79% N_2) is sent to a distillation column where it is separated as stream 1 (60% N_2, 40% O_2) leaving as bottom and stream 2 (96% N_2, 4% O_2) leaving at the top. Both streams are then fed to another column. Nitrogen-rich product (98% N_2, 2% O_2) is withdrawn as distillate and oxygen-rich product (99.5% O_2, 0.5% N_2) is withdrawn as the bottom product. Refer to Figure 9.37.

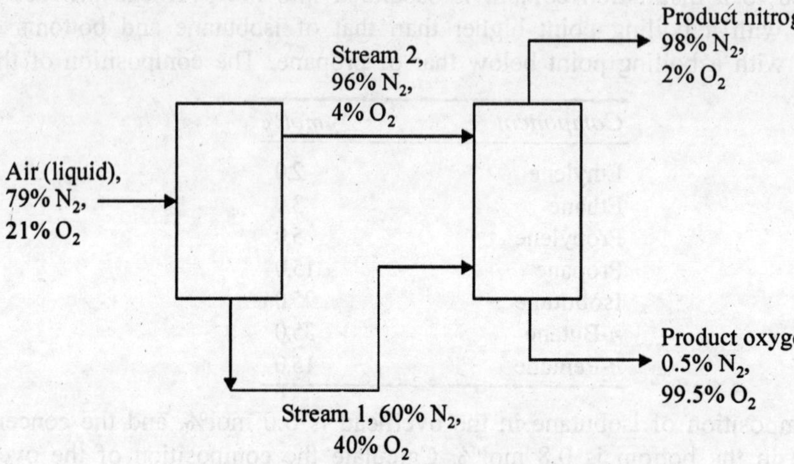

Figure 9.37 Air separation in Exercise 9.92.

Calculate the following:

(a) The percent recovery of oxygen in the oxygen-rich product
(b) The flow rate of stream 1 in mol/s for the production of oxygen-rich product at a rate of 3.2 kg/s.

10 Material Balance with Chemical Reaction

We have already seen that the general material balance equation can be written in the following form:

$$\begin{bmatrix}\text{material input}\\\text{to the system}\end{bmatrix} - \begin{bmatrix}\text{material output}\\\text{from the system}\end{bmatrix} + \begin{bmatrix}\text{material generation}\\\text{within the system}\end{bmatrix} - \begin{bmatrix}\text{material consumption}\\\text{within the system}\end{bmatrix}$$

$$= \begin{bmatrix}\text{accumulation of}\\\text{material within}\\\text{the system}\end{bmatrix} \tag{10.1}$$

In the absence of generation and consumption of materials, the above equation reduces to

$$\text{input} - \text{output} = \text{accumulation} \tag{10.2}$$

Under steady state, there is no accumulation and the terms on the right-hand sides of both the above equations reduce to zero. For processes with chemical conversion, Eq. (10.2) is not universally valid. Consider, for example, a process in which 1 kmol (100.09 kg) $CaCO_3$ is subjected to calcination reaction in a kiln. The following reaction takes place.

$$CaCO_3 \rightarrow CaO + CO_2$$

Assume that only 90% of the feed is converted. Then the quantities of the reactants and products in mass and molar units are given as follows:

Material	*Input*		*Output*	
	Mass (kg)	*Moles* (kmol)	*Mass* (kg)	*Moles* (kmol)
$CaCO_3$	100.09	1	10.01	0.1
CaO	0	0	50.47	0.9
CO_2	0	0	39.61	0.9
Total	100.09	1	100.09	1.9

We can see that Eq. (10.2) is satisfied for the total material balance written on the mass basis, but is not satisfied if the balance is considered on molar units. When writing component balances for individual chemical species, Eq. (10.2) is not applicable either on a mole basis or on a mass basis. In this case the balance equation Eq. (10.1) should be used taking the generation and consumption terms into account. However, Eq. (10.2) is still valid on a mole or mass basis if the balance is written for atomic species. Thus, the amounts of carbon entering and leaving the calciner are the same irrespective of whether the amount is expressed on a mole basis or mass basis.

10.1 COMBUSTION OF SOLID, LIQUID AND GASEOUS FUELS

The consumption of energy by chemical process industries is quite large in comparison with other industries. At present almost all the process energy requirement is met by the combustion of fossil fuels. The fossil fuels consist chiefly of carbon and hydrogen, with small amounts of oxygen, nitrogen, sulphur, mineral matter and moisture. The composition of the fuels varies widely depending upon the type and source of the fuels.

During combustion the chemical energy stored in the fuels is released as heat energy. The combustion process is the rapid chemical reaction of oxygen with materials and is used generally for the purpose of liberating heat energy. The combustion of coal, fuel oil or natural gas is the main source of heat energy in process industries. The combustion process is generally carried out in furnaces and the heat liberated may be utilized for the production of steam for power generation and for heating purposes in the plant. Combustion calculations are carried out to relate quantities and composition of the fuel to those of the flue gas. In the following paragraphs some special terms used with reference to combustion calculations are discussed.

Theoretical oxygen is the oxygen required to burn all carbon in the fuel to carbon dioxide and all hydrogen to water. Oxygen required for combustion is supplied through air. To ensure complete combustion, usually more air is supplied than what is needed theoretically. If there is no adequate supply of air, incomplete combustion results, and carbon monoxide and free hydrogen will be present in the flue gases. *Excess oxygen* is the oxygen furnished through air in excess of the theoretical requirement of oxygen for the combustion. It is usually expressed as *percent excess oxygen*. Percent excess oxygen and percent excess air are the same. If oxygen is already present in the fuel, the percent excess oxygen is based on the *net oxygen*. Net oxygen is the total oxygen required for complete combustion minus the oxygen present in the fuel. Though a large percent excess air is preferred from the point of view of complete combustion, excess air supply is not without some problems. A large volume of gas in the furnace reduces the temperature attained on combustion and increases the loss of heat with the flue gas leaving the furnace.

10.1.1 Orsat Analysis

The Orsat analysis lists the composition of the flue gas on a dry basis. The orsat apparatus measures the amount of CO_2, CO and O_2 in a given sample of flue gas. The principle of measurement can be explained with the help of Figure 10.1.

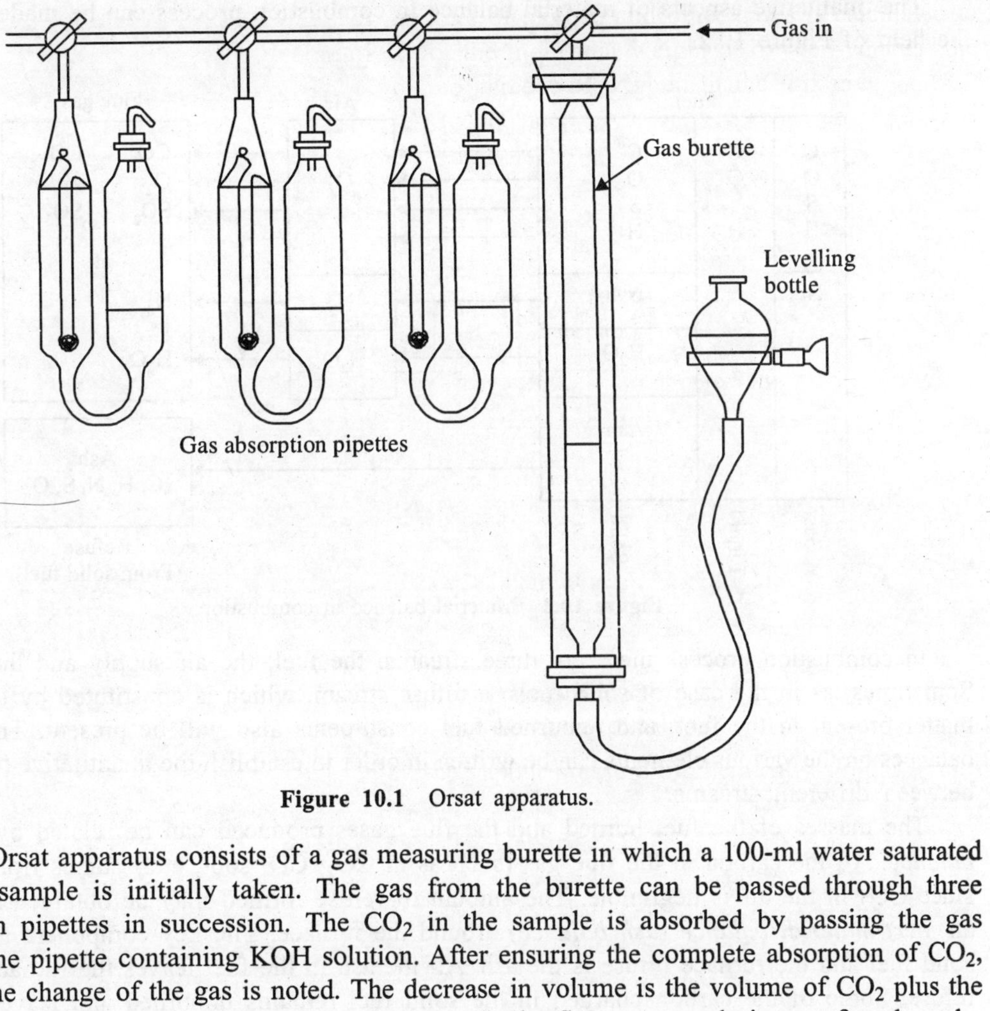

Figure 10.1 Orsat apparatus.

The Orsat apparatus consists of a gas measuring burette in which a 100-ml water saturated flue gas sample is initially taken. The gas from the burette can be passed through three absorption pipettes in succession. The CO_2 in the sample is absorbed by passing the gas through the pipette containing KOH solution. After ensuring the complete absorption of CO_2, the volume change of the gas is noted. The decrease in volume is the volume of CO_2 plus the water vapour it contained. The amount of oxygen in the flue gas sample is next found out by passing the gas through the second pipette which contains an oxygen-absorbing liquid such as alkaline pyrogallate solution and by measuring the decrease in volume. Similarly, the CO content is measured by absorbing the gas in cuprous chloride solution. The decrease in volume in each case measures the volume of CO_2, O_2 and CO respectively plus the water vapour associated with each of these gases. The ratio of the amount of water vapour associated with each gas to the amount of gas absorbed in each case is constant. This means that the decrease in volume in each case is equal to the volume percent of the gas on a dry basis. For example, assume that 100 mL of the gas sample contained 50 mL of CO_2, 10 mL of CO, 30 mL of O_2 and 10 mL of H_2O vapour. On absorbing CO_2, the decrease in volume would be $50 + 50 \times (10/90) = 55.56$ mL which is the percent of CO_2 in the gas on a water-free basis. Similarly, the concentrations of O_2 and CO are respectively 33.33% and 11.11%. The percent of nitrogen is found out by subtracting from 100 the sum of the percent concentrations of CO_2, CO and O_2.

The qualitative aspects of material balance in combustion process can be made clear with the help of Figure 10.2.

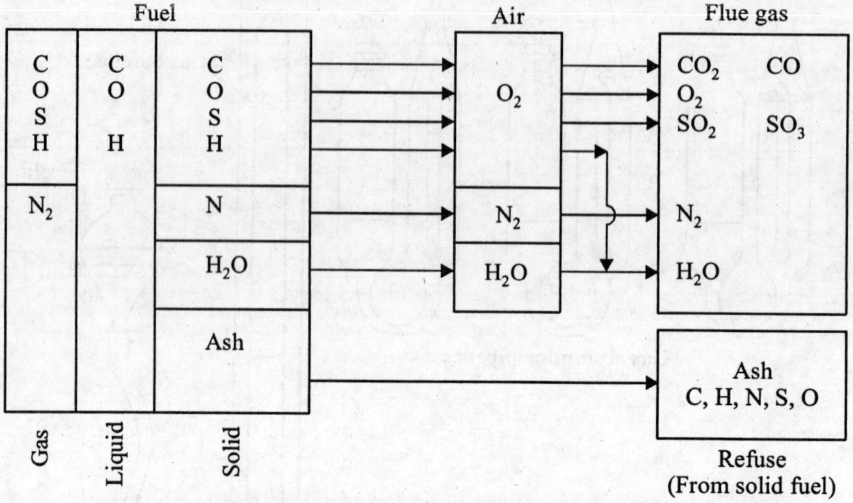

Figure 10.2 Material balance in combustion.

In combustion process there are three streams: the fuel, the air supply and the flue gas. Sometimes, as in the case of solid fuels, a refuse stream, which is constituted by the mineral matter present in the fuel, and unburned fuel constituents also will be present. The material balances on the various elements can be written in order to establish the quantitative relationship between different streams.

The masses of the fuel burned and the flue gases produced can be related by a *carbon balance*. All the carbon in the flue gas (such as in CO_2, CO, soot, etc.) comes from the fuel, since CO_2 in the air is negligible. The amount of refuse formed may be obtained by making an *inert material balance* (*ash balance*) around the furnace. The key component relating the solid fuel and the furnace refuse is the ash. All the ash in the fuel leaves the furnace with the refuse. Some of the carbon charged in the solid fuel remains unburned and leaves with the refuse. An ash balance and a refuse analysis would be necessary to determine the fraction of carbon charged that is unburned. The nitrogen in the fuel plus that in the air equals the nitrogen in the flue gas. Thus the *nitrogen balance* can be used to find out the amount of air that was supplied per unit mass of the flue gas. On combustion, the hydrogen in the fuel becomes water and it leaves with the flue gas. The total water present in the flue gas is made up of water produced by the combustion of hydrogen with oxygen plus the moisture present in the fuel and air. The *oxygen balance* can thus be made to determine the amount of hydrogen in the fuel. This gives

moles of oxygen used for burning H_2 = moles of O_2 in the fuel and air – moles of oxygen in the dry flue gas as CO_2, SO_2, CO, O_2, etc.

Knowing that one mole of oxygen is equivalent to two moles of hydrogen, we can determine the moles of hydrogen in the fuel that have combined with oxygen to form water. Since the flue gas analysis is reported on a dry basis, it is necessary that the amount of water in the flue gas be known to find the volume of wet flue gas. This is done by taking a *water balance* around the furnace, i.e.

mass of water in fuel and air + mass of water formed by the combustion of H_2 in the fuel

= mass of water in the wet flue gas

10.1.2 Proximate and Ultimate Analysis of Coal

The proximate analysis of coal is an empirical procedure for analysis of coal in order to list the composition in terms of the following items, all expressed as percentages by weight of coal:

- *Moisture:* The loss of weight when coal is heated in an oven at 105°C.
- *Volatile combustible matter:* The loss in weight when the coal sample is heated in a covered crucible for about 7 minutes at 950°C minus the weight of the moisture.
- *Ash:* The weight of residue obtained when the sample of coal is subjected to complete combustion in a muffle furnace at 700°C–750°C.
- *Fixed carbon:* Calculated as 100 minus the sum of the moisture, volatile matter and ash already obtained, i.e.

$$FC = 100 - (M + VCM + A) \tag{10.3}$$

where *FC*, *M*, *VCM* and *A* are the percentages of fixed carbon, moisture, volatile combustible matter and ash respectively. Due to the environmental problems associated with the combustion of sulphur, it is now required to report the percentage by weight of sulphur also in the proximate analysis.

The ultimate analysis of coal gives the composition as the percentage by weight of the various elements such as C, H, N, O, S, etc. and the ash. The ultimate analysis provides the necessary data required for material balance in the combustion of coal. The analysis is carried out by elaborate and standard chemical techniques. The carbon dioxide, water and sulphur dioxide formed by burning coal in a sufficient supply of oxygen determine respectively the amounts of carbon, hydrogen and sulphur in coal. Carbon dioxide and water are absorbed in suitable solvents and sulphur dioxide in lead chromate. The nitrogen in the coal is determined by the Kjeldahl method. The percentage of oxygen in the coal is found out by subtracting the percentages of all other constituents from 100.

The hydrogen reported in the ultimate analysis includes the hydrogen in the volatile matter and the hydrogen in the moisture present in coal. The carbon reported in the ultimate analysis includes the fixed carbon and the carbon present in the volatile matter. The volatile matter contains complex organic substances made up of C, H, O, N and S. The fixed carbon is the combustible residue of tar and coke resulting from the pyrolysis reaction at 950°C. Oxygen and total hydrogen may be reported in the ultimate analysis, or the oxygen may be reported as equivalent "combined water" in which case, the remaining hydrogen is the available hydrogen or net hydrogen. The available hydrogen is the only hydrogen that requires oxygen from the air for combustion.

EXAMPLE 10.1 Propane is burned with excess air to ensure complete combustion. If 55 kg of CO_2 and 15 kg of CO are obtained when propane is completely burned with 500 kg air, determine the following:

(a) The mass of propane burnt (in kilograms)
(b) The percent excess air
(c) The composition of flue gas

Solution *Basis:* 55 kg of CO_2 and 15 kg of CO obtained by burning propane with 500 kg of air

The combustion of propane in this example is schematically represented in Figure 10.3.

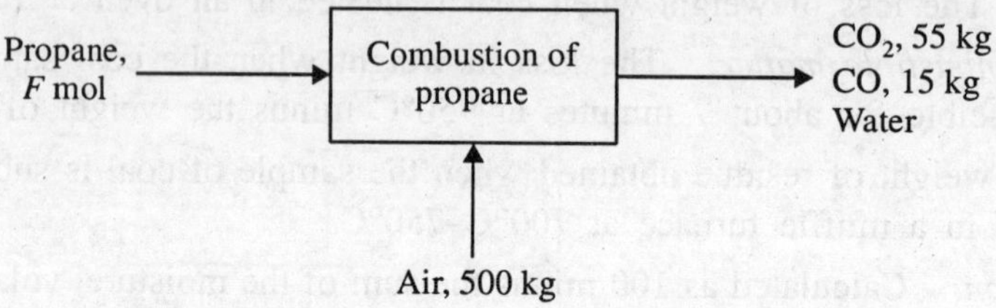

Figure 10.3 Combustion of propane (Example 10.1).

The molecular weight of CO_2 is 44 and the molecular weight of CO is 28. Therefore,

$$\text{Moles of } CO_2 \text{ in the flue gas} = \frac{55}{44} = 1.25 \text{ kmol}$$

$$\text{Moles of CO in the flue gas} = \frac{15}{28} = 0.5357 \text{ kmol}$$

$$\text{Air supplied} = \frac{500}{29} = 17.24 \text{ kmol}$$

(a) Let F be the moles of propane burned. It can be calculated by a carbon balance:

moles of carbon in the inlet gas = moles of carbon in the exit gases

Since one mole of propane contains three moles of carbon, the carbon balance gives

$$3F = 1.25 + 0.5357 \tag{A}$$

Therefore, the moles of propane burned $F = 0.5952$ kmol.

$$\text{Mass of propane burned} = 0.5952 \times \text{molecular weight of propane}$$
$$= 0.5952 \times 44.064 = 26.23 \text{ kg}$$

(b) One mole propane requires 5 moles of oxygen for complete combustion:

$$C_3H_8 + 5O_2 \rightarrow 3CO_2 + 4H_2O \tag{B}$$

Theoretical oxygen requirement is

$$0.5952 \times 5 = 2.976 \text{ kmol}$$

Theoretical requirement of air is

$$\frac{100}{21} \times 2.976 = 14.17 \text{ kmol}$$

$$\text{Percent excess of air} = \frac{\text{excess air}}{\text{thoretical air}} \times 100 = \frac{\text{air supplied} - \text{theoretical air}}{\text{theoretical air}} \times 100$$

$$= \frac{17.24 - 14.17}{14.17} \times 100 = 21.67\%$$

(c) The products of combustion contains CO_2, CO, O_2, N_2, and H_2O vapour produced on combustion. The reactions that occur are as follows:

$$C_3H_8 + 5O_2 \rightarrow 3CO_2 + 4H_2O \qquad \text{(C)}$$

$$C_3H_8 + \tfrac{7}{2}O_2 \rightarrow 3CO + 4H_2O \qquad \text{(D)}$$

When 0.5952 kmol propane is burnt, 0.5952 × 4 = 2.3808 kmol water is produced. The moles of oxygen present in the combustion gases can be determined by an oxygen balance:

O_2 in the air supplied = O_2 present in the flue gas as CO_2, CO and H_2O + unburned oxygen

That is,

$$17.24 \times \frac{21}{100} = 1.25 + 0.5357 \times 0.5 + 2.3808 \times 0.5 + \text{unburned } O_2$$

Therefore, the amount of free oxygen present in the flue gas is

$$17.24 \times \frac{21}{100} - (1.25 + 0.5357 \times 0.5 + 2.3808 \times 0.5) = 0.9122 \text{ kmol}$$

[Alternatively the moles of oxygen in the flue gas can be obtained by considering the reaction in steps. With every 3 mol CO_2 formed according to reaction *C*, 5 mol oxygen is consumed, and with every 3 mol CO formed according to reaction *D* 3.5 mol oxygen is consumed. Thus,

$$\text{moles of oxygen consumed by reaction } C = 1.25 \times \frac{5}{3} = 2.0833 \text{ kmol}$$

Therefore,

$$\text{moles of oxygen consumed by reaction } D = 0.5357 \times \frac{3.5}{3} = 0.6250 \text{ kmol}$$

$$\text{total moles of oxygen consumed} = 2.0823 + 0.6250 = 2.7083 \text{ kmol}$$

$$\text{moles of oxygen supplied} = 17.24 \times 0.21 = 3.6204 \text{ kmol}$$

Moles of free oxygen present in the flue gas by an oxygen balance is equal to the moles of oxygen supplied minus the moles of oxygen consumed. Therefore, moles of free oxygen = 3.6204 – 2.7083 = 0.9121 kmol]

The entire nitrogen in the air supplied will appear in the flue gas. Therefore, flue gas contains 17.24 × 0.79 = 13.6196 kmol. The composition of the flue gas can now be calculated as shown in the table below.

Constituent	*The number of moles in the flue gas*	*Percent composition, (mol%)*
CO_2	1.25	$\frac{1.25}{18.6983} \times 100 = 6.69\%$
CO	0.5357	$\frac{0.5357}{18.6983} \times 100 = 2.87\%$
O_2	0.9122	$\frac{0.9122}{18.6983} \times 100 = 4.88\%$
N_2	13.6196	$\frac{13.6196}{18.6983} \times 100 = 72.84\%$
H_2O	2.3808	$\frac{2.3808}{18.6983} \times 100 = 12.72\%$
Total	18.6983	100

EXAMPLE 10.2 Hydrogen-free coke containing 85% (weight) carbon and the rest inert materials is burned in a furnace. It is found that during combustion 5% of the coke charged is lost unburned. The flue gas analysis shows 14.84% CO_2, 1.65% CO, 5.16% O_2 and 78.35% N_2. The flue gas leaves the furnace at 500 K and 100 kPa. Calculate the following:

(a) The percent excess air on the basis of complete combustion of coke
(b) The weight of air supplied per kg of coke charged
(c) The volume of flue gas per kg of coke charged
(d) The composition of the refuse from the furnace

Solution *Basis:* 100 kmol flue gas

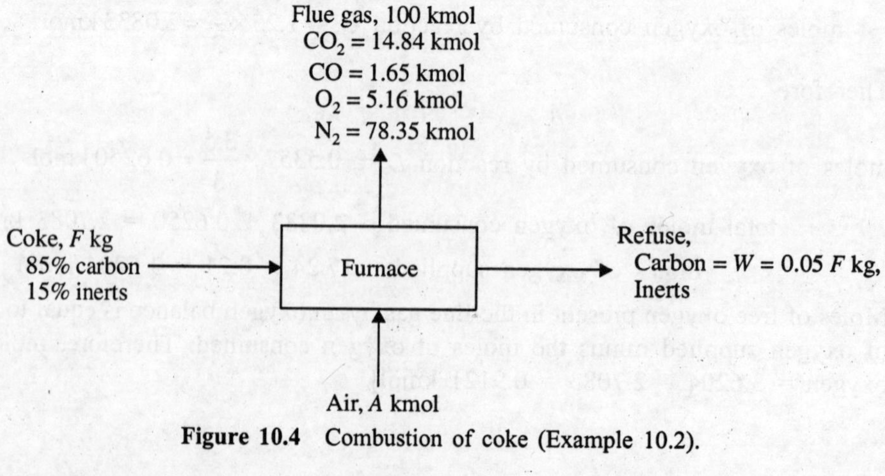

Figure 10.4 Combustion of coke (Example 10.2).

Let F kg be the mass of coke charged and W kg be the mass of coke left unburned per 100 kmol flue gas produced. $W = 0.05\ F$.

The amount of coke charged to the burner to obtain 100 kmol flue gas can be obtained by taking a carbon balance.

Carbon present in the flue gas = carbon in CO_2 + carbon in CO

$$= 14.84 + 1.65 = 16.49 \text{ kmol} = 16.49 \times 12 = 197.88 \text{ kg}$$

Carbon in the coke charged = $F \times 0.85$ kg

Carbon in the unburned coke = $W \times 0.85 = 0.05F \times 0.85$

Carbon balance:

$$F \times 0.85 = 0.05F \times 0.85 + 197.88$$

Solving the above equation, we get

$$F = \frac{197.88}{0.95 \times 0.85} = 245.05 \text{ kg}$$

(a) Let A kmol air is supplied for combustion. Since the nitrogen in the flue gas is coming from the air supplied, a nitrogen balance gives

$$A \times 0.79 = 78.35$$

or

$$A = 99.177 \text{ kmol}$$

Thus,

$$\text{Oxygen supplied} = A \times 0.21 = 99.177 \times 0.21 = 20.827 \text{ kmol}$$

Theoretical oxygen requirement is calculated as the moles of oxygen required for the complete conversion of carbon in the coke charged to CO_2. That is,

$$\text{theoretical requirement of oxygen} = \frac{245.05 \times 0.85}{12} = 17.358 \text{ kmol}$$

$$\text{Percent excess of oxygen} = \frac{\text{excess oxygen}}{\text{theoretical oxygen}} \times 100$$

$$= \frac{20.827 - 17.358}{17.358} \times 100 = 19.99\%$$

Therefore,

$$\text{Percent excess air} = \text{percent excess oxygen} = 19.99\%$$

(b) Air supplied = 99.177 kmol = 99.177 × 29 = 2876.133 kg

$$\text{Coke charged} = 245.05 \text{ kg}$$

$$\text{Air supplied per kg of coke charged} = \frac{2876.133}{245.05} = 11.74 \text{ kg}$$

(c) For 100 kmol flue gas, 245.05 kg of coke is charged. Therefore, the volume of flue gas at 500 K and 100 kPa is

$$\frac{100}{245.05}\times 22.414\times\frac{101.325}{100}\times\frac{500}{273.13}=16.96\ \text{m}^3$$

(d) The refuse from the furnace is made up of inerts in the coke charged and the unburned carbon. Therefore, the total weight of refuse is:

$$\text{carbon in the refuse} = F \times 0.05 \times 0.85 = 10.41\ \text{kg}$$

$$\text{inerts in the refuse} = F \times 0.15 = 36.76\ \text{kg}$$

Therefore, the total weight of refuse is

$$36.76 + 10.41 = 47.17\ \text{kg}$$

Composition of refuse is

$$\text{carbon} = \frac{10.41}{47.17}\times 100 = 22.07\%$$

$$\text{inerts} = \frac{36.76}{47.17}\times 100 = 77.93\%$$

EXAMPLE 10.3 A fuel oil containing 70% carbon by weight and the rest combustible hydrogen and moisture is burned with excess air. The flue gas analyzed 9% CO_2, 2% CO, 3% O_2 and 86% N_2. Determine the following:

(a) The percentage of excess air
(b) The ratio of carbon to combustible hydrogen in the fuel on a weight basis
(c) The ratio of carbon to total hydrogen in the fuel on a weight basis
(d) The percentages of combustible hydrogen and moisture in the fuel
(e) The mass of moisture present in the flue gas per kg of oil burned

Solution *Basis:* 100 kmol flue gas
Carbon content in the flue gas = carbon in CO_2 and CO = 11 kmol
Let F kg be the mass of fuel burned and A kmol be the amount of air supplied.

(a) Nitrogen in the flue gas = 86 kmol
Since the fuel contains no nitrogen, the nitrogen present in the flue gas comes from air. The nitrogen balance is

$$A\times 0.79 = 86 \Rightarrow A = \frac{86}{0.79} = 108.86\ \text{kmol}$$

Oxygen in the air supplied = 108.86 × 0.21 = 22.86 kmol
If CO in the flue gas were completely converted to CO_2, the moles of oxygen present in the flue gas would be 3 – 1 = 2 kmol.
That is, excess oxygen is 2 kmol.

$$\text{Percent excess oxygen} = \frac{\text{oxygen excess}}{\text{theoretical oxygen}} \times 100$$

$$= \frac{2}{22.86 - 2} \times 100 = 9.59\%$$

(b) The combustible hydrogen in the fuel is determined from the quantity of oxygen that is consumed by the net hydrogen in the formation of water. Of the 22.86 moles of oxygen supplied, the oxygen that is accounted in the flue gas as CO_2, CO or excess oxygen is only 9 + 1 + 3 = 13 kmol. The difference has been utilized by the net hydrogen in the formation of water. That is, the oxygen balance is

Oxygen in the air supplied = oxygen accounted in the flue gas as CO_2, CO or excess oxygen + oxygen consumed by the net hydrogen

The amount of oxygen consumed by net hydrogen is

$$22.86 - 13 = 9.86 \text{ kmol}$$

Therefore, the net hydrogen burned is

$$2 \times 9.86 = 19.72 \text{ kmol} = 39.76 \text{ kg}$$

The mass of carbon burned is

$$9 + 2 = 11 \text{ kmol} = 132 \text{ kg}$$

The ratio of carbon to net hydrogen in the fuel is

$$\frac{132}{39.76} = 3.32:1$$

(c) For 1 kg of net hydrogen in the fuel, carbon present is 3.32 kg. Let x kg be the mass of moisture present in the fuel for every kilogram of net hydrogen present. It is given that the weight percentage of carbon in the fuel is 70%. Therefore,

$$0.7 = \frac{3.32}{(1 + 3.32 + x)}$$

which on solving gives x = 0.423 kg.
The mass of hydrogen present in 0.423 kg of moisture is

$$0.423 \times \frac{2.016}{18.016} = 0.047 \text{ kg}$$

Therefore,

$$\text{Total hydrogen present} = 0.047 + 1 = 1.047 \text{ kg}$$

The ratio of carbon to total hydrogen is

$$\frac{3.32}{1.047} = 3.17:1$$

(d) Carbon = 3.32 kg; combustible hydrogen = 1.0 kg and moisture = 0.423 kg. The percentage composition by weight is

$$\text{net hydrogen} = \frac{1.00}{(1+3.32+0.423)}\times 100 = \frac{1.00}{4.743}\times 100 = 21.08\%$$

$$\text{moisture} = \frac{0.423}{4.743}\times 100 = 8.92\%$$

(e) Since the fuel contains 70% carbon, 21.08% net hydrogen and 8.92% moisture, the products of combustion when 100 kg of fuel is burned will contain

$$8.92 + \frac{18.016}{2.016}\times 21.08 = 197.3 \text{ kg}$$

The mass of moisture in the flue gas per kg of fuel burned = 1.973 kg.

EXAMPLE 10.4 A producer gas contains 9.2% CO_2, 21.3% CO, 18% H_2, 2.5% CH_4, and the rest N_2. It is burned with an excess supply of air. The flue gas analysed 9.05% CO_2, 1.34% CO, 9.98% O_2 and 79.63% N_2. Determine the following:

(a) The volumetric ratio of air supplied to the fuel burned
(b) The percent excess air supplied
(c) The percent of nitrogen in the flue gas that came from the fuel.

Solution *Basis:* 100 moles of flue gas

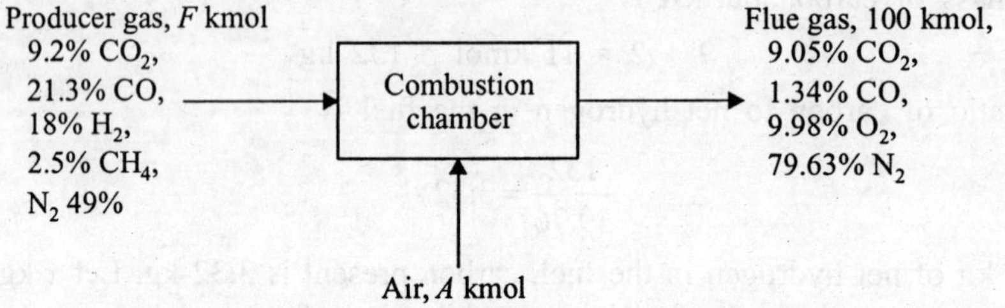

Figure 10.5 Combustion of producer gas (Example 10.4).

Let F moles of fuel be burned with A moles of air to obtain 100 moles of dry flue gas.

Carbon balance:

carbon in the fuel burned = carbon in the flue gas as CO_2 and CO

$$0.092F + 0.213F + 0.025F = 9.05 + 1.34 = 10.39 \text{ kmol}$$

Therefore, the moles of fuel burned to produce 100 kmol flue gas, F = 31.4848 kmol.

Air supplied for the combustion may be obtained by a nitrogen balance. Nitrogen in the flue gas is the sum of the nitrogen present in the fuel and that in the air. The mole percentage of nitrogen in the producer gas = 100 – 21.3 – 9.2 – 2.5 – 18.0 = 49%.

Nitrogen balance:

nitrogen from the fuel + nitrogen in the air = nitrogen in the flue gas

$$31.4848 \times 0.49 + A \times 0.79 = 79.63$$

Solving this we get A = 81.2689 kmol.

(a) The molar ratio of air to fuel is $= \dfrac{81.2689}{31.4848} = 2.58$, which is the same as the volumetric ratio of air to fuel.

(b) The percent excess of oxygen supplied can be calculated as

$$\text{percent excess} = \frac{\text{oxygen excess}}{\text{theoretical requirement}} \times 100$$

The flue gases contain 9.98 kmol oxygen. The amount of oxygen that would have been present in the flue gas had the combustion were complete would be less than this by the amount of oxygen required by CO present in the flue gas for its combustion to CO_2. That is,

$$\text{excess oxygen} = 9.98 - 0.5 \times 1.34 = 9.31 \text{ kmol}$$

$$\text{Theoretical requirement of oxygen} = \text{oxygen supplied} - \text{oxygen excess}$$

$$= 81.2689 \times 0.21 - 9.31 = 7.76 \text{ kmol}$$

$$\text{Percent excess of air} = \text{percent excess of oxygen}$$

$$= \frac{9.31}{7.76} \times 100 = 120\%$$

(c) 100 kmol of flue gas contains 79.63 kmol nitrogen. 100 kmol fuel results from the combustion of 31.4848 kmol producer gas containing $31.4848 \times \dfrac{49}{100} = 15.43$ kmol nitrogen.

The percent of nitrogen in the flue gas that came from the fuel is

$$\frac{15.43}{79.63} \times 100 = 19.38\%$$

Note: Consider the calculation of percent excess air in part (b). Instead of calculating the theoretical requirement of oxygen as the difference between the oxygen supplied and oxygen present in excess, the theoretical requirement of oxygen can be computed as the oxygen required for complete combustion of the various constituents of the fuel based on the fuel analysis. Ideally both calculations should give the same result. Here, let us calculate the oxygen required by 31.4848 kmol producer gas. The producer gas contains 21.3% CO, 18% H_2 and 2.5% CH_4.

$$CO + \frac{1}{2}O_2 \rightarrow CO_2$$

$$CH_4 + 2O_2 \rightarrow CO_2 + 2H_2O$$

$$H_2 + \frac{1}{2}O_2 \rightarrow H_2O$$

$$\text{Oxygen required by CO} = 0.5 \times 21.3 = 10.65 \text{ kmol}$$

$$\text{Oxygen required by methane} = 2 \times 2.5 = 5.0 \text{ kmol}$$

$$\text{Oxygen required by hydrogen} = 0.5 \times 18 = 9 \text{ kmol}$$

Theoretical requirement of oxygen is

$$\frac{(10.65 + 5.0 + 9.0)}{100} \times 31.4848 = 7.76 \text{ kmol}$$

The value obtained is equal to the theoretical requirement already calculated in part (b) on the basis of the flue gas analysis. This will rarely be the case in many practical situations. Though the composition of the products of combustion is a unique function of the composition of the fuel and the air-fuel ratio employed, since the flue gas analysis and the fuel analysis are done by two independent measurements, these data are rarely consistent. Any inconsistency in these measurements may lead to difference in the values calculated for theoretical air requirement by different methods. For example, try the following problem:

Producer gas containing 5.3% CO_2, 21.1% CO, 15.3% H_2, 1.0% CH_4, and 57.3% N_2 is burned in a furnace in an excess supply of air. The flue gas contains 14.1% CO_2, 2.3% O_2 and 83.6% N_2. Find the percent excess air.

Calculate the percent excess air by the following three different formulae:

$$\text{Percent excess} = \frac{\text{excess oxygen}}{\text{theoretical oxygen}} \times 100$$

$$\text{Percent excess} = \frac{\text{excess oxygen}}{\text{oxygen supplied} - \text{excess oxygen}} \times 100$$

$$\text{Percent excess} = \frac{\text{oxygen supplied} - \text{theoretical oxygen}}{\text{theoretical oxygen}} \times 100$$

and you will see that the percent excess air calculated by these different formulae are not the same.

EXAMPLE 10.5 The following data were obtained in a test on coal fired steam generator. The ultimate analysis of coal: C, 80.5%; H, 4.6%; O, 5.0%; N, 1.1%; ash, 8.8%. No carbon is lost in the refuse. The Orsat analysis of flue gas: CO_2, 16.4%; O_2, 2.3%; CO, 0.4%; N_2, 80.9%. Calculate

(a) The weight of dry gaseous products formed per 100 kg of coal fired

(b) The percent excess air supplied for combustion

Solution *Basis:* 100 kmol flue gas

(a) Let W be the mass of coal burned to produce 100 kmol dry flue gas. Taking a carbon balance, we get

$$W \times 0.805 = (16.4 + 0.4) \times 12$$

This is solved to get W = 250.435 kg.

The mass of 100 kmol flue gas is calculated as below.

Component	*Moles* (kmol)	*Mol. wt.*	*Weight* (kg)
CO_2	16.4	44	721.6
O_2	2.3	32	73.6
CO	0.4	28	11.2
N_2	80.9	28.014	2266.3
Total			3072.7 kg

Therefore, the mass of dry flue gas produced by the combustion of 100 kg coal

$$= \frac{3072.7}{250.435} \times 100 = 1226.95 \text{ kg}$$

(b) Take a nitrogen balance:

nitrogen from coal + nitrogen from air = nitrogen in the flue gas

$$250.435 \times 0.011 + 28.014x = 2266.3$$

where x is the moles of nitrogen in the air supplied. Solving the equation, we get

$$x = 80.80 \text{ kmol}$$

The amount of oxygen in the air supplied is

$$80.80 \times \frac{21}{79} = 21.48 \text{ kmol}$$

The amount of oxygen required for complete combustion is

$$\frac{250.435}{100} \times \left(\frac{80.5}{12} + \frac{4.6}{2.016} \times \frac{1}{2} - \frac{5}{32} \right) = 19.2659 \text{ kmol}$$

Percent excess air is

$$\frac{21.48 - 19.27}{19.27} \times 100 = 11.47\%$$

EXAMPLE 10.6 A furnace is fired with a coal having the following analysis:

Ultimate analysis: C, 63%; H, 12%; O, 16% and ash, 9%.

Proximate analysis: fixed carbon, 39%; volatiles, 42%; moisture, 10% and ash, 9%.

1000 cubic metres of dry air at STP is supplied per 100 kg of coal fired. The dry refuse from this furnace contains 10.8% fixed carbon, 9.0% volatiles, and 80.2% ash. Only 80% of carbon that was burnt has gone to CO_2, the rest to CO. Find the following:

(a) The composition of the volatile matter reported in the proximate analysis of coal
(b) The mass and composition of the refuse
(c) The Orsat analysis of the flue gas
(d) The percent excess air supplied
(e) The mass of water vapour present per 100 moles of dry flue gas.

Solution *Basis:* 100 kg coal charged

(a) 100 kg of coal contains 63 kg of carbon of which 39 kg is fixed carbon. Therefore, the volatile matter contains 63 – 39 = 24 kg of carbon.

Coal contains 12 kg of hydrogen of which $10 \times \frac{2.016}{18.016} = 1.12$ kg of hydrogen is present as moisture. Therefore, hydrogen in the volatile matter is 12 – 1.12 = 10.88 kg of hydrogen. Oxygen present in 10 kg of moisture is = 10 – 1.12 = 8.88 kg. Therefore, oxygen present as volatile matter = 16 – 8.88 = 7.12 kg.

The composition of volatile matter is:

Constituent	*Mass* (kg)	*Weight* %
Carbon	24.00	$\dfrac{24}{42}\times 100 = 57.14\%$
Hydrogen	10.88	$\dfrac{10.88}{42}\times 100 = 25.91\%$
Oxygen	7.12	$\dfrac{7.12}{42}\times 100 = 16.95\%$
Total	42.00	

(b) An ash balance on the furnace reveals that all the ash present in the coal appears in the furnace refuse. Let W be the weight of refuse formed. 80.2% of the refuse is ash. Then, the ash balance is:

$$9.00 = 0.802W$$

or

$$W = 11.22 \text{ kg}$$

The weight of refuse is 11.22 kg of which ash is 9 kg, volatile matter is $11.22\times\dfrac{9.0}{100} = 1.01$ kg, and fixed carbon is $11.22\times\dfrac{10.8}{100} = 1.21$ kg. Since the volatile matter is 57.14% carbon, the total carbon present in the refuse is $1.21 + 1.01 \times \dfrac{57.14}{100}$ = 1.79 kg. Similarly, the total hydrogen present in the refuse is $1.01\times\dfrac{25.91}{100} = 0.26$ kg and the total oxygen is $1.01\times\dfrac{16.95}{100} = 0.17$ kg.

Thus the refuse analysis is

Constituent	*Mass* (kg)	*Weight* %
Carbon	1.79	15.95
Hydrogen	0.26	2.32
Oxygen	0.17	1.52
Ash	9.00	80.21
Total	11.22	

(c) The coal burned = 100 – mass of refuse = 100 – 11.22 = 88.78 kg.
The coal burned contains carbon, hydrogen and oxygen the amounts of which are calculated as below:

$$\text{Carbon} = \text{carbon in the feed} - \text{carbon in refuse}$$
$$= 63 - 1.79 = 61.21 \text{ kg} = 5.10 \text{ kmol}$$

$$\text{Hydrogen} = \text{hydrogen in feed} - \text{hydrogen in refuse}$$
$$= 12 - 0.26 = 11.74 \text{ kg} = 5.82 \text{ kmol}$$
$$\text{Oxygen} = \text{oxygen in feed} - \text{oxygen in refuse}$$
$$= 16 - 0.17 = 15.83 \text{ kg} = 0.495 \text{ kmol}$$

Flue gas:

The flue gas contains CO_2, CO, H_2O, O_2 and N_2.
80% of the carbon burned forms CO_2 and 20% forms CO. Therefore,

$$CO_2 \text{ formed} = 0.8 \times 5.10 = 4.08 \text{ kmol}$$
$$CO \text{ formed} = 0.2 \times 5.10 = 1.02 \text{ kmol}$$

$$\text{Air supplied} = 1000 \text{ cubic metres of dry air at STP}$$
$$= \frac{1000}{22.414} = 44.615 \text{ kmol}$$

This consists of $44.615 \times 0.79 = 35.25$ kmol N_2 and $44.615 \times 0.21 = 9.37$ kmol oxygen.

$$\text{Nitrogen in the flue gas} = \text{nitrogen in the air} = 35.25 \text{ kmol}$$

$$\text{Hydrogen in the coal burned} = 12 - 0.26 = 11.74 \text{ kg} = 5.82 \text{ kmol.}$$

Therefore, the flue gas contains 5.82 kmol water. Oxygen present in the flue gas can be determined by an oxygen balance.

$$\text{oxygen in the coal charged} + \text{oxygen in the air} = \text{oxygen in the refuse} + \text{oxygen present in the flue gas as } CO_2, CO, \text{ and } H_2O + \text{free } O_2$$
$$\text{oxygen in the coal charged} = 16 \text{ kg} = 0.5 \text{ kmol}$$
$$\text{oxygen in the refuse} = 0.17 \text{ kg} = 0.005 \text{ kmol}$$

$$\text{oxygen in } (CO_2 + CO + H_2O) = 4.08 + 0.5 \times 1.02 + 0.5 \times 5.82 = 7.50 \text{ kmol}$$

Therefore, the oxygen balance yields

$$0.5 + 9.37 = 0.005 + 7.50 + \text{free oxygen}$$

Therefore, the amount of free oxygen is 2.36 kmol.
Flue gas analysis on a dry basis is given as follows:

Component	*Moles*	*mol %*
CO_2	4.08	9.55
CO	1.02	2.39
O_2	2.36	5.53
N_2	35.25	82.53
Total	42.71	

(d) Oxygen supplied = 9.37 kmol

Oxygen required for complete combustion = oxygen required by carbon in the coal + oxygen required by hydrogen in the coal − oxygen present in the coal

Theoretical requirement of oxygen is

$$\frac{63}{12}+\frac{12}{2.016}\times 0.5-\frac{16}{32}=7.73 \text{ kmol}$$

Excess oxygen = oxygen supplied – theoretical requirement

9.37 – 7.73 = 1.64 kmol

$$\text{Percent excess} = \frac{\text{excess}}{\text{theoretical oxygen}}\times 100$$

$$=\frac{1.64}{7.73}\times 100 = 21.22\%$$

Note: Excess oxygen is the free oxygen present in the flue gas if the fuel is completely burned. Here it is equal to the free oxygen present in the flue gas less the oxygen required by CO in the flue gas for converting it into CO_2 and the oxygen required by the combustible matter in the refuse. The refuse contains 1.79 kg (0.149 kmol) of carbon, 0.26 kg (0.129 kmol) of hydrogen and 0.17 kg (0.005 kmol) of oxygen. Therefore, the material lost with the refuse requires 0.149 + 0.5 × 0.129 – 0.005 = 0.2085 for complete combustion. Therefore, the amount of excess oxygen is

$$2.36 - 0.5 \times 1.02 - 0.2085 = 1.64 \text{ kmol}$$

$$\text{Percent excess} = \frac{1.64}{9.37-1.64}\times 100 = 21.22\%$$

(e) Hydrogen in the coal burned = 12 – 0.26 = 11.74 kg = 5.82 kmol. Therefore, the flue gas contains 5.82 kmol (104.85 kg) water. It means 104.85 kg of water vapour is associated with 42.71 kmol dry flue gas. Thus, there is $\frac{104.85}{42.71} \times 100 = 254.49$ kg of water vapour per 100 kmol dry flue gas.

10.2 OXIDATION OF SULPHUR COMPOUNDS AND RELATED PROCESSES

Sulphur is one of the raw materials for the production of sulphuric acid. It is available as elemental sulphur or as sulphide ores such as pyrites. Combustion of sulphur or sulphide ores is carried out for producing sulphur dioxide. Some sulphur trioxide is also formed during combustion. It is usually impossible to eliminate completely the formation of sulphur trioxide. Usually, about 2–10 % of the sulphur burned is converted to trioxide.

$$S + O_2 \rightarrow SO_2$$

$$2S + 3O_2 \rightarrow 2SO_3$$

In the combustion gas analysis, the SO_3 content will not be given since it is removed from the gas as H_2SO_4 before the analysis. However, its concentration in the gas can be determined from the moles of oxygen supplied for combustion, which in turn is calculated from the nitrogen content of the gas, and the oxygen that is accounted for in the SO_3-free gas. The difference between the oxygen supplied and the oxygen accounted for in the product gas is the oxygen that is consumed for the formation of SO_3.

When iron pyrites is burned in air the following reactions occur:

$$4FeS_2 + 11O_2 \rightarrow 2Fe_2O_3 + 8SO_2$$
$$4FeS_2 + 15O_2 \rightarrow 2Fe_2O_3 + 8SO_3$$

Since the percent of SO_3 is not reported in the burner gas analysis, it has to be determined by the quantity of oxygen unaccounted for in the analysis. The oxygen equivalent of nitrogen in the burner gas gives the amount of oxygen supplied through air. Subtracting from this the oxygen consumed for the formation of SO_2 and the excess oxygen present in the burner gas, we get the oxygen that is unaccounted. The moles of SO_3 formed with this oxygen can be calculated. It is probable that some of the SO_3 that is formed gets adsorbed in the solid refuse known as *cinder* formed on combustion. Analysis of the cinder is necessary to determine the fraction of the SO_3 formed that is adsorbed in it. The cinder is made up of Fe_2O_3 formed during the reaction, the unburned sulphides, and the incombustible materials known as *gangue.*

Assume that the burner gas analyzes 5% SO_2 and 10% oxygen, the rest being nitrogen. The number of moles of oxygen supplied is obtained from the moles of nitrogen in the burner gas. For a basis of 100 moles of the burner gas, the number of moles of oxygen supplied is $85 \times \dfrac{21}{79} = 22.595$ mol. The amount of oxygen consumed for the formation of 5 moles of SO_2 is, according to the chemical reaction given above, $5 \times \dfrac{11}{8} = 6.875$ mol. Therefore, the amount of oxygen unaccounted for is 22.595 – 6.875 – 10 = 5.72 mol. Since according to the reaction equation, 15 moles of O_2 is required for the formation of 8 moles of SO_3, the moles of SO_3 formed is $5.72 \times \dfrac{8}{15} = 3.05$ mol. Not all this SO_3 may be present in the burner gas. A part may get adsorbed on the cinder. The quantity adsorbed is to be measured by analysis of the cinder, and by subtracting this quantity from the total SO_3 produced, the SO_3 in the burner gas can be determined.

In the contact process for the manufacture of sulphuric acid, SO_2 is converted to SO_3 in series of vessels known as *converters.*

$$SO_2 + \frac{1}{2}O_2 \rightarrow SO_3$$

For completing the conversion of SO_2 to SO_3, secondary air supply may be necessary in the converters. The SO_3 leaving the converters is absorbed in concentrated sulphuric acid by admitting the gas at the bottom of an absorption column packed with ceramic packings. 98.5 to 99% acid is used as absorption solvent, and it is fed at the top and flows counter-current to the gas. Usually a 20% *oleum* product is obtained as the concentrated solution leaving the absorption tower. (Oleum is SO_3 dissolved in H_2SO_4. One hundred kilograms of 20% oleum contains 20 kg SO_3 dissolved in 80 kg of H_2SO_4.)

EXAMPLE 10.7 Find the Orsat analysis of the burner gas when pure sulphur is burned with 20% excess air. Of the sulphur burned, 5% is converted to SO_3 and the rest to SO_2.

Solution *Basis:* 1 kg-atom sulphur burned

$$S + O_2 \rightarrow SO_2$$

Theoretical requirement of oxygen = 1 kmol
Oxygen supplied = 1.2 kmol

$$\text{Nitrogen in the air supplied} = 1.2 \times \frac{79}{21} = 4.514 \text{ kmol}$$

Since only 95% is converted to SO_2, SO_2 produced is 0.95 kmol.
Therefore,

$$SO_3 \text{ produced} = 0.05 \text{ kmol}$$

$$\text{Oxygen consumed} = 0.95 + \frac{3}{2} \times 0.05 = 1.025 \text{ kmol}$$

Therefore,

$$\text{Oxygen remaining} = 1.2 - 1.025 = 0.175 \text{ kmol}$$

The burner gas contains:

Component	kmol	*Mole%*	*Mole% on SO_3-free basis*
SO_2	0.95	$\frac{0.95}{5.689} \times 100 = 16.70$	$\frac{0.95}{5.689 - 0.05} \times 100 = 16.85$
SO_3	0.05	$\frac{0.05}{5.689} \times 100 = 0.88$	
O_2	0.175	$\frac{0.175}{5.689} \times 100 = 3.08$	$\frac{0.175}{5.689 - 0.05} \times 100 = 3.10$
N_2	4.514	$\frac{4.514}{5.689} \times 100 = 79.34$	$\frac{4.514}{5.689 - 0.05} \times 100 = 80.05$
Total	5.689		

The results in the last column gives the Orsat analysis of the burner gas.

EXAMPLE 10.8 In a sulphuric acid plant, pyrites containing 50% (weight) sulphur is burned to give SO_2 which is subsequently converted to SO_3 in a converter. The analysis of the burner gas shows 9.5% SO_2 and 7.0% oxygen. The cinder is analysed and it is found that it contains 2.5% sulphur as SO_3. Assuming that all the sulphur in the feed is burned, calculate the following:

(a) The weight of pyrites burned per 100 kmol SO_3-free burner gas
(b) The weight of cinder obtained per 100 kmol SO_3-free burner gas
(c) The percent of the sulphur in the charge that is lost in the cinder
(d) The percent of sulphur charged that is there in the burner gas as SO_3

Solution *Basis:* 100 kmol SO_3-free burner gas
The amount of nitrogen present in the burner gas is

$$100 - (9.5 + 7.0) = 83.5 \text{ kmol}$$

Since the nitrogen in the burner gas is that present in the air supplied for combustion, the amount of oxygen supplied is

$$83.5\times\frac{21}{79}=22.20 \text{ kmol}$$

$$4FeS_2 + 11O_2 \rightarrow 2Fe_2O_3 + 8SO_2$$

$$4FeS_2 + 15O_2 \rightarrow 2Fe_2O_3 + 8SO_3$$

With every 8 kmol SO_2 in the burner gas, 2 kmol Fe_2O_3 is produced accounting for 3 mol of oxygen. Therefore the total moles of oxygen accounted for = Moles of oxygen in the burner gas + Moles of oxygen in the cinder as Fe_2O_3. This is equal to

$$9.5 + 7.0 + 9.5 \times \frac{3}{8} = 20.06 \text{ kmol}$$

The amount of oxygen unaccounted = 22.20 – 20.06 = 2.14 kmol.

2.14 kmol oxygen is consumed by sulphur in the pyrites for the formation of SO_3. As 15 mol oxygen is required for 8 mol SO_3, the total SO_3 formed is

$$2.14\times\frac{8}{15}=1.14 \text{ kmol}$$

(a) Total S burned = 9.5 + 1.14 = 10.64 kmol = 10.64 × 32.064 = 341.16 kg
This constitutes 50% of the pyrites burned. The total pyrites burned is

$$\frac{341.16}{0.5}=682.32 \text{ kg}$$

(b) 10.64 kmol sulphur is equivalent to $\frac{10.64}{2}$ = 5.32 kmol FeS_2. Since the molecular weight of FeS_2 is 119.975, the mass of FeS_2 present in the ore is

$$5.32 \times 119.975 = 638.27 \text{ kg}$$

Weight of gangue = weight of pyrites burned – weight of FeS_2 in the ore

$$= 682.32 - 638.27 = 44.05 \text{ kg}$$

5.32 kmol FeS_2 on combustion will produce

$$5.32 \times \frac{1}{2} = 2.66 \text{ kmol } Fe_2O_3$$

Therefore, the mass of Fe_2O_3 in the cinder is

$$2.66 \times 159.694 = 424.79 \text{ kg}$$

The cinder contains 2.5% sulphur as SO_3. That is, 100 kg of cinder contains

$$\frac{2.5}{32.064} = 0.078 \text{ kmol S or } 0.078 \text{ kmol } SO_3.$$

SO_3 present in 100 kg of cinder is

$$0.078 \times 80.064 = 6.245 \text{ kg}$$

The Fe_2O_3 and gangue together constitutes 100 – 6.245 = 93.755% of the total weight of cinder.

Let x be the weight of cinder. Then,

$$0.937\,55x = 424.79 + 44.05 \quad \text{or} \quad x = 500.07 \text{ kg}$$

Hence, the weight of cinder produced = 500.07 kg.

(c) Since 100 kg of cinder contains 0.078 kmol sulphur, the total amount of sulphur lost in the cinder is

$$500.07 \times \frac{0.078}{100} = 0.3901 \text{ kmol sulphur.}$$

The total amount of sulphur burned is 10.64 kmol.
Percent of total S burned that is lost in the cinder is

$$\frac{0.3901}{10.64} \times 100 = 3.67\%$$

(d) The mass of SO_3 present in the cinder is

$$500.07 \times \frac{6.245}{100} = 31.23 \text{ kg}$$

31.23 kg of SO_3 is equivalent to 0.3904 kmol SO_3. That is 0.3904 kmol SO_3 gets adsorbed in the cinder. Since the total amount of SO_3 formed is 1.14 kmol, the amount of SO_3 present in the burner gas is

$$1.14 - 0.3904 = 0.7496 \text{ kmol}$$

Therefore, the percent of S charged that is present as SO_3 in the burner gas is

$$\frac{0.7496}{10.64} \times 100 = 7.05\%$$

EXAMPLE 10.9 One hundred kilograms per hour of sulphur of 95% purity is burned in dry air in a sulphur burner. The gas leaving the burner containing 15% SO_2, 5% O_2 and 80% N_2 enters a converter after mixing with secondary air. The gas leaving the converter has the following analysis: SO_2, 4.5%; O_2, 7.5% and N_2, 88.0%. The converter gas is passed through an absorber so that 95% of the SO_3 in the gas is absorbed in 98% H_2SO_4 to produce 100% sulphuric acid. Determine the following:

(a) The volume of secondary air (in m^3) supplied per hour at 100 kPa and 300 K
(b) The mass of 98% H_2SO_4 (in kilograms) supplied per hour
(c) The mass of 100% acid (in kilograms) produced per hour

Solution *Basis:* 100 kmol SO_3-free converter gas

Figure 10.6 schematically represents the production of sulphuric acid in this example. The amount of oxygen in the air supplied is

$$88.0 \times \frac{21}{79} = 23.39 \text{ kmol}$$

The amount of oxygen accounted in the converter gas is 12.0 kmol.
The amount of oxygen consumed for the formation of SO_3 is

$$23.39 - 12.0 = 11.39 \text{ kmol}$$

Therefore, the amount of SO_3 formed is

$$\frac{11.39}{1.5} = 7.59 \text{ kmol}$$

The total amount of sulphur present in the gases leaving the converter is 7.59 + 4.5 = 12.09 kmol.

Or, 12.09 kmol sulphur on combustion produces 100 kmol SO_3-free converter gas.

Basis: 100 kmol SO_3-free burner gas

The amount of oxygen accounted for in the burner gas = 15.0 + 5.0 = 20.0 kmol. Since the burner gas contains 80 kmol nitrogen, the amount of oxygen in the air supplied is

$$80 \times \frac{21}{79} = 21.2658 \text{ kmol}$$

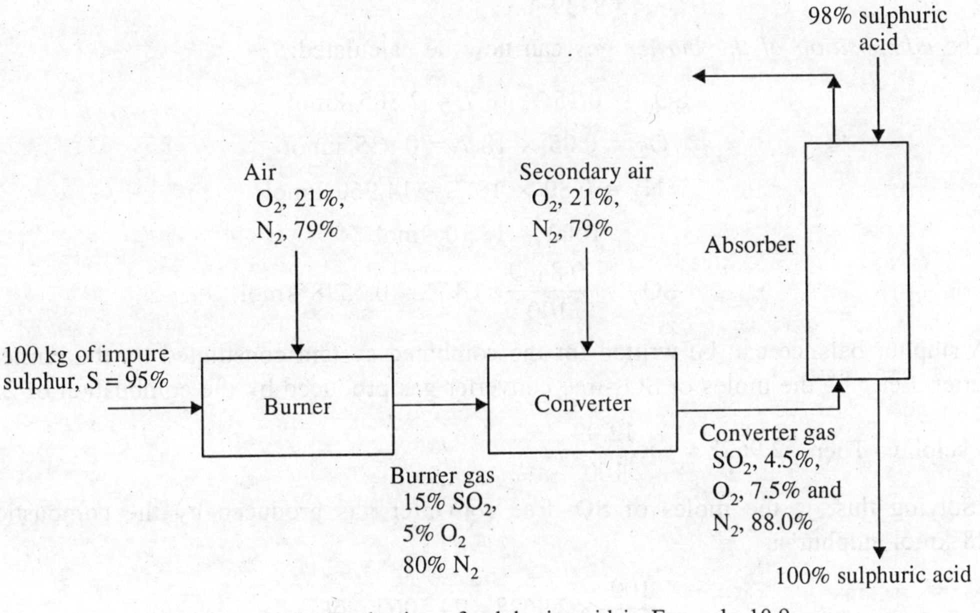

Figure 10.6 Production of sulphuric acid in Example 10.9.

Therefore, the amount of oxygen consumed forming SO_3 is

$$21.2658 - 20 = 1.2658 \text{ kmol}$$

The amount of SO_3 present in the gases leaving the burner is

$$1.2658 \times \frac{1}{1.5} = 0.8439 \text{ kmol}$$

Therefore, the total sulphur burned to produce 100 kmol SO_3-free burner gas is

$$15.0 + 0.8439 = 15.8439 \text{ kmol}$$

Basis: 100 kg impure sulphur charged

The amount of sulphur burned is

$$95 \text{ kg} = \frac{95}{32.064} = 2.9628 \text{ kmol}$$

Let x kmol SO_3-free burner gas be produced per 100 kg of sulphur charged. Since the total moles of S present in the gas leaving the burner is 15 + 0.8439 = 15.8439 kmol, a sulphur balance can be written for the burner, i.e.

$$2.9628 = x \times \frac{15.8439}{100}$$

Therefore, the moles of SO_3-free burner gas produced by the combustion of 2.9628 kmol sulphur is

$$x = \frac{100}{15.8439} \times 2.9628 = 18.70 \text{ kmol}$$

The *composition of the burner gas* can now be calculated:

$$SO_2 = 0.15 \times 18.7 = 2.805 \text{ kmol}$$

$$O_2 = 0.05 \times 18.7 = 0.935 \text{ kmol}$$

$$N_2 = 0.80 \times 18.7 = 14.960 \text{ kmol}$$

$$\text{Total} = 18.70 \text{ kmol}$$

$$SO_3 = \frac{0.8439}{100} \times 18.7 = 0.1578 \text{ kmol}$$

A sulphur balance can be written for the combined system constituted by the burner and converter. Let y be the moles of SO_3-free converter gas produced by the combustion of 2.9628 kmol sulphur. Then, $2.9628 = y \times \frac{12.09}{100}$.

Solving this, y, the moles of SO_3-free converter gas produced by the combustion of 2.9628 kmol sulphur is

$$\frac{100}{12.09} \times 2.9628 = 24.506 \text{ kmol}$$

The converter gas contains:

$$SO_2 = 0.045 \times 24.506 = 1.1028 \text{ kmol}$$

$$O_2 = 0.075 \times 24.506 = 1.838 \text{ kmol}$$

$$N_2 = 0.88 \times 24.506 = 21.5653 \text{ kmol}$$

$$\text{Total} = 24.506 \text{ kmol}$$

Therefore,

$$SO_3 = \frac{7.59}{100} \times 24.506 = 1.86 \text{ kmol}$$

Now, refer to Figure 10.7, which schematically shows the material balance.

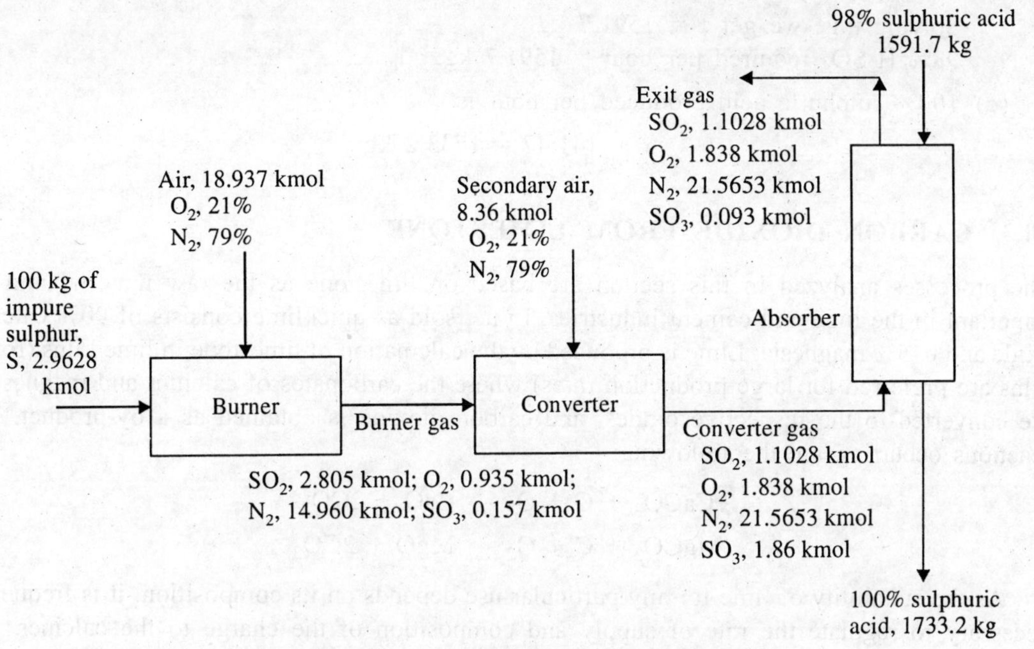

Figure 10.7 Material balance in Example 10.9.

(a) The converter gas contains 21.5653 kmol nitrogen and the burner gas contains 14.960 kmol nitrogen. The difference must be the amount of nitrogen coming from the secondary air supplied. Therefore, the amount of secondary air supplied is

$$(21.5653 - 14.960) \times \frac{100}{79} = 8.36 \text{ kmol}$$

At 100 kPa and 300 K, this corresponds to

$$8.36 \times 22.414 \times \frac{101.3}{100} \times \frac{300}{273.15} = 208.5 \text{ m}^3/\text{h}$$

(b) Let x kg 98% sulphuric acid is supplied for absorption.

$$SO_3 \text{ absorbed} = 0.95 \times 1.86 \times 80.064 = 141.47 \text{ kg}$$

Then amount of 100% H_2SO_4 produced is

$$(x + 141.47) \text{ kg}$$

Consider an SO_3 balance for the absorber.

SO_3 in the converter gas + SO_3 in the 98% sulphuric acid
= unabsorbed SO_3 in the exit gas + SO_3 in the 100% H_2SO_4

or

SO_3 absorbed + SO_3 in the 98% sulphuric acid = SO_3 in 100% H_2SO_4

$$141.47 + 0.98x \times \frac{80.064}{98.08} = (x + 141.47) \times \frac{80.064}{98.08}$$

Solving this, we get x = 1591.7 kg.
98% H_2SO_4 required per hour = 1591.7 kg.

(c) 100% sulphuric acid produced per hour is

$$x + 141.47 = 1733.2 \text{ kg}$$

10.3 CARBON DIOXIDE FROM LIMESTONE

The processes analyzed in this section are based on limestone as the raw material and are important in the lime and cement industries. Lime, sold as quicklime consists of 90% calcium oxide and 0–5% magnesia. Lime is produced by the calcination of limestone in lime kilns (rotary kilns are preferred for large production rates) where the carbonates of calcium and magnesium are converted to the respective oxides, and carbon dioxide is obtained as a by-product. The reactions occurring are the following:

$$CaCO_3 + C + O_2 \rightarrow CaO + 2CO_2$$

$$MgCO_3 + C + O_2 \rightarrow MgO + 2CO_2$$

As the suitability of lime for any particular use depends on its composition, it is frequently necessary to regulate the rate of supply and composition of the charge to the calciner. The material balance calculations are necessary for achieving this as well as for estimating the amount of CO_2 obtained as the by-product. Carbon dioxide is used in food industries as a refrigerant and in carbonated beverages. Also, it is an important fire-extinguisher and a raw material for many products.

The amount of lime produced per unit weight of fuel charged is known as the *fuel ratio*, which can be readily evaluated from the analysis of kiln gas provided the analysis of fuel and limestone are known. If the fuel ratio is specified, the composition of the product can be estimated by the material balance. The following examples illustrate the calculations generally encountered in the calcination of limestone.

EXAMPLE 10.10 Limestone mixed with coke is being burnt in a kiln. An average analysis of the limestone is $CaCO_3$ = 84.5%, $MgCO_3$ = 11.5% and the rest inerts. The coke contains 76.0% carbon, 21% ash and 3% moisture. The calcination of $CaCO_3$ is only 95% complete and that of $MgCO_3$ is 90%. The carbon in the coke is completely burnt to CO_2. The kiln is fed with one kg of coke per 5 kg of limestone. Calculate the weight percent of CaO in the product leaving the kiln. Assume that the moisture in the feed is completely vaporized.

Solution *Basis:* A charge consisting of 5 kg of limestone and 1 kg of coke
The limestone charged as feed consists of the following constituents:

$$CaCO_3 = 5 \times 0.845 = 4.225 \text{ kg} = \frac{4.225}{100.09} = 0.0422 \text{ kmol}$$

$$MgCO_3 = 5 \times 0.115 = 0.575 \text{ kg} = \frac{0.575}{84.312} = 6.82 \times 10^{-3} \text{ kmol}$$

$$\text{Inerts} = 5 \times 0.04 = 0.2 \text{ kg}$$

The coke charged contains the following constituents:

$$\text{Carbon} = 0.76 \text{ kg} = \frac{0.76}{12} = 0.0633 \text{ kmol}$$

$$\text{Water} = 0.03 \text{ kg} = \frac{0.03}{18.016} = 1.665 \times 10^{-3} \text{ kmol}$$

$$\text{Ash} = 0.21 \text{ kg}$$

The reactions occurring are as follows:

$$CaCO_3 + C + O_2 \rightarrow CaO + 2CO_2$$

$$MgCO_3 + C + O_2 \rightarrow MgO + 2CO_2$$

The product contains CaO, MgO, unconverted $CaCO_3$ and $MgCO_3$, inerts and ash

$$CaO = 0.95 \times 0.0422 = 0.0401 \text{ kmol} = 0.0401 \times 56.08 = 2.2482 \text{ kg}$$

$$MgO = 0.90 \times 6.82 \times 10^{-3} = 6.138 \times 10^{-3} \text{ kmol} = 6.138 \times 10^{-3} \times 40.312 = 0.2474 \text{ kg}$$

$$CaCO_3 = 0.05 \times 4.225 = 0.2113 \text{ kg}$$

$$MgCO_3 = 0.10 \times 0.575 = 0.0575 \text{ kg}$$

$$\text{Inerts} = 0.2 \text{ kg}$$

$$\text{Ash} = 0.21 \text{ kg}$$

The total weight of product is

$$2.2482 + 0.2474 + 0.2113 + 0.0575 + 0.2 + 0.21 = 3.1744 \text{ kg}$$

The weight percent of CaO in the product leaving the kiln is

$$\frac{2.2482}{3.1744} \times 100 = 70.82\%$$

EXAMPLE 10.11 Pure CO_2 may be prepared by treating limestone with aqueous H_2SO_4. The limestone used contained $CaCO_3$ and $MgCO_3$, the remainder being inert insoluble materials. The acid used contained 12% H_2SO_4 by weight. The residue from the process had the following composition—$CaSO_4$: 9.00%, $MgSO_4$: 5.00%, H_2SO_4: 1.2%, inerts: 0.50%, CO_2: 0.2% and H_2O: 84.10%. During the process, the mass was warmed, and CO_2 and H_2O vapour were removed. Calculate the following:

(a) The analysis of limestone used
(b) The percentage excess of acid used
(c) The mass of water vaporized and removed per 100 kg of limestone
(d) The mass of CO_2 obtained per 100 kg of limestone

Solution *Basis:* 100 kg of residue
Molecular weight of $CaSO_4$ = 136.144 and molecular weight of $MgSO_4$ = 120.376
The amount of $CaSO_4$ present in the residue is

$$9.00 \text{ kg} = \frac{9.00}{136.144} = 0.0661 \text{ kmol}$$

The amount of $MgSO_4$ present in the residue is

$$5.00 \text{ kg} = \frac{5.00}{120.376} = 0.0415 \text{ kmol}$$

(a)
$$CaCO_3 + H_2SO_4 \rightarrow CaSO_4 + H_2O + CO_2$$
$$MgCO_3 + H_2SO_4 \rightarrow MgSO_4 + H_2O + CO_2$$

The mass of $CaCO_3$ burned to produce 0.0661 kmol $CaSO_4$ is

$$0.0661 \text{ kmol} = 0.0661 \times 100.08 = 6.6153 \text{ kg}$$

The mass of $MgCO_3$ burned to produce 0.0415 kmol $MgSO_4$ is

$$0.0415 \text{ kmol} = 0.0415 \times 84.312 = 3.4989 \text{ kg}$$

Therefore, the limestone used has the following analysis:

Constituent	*Mass* (kg)	*Weight* %
$CaCO_3$	6.6153	$\frac{6.6153}{10.6142} \times 100 = 62.32$
$MgCO_3$	3.4989	$\frac{3.4989}{10.6142} \times 100 = 32.96$
Inerts	0.50	$(100 - 62.32 - 32.96) = 4.72$
Total	10.6142	

(b) The amount of H_2SO_4 required by $CaCO_3$ and $MgCO_3$ for complete reaction is

$$(0.0661 + 0.0415) = 0.1076 \text{ kmol} = 10.5534 \text{ kg}$$

The mass of excess H_2SO_4 present in the refuse is 1.2 kg.

$$\text{Percent excess acid supplied} = \frac{1.2}{10.5534} \times 100 = 11.37\%$$

(c) Total sulphuric acid supplied = 10.5534 + 1.2 = 11.7534 kg

Since the acid used is 12% sulphuric acid, the mass of water entering the process is

$$\frac{11.7534}{0.12} \times 0.88 = 86.1916 \text{ kg}$$

The amount of water produced in the reaction is

$$0.1076 \times 18.016 = 1.9385 \text{ kg}$$

$$\text{Total water present} = 86.1916 + 1.9385 = 88.1301 \text{ kg}$$

Since 100 kg of residue contains 84.10 kg of water, the mass of water vaporized is

$$88.1301 - 84.10 = 4.0301 \text{ kg}$$

This is the amount of water vaporized per 10.6142 kg of limestone.
The amount of water vaporized per 100 kg of limestone is

$$\frac{4.0301}{10.6142} \times 100 = 37.97 \text{ kg}$$

(d) The amount of CO_2 produced per 100 kg of residue is

$$0.1076 \text{ kmol} = 0.1076 \times 44 = 4.7344 \text{ kg}$$

The amount of CO_2 released per 100 kg of residue is

$$4.7344 - 0.20 = 4.5344 \text{ kg}$$

Therefore, the amount of CO_2 released per 100 kg of limestone is

$$\frac{4.5344}{10.6142} \times 100 = 42.72 \text{ kg}$$

10.4 REACTIONS INVOLVING PHOSPHORUS AND PHOSPHORUS COMPOUNDS

Phosphorus exists in nature as phosphate rock. Phosphate rock is mixed with coke and is reduced to elemental phosphorus in an electric furnace. The liquid yellow phosphorus (P_4) formed can be converted to red phosphorus by heating.

$$2Ca_3(PO_4)_2 + 10C + 6SiO_2 \rightarrow P_4 \text{ (yellow grade)} + 6CaSiO_3 + 10CO$$

$$P_4 \text{ (yellow grade)} \xrightarrow{523\text{–}723 \text{ K}} P_4 \text{ (red grade)}$$

In India almost all the phosphate rock processing is done for the production of phosphoric acid, which in turn is converted to calcium and ammonium phosphate for using as fertilizers.

$$Ca_3(PO_4)_2 + 3H_2SO_4 + 6H_2O \rightarrow 2H_3PO_4 + 3(CaSO_4{\cdot}2H_2O)$$

Pyrophosphoric acid ($H_4P_2O_7$) can be produced by heating the orthophosphoric acid (H_3PO_4) to 523 K. Hydrochloric acid leaching may also be used for the production of phosphoric acid.

$$Ca_3(PO_4)_2 + 6HCl \rightarrow 2H_3PO_4 + 3CaCl_2$$

Other phosphorus compounds of industrial importance are the different grades of phosphates of calcium, ammonium and sodium. Based on the method of production from phosphate rock, there are two grades for calcium phosphate; they are superphosphate and triple superphosphate, both used as fertilizers.

$$[Ca_3(PO_4)_2]_3{\cdot}CaF_2 + 7H_2SO_4(aq) \rightarrow 3CaH_4(PO_4)_2{\cdot}7CaSO_4 \text{ (superphosphate)} + 2HF$$

$$[Ca_3(PO_4)_2]_3{\cdot}CaF_2 + 14H_3PO_4 \rightarrow 10CaH_4(PO_4)_2 \text{ (triple superphosphate)} + 2HF$$

The basic reaction in the manufacture of ammonium phosphate is the neutralization of phosphoric acid with ammonia.

$$NH_3 + H_3PO_4 \rightarrow NH_4H_2PO_4 \text{ (monoammonium phosphate)}$$

$$NH_3 + NH_4H_2PO_4 \rightarrow (NH_4)_2HPO_4 \text{ (diammonium phosphate)}$$

Sodium phosphates are manufactured from pure phosphoric acid. The orthophosphates include monosodium phosphate (NaH_2PO_4), disodium phosphate (Na_2HPO_4) and trisodium phosphate (Na_3PO_4). These salts are used in water treatment, baking powder, fire proofing, detergent and cleaners.

The calculations in processes involving phosphorus and phosphorus compounds are illustrated with the help of Examples 10.12 and 10.13.

EXAMPLE 10.12 Trisodium phosphate (TSP) is an important water softening compound. The production of TSP is carried out in a two-step process.

$$Na_2CO_3\ (s) + H_3PO_4\ (aq) \rightarrow Na_2HPO_4\ (aq) + H_2O\ (l) + CO_2\ (g)$$

$$Na_2HPO_4\ (aq) + NaOH\ (s) \rightarrow Na_3PO_4\ (aq) + H_2O\ (l)$$

In the first step, a dilute phosphoric acid containing 1.25% H_3PO_4 is neutralized by reacting with a stoichiometric quantity of powdered dry soda ash to form a solution of disodium phosphate. In the second step, the neutralization is completed by the addition of sufficient solid caustic soda to produce a solution of TSP. The hydrated salt of trisodium phosphate ($Na_3PO_4{\cdot}12H_2O$) is produced from the solution by evaporation and subsequent crystallization. For a basis of 1000 kg of dilute phosphoric acid, calculate

(a) The maximum weight of TSP that can be produced assuming 100% yield in both the reactions

(b) The volume of CO_2 at 310 K and 100 kPa saturated with water vapour given that the vapour pressure of water at 310 K is 6.27 kPa

Solution *Basis:* 1000 kg of dilute phosphoric acid treated.
Molecular weight of phosphoric acid = 97.998.
The amount of H_3PO_4 present in the dilute acid is

$$1000 \times \frac{1.25}{100} = 12.5\ \text{kg} = \frac{12.5}{97.998} = 0.1276\ \text{kmol}$$

(a) Since one mole of phosphoric acid gives one mole of trisodium phosphate, moles of trisodium phosphate produced = 0.1276 kmol.

Molecular weight of TSP ($Na_3PO_4{\cdot}12H_2O$) = 380.166

Maximum weight of TSP obtained = 0.1276 × 380.166 = 48.51 kg

(b) Moles of CO_2 produced = 0.1276 kmol.
Since the CO_2 produced is saturated with water vapour and the vapour pressure of water at 310 K is 6.27 kPa, moles of water vapour present with 0.1276 kmol CO_2 is

$$0.1276 \times \frac{6.27}{(100 - 6.27)} = 8.5357 \times 10^{-3}\ \text{kmol}$$

Total moles of wet CO_2 produced = 0.1276 + 0.0085 = 0.1361 kmol

$$\text{Volume of } CO_2 = 0.1361 \times \frac{101.3}{100} \times \frac{310}{273.15} \times 22.414 = 3.51\ \text{m}^3$$

EXAMPLE 10.13 Sodium phosphate is produced by the reaction of soda ash with phosphoric acid:

$$H_3PO_4 + Na_2CO_3 \rightarrow Na_2HPO_4 + CO_2 + H_2O$$

$$Na_2HPO_4 + NaOH \rightarrow Na_3PO_4 + H_2O$$

The first reaction is carried out by treating an aqueous solution of soda ash with commercial phosphoric acid containing 85% acid. The disodium phosphate solution is then treated with 50% caustic lye to produce a 20% solution of trisodium phosphate in water. To produce 1000 kg of trisodium phosphate solution, calculate

(a) The quantity and composition of aqueous soda ash solution

(b) The weight ratio in which the soda ash solution and commercial phosphoric acid are to be mixed

Solution *Basis:* 1000 kg 20% solution of trisodium phosphate in water

$$Na_3PO_4 \text{ produced} = 1000 \times 0.2 = 200 \text{ kg}$$

$$= \frac{200}{163.974} = 1.2197 \text{ kmol}$$

(a) The amount of soda ash required for the reaction is

$$1.2197 \text{ kmol} = 1.2197 \times 106 = 129.29 \text{ kg}$$

The amount of phosphoric acid reacted is

$$1.2197 \text{ kmol} = 1.2197 \times 97.998 = 119.53 \text{ kg}$$

The amount of NaOH reacted is

$$1.2197 \text{ kmol} = 1.2197 \times 40.008 = 48.80 \text{ kg}$$

Phosphoric acid is available as 85% solution in water, NaOH is present as 50% solution in water and soda ash is supplied as an aqueous solution. Let x kg water be present in the soda ash solution. Then we can write the water balance for the process:

water present in the soda ash solution + water present in the aqueous NaOH solution + water present in the phosphoric acid solution + water produced = water present in the trisodium phosphate.

$$x + \frac{48.80}{0.5} \times 0.5 + \frac{119.53}{0.85} \times 0.15 + 1.2197 \times 2 \times 18.016 = 800$$

Solving the above equation, we get $x = 686.16$ kg.

Amount of aqueous soda ash solution = 129.29 + 686.16 = 815.45 kg

$$\text{Concentration of soda ash solution} = \frac{129.29}{815.45} \times 100 = 15.86\%$$

(b) Since the phosphoric acid reacted is 119.53 kg, the weight of commercial acid mixed is $\frac{119.53}{0.85} = 140.62 \text{ kg}$. The weight of soda ash solution mixed is 815.45 kg.

$$\frac{\text{weight of soda ash solution}}{\text{weight of commercial phosphoric acid}} = \frac{815.45}{140.62} = \frac{5.8}{1}$$

10.5 RECOVERY OF METALS AND NON-METALS FROM ORES

Chemical engineering techniques find wide applications in metallurgical industies manufacturing iron and steel, aluminium, copper, lead and zinc, etc. Iron and steel industry is the most prominent among the metallurgical industires. Hematite (Fe_2O_3), magnetite (Fe_3O_4), pyrites (FeS_2), and siderite ($FeCO_3$) are the common ores of iron. Hematite is the principal ore used for the producton of iron and steel.

The first step in the manufacture of steel is the production of pig iron in a blast furnace. The charge consisting of the ore, coke and the limestone *flux* is introduced into the blast furnace. Hot air blast is introduced at the bottom of the furnace. Air reacts with coke producing CO and heat, and iron oxide gets reduced to molten pig iron which is collected at the bottom of the furnace. Though many different reactions occur in various regions of the furnace, the essential process occurring in the furnace can be represented by the following chemical reaction:

$$Fe_2O_3 + 3C \rightarrow 2Fe + 3CO$$

At the highest temperature in the furnace, a part of the silica present is reduced to the metallic state and is dissolved in the molten iron while the remaining silica and alumina in the charge reacts with the metallic bases present in the flux and a fusible *slag* consisting of complex silicates and aluminates of calcium, magnesium and iron are formed. Thus, the products of a blast furnace operation are the pig iron, slag and blast furnace gas, the latter being a mixture of nitrogen, CO, and water vapour with small amounts of hydrogen and methane. The slag consists of the lime, magnesia, and alumina and alkalies originally present in the charge together with some silica and manganese oxides.

Example 10.14 illustrates a simple material balance calculation for a blast furnace operation in which the amount of flux required for producing a given quantity of pig iron is calculated knowing the composition of the ore, the flux and the coke. The methdos of calculations are applicable to similar reactions in the recovery of other metals.

EXAMPLE 10.14 For producing 1000 kg of pig iron of composition 95% Fe, 4% C and 1% Si, iron ore containing 85.0% Fe_2O_3 and the rest SiO_2 is reduced in a blast furnace using 1000 kg of coke containing 90% C and 10% SiO_2. The flux used contains 90% $CaCO_3$, 5% $MgCO_3$ and 5% SiO_2. If it is desired that the slag contains 40% (CaO + MgO), how much flux is required for producing 1000 kg of pig iron? Assume that no iron is present in the slag.

Solution *Basis:* 1000 kg of pig iron produced in the blast furnace

Figure 10.8 schematically shows the production of pig iron.

Let x be the weight of iron ore charged, y be the amount of flux added and z be the weight of slag produced, all expressed in kilograms.

Assuming that no iron is present in the slag, the iron balance gives

$$x \times 0.85 \times \frac{111.694}{159.694} = 950$$

Solving this, we get the weight of iron ore charged, x = 1597.95 kg.

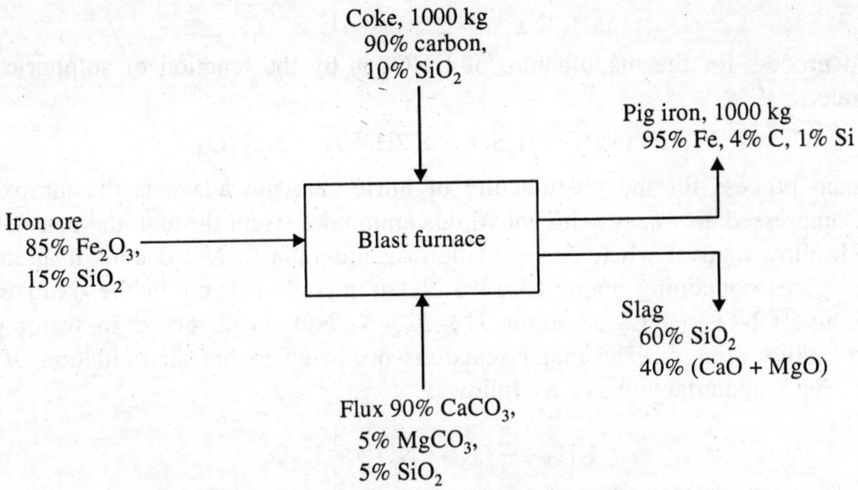

Figure 10.8 Production of pig iron in Example 10.14.

Silicon balance:

$$1597.95 \times \frac{15}{100} \times \frac{28.086}{60.086} + 1000 \times \frac{10}{100} \times \frac{28.086}{60.086} + y \times \frac{5}{100} \times \frac{28.086}{60.086}$$

$$= 10 + z \times \frac{60}{100} \times \frac{28.086}{60.086} \quad \text{(A)}$$

(CaO + MgO) balance:

$$y \times \left(\frac{90}{100} \times \frac{56.08}{100.08} + \frac{5}{100} \times \frac{40.305}{84.305} \right) = z \times \frac{40}{100}$$

which can be simplified as $z = 1.3205\, y$ (B)

Substituting (B) into (A) and simplifying, we get $y = 428.9$ kg.

Thus, the amount of flux required to produce 1000 kg of pig iron = 428.9 kg.

10.6 NITROGEN, AMMONIA AND NITRIC ACID

In this section we discuss through example problems the material balance calculations in processes such as the synthesis of ammonia, air oxidation of ammonia to form nitric oxide, the manufacture of urea from ammonia and carbon dioxide, the manufacture of nitric acid by the oxidation of ammonia and by the reaction of nitrates with sulphuric acid, etc.

Being the major constituent of air, nitrogen is the most abundant raw material which is commercially utilized for the production of ammonia. Ammonia serves as the source for nitrogen fertilizers in the form of liquid and aqueous ammonia solutions, ammonium salts and urea. It also serves as the source of nitric acid and many nitrogen compounds.

Ammonia is produced by the high-pressure catalytic reaction of nitrogen and hydrogen. Synthesis gas which is a mixture of nitrogen and hydrogen in the stoichiometric ratio (1:3) compressed to about 100–1000 bar is converted to ammonia in the presence of promoted iron catalyst at 773–823 K.

$$N_2 + 3H_2 \rightarrow 2NH_3$$

The old process for the manufacture of HNO_3 is by the reaction of sulphuric acid with sodium nitrate.

$$2NaNO_3 + H_2SO_4 \rightarrow 2HNO_3 + Na_2SO_4$$

The prominant process for the manufacture of nitric acid nowadays is the air oxidation of ammonia. Compressed air mixed with anhydrous ammonia is sent through the converter packed with a Pt-Rh alloy catalyst where the oxidation of ammonia to NO occurs at about 1073 K. The product gases containing about 10–12% NO is mixed with air in the oxidizer-absorber system to convert NO to NO_2 at about 313–323 K. NO_2 is absorbed in water producing HNO_3 of 57–60% strength. The major reactions occurring in the air oxidation of ammonia process for HNO_3 manufacture are as follows:

$$NH_3 + \frac{5}{4}O_2 \rightarrow NO + \frac{3}{2}H_2O$$

$$2NO + O_2 \rightarrow 2NO_2$$

$$3NO_2 + H_2O \rightarrow 2HNO_3 + NO$$

$$2NO_2 \rightarrow N_2O_4$$

Among the nitrogen fertilizers, the most important one is urea. In the ammonium carbamate decomposition process for the manufacture of urea, ammonia and carbon dioxide are compressed to 100–200 bar and reacted at 445–465 K. Urea is formed from the ammonium carbamate solution by dehydration.

$$CO_2 + 2NH_3 \rightarrow NH_4{\cdot}COO{\cdot}NH_2$$

$$NH_4{\cdot}COO{\cdot}NH_2 \rightarrow NH_2{\cdot}CO{\cdot}NH_2 + H_2O$$

EXAMPLE 10.15 Nitric oxide is produced by the air oxidation of ammonia:

$$4NH_3 + 5O_2 \rightarrow 4NO + 6H_2O$$

The following side-reaction also occurs:

$$4NH_3 + 3O_2 \rightarrow 2N_2 + 6H_2O$$

The product gases are scrubbed to remove all water and 80% of NO produced. The gases leaving the scrubber have the following analysis: O_2 = 5.6%, N_2 = 92.0%, NO = 2.4%. Determine the following:

(a) The percent excess air used
(b) The fraction of the ammonia fed that is taking part in the side reaction

Solution *Basis:* 100 mol gas leaving the scrubber

The composition of the gas:

$$NO = 2.40 \text{ mol}$$
$$N_2 = 92.00 \text{ mol}$$
$$O_2 = 5.60 \text{ mol}$$

The amount of NO present in the gas is 2.4 mol which is 20% of the total NO produced.

Therefore, the amount of NO produced by the reaction is $\frac{2.4}{0.2} = 12$ mol.

Let x mol nitrogen is produced by the reaction. Then, the amount of nitrogen present in the air supplied is

$$92 - x \text{ mol} \tag{A}$$

Since 4 moles of NO formed in the reaction accounts for 5 moles of oxygen and 2 moles of nitrogen accounts for 3 moles of oxygen, the total oxygen used up in the reaction is

$$12 \times \frac{5}{4} + x \times \frac{3}{2} = 15 + 1.5x \text{ mol}$$

Since 5.6 mol oxygen is present in the products, the total oxygen supplied is

$$15 + 1.5x + 5.6 = 20.6 + 1.5x$$

The amount of nitrogen associated with the oxygen supplied is

$$(20.6 + 1.5x) \times \frac{79}{21} = 77.495 + 5.643x \tag{B}$$

Comparing Eqs. (A) and (B), we have

$$77.495 + 5.643x = 92 - x$$

On solving this equation, we get

$$x = 2.1835$$

Or the amount of nitrogen produced by the reaction = 2.1835 mol.

(a) 12 mol NO in the products requires reaction of 12 mol ammonia and formation of 2.1835 mol nitrogen requires reaction with 2 × 2.1835 = 4.367 mol ammonia, the total ammonia reacted is

$$12 + 4.367 = 16.367 \text{ mol}$$

The theoretical requirement of oxygen is the oxygen required by ammonia for its complete conversion to NO according to the first reaction is

$$16.367 \times \frac{5}{4} = 20.4588 \text{ mol}$$

$$\text{Oxygen supplied} = 20.6 + 1.5x = 23.8753 \text{ mol}$$

$$\text{Percent excess oxygen} = \frac{23.8753 - 20.4588}{20.4588} \times 100 = 16.7\%$$

(b) Ammonia fed = 16.367 mol

The amount of ammonia converted to N_2

$$2 \times 2.1835 = 4.367 \text{ mol}$$

Therefore, the fraction of ammonia that is taking part in the side reaction is

$$\frac{4.367}{16.367} = 0.267$$

(*Note:* Solve this problem taking 1 mole NH_3 admitted as a basis. Let x be the number of moles of ammonia that take part in the side reaction.)

EXAMPLE 10.16 In the manufacture of nitric acid, pure dry sodium nitrate is treated with 95% (weight) sulphuric acid.

$$2NaNO_3 + H_2SO_4 \rightarrow 2HNO_3 + Na_2SO_4$$

Acid is supplied in excess to ensure that the resulting nitre cake is liquid containing 35% sulphuric acid by weight. The cake contains 1.5% water by weight. In addition 2% of the nitric acid formed will remain with the cake. For a basis of 100 kg of sodium nitrate treated, calculate the following:

(a) The weight and composition of the nitre cake
(b) The weight of sulphuric acid to be used
(c) The weight of nitric acid product obtained
(d) The water vapour that is distilled from the nitre cake

Solution *Basis:* 100 kg of sodium nitrate reacted

$$\text{Moles of } NaNO_3 \text{ reacted} = \frac{100}{85} = 1.1765 \text{ kmol}$$

$$\text{Mass of } H_2SO_4 \text{ consumed} = \frac{1.1765}{2} \times 98.08 = 57.6956 \text{ kg}$$

$$\text{Mass of } HNO_3 \text{ produced} = 1.1765 \text{ kmol} = 1.1765 \times 63.008 = 74.1289 \text{ kg}$$

$$\text{Mass of } Na_2SO_4 \text{ produced} = \frac{1.1765}{2} \times 142.064 = 83.569 \text{ kg}$$

(a) Nitric acid remaining with the cake = 0.02 × 74.1289 = 1.4826 kg
Nitre cake contains 83.569 kg of Na_2SO_4, 1.4826 kg of nitric acid which together will constitute 100 – 35.0 – 1.5 = 63.5% of the total weight of cake. Therefore, the total weight of cake is

$$\frac{(83.569 + 1.4826)}{0.635} = 133.94 \text{ kg}$$

The composition of the cake:

Na_2SO_4 =	83.569 kg	(62.39%)
H_2SO_4 =	46.879 kg	(35.00%)
HNO_3 =	1.483 kg	(1.11%)
Water =	2.009 kg	(1.50%)
Total =	133. 94 kg	

(b) The mass of H_2SO_4 consumed = 57.6956 kg
The mass of H_2SO_4 present in the cake = 46.879 kg
The total H_2SO_4 required = 104.5746 kg

The weight of 95% sulphuric acid is

$$\frac{104.5746}{0.95} = 110.08 \text{ kg}$$

(c) Since 2% of the nitric acid formed is lost with the cake, the mass of nitric acid product is

$$0.98 \times 74.1289 = 72.65 \text{ kg}$$

(d) The amount of water associated with sulphuric acid supplied = 110.08 × 0.05 = 5.504 kg. Out of this, 2.009 kg of water remains with the cake. Therefore, the water vapour that is distilled off from the cake is

$$5.504 - 2.009 = 3.495 \text{ kg}$$

Figure 10.9 schematically represents the material balance of this example.

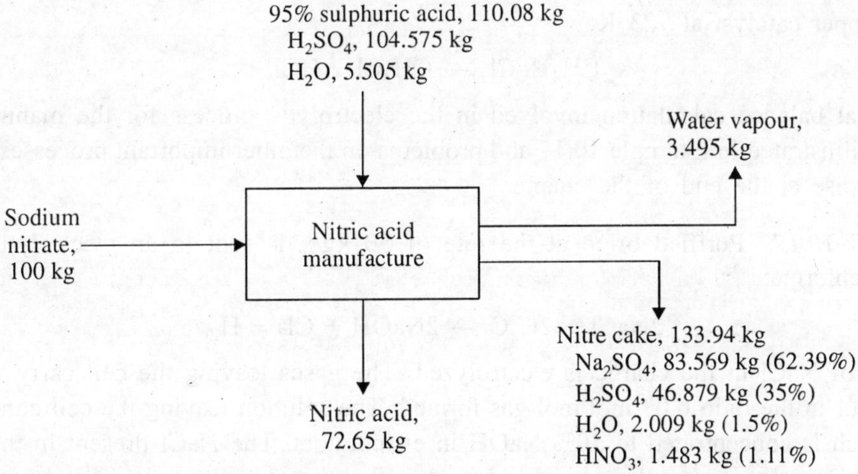

Figure 10.9 Material balance in nitric acid production of Example 10.16.

10.7 CHLORINE, HYDROCHLORIC ACID AND CHLORINATION

Electrolysis of brine results in the production of two industrially important chemicals: chlroine and caustic soda (NaOH).

$$2NaCl + 2H_2O \rightarrow 2NaOH + Cl_2 + H_2$$

Diaphragm electrolytic cell produces 10–12% NaOH solution whereas mercury electrolytic cell gives concentrated NaOH of strength 70%.

In the Deacon process for the manufacture of chlorine by the air oxidation of hydrochloric acid, a dry mixture of hydrochloric acid and air is passed over a heated Fe_2O_3 catalyst. Acid is oxidized in the presence of air, which is supplied in excess of that theoretically required:

$$4HCl + O_2 \rightarrow 2H_2O + 2Cl_2$$

Chlorine is industrially very important chemical. Its uses include production of chlorinated solvents, insecticides (DDT, BHC), plastics (PVC), refrigerants (methyl chloride), hydrochloric acid, etc. It is also used in pulp and paper bleaching, textile bleaching and water treatment.

Some of the important chlorination reactions are as follows:

(a) Oxychlorination of ethylene to ethylene dichloride:

$$2C_2H_4 + 4HCl + O_2 \rightarrow 2C_2H_4Cl_2 + 2H_2O$$

(b) Chlorination of glacial acetic acid to mono and dichloroethylene:

$$CH_3COOH + Cl_2 \rightarrow CH_2ClCOOH + HCl$$

$$CH_2ClCOOH + Cl_2 \rightarrow CHCl_2COOH + HCl$$

(c) Catalytic vapour phase chlorination of carbon disulphide to carbon tetrachloride:

$$CS_2 + 3Cl_2 \rightarrow CCl_4 + S_2Cl_2$$

(d) Manufacture of methyl chloride by the chlorination of methane in the presence of a copper catalyst at 723 K:

$$CH_4 + Cl_2 \rightarrow CH_3Cl + HCl$$

The material balance calculation involved in the electrolytic process for the manufacutre of chlorine is illustrated in Example 10.17 and problems on the other important processes are given in the exercise at the end of the chapter.

EXAMPLE 10.17 Purified brine at the rate of 50 kg/h is sent to an electrolytic cell for producing chlorine:

$$2NaCl + 2H_2O \rightarrow 2NaOH + Cl_2 + H_2$$

Only 50% of NaCl in the charge is electrolyzed. The gases leaving the cell carry with them water vapour in the ratio 0.03 mol/mol gas formed. The solution leaving the cell contains 10% NaOH which is concentrated to 50% NaOH in evaporators. The NaCl present in the solution fed to the evaporator is crystallized and removed so that the concentrate leaving the evaporators contain only 1.5% NaCl. Calculate the following:

(a) The rate of production of 50% NaOH in kg/h
(b) The rate of production of chlorine and hydrogen gas
(c) The rate at which NaCl is crystallized in kg/h
(d) The rate of evaporation of water in kg/h

Solution *Basis:* 50 kg/h of brine charged

Let the brine contain x kg of NaCl and the rest water. Therefore, the mass of water present $= 50 - x$ kg.

The amount of NaCl reacted is

$$0.5x \text{ kg} = \frac{0.5x}{58.45} \text{ kmol}$$

$$2NaCl + 2H_2O \rightarrow 2NaOH + Cl_2 + H_2$$

The mass of water reacted is

$$\frac{0.5x}{58.45} \times 18.016 = 0.1541x \text{ kg}$$

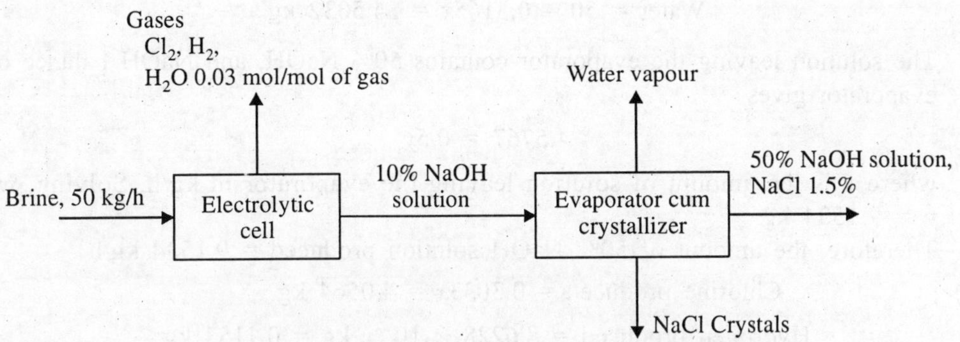

Figure 10.10 Electrolysis of brine in Example 10.17.

Gases produced:

$$\text{Chlorine} = \frac{0.5x}{58.45} \times \frac{1}{2} = 4.2772 \times 10^{-3}\, x \text{ kmol}$$

$$= 0.3033x \text{ kg}$$

$$\text{Hydrogen} = \frac{0.5x}{58.45} \times \frac{1}{2} = 4.2772 \times 10^{-3}\, x \text{ kmol}$$

$$= 8.6228 \times 10^{-3}x \text{ kg}$$

Mass of water vapour present in the gas

$$0.03 \times 2(4.2772 \times 10^{-3})x = 2.5663 \times 10^{-4}x \text{ kmol} = 4.6235 \times 10^{-3}x \text{ kg}$$

Solution leaving the cell:

$$\text{NaCl} = 0.5x \text{ kg}$$

$$\text{NaOH} = \frac{0.5x}{58.45} \times 40.008 = 0.3422x \text{ kg}$$

Water = water in the brine charge – water reacted – water present in the gas

$$= 50 - x - 0.1541x - 4.6235 \times 10^{-3}x$$

$$= 50 - 1.1587x \text{ kg}$$

$$\text{Total weight of solution} = 0.5x + 0.3422x + 50 - 1.1587x$$

$$= 50 - 0.3165x \text{ kg}$$

Since NaOH in the solution leaving the cell constitutes 10% of the total weight of solution, we have the following relationship:

$$0.3422x = 0.10(50 - 0.3165x)$$

The above equation gives, $x = 13.3743$ kg.

(a) The solution leaving the cell is taken to the evaporator. This solution contains

$$\text{NaOH} = 0.3422x = 4.5767 \text{ kg}$$

$$\text{NaCl} = 0.5x = 6.6872 \text{ kg}$$

$$\text{Water} = 50 - 0.3165x = 34.5032 \text{ kg}$$

The solution leaving the evaporator contains 50% NaOH, and NaOH balance on the evaporator gives

$$4.5767 = 0.5y$$

where y is the amount of solution leaving the evaporator in kg/h. Solving, we get $y = 9.1534$ kg.

Therefore, the amount of 50% NaOH solution produced = 9.1534 kg/h

(b) Chlorine produced = $0.3033x = 4.0564$ kg

Hydrogen produced = $8.6228 \times 10^{-3}x$ kg = 0.1153 kg

(c) NaCl in the solution leaving the evaporator is only 1.5%, the amount of NaCl present in the solution is

$$0.015 \times 9.1534 \text{ kg} = 0.1373 \text{ kg}$$

Since the amount of NaCl entering the evaporator is 6.6872 kg, the amount of NaCl crystallized is

$$6.6872 - 0.1373 = 6.5499 \text{ kg/h}$$

(d) The weight of water in the concentrated solution leaving the evaporator is

total weight of solution – weight of NaOH – weight of NaCl

$$= 9.1534 - 4.5767 - 0.1373 = 4.4394 \text{ kg}$$

The amount of water entering the evaporator is 34.5032 kg.

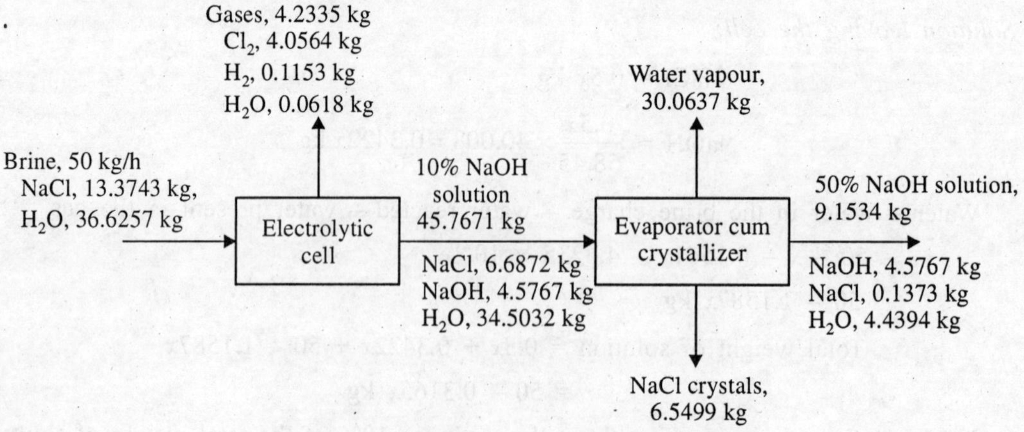

Figure 10.11 Material balance in Example 10.17.

Therefore, a water balance on the evaporator gives:

$$34.5032 = 4.4394 + \text{water evaporated and removed in the evaporator}$$

Therefore, the weight of water evaporated is

$$34.5032 - 4.4394 = 30.0637 \text{ kg}$$

Figure 10.11 shows the material balance in Example 10.17.

10.8 HYDROGENATION, HYDRATION AND OXIDATION

In addition to the applications discussed so far, the material balance will be required to find answers to such questions as:

- How much hydrogen is required for the hydrogenation of an unsaturated edible oil?
- What are the quantity and composition of the product gases that leave the reaction chamber where the vapour-phase oxidation of methanol to formaldehyde takes place in presence of air?
- At what rate are hydrocarbon fuels produced when coal is hydrogenated in a supply of air and steam? What should be the supply of air and steam for the reaction of a given quantity of coal?
- Given the analysis of the gas leaving the reaction chamber where ethylene is subjected to oxidation, what is the percent conversion of ethylene and what is the percent yield of ethylene oxide?

The material balance problems of the above type are encountered in processes employing the hydrogenation, hydration and oxidation reactions. Hydrogenation reactions involve the addition of hydrogen to a compound. Generally hydrogenation is used to improve existing products, to convert complex raw materials into valuable chemicals, and to transform solid fuels into liquid fuels. The commercial applications of hydrogenation include hydrogenation of oils, high-pressure hydrogenation of carbon monoxide to synthetic methanol and higher alcohols, hydrogenation of fatty acids to alcohols for the production of detergents, etc. The unit process of hydrolysis is a double decomposition reaction between a compound and water. The saponification of fats, the inversion of sugars, the conversion of starch, the saccharification of cellulose, the breakdown of protein are examples of hydrolysis. Oxidation is one of the most powerful tools used in the synthesis of chemical compounds.

The material balance calculation generally encountered in the various processes outlined in this section is illustrated for the process of oxidation of methanol to formaldehyde using air in the following example. Other processes can be seen in exercise problems 10.101 through 10.110.

EXAMPLE 10.18 Formaldehyde is made by the oxidation of methanol with air. The analysis of the exit gas from the reactor shows 64.49% N_2, 13.88% O_2, 5.31% H_2O, 11.02% CH_3OH, 4.08% HCHO and 1.22% HCOOH. Calculate the following:

(a) The percent conversion of formaldehyde

(b) The ratio of air to methanol in the feed

Solution *Basis:* 100 mol reactor exit gas

The following reactions take place in the reactor.

$$CH_3OH + O_2 \rightarrow HCOOH + H_2O$$

$$CH_3OH + \frac{1}{2}O_2 \rightarrow HCHO + H_2O$$

(a) Let x moles of methanol reacted to produce 100 mol gas. A carbon balance on the reactor gives

$$x = 11.02 + 4.08 + 1.22 = 16.32 \text{ mol}$$

The amount of methanol reacted = 16.32 mol.

The amount of methanol converted to formaldehyde = 4.08 mol.

$$\text{Percent conversion of formaldehyde} = \frac{4.08}{16.32} \times 100 = 25\%$$

(b) Air supplied can be determined by a nitrogen balance.

The amount of air supplied is

$$64.49 \times \frac{100}{79} = 81.63 \text{ mol}$$

The ratio of air to methanol in the feed is

$$\frac{81.63}{16.32} = 5.00$$

10.9 PROCESSES IN BIOLOGICAL, ENVIRONMENTAL AND ENERGY SYSTEMS

The principles of process calculations that we have applied to conventional processes can be readily extended to solve material and energy balance problems dealing with biotechnology, energy conservation, environmental pollution abatement, safety and health hazards. Though a detailed discussion is beyond the scope of this book, some illustrative problems with a brief discussion of related terms not covered so far are given here.

Stoichiometry of bioprocesses provides information on substrate conversion into product and cell mass production from substrate. The term *substrate* refers to medium such as glucose on which the biomass (bacteria, fungi, etc.) are expected to grow either on the surface or in solution. The following simple equation can be used to represent the process:

$$\text{Cells + medium} + O_2 \rightarrow \text{more cells + extracellular product} + CO_2 + H_2O \qquad (10.4)$$

As nitrogen sources, compounds such as ammonia, ammonium, urea, nitrate, secondary treatment wastes are used for biomass growth. Products may contain cellular as well as extracellular products. Cellular products are the biomasses that are produced in the reaction(s). Growth takes place throughout the biomass, not restricted to the surface. A product may occur jointly with biomass growth that probably will have to be separated from the biomass. Various metabolites such as acetates, citrates, glycerol, and so on as well as CO_2 and H_2O constitute the extracellular products.

The following equivalent reaction can also be written to represent the general chemical equation for bioprocesses:

$$C_wH_xO_yN_z + aO_2 + bH_gO_hN_i \rightarrow cCH_\alpha O_\beta N_\delta + dCO_2 + eH_2O + fC_jH_kO_lN_m \quad (10.5)$$

(Substrate) (Nitrogen source) (Dry biomass) (Extracellular product)

where a, b, α, β, δ, x, y, z depend on the type of cell involved. a, b, c, d, e and f are stoichiometric coefficients. In anaerobic processes, second term on the left side is not present. A cell contains elements other than C, H, O and N, but the minor components such as P, S, K, Ca, Mg, Cl and Fe require so little O_2 that they are often treated as ash, a non-reacting compound. The cells are generally represented by the molecular formula $CH_\alpha O_\beta N_\delta$. In the absence of sufficient information on the cell composition, an approximate cell composition of $CH_{1.8}O_{0.5}N_{0.2}$ is accepted. The stoichiometric equation is generally normalized to 1 mole of carbon source compound. The values of stoichiometric coefficients are required to predict the yield of product and biomass, the oxygen demand for the process, etc.

10.9.1 Yield and Yield Coefficients

Stoichiometric or *theoretical yield* is defined as the total mass or moles of product formed per unit mass or moles of reactant used to form that particular product. *Observed* or *apparent yield* is the mass or moles of product present per unit total mass or moles of reactant consumed. Yield coefficients are denoted by Y and several such coefficients are used. Thus,

$Y_{X/S}$ = mass or moles of biomass produced per unit mass or moles of substrate consumed
$Y_{P/S}$ = mass or moles of product formed per unit mass or moles of substrate consumed
$Y_{P/X}$ = mass or moles of product formed per unit mass or moles of biomass
$Y_{X/O}$ = mass or moles of biomass formed per unit mass or moles of oxygen consumed
$Y_{CO_2/S}$ = mass or moles of CO_2 formed per unit mass or moles of substrate consumed

10.9.2 Elemental Balance

Elemental balance helps in determining the stoichiometric coefficients. For illustration, consider a process represented by the following equation:

$$C_wH_xO_yN_z + aO_2 + bH_gO_hN_i \rightarrow cCH_\alpha O_\beta N_\delta + dCO_2 + eH_2O + fC_jH_kO_lN_m$$

(Substrate) (Nitrogen source) (Dry biomass) (Extracellular product)

C balance $$w = c + d + fj \quad (10.6)$$

H balance $$x + bg = c\alpha + 2e + fk \quad (10.7)$$

O balance $$y + 2a + bh = c\beta + 2d + e + fl \quad (10.8)$$

N balance $$z + bi = c\delta + fm \quad (10.9)$$

As the elemental balance generally fails to provide sufficient data, additional informations such as respiratory coefficient, degree of reduction, etc., are required to determine stoichiometric coefficients. The values of c and f ($f = 0$ when no extracellular products are formed) are sometimes obtained by measurements.

$$c = \frac{\text{moles of biomass produced}}{\text{moles of substrate consumed}}$$

$$f = \frac{\text{moles of extracellular product}}{\text{moles of substrate consumed}}$$

10.9.3 Respiratory Quotient

Respiratory coefficient (RQ) is defined as the number of moles of CO_2 formed from one mole of O_2 consumed.

$$RQ = \frac{d}{a} \tag{10.10}$$

Respiratory coefficient provides information on the metabolic state of the cell. A high RQ means that much CO_2 is produced, and hence, the metabolism operates at high efficiency. Equations (10.6) through (10.10) are solved for five unknowns *a, b, c, d* and *e*.

10.9.4 Degree of Reduction

Degree of reduction (γ) is defined as the number of equivalents of available electrons per gram atom of carbon. Available electrons are those that would be transferred to oxygen upon oxidation of a compound to CO_2, H_2O, NH_3. Degree of reduction provides another independent equation to find the stoichiometric coefficients. Degree of reduction (γ) for C = 4, H = 1, N = –3, O = –2, P = 5, S = 6.

For example, the degree of reduction of ethanol is evaluated as follows:

$$C_2H_5OH: [2(4) + 6(1) + 1(-2)]/2 = 12/2 = 6$$

In the absence of extracellular products formed during the reaction, we can write

$$w\gamma_S - 4a = c\gamma_B \tag{10.11}$$

where γ_B and γ_S are the degree of reduction of substrate and biomass, respectively.

$$a = \frac{1}{4}(w\gamma_S - c\gamma_B) \tag{10.12}$$

Equation (10.12) can also be written as

$$\frac{4a}{w\gamma_S} + \frac{c\gamma_B}{w\gamma_S} = 1 \tag{10.13}$$

$\frac{4a}{w\gamma_S}$ = number of electrons transferred to O_2

$\frac{c\gamma_B}{w\gamma_S}$ = number of electrons transferred to biomass

Maximum value of stoichiometric coefficient *c* (all electrons are used for biomass synthesis) is

$$C_{\text{max}} = \frac{w\gamma_S}{c\gamma_B}$$

When a product is also formed,

$$f_{\text{max}} = \frac{w\gamma_S}{j\gamma_P}$$

where γ_P is the degree of reduction of the product.

10.9.5 Oxygen Requirement

A number of empirical measures of oxygen usage in biosystems are employed to measure oxygen demand. The most important parameters used in biological processes and environmental systems are the chemical oxygen demand (COD), total oxygen demand (TOD), biochemical oxygen demand (BOD) and theoretical oxygen demand (ThOD). The definitions are given below.

Chemical oxygen demand (COD): Since there are many organic compounds that are difficult or impossible to decompose biologically, the parameter COD has been defined indicating the amount of oxygen required for oxidizing all organics completely. A chemical oxidant is added to the sample, the consumption of which is then determined. The internationally dominant method today is the 'dichromate' method, which is characterized by the acidification of the sample with sulphuric acid and the addition of silver nitrate. The COD measures all organic carbon with the exception of certain aromatics which are not completely oxidized in the reaction. Substances such as sulphides, sulphites and ferrous iron are also oxidized and reported as COD. Nitrogen in ammonia is not oxidized.

Total oxygen demand (TOD): It is the quantity of oxygen required to completely oxidize all the organic and inorganic compounds present in the sample and is based on the same principle as COD. To define TOD, the sample is thermally oxidized in an oven and the emerging oxygen is measured directly in the carrier gas. The amount of oxygen is reported as milligrams of oxygen per litre containing the sample, or as grams of oxygen per gram of sample.

Biochemical oxygen demand (BOD): The quantity of oxygen required by microorganisms to oxidize the organic compounds in the sample is determined by a BOD test. The test is carried out for five days at 20°C. The quantity of oxygen used is determined by the difference in the dissolved oxygen in the sample in the beginning of the test and at the end of the test. BOD is expressed as grams oxygen per gram sample. It is used mainly for evaluating water and waste water quality.

Theoretical oxygen demand (ThOD): It is the quantity of oxygen required according to a balanced chemical reaction equation to oxidize the reactants to CO_2, H_2O, and the highest oxidation state of other products, ignoring any extracellular product from a sample ($f = 0$). The highest state of nitrogen compounds is nitrates.

EXAMPLE 10.19 Production of single cell proteins from hexadecane is given by the following equation. If respiratory quotient (RQ) is 0.4, determine the stoichiometric coefficients.

$$C_{16}H_{34} + aO_2 + bNH_3 \rightarrow cCH_{1.66}O_{0.27}N_{0.2} \text{ (Biomass)} + dCO_2 + eH_2O$$

Solution

By definition, $$RQ = d/a = 0.40 \quad \text{(A)}$$

C balance: $$16 = c + d \quad \text{(B)}$$

H balance: $$34 + 3b = 1.66c + 2e \quad \text{(C)}$$

O balance: $$2a = 0.27c + 2d + e \quad \text{(D)}$$

N balance: $$b = 0.2c \quad \text{(E)}$$

On solving,

$$a = 11.73,\ b = 2.26,\ c = 11.31,\ d = 4.69,\ e = 11.03$$

EXAMPLE 10.20 Ammonia is used as the nitrogen source for the production of biomass *Klebsiella aerogenes* ($CH_{1.73}O_{0.43}N_{0.24}$) from substrate glycerol aerobically. 0.5 g biomass is produced for each gram of glycerol consumed. The biomass contains 6% ash. No extracellular products are formed. Determine the oxygen requirement for this process.

Solution Molecular weight (MW) of glycerol = 92.064

Molecular weight (MW) of biomass *Klebsiella aerogenes* $CH_{1.73}O_{0.43}N_{0.24}$ = 23.99

Molecular weight (MW) with ash = 23.99/0.94 = 25.52

Degree of reduction for substrate glycerol ($C_3H_8O_3$),

$$\gamma_S = \frac{3\times4+8\times1+3\times(-2)}{3} = 4.67$$

Degree of reduction for biomass $CH_{1.73}O_{0.43}N_{0.24}$,

$$\gamma_B = \frac{4\times1+1\times1.73-2\times0.43-3\times0.24}{1} = 4.15$$

Number of *C* atoms in glycerol = 3

Yield of biomass Y_{xs} = 0.5 g/g of substrate

The general equation is

$$\underset{\text{(Substrate)}}{C_wH_xO_yN_z} + aO_2 + \underset{\text{(Nitrogen source)}}{bH_gO_hN_i} \rightarrow \underset{\text{(Dry biomass)}}{cCH_\alpha O_\beta N_\delta} + dCO_2 + eH_2O$$

$Y_{X/S}$ = mass of biomass produced per unit mass of substrate consumed

$$Y_{X/S} = \frac{\text{Mass of cells}}{\text{Mass of substrate}} = \frac{c\times\text{MW of cells}}{\text{MW of substrate}} = 0.5$$

$$\text{Coefficient } c = Y_{X/S}\times\frac{\text{MW of substrate}}{\text{MW of cells}} = 0.5\times\frac{92.064}{25.52} = 1.80$$

Oxygen requirement is

$$a = \frac{w\times\gamma_s - c\times\gamma_B}{4} = \frac{3\times4.67 - 1.80\times4.15}{4} = 1.63 \text{ mol } O_2 \text{ per mol substrate}$$

On mass basis, oxygen demand = 1.63 × 32/92.064 = 0.567 g oxygen/g substrate.

EXAMPLE 10.21 Anaerobic digestion of volatile acids by methane bacteria is represented by the equation given below:

$$CH_3COOH + bNH_3 \rightarrow cCH_{1.4}O_{0.4}N_{0.2} \text{ (Biomass)} + dCO_2 + eH_2O + fCH_4$$

The composition of methane bacteria is $CH_{1.4}O_{0.4}N_{0.2}$. For each gram acetic acid consumed, 0.60 g CO_2 is evolved. How does the yield of methane under these conditions compare with the maximum possible yield?

Solution In anaerobic digestion, no oxygen participates in the reaction.

Molecular weight of biomass (methane bacteria) $CH_{1.4}O_{0.4}N_{0.2}$ = 22.6

Degree of reduction of biomass, $\gamma_B = 4\times1 + 1\times1.4 + (-2)\times0.4 + (-3)\times0.2 = 4$

Degree of reduction of substrate, CH_3COOH, $\gamma_S = \dfrac{4\times2 + 1\times4 + (-2)\times2}{2} = 4$

Degree of reduction of CH_4, $\gamma_P = 4\times1 + 1\times4 = 8$

0.60 gram CO_2 is produced from one gram acetic acid.

That is, $\dfrac{0.60/44}{1/60.032} = 0.82 \text{ mol } CO_2\text{/mol acetic acid}$

So, d = 0.82.

From elemental balance, the stoichiometric coefficients are determined. The general equation is

$$\underset{\text{(Substrate)}}{C_wH_xO_yN_z} + aO_2 + \underset{\text{(Nitrogen source)}}{bH_gO_hN_i} \rightarrow \underset{\text{(Dry biomass)}}{cCH_\alpha O_\beta N_\delta} + dCO_2 + eH_2O + \underset{\text{(Extracellular product)}}{fC_jH_kO_lN_m}$$

Here, the extracellular product is CH_4.

N balance:	$b = 0.2c$	(A)
C balance:	$2 = c + d + f$	(B)
O balance:	$2 = 0.4c + 2d + e$	(C)
H balance:	$4 + 3b = 1.4c + 2e + 4f$	(D)

On solving, we get f = 0.82.
So, product yield = 0.82 mol/mol substrate

$$f_{max} = \frac{w\gamma_s}{f\gamma_P} = \frac{2\times4}{1\times8} = 1$$

Comparing the values of f and f_{max}, $f = 0.82\, f_{max}$.

EXAMPLE 10.22 A bioreactor was charged with 5000 kg/h of an aqueous solution of fermented grain containing 15% by weight glucose ($C_6H_{12}O_6$). Yeast digests the glucose to form ethanol and acrylic acid (C_2H_3COOH).

$$C_6H_{12}O_6 \rightarrow 2C_2H_5OH + 2CO_2$$

$$C_6H_{12}O_6 \rightarrow 2C_2H_3COOH + 2H_2O$$

If 150 kg CO_2 is produced and 110 kg unreacted glucose remains in the broth, determine the percent composition by weight of the products in the broth.

Solution Basis: 5000 kg/h 15% glucose solution

Glucose charged = 5000 × 0.15/180.096 = 4.164 kmol

CO_2 produced = 150/44 = 3.41 kmol

Glucose reacted to produce CO_2 and ethanol = 3.41/2 = 1.705 kmol

Ethanol produced = 1.705 × 2 × 46.048 = 157.02 kg

Unreacted glucose = 110/180.096 = 0.61 kmol

Glucose reacted to produce acrylic acid = 4.164 – 1.705 – 0.61 = 1.849 kmol

Acrylic acid produced = 1.849 × 2 × 72.032 = 266.37 kg

Water remaining = 5000 × 0.85 + 1.849 × 2 × 18.016 = 4316.61 kg

The product composition is as follows:

Component	*Mass,* kg	*Weight %*
Ethanol	157.02	(157.02/4850) × 100 = 3.24
Acrylic acid	266.37	(266.37/4850) × 100 = 5.49
Glucose	110.00	(110.00/4850) × 100 = 2.27
Water	4316.61	(4316.61/4850) × 100 = 89.00
	4850.00	

EXAMPLE 10.23 In an experimental investigation on renewable energy resources, rice husks were subjected to pyrolysis. After pyrolysis, the gas analyzed 6.4% CO_2, 0.1% O_2, 39% CO, 51.8% H_2, 0.6% CH_4 and 2.1% N_2. It entered a combustion chamber at 300 K and a pressure of 120 kPa and was burned with 50% excess dry air at 295 K and 101.3 kPa. 10% of the CO remained unconverted. For each cubic metre of gas entering the combustion chamber, determine:

(a) Air supplied in cubic metres
(b) Composition of product gas
(c) Cubic metres of product gas at STP

Solution Basis: 100 kmol of pyrolysis gas entered the combustion chamber.

6.4 kmol CO_2, 0.1 kmol O_2, 39 kmol CO, 51.8 kmol H_2, 0.6 kmol CH_4, 2.1 kmol N_2

Reactions requiring O_2 are:

$$CO + 1/2O_2 \rightarrow CO_2$$

$$H_2 + 1/2O_2 \rightarrow H_2O$$

$$CH_4 + 2O_2 \rightarrow CO_2 + 2H_2O$$

(a) Oxygen required $= \frac{39}{2} + \frac{51.8}{2} + 2 \times 0.6 - 0.1 = 46.5$ kmol

Oxygen supplied = 50% excess = 46.5 × 1.5 = 69.75 kmol

Air supplied $= \frac{69.75}{0.21} = 332.14$ kmol

Volume of air entered at 295 K and 101.3 kPa,

$$V = \frac{nRT}{P} = \frac{332.14 \times 8.314 \times 295}{101.3} = 8041.7 \text{ m}^3$$

Volume of pyrolysis gas entered at 120 kPa and 300 K =

$$\frac{nRT}{P} = \frac{100 \times 8.314 \times 300}{120} = 2078.5 \text{ m}^3$$

Air supplied per cubic metre of entering gas $= \frac{8041.7}{2078.5} = 3.87 \text{ m}^3$

(b) The composition of product gas is as follows:

Component	kmol	*Mol %*
CO_2	6.4 + 35.1 + 0.6 = 42.1	10.8
CO	3.9	1.0
N_2	2.1 + 262.4 = 264.5	68.1
O_2	69.75 + 0.1 − ((35.1/2) + (51.8/2) + 1.2) = 25.2	6.5
H_2O	51.8 + 1.2 = 53	13.6
Total	388.7	100

(c) Total moles of product gases = 388.7 kmol

Volume of the product gas at STP, $V = 388.7 \times 22.414 = 8712.3 \text{ m}^3$

EXAMPLE 10.24 Fuel cell can be considered as the energy supplier for the future. A fuel cell is an open system into which fuel and air are fed and produces electricity and waste products. In a fuel cell using methane as fuel, CO_2 and water are the products along with electricity. Special membranes and catalysts are used to promote oxidation of methane with air. If air and methane are fed to the fuel cell in the molar ratio 10:1, determine the composition of the products.

Solution Basis: Methane and air admitted in the ratio 1:10.

$$CH_4 + 2O_2 \rightarrow CO_2 + 2H_2O$$

Moles of oxygen = 10 × 0.21 = 2.1
Moles of nitrogen = 7.9

Component	*Moles*	*Mole %*
CH_4	0	0
O_2	2.1 – 2 = 0.1	0.91
N_2	7.9	71.82
CO_2	1.0	9.09
H_2O	2.0	18.18
Total	11.0	100.0

EXAMPLE 10.25 Air oxidation and chlorination are two processes employed in waste water treatment for odour removal. But, chlorination using HOCl solution is preferred as it not only destroys H_2S and other odourous compounds but also retards the growth of bacteria. 0.1 cubic metres of waste water contains 63 ppm H_2S. If a 5% solution of HOCl in water is used, determine the quantity required. The reaction is

$$HOCl + H_2S \rightarrow S + HCl + H_2O$$

Solution Basis: 0.1 m^3 solution contains 63 ppm H_2S.

$$63 \text{ ppm} = 63 \text{ mg/L} = 63\times10^{-3}\text{ g} = \frac{63\times10^{-3}}{34.081}\times10^3 = 1.849 \text{ mol/m}^3$$

1 mol H_2S requires 1 mol HOCl.

HOCl required = $1.849\times0.1 = 0.1849$ mol

HOCl supplied = 2×0.1849 mol $= 0.3698$ mol

Weight of HOCl = $0.3698\times52.461 = 19.4$ g

Weight of 5% solution = 19.4/0.05 = 388 g (= 0.388 kg)

EXAMPLE 10.26 Activated sludge process is one of the processes for sewage treatment. After treatment, the sludge produced must be dewatered and disposed. Sludge incineration is one of the options for sludge disposal. In a typical treatment plant, the dried sludge with the composition C = 40%, S = 32%, H_2 = 4% and O_2 = 24% is mixed with fuel oil and burned in a furnace with air. The effluent gases leaving the furnace analyzed SO_2 = 1.52%, CO_2 = 10.14%, O_2 = 4.65%, N_2 = 81.67% and CO = 2.02%. Determine:

(a) The weight percent of carbon and hydrogen in the fuel oil
(b) Ratio of weights of sludge to fuel oil fed to the furnace

Solution Basis: 100 kmol product gas

N_2 comes from air only. N_2 = 81.67 kmol

O_2 supplied = 81.67 × 21/79 = 21.71 kmol

C in product gas = 10.14 + 2.02 = 12.16 kmol

S in product gas = 1.52 kmol

All S comes from sludge. Weight of S = 1.52 × 32.065 = 48.74 kg

48.74 kg = 32% of sludge.

Sludge burned = 48.74/0.32 = 152.31 kg

$C = 40\% = 152.31 \times 0.4 = 60.92$ kg

$H_2 = 4\% = 0.04 \times 152.31 = 6.092$ kg

$O_2 = 24\% = 0.24 \times 152.31 = 36.55$ kg

Total oxygen present = 36.55/32 + 21.27 = 22.41 kmol

Oxygen present in flue gas = 1.52 + 10.14 + 4.65 + 2.02/2 = 17.32 kmol

So, oxygen used by hydrogen = 22.41 – 17.32 = 5.04 kmol

H_2 from sludge = 6.092/2.016 = 3.022 kmol

H_2 from fuel oil = 5.09 – 3.022 = 2.068 kmol = 4.169 kg

Carbon in the product gas = 12.16 kmol = 145.92 kg

Carbon from sludge = 60.92 kg

So, carbon from fuel oil = 145.92 – 60.92 = 85 kg

(a) Weight percent of carbon and hydrogen in fuel oil
Fuel oil, C = 85 kg = 95.33%
H = 4.169 kg = 4.67 %

(b) Weight of dry sludge = 152.31 kg
Weight of oil = 89.169 kg
Weight ratio of sludge to oil = 152.31/89.169 = 1.71

EXERCISES

Combustion of solid, liquid and gaseous fuels

10.1 A synthesis gas analyzing 6.0% CO_2, 0.5% O_2, 40.0% CO, 50% H_2 and the rest N_2 is burned with 50% excess air. What is the composition of the flue gas?

10.2 A gas mixture consisting of 80% ethane and 20% oxygen is burned in an engine with 200% excess air. 80% of the ethane goes to CO_2, 10% to CO and 10% remains unburned. Calculate the composition of the exhaust gases on

(a) a wet basis
(b) a dry basis

10.3 Pure propane is burned with as much quantities of air as is required to limit the carbon dioxide content in the flue gas to 8%. How much excess air is to be supplied?

10.4 One kilomole methane is completely burnt to form CO_2 and water vapour using 10% excess air than that required for complete combustion. The resulting gas mixture is treated to remove all the water content.

(a) What will be the volume of dry gas leaving the burner at 110.3 kPa and 275 K?
(b) How many kilograms of water are removed from the products of combustion?

10.5 Pure methane is completely burned with air. The gas leaving the burner which contains no oxygen is passed through a cooler where some of the water is removed by condensation. The mole fraction of nitrogen in the gas leaving the cooler was 0.8335. Calculate the following:

(a) The analysis of the gas leaving the cooler
(b) Water condensed in kg per kmol of methane burned
(c) The partial pressure of water in the gas leaving the cooler at 325 K and 1.5 bar

10.6 A mixture of ethane (C_2H_6) and ethene (C_2H_4) occupies 40 L at 1 atm and 400 K. The mixture reacts completely with 130 g of O_2 to produce CO_2 and H_2O. Assuming ideal gas behaviour, calculate the mole fractions of C_2H_4 and C_2H_6 in the mixture.

10.7 Pure methane is completely burned with air. The outlet gases from the burner which contains no oxygen, are passed through a cooler, where some of the water is removed by condensation. The gases leaving the cooler have a nitrogen mole fraction of 0.75. Calculate the following:

(a) The analysis of the gases leaving the cooler
(b) The mass of water condensed per one kilogram of methane burned
(c) The average molecular weight of the gases leaving the cooler

10.8 Calculate the flue gas analysis on the dry basis for the following furnace operation. The fuel contains 85% carbon and the rest hydrogen on a weight basis. 40% excess air is used. It is found that all H_2 is converted to H_2O and 95% carbon to carbon dioxide and the rest to carbon monoxide.

10.9 Producer gas has the following composition by volume: carbon monoxide = 25.0%, carbon dioxide = 4.0%, oxygen = 3.0% and nitrogen = 68.0%.

(a) Determine the volume of the gas at 1 bar and 290 K per kg of carbon.
(b) 100 m^3 of the gas at 1 bar and 290 K is to be burned with 20% excess air than that is theoretically required for complete combustion. What volume of air is required at 1 bar and 290 K?
(c) For the conditions in part (b), what would be the composition of the gas after complete combustion?

10.10 A producer gas contains 28% CO, 3.5% CO_2, 0.5% O_2 and 68% N_2. 100 kg of this gas is burned with 20% excess air. If the combustion is only 90% complete, determine the following:

(a) The composition of the flue gas
(b) The weight of the gaseous products

10.11 A gas mixture analyzing 6% pentane, 10% butane, 15% propane, 10% ethane, 55% methane and 4% nitrogen by volume is burned at a rate of 100 m^3 per hour at 350 kPa and 300 K with 12% excess air. Air is supplied at 100 kPa and 300 K. It is found that the flue gas which leaves at 105 kPa and 800 K contains CO_2 and CO in the mole ratio 18:1. Determine the following:

(a) The rate of air supply (m^3 per hour)
(b) The Orsat analysis of flue gas
(c) The flue gas produced (m^3 per hour)

10.12 A fuel oil analyzing 85% (weight) C and 15% (weight) H is oxidized with 50% excess air. For the oxidation of 100 kg of oil, calculate the following:

(a) The air requirement in kilograms
(b) The composition of the flue gas
(c) The average molecular weight of the flue gas

10.13 A fuel gas on a dry basis contains 80% methane, 6% ethane, 4% propane, 2% oxygen and 8% nitrogen. The fuel saturated with water vapour at 300 K and 101.3 kPa and measuring 100 m^3 is burned with 50% excess air supplied at 305 K and 101.3 kPa with a relative saturation of 60%. The vapour pressure of water is given by the Antoine equation:

$$\ln P^S = 16.26205 - \frac{3799.887}{T - 46.854}$$

where P is in kPa and temperature is in K. Calculate the following:

(a) The analysis of the gas leaving the combustion chamber
(b) The volume of the gas (in m^3) leaving the furnace at 550 K and 101.3 kPa assuming complete combustion
(c) The dew point of the flue gas.

10.14 A liquid fuel analyzing 88% (weight) C and 12% (weight) H_2 when burned with excess air gives a flue gas of the following composition on a dry basis: CO_2 = 13.5%, O_2 = 3.5% and N_2 = 83.0%. Calculate the following:

(a) The percent excess air used
(b) The volume of dry flue gases at STP per 100 kg of the fuel burned

10.15 The Orsat analysis of the flue gas produced by the combustion of a hydrocarbon fuel shows 10.2% CO_2, 1.0%CO, 8.4% O_2 and 80.4% N_2. What is the atomic ratio of H to C in the fuel?

10.16 The Orsat analysis of the flue gas resulting from the combustion of a pure hydrocarbon fuel oil is found to be: CO_2 = 14.6%, CO = 2.0%, O_2 = 2.8%, N_2 = 80.6%. Calculate the composition of the fuel in weight percent.

10.17 A hydrocarbon fuel is burned with dry air. The combustion gases leave the furnace at 750 K and 100 kPa. The flue gas analysis shows CO_2 = 12.0%, CO =1.0%, O_2 = 4.0% and N_2 = 83%. What is the dew point of the stack gas? The vapour pressure of water is given by the Antoine equation:

$$\ln P^S = 16.26205 - \frac{3799.887}{T - 46.854}$$

where P is in kPa and temperature is in K.

10.18 Pure propane is burnt in an excess of air to give the following analysis of combustion products in volume percent: CO_2 = 5.0, CO = 3.5, H_2O = 11.4, O_2 = 7.0, N_2 = 73.1. Calculate the percent of excess air used.

10.19 The Orsat analysis of flue gas produced by the combustion of pure methane in excess of dry air is 8.17% CO_2, 0.96% CO, 5.04% O_2 and 85.83% N_2. Calculate the percent excess air used for combustion.

10.20 The Orsat analysis of a flue gas produced by the combustion of a pure hydrocarbon fuel with an excess of dry air is found to be 8.5% CO_2, 1.0% CO, 5.0% O_2 and 85.5% N_2. Calculate

(a) the percent excess air used for combustion and
(b) the weight ratio of C:H in the fuel.

10.21 A gas containing only CH_4 and N_2 is burned with air yielding a flue gas that has an Orsat analysis of CO_2 = 8.7%, CO = 1.0%, O_2 = 2.0% and N_2 = 88.3%. Calculate the following:

(a) The percent composition of the fuel
(b) The percent excess air used

10.22 The exhaust gases from a diesel engine using a high grade fuel oil show 9.76% CO_2, 8.18% O_2, and the rest N_2. Calculate the following:

(a) The H/C weight ratio in the fuel
(b) The amount of dry air supplied per kilogram of fuel burnt
(c) The total amount of exhaust gases in kilograms per kilogram of fuel burnt.

10.23 A solid fuel has the following analysis: H = 5.0%, S = 4.0%, C = 65%, O = 10% and inerts = 16%.

The fuel is burned with 20% excess air. If only 80% of the carbon burned gets converted to CO_2, 15% is converted to CO and 5% is left behind as soot, determine the composition of the combustion gases.

10.24 The flue gas produced by the combustion of pure carbon with excess air contains N_2 and O_2 in the ratio 7.18:1 by mole and CO_2 and CO in the ratio 2:1, what is the percent excess air used? No other compounds are present in the flue gas.

10.25 Coal containing 80% carbon and 6% ash by weight when burned leaves a cinder which contains 90% ash and 10% carbon. If 100 kg of coal is charged, calculate the following:

(a) The weight of cinder produced
(b) The per cent fuel value wasted

10.26 The combustible matter in a certain sample of coal is found to be 80% C and 20% H by weight. 50 kg of coal is burned with 1000 kg of air. The Orsat analysis of the combustion gas showed CO_2 to CO in the ratio 3:2. Determine (a) the percent excess air and (b) the total moles of flue gas produced.

10.27 When coal containing 74% C, 14.9% H and 11.1% ash is burned, it gives a flue gas containing 12.5% CO_2, 1.0% CO, 1.5% O_2 and 85% N_2 on a dry basis. Determine the following:

(a) The mass of coal fired in kilograms per kilomole of dry flue gas
(b) The percent excess air used
(c) The amount of air supplied in kilograms per kg of coal burned

10.28 A gaseous fuel made up of methane and ethane is burned with enriched air containing 50% oxygen and 50% nitrogen. The flue gas analysis on a dry basis is 25% CO_2, 15% O_2 and 60% N_2. Determine the following:

(a) The mole percent of methane in the fuel
(b) The moles of air used per mole of fuel

10.29 A fuel gas consisting entirely of methane and ethane is burned with air to yield a flue gas whose Orsat analysis is 8.68% CO_2, 6.44% O_2 and 84.88% N_2. Calculate the following:

(a) The analysis of the fuel in mole percent
(b) The percent excess air

10.30 A furnace is fired with natural gas consisting of 95% methane and 5% nitrogen. The flue gas analysis shows 9.1% CO_2, 0.2% CO, 4.6% O_2, 86.1% N_2. Calculate the percentage excess air supplied.

10.31 A pure hydrocarbon gas is burned with air and the Orsat analysis of the flue gas gave the following results: initial volume = 100 mL, volume after caustic absorption = 92.6 mL, volume after pyrogallol absorption = 87.1 mL, volume after cuprous chloride absorption = 83.4 mL. On the basis of this analysis, determine the following:

(a) The percent excess air
(b) The atomic ratio of carbon to hydrogen in the fuel
(c) The volume of dry flue gas (in m^3) at STP formed per kg fuel burned

10.32 A relatively pure saturated hydrocarbon gas is burnt with air in a small furnace. The Orsat analysis of the sample of flue gas gives the following data: Initial volume of the sample = 100 mL, volume after caustic absorption = 93.21 mL, volume after pyrogallol absorption = 85.74 mL, volume after cuprous chloride absorption = 83.48 mL. Determine the following:

(a) The percent excess air
(b) The atomic ratio H/C in the fuel and the molecular formula of the fuel
(c) The volume (in cubic metres) at STP of flue gas produced by the combustion of one kg of fuel

10.33 A furnace is supplied with a gaseous mixture of ethanol, acetic acid and nitrogen containing 15% nitrogen by volume at 400 K and 90 kPa. It is burned in excess of air and the flue gas contains 12.5% CO_2, 1.7% CO, 3.2% O_2 and 82.6% N_2. The flue gas leaves the furnace at 550 K and 100 kPa. Calculate the following:

(a) The percent excess air
(b) The percentage ethanol in the fuel by volume
(c) The moles of air per mole of fuel used
(d) The volume of flue gas in cubic metres per 100 m^3 of fuel burned.

10.34 A producer gas analyzing 2.0% CO_2, 30.0% CO, 2.0% O_2, 8.00% H_2 and 58.0% N_2 is burned with 20% excess air. Due to an air leak the flue gas gets diluted and the Orsat analysis (with water as the confining fluid) of the diluted flue gas is found to be 10.71% CO_2, 3.57% CO, 7.99% O_2 and 77.73% N_2. Calculate the following:

(a) The moles of air leaked into the flue gas per 100 kg fuel burned
(b) The analysis of the dry flue gas before dilution

10.35 The flue gas leaving two boilers, one using natural gas analyzing 96% methane and 4% CO_2 and the other using hydrocarbon oil containing C and H in the mole ratio 1:0.9 has the following analysis: CO_2 = 10.0%, O_2 = 4.5% and N_2 = 85.5%. What percent of the total carbon burned comes from the natural gas?

10.36 One hundred kg per hour of a pure hydrocarbon gas C_nH_{2n+2} at 295 K is burnt in a furnace. The flue gas produced analyzed 9.57% CO_2, 6.38% O_2 and the rest nitrogen. The flue gas leaves the furnace at 700 K and 100 kPa. Calculate the following:

(a) The atomic ratio H/C, and from this the formula of the fuel
(b) The volumetric ratio of flue gas to fuel gas
(c) The amount of flue gas produced in cubic metres per minute

10.37 A fuel gas is made up of 20% propane and 80% C_nH_{2n+2}, where n is a whole number. It is burned with an oxygen-enriched air stream containing 50% oxygen and 50% nitrogen. The flue gas analyzed 29.72% CO_2, 3.49% CO, 7.17% O_2 and 59.62% N_2. Determine the following:

(a) The percent excess air used
(b) The empirical formula of C_nH_{2n+2}
(c) The rate of air supplied in m^3/h at 400 K and 350 kPa if the fuel is burned at a rate of 1000 cubic metres per hour at 300 K and 110 kPa
(d) The moisture carried by the flue gas in kg/h.

10.38 The Orsat analysis of the flue gas from an oil-fired furnace is CO_2: 8%, CO: 3.0%, O_2: 4%, and N_2: 85%. An analysis indicates that the oil contains 78% by weight carbon, the remainder being combustible hydrogen and moisture. Air enters at 300 K and 1.013 bar. Assume air to be dry. Calculate the following:

(a) The percentage excess air used
(b) The carbon-hydrogen weight ratio of fuel oil
(c) The volume of air used per kg of oil fired
(d) The mass of moisture (kilograms) in the flue gas per kg of oil fired

10.39 A furnace is burning fuel oil having the following analysis on a weight basis: 85.0% C, 14.0% H and 1.0% S. Dry air is used for combustion and is supplied in 20% excess. 10 percent of the carbon burns to CO and S is converted completely to SO_2. It has come to light that there is an air leak at some point in the base of the stack. The air is at 300 K and 100 kPa with 70% relative saturation. The Orsat analysis of the stack gases above this point indicates concentration of CO_2 and SO_2 together as 8%. What percent of the stack gas is air leaked into it? The vapour pressure of water at 300 K is 3.56 kPa.

10.40 Coal having the following analysis on a dry basis and containing 3.9% moisture on a dry basis is burned with excess air.
C = 83.05%, H = 4.45%, O = 3.36%, N = 1.08%, S = 0.70% and ash = 7.36%.
Air used contains 0.0048 kg of water vapour per kg of dry air. The Orsat analysis of

the combustion gases gave the following result: $CO_2 + SO_2 = 15.4\%$, $CO = 0.0$, $O_2 = 4.0\%$ and $N_2 = 80.6\%$. The refuse analyzed 14.0% unburned ash-free coal and the rest ash. Check the consistency of the given data and calculate the percent excess air used.

10.41 Coal having the following analysis on a dry basis and containing 3.9% moisture is burned with excess air.

$C = 65.4\%$, $H = 5.3\%$, $O = 18.5\%$, $N = 1.1\%$, $S = 0.60\%$ and ash = 9.1%.

The Orsat analysis of the combustion gases gave the following result: $CO_2 + SO_2 = 13.0\%$, $CO = 0.76$, $O_2 = 6.17\%$, $H_2 = 0.87\%$ and $N_2 = 79.2\%$. Determine the following:

(a) The mass of coal fired per 100 kmol of dry flue gas analysed.
(b) The ratio of the moles of total combustion gases to the moles of dry air supplied
(c) The total moles of water vapour in the stack gas per 100 kg of coal if the air is dry
(d) The percent excess air

10.42 A furnace is fired with a coal analyzing 70% carbon, 5% hydrogen, 12% combined moisture, 5% free moisture and 8% ash. The relative saturation of air with water vapour is 60% at 295 K and 101.3 kPa. The combustion of coal is complete and the refuse contains no combustible matter. The flue gas leaves at 875 K and 100 kPa and has the following analysis on a dry basis: $CO_2 = 9\%$, $CO = 2\%$, $O_2 = 7\%$ and $N_2 = 82\%$. The vapour pressure of water at 295 K is 2.64 kPa. Calculate the following:

(a) The percent excess air
(b) The volume of flue gas (m^3) formed per kg of coal
(c) The volume of air (m^3) supplied per kg of coal.

10.43 Hydrogen-free coke containing 85% (weight) carbon and the rest inert materials is burned in a furnace. It is found that during combustion, 5% of the coke charged is lost unburned. The flue gas analysis shows 15.0% CO_2, 2.0% CO, 4.5% O_2 and 78.5% N_2. The flue gas leaves the furnace at 500 K and 100 kPa. Calculate the following:

(a) The percent excess air assuming complete combustion of coke.
(b) The weight of air supplied per kg of coke charged
(c) The volume of flue gas per kg of coke charged
(d) The composition of the refuse from the furnace

10.44 100 kg/h of coal containing 75.0% C, 4.4% H, 8.5% O, 1.2% S, 1.4% N and 9.5% ash is burnt with 20% excess air which contains 0.01 kg of water per kg of dry air. During combustion, 90% of the carbon is burnt to CO_2 and the rest to CO; and all sulphur is oxidized to SO_2. The refuse produced is dry and contains only ash. Calculate the following:

(a) Wet air supplied in kg/h
(b) Wet furnace gas in kmol/h
(c) Analysis of wet furnace gas

10.45 One hundred kilograms of coal having the analysis 70% C, 5% H, 1.0% N, 9.0% O, 2% moisture, and 13.0% ash is contacted with 75 kg of steam in a producer gas plant. Dry air is also supplied to the unit. The producer gas analyzed 13.11% CO_2, 16.38% CO, 4.37% CH_4, 18.08% H_2 and 48.06% N_2. The producer gas mixed with undecomposed steam leaves the gas producer. Tar formed in the unit is found to be 10% of the weight of coal charged and contained 10% (weight) hydrogen and the rest carbon. Refuse contains 8% C, and the rest ash. Calculate the following:

(a) The amount of dry air supplied, in kilomoles
(b) The percentage of steam fed to the gas producer that is undecomposed
(c) The volume of wet producer gas (in m^3) obtained at STP

10.46 Producer gas is obtained by burning coke in a restricted supply of air so that more CO is produced than CO_2. A producer gas having the CO:CO_2 mole ratio of 5:1 is to be prepared from the coke containing 75% carbon and 25% ash. The solid residue after combustion contains 5% unburned carbon. Calculate the following:

(a) The moles of gas produced per 1000 kg of coke burnt
(b) The moles of air supplied per 1000 kg of coke burnt
(c) The percentage of carbon lost in the ash

10.47 The ultimate analysis of coal is 78% carbon, 6% hydrogen, 7% oxygen, 3% sulphur, 2% nitrogen and 4% ash. The proximate analysis is 65% FC, 4% moisture, 27% VM and 4% ash. Calculate the masses of the combined water and net hydrogen in the coal.

10.48 During a test on a coal-fired steam generator, the following data were obtained: *Ultimate analysis of coal*: carbon 80.5%, hydrogen 4.6%, oxygen 5.0%, nitrogen 1.1%, sulphur 1.5% and ash 7.3%. *Data on the refuse from ash pit*: ash content 96% and carbon 4%. *Orsat analysis of flue gas*: CO_2 16.4%, O_2 2.3%, CO 0.4% and N_2 80.9%. Calculate the following:

(a) The weight of dry gaseous products formed per 100 kg of coal fired
(b) The percent excess air supplied for the combustion

10.49 Coal containing 72% carbon, 15% ash and the remainder being net hydrogen and moisture is completely burnt in dry air. The Orsat analysis of the flue gas is 9% CO_2, 4% CO, 8% O_2 and the rest N_2. Calculate the following:

(a) The net hydrogen content in the fuel
(b) The percent excess air used
(c) The volume of wet flue gas per kg of coal at STP

10.50 Coal with the following ultimate analysis is burned in boiler: carbon 66%, hydrogen 4%, nitrogen 1%, combined water 6%, free water 4% and ash 19%. The refuse drawn from the ash pit contains 12.6% carbon and the rest ash. The flue gas analysis gives CO_2 11.51%, O_2 7.64%, CO 0.09% and N_2 80.76%. The humidity of dry air is 0.012 kg/kg dry air. For 100 kg of coal fired, determine the following:

(a) The percent excess air supplied
(b) The mass of wet air supplied
(c) The weight of moisture in gaseous products
(d) The volume of total flue gases at STP

10.51 A furnace is fired with coal having the following proximate analysis: 2.9% moisture, 33.8% volatile matter, 53.1% fixed carbon and the rest ash. Only incomplete data are available for the ultimate analysis of coal. These are 1.1% sulphur and 73.8% carbon. The dry refuse from the furnace analyzed as 3.1% volatile matter, 18.0% fixed carbon, and 78.9% ash. The Orsat analysis of the flue gas is CO_2 = 12.1%, CO = 0.2%, O_2 = 7.2% and N_2 = 80.5%. Air enters the furnace at 293 K with 60% humidity at 101.3 kPa. For 100 kg of coal charged, determine the following:

(a) The total carbon content in the refuse
(b) The weight of dry flue gases
(c) The volume of wet air admitted
(d) The available hydrogen in the coal
(e) The complete ultimate analysis of coal

10.52 A furnace is fired with coal of the following analysis: carbon 60%, hydrogen 14%, oxygen 16%, ash 10% and negligible nitrogen and sulphur. The proximate analysis is 40% FC, 41% VM, 9% moisture and 10% ash. The refuse has the following analysis: 40% FC, 60% ash. It is dry. 1200 m^3 of air at 100 kPa, 295 K with a partial pressure of water of 1350 Pa is supplied per 100 kg of coal. The total pressure of the wet flue gas is 100 kPa. The ratio of mol CO_2 to mol CO in the flue gas is 4:1. Determine the following:

(a) The percent excess air
(b) Orsat analysis of flue gas
(c) Partial pressure of water vapour in the hot wet flue gas

Oxidation of sulphur compounds and related processes

10.53 The analysis of a gas entering the secondary converter in a contact sulphuric acid plant is 5% SO_2, 15% O_2 and 80% N_2. The gas leaving the converter contains 1.06% SO_2 on an SO_3-free basis. Calculate the percentage of SO_2 converted.

10.54 Pure sulphur is burned with excess air to give sulphur trioxide. Due to incomplete oxidation some sulphur dioxide is also formed. The analysis of the burner gases showed 8.2% SO_3, 0.91% SO_2, 8.65% O_2 and 82.24% N_2. Determine the following:

(a) The percent excess air supplied
(b) The volume of burner gases at 1.2 bar and 600 K per kg of sulphur burned
(c) The volume of air supplied at 1.013 bar and 288 K per kg of sulphur burned.

10.55 The gases leaving a sulphur burner has the following analysis: 9.8% SO_2, 8.5% O_2 and 81.7% N_2. The gas is sent to a converter where SO_2 is converted to SO_3. The gas leaving the converter is found to contain 0.5% SO_2, 4.44% O_2 and the rest nitrogen. What percent of the SO_2 entering the converter is oxidized to SO_3?

10.56 In the manufacture of sulphuric acid from elemental sulphur, the sulphur is fed to the burner where it is burned with 100% excess air to produce sulphur dioxide. The conversion of sulphur to sulphur dioxide is only 90%. The hot combustion gases from the burner are taken to a converter where SO_2 is converted to SO_3. This conversion is 95% complete. Calculate the following:

(a) The amount of air supplied to the burner per 100 kg of sulphur burned
(b) The composition of the burner gas
(c) The composition of the converter gas

10.57 Iron pyrites (FeS_2) is burnt with air 100% in excess of that required to oxidize all iron to Fe_2O_3 and all sulphur to sulphur dioxide. Calculate the composition of the exit gases, if 80% of sulphur is oxidized to sulphur dioxide and the rest to sulphur trioxide. All iron is oxidized to Fe_2O_3.

10.58 The analysis of a gas entering the secondary converter of a contact sulphuric acid plant is 4% SO_2, 13% O_2 and 83% N_2. In the converter, SO_2 is oxidized to SO_3. The gases leaving the converter contains 0.45% SO_2 on an SO_3-free basis. Calculate the percent conversion of SO_2.

10.59 The Orsat analysis of a gas leaving a pyrites burner shows 10% SO_2 and 5% O_2 and the rest N_2. What percent of the sulphur burned is oxidized to SO_3?

10.60 In the combustion of iron pyrites containing 85% FeS_2 and 15% gangue, some FeS_2 is lost in the cinder unburned. If the cinder carries 1.5% sulphur as FeS_2, how many kilograms of FeS_2 are lost in the cinder per 100 kg of pyrites charged?

10.61 In the combustion of iron pyrites containing 85% FeS_2 and 15% gangue, a portion of the SO_3 formed during the combustion gets adsorbed in the cinder. If the cinder carries 1.5% sulphur as SO_3, what mass of SO_3 is adsorbed by the cinder per 100 kg of pyrites charged?

10.62 The combustion of iron pyrites containing 90% FeS_2 and 10% gangue is found to produce a burner gas having the Orsat analysis of 10% SO_2 and 5% O_2 and the rest nitrogen. Determine the volume of gas (in m^3) leaving the burner at 101.3 kPa and 300 K per 100 kg of pyrites charged.

10.63 A mixture of pyrites and zinc sulphide ore is burned in a burner. The mixture contains 75% pyrites and 25% zinc sulphide ore. The pyrites yield 90% FeS_2 and the rest gangue. The zinc sulphide ore contains 70% ZnS and the rest inerts. A sample of cinder yields 4.0% S. 70% of the sulphur in the cinder is in the form of SO_3 absorbed in it, and the rest is unoxidized FeS_2. Based on 100 kg of mixed charge, calculate the following:

(a) The amount of cinder formed and its analysis
(b) The percentage of the sulphur left in the cinder based on the total sulphur charged

10.64 In the manufacture of sulphuric acid by the chamber process, iron pyrites having the composition 75% FeS_2 and 25% gangue is burned at 773 K using air. The reaction is

$$4FeS_2 + 11O_2 \rightarrow 2Fe_2O_3 + 8SO_2$$

The gases leaving the burner contains 9.3% SO_2, 8.95% O_2 and 81.75% N_2. For 100 kg of pyrites charged, calculate the following:

(a) The volume of air (in m^3) supplied at 300 K and 100 kPa
(b) The percent excess air used
(c) The volume of gas (in m^3) leaving the burner at 773 K and 100 kPa

10.65 In a contact sulphuric acid plant, pyrites is burned in dry air to produce a gas analyzing 10% SO_2 and 7% O_2. The cinder carries 2% by weight sulphur as SO_3. The burner gas at a rate of 2000 m^3/h at 365 K and 100 kPa is fed to an absorber to absorb the entire SO_3 present in it. 70% H_2SO_4 is used for the absorption at a rate of 25 kg/h and produces 90% H_2SO_4. Determine the percentage of FeS_2 in the pyrites charged.

10.66 One hundred kilograms of pyrites containing 32% S is mixed with 10 kg of pure sulphur and is burned in excess air. A part of sulphur is converted to SO_3 and the remaining part to SO_2. But the sulphur trioxide in the combustion gases will not be detected in the Orsat analysis. The Orsat showed 13.5% SO_2, 3.0% O_2 and 83.5% N_2. Assuming that the entire SO_3 in the combustion gases comes from the oxidation of pure sulphur, what is the fraction of the sulphur charged that is converted to SO_3?

10.67 Impure FeS_2 is burned to give a burner gas containing 8.0% SO_2, 10.2% O_2 and 81.8% N_2.

$$FeS_2 + \frac{15}{4}O_2 \rightarrow \frac{1}{2}Fe_2O_3 + 2SO_3$$

$$FeS_2 + \frac{11}{4}O_2 \rightarrow \frac{1}{2}Fe_2O_3 + 2SO_2$$

$$Fe_2O_3 + 3SO_3 \rightarrow Fe_2(SO_4)_3$$

The pyrites contains 48% (weight) sulphur and the cinder contains 2% (weight) sulphur as $Fe_2(SO_4)_3$ and no FeS_2. For a basis of 100 kg of pyrites burned, calculate the following:

(a) The weight of cinder
(b) The moles of burner gas constituents
(c) The percentage excess air based on conversion to SO_3
(d) The percentage conversion of S to SO_3.

10.68 A mixture of pure sulphur and pyrites analyzing 85% FeS_2 and 15% gangue is burnt in a standard pyrites burner. The burner gas contains 10% SO_2, 7% O_2, and 83% N_2 on a SO_3-free basis and contains 1 mole of SO_3 per 100 moles of SO_3-free burner gas. The cinder contains 2% S as SO_3. Calculate the percentage of FeS_2 in the charge.

10.69 In the contact process for the manufacture of H_2SO_4, SO_3 is obtained by burning iron pyrites (FeS_2) in air. The air supplied is 40% in excess of that required for the complete conversion of all S to SO_3. However, in the burner, only 40% of the sulphur is converted to sulphur trioxide and the remainder is converted to SO_2. Iron is oxidized to Fe_2O_3. Of the pyrites charged, 15% is lost by falling through a grate. The burner gas is taken to a converter, where 95% of SO_2 present is converted to SO_3. For a basis of 100 kg pyrites charged, calculate the following:

(a) The weight of air supplied, kg
(b) The burner gas analysis, wt%
(c) The total weight of SO_3 produced in kilograms
(d) The converter gas analysis, vol%
(e) The weight of cinder in kilograms

10.70 In the contact process for the manufacture of sulphuric acid, iron pyrites (FeS_2) is burned with air so that iron is oxidized as Fe_2O_3 which leaves with the cinder and S is oxidized to its oxides. It is found that in the burner 40% of sulphur is converted to SO_3 and the rest to SO_2. The latter is further oxidized to SO_3 in the catalytic converter. This conversion may be assumed to be 96% complete. About 10% of the pyrites charged is lost unburned and leaves with the cinder. The air supplied to the burner is 40% in excess of that required for the conversion of all S to SO_3.

(a) How many kilograms of air are supplied per kg of pyrites charged?
(b) What is the percent composition on a weight basis of the gases leaving the burner?
(c) What mass of SO_3 is obtained per 100 kg of pyrites charge?
(d) What is the degree of conversion of S in the charge to SO_3?

10.71 Dry pyrites containing 48% sulphur is burned in a sulphuric acid plant. The cinder carries 2% sulphur as SO_3. The burner gas analysis shows 8.25% SO_2, 7.85% O_2 and the rest N_2. The burner gas is passed through a converter where SO_2 is oxidized to SO_3. The converter gas is admitted into an absorber to absorb the SO_3 in 70% H_2SO_4. The gas leaving the absorber is free of SO_3 and contains 1.09% SO_2, 4.73% O_2 and 94.18% N_2, and the acid produced is 100% H_2SO_4. No secondary air supply is used in the converter. For 100 kg of pyrites charged, determine the following:

(a) Moles of SO_2 in the burner gas
(b) Mole percent of SO_3 in the converter gas
(c) The amount of 100% acid produced in the absorber

10.72 Sulphuric acid can be produced as a by-product in the manufacture of zinc from sulphide ores. The ore analyzes 65% ZnS and 35% inert impurities by weight. The ore is burned in a furnace and the resulting SO_2 is converted to SO_3 in a catalytic reactor. The SO_3 is then absorbed by water to give sulphuric acid, the final product being 2% water and 98% H_2SO_4 by weight. A total of 99% of sulphur in the ore is recovered in the acid. The chemical reactions are:

$$ZnS + 1.5O_2 \rightarrow ZnO + SO_2$$
$$SO_2 + 0.5O_2 \rightarrow SO_3$$
$$SO_3 + H_2O \rightarrow H_2SO_4$$

Compute

(a) The quantity of the product sulphuric acid produced in a zinc plant that processes 2×10^5 kg/day of ore
(b) The quantity of water required per day

10.73 In a sulphuric acid plant, 100 kg/h of dry pyrites containing 90% FeS_2 and 10% inert material is burned with dry air in the burner. Only 85% of FeS_2 is oxidized. In the burner, 90% of the sulphur burnt is oxidized to SO_2 and the balance to SO_3. Dry air supplied is 50% in excess of that required for the complete conversion of S to SO_3. Burner gas is fed to a converter where 95% of the SO_2 is converted to SO_3 using the oxygen present in the burner gas. All the SO_3 produced is absorbed in 70% H_2SO_4 fed at a rate of 100 kg/h in an absorption tower. No SO_3 is adsorbed on the cinder. Determine the following:

(a) The percentage of sulphur in cinder
(b) The amount of sulphuric acid produced in kg/h
(c) The concentration of the acid product
(d) The amount of gas leaving the absorber in kg/h
(e) The composition of the gas leaving the absorber

Carbon dioxide from limestone

10.74 A limestone analyses 92.9% $CaCO_3$, 5.4% $MgCO_3$ and 1.7% insolubles.

(a) How many kilograms of CaO can be made from 1000 kg of this limestone?
(b) How many kilograms of CO_2 can be recovered per kg of limestone?
(c) How many kilograms limestone are needed to make 1000 kg of lime?

10.75 Carbon dioxide is produced by treating limestone with sulphuric acid. The analysis of the residue leaving the process showed 9.0% $CaSO_4$, 5.5% $MgSO_4$, 1.0% H_2SO_4, 0.5% inerts and 0.1% CO_2 and 83.9% water. The limestone charged contains calcium carbonate, magnesium carbonate and inert insoluble matter. The CO_2 and water vapour produced are removed. Determine the following:

(a) The percent composition of the limestone used
(b) The percent excess acid used
(c) The quantity of CO_2 recovered per 100 kg of limestone charged

10.76 The calcinations of limestone can be represented by the following reactions:

$$CaCO_3 \rightarrow CaO + CO_2$$

$$MgCO_3 \rightarrow MgO + CO_2$$

In a certain operation 100 kg of limestone containing only calcium and magnesium carbonates and no inert materials is calcined producing 30 m^3 of CO_2 at 348.32 K and atmospheric pressure. What is the composition of the limestone in weight percent?

10.77 Carbon dioxide is produced by treating limestone ore with 15.0% sulphuric acid. 1000 kg of ore consisting of $CaCO_3$, $MgCO_3$ and inert materials is mixed with an excess of acid:

$$CaCO_3 + H_2SO_4 \rightarrow CaSO_4 + CO_2 + H_2O$$

$$MgCO_3 + H_2SO_4 \rightarrow MgSO_4 + CO_2 + H_2O$$

The liquid product containing 6.37% $MgSO_4$, 1.25% H_2SO_4 and the rest water is treated with pure $CaCO_3$ to neutralize the free acid. The solid residue consisting of $CaSO_4$ and inerts is separated from the liquid by filtration. If the filter-cake analyzed 1 kg of inert per 20 kg of $CaSO_4$, determine the following:

(a) The composition of the limestone used
(b) The moles of CO_2 produced
(c) The weight of $CaCO_3$ required to neutralize the excess acid
(d) The percent excess of acid used

10.78 A lime furnace burns 100 kg of coal containing 80% C, 5% H, 2% N, 1.0% S and 12% ash and limestone containing 82.0% $CaCO_3$, 7.0% $MgCO_3$ and 11.0% inert material with excess air. The average stack gas analysis is 14.42% CO_2, 6.03% O_2, and 79.55% N_2. CO_2 percentage includes SO_2 also. Calculate the following:

(a) The percent excess air used for combustion
(b) The fuel ratio on a weight basis
(c) The amount of limestone supplied, in kilograms
(d) The amount of total lime produced, in kilograms

Reactions involving phosphorus and phosphorus compounds

10.79 In the manufacture of phosphorus, calcium phosphate is mixed with sand and charcoal, and heated in an electric furnace. The silica used is 15% in excess of that theoretically required to convert all the phosphorus to silicate. Charcoal used is 50% in excess of that required to combine with oxygen in the phosphorus pentoxide to form carbon monoxide. It is observed that the conversion to silicate is 85% complete and reduction of the oxide to phosphorus is only 70% complete. Determine the following:

(a) The ratio in which phosphate, sand and charcoal are mixed before they are sent to the furnace
(b) The quantity of phosphorus produced per 100 kg of the above mixture

10.80 The off-gases from the phosphate reduction furnace analyzes 8.0% P_4, 85.0% CO, and 7.0% N_2 on a mole basis. This gas is burned with air under such conditions that the phosphorus is selectively oxidized. The oxides of phosphorus (P_4O_6 and P_4O_{10}) present in the flue gas are precipitated by cooling and separated. After separation, the gases analyzed 0.8% CO_2, 23.0% CO, 68.0% N_2 and 8.2% O_2. Calculate the following:

(a) The percent of CO entering the burner that is converted to CO_2
(b) The percent of P_4 that is oxidized to P_4O_{10}.

10.81 Phosphorus is produced by reacting calcium phosphate with silica and carbon in an electric furnace:

$$Ca_3(PO_4)_2 + 3SiO_2 \rightarrow 3CaSiO_3 + P_2O_5$$
$$2P_2O_5 + 5C \rightarrow 4P + 5CO_2$$

Calcium silicate forms a liquid slag, whereas phosphorus passes off as vapour along with CO_2 and is condensed. Phosphate rock containing 80% $Ca_3(PO_4)_2$ and the rest inert impurities is charged with 25% excess silica and 50% excess carbon. The first reaction is 80% complete, whereas the second is 90% complete. Calculate the following on a basis of per 1000 kg of the rock charged to the furnace:

(a) The weight of phosphorus produced
(b) The weight of calcium silicate produced
(c) The percent recovery of phosphorus

Recovery of metals and non-metals from ores

10.82 One hundred kilograms of flurospar containing 75% by weight CaF_2 and 25% inert materials is treated with concentrated sulphuric acid to produce hydrofluoric acid (HF).

$$CaF_2 + H_2SO_4 \rightarrow CaSO_4 + 2HF$$

Pure sulphuric acid is used in 30 percent excess of that theoretically required. 95% of the hydrofluoric acid produced in the reaction chamber leaves as gas while the remaining 5% leaves with the solid residue which consists of $CaSO_4$ produced in the reaction, the unreacted sulphuric acid and the inert materials.

(a) Determine the amount of solid residue formed.
(b) Express the concentration in weight percent.

10.83 Copper is extracted from its ore having the following analysis: CuS = 10%, FeS_2 = 30%, inerts = 60%. The ore is first crushed and subjected to flotation during which 70% of the inert materials are eliminated. The resulting concentrate has the following analysis: CuS = 16%, FeS_2 = 48%, inerts = 36%. This is then roasted where the following reactions occur:

$$2CuS + \frac{5}{2}O_2 \rightarrow Cu_2O + 2SO_2$$

$$2FeS_2 + \frac{11}{2}O_2 \rightarrow Fe_2O_3 + 4SO_2$$

The Cu_2O is then reduced to metallic copper. If 1000 kg of copper ore is originally charged, how many kilograms of the metal can be obtained if there is 3% loss during the reduction of the oxide to the metal.

10.84 For producing 1000 kg of pig iron of composition 94% Fe, 5% C and 1% Si, iron ore containing 89.6% Fe_2O_3 and the rest SiO_2 is reduced in a blast furnace using 1000 kg of coke containing 90% C and 10% SiO_2. The flux used contains 94% $CaCO_3$, 3% $MgCO_3$ and 3% SiO_2. If it is desired that the slag contains 40% (CaO + MgO), how much flux is required for producing 1000 kg of pig iron? Assume that no iron is present in the slag.

10.85 Powdered bauxite is reacted with dilute sulphuric acid to produce aluminium sulphate. In a typical operation, 1000 kg of bauxite containing 50% Al_2O_3 is treated with 1700 kg of 75% sulphuric acid. The reacted mass is filtered to remove solid residue. The filtrate is concentrated to crystallize 2000 kg of $Al_2(SO_4)_3 \cdot 9H_2O$. Calculate the following:

(a) The percent excess reactant used
(b) The degree of completion of the reaction
(c) The weight and composition of solid residue produced.

Nitrogen, ammonia and nitric acid

10.86 Synthesis gas consists of nitrogen and hydrogen in the ratio 1:3. It is prepared by mixing air with pure hydrogen. Oxygen is removed as water formed in the reaction between hydrogen and oxygen. For treating 100 moles of air containing 21% oxygen, 78% nitrogen and 1% argon, determine (a) moles of hydrogen added (b) weight of water condensed in kg.

10.87 A plant produces 600 tons/day of ammonia. Nitrogen and hydrogen in the mole ratio 1:4 is sent to the converter. The gases leaving the converter contains these gases in the ratio 1:4.25. Calculate the volume of gases measured at 773 K and 100 kPa that is admitted to the reactor per day.

10.88 In the fixation of nitrogen by the arc process, air is passed through an electric arc. Some of the nitrogen is oxidized to NO which on cooling is oxidized to NO_2. Of the NO_2 formed, 66% will get associated to N_2O_4 at 300 K. The gases are then passed into absorption towers where HNO_3 is formed by the following reaction:

$$H_2O + 3NO_2 \rightarrow NO + 2HNO_3$$

The NO liberated in this reaction will be reoxidized in part and form more nitric acid. In the operation of such a plant, it is found possible to produce gases from the arc furnace in which the nitric oxide is 2% by volume while hot. The gases are cooled to 300 K at a pressure of 1.013 bar before being sent to the absorber. Calculate the following:

(a) The complete analysis by volume of the hot gases leaving the furnace assuming that the air entering the furnace was of average composition
(b) The partial pressures of NO_2 and N_2O_4 in the gases entering the absorption tower
(c) The weight of HNO_3 formed by 1000 m^3 of gas entering the absorption system if the conversion to HNO_3 of the combined N_2 in the furnace gases is 85% complete

10.89 HNO_3 is produced by treating $NaNO_3$ with 95% H_2SO_4. The resulting nitre cake contains 34% H_2SO_4 and 1.5% H_2O. 2% of the HNO_3 formed remains in the cake. The reaction is complete. Calculate the following:

(a) The weight of nitre cake per 100 kg of $NaNO_3$
(b) The composition of nitre cake
(c) The weight of aqueous H_2SO_4 used
(d) The weight of HNO_3 and H_2O distilled from the nitre cake for 100 kg of $NaNO_3$

10.90 Urea is produced synthetically by reacting ammonia and carbon dioxide to form ammonium carbamate which is then decomposed to urea and water:

$$2NH_3 + CO_2 \rightarrow NH_2COONH_4$$

$$NH_2COONH_4 \rightarrow NH_2CONH_2 + H_2O$$

The conversion of ammonia is only 60%. If 1000 kg of urea is to be produced, determine the following:

(a) The volume of ammonia fed at STP
(b) The amount of water produced

10.91 In the nitric acid manufacture 100 m^3/h of ammonia at 300 K and 101.3 kPa is mixed with air and passed over a catalyst in a converter and the following reaction takes place:

$$4NH_3 + 5O_2 \rightarrow 6H_2O + 4NO$$

The gases are then passed into an oxidizing tower where the oxidation is completed according to the following reaction:

$$2NO + O_2 \rightarrow 2NO_2$$

The gases from the oxidizing tower are passed into a cooling tower where NO_2 is absorbed in 150 kg/h of water and forms nitric acid as per reaction:

$$3NO_2 + H_2O \rightarrow 2HNO_3 + NO$$

The overall reaction in the process is given by

$$NH_3 + 2O_2 \rightarrow HNO_3 + H_2O$$

Oxygen present in the air supplied is 20% in excess of that required for complete oxidation of the ammonia to nitric acid and water. Assume that (1) the degree of completion of the reaction in the converter is 85% and no other decompositions take place, (2) 90% of the nitric oxide entering the oxidizing tower is oxidized to nitrogen peroxide, and (3) the cooling tower exit gas contains no water vapour and all NO_2 is absorbed in water. Calculate the following:

(a) The amount of air to be used at 300 K, 101.3 kPa in m^3/h
(b) The moles of gases leaving the converter and its composition
(c) The moles of gases leaving the oxidiser and its composition
(d) The amount in kg/h and concentration by weight percent of nitric acid produced

Chlorine, hydrochloric acid and chlorination reactions

10.92 Hydrochloric acid is oxidized to chlorine by air in the presence of a catalyst. Air is supplied 30% in excess of the theoretical requirement and the oxidation of the acid is 60% complete. Calculate the following:

(a) The ratio of acid to air on a weight basis
(b) The composition of the gases leaving the reactor on a weight basis

10.93 Chlorine is produced by the oxidation of hydrogen chloride gas with air

$$4HCl + O_2 \rightarrow 2Cl_2 + 2H_2O$$

The reaction is carried out at 1.2 × 10^5 N/m^2 and 400 K. 50 percent excess air is used and the reaction is only 80 percent complete. Calculate

(a) The volume of air admitted per 100 m^3 of HCl if both air and HCl enter the reactor at 1.0 bar and 290 K
(b) The volume of gases leaving the reactor per 100 m^3 of HCl entering
(c) The weight of chlorine produced per 100 m^3 of HCl entering
(d) The percent composition by volume of the exit gas on a dry basis

10.94 In the Deacon process for the manufacture of chlorine, a dry mixture of hydrochloric acid and air is passed over a heated catalyst. Acid is oxidized in the presence of air, which is supplied 20% in excess of that theoretically required:

$$4HCl + O_2 \rightarrow 2H_2O + 2Cl_2$$

Calculate the following for the oxidation of 1 kg of acid:

(a) The amount of air supplied, in kilograms
(b) If 60% conversion is achieved, the composition of the product gases expressed in weight percent
(c) The dew point of the exit gas stream which is at a pressure of 100 kPa and temperature of 450 K

The vapour pressure of water is given by the Antoine equation:

$$\ln P^S = 16.26205 - \frac{3799.887}{T - 46.854}$$

where pressure is in kPa and temperature is in K.

10.95 Carbon tetrachloride is made by chlorination of carbon disulphide in the following two steps:

$$CS_2 + 3Cl_2 \rightarrow CCl_4 + S_2Cl_2$$

$$CS_2 + 2S_2Cl_2 \rightarrow CCl_4 + 6S$$

The reactants are admitted in stoichiometric ratio. Determine the mass of carbon tetrachloride and sulphur produced from 1000 kg of sulphide if

(a) The conversion in both the reactions are 100%
(b) The yield is 85% in the first reaction and 90% in the second reaction.

10.96 Catalytic vapour-phase chlorination of CS_2 gives carbon tetrachloride:

$$CS_2 + 3Cl_2 \rightarrow CCl_4 + S_2Cl_2$$

The products leaving the converter analyzed 26.0% CCl_4, 26.0% S_2Cl_2, 17.5% CS_2 and 30.5% Cl_2. Determine the following:

(a) The excess reactant and the percent excess
(b) The percent conversion
(c) The quantity of CCl_4 obtained (in kilograms) per 100 kg of chlorine admitted

10.97 Ethylene dichloride is manufactured by the oxychlorination of ethylene:

$$2C_2H_4 + 4HCl + O_2 \rightarrow 2C_2H_4Cl_2 + 2H_2O$$

Though ethylene and air are supplied respectively 5% and 10% in excess over those required for the complete conversion of hydrogen chloride, the conversion attained is found to be only 90%. For 500 kmol of HCl supplied, calculate the following:

(a) The moles of the reactant and products
(b) The mass of the reactants and products
(c) The composition of the reactant stream in weight percent
(d) The composition of the product stream in weight percent.

10.98 Monochloroacetic acid (MCA) is produced at a rate of 5000 kg/h by chlorination of glacial acetic acid with gaseous chlorine.

$$CH_3COOH + Cl_2 \rightarrow CH_2ClCOOH + HCl$$

Chlorine is supplied at a rate of 4500 kg/h. It is found that 300 kg/h of dichloroacetic acid (DCA) also is formed according to

$$CH_2ClCOOH + Cl_2 \rightarrow CHCl_2COOH + HCl$$

Acetic acid is supplied in excess to minimize DCA production. Determine the conversion and yield of MCA.

10.99 Hydrochloric acid is made by the action of NaCl with H_2SO_4. The following reactions occur in series:

$$NaCl + H_2SO_4 \rightarrow NaHSO_4 + HCl$$

$$NaHSO_4 + NaCl \rightarrow Na_2SO_4 + HCl$$

One thousand kilograms of pure NaCl is treated with aqueous sulphuric acid solution of strength 75% H_2SO_4. The solid cake obtained analyzed 91.5% Na_2SO_4, 4.8% $NaHSO_4$, 2.0% NaCl, 1.3% H_2O and 0.4% HCl. Determine the following:

(a) The degree of completion of the first reaction
(b) The degree of completion of the conversion to Na_2SO_4
(c) The quantity of acid added
(d) The quantity of cake obtained
(e) The weight of gases if the gases driven off are HCl and water vapour

9.100 For purification of silicon from the metallurgical grade to the semiconductor grade, 100 kg of the metal is reacted with hydrogen chloride gas at 573 K. From several polychlorinated silanes that are formed in the reaction, trichlorosilane ($HSiCl_3$), which is a liquid at room temperature, is separated by fractional distillation. Determine the quantity of pure trichlorosilane separated in the distillation unit if the gases leaving analyzed 21.0% H_2SiCl_2, 14.0% $SiCl_4$ and 65% H_2.

Hydrogenation, hydration and oxidation

10.101 Coal is hydrogenated in a fluidized bed reactor in the presence of air and steam at 675 K in order to produce gaseous hydrocarbon fuels. 2000 kg per hour of coal containing 10% inert material by weight is charged into the reactor from the top, and air and steam admitted from the bottom at a rate of 25 000 kg/h and 3500 kg/h respectively. Assuming the complete conversion of coal, determine the flow rate of exit gases in kg/h.

10.102 An organic ester of formula $C_{19}H_{36}O_2$ is to be hydrogenated at a rate of 100 kg/h to yield $C_{19}H_{38}O_2$. The hydrogen required for the plant, which runs continuously, is available as 50 L cylinders in which the gas is contained at 70 bar and 300 K. How many cylinders should the company order per week?

10.103 Formaldehyde is made by the oxidation of methanol with air. The analysis of the exit gas from the reactor shows 63.1 % N_2, 13.4% O_2, 5.9% H_2O, 12.3% CH_3OH, 4.1% HCHO and 1.2% HCOOH. Calculate the following:

(a) The conversion per pass
(b) The ratio of air to methanol in the feed

10.104 In a propylene plant, 100 kg/h of pure propane is fed to a catalytic reactor where it is converted to a gas containing 25% propylene, 45% propane, and 30% hydrogen by volume. This reactor gas is separated in a separator unit into three streams: (1) light gas containing 1.2% propane, 0.8% propylene and all hydrogen formed in the reactor, (2) desired product containing 99% propylene and 1% propane, and (3) bottom product containing 2% propylene and 98% propane. All the bottom product is returned back to the reactor. Calculate the following:

(a) The amount of product obtained in kg/h
(b) The amount of propane recycled, kg/h
(c) The amount of propane in light gas, kg/h

10.105 Formaldehyde is produced by the gas phase oxidation of methanol with air over a catalyst

$$CH_3OH + \frac{1}{2}O_2 \rightarrow HCHO + H_2O$$

100 m^3 of methanol vapour at 1.013×10^5 N/m^2 and 550 K is to be treated. If 10% excess air is supplied and the reaction is only 80% complete, calculate the following:

(a) The composition of the product gas
(b) The volume of product gases at 1.5×10^5 N/m^2 and 800 K

10.106 Ethylene oxide is produced by burning ethylene gas with air in the presence of a catalyst. In the reaction, a substantial portion of ethylene is converted to ethylene oxide, a small fraction is completely oxidized to CO_2, and water and some gases remain unconverted. The ethylene oxide in the product gas is recovered by absorption. The Orsat analysis of the gases leaving the absorber is: 8.77% CO_2, 4.38% O_2 and 4.38% ethylene. Calculate the percent of the ethylene fed to the reactor that is converted to the oxide.

10.107 Ethylene glycol $(CH_2OH)_2$ is produced by air oxidation of ethylene to ethylene oxide and subsequent absorption of the oxide in water.

$$C_2H_4 + \frac{1}{2}O_2 \rightarrow C_2H_4O$$

$$C_2H_4O + H_2O \rightarrow (CH_2OH)_2$$

A portion of the ethylene is oxidized to CO_2:

$$C_2H_4 + 3O_2 \rightarrow 2CO_2 + 2H_2O$$

100 kmol of a gas mixture containing 10% ethylene and 90% air is fed to the oxidizer. The gases leaving the reactor are absorbed in water fed at the rate of 10 moles per

100 moles of the original gas mixture. The off-gases leaving the absorber gave the following analysis on a dry basis: C_2H_4 2.31%, CO_2 6.94% and the rest oxygen and nitrogen. The gases leave the absorber at 101.325 kPa saturated with water vapour at 306 K. (The vapour pressure of water at 306 K is 5 kPa.) Determine the following:

(a) The moles of ethylene glycol produced
(b) The percent conversion of ethylene to glycol
(c) The moles of dry gases leaving the absorber
(d) The weight percent of ethylene glycol in the glycol water solution

10.108 Acetylene is produced industrially by reacting calcium carbide and water:

$$CaC_2 + H_2O \rightarrow CaO + C_2H_2$$

Calcium carbide containing 8% inert impurities is reacted with water so that the reaction goes to completion. For producing 1000 kg of acetylene, calculate the following:

(a) The weight of impure carbide needed
(b) The weight of solid residue after the reaction
(c) The weight ratio of acetylene to water vapour if the acetylene produced is saturated with water vapour at 306 K and 101.3 kPa. The vapour pressure of water at 306 K is 5 kPa.

10.109 Ethanol is produced by the hydration of ethylene:

$$C_2H_4 + H_2O \rightarrow C_2H_5OH$$

Some of the product is converted to ethyl ether by the side reaction

$$2C_2H_5OH \rightarrow (C_2H_5)_2O + H_2O$$

The feed to the reactor contains 25% C_2H_4, 65% H_2O, and the balance inert materials on a mole basis. An ethylene conversion of 10% and ethanol yield of 90% is achieved. Compute the composition of the reactor output stream.

10.110 A pilot plant reactor was charged with 50 kg of naphthalene and 200 kg (98% by weight) of sulphuric acid. The reaction was carried out for 3 hours at 430 K. The reaction goes to completion:

$$C_{10}H_8 + H_2SO_4 \rightarrow C_{10}H_7(SO_3H) + H_2O$$

$$C_{10}H_8 + 2H_2SO_4 \rightarrow C_{10}H_6(SO_3H)_2 + 2H_2O$$

The product distribution by weight was found to be 18% monosulphonate and 82% disulphonate of naphthalene. Calculate

(a) The quantities of monosulphonate and disulphonate products
(b) The complete analysis of the product

11 Bypass, Recycle and Purge Operations

Chemical processing would have been a smooth affair, if the desired changes could be brought about by taking the input streams in a single pass through the series of equipment that constitute the process unit. But, this is not the case in many situations where the single pass operations usually fail to provide the desired change. Chemical reactions generally never go to completion when sent through the system in a single pass. It becomes necessary to return back the unconverted streams and admit them to the process units again along with the feed admitted afresh. Returning the streams back ensures higher conversion and higher yield of the products. Reuse of materials in such recycle loops leads to reduction in cost and improvement in efficiency of the process. It is also true that in some cases, it may not be necessary to admit the entire input streams to certain stages of the process. Such streams that are allowed to skip one or more stages and go directly to another downstream stage are called *bypass streams*. The desired change can be brought about by admitting the remaining part of the input streams to the process unit that was skipped by the major part of the input stream. The processed stream is then mixed with the bypass stream downstream from the unit in such proportions as demanded by the process. Purging operations are performed when there is accumulation of undesired materials within the recycle loop. By bleeding off a small fraction of the streams, the undesired impurity can be maintained within acceptable limits. In this chapter, we will discuss these three special streams—recycle, bypass and purge—and explore the material balance calculations involving them in detail.

11.1 BYPASS

Bypass is employed mainly for effecting relatively small changes in a process stream. This is achieved by introducing large changes in a small portion of the original stream and diverting the other portion without passing it through the unit accomplishing this change. The portion so diverted is known as *bypass stream*. The stream that is passed through the process unit is later mixed with the bypassed stream to get the final desired change in the combined stream.

For example, consider the air-conditioning practised in process industries to produce air of desired humidity. Air can be humidified or dehumidified by passing it through a spray chamber where water is sprayed into the air stream. An accurate control of the final humidity is possible by splitting the air stream into two parallel streams and subjecting one stream to the humidification or dehumidification operation described and bypassing the other stream around the humidifier or dehumidifier. The desired final mixture is obtained by combining the bypass stream with the conditioned air (refer to Figure 11.1).

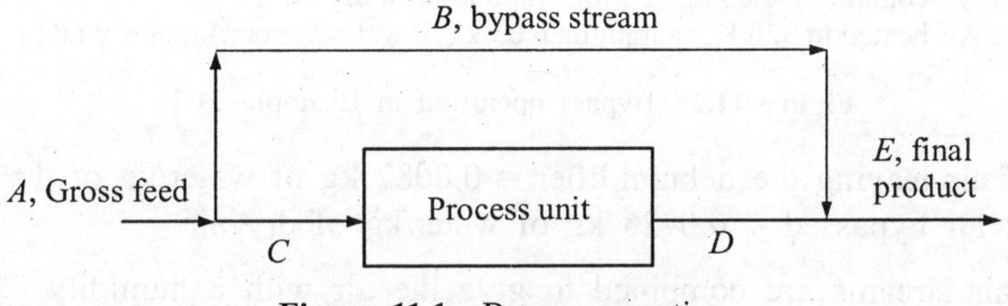

Figure 11.1 Bypass operation.

A portion of the gross feed stream (*A*) is diverted before entering the process unit and constitutes the bypass stream (*B*). The remaining portion (*C*) is passed through the unit and subjected to the necessary operation or process. The streams *A*, *B* and *C* have the same composition as well as other properties. The product stream leaving the processing unit (*D*) combines with the bypass stream and forms the final product (*E*) of the desired properties.

Bypass is desired when we need a product with properties in-between the untreated process stream and the process outlet product. For example, in a juice concentration process, the dehydration process runs most effectively by removing more water than is actually desired. A small portion is subjected to dehydration, which is later mixed with the major portion of the unprocessed feed in a bypass loop.

EXAMPLE 11.1 Air at 320 K saturated with water vapour is dehumidified by cooling to 285 K and by condensation of water vapour. Air leaving the dehumidifier saturated at 285 K is mixed with a part of the original air which has bypassed the dehumidifier. The resulting air stream is reheated to 320 K. It is desired that the final air contains water vapour not more than 0.03 kg per kg of dry air. Calculate:

(a) The dry air bypassed (kg) per each kg of dry air sent through the dehumidifier
(b) The water vapour condensed in the dehumidifier (kg) per 100 m^3 of air sent through it
(c) The volume of final air obtained per 100 m^3 of air passed through the dehumidifier.

The total pressure is atmospheric and the vapour pressures of water are 1.4 kPa at 285 K and 10.6 kPa at 320 K.

Solution *Basis:* 1 kg of dry air passed is through the dehumidifier.
Let *x* kg of dry air be present in the air bypassed.

$$\text{Saturation humidity of air at 320 K} = \frac{10.6}{101.3 - 10.6} \times \frac{18}{29} = 0.0725 \text{ kg of water/kg of dry air}$$

$$\text{Saturation humidity at 285 K} = \frac{1.4}{101.3 - 1.4} \times \frac{18}{29} = 0.0087 \text{ kg of water/kg of dry air}$$

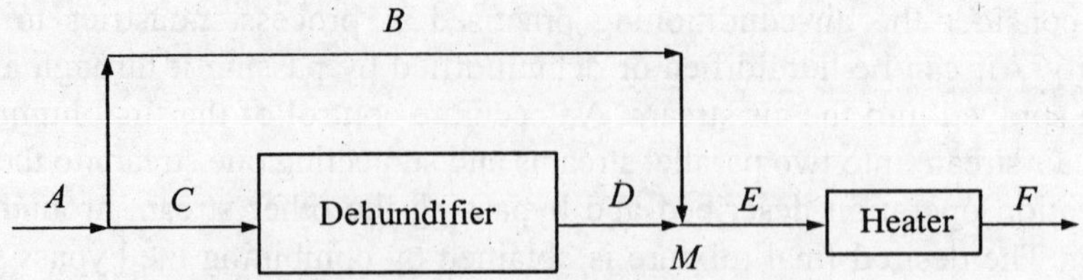

A, *B*, *C* : Saturated air at 320 K, D: Saturated air at 285 K
E : Air containing 0.03 kg of water vapour/kg of dry air
F : Air heated to 320 K, containing 0.03 kg of water vapour/kg of dry air

Figure 11.2 Bypass operation in Example 11.1.

Humidity of air leaving the dehumidifier = 0.0087 kg of water/kg of dry air
Humidity of air bypassed = 0.0725 kg of water/kg of dry air

These two air streams are combined to give the air with a humidity 0.03 kg of water vapour per kg of dry air at point *M*. Take a moisture balance at point *M*. We get

$$1 \times 0.0087 + x \times 0.0725 = (1 + x) \times 0.03$$

On solving, we get $x = 0.5012$.

(a) 0.5012 kg of dry air is bypassed per kg of dry air sent through the dehumidifier.

(b) For each kilogram of dry air passed through the dehumidifier, water condensed = 0.0725 – 0.0087 = 0.0638 kg.

$$1 \text{ kg of dry air} = 1 + 0.0725 = 1.0725 \text{ kg of wet air} = \frac{1}{29} + \frac{0.0725}{18} = 0.0385 \text{ kmol of wet air}$$

$$0.0385 \text{ kmol of air} = 0.0385 \times 22.414 = 0.8632 \text{ m}^3 \text{ at STP}$$

$$0.8632 \text{ m}^3 \text{ at STP} = 0.8632 \times \frac{1.01325}{1.013} \times \frac{320}{273.15} = 1.0115 \text{ m}^3 \text{ at 101.3 kPa and 320 K}$$

For 1.0115 m^3 of air sent through the dehumidifier, 0.0638 kg of water vapour is condensed.

$$\text{For 100 m}^3 \text{ of air, amount of water vapour condensed} = \frac{0.0638}{1.0115} \times 100 = 6.3075 \text{ kg.}$$

(c) 0.5012 kg of dry air is bypassed per 1 kg dry air passed through the dehumidifier. Hence, for each kilogram of dry air, the final air obtained is 1.5012 kg of dry air. Since 1 kg of dry air along with the accompanying water vapour has a volume of 1.0115 m^3, the kilogram of dry air in the final air is

$$\frac{1.5012}{1.0115} \times 100 = 148.41 \text{ kg}$$

$$\text{Moles of dry air produced} = \frac{148.41}{29} = 5.1177 \text{ kmol}$$

Moisture content in the final air $= 0.03\ \dfrac{\text{kg of water}}{\text{kg of dry air}} = 0.03 \times \dfrac{29}{18} = 0.0483\ \dfrac{\text{kmol of water}}{\text{kmol of dry air}}$

It means that 1.0483 kmol of wet air is present per each kilomole of dry air. Therefore, total moles of final wet air $= 5.1177 \times \dfrac{1.0483}{1} = 5.3651\,\text{kmol}$. At 320 K and 101.3 kPa, this has a volume of $5.3651 \times 22.414 \times \dfrac{101.325}{101.3} \times \dfrac{320}{273.15} = 140.91\ \text{m}^3$

The volume of final air per 100 m^3 of air passed through the dehumidifier = 140.91 m^3.

11.2 RECYCLE

A system with recycle is one in which a stream leaving a processing unit is partly returned to the same unit for reprocessing or reusing. The recycling of the fluid stream in chemical processing is a common practice to increase yields, to increase the purity, to enrich a product, to conserve heat, or to improve operations. Recycling may also be used to recover an expensive catalyst, reuse a lubricant or refrigerant, maintain feed concentration below certain levels by diluting the process streams either to control flow or to control the rate of reaction. A common example is the drying operation where a portion of the hot air leaving the dryer is recycled to conserve heat or to reduce the rate of drying by controlling humidity of the air entering the dryer. In a fractionating column, a part of the distillate is returned to the column as reflux to enrich the product.

The recycling operation is shown in Figure 11.3. The product stream *D* is split into two streams and one is recycled (*B*). The other stream is removed as the net product (*E*) from the operation. The physical properties and chemical constitution of streams *D*, *E* and *B* may be same or different depending on the operations that the stream *D* has undergone before it is separated into streams *E* and *B*. However, in majority of cases, the process effluent (*D*), process product (*E*) and the recycle (*B*) have identical compositions. The recycled stream combines with the fresh feed (*A*) and enters the unit as gross feed (*C*). The ratio of the quantity of a substance in the recycled stream to the quantity of the same substance in the fresh feed is known as *recycle ratio*.

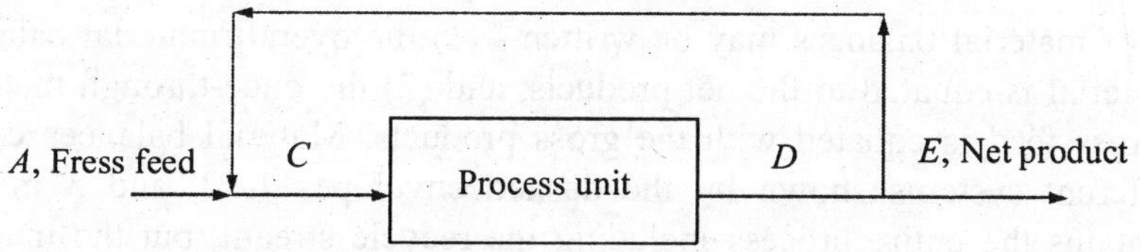

Figure 11.3 Recycle operation.

In processes involving chemical reactions, recycle is generally employed to increase the yield. Many organic reactions do not give 100% yield of the desired product. In cases where the yield is low, the unconverted reactants are separated from the product and sent back to the reactor as recycle stream and joined with a stream of reactants that enter the reactor. In ammonia synthesis, the gas mixture leaving the converter after recovery of ammonia is recycled through the converter. Recycle process involving chemical reactions can be schematically represented as shown in Figure 11.4.

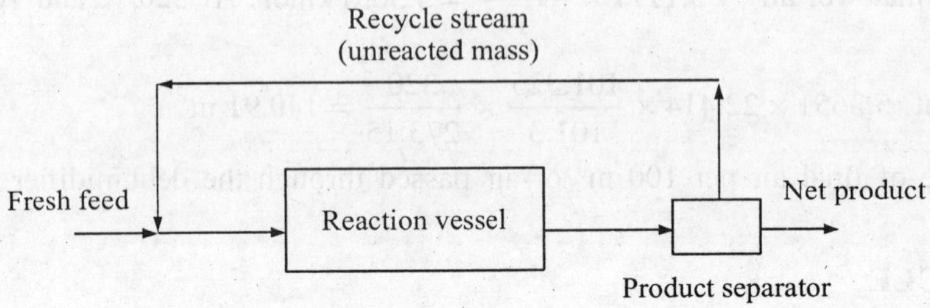

Figure 11.4 Recycle in processes with chemical reaction.

Recycle process necessitates the specification of two types of conversions—the *overall conversion* and the *single pass conversion*. The overall conversion is the fraction of the reactant in the fresh feed that is reacted considering the overall process.

$$\text{Overall conversion} = \frac{\begin{matrix}\text{moles of reactant}\\ \text{in the fresh feed}\end{matrix} - \begin{matrix}\text{moles of reactant in the product}\\ \text{from the overall process}\end{matrix}}{\text{moles of reactant in the fresh feed}}$$

$$= \frac{\text{moles of reactant in stream } A - \text{moles of reactant in stream } E}{\text{moles of reactant in stream } A}$$

The single pass conversion is the fraction of the reactants that actually enter the reactor in a single pass that is converted.

$$\text{Single pass conversion} = \frac{\begin{matrix}\text{moles of reactant fed}\\ \text{into the reactor}\end{matrix} - \begin{matrix}\text{moles of reactant}\\ \text{leaving the reactor}\end{matrix}}{\text{moles of reactant fed into the reactor}}$$

$$= \frac{\text{moles of reactant in stream } C - \text{moles of reactant in stream } D}{\text{moles of reactant in stream } C}$$

Two types of material balances may be written—(1) the overall material balance in which the net feed material is equated to the net products, and (2) the once-through material balance, in which the gross feed is equated with the gross products. Material balances can be written for several different systems shown by the dashed envelopes 1, 2, and 3 in Figure 11.5. Envelope 1 contains the entire process including the recycle stream, but the material balance over this gives no insight into the recycle stream. However, the balances written over envelopes 2 and 3 contain recycle streams. Consider a chemical reaction $A + B \rightarrow C$ being carried out in a unit shown below, where the single pass conversion of A is 25%. Assume that the reactants are present in stoichiometric proportions in the fresh feed. What is the amount recycled per 100 moles of fresh feed?

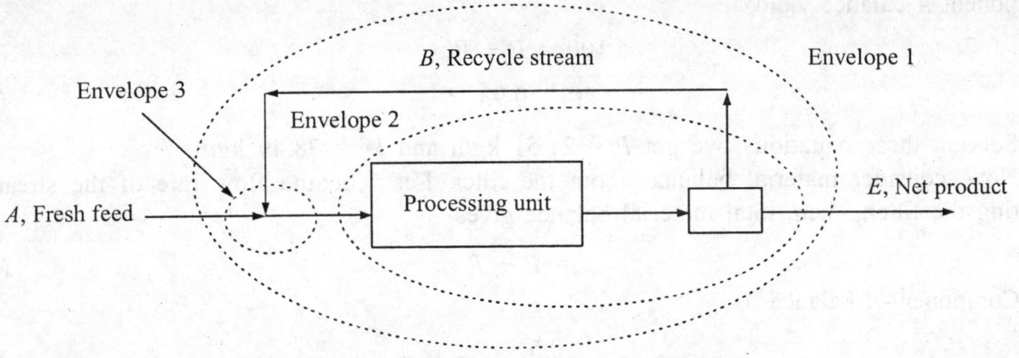

Figure 11.5 Material balance in recycle operation.

Taking an A-balance or $A + B$ balance for the overall unit (envelope1) does not give us any information on the recycle. Consider envelope 2 that encompasses the reactor and the separator in which the recycle is separated from the reaction products, and take a material balance for $A + B$ assuming that R moles are recycled per 100 moles of fresh feed. For steady-state material balance for processes with chemical reaction, the general material balance can be written as

material in = material out + material consumed due to chemical reaction

Since the moles entering the reactor is $100 + R$, we get

$$100 + R = (100 + R)0.25 + R$$

which gives $R = 300$ moles.

EXAMPLE 11.2 Final purification stage in the preparation of a pharmaceutical product A from natural sources requires centrifuging and continuous filtration as depicted in Figure 11.6. Determine the flow rate of the recycle stream in kg/h.

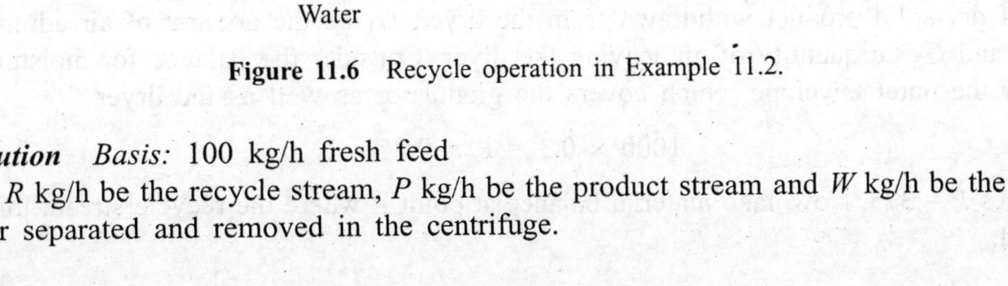

Figure 11.6 Recycle operation in Example 11.2.

Solution *Basis:* 100 kg/h fresh feed

Let R kg/h be the recycle stream, P kg/h be the product stream and W kg/h be the amount of water separated and removed in the centrifuge.

Considering the envelope which contains the centrifuge and the filter, the total material and component-A balance yield

$$100 = P + W \tag{A}$$

$$20 = 0.93\ P \tag{B}$$

Solving these equations, we get P = 21.51 kg/h and W = 78.49 kg/h.

Now, consider material balance about the filter. Let x be the flow rate of the streams entering the filter. Then, total material balance gives

$$x = P + R \tag{C}$$

Component-A balance

$$0.65x = \frac{0.5}{1.5} R + 0.93P \tag{D}$$

Substitute P = 21.51 in Eqs. (C) and (D) and solve the equations simultaneously. The result is R = 19.

The amount of recycle stream = 19 kg/h.

EXAMPLE 11.3 A wet solid containing 70% water is mixed with recycled dry solid to reduce the water content to 50% before being admitted into the granulator. The solid leaving the granulator is fed to a drier where it is brought into contact with dry air initially containing 0.25% water by weight. In the drier, the air picks up moisture and leaves with a moisture content of 5%. The solids leaving the drier contain 20% water. A portion of this solid is recycled.

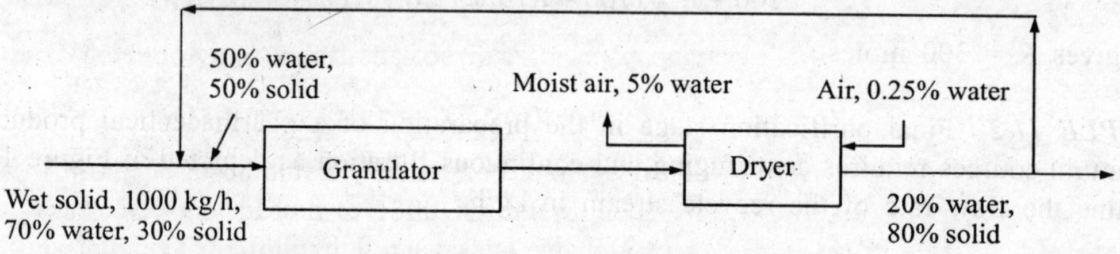

Figure 11.7 Recycle operation in Example 11.3.

For 1000 kg/h of wet solid sent to the granulator as fresh feed, determine:

(a) The amount of solid recycled

(b) The circulation rate of air in the drier on a dry basis.

Solution *Basis:* 1000 kg/h of fresh solid feed containing 70% water. Let R be the amount of recycle stream, S be the amount of combined stream entering the granulator, P be the amount of dry solid product withdrawn from the dryer, G_1 be the amount of air admitted to the dryer and G_2 be quantity of air leaving the dryer. Consider the balance for moisture-free solid over the outer envelope which covers the granulator as well as the dryer.

$$1000 \times 0.3 = \text{P} \times 0.8$$

which gives P = 375. Now, take material balance at point A where the recycle stream joins the fresh feed.

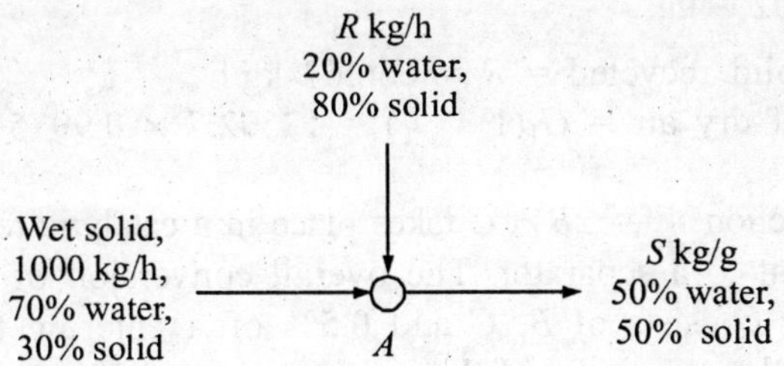

Figure 11.8 Material balance around point *A*.

Total material balance,

$$1000 + R = S$$

Water balance,

$$1000 \times 0.70 + R \times 0.20 = S \times 0.5$$

Solution of the above equations give R = 666.67 and S = 1666.67.

Refer to Figure 11.9 which shows the quantity and composition of material entering and leaving the dryer. The solids entering the dryer, S = 1666.67 kg, solids leaving = $P + R$ = 1041.67 kg.

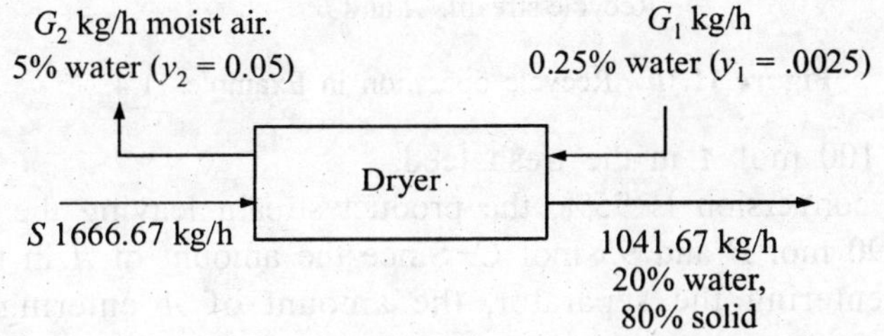

Figure 11.9 Material balance around the dryer in Example 11.3.

Since all the dry air entering the dryer is present in the air stream leaving the dryer, the material balance for the dry air is simply

$$G_1(1 - y_1) = G_2(1 - y_2)$$

Here, y_1 = 0.0025 and y_2 = 0.05 so that Eq. (9.58) becomes

$$G_2 = G_1 \frac{0.9975}{0.95} = 1.05 G_1 \tag{A}$$

Finally, we can take a water balance around the drier, which is

$$S \times 0.5 + G_1 y_1 = G_2 y_2 + (P + R) \times 0.20 \tag{B}$$

Using Eq. (A) and substituting known values into Eq. (B), we get

$$1666.67 \times 0.5 + 0.0025 G_1 = 1.05 G_1 \times 0.05 + 1041 \times 0.20$$

Therefore, $G_1 = 12502.7$ kg.

(a) The amount of solid recycled = R = 666.67 kg/h

(b) Circulation rate of dry air = $G_1(1 - y_1) = 12502.7 \times 0.9975 = 12500$ kg/h

EXAMPLE 11.4 The reaction $A \rightarrow 2B + C$ takes place in a catalytic reactor (see Figure 11.10). The reactor effluent is sent to a separator. The overall conversion of A is 95%. The product stream from the separator consists of B, C and 0.5% of A entering the separator, while the recycle stream consists of the remainder of the unreacted A and 1% of B entering the separator. Calculate the

(a) Single pass conversion of A in the reactor

(b) Molar ratio of recycle to feed.

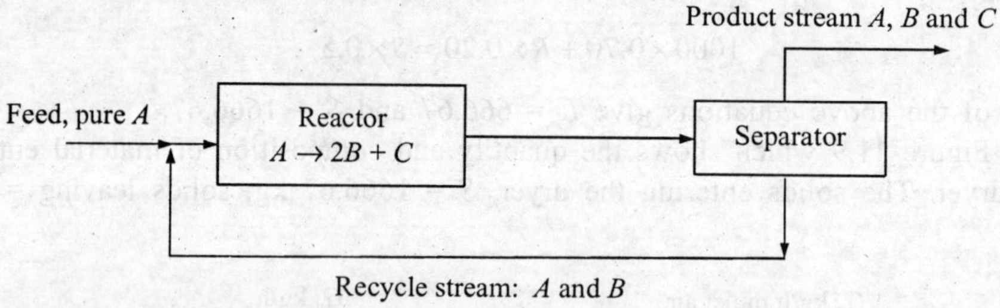

Figure 11.10 Recycle operation in Example 11.4.

Solution *Basis:* 100 mol A in the fresh feed.

Since the overall conversion is 95%, the product stream leaving the separator contains 5 mol A, $95 \times 2 = 190$ mol B and 95 mol C. Since the amount of A in the product stream is only 0.5% of A entering the separator, the amount of A entering the separator is $5 \times \dfrac{100}{0.5} = 1000$ mol. 99% of the B entering the separator should leave with the product stream. Then, the amount of B entering the separator is $190 \times \dfrac{100}{99} = 191.92$ mol. Since no C is present in the recycle stream, C entering the separator is 95 mol, same as the C leaving in the product stream.

(a) Now, the stream leaving the reactor consists of A = 1000 mol, B = 191.92 mol and C = 95 mol. The recycle stream consists of $A = 1000 - 5 = 995$ mol and $B = 191.92 - 190 = 1.92$ mol.

The combined feed entering the reactor contains $A = 100 + 995 = 1095$ mol and B = 1.92 mol. Since A entering the reactor is 1095 mol and A leaving the reactor is 1000 mol, single pass conversion of $A = \dfrac{(1095 - 1000)}{1095} \times 100 = 8.68\%$

(b) Moles recycled = 995.00 + 1.92 = 996.92 mol. Moles of fresh feed = 100. Therefore,

ratio of recycle to fresh feed $= \dfrac{996.92}{100} = 9.97.$

EXAMPLE 11.5 Refined sugar is converted to glucose and fructose according to

$$\underset{\text{Sucrose}}{C_{12}H_{22}O_{11}} + H_2O \rightarrow \underset{d\text{-Glucose}}{C_6H_{12}O_6} + \underset{d\text{-Fructose}}{C_6H_{12}O_6}$$

The conversion of sucrose in one pass through the reactor is found to be 71.7%. A part of the product leaving the reactor is recycled in such quantities that after mixing the recycle stream with the fresh feed, the combined concentration of glucose and fructose in the stream entering the reactor is 4%. For 100 kg sucrose solution containing 25% sucrose and 75% water charged to the reactor as fresh feed, calculate:

(a) The recycle flow
(b) The combined concentration of fructose and glucose in the recycle stream.

Solution *Basis:* 100 kg sucrose solution admitted as fresh feed

Refer to Figure 11.11. Let R kg of reactor exit be recycled and joins with the fresh feed at point 1 before entering the reactor. Let x be the weight fraction of sucrose and y be the weight fraction of inversion sugar (glucose and fructose) in the recycle stream. It is given that the weight fraction of glucose and fructose in the combined stream is 0.04. Let z be the weight fraction of sucrose in the combined stream entering the reactor.

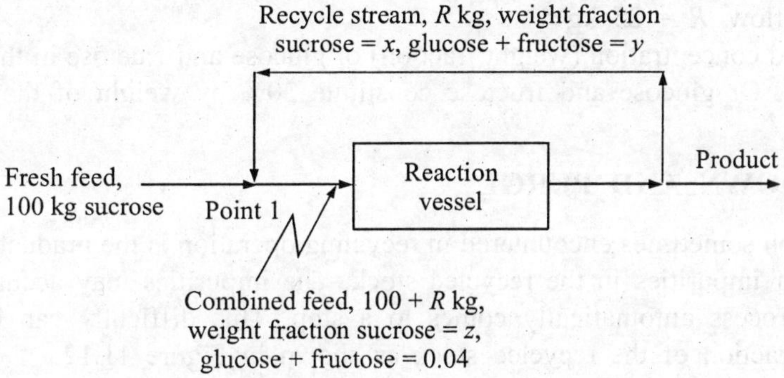

Figure 11.11 Recycle in processes in Example 11.5.

We can write a sucrose balance at point 1 as

$$25 + Rx = (100 + R)z \tag{A}$$

Similarly, glucose + fructose balance yields

$$Ry = (100 + R)\ 0.04 \tag{B}$$

A sucrose balance around the reactor can be written as follows:

Sucrose entering the reactor = sucrose consumed in the reaction + sucrose leaving the reactor

$$(100 + R)z = 0.717(100 + R)z + (100 + R)x \tag{C}$$

Note that the total material intering the reactor is equal to the total material leaving the reactor, both being equal to (100 + R) kg. Equation (C) can be simplified as

$$x = 0.283\,z \tag{D}$$

In Eqs. (C) and (D), the amount of sucrose converted to fructose and glucose = 0.717 (100 + R)z kg. Then, the glucose and fructose formed in the reactor is $0.717(100 + R)z \times \dfrac{360.192}{342.176}$ kg. A glucose + fructose balance around the reactor is written by equating the sum of the amount in the stream entering the reactor and the amount formed by the reaction to the amount in the stream leaving the reactor.

$$(100 + R)0.04 + 0.717(100 + R)\,z \times \frac{360.192}{342.176} = (100 + R)\,y \tag{E}$$

This equation can be simplified to

$$y = 0.04 + 0.7548z \tag{F}$$

Equations (A), (B), (D) and (F) are solved simultaneously and the following values are obtained.

$$x = 0.06, \quad y = 0.20, \quad z = 0.212, \quad R = 25$$

(a) Recycle flow, R = 25 kg

(b) Combined concentration (weight fraction) of glucose and fructose in the recycle stream, y = 0.20. Or glucose and fructose constitute 20% by weight of the recycle stream.

11.3 BLOWDOWN AND PURGE

A major limitation sometimes encountered in recycling operation is the gradual accumulation of inert materials or impurities in the recycled stock. The impurities may accumulate to such a level that the process automatically comes to a stop. This difficulty can be overcome by bleeding off a fraction of the recycled stock as shown in Figure 11.12.

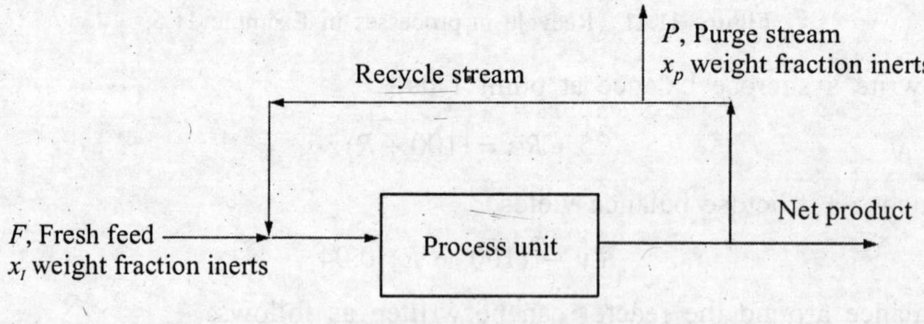

Figure 11.12 Purging operation.

Similarly, in processes involving chemical reaction employing recycle, if a small amount of a non-reactive material is present in the feed, it may be necessary to remove the non-reactive material in a purge stream to prevent its building up above the maximum tolerable value. In the synthesis of ammonia from atmospheric nitrogen and hydrogen, the percentage conversion of a 1:3 mixture is 25% in a single pass through the reactor. The ammonia formed is removed by cooling and condensation under high pressure and the unconverted nitrogen and hydrogen are re-circulated to the reactor. The methane present in the feed stream through the methanation reaction of hydrocarbon feed stock and argon coming from the atmospheric nitrogen enter the converter with the synthesis gas and accumulate in the gas stream entering the converter. The build-up of methane and argon beyond the permissible upper limit can be prevented by purging off a fraction of the recycled stream (refer to Figure 11.13). In the manufacture of methanol from synthesis gas also, purging is necessary to prevent accumulation of methane.

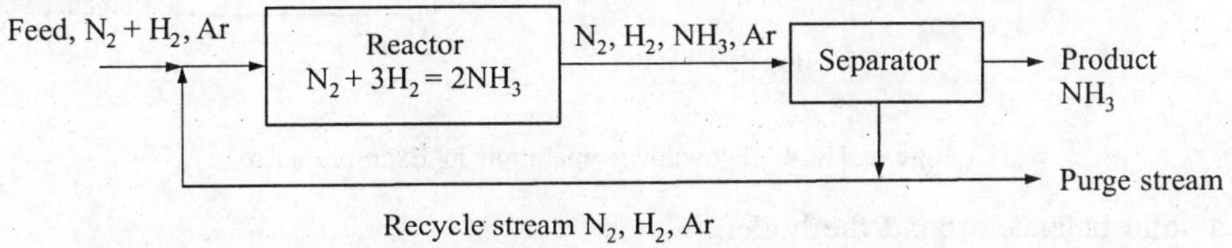

Figure 11.13 Purging operation in process with chemical reaction.

In the electrolytic refining of copper employing the recycle of electrolyte, the accumulation of impurities is prevented by bleeding off a portion of the electrolyte and replacing the same with fresh electrolyte. In the steam boiler using feed water containing dissolved solids, the solids would accumulate in the boiler, as steam is removed. A certain amount of boiler contents are periodically removed as "blowdown" which, in principle, is analogous to purging operation.

If the impurities or inert material is removed only through the purge stream, then the amount of inerts entering with the fresh feed should be equal to the amount of inerts in the purge. Let P be the amount of purge, and x_P be the composition of the inert in it. Then,

$$Px_P = Fx_I$$

where F is the quantity of fresh feed and x_I is its inert concentration. Thus, the quantity purged off can be evaluated as

$$\text{Purge} = \frac{\text{feed rate} \times \text{mole fraction of inerts in the feed}}{\text{mole fraction of inerts in the recycle}}$$

EXAMPLE 11.6 One half of the high-pressure steam after being utilized for power generation is sent to the plant as process steam. The other half is returned to the boiler as condensate carrying 50 ppm solids.

To keep the solid level in the boiler below 1600 ppm solids, a part of the boiler water is blown down continuously. The fresh boiler feed water is found to contain 500 ppm solids. If steam produced is free of solids, calculate the weight ratio of feed water to the blowdown water.

Solution Refer to Figure 11.14. Let F be the feed water rate, B be the rate of blowdown, S be the high-pressure steam produced and P be the rate at which the process steam leaves the power generation unit. P is also the rate at which condensate is returned to the boiler. The total material balance over the overall unit is

$$F = B + P \tag{A}$$

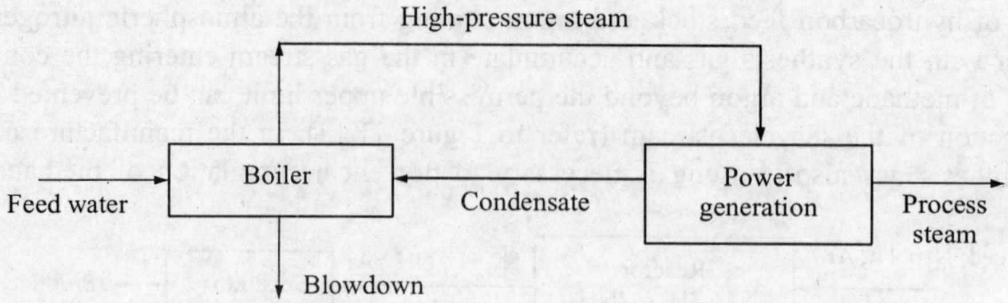

Figure 11.14 Blowdown operation in Example 11.6.

Take a solid balance around the boiler.

$$F \times \frac{500}{10^6} + P \times \frac{50}{10^6} = B \times \frac{1600}{10^6}$$

which can be simplified as

$$5F + 0.5P = 16B \tag{B}$$

Eliminating P from Eq. (B) using Eq. (A), we get

$$5F + 0.5(F - B) = 16B \quad \Rightarrow \quad 5.5F = 16.5B$$

which means $\dfrac{F}{B} = \dfrac{16.5}{5.5} = 3$

Ratio of feed water to blowdown = 3.

EXAMPLE 11.7 For the reaction $A \rightarrow B$, the process flow diagram is shown in Figure 11.15. The fresh feed of A contains 0.5% of inerts by volume. 60% conversion of A per pass is obtained. The concentration of inerts going into the reactor at (1) must be held at 2% by volume. All streams are ideal gases and the process is at steady state.

(a) How many moles need to be recycled per mole of total feed to the reactor at (1)?
(b) How many moles need to be purged?
(c) What is the overall conversion of A?

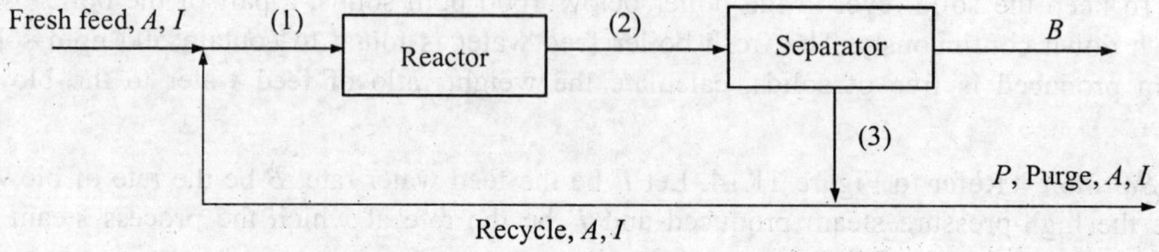

Figure 11.15 Purging operation in Example 11.7.

Solution *Basis:* 1 mol combined feed entering the reactor at point (1).

Let F be the moles of fresh feed introduced (see Figure 11.15). Composition of the streams at points (1), (2) and (3) can be found out first. At point (1), the inert content is 2%. Therefore, A at point (1) is 0.98 mol, inerts = 0.02 mol. Since single pass conversion is 60%, stream at point (2) contains $A = 0.40 \times 0.98 = 0.392$ mol, $B = 0.60 \times 0.98 = 0.588$ mol, inerts = 0.02 mol. Since B is completely removed from the separator, stream at point (3) contains 0.392 mol A and 0.02 mol inerts. Let R be the moles of the gas recycled and P be the moles purged. Then,

$$R + P = 0.392 + 0.02 = 0.412 \tag{A}$$

Also, the composition of R and P are the same as stream (3). Inert content in the recycle stream is $= \dfrac{0.02}{(0.392 + 0.02)} \times 100 = 4.85\%$. The purge stream also has the same inert content.

Taking a component – A balance over the combined system, we get

A in the fresh feed = A leaving with the purge stream + A consumed in the reaction

Number of moles of A consumed in the reaction is the same as the number of moles of product B leaving the separator, which is equal to 0.558 mol. Therefore,

$$F \times 0.995 = P \times 0.9515 + 0.588 \quad \Rightarrow \quad P = 1.0457F - 0.618 \tag{B}$$

An inert balance at the point where fresh feed is mixed with the recycle,

$$F \times 0.005 + R \times 0.0485 = 1 \times 0.02 \tag{C}$$

Equations (A), (B) and (C) can be solved to get the following:

$$F = 0.6552 \text{ mol}, \; P = 0.0671 \text{ mol}, \; R = 0.3448 \text{ mol}$$

(a) Moles of recycle stream = 0.3448 mol
(b) Moles of purge stream = 0.0671 mol
(c) Moles of A converted = 0.588 mol

Moles of A entering in the fresh feed = 0.995 × 0.6552 = 0.6519 mol

$$\text{Overall conversion of } A = \frac{0.588}{0.6519} \times 100 = 90.2\%$$

EXAMPLE 11.8 Carbon monoxide and hydrogen react to give methanol

$$CO + 2H_2 \rightarrow CH_3OH$$

The conversion of CO entering the reactor is only 20%. A feed stream consisting of 33% CO, 66.5% H_2, and 0.5% CH_4 is mixed with a recycle stream and fed to a reactor. Methanol leaving the reactor is separated and the unconverted gases are recycled. To prevent the accumulation of CH_4 and keep its concentration in the recycle stream at 3%, a portion of the recycled stream is blown off. For 100 moles of fresh feed, determine:

(a) The moles of recycle stream
(b) The moles of purge stream
(c) The composition of the purge stream
(d) The moles of methanol produced.

Solution *Basis:* 100 moles of fresh feed.

Mole fractions of the various components of the fresh feed are: CO = 0.33, H_2 = 0.665 and CH_4 = 0.005. Let *R* be the moles of recycle stream and *P* be the moles purged. Let *x* be the mole fraction of CO in the recycle stream. Then, the composition of the recycle stream in mole fraction of the components is CH_4 = 0.03, CO = *x*, H_2 = 0.97 – *x*; the purge and recycle streams have the same composition (refer to Figure 11.16).

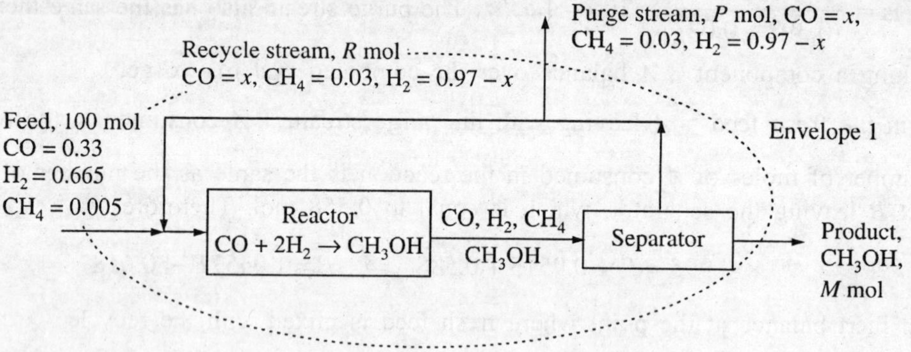

Figure 11.16 Purging operation in Example 11.8.

Taking a methane balance over the entire system (envelope 1), we get

$$0.5 = 0.03P \quad \Rightarrow \quad P = 16.67 \text{ mol}$$

Taking carbon balance,

$$33.5 = M + P(0.03 + x)$$

Hydrogen balance gives

$$66.5 + 2 \times 0.5 = 2M + P(2 \times 0.03 + 0.97 - x)$$

Substituting *P* = 16.67 mol into the above equations give

$$M + 16.67x = 33.0 \tag{A}$$

and

$$2M - 16.67x = 50.33 \tag{B}$$

Solving Eqs. (A) and (B) simultaneously, we get *M* = 27.78 and *x* = 0.3134.

Since the single pass conversion of CO is 20%, moles of methanol produced will be equal to the moles of CO converted per pass. That is, a methanol balance over the reactor is

$$\text{methanol entering} + \text{methanol produced} = \text{methanol leaving}$$

Since no methanol is present in the stream entering the reactor, we have

$$(33 + Rx) \times 0.2 = M$$

Substituting the values of M and x in this equation, we get $R = 337.91$

(a) Moles of recycle stream, $R = 337.91$ mol
(b) Moles of purge stream, $P = 16.67$ mol
(c) The mole fraction of $CH_4 = 0.03$, mole fraction of $CO = x = 0.3134$; mole fraction of $H_2 = 0.97 - x = 0.6566$. Therefore, the purge stream consists of 3% CH_4, 31.34% CO and 65.66% H_2.
(d) Methanol produced, $M = 27.78$ mol

EXERCISES

Bypass

11.1 For using brackish water for irrigation, it is to be desalinated by evaporation. A fraction of the feed water is sent through an evaporator and the other part is bypassed. The fraction bypassed and the pure water leaving the evaporator are mixed together to give the desalinated water of the desired quality. If the feed water has 500 ppm (parts per million) of salt, and the salt content in the water used for irrigation is to be limited to a maximum of 50 ppm, determine the fraction of the feed water bypassed.

11.2 It is desired that the concentration of *i*-pentane in the feed to a natural gasoline plant should not be more than 10% by mole. This is prepared by passing a fraction of a hydrocarbon feed stream made up of *n*-pentane and *i*-pentane in the mole ratio 4:1 through an *i*-pentane recovery unit as a side-stream. The stream leaving this tower is free of *i*-pentane. It is mixed with the fraction of the original mixture that is not passed through the recovery tower to get the feed stock to the natural gasoline plant. What percent of the original hydrocarbon mixture should be treated for removing the isopentane?

11.3 A process stream contains 4%(weight) salt and the rest water. This is prepared by passing a part of pure water stream through a saturator containing the salt. The solution leaving the saturator containing 20% salt is mixed with the pure water bypassed to get the process stream. What fraction of the pure water available is to be passed through the saturator?

11.4 The maximum allowable limit of impurities in the effluent from a processing unit is 100 ppm, whereas the discharge from the plant contains 500 ppm impurities. To reduce the level of impurities to the allowable limit, a fraction of the effluent is sent through a treatment plant which brings down the impurities to 10 ppm and mixes the treated effluent with the fraction of the effluent that is bypassed. Determine the fraction of the effluent that is bypassed.

11.5 Air at 313 K saturated with water vapour is dehumidified by cooling and condensation of water vapour at 286K and by consequent condensation of water vapour. Air leaving the dehumidifier saturated at 286 K is mixed with a part of the original air which is bypassed through the dehumidifier. The resulting air stream is reheated to 313 K. It is desired that the final air contains water vapour not more than 0.02 kg/kg of dry air. Calculate:

(a) The mass of dry air bypassed (kg) per each kg of dry air sent through the dehumidifier
(b) The mass of water vapour condensed (kg) in the dehumidifier per 100 m^3 of air sent through it
(c) The volume of final air obtained per 100 m^3 of air passed through the dehumidifier.

The total pressure is atmospheric and the vapour pressures of water are 1.5 kPa at 286 K and 7.5 kPa at 313 K.

11.6 In a process for concentrating 1000 kg freshly extracted orange juice containing 15% solids, the juice is strained, yielding 750 kg of strained juice and 250 kg of pulpy juice. The strained juice is concentrated in a vacuum evaporator to give an evaporated juice of 60% solids. The 250 kg of pulpy juice is bypassed around the evaporator and mixed with the evaporated juice in a mixer to improve the flavour. The final concentrated juice contains 40% solids. Calculate:

(a) The mass of final concentrated juice (kg)
(b) The concentration of solids in the strained juice
(c) The concentration of solids in the pulpy juice.

Recycle and purge involving no chemical reaction

11.7 Dry salt is to be produced at a rate of 20000 kg/h by evaporating water from a feed containing 20% NaCl. The brine (27% NaCl) leaving the evaporator may either be recycled to the evaporator or may be discarded. If the brine produced is 25% of the weight of dry salt, calculate:

(a) The feed rate if the brine is recycled
(b) The percent excess of the feed rate if the brine is discarded over that required when the brine is recycled.

11.8 Ten thousand kilograms of an equimolar mixture of benzene and toluene is subjected to fractionation (distillation) so as to produce a distillate having the composition of 95% benzene and the bottom product with a composition of 96% toluene. The column is provided with a condenser in which the vapour product leaving the top is totally condensed. It was found that the vapour flow rate was 8000 kg/h. A part of the condensate is recycled to the column as reflux and the other part is removed as the distillate product. The vapour, reflux and the distillate have the same composition. All compositions are on mole basis. Determine the reflux ratio of the column if it is defined as the quantity of condensate recycled to the quantity withdrawn as distillate product.

11.9 For the preparation of potassium nitrate, 10000 kg/h of a 20% KNO_3 solution is mixed with a recycle stream and sent to an evaporator. The rate of evaporation is 1.25 times the rate of introduction of recycle stream. The concentrated solution leaving the evaporator contains 50% KNO_3. This is admitted to the crystallizer which yields crystals containing 4% water. At the crystallization temperature, the solubility is 60 kg/100 kg of water. Major part of the mother liquor leaving the evaporator is returned to the evaporator as recycle. Calculate:

(a) The concentration of KNO_3 in the stream entering the evaporator
(b) kg/h of recycle stream
(c) The rate of production of crystals.

11.10 In a wood drier, the hot air must contain at least 2.5% water to prevent the wood from drying too rapidly and splitting or warping. The original fresh air fed contains 1% water. Wood is dried from 25% water to 5% water. The wet air leaving the drier contains 4% water. Calculate the amount of wet air that must be returned to the drier if 1000 kg per hour of wet wood is dried. All moisture contents are on wet basis.

11.11 Final purification stage in the preparation of vitamins from natural sources requires centrifuging and continuous filtration, as depicted in Figure 11.17.

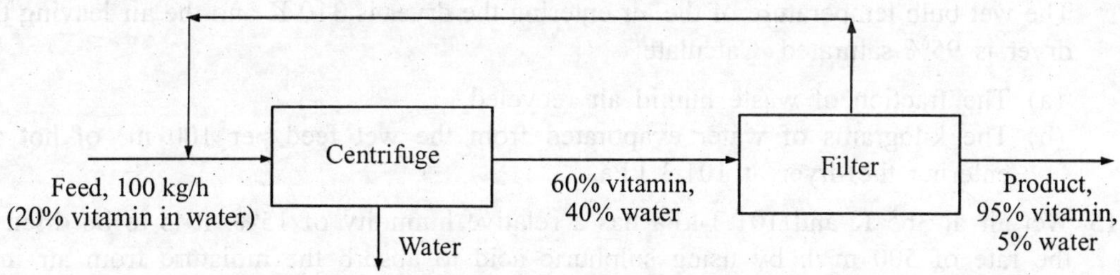

Figure 11.17 Recycle operation in Exercise 11.11.

Determine the flow rate of the recycle stream in kg/h.

11.12 Fresh atmospheric air mixed with recycled air is heated to 349 K admitted to a drier at 101.3 kPa at a rate of 100 m^3/min. The relative humidity of air entering the drier is 7.5%. Atmospheric air is at 302 K and has a relative humidity of 30%. The air leaves the drier at 306 K and 90% relative humidity, a portion of which is recycled and the other portion is discarded. The vapour pressures of water are 4 kPa at 302 K, 5 kPa at 306 K and 40 kPa at 349 K. Calculate:

(a) The ratio of wet recycle stream to wet fresh stream
(b) The mass of dry air (kg) to be fed to the drier to remove 1 kg water in the drier
(c) The rate at which water is removed in kg/min.

11.13 Wet solid containing 75% water is mixed with recycled dry solid to reduce the water content to 50% before being admitted into the granulator. The solid leaving the granulator is fed to a drier where it is brought into contact with dry air initially containing 0.3% water by weight. In the drier, the air picks up moisture and leaves with a moisture content of 6%. The solids leaving the drier contains 15% water. A portion of this solid is recycled.

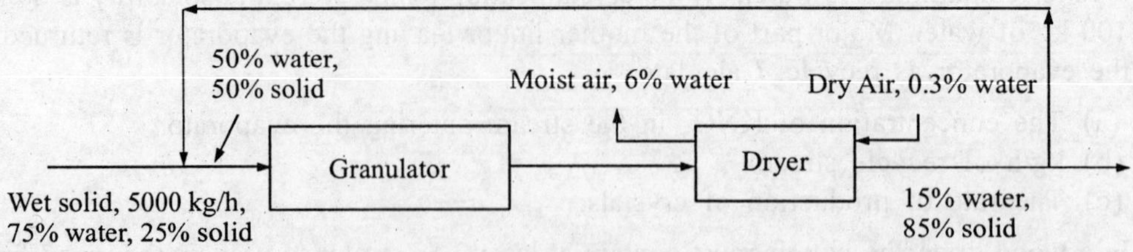

Figure 11.18 Recycle operation in Exercise 11.13.

For 5000 kg/h of wet solid sent to the granulator as fresh feed, determine:

(a) The amount of solid recycled

(b) The circulation rate of air in the drier on a dry basis.

11.14 Fresh air with a dew point of 290 K is used in an adiabatic dryer which recycles a portion of the wet air. The combined mixture of fresh air and recycled air is passed through a heater which raises its temperature to 365 K before admitted to the dryer. The wet bulb temperature of the air entering the dryer is 310 K and the air leaving the dryer is 95% saturated. Calculate:

(a) The fraction of waste humid air recycled

(b) The kilograms of water evaporated from the wet feed per 100 m^3 of hot air entering the dryer at 101.3 kPa.

11.15 Wet air at 365 K and 101.3 kPa has a relative humidity of 15%. It is to be dried at the rate of 500 m^3/h by using sulphuric acid to absorb the moisture from air in a packed tower. The dried air will leave the tower at 327 K and 101.3 kPa and its relative humidity will be 10%. The spent acid (65% H_2SO_4) is partly withdrawn and partly recirculated after making up with fresh 98% H_2SO_4 so that the acid entering the tower has a concentration of 75% H_2SO_4. Calculate:

(a) Cubic meters of wet air leaving the scrubber per hour

(b) Kilograms of make up acid per hour

(c) Kilograms of acid entering the tower per hour.

Vapour pressure of water at 327 K = 15 kPa and at 365 K = 75 kPa.

11.16 A feed consisting of 30% benzene, 40% toluene, and 30% xylene is separated into a distillate containing 93% benzene 5.5% toluene and 1.5% xylene and a bottom product containing 2.0% benzene. All percentages are on mole basis. The reflux ratio is 3.0 and the feed rate is 100 kmol/h. Calculate:

(a) Composition of the residue

(b) Masses of distillate and residue

(c) Recovery of xylene plus toluene in the residue

(d) Mass of vapour condensed in the condenser.

11.17 It is desired to produce 1000 kg/h of $Na_3PO_4.12H_2O$ crystals from a feed solution containing 6% (weight) Na_3PO_4 and traces of impurity. The original solution is first evaporated in an evaporator to a 40% (weight) Na_3PO_4 solution and then cooled to

293 K in a crystallizer, where the hydrated crystals and mother liquor solution are removed. One out of every 10 kg of mother liquor is discarded to waste to get rid of the impurities, and the remaining mother liquor is recycled to the evaporator. The solubility of Na_3PO_4 at 293 K is 9.9% (weight). Calculate:

(a) Feed rate (kg/h)
(b) Rate of evaporation of water (kg/h)
(c) Rate at which the mother liquor is recycled (kg/h)
(d) Concentration of the solution entering the evaporator

11.18 A solution containing 10% NaCl, 5% KCl and 85% water is mixed with a recycle stream and sent to the evaporator. In the evaporator is removed as vapour and NaCl is removed as crystals. The concentrated solution leaving the evaporator (NaCl = 17.0%, KCl = 22%, H_2O = 61%) is sent to a crystallizer from which KCl and NaCl crystals are removed. The mother liquor leaving the crystallizer (NaCl = 15.86%, H_2O = 84.14%) is sent back to the evaporator. Calculate the quantity and composition of the combined stream entering the evaporator if the feed rate is 20000 kg/h.

11.19 A distillation column separates 10000 kg/h of a 55% benzene and 45% toluene mixture. The product recovered from the condenser at the top of the column contains 97% benzene and the bottoms from the column contains 98% toluene. The vapour stream entering the condenser from the top of the column is 9000 kg/h. A portion of the product is returned to the column as reflux, and the rest is withdrawn for use elsewhere. Assume that the composition of the streams at the top of the column, the product withdrawn and the liquid reflux are identical. All compositions are on weight basis. Find the ratio of the amount refluxed to the product withdrawn.

11.20 A solid material containing 12% moisture is dried so that it contains 5% water by blowing fresh warm air mixed with recycled air over the solid in the drier. The inlet fresh air has a humidity of 0.01 kg of water/kg of dry air, the air from the drier that is recycled has a humidity of 0.1 kg of water/kg of dry air, and the mixed air to the drier 0.04 kg of water/kg of dry air. For a feed of 1000 kg solid/h fed to the drier, calculate:

(a) kg/h of dried product
(b) kg of dry air/h in the fresh air
(c) kg of dry air/h in the recycled air

11.21 Fresh air containing 0.01 kg of water/kg of dry air is mixed with recycled air containing 0.15 kg of water/kg of dry air and is blown over a wet solid in a drier. In a certain operation, the wet solid contains 15% (weight) moisture and it is to be dried to a final moisture content of 5% by evaporating moisture into the air blown over it. The fresh air and recycled air are mixed in such proportions that the air blown over the solid contains 0.03 kg of moisture/kg of dry air. A part of the air leaving the drier which contains 0.1 kg of water is recycled. For 100 kg of wet material charged, determine:

(a) The ratio of dry air in the recycled air to that in the fresh air
(b) The quantity of dry air in the fresh air feed
(c) The amount of dry air recycled.

11.22 A solution containing 10% NaCl, 3% KCl and 87% water is fed to the process shown in Figure 11.19 at the rate of 20000 kg/h. The composition of the evaporator product *P* is NaCl = 16.8, KCl = 21.6, H_2O = 61.6. Calculate:

(a) Flow rate of stream *P*
(b) Flow rate of stream *R*
(c) Rate of evaporation
(d) Quantity and composition of stream entering the evaporator

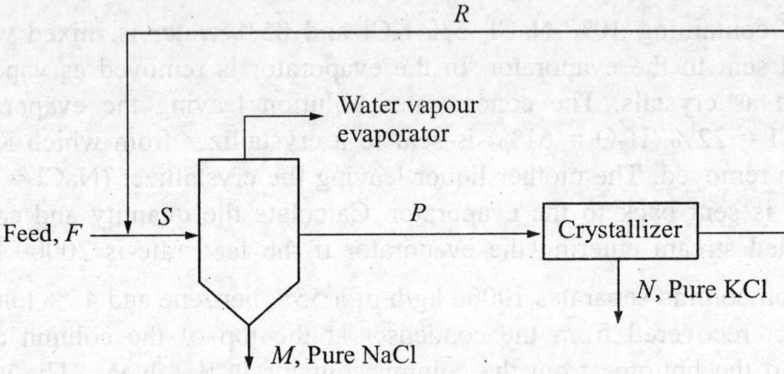

Figure 11.19 Recycle operation in Exercise 11.22.

11.23 A wet organic pigment containing 40% CCl_4 is to be dried to 5%. The drier can handle only 200 kg/h of dry solid. The air used for drying is obtained by mixing fresh air with recycled air. The fresh air is essentially free of CCl_4 and it enters the dryer at 373 K and 101.3 kPa with a dew point 288 K for CCl_4. The air leaving the drier has a CCl_4 dew point of 298 K. Calculate:

(a) The fresh air to be supplied to the drier (kg/h)
(b) The recycle ratio, CCl_4-free basis

The vapour pressure of CCl_4 at 288 K = 7.60 kPa and at 298 K = 15.3 kPa.

11.24 Cloth is dried in a stenter (hot-air dryer) in a textile mill. In this machine, fresh air first mixes with recirculated air. The mixture then passes through a heater. Hot air is jetted over the cloth to evaporate the mixture. A major portion of the jetted air is recirculated while the remaining small portion is exhausted. The operation is shown schematically in Figure 11.20.

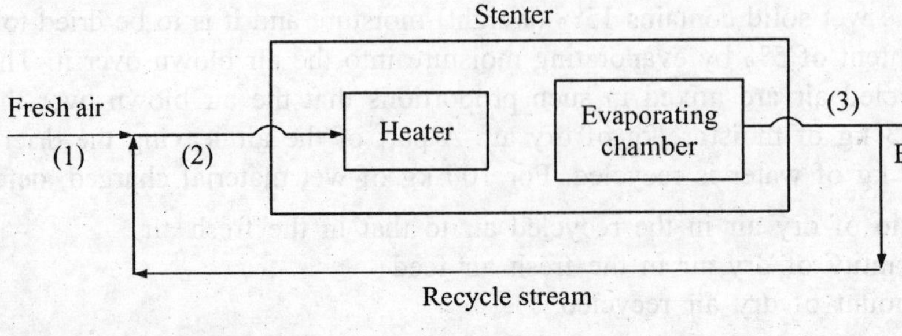

Figure 11.20 Cloth drying process in Exercise 11.24.

Test data on a particular run of a stenter are given below:

Cloth details: width: 1.1 m, density: 0.085 kg/m^2 (dry basis), speed: 60 m/min, inlet moisture: 83% (dry basis), outlet moisture: 10% (dry basis).

Air conditions: Moisture of air at point (1): 0.02 kg/kg of dry air, Moisture of air at point (2): 0.09 kg/kg of dry air, Moisture of air at (3): 0.10 kg/kg of dry air.

Calculate:

(a) The rate of evaporation in the stenter
(b) The mass flow rate of fresh air
(c) The volumetric flow rate of fresh air (m^3/min), if it is supplied at 300 K and 100 kPa
(d) The mass flow rate of recirculating dry air.

11.25 In a process producing KNO_3 salt, 1000 kg/h of a feed solution containing 10% KNO_3 is fed to an evaporator which evaporates some water to produce a 50% KNO_3 solution. This is then fed to a crystalliser, where crystals containing 95% KNO_3 are removed. The saturated solution containing 35% KNO_3 is recycled to the evaporator. Calculate:

(a) The amount of recycle stream (kg/h)
(b) The amount of crystals (kg/h)
(c) The quantity of water evaporated (kg/h)

11.26 Schematic diagram of a process is shown in Figure 11.21.

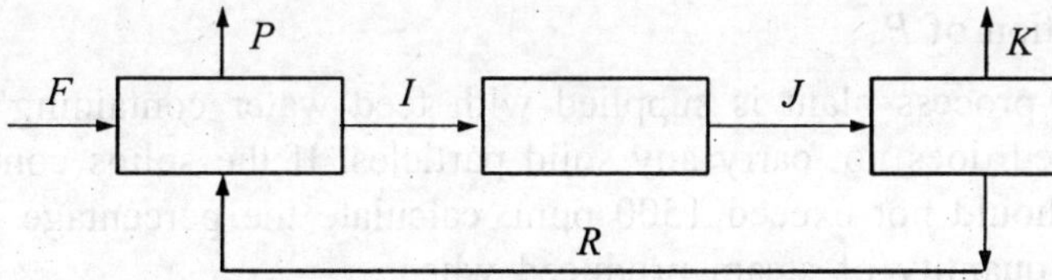

Figure 11.21 Recycle operation in Exercise 11.26.

Composition of various streams are as follows:

Component	*Mol % F*	*Mol % I*	*Mol % J*	*Mol % K*
A	15	4	7	60
B	20	29.4	27	6.5
C	50	54	44.4	16.5
D	15	12.6	21.6	17

Given that 80% of *A* in *J* is present in *K* and the ratio of *F* to *P* is 1.63, calculate:

(a) *R*/*J*
(b) Composition of *P*.

11.27 Schematic flow sheet of a process is shown in Figure 11.22.

The following table gives the composition of various streams in mole%.

Component	*Mol % F*	*Mol % L*	*Mol % M*	*Mol % N*
A	20	30	27.6	6
B	50	50	40.4	14
C	15	5	8	60
D	15	15	24	20

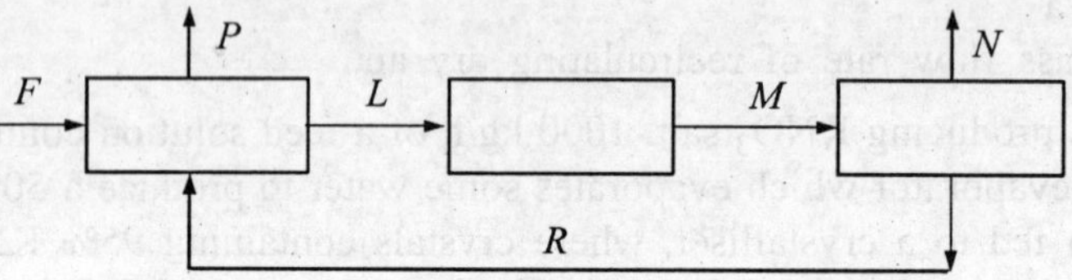

Figure 11.22 Recycle operation in Exercise 11.27.

It is known that stream *N* contains 75% of *C* present in stream *M*. The ratio of *F*/*P* is 5:3 on mole basis. Calculate:

(a) *R*/*M*
(b) Composition of *R*
(c) Composition of *P*.

11.28 A boiler in a process plant is supplied with feed water containing 250 ppm solids. Steam produced does not carry any solid particles. If the solids content in the boiler at any time should not exceed 1500 ppm, calculate the percentage boiler blowdown based on the quantity of steam produced when

(a) The entire feed to the boiler is fresh feed water
(b) 50% of the feed entering the boiler is constituted by the condensate recycled from the boiler.

11.29 A boiler feed water contains 2000 ppm dissolved solids. The permissible limit of dissolved solids in the water in the boiler is 1 part per 100 parts. What percent of the feed water is to be blowndown?

11.30 High-pressure steam generated in a plant is sent to the power generation unit, 50% of which is further utilized in the plant as process stream and the remaining 50% is sent back to the boiler as condensate. The condensate carries with it dissolved solids to the extent of 50 ppm. To keep the solid content in the boiler water within 1500 ppm, a part of the boiler water is blowndown continuously. To make up for the blowdown as well as for the water lost as the process steam, feed water containing 450 ppm solids is fed to the boiler. Calculate the feed to blowdown ratio.

11.31 Seawater is desalinated by reverse osmosis, as shown in Figure 11.23. All compositions are on mass basis. Calculate *R*/*E*.

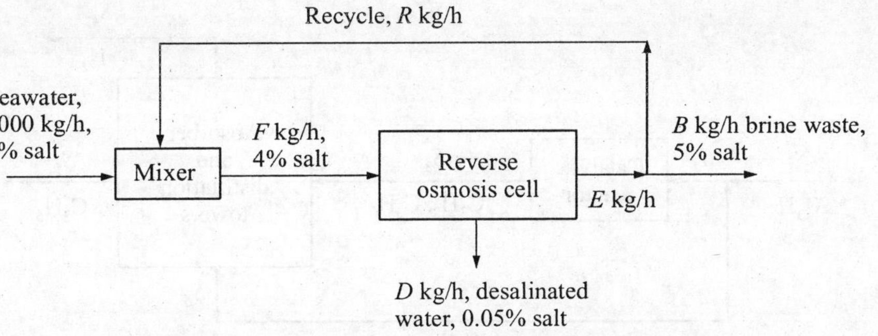

Figure 11.23 Reverse osmosis process in Exercise 11.31.

11.32 Potable water containing not more than 500 ppm dissolved salt is made by desalinization through reverse osmosis of seawater which contains 3.0% salt. Fresh seawater is admitted at a rate of 1000 kg/h. Potable water is withdrawn from the reverse osmosis cell as product while a fraction of the brine that leaves the cell and containing 5.25% salt is recycled. The concentration of salt in the stream entering the cell after mixing the recycle stream with fresh seawater is 4.0%. Determine:

(a) The rate at which brine is removed from the plant (kg/h)
(b) The rate at which potable water is produced (kg/h)
(c) The fraction of the brine leaving the cell that is recycled.

Recycle and purge involving chemical reactions

11.33 Methanol is converted to formaldehyde according to:

$$2CH_3OH + O_2 \rightarrow 2HCHO + 2H_2O$$

$$CH_3OH \rightarrow HCHO + H_2$$

A mixture containing 80% methanol and 20% oxygen is supplied as fresh feed in a formalin plant. The conversion of methanol per pass is 50% and all the oxygen is converted. The exit stream from the reactor contains unconverted methanol, hydrogen, water and formaldehyde. All the formaldehyde and water are removed from the separator. The hydrogen is recovered from the recycle stream before it is sent to the reactor. For a basis of 100 kmol methanol entering as fresh feed, determine:

(a) The mass of formaldehyde produced (kg)
(b) The mass of hydrogen produced (kg)
(c) The mass of methanol recycled (kg).

11.34 The process schematic of a propane dehydrogenation plant is shown below in Figure 11.24. It is desired to set up a simplified version of the material balance for this plant. Assume that the only reaction is the dehydrogenation of propane to propylene; there are no side reactions. The yield of propylene per pass is 30%. Assume that the amount of carbon formed on the catalyst is negligible. The product flow rate (stream S_5) is 60 kmol/h. Calculate the flow rates of all the other streams. Notice that all streams except stream S_3 are pure.

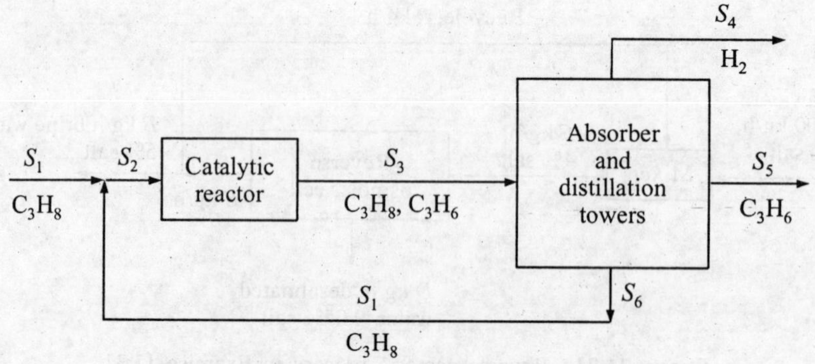

Figure 11.24 Propane dehydrogenation, Exercise 11.34.

11.35 Methanol is produced by the reaction of CO with H_2.

$$CO + 2H_2 \rightarrow CH_3OH$$

Only 15% of carbon monoxide entering the reactor is converted to methanol. The methanol formed is condensed and recovered completely. The unreacted CO and H_2 are recycled back to the reactor. The feed will contain H_2 and CO in the ratio 2:1. For 3200 kg/h of methanol produced, calculate

(a) The amount of fresh feed (kmol/h)
(b) The amount of recycle gas (kmol/h).

11.36 It is proposed to produce acetaldehyde by oxidation of ethanol in the gas phase.

$$C_2H_5OH(g) + \frac{1}{2}O_2(g) \rightarrow CH_3CHO(g) + H_2O(g)$$

the ratio of air to ethanol in the fresh feed (before it is mixed with recycle stream) is 10 to 1 by mole. The conversion of ethanol on a single pass through the reactor is 25%. The unreacted ethanol is completely separated from the reaction products and recycled.

(a) What is the ratio of recycle stream to the fresh feed stream?
(b) What is the composition of the outlet stream from the reactor in mass percent and mole percent?

11.37 Glucose ($C_6H_{12}O_6$) is enzymatically converted to fructose ($C_6H_{12}O_6$) in a fixed-bed reactor. Only a portion of the exit stream from the reactor is removed as product, the other portion being recycled to the reactor. The ratio of the moles of product to recycle stream is 8.5. The recycle stream is mixed with the fresh feed and the combined stream containing 5% fructose is admitted to the reactor. The fresh feed is a 60% glucose solution in water. What is the percent conversion of glucose to fructose on one pass through the reactor?

11.38 Limestone containing 95 weight % $CaCO_3$ and 5% SiO_2 is calcined in a kiln.

$$CaCO_3 \rightarrow CaO + CO_2$$

The heat requirement in the kiln is met by burning coke which is essentially carbon in a furnace and by admitting the hot flue gases into the kiln. The flue gas analysis at the furnace exit shows 8% CO_2. This is mixed with a fraction of the gases leaving the kiln so that CO_2 content at the kiln inlet is 10% CO_2. If the gases leaving the kiln analyzed 12% CO_2, determine for 100 kg coke burned,

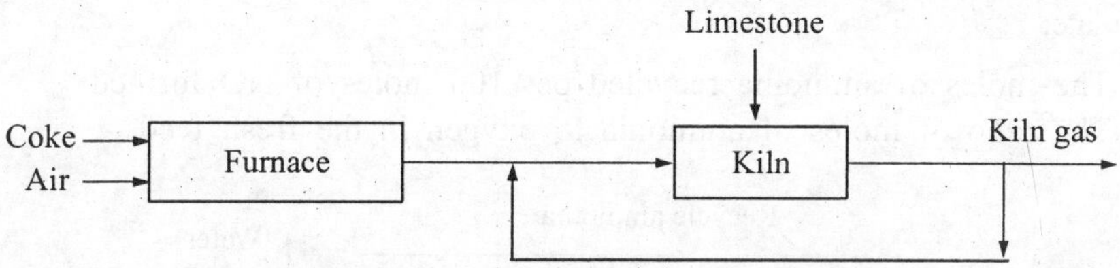

Figure 11.25 Calcination of lime, Exercise 11.38.

(a) The mass of lime produced (kg) and its percent purity.
(b) Recycle ratio
(c) Moles of stack gases.

11.39 Methanol vapour can be converted into formaldehyde by the following reaction scheme:

$$CH_3OH + \frac{1}{2}O_2 \rightarrow HCHO + H_2O$$

$$CH_3OH \rightarrow HCHO + H_2$$

The fresh feed to the process is 0.5 k mol/h of O_2 and an excess methanol. All of the oxygen reacts in the reactor. Formaldehyde and water are removed from the product stream first, after which hydrogen is removed from the recycled methanol. The recycle flow rate of methanol is 1 kmol/h. The ratio of methanol reacting by decomposition to that by oxidation is 3. Calculate:

(a) The fresh feed rate of methanol
(b) The single pass conversion of methanol
(c) The hydrogen produced.

11.40 The chlorination of methane to form methyl chloride results in the formation of some polychlorinated methane also. A mixture of methane and chlorine in the mole ratio 10:1 is sent to a reactor. The products analyze 92.2% CH_4, 6.2% CH_3Cl, 1.0%CH_2Cl_2, 0.4% $CHCl_3$, 0.1% CCl_4 on an HCl-free basis. The mixture is scrubbed with water to remove HCl gases. The chlorinated products in the gas leaving the scrubber are condensed and removed from the excess methane which is recycled. Additional chlorine is added to keep the methane: chlorine ratio in the feed to the reactor at 10:1. For a basis of 100 kg of chlorinated products per hour, determine:

(a) Moles of methane and chlorine admitted as fresh feed.
(b) The mass of methane (kg) recycled per hour
(c) The percent excess chlorine in the fresh feed based on conversion to methyl chloride
(d) The cubic metres of gas leaving the separator at STP.

11.41 In an attempt to provide a means of generating NO cheaply, gaseous ammonia is burned with O_2. It is desired that oxygen entering the reactor is 20% in excess of that theoretically required. The reaction is 70% complete. The nitric oxide is separated from the unreacted NH_3, and the latter is recycled, as shown in Figure 11.26.

$$4NH_3 + 5O_2 \rightarrow 4NO + 6H_2O$$

Calculate:

(a) The moles of ammonia recycled per 100 moles of NO formed
(b) The ratio of moles of ammonia to oxygen in the fresh feed.

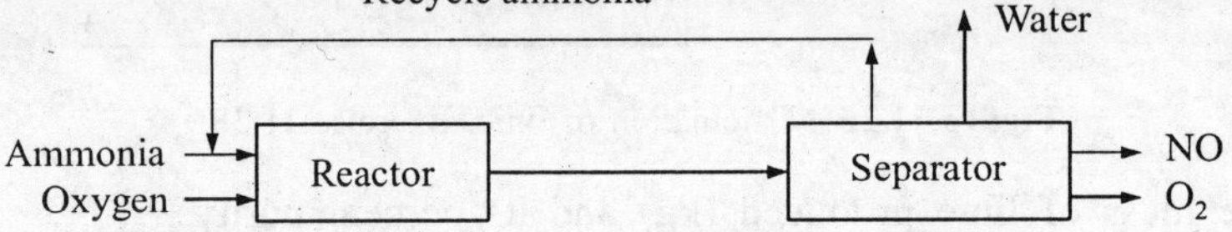

Figure 11.26 Ammonia oxidation, Exercise 11.41.

11.42 Figure 11.27 shows the three main steps in the production of vinyl chloride from ethylene—chlorination (A), oxyhydrochlorination (B) and pyrolysis (C).

The reactions occurring are:

Chlorination: (The yield on ethylene is 98%)

$$C_2H_4 + Cl_2 \rightarrow C_2H_4Cl_2$$

Oxyhydrochlorination: (Yield on ethylene is 95%, on HCl 90%)

$$C_2H_4 + 2HCl + \frac{1}{2}O_2 \rightarrow C_2H_4Cl_2 + H_2O$$

Pyrolysis: (Yield on dichloroethylene is 99%, on HCl 99.5%)

$$C_2H_4Cl_2 \rightarrow C_2H_3Cl + HCl$$

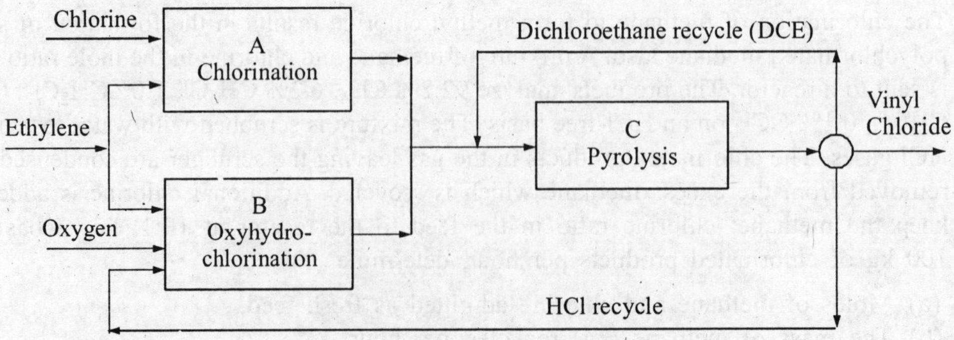

Figure 11.27 Vinyl chloride production from ethylene, Exercise 11.42.

The HCl from the pyrolysis step is recycled to the oxyhydrochlorination step. The flow of ethylene to the chlorination and oxyhydrochlorination reactors is adjusted so that the production of HCl is in balance with the requirement. The conversion in the pyrolysis reaction is limited to 55%, and the unreacted dichloroethane is separated and recycled. For the production of 25000 kg/h of vinyl chloride, determine (a) the flow of ethylene to each reactor and (b) the total flow of dichloroethane to the pyrolysis reactor.

11.43 Methanol is produced by the reaction of CO with H_2.

$$CO + 2H_2 \rightarrow CH_3OH \tag{A}$$

The side reaction is

$$CO + 3H_2 \rightarrow CH_4 + H_2O \tag{B}$$

The conversion per pass is 15%. Of this amount, 87.5% is assumed to react via equation (A) and 12.5% via equation (B). The stream leaving the reactor passes through a condenser and a separator. The CO and H_2 leaving these units are recycled. The ratio of H to CO in the recycle is 2:1. The CH_4 leaves as a gas and the liquid mixture of methanol and water passes to a distillation column for the concentration of methanol. The flow diagram is given in Figure 11.28.

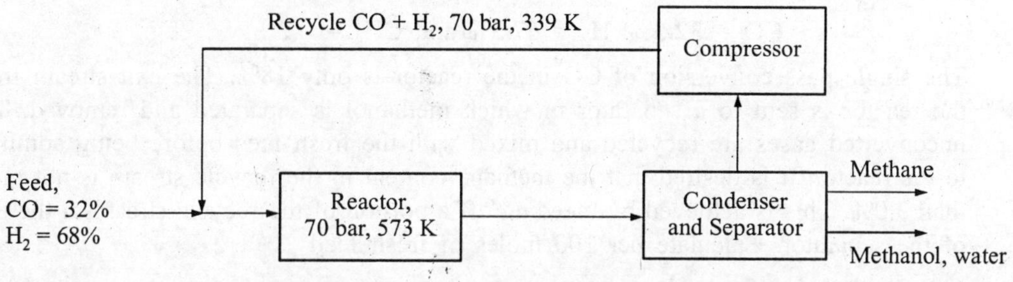

Figure 11.28 Methanol synthesis, Exercise 11.43.

For 100 moles of fresh feed, determine:

(a) Moles of gas recycled
(b) Composition of the stream leaving the reactor
(c) Moles of methanol produced

11.44 In the production of ammonia from hydrogen and nitrogen, the conversion based on either raw material is limited to 15%. The ammonia produced is condensed and separated. The uncondensed gases are recycled. The argon in the recycle is to be limited to 5%. If the feed contains 0.2% argon, calculate the purge required.

11.45 *o*-Xylene is oxidized to phthalic anhydride by oxygen according to:

$$C_8H_{10} + 3O_2 \rightarrow C_8H_4O_3 + 3H_2O$$

A mixture of *o*-xylene and air containing 6% xylene is sent to a fixed-bed reactor at 625 K along with a recycle stream coming from the condenser in which the phthalic anhydride and water produced in the reactor are separated and removed. The recycle is provided with a purge and the purge stream analyzed 78% N_2. It is found that for every mole of gas mixture entering as fresh feed, 1.2 moles leave the reactor. If 60% conversion is achieved in a single pass and 100 kmol of a mixture of xylene and air is admitted as fresh feed, determine: (a) the rate of production of pthalic anhydride (kg/h) and (b) xylene produced (kg/h).

11.46 A 1:3 nitrogen-hydrogen mixture is fed to a converter resulting in 20% conversion to ammonia. After complete separation of ammonia, the remaining unconverted gases are recycled to the converter. The initial reaction mixture contains 0.2% by volume argon. If the limit of argon in the reactor is 5% by volume of the N_2–H_2 mixture in the reactor, estimate:

(a) Fraction of the recycle that is purged
(b) Moles of ammonia produced per 100 mole feed
(c) Overall conversion of ammonia.

11.47 Methanol is made by the reaction of CO with H_2 according to the reaction

$$CO + 2H_2 \rightarrow CH_3OH$$

The gas mixture entering the reactor is contaminated by small amounts of methane. The composition of the gas mixture is as follows:

$$CO = 32.5\%,\ H_2 = 67.3\% \text{ and } CH_4 = 0.2\%$$

The single pass conversion of CO in the reactor is only 18%. The exit stream from the reactor is sent to a separator in which methanol is separated and removed. The unconverted gases are recycled and mixed with the fresh feed before being admitted to the reactor. It is desired that the methane content in the recycle stream is not more than 3.0%. This is achieved by bleeding off a portion of the recycle stream at the exit of the separator. Calculate per 100 moles of fresh feed

(a) The moles of recycle stream
(b) The moles of purge stream
(c) Composition of the purge stream
(d) Moles of methanol produced.

11.48 Refer to Figure 11.29. Ethylene is oxidized to ethylene oxide by the reaction:

$$C_2H_4 + \frac{1}{2}O_2 \rightarrow CH_2CH_2O$$

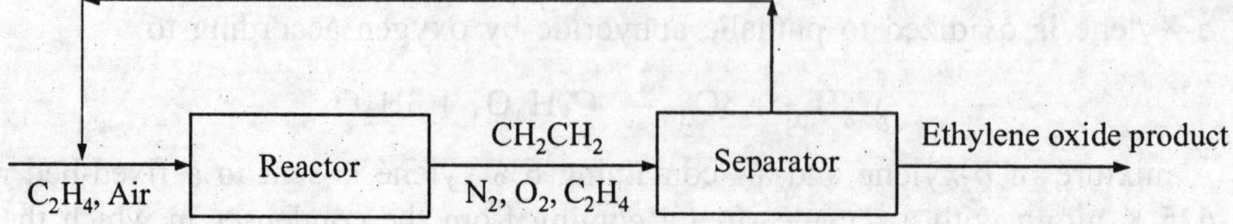

Figure 11.29 Oxidation of ethylene, Exercise 11.48.

Ethylene to air ratio in the stream entering the reactor is to be 1:10. The conversion of ethylene is 25% per pass. Ethylene oxide is separated and the unconverted gas is recycled. To maintain the ethylene-air ratio in the reactor, 25% by volume of the recycle stream is purged off. Also, pure oxygen is supplied in addition to the ethylene-air mixture supplied as fresh feed. For producing 100 kg of ethylene oxide per hour, calculate:

(a) Ethylene-air ratio in the fresh feed
(b) Moles of gas purged off
(c) kg/h of ethylene to be supplied as fresh feed
(d) kg/h of pure oxygen supplied.

11.49 In the Haber process for the manufacture of ammonia, synthesis gas which analyzes 75.2% H_2, 24.5% N_2 and 0.3% Ar is sent to the converter after mixing with recycle stream. The following reaction takes place in the converter:

$$N_2 + 3H_2 \rightarrow 2NH_3$$

To avoid building up of argon in the recycle stream, which will adversely affect the equilibrium conversion after a certain concentration, a small purge is provided in the recycle. The gas mixture entering the converter contains 79.5% H_2. Ammonia is separated from the converter gases in a separator and the unreacted gases are recycled. The gases leaving the separator are free from ammonia and contain 80.0% H_2 and the product ammonia contains no dissolved gas. For a basis of 100 moles of fresh feed, calculate:

(a) The moles of recycle stream
(b) The moles of purge stream
(c) The percent conversion of hydrogen per pass.

11.50 A synthesis gas consisting of 24.3% N_2, 75.2% H_2, and 0.5% Ar is fed to an ammonia converter mixed with recycled stream. The single pass conversion of nitrogen is 10%. Unconverted gases leaving the reactor is separated and returned to the reactor along with fresh feed. In order to keep the concentration of Ar in the recycled stream below 1%, a portion of the recycled stream is purged off. For 100 kg of fresh synthesis gas, determine:

(a) Moles of recycle stream
(b) Moles of purge stream
(c) Ammonia produced (kmol).

11.51 A gas mixture containing nitrogen and hydrogen in the ratio of 1:3 by volume is sent to the ammonia converter where its conversion to NH_3 is 25% complete. The ammonia formed is condensed and separated by passing the gases leaving the converter through a condenser. The unconverted gases are recycled. To prevent the accumulation of argon, a portion of the recycle stream is purged off. The fresh feed contains 0.003 mol argon per mole of N_2–H_2 mixture, and the level of argon in the reactor should not exceed 0.05 mol per mol of N_2–H_2 mixture, determine:

(a) The fraction of the recycle stream that is purged off
(b) The concentration of Ar in the recycle stream
(c) Moles of N_2–H_2 mixture entering the reactor per 100 moles of gas admitted as fresh feed.

11.52 Benzene is produced from toluene as per the following reaction:

$$C_7H_8 + H_2 \rightarrow C_6H_6 + CH_4$$

The following side reaction is also found to occur:

$$2C_7H_8 + H_2 \rightarrow C_{12}H_{10} + 2CH_4$$

The reaction is carried out according to Figure 11.30.

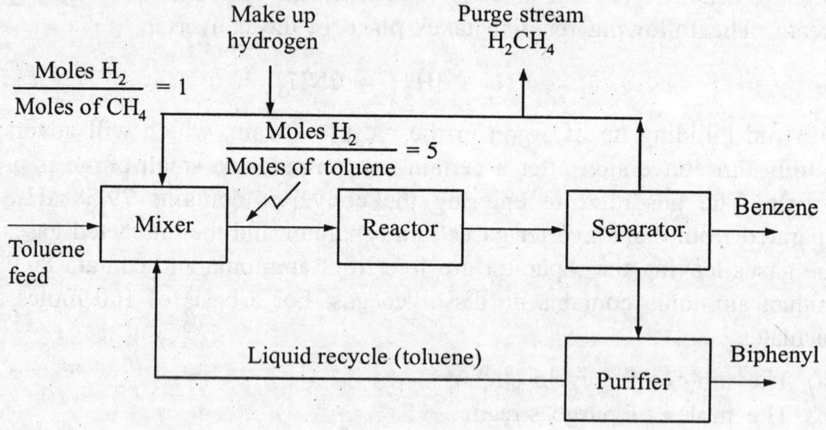

Figure 11.30 Production of benzene from toluene, Exercise 11.52.

It is known that conversion of toluene per pass is 75%. And 90% of the toluene reacted is converted to benzene. For 100 moles of toluene entering the reactor, calculate:
(a) Moles of gas purged
(b) Composition of the purge stream
(c) Moles of makeup hydrogen

11.53 N_2 and H_2 in the ratio of 1:3 constitute the synthesis gas for the manufacture of ammonia. Synthesis gas is obtained by burning excess hydrogen in air. Since air contains 0.95% Ar, the gas produced will be contaminated with Ar. The gas is fed to a converter after mixing with recycle stream. The conversion of the gas to ammonia is only 25% complete. The unconverted gases are separated from the product ammonia and are recycled. The combined stream consisting of fresh synthesis gas and the recycle stream should not contain Ar above 5%. To keep the level of Ar below this, a portion of recycle is purged off. For 100 kmol of hydrogen in the fresh feed, determine:
(a) The moles of ammonia produced
(b) The percent of hydrogen in the fresh feed that is converted to ammonia
(c) The moles of gas purged off.

12 Energy Balance
Thermophysics

Chemical process industries in general are huge consumers of energy. Efficient and proper conversion of the scarce resources into energy and the proper utilization of the energy so obtained in the operation of the plant are naturally two major concerns in chemical engineering. So, the engineer should be competent to identify cheap sources of fuel, choose proper operating conditions to reduce loss of energy, design appropriate waste heat recovery systems, etc. An energy balance helps the engineer in all these tasks. Energy balances are carried out in the operation and the process design stages of chemical process industries. The energy requirements for heating and cooling or the power requirement of a process are determined by carrying out energy balance during the process design stage. An energy balance in plant operation helps us to identify the pattern of energy consumption and suggest possible means for the conservation of energy.

The energy balance principle can be made clear by considering the theory behind heat exchangers, which is one of the most common and widely used process equipment. Almost all process industries employ heat exchangers in one form or another, either for heating or cooling of process streams or for the recovery of waste heat. A schematic sketch of a simple shell-and-tube heat exchanger is shown in Figure 12.1.

A heat exchanger is typically a device that performs the transfer of heat between two fluids. The shell-and-tube type heat exchanger consists of a large number of tubes welded into fixed tube plates at each end. The tube bundle is housed in a cylindrical chamber known as *shell*. Baffles are placed in the shell to ensure better transfer of heat. One fluid is passed through the inside of the tubes and the other fluid outside the tubes. The former is called *tube-side fluid* and the latter the *shell-side fluid*. In a parallel flow exchanger, both fluids enter at one end and flows in the same direction throughout the length of the exchanger. In a counter-current exchanger, the fluids are admitted at the opposite ends and flow in the opposite directions. Heat is transferred from the hot fluid to the cold fluid. The hot fluid gets cooled, and the cold fluid gets heated in the exchanger. The heat given out by the hot fluid is equal to the heat absorbed by the cold fluid, if heat losses are negligible. This heat balance forms the basis for the calculation of the exit temperature of the fluids, the flow rates of streams and the size of the exchanger.

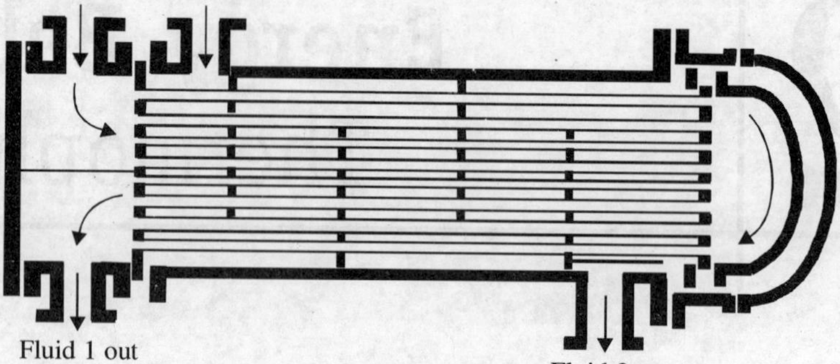

Figure 12.1 Shell-and-tube heat exchanger.

The heat balance is just a simple form of the general energy balance. We present here and in the next chapter the basic principles of energy balance and illustrate the use and application of energy balance through practical problems. Just as material balance is based on the law of conservation of mass, energy balance is based on the law of conservation of energy, also known as the *first law of thermodynamics.*

12.1 LAW OF CONSERVATION OF ENERGY

The law of conservation of energy states that *although energy may be converted from one form to another, it cannot be created or destroyed.* Whenever a quantity of one form of energy is produced, an exactly equivalent amount of another kind must be used up. Therefore, it is impossible to construct a perpetual motion machine (PMM) of the first kind that would produce mechanical work continuously without drawing energy from an external source and without itself undergoing a change. In fact, the failure to construct such a machine formed the basis for the law of conservation of energy (H. Von Helmholtz, 1847). When the law of conservation of energy was proposed, heat was not considered to be a form of energy. It was a series of experiments conducted by Joule between 1843 and 1847 that established heat as a form of energy.

Joule's classical experiment established a quantitative equivalence between heat and work. This was achieved by doing mechanical work on a known quantity of water taken in an insulated container by the following different methods:

1. Converting mechanical work into electricity and passing the resulting electric current through water
2. Compressing a gas confined in a cylinder immersed in water mechanically
3. Agitating the water by means of a paddle wheel
4. Rubbing two metal pieces immersed in water against each other

In all the above methods, mechanical work was converted to heat which was measured by a rise in temperature. Irrespective of the way in which mechanical work was converted to heat energy, it was found that an exact proportionality existed between the amount of work expended on the water and the rise in temperature. Thus, Joule's experiments established a quantitative relationship between heat and work thus rejecting the earlier theory that treated heat as a substance. The recognition that heat and later, internal energy as forms of energy led to their inclusion in the law of conservation of energy along with work, potential energy and kinetic energy, and the law came to be known as the first law of thermodynamics. Before we embark on the general mathematical treatment of the first law of thermodynamics and formulate energy balance equations applicable for different situations, we should be familiar with the various forms of energy that appear in an energy balance. So, we defer the detailed mathematical treatment of energy balance to section 12.8.

12.2 COMPONENTS OF ENERGY BALANCE EQUATIONS

12.2.1 Heat and Work

Work: Energy is expended in the form of work when a force acts through a distance. Thus,

$$dW = F\,dZ \tag{12.1}$$

where W is the work done, F is the force acting and Z is the displacement. The unit of work in the SI system is N m (newton-metre) or J (joule).

Let us consider the expansion or compression work in a cylinder, an important quantity in engineering calculations. Assume that a gas is confined in a cylinder and let the pressure of the gas be P and volume V. If the surface area of the piston exposed to the gas is A, the force acting on the piston is

$$F = PA \tag{12.2}$$

The displacement of the piston in the direction of the force dZ is related to the change in volume dV of the gas as

$$dZ = \frac{dV}{A} \tag{12.3}$$

Substituting Eqs. (12.2) and (12.3) in Eq. (12.1), we get

$$dW = P\,dV \tag{12.4}$$

If the volume of the gas changes from the initial value V_1 to the final value V_2, Eq. (12.4) may be readily integrated to get the work done on the face of the piston:

$$W = \int_{V_1}^{V_2} P\,dV \tag{12.5}$$

The pressure–volume history of the gas undergoing the change in state is illustrated on the P-V diagram of Figure 12.2. The integral of Eq. (12.5) is given by the area under the curve between the limits V_1 and V_2.

The area, and hence the work done in the compression or expansion of the gas depend on the shape of the PV-curve, thus establishing that the work done in a process is a path function.

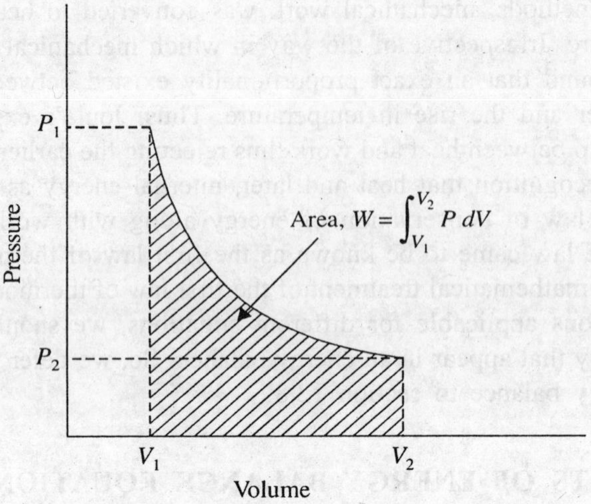

Figure 12.2 Work of expansion on *PV*-diagram.

Power: The power is defined as the rate of doing work. Its unit in the SI system is J/s, commonly designated as W (watts). In engineering calculations, power is sometimes expressed as horsepower (hp). 1 hp = 745.7 W.

Heat: Heat is that quantity which is transferred between bodies due to the difference in temperatures existing between them. In the early years of the development of the science of thermodynamics, this quantity was thought of as a substance called *caloric*. Heat is now recognised as a form of energy that cannot be stored as such within the system. Heat is manifested only during a change of state of the system; it is energy in transit, like work. Heat exchanged in a process depends on the way in which the process is carried out, as pointed out earlier. Therefore, just as work, heat is a path function. And like work, it is expressed in J. Two other units used for heat are the *calorie* and the *BTU* (the British Thermal Unit):

$$1 \text{ calorie} = 4.1868 \text{ J}; \; 1 \text{ BTU} = 1055.04 \text{ J}$$

12.2.2 Kinetic Energy, Potential Energy and Flow Energy

Kinetic and potential energy: Energy is a quantity that can be stored within the system and can be exchanged between the system and the surroundings. The exchange of energy occurs either as heat or as work. Heat and work are called energy in transit; they cannot be stored within the system. The energy stored within the system due to its position above some arbitrary reference plane is referred to as its *potential energy* (PE). If a body of mass m is at an elevation z above the ground, the potential energy of the mass is

$$\text{PE} = mgz \tag{12.6}$$

where g is the acceleration due to gravity. The energy possessed by the body by virtue of its motion is called its *kinetic energy* (KE). If a body of mass m is moving at a velocity u, the kinetic energy of the body is

$$\text{KE} = \frac{1}{2} mu^2 \tag{12.7}$$

Flow energy: A fluid element in a flow system possesses an additional energy which is known as the *flow energy*. Consider a fluid entering a tube with a pressure P, and let its specific volume be V. A mass of fluid entering the system is being pushed by the fluid behind it by a certain force. This force multiplied by the distance through which the force acts is equal to the *Flow energy* or *entrance work*. The force acting on the fluid equals the pressure times the cross-sectional area of the tube. Since a unit mass of the fluid at pressure P occupies a volume V, the distance through which the force acts in order to move the fluid into the system is equal to V/A, where A is the cross-sectional area of the tube. This work which is equal to PV for unit mass of the fluid is done by the flowing fluid and therefore is called "flow energy."

EXAMPLE 12.1 A body of mass 75 kg is lifted through a distance of 10 m. What is the work done? If it takes 2.5 minutes for lifting the mass, calculate the power. (Take g = 9.81 m/s^2)

Solution Work is done when the force is undergoing a displacement. Force is measured as mass times acceleration.

Here,

$$\text{force} = mg = 75 \times 9.81 = 735.75 \text{ kg m/s}^2 = 735.75 \text{ N}$$

$$\text{Work done} = F \times \Delta Z$$

$$= 735.75 \times 10 = 7357.5 \text{ N m } (= 7357.5 \text{ J})$$

Power is the work done in unit time.

$$\text{Power required} = \frac{7357.5}{2.5 \times 60} = 49.05 \text{ J/s } (= 49.05 \text{ W})$$

EXAMPLE 12.2 The potential energy of a body of mass 10.0 kg is 1.5 kJ. What is the height of the body from the ground? If a body of mass 10 kg is moving at a velocity of 50 m/s, what is its kinetic energy?

Solution The potential energy (PE) is given by Eq. (12.6), i.e.

$$\text{PE} = mgz$$

$$1.5 \times 10^3 = 10 \times 9.8067 \times z$$

Thus, the elevation z = 15.3 m.

Kinetic energy (KE) is given by Eq. (12.7), i.e.

$$\text{KE} = \frac{1}{2} mu^2$$

$$= \frac{1}{2} \times 10 \times (50)^2 = 12\,500 \text{ N m} = 12.5 \text{ kJ}$$

EXAMPLE 12.3 A gas confined in a cylinder is maintained at a constant pressure by means of a weight placed on the piston. The piston is frictionless. The diameter of the piston is 100 mm and the mass of the piston and the weight together is 50 kg. The atmospheric pressure is 1.0133 bar (take g = 9.8 m/s^2). Determine

(a) The pressure of the gas
(b) The work done by the gas in J if the gas is allowed to expand pushing up the piston and the weight by 500 mm

Solution

(a) Force exerted by the atmosphere

$$= \text{pressure} \times \text{area}$$

$$= 1.0133 \times 10^5 \times \frac{\pi}{4} \times (100 \times 10^{-3})^2 = 795.84 \text{ N}$$

Force exerted by the piston and weight is

$$m \times g = 50 \times 9.81 = 490.5 \text{ N}$$

Total force acting on the gas is

$$795.84 + 490.5 = 1286.34 \text{ N}$$

$$\text{Pressure} = \text{force/area} = \frac{1286.34}{(\pi/4) \times (100 \times 10^{-3})^2} = 1.6378 \times 10^5 \text{ N/m}^2 \ (= 1.6378 \text{ bar})$$

(b) Work done = force × displacement = 1286.34 × (500 × 10^{-3}) = 643.17 J
(*Note:* The work done can be calculated using Eq. (12.5). The change in volume due to the expansion of gas is

$$\Delta V = \frac{\pi}{4} \times (100 \times 10^{-3})^2 \times 500 \times 10^{-3} = 3.927 \times 10^{-3} \text{ m}^3$$

The work done against a constant external pressure by the gas inside the cylinder is

$$W = \int P\, dV = P\Delta V$$

$$= 1.6378 \times 10^5 \times 3.927 \times 10^{-3} = 643.16 \text{ J})$$

EXAMPLE 12.4 Liquid benzene of specific gravity 0.879 flows at a rate of 5 m^3/h with a pressure differential of 3500 kPa. What is the power requirement of the pump in *W*?

Solution Energy input by the pump per unit mass of fluid is $\frac{\Delta P}{\rho}$ where ΔP is the pressure differential against which the fluid is being pumped. The power requirement of the pump is calculated as the energy to be supplied in unit time. This is obtained by multiplying $\frac{\Delta P}{\rho}$ by m where m is the mass flow rate of the fluid. Here,

$$\Delta P = 3500 \times 10^3 \text{ N/m}^2$$

$$\rho = 879 \text{ kg/m}^3$$

$$m = \frac{5 \times 879}{3600} \text{ kg/s}$$

Therefore, the power requirement of the pump is

$$W = \frac{\Delta Pm}{\rho} = \frac{3500 \times 10^3 \times 5}{3600} = 4861 \text{ W}$$

12.2.3 Internal Energy

Heat and work represent energy in transit. Energy exchange between the system and the surroundings occurs either as heat or as work. Heat and work are thus external effects of changes occurring within the system or are external manifestations of stored energy. What happens to the state of the system after heat is supplied to it and before work is extracted from it? The system undergoes a change because energy gets stored within it. The energy stored in the system by virtue of the configuration and motion of the molecules constituting the system is called its *internal energy*. It is a definite property of the system and is denoted by U.

The molecules constituting the system possess kinetic energy of translation, rotation and vibration. They also possess potential energy due to the forces of attraction existing among them. These molecular potential and kinetic energies contribute to the internal energy of the system. These are different from the energy due to the mass motion of the system as a whole (the kinetic energy), and that due to its external position in a gravitational, electrical or magnetic field (the potential energy) which we have defined earlier. The addition of heat to the system results in the increase of its molecular kinetic energy and thus increases the internal energy of the system.

A given system under a given set of conditions has a definite internal energy. In a cyclic process the internal energy of the system after a series of changes remains the same as it was before. The heat and work involved in a non-cyclic process are not equal. Consequently, during a non-cyclic process some energy gets stored in the system or some stored energy gets removed from the system. These changes in the stored energy are measured as the change in the internal energy of the system. The change in internal energy ΔU in a closed system is equal to the heat added to the system (Q) minus the work done by the system (W).

$$\Delta U = Q - W \tag{12.8}$$

Equation (12.8) is the mathematical statement of the first law of thermodynamics for a closed system.

The internal energy which a system possesses by virtue of the molecular configuration and motion of molecules is a property of the system and is solely determined by the thermodynamic state and not by the path by which that state was arrived at. That is to say, internal energy is a state function. As is obvious, the kinetic energy and potential energy are not thermodynamic properties of the system. They do not change with change in the temperature or pressure of the body. It should be borne in mind that it is impossible to measure the total internal energy of a substance and therefore absolute values of internal energy are not known.

12.2.4 Enthalpy

For a system kept at constant volume in which no work of expansion or any other kind of work is done, the entire heat supplied goes to change its internal energy. When the system is free to change its volume against a constant external pressure, the change in internal energy is no longer equal to the energy supplied as heat. In effect, a part of the energy supplied is utilized by the system for occupying a new volume; the energy thus utilized is equal to the work required to 'push' the surroundings against a constant pressure. Consequently, $\Delta U < Q$. However, the heat supplied at constant pressure can be measured as the change in another thermodynamic property of the system which we call the *enthalpy*. Enthalpy is denoted by H and is perhaps, the most widely used of all thermodynamic functions. It is defined as

$$H = U + PV \tag{12.9}$$

Here U is the internal energy of the system, P the absolute pressure and V the volume of the system. Since U, P and V are all state functions, any combination of them also must be a state function. Enthalpy is, therefore, a state function. Enthalpy may be treated as 'total energy' because it includes both the intrinsic energy it possesses (U) and the energy due to the expansion possibilities of the system or the energy that the system possesses because of it occupying a space (PV).

In differential form Eq. (12.9) can be written as

$$dH = dU + d(PV) \tag{12.10}$$

which on integration gives

$$\Delta H = \Delta U + \Delta(PV) \tag{12.11}$$

Equation (12.11) is applicable for any finite change occurring in the system.

We can show that the change in enthalpy for a reversible, non-flow process at constant pressure is equal to the heat supplied. Equation (12.10) can be expanded as

$$dH = dU + P\,dV + V\,dP \tag{12.12}$$

Substituting Eq. (12.8) into Eq. (12.12) we get

$$dH = dQ - dW + P\,dV + V\,dP$$

Since $dW = P\,dV$ and $V\,dP = 0$ for constant pressure process, the above equation simplifies to

$$dH = dQ \quad \text{(for constant pressure process)} \tag{12.13}$$

That is, when a system is heated at constant pressure, and only expansion work can occur, the change in enthalpy is equal to the heat supplied. On the other hand, for a process occurring at constant volume, work of expansion is zero, and Eq. (12.8) reveals that the change in internal energy is equal to the heat supplied.

$$dU = dQ \quad \text{(for constant volume process)} \tag{12.14}$$

EXAMPLE 12.5 A spherical storage tank of 3 m in diameter is half-filled with 12 500 kg of an organic liquid at 7000 kPa. If the total internal energy in the tank is 5.3×10^6 kJ, what is the specific enthalpy of the fluid in the tank?

Solution The internal energy U of the fluid referred to some arbitrary reference state is 5.3×10^6 kJ. The enthalpy of the fluid is given by Eq. (12.9), i.e.

$$H = U + PV$$

$$P = 7000 \text{ kPa } (= 7 \times 10^6 \text{ N/m}^2)$$

$$\text{Volume of the tank} = \frac{4}{3}\pi r^3 = \frac{4}{3}\pi(1.5)^3 = 14.137 \text{ m}^3$$

$$\text{Volume occupied by the liquid, } V = 0.5 \times 14.137 = 7.0685 \text{ m}^3$$

Thus
$$H = U + PV = 5.3 \times 10^9 + 7 \times 10^6 \times 7.0685$$
$$= 5.3495 \times 10^9 \text{ J } (= 5.3495 \times 10^6 \text{ kJ})$$
$$\text{Specific enthalpy} = 5.3495 \times 10^6/12\,500 = 427.96 \text{ kJ/kg}$$

EXAMPLE 12.6 Calculate ΔU and ΔH in kJ for 1 kmol water as it is vaporized at a constant temperature of 373 K and a constant pressure of 101.3 kPa. The specific volumes of liquid and vapour at these conditions are 1.04×10^{-3} and 1.675 m^3/kmol respectively. 1030 kJ of heat is added to water for this change.

Solution During vaporization the specific volume of water increases at constant pressure, doing work of expansion which is equal to $P\Delta V$.

The expansion work done by the system against 101.3 kPa pressure is

$$W = P\Delta V = 101.3 \times 10^3(1.675 - 1.04 \times 10^{-3}) \times 10^{-3} = 169.57 \text{ kJ/kmol}$$
$$Q = 1.03 \times 10^3 \text{ kJ/kmol}$$

By Eq. (12.8)
$$\Delta U = Q - W = 1.03 \times 10^3 - 169.57 = 860.43 \text{ kJ/kmol}$$

By Eq. (12.11)
$$\Delta H = \Delta U + \Delta(PV)$$

For the constant pressure process, this becomes
$$\Delta H = \Delta U + P\Delta V = 860.43 + 169.57 = 1.03 \times 10^3 \text{ kJ/kmol}$$
which is the same as the heat added.

EXAMPLE 12.7 The contents in a stirred tank reactor are being agitated by means of a 2-hp stirrer. The heat generated due to stirring is dissipated to the surroundings at a rate of 3000 kJ/h. Determine the change in internal energy.

Solution Work W is positive when work is done by the system. Here, work is done on the system by means of the stirrer. Therefore, the work done W is negative.

$$W = -2 \text{ hp} = -1491.4 \text{ W } (= -1491.4 \text{ J/s})$$

Similarly, the heat transferred is positive when heat is transferred to the system. Here, heat is transferred to the surroundings. Therefore, Q is negative.

$$Q = -3000 \text{ kJ/h} = -\frac{3000 \times 10^3}{3600} = -833.33 \text{ J/s}$$

By Eq. (12.8),
$$\Delta U = Q - W = -833.33 - (-1491.4) = 658.07 \text{ J/s}$$

The internal energy of the system increases by 658.07 J in one second.

EXAMPLE 12.8 Iron reacts with hydrochloric acid at 300 K according to

$$Fe(s) + 2HCl(aq) \rightarrow FeCl_2(aq) + H_2(g)$$

Determine the work done when 1 kg of iron filings reacts with HCl in a closed vessel maintained at a constant pressure. Assume that hydrogen behaves as an ideal gas.

Solution The atomic weight of iron is 55.845.

The number of moles of iron reacted is

$$\frac{1\times 10^3}{55.845} = 17.906 \text{ mol}$$

Since one g-atom of Fe produces 1 mol hydrogen, the number of moles of hydrogen produced = 17.906 mol.

The work done during the process is $P\Delta V$. Since the volumes of solids and liquids are negligible in comparison with the volume of gases, the volume change during the reaction is the volume occupied by the hydrogen produced during the reaction. Since hydrogen behaves as an ideal gas, the volume of hydrogen is $\frac{(\Delta n)RT}{P}$. Therefore, the work done is

$$P\Delta V = (\Delta n)RT = 17.906 \times 8.314 \times 300$$

$$= 44\,661.15 \text{ J} = 44.661 \text{ kJ}$$

12.3 HEAT CAPACITIES

The *heat capacity* of a substance is the quantity of heat to be supplied to cause a temperature rise of one degree. Thus,

$$dQ = C\,dT$$

where C is known as heat capacity of the substance. The heat capacity of the unit mass of a substance is also known as *specific heat* of the substance. Heat capacity depends on the way in which the heat is supplied. When heat is supplied to a system at constant volume, the system is unable to do any work and the quantity of heat required is given by

$$dQ = C_V\,dT \quad \text{(constant volume)} \tag{12.15}$$

where C_V is known as the heat capacity at constant volume. Thus,

$$C_V = \left(\frac{\partial Q}{\partial T}\right)_V \tag{12.16}$$

According to Eq. (12.8), $dU = dQ$ for a constant volume process, so that we can write Eq. (12.16) as

$$C_V = \left(\frac{\partial U}{\partial T}\right)_V \tag{12.17}$$

If heat is supplied to a substance at constant pressure, it is free to expand doing work against the constant pressure. A part of the heat supplied to the system is utilized for the work of expansion and more heat will be required to raise the temperature than that required in a constant volume process for the same temperature change. The amount of heat required is related to the temperature rise as

$$dQ = C_P\, dT \qquad \text{(constant pressure)} \tag{12.18}$$

where C_P is called the heat capacity at constant pressure. Thus,

$$C_P = \left(\frac{\partial Q}{\partial T}\right)_P \tag{12.19}$$

For a constant pressure process, Eq. (12.8) becomes

$$dU = dQ - P\, dV$$

or

$$dQ = dU + P\, dV = dH \qquad \text{(constant pressure)}$$

Equation (12.19) may be written as

$$C_P = \left(\frac{\partial H}{\partial T}\right)_P \tag{12.20}$$

Since for a given temperature change, the heat required is more in a constant pressure process than that in a constant volume process, generally $C_P > C_V$. However, the difference between C_P and C_V for solids and liquids are very small compared to that for gases, as the change in volume of solids and liquids during heating is not very significant.

Equations (12.17) and (12.20) give

$$dU = C_V\, dT \qquad \text{(constant volume)} \tag{12.21}$$

$$dH = C_P\, dT \qquad \text{(constant pressure)} \tag{12.22}$$

The change in enthalpy in a constant pressure process or a process in which the initial and final pressures are the same, can be evaluated by Eq. (12.22) whereas Eq. (12.18) determines the heat supplied in a true isobaric process only.

Generally, the heat capacity of a substance is not constant; but varies with temperature. For convenience in engineering calculations, it is usually expressed by a polynomial equation such as

$$C_P = a + bT + cT^2 \tag{12.23}$$

$$C_P = a + bT + cT^{-2} \tag{12.24}$$

where a, b and c are constants which are determined by standard methods for curve fitting. Table 12.1 gives the heat capacities of different gases at constant pressure.

The enthalpy change in a constant pressure process occurring between temperature limits T_1 and T_2 can be evaluated by integrating Eq. (12.22).

$$\Delta H = \int_{T_1}^{T_2} C_P\, dT = \int_{T_1}^{T_2} (a + bT + cT^2)\, dT$$

If the heat capacity versus temperature data is available as an experimentally measured result and not expressed in the form of a mathematical expression such as the ones presented above, the enthalpy change is to be determined by graphical integration. This is done by plotting C_P as the ordinate against temperature as the abscissa (Figure 12.3).

Table 12.1 Heat capacities of gases at constant pressure (Pressure = 0)
$C_P = a + bT + cT^2$ (C_P in kJ/kmol K and T in K)
Valid in the range of 300 K to 1500 K

Gas	*a*	$b \times 10^3$	$c \times 10^6$
Air	26.737	7.377	−1.1120
Ammonia	24.786	37.526	−7.3855
Carbon dioxide	26.540	42.454	−14.2979
Carbon monoxide	26.586	7.582	−1.1200
Chlorine	32.042	9.299	−3.6563
Ethane	9.722	159.266	−45.9292
Ethylene	12.640	118.110	−35.7427
Hydrogen	29.082	−0.821	1.9917
Hydrogen chloride	28.194	1.717	1.5127
Methane	13.415	77.079	−18.7569
Nitric oxide	26.963	8.663	−1.7610
Nitrogen	27.034	5.815	−0.2889
Oxygen	25.611	13.260	−4.2077
Sulphur dioxide	29.077	41.910	−15.8847
Sulphur trioxide	31.208	80.093	−27.7501

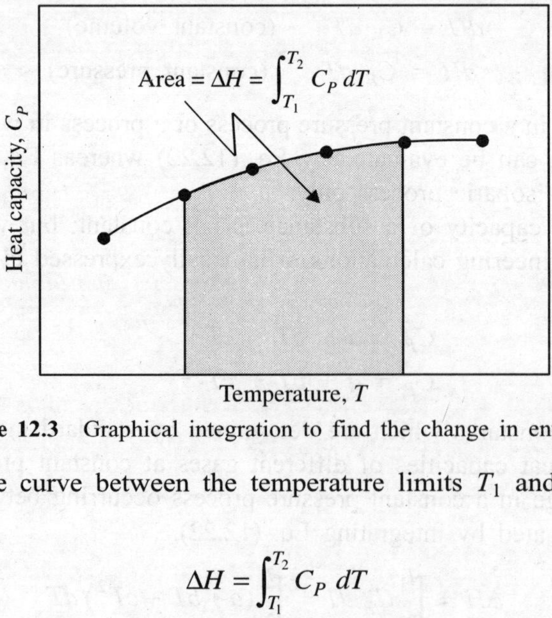

Figure 12.3 Graphical integration to find the change in enthalpy.

The area under the curve between the temperature limits T_1 and T_2 gives the enthalpy change, i.e.

$$\Delta H = \int_{T_1}^{T_2} C_P\, dT$$

EXAMPLE 12.9 1000 kg/h of a thermic fluid, to be used as a heat transfer medium, is being indirectly heated in a heater from 380 K to 550 K. Calculate the heat load on the heater in kW. The heat capacity equation for the thermic fluid is

$$C_P = 1.436 + 2.18 \times 10^{-3}\ T$$

where C_P is in kJ/kg K and T is in K.

Solution The amount of material being heated in unit time is

$$\frac{1000}{3600} = 0.278 \text{ kg/s}$$

The heat required in unit time for heating the fluid from 380 K to 550 K is

$$m\int_{T_1}^{T_3} C_P\, dT = 0.278\int_{380}^{550} (1.436 + 2.18\times10^{-3}\ T)\, dT$$

$$= 115.77 \text{ kJ/s } (= 115.77 \text{ kW})$$

EXAMPLE 12.10 The heat capacity of carbon dioxide is given by the following relation

$$C_P = 26.540 + 42.454 \times 10^{-3}\ T - 14.298 \times 10^{-6}\ T^2$$

where C_P is in kJ/kmol K and T is in K.

(a) How much heat is required to heat 1 kg of CO_2 from 300 K to 1000 K?
(b) Obtain the relation expressing the heat capacity in kcal/kmol °C and temperature in °C.
(c) Obtain the relationship giving heat capacity in Btu/lb-mol °F and temperature in °F.

Solution

(a) The heat requirement can be computed using the equation

$$dQ = C_P\, dT \qquad \text{(constant pressure)}$$

which on integration gives

$$Q = m\int_{T_1}^{T_2} C_P\, dT$$

Here m = 1/44 kmol. Therefore,

$$Q = \frac{1}{44}\int_{300}^{1000} (26.540 + 42.454\times10^{-3}\ T - 14.298\times10^{-6}\ T^2)\, dT$$

$$= 755.86 \text{ kJ}$$

(b) Let us use the given equation to evaluate the heat capacity at temperature t °C. The specific heat at this temperature is evaluated by the given equation as

$$C_P = 26.540 + 42.454 \times 10^{-3}\ (t + 273.15) - 14.298 \times 10^{-6}\ (t + 273.15)^2$$

$$= 37.068 + 34.643 \times 10^{-3} t - 14.298 \times 10^{-6}\ t^2$$

in the above equation t is in °C and the specific heat is given in kJ/kmol K. By dividing the equation by 4.1868, the specific heat evaluated is in kcal/kmol K. Hence the desired equation is

$$C_P = 8.854 + 8.274 \times 10^{-3}\ t - 3.415 \times 10^{-6}\ t^2$$

By replacing t in °C in the above equation by $(t' - 32)/1.8$ the equation becomes

$$C_P = 8.7058 + 4.6642 \times 10^{-3}\, t' - 1.0540 \times 10^{-6}\, t'^2$$

In the above equation t' is in °F and C_P is in kcal/kmol K. Since 1 Btu/lb-mol °F = 1 kcal/kmol °C, the above equation gives C_P in Btu/lb-mol °F as well, when the temperature is in °F.

EXAMPLE 12.11 The heat capacity of chlorine at 101.3 kPa varies with temperature as follows:

T (K)	273	373	473	573	673	773	873	973	1073	1173	1273
C_p (kJ/kmol K)	33.6	35.1	36.0	36.6	37.0	37.3	37.5	37.6	37.7	37.8	37.9

Calculate the enthalpy change when chlorine gas is heated from 273K to 1273K at 101.3 kPa.

Solution At constant pressure, the enthalpy change is given by

$$\Delta H = \int_{T_1}^{T_2} C_P\, dT$$

Since the exact functional relationship is not available as a mathematical equation, the integration is carried out graphically. For this purpose, a graph is plotted taking C_P as the ordinate against T as the abscissa. The area under the curve between the limits 273 K and 1273 K gives $\int_{273}^{1273} C_P\, dT$. Refer to Figure 12.4.

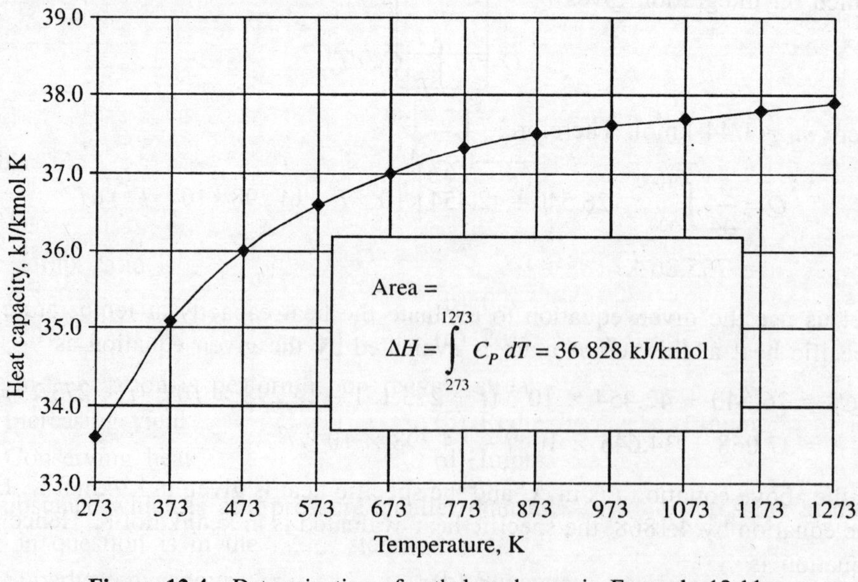

Figure 12.4 Determination of enthalpy change in Example 12.11.

The area is found to be 36 828 kJ/kmol. Therefore,

$$\Delta H = 36\,828 \text{ kJ/kmol}$$

12.3.1 Mean Heat Capacity

For calculating the heat interaction and enthalpy changes in processes involving gases, it is sometimes convenient to use a mean or average heat capacity in the temperature range of interest. The mean heat capacity at constant pressure may be denoted by C_{Pm} and can be obtained as

$$C_{Pm} = \frac{Q}{(T_2 - T_1)} = \frac{\int_{T_1}^{T_2} C_P \, dt}{(T_2 - T_1)} \tag{12.25}$$

If C_P is expressed as a power series such as the one given by Eq. (12.23), then

$$C_{Pm} = \frac{1}{(T_2 - T_1)} \int_{T_1}^{T_2} (a + bT + cT^2)\, dt$$

$$= \frac{1}{(T_2 - T_1)} \left[a(T_2 - T_1) + \frac{b}{2}(T_2^2 - T_1^2) + \frac{c}{3}(T_2^3 - T_1^3) \right] \tag{12.26}$$

Mean heat capacities at various temperatures are evaluated and are available in standard references. These values are evaluated between the given temperature and a reference temperature (T_R). Therefore, in the calculation of heat requirement or the enthalpy changes, their use is limited to temperatures lying between these two values. To calculate the heat required to heat a gas from temperature T_1 to temperature T_2, and if T_R is outside these limits, it is necessary to calculate the quantity of heat required for heating from the reference temperature to these temperatures separately and subtract one quantity from the other as detailed below:

$$Q = Q_1 - Q_2 = C_{Pm,1}\,(T_1 - T_R) - C_{Pm,2}\,(T_2 - T_R) \tag{12.27}$$

where $C_{Pm,1}$ and $C_{Pm,2}$ are the mean specific heat values at temperatures T_1 and T_2 respectively.

EXAMPLE 12.12 The molal heat capacity of CO is given by

$$C_P = 26.586 + 7.582 \times 10^{-3}\ T - 1.12 \times 10^{-6}\ T^2$$

where C_P is in kJ/kmol K and T is in K.

(a) Calculate the mean molal heat capacity in the temperature range of 500–1000 K.
(b) CO enters a heat exchanger at a rate of 500 cubic metres per hour at STP. Calculate the heat to be supplied to the gas to raise its temperature from 500 to 1000 K.
(c) CO is to be heated from 500 to 1500 K. What percent error is expected if the heat requirement is calculated using the mean heat capacity value determined in part (a)?

Solution

(a) Mean molal heat capacity

$$C_{Pm} = \frac{1}{(T_2 - T_1)} \int_{T_1}^{T_2} (a + bT + cT^2)\, dt$$

$$= \frac{1}{(T_2 - T_1)} \left[a(T_2 - T_1) + \frac{b}{2}(T_2^2 - T_1^2) + \frac{c}{3}(T_2^3 - T_1^3) \right]$$

$$= \frac{1}{1000-500}\left[26.586(1000-500)+\frac{7.582\times10^{-3}}{2}\times(1000^2-500^2)-\frac{1.12\times10^{-6}}{3}(1000^3-500^3)\right]$$

$$= 31.293 \text{ kJ/kmol K}$$

(b) Moles of gas to be heated = 500/22.4 = 22.32 kmol/h

$$\text{Heat to be supplied} = m \times C_{Pm} \times (T_2 - T_1)$$

$$= 22.32 \times 31.293 \times 500 = 3.4923 \times 10^5 \text{ kJ/h}$$

(c) The mean heat capacity calculated in part (a) is valid for the temperature range of 500–1000 K. The heat required for heating from 500 to 1500 K calculated using this value will not be accurate. For this calculation, we must use the mean heat capacity valid for the temperature range of 500–1500 K.
Heat requirement calculated using the mean heat capacity calculated in part (a) is

$$m \times 31.293 \times (1500 - 500) = 31293\ m \text{ kJ}$$

Actual heat requirement is

$$m\int_{500}^{1500}(26.586+7.582\times10^{-3}\,T-1.12\times10^{-6}\,T^2)\,dT = 32955\,m\text{ kJ}$$

$$\text{Percent error} = \frac{(32955-31293)}{32955}\times100 = 5.0\%$$

EXAMPLE 12.13 Calculate enthalpy change for one mole of a gas when it is heated from 400 K to 1500 K at 1 atm pressure given that the mean specific heat of the gas at the reference temperature of 273 K are 35 kJ/kmol K at 400 K and 50 kJ/kmol K at 1500 K.

Solution
$$\Delta H_{400\text{–}1500\text{ K}} = \Delta H_{1500\text{–}273} - \Delta H_{400\text{–}273\text{ K}}$$

$$= C_{Pm,1}\,(T_2 - T_R) - C_{Pm,2}\,(T_1 - T_R)$$

$$= 50(1500 - 273) - 35(400 - 273) = 56905 \text{ kJ/kmol}$$

12.3.2 Heat Capacity of Mixtures of Gases

The heat capacity of an ideal gas mixture can be evaluated as the sum of the heat capacities of the individual constituents each weighted according to the mole fraction in the mixture. Thus

$$C_{P,\text{mixture}} = \sum_{i=1}^{n} y_i C_{P,i} \tag{12.28}$$

where y_i represents the mole fraction of component i in the mixture of n components and $C_{P,i}$ is its specific heat in the pure state. If the heat capacity can be expressed as a quadratic function of temperature, i.e.

$$C_P = \alpha + \beta T + \gamma T^2 \tag{12.29}$$

the heat capacity of the mixture is

$$C_{P,\text{mixture}} = \sum_{i=1}^{n} y_i C_{P,i}$$

$$= \Sigma(y_i\alpha_i) + \Sigma(y_i\beta_i)T + \Sigma(y_i\gamma_i)T^2 \tag{12.30}$$

EXAMPLE 12.14 Combustion of solid wastes produces a flue gas of the following analysis: CO_2 = 9.0%, CO = 2.00%, O_2 = 7.00% and N_2 = 82.0%. Find the difference in enthalpies for this gas between the bottom and top of the stack if the temperature of the gas at the bottom is 600 K and that at the top is 375 K. The heat capacities of the gas are:

CO $\quad C_P = 26.586 + 7.582 \times 10^{-3}\ T - 1.12 \times 10^{-6}\ T^2$

CO_2 $\quad C_P = 26.540 + 42.454 \times 10^{-3}\ T - 14.298 \times 10^{-6}\ T^2$

O_2 $\quad C_P = 25.74 + 12.987 \times 10^{-3}\ T - 3.864 \times 10^{-6}\ T^2$

N_2 $\quad C_P = 27.03 + 5.815 \times 10^{-3}\ T - 0.289 \times 10^{-6}\ T^2$

C_p is in kJ/kmol K and T is in K.

Solution The heat capacity of the mixture is

$$C_{P,\text{mixture}} = \sum_{i=1}^{n} y_i C_{P,i}$$

$$= \Sigma(y_i\alpha_i) + \Sigma(y_i\beta_i)T + \Sigma(y_i\gamma_i)T^2$$

$$\Sigma(y_i\alpha_i) = 0.09 \times 26.540 + 0.02 \times 26.586 + 0.07 \times 25.74 + 0.82 \times 27.03$$

$$= 26.8867$$

$$\Sigma(y_i\beta_i) = (0.09 \times 42.454 + 0.02 \times 7.582 + 0.07 \times 12.987 + 0.82 \times 5.815) \times 10^{-3}$$

$$= 9.6499 \times 10^{-3}$$

$$\Sigma(y_i\gamma_i) = -(0.09 \times 14.298 + 0.02 \times 1.12 + 0.07 \times 3.864 + 0.82 \times 0.289) \times 10^{-6}$$

$$= -1.8167 \times 10^{-6}$$

Therefore,

$$C_{P,\text{mixture}} = 26.8867 + 9.6499 \times 10^{-3}\ T - 1.8167 \times 10^{-6}\ T^2$$

Enthalpy difference between the bottom and top of the stack is

$$\Delta H = \int_{600}^{375} (26.8867 + 9.6499 \times 10^{-3}\ T - 1.8167 \times 10^{-6}\ T^2)\, dT = -6656.29 \text{ kJ/kmol}$$

The enthalpy of the gas at the top of the stack is less than that at the bottom by 6656.29 kJ per kmol of gas.

12.4 PREDICTION OF HEAT CAPACITIES OF SOLIDS AND LIQUIDS

12.4.1 Heat Capacity of Solids

It has been proposed (*Dulong* and *Petit*) that the constant volume atomic heat capacities of crystalline solid elements are almost constant and equal to 6.2 cal/g-atom K (25.96 J/g-atom K). Though this observation was based on a very few experimental evidence then available, this rule applies satisfactorily to elements having atomic weights above 40 at room temperature. For low-atomic weight elements such as carbon, boron, silicon, phosphorous, sulphur, etc., the atomic heat capacities at room temperature are lower than 25.96 J/g-atom K. On increasing the temperature, the atomic heat capacities of even these elements approach the value of 25.96. The atomic heat capacities of all solid elements decrease greatly with decrease in temperature, and finally for crystalline solids at absolute zero of temperature, the heat capacity approaches zero. For accurate heat balance calculations, therefore, experimental data on the specific heats of solids should be used. Table 12.2 adapted from Chemical Engineer's Handbook gives constant pressure atomic heat capacities of certain elements in the crystalline state at 300 K.

Table 12.2 Atomic heat capacities of elements at 300 K, J/g-atom K

Element	*Heat capacity,* C_P *(*J/g-atom K *at* 300 K*)*
Aluminium	24.141
Antimony	25.305
Arsenic	24.585
Bismuth	25.791
Cadmium	25.958
Calcium	26.415
Carbon, graphite	9.039
Carbon, diamond	6.833
Chromium	23.969
Cobalt	25.619
Copper	24.614
Iron	25.305
Lead	26.695
Lithium	25.456
Magnesium	25.476
Manganese	25.125
Nickel	25.874
Phosphorous, yellow	23.027
Phosphorous, red	23.488
Potassium	28.910
Silicon	20.109
Sodium	27.708
Sulphur, rhombic	25.237
Sulphur, monoclinic	23.865
Tin	27.172
Titanium	18.593

Kopp's rule: When experimental data are lacking, Kopp's rule may be used for making a rough estimation of the heat capacity of solids. Kopp's rule predicts that the heat capacity of a solid compound at room temperature is approximately equal to the sum of the heat capacities of the individual elements. The atomic heat capacity values of the elements given in Table 12.3 may be used for estimating the heat capacity of compounds using this rule.

Table 12.3 Atomic heat capacity values for use in Kopp's rule (Heat capacity at 293 K, J/g-atom K)

Element	*Solids*	*Liquids*
Carbon	7.5	11.7
Hydrogen	9.6	18.0
Boron	11.3	19.7
Silicon	15.9	24.3
Oxygen	16.8	25.1
Fluorine	20.9	29.3
Phosphorus or sulphur	22.6	31.0
All others	25.96	33.5

According to Kopp's rule, the heat capacity of potassium permanganate may be estimated as equal to the heat capacity of potassium + heat capacity of manganese + 4 times the heat capacity of oxygen. This is equal to 25.96 + 25.96 + 4 × 16.8 = 119.12 J/mol K. This compares well with the experimental value of 117.2 J/mol K. The deviation is less than 2%. For $CaSO_4 \cdot 2H_2O$, the experimental heat capacity value is 195.9 J/mol K, whereas the value estimated using Kopp's rule is 25.96 + 22.60 + 6 × 16.80 + 4 × 9.60 = 187.80 J/mol K. The deviation in this case is about 4%.

Generally the heat capacities of compounds are higher in the liquid than in the solid state. The heat capacity of a *heterogeneous mixture* of solids is equal to the sum of the heat capacities of the component parts.

12.4.2 Heat Capacities of Liquids

In the absence of experimental data on heat capacities of liquids, Kopp's rule may be used to evaluate heat capacities of liquids at room temperature with a modified set of atomic heat capacities of constituent elements. These modified heat capacity values are given in Table 12.3.

The heat capacities of most liquids increase with increase in temperature. An empirical equation of the form $C_P = a + bT + cT^2 + dT^3$ usually represents heat capacity data of many liquids satisfactorily. Table 12.4 gives the constants a, b, c and d for use in this equation for a number of liquids (C_P is in J/mol K and T in K). For petroleum fractions heat capacity can be correlated to the specific gravity in API scale.

Heat capacities of *heterogeneous mixtures of liquids* can be computed as the weighted average of the heat capacities of the constituent pure liquids. This is found to be approximately valid even for miscible systems made up of chemically similar materials and for systems with small heats of mixing (ideal solutions). Thus for hydrocarbon mixtures, heat capacity may be calculated as

Table 12.4 Constants in the heat capacity equation for liquids

$$C_P = a + bT + cT^2 + dT^3$$
(C_p in J/mol K and T in K)

Liquid	*Temperature range* (K)	*a*	$b \times 10^2$	$c \times 10^5$	$d \times 10^9$
Acetaldehyde	150–324	16.8842	81.0208	−308.085	4425.9
Acetic acid	290–421	−36.0814	60.4681	−39.3957	−561.602
Aniline		−13.6683	93.1971	−160.401	1367.15
Benzene	279–350	−7.27329	77.054	−164.82	1897.9
Carbon disulphide		17.4151	55.4537	−172.346	15501.9
Carbon tetrachloride	273–343	12.285	0.01095	−318.26	3425.2
Ethyl acetate		4.2905	93.4378	−264.0	3342.58
Ethyl alcohol	250–400	−325.137	0.041379	−1403.1	1.7035×10^4
Ethylene glycol		31.0224	110.034	−284.571	2889.21
Formaldehyde		25.099	79.3671	−382.691	6104.92
Hydrogen peroxide	273–698	63.2314	17.07533	−38.99272	372.4104
Methyl alcohol	273–400	−259.25	0.03358	−11639	1.4052×10^4
n-Hexane	273–400	31.421	0.97606	−235.37	3092.7
Nitric acid	298	110.0			
n-Pentane	270–350	33.24	192.41	−236.87	17944
Phenol		−36.1614	115.354	−212.291	1741.83
Styrene		−38.0191	119.721	−219.565	1933.12
Sulphuric acid	283–318	−4119.31	15.59		
Toluene	270–370	1.8083	81.222	−151.27	1630
Water	273–373	18.2964	47.212	−133.88	1314.2

$$C_P = \sum w_i C_{P,i} \tag{12.31}$$

where w_i is the weight fraction of the constituent i and $C_{P,i}$ is its heat capacity at constant pressure.

Heat capacities of *aqueous solutions*, in the absence of experimental data may be taken as equal to the heat capacity of water at the same temperature. Thus, one kilogram of a 20% aqueous solution contains 0.8 kg of water and therefore the heat capacity of the solution is roughly equal to $0.8 \times 4.1868 = 3.35$ kJ/kg K.

EXAMPLE 12.15 Determine the heat capacity of $Na_2SO_4 \cdot 10H_2O$ at room temperature using Kopp's rule. The atomic heat capacities of elements (J/g-atom K) are 26.04 for Na, 22.6 for S, 16.8 for O, and 9.6 for H. Compare the result with the experimental value of the heat capacity of 592.2 J/mol K).

Solution According to Kopp's rule, the heat capacity of $Na_2SO_4 \cdot 10H_2O$ is 2 times the atomic heat capacity of sodium + atomic heat capacity of sulphur + 14 times the heat capacity of oxygen + 20 times the atomic heat capacity of hydrogen.

Therefore,

$$\text{heat capacity of } Na_2SO_4{\cdot}10H_2O = 2 \times 26.04 + 22.6 + 14 \times 16.8 + 20 \times 9.6$$
$$= 501.9 \text{ kJ/mol K}$$

The heat capacity predicted using Kopp's rule deviates from the experimental value by

$$\frac{592.2 - 501.9}{592.2} \times 100 = 15.25\%$$

12.5 ENTHALPY CHANGE OF PHASE CHANGES

Consider pure water at atmospheric pressure. At temperatures below 273 K, water cannot exist in the liquid state; it will be in the solid state called ice. When ice is heated at constant pressure the temperature rises to 273 K, the melting point of water at atmospheric pressure. At this temperature ice melts, the temperature remains constant at 273 K till the solid is completely converted to liquid water. We see that a phase change occurs from the solid state to the liquid state. This change of state from the solid to the liquid occurs at constant temperature and constant pressure. The enthalpy change accompanying this process is known as *heat of fusion* or *latent heat of fusion.* On further heating the temperature of water increases until it becomes equal to 373 K, the normal boiling point of water. The heat absorbed by water that resulted in its temperature rise is its *sensible heat.* On further heating, the heat supplied is not utilized as sensible heat for a temperature rise, rather, the heat supplied is used for a change of phase from the liquid state to the vapour state. The heat required to change the state of unit mass of a substance from the liquid to the vapour state is known as the *heat of vaporization* or the *latent heat of vaporization*. The heat given out when unit mass of vapour condenses to liquid is known as the *heat of condensation.* The heat of vaporization is numerically equal to the heat of condensation. For water this is 2256.1 kJ/kg at standard atmospheric pressure. Besides the heat of vaporization and the heat of fusion, enthalpy change accompanying other phase changes such as change from one solid form to another and change from solid phase to gas phase (sublimation) are also important.

EXAMPLE 12.16 Enthalpy of steam at 75 kPa and 573 K is 3075 kJ/kg referred to liquid water at 273 K. If the mean heat capacity of liquid water and water vapour are 4.2 kJ/kg K and 1.97 kJ/kg K respectively, calculate the heat of vaporization of water at 75 kPa. The saturation temperature of water at 75 kPa is 365 K.

Solution Since the boiling point of water at 75 kPa is 365 K, the vapour at 573 K and 75 kPa is superheated steam. The process of converting liquid water at 75 kPa and 273 K to superheated steam at 75 kPa and 573 K can be assumed to take place along the following route (Figure 12.5):

Step 1: Water is heated from 273 K to 365 K at constant pressure. Enthalpy change, ΔH_1 is equal to the heat required to raise the temperature of water from 273 to 365 K = $mC_P\Delta T$ = 1 × 4.2 × (365 – 273) = 386.4 kJ/kg

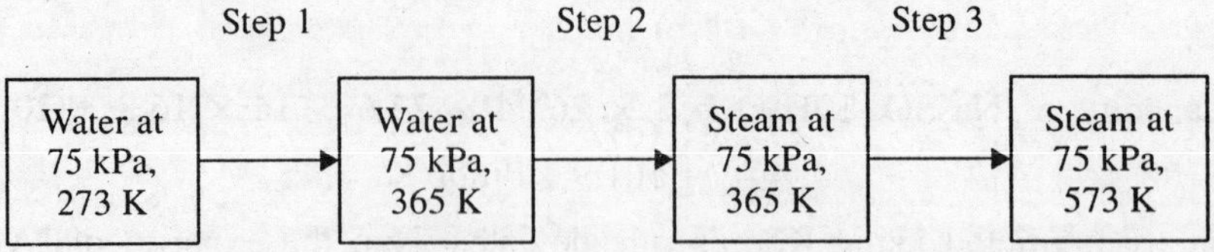

Figure 12.5 Process of converting liquid water to superheated steam (Example 12.16).

Step 2: The liquid is vaporized at constant pressure and constant temperature. The enthalpy change for this process is ΔH_2 and is equal to the heat of vaporization at 75 kPa, ΔH_v.

Step 3: The saturated vapour at 365 K is heated to 573 K at constant pressure. The enthalpy change for this process,

$$\Delta H_3 = mC_P\Delta T = 1 \times 1.97 \times (573 - 365) = 409.76 \text{ kJ/kg}$$

Assuming that the enthalpy of liquid water at 273 K to be zero, then the enthalpy of the superheated steam in the final state is

$$0 + \Delta H_1 + \Delta H_V + \Delta H_3$$

Therefore,

$$\Delta H_1 + \Delta H_V + \Delta H_3 = 3075 \text{ kJ/kg}$$

or

$$\Delta H_V = 3075 - \Delta H_1 - \Delta H_3$$

$$= 3075 - 386.4 - 409.76 = 2278.84 \text{ kJ/kg}$$

Therefore, the heat of vaporization of water at 75 kPa is 2278.84 kJ/kg.

EXAMPLE 12.17 One kg of water is heated from 250 K to 400 K at one standard atmosphere pressure. How much heat is required for this? The mean heat capacity of ice between 250 K and 273 K is 2.037 kJ/kg K, the mean heat capacity of water between 273 K and 373 K is 75.726 kJ/kmol K and the heat capacity of water vapour (kJ/kmol K) is

$$C_P = 30.475 + 9.652 \times 10^{-3}\, T + 1.189 \times 10^{-6}\, T^2$$

where T is in K. The latent heat of fusion and vaporization of water are, respectively, 6012 kJ/kmol and 40608 kJ/kmol.

Solution The heat required ($Q = \Delta H$) for the proposed change is evaluated as the sum of the following quantities:

1. Sensible heat for raising the temperature of ice from 250 K to 273.15 K, ΔH_1
2. Latent heat of fusion of ice, ΔH_f
3. Sensible heat of raising the temperature of water from 273.15 K to 373.15 K, ΔH_2
4. Latent heat of vaporization of water, ΔH_v
5. Sensible heat of raising the temperature of water vapour from 373.15 K to 400 K, ΔH_3

$$\Delta H_1 = 2.037 \times (273.15 - 250) = 47.16 \text{ kJ}$$

$$\Delta H_f = \frac{6012}{18.016} = 333.70 \text{ kJ}$$

$$\Delta H_2 = \frac{75.726}{18.016} \times (373.15 - 273.15) = 420.3 \text{ kJ}$$

$$\Delta H_v = \frac{40,608}{18.016} = 2254 \text{ kJ}$$

$$\Delta H_3 = \frac{1}{18.016} \int_{373.15}^{400} (30.475 + 9.652 \times 10^{-3} T + 1.189 \times 10^{-6} T^2)\, dT = 51.24 \text{ kJ}$$

$$Q = \Delta H = \Delta H_1 + \Delta H_f + \Delta H_2 + \Delta H_v + \Delta H_3 = 3106.4 \text{ kJ}$$

EXAMPLE 12.18 A high boiling organic liquid at 650 K is mixed with carbon tetrachloride at 295 K in the weight ratio 1:1. The pressure is one standard atmosphere. What will be the equilibrium temperature of the mixture? The heat capacity of the organic liquid is given by the relation:

$$C_P = 0.16 + 4.78 \times 10^{-3}\ T$$

Assume that the heat capacity of CCl_4 liquid varies linearly with temperature as

$$C_P = 0.7935 + 1.298 \times 10^{-4}\ T$$

In the above relations C_P is in kJ/kg K and T is in K. The boiling point of carbon tetrachloride is 349.9 K and the heat of vaporization is 195 kJ/kg. The mean heat capacity of CCl_4 vapour is 0.4693 kJ/kg K.

Solution Suppose that the heat given out by cooling of the organic liquid is absorbed by CCl_4 to raise its temperature and that there is no phase change. Let T be the final temperature attained. Then

$$\int_T^{650} (0.16 + 4.78 \times 10^{-3})\, dT = \int_{295}^{T} (0.7935 + 1.298 \times 10^{-4}\ T)\, dT$$

This relation yields a quadratic equation in T

$$2.4549 \times 10^{-3}\ T^2 + 0.9535T - 1353.51 = 0$$

which is solved to give $T = 573.3$ K. This temperature is above the boiling point of CCl_4. Therefore, the CCl_4 liquid gets converted to the vapour state using the heat given out by the organic liquid. Let T be the final temperature attained which is above 349.9 K, the boiling point of CCl_4. To find T, the heat balance is written as follows:

$$\int_T^{650} (0.16 + 4.78 \times 10^{-3})\, dT = \int_{295}^{349.9} (0.7935 + 1.298 \times 10^{-4}\ T)\, dT + \Delta H_v + \int_{349.9}^{T} 0.4693\, dT$$

Substitute $\Delta H_v = 195$ kJ/kg in the above equation. The resulting equation is quadratic in T which is solved for T. The temperature attained by the system $T = 540.1$ K.

12.5.1 Estimation of Heat of Fusion

No accurate methods are available for the estimation of heat of fusion, ΔH_f. The Clausius–Clapeyron equation is applicable, but its use to calculate the heat of fusion requires data on the variation of melting point with pressure, which is seldom available. A rough estimate of heat of fusion is made using the following approximation:

$$\frac{\Delta H_f}{T_f} = \text{constant} \tag{12.32}$$

where ΔH_f is in J/mol and T_f is the melting point in K. The values of the constant lie between 8.4 and 12.6 for elements, between 21 and 29.4 for inorganic compounds, and between 37.8 and 46.2 for organic compounds.

12.5.2 Estimation of Heat of Vaporization

The heat of vaporization or the latent heat of vaporization is the difference between the enthalpy of saturated vapour and the enthalpy of saturated liquid at a given pressure. The latent heat of vaporization decreases steadily with increase in temperature and is zero at the critical point. The heat of vaporization at the normal boiling point of a few compounds are listed in Table 12.5.

Table 12.5 Heat of vaporization at normal boiling point (kJ/mol)

Liquid	*Normal boiling point* (K)	ΔH_v (kJ/mol)
Acetic acid	390.4	24.4
Acetone	329.2	30.2
Ammonia	239.73	23.35
Benzene	353.26	30.76
Carbon disulphide	319.41	26.8
Carbon tetrachloride	349.9	30.0
Cyclohexane	353.90	30.1
Diethyl ether	307.76	26.05
Ethyl alcohol	351.7	38.6
Ethylene glycol	470.4	56.9
Formaldehyde	253.9	24.5
Hydrogen chloride	188.11	16.15
Methyl alcohol	337.9	35.3
Methyl cyclohexane	374.10	31.7
n-Heptane	371.59	31.69
n-Hexane	341.9	28.85
Nitric acid	359	30.30
n-Pentane	309.23	25.77
Toluene	383.78	33.5
Water	373.16	40.65

There are several techniques for estimating enthalpy of vaporization, the one based on the Clapeyron equation being an exact thermodynamic relationship.

Clapeyron equation: The Clapeyron equation predicts the dependence of vapour pressure on the temperature. It is given by

$$\frac{dP^S}{dT} = \frac{\Delta H_v}{T\Delta V} \tag{12.33}$$

where ΔH_v is the latent heat of vaporization, and ΔV is the volume change accompanying vaporization. The left-hand side of Eq. (12.33) is the slope of the vapour pressure–temperature plot. Experimental vapour pressure data or a correlation of vapour pressure with temperature will provide this value. The (dP^S/dT) at a given temperature and the volume change occurring on vaporization are substituted into Eq. (12.33) to calculate the latent heat of vaporization. If the temperature is not too near the critical point, the volume of the liquid is small in comparison with the volume of the vapour. The volume change accompanying vaporization $\Delta V = V_G - V_L$ is therefore approximately equal to V_G, the molar volume of vapour. Equation (12.33) now becomes

$$\frac{dP^S}{dT} = \frac{\Delta H_v}{TV_G} \tag{12.34}$$

The vapour pressure in regions well below the critical point is relatively small so that the vapour can be assumed to behave as an ideal gas. The molar volume V_G can now be replaced by RT/P^S so that Eq. (12.34) becomes

$$\frac{dP^S}{dT} = P^S \frac{\Delta H_v}{RT^2} \tag{12.35}$$

or

$$\frac{dP^S}{P^S} = \frac{\Delta H_v}{R} \frac{dT}{T^2} \tag{12.36}$$

Equation (12.36) can be rewritten as

$$d \ln P^S = -\frac{\Delta H_v}{R} d\left(\frac{1}{T}\right) \tag{12.37}$$

or

$$\frac{d \ln P^S}{d\left(\frac{1}{T}\right)} = -\frac{\Delta H_v}{R} \tag{12.38}$$

The slope of ln P^S versus $(1/T)$ at any temperature gives ΔH_v at that temperature. The integration of Eq. (12.38) assuming that the heat of vaporization remains constant gives the Clausius–Clapeyron equation which after substituting the limits takes the following form in the temperature range of T_1 to T_2.

$$\ln \frac{P_2^S}{P_1^S} = \frac{\Delta H_v}{R}\left(\frac{1}{T_1} - \frac{1}{T_2}\right) \tag{12.39}$$

where P_1^S and P_2^S are the saturation pressures at temperatures T_1 and T_2 respectively. Over small temperature intervals, the heat of vaporization remains essentially constant and may be estimated using Eq. (12.39).

The Clapeyron equation in the reduced form can be combined with the Antoine equation to yield the following relation for heat of vaporization in terms of Antoine constants.

$$\frac{\Delta H_v}{ZRT_C} = \frac{B}{T_C}\left[\frac{T_r}{T_r - (C/T_C)}\right]^2 \tag{12.40}$$

where B and C are Antoine constants.

Watson equation: Watson proposed [K.M. Watson, *Ind. Eng. Chem.*, 23, 360(1931), 35,398 (1943)] the following empirical equation for calculating latent heat of vaporization in terms of latent heat of vaporisation at the normal boiling point.

$$\frac{\Delta H_{v2}}{\Delta H_{v1}} = \left(\frac{1 - T_{r2}}{1 - T_{r1}}\right)^{0.38} \tag{12.41}$$

where ΔH_{v1} and ΔH_{v2} are the heat of vaporization at temperature T_1 and T_2 respectively, and T_{r1} and T_{r2} are the respective reduced temperatures.

Trouton's rule: It is observed that the ratio of the molal heat of vaporization at the normal boiling point to the boiling point in absolute units is constant, or

$$\frac{\Delta H_{vb}}{T_b} = K \tag{12.42}$$

where $\Delta H_{v\,b}$ is the heat of vaporization at the normal boiling point T_b. K is known as *Trouton's ratio*. This ratio is approximately equal to 87.9 J/mol K (=21 cal/mol K). For non-polar liquids, this ratio increases slightly with increase in the normal boiling point. For polar liquids, this ratio is much greater than 87.9 J/mol K. For water and lower alcohols, this value is found to be 108.9 J/mol K (=26 cal/mol K).

Kistyakowsky equation: The Kistyakowsky equation is an exact thermodynamic equation that can be utilized for estimating the enthalpy of vaporization of non-polar liquids at their normal boiling points. Expressing the heat of vaporization at the normal boiling point in J/mol and temperature in K, the equation takes the following form:

$$\frac{\Delta H_{vb}}{T_b} = 36.63 + 8.31 \ln T_b \tag{12.43}$$

The Kistyakowsky equation is not applicable to polar liquids.

Reference substance plots: Two such plots are in common use. the equal-pressure reference-substance plots and the equal-temperature reference-substance plots. Duhring's plot is an equal-pressure reference-substance plot. In this plot the temperature of a substance is plotted against the temperature of a reference substance (usually water) having equal vapour pressure. The slope of the resulting straight line is equal to $\dfrac{\Delta H_v}{\Delta H_{vR}}\left(\dfrac{T_R}{T}\right)^2$ where ΔH_v is the latent heat of vaporization at temperature T and ΔH_{vR} is the latent heat of vaporisation of the reference substance at temperature T_R where T_R is the temperature at which the reference substance has the same vapour pressure as the substance in question. On the other hand, in equal temperature reference-substance plots, the logarithm of the vapour pressure of the substance is plotted against the logarithm of the vapour pressure of a reference substance at the same temperature. The slope of the resulting straight line is $\dfrac{\Delta H_v}{\Delta H_{vR}}$. (A detailed discussion of these plots is given in Chapter 6.)

EXAMPLE 12.19 The vapour pressure of water at 363 K and 373 K are respectively 70.11 kPa and 101.3 kPa. Estimate the mean heat of vaporization of water in this temperature range.

Solution Refer to Eq. (12.39). T_1 = 363 K, T_2 = 373 K, P_1^S = 70.11 kPa, P_2^S = 101.3 kPa, R = 8.314 kJ/kmol K, ΔH_v = ? Substituting the values in Eq. (12.39), we get

$$\ln \frac{101.3}{70.11} = \frac{\Delta H_v}{8.314}\left(\frac{1}{363} - \frac{1}{373}\right)$$

or

$$\Delta H_v = 41\,428 \text{ J/mol } (= 2302 \text{ kJ/kg})$$

From steam tables, the heat of vaporization of water at 368 K is 2270 kJ/kg, indicating that the value estimated is higher than this by 1.4%.

EXAMPLE 12.20 Estimate the heat of vaporization of methyl chloride at 243.15 K from the Clapeyron equation assuming ideal behaviour for the vapour phase. The vapour pressure (kPa) is given by

$$\ln P^S = 14.2410 - \frac{2137.72}{T - 26.72}$$

where T is in K.

Solution Equation (12.36) can be written as

$$\frac{d \ln P^S}{dT} = \frac{\Delta H_v}{RT^2} \tag{A}$$

Differentiating the given relationship for vapour pressure with respect to T, we get

$$\frac{d \ln P^S}{dT} = \frac{2137.72}{(T - 26.72)^2} \tag{B}$$

Comparing Eq. (A) with Eq. (B), we see that

$$\frac{\Delta H_v}{RT^2} = \frac{2137.72}{(T - 26.72)^2}$$

or

$$\Delta H_v = \frac{2137.72\, RT^2}{(T - 26.72)^2} \tag{C}$$

To find the heat of vaporization at 243.15 K, substitute T = 243.15 K into Eq. (C). Thus,

$$\Delta H_v = \frac{2137.72 \times 8.314 \times (243.15)^2}{(243.15 - 26.72)^2} = 22\,432.33 \frac{\text{kJ}}{\text{kmol}}$$

EXAMPLE 12.21 Using the Watson equation, find the latent heat of vaporization of water at 473 K given that the latent heat of vaporization at 373 K is 2256 kJ/kg. The critical temperature of water is 647 K.

Solution The following values in the Watson Equation are known:

$$\Delta H_{v1} = 2256 \text{ kJ/kg}, \quad T_1 = 373 \text{ K}, \quad T_{r1} = \frac{T_1}{T_C} = \frac{373}{647} = 0.5765,$$

$$T_2 = 473 \text{ K}, \quad T_{r2} = \frac{T_2}{T_C} = \frac{473}{647} = 0.7311$$

Substitute these values in the Watson equation

$$\frac{\Delta H_{v2}}{\Delta H_{v1}} = \left(\frac{1 - T_{r2}}{1 - T_{r1}}\right)^{0.38}$$

$$\frac{\Delta H_{v2}}{2256} = \left(\frac{1 - 0.7311}{1 - 0.5765}\right)^{0.38} = 0.8415$$

Therefore, $\Delta H_{v2} = 0.8415 \times 2256 = 1898 \text{ kJ/kg}$

So, the latent heat of vaporization of water at 473 K = 1898 kJ/kg.

EXAMPLE 12.22 Heat capacity of liquid benzene is given below:

T (K)	293	323
C_P (J/mol K)	131.05	138.04

(a) Fit the data into an equation of the form

$$C_P = a + bT$$

(b) Calculate the heat required to convert 100 kg of liquid benzene from 293.15 K into saturated vapour at the normal boiling point of 353.25 K. The latent heat of vaporization may be estimated using the Kistyakowsky equation.

Solution

(a) If the variation of heat capacity of benzene with temperature is represented by $C_P = a + bT$, then

$$a + 293b = 131.05$$

$$a + 323b = 138.04$$

Solving these two equations simultaneously, we get $a = 62.781$ and $b = 0.233$
The desired equation for heat capacity of benzene is

$$C_P = 62.781 + 0.233T$$

(b) Kistyakowsky equation is

$$\frac{\Delta H_{vb}}{T_b} = 36.63 + 8.31 \ln T_b$$

Substituting $T_b = 353.25$ K into the above equation, we get $\Delta H_{vb} = 30163$ J/mol.

The total heat required for the desired process is the sum of (1) heat required to heat the liquid from 293.15 K to 353.25 K and (2) the heat of vaporization, i.e.

$$\text{Heat required} = m\int_{T_1}^{T_2} C_P\, dT + m\Delta H_v$$

$$= \frac{100\times10^3}{78.048}\int_{293.15}^{353.25}(62.781+0.233T)\,dT + \frac{100\times10^3}{78.048}\times 30163$$

$$= 1.0633 \times 10^7 + 3.8647 \times 10^7 = 4.928 \times 10^7 \text{ J}$$

12.6 STEAM TABLES

Steam tables provide a tabular representation of the thermodynamic properties of water in the saturated liquid state or in the saturated or superheated vapour state. Usually steam tables are given in two parts (Refer to steam tables given in Appendix A). The first one gives the thermodynamic properties of saturated steam and the second one gives the thermodynamic properties of superheated steam. The *saturated steam tables* give the enthalpy, entropy, and specific volume of saturated water and saturated steam as functions of saturation temperature and/or saturation pressure. The phase rule predicts that only one property, either the temperature or the pressure is sufficient to define the state of the saturated water–water vapour system in equilibrium. The tables give these properties in the temperature range of 273.15 K to 647.30 K (0°C to 374.15°C) the temperature range in which the liquid and vapour can coexist in equilibrium. The changes in these property values when the unit weight of liquid water is vaporized is also provided in the saturated steam tables. For example, let us consider the thermodynamic properties of saturated liquid and saturated vapour at 200 kPa. Refer to the saturated steam table (Table A.3) given in Appendix A. The table gives the following values in the horizontal row corresponding to 200 kPa:

Saturation temperature, T_s = 393.38 K
Specific volume of saturated water, $V_l = 1.061 \times 10^{-3}$ m³/kg
Specific volume of saturated steam, $V_g = 885.7 \times 10^{-3}$ m³/kg
Internal energy of saturated liquid, U_l = 504.49 kJ/kg
Internal energy of saturated vapour, U_g = 2529.50 kJ/kg
Enthalpy of saturated liquid, H_l = 504.70 kJ/kg
Enthalpy of saturated vapour, H_g = 2706.70 kJ/kg
Entropy of saturated liquid, S_l = 1.5301 kJ/kg K
Entropy of saturated vapour, S_g = 7.1271 kJ/kg K

The volume change accompanying the vaporization and the internal energy, enthalpy and entropy of vaporization can be obtained as the difference between the respective property values of saturated vapour and saturated liquid. Thus enthalpy of vaporization at 200 kPa is

$$2706.70 - 504.70 = 2202 \text{ kJ/kg}$$

These values are also given in some steam tables.

In a system made up of saturated liquid and saturated vapour, the weight fraction of the mixture that is dry steam is called the *quality* or *dryness fraction* of steam. Denoting the fraction of dry steam or the quality of the mixture by y, the specific volume and enthalpy of wet steam can be determined by the following relationships:

$$V = (1 - y)V_l + yV_g \tag{12.44}$$

$$H = (1 - y)H_l + yH_g \tag{12.45}$$

In the *superheat tables* (Table A.4), the specific volume, internal energy, enthalpy and entropy of steam at a particular pressure are given for various degrees of superheat. In addition, these tables also give the values of these properties for saturated liquid and saturated vapour at that pressure.

EXAMPLE 12.23 The quality of wet steam is determined using steam calorimeters. The wet steam is throttled in a steam calorimeter, and the enthalpy of the resulting super heated steam is measured by noting its temperature and pressure. Throttling is a constant enthalpy process. If wet steam at 700 kPa enters the throttling calorimeter and the discharged steam is at 10 kPa and 358.15 K. Using steam tables find the quality of steam.

Solution The enthalpy of superheated steam at 10 kPa and 358.15 K is determined from steam tables. From Table A.4:

$$\text{At 10 kPa, 323.15 K, } H = 2592.6 \text{ kJ/kg}$$

$$\text{At 10 kPa, 373.15 K, } H = 2687.5 \text{ kJ/kg}$$

The enthalpy of superheated steam at 10 kPa and 358.15 K is found by interpolation.

$$H = 2592.6 + \frac{2687.5 - 2592.6}{373.15 - 323.15}(358.15 - 323.15) = 2659 \text{ kJ/kg}$$

Since throttling is a constant enthalpy process, the enthalpy of the original wet steam at 700 kPa is also 2659 kJ/kg. At 700 kPa, enthalpies of saturated liquid and saturated vapours are 697.061 and 2762.0 kJ/kg respectively. The enthalpy of the mixture is

$$H = (1 - y)H_l + yH_g = 697.061(1 - y) + 2762.0y$$

Since $H = 2659$ kJ/kg, we have

$$697.061(1 - y) \times 2762.0 + y = 2659$$

which is solved to give $y = 0.9501$ or the quality or the dryness fraction $= 0.9501$.

The wet mixture contains 95% dry saturated steam and 5% moisture.

EXAMPLE 12.24 A large main contains 100 kg of steam at 3500 kPa and 673.15 K. It is closed at both ends. (a) Find the temperature and pressure in the main after half the steam has condensed. (b) How much heat is removed from the system during this process?

Solution

(a) The specific volume of superheated steam at 3500 kPa and 673.15 K is, from the steam tables (Table A.4), 0.084 53 m^3/kg. When half the vapour is condensed, the volume occupied by the 100 kg of the original superheated steam is now being occupied by 50 kg of vapour and 50 kg of condensate. Since the volume of liquid is negligible in comparison with the volume of vapour, the specific volume of the saturated vapour is

$$V = \frac{100 \times 0.084\,53}{50} = 0.1690 \text{ m}^3/\text{kg}$$

Referring to saturated steam tables, we see that the saturated steam having this specific volume for the vapour is at 1158 kPa and the saturation temperature is 459.5 K. (*Note:* If we consider the volume of liquid also, the temperature and pressure is to be determined using the following relation:

$$50(V_l + V_g) = 100 \times 0.0845$$

But V_l is very small in comparison with V_g, and therefore no appreciable error is introduced by the assumption that $V \approx V_g$).

(b) The heat liberated in the process is obtained from the relation

$$\Delta U = Q - W$$

Since this is a constant volume process, $W = 0$; therefore,

$$Q = \Delta U = U_2 - U_1$$

where U_2 is the internal energy of the system in the final state and U_1 is the internal energy in the initial state.

The internal energy of superheated steam at 3500 kPa and 673.15 K = 2928.4 kJ/kg (from Table A.4)

$$U_1 = 100 \times 2928.4 = 292\,840 \text{ kJ}$$

The internal energies of saturated vapour and liquid at 1158 kPa (459.5 K) are also read from saturated steam tables,

$$U_l = 790 \text{ kJ/kg}$$

$$U_g = 2585.9 \text{ kJ/kg}$$

$$U_2 = 50 \times 790 + 50 \times 2585.9 = 168\,795 \text{ kJ}$$

$$Q = \Delta U = U_2 - U_1 = 168\,795 - 292\,840 = -124\,045 \text{ kJ}$$

The amount of heat removed from the system during this process is 124 045 kJ.

12.7 HEAT OF MIXING

The environment of a molecule in a solution is different from that in the pure state. Since the energy of interaction between like molecules is different from that between unlike molecules, the energy of a solution is different from the sum of the energies of its constituents. This difference between the energy of the solution and the energy of the constituents leads to the absorption or evolution of heat during the mixing process. The heat of mixing (or the enthalpy change of mixing) is the enthalpy change when pure species are mixed at constant pressure and temperature to form one mole (or unit mass) of solution. For binary mixtures,

$$\Delta H = H - (x_1 H_1 + x_2 H_2) \tag{12.46}$$

Here x_1 and x_2 are the mole fractions of components 1 and 2 in the solution. Knowing the enthalpies of the pure constituents H_1 and H_2, and the heat of mixing at the given concentration, the enthalpy of the solution can be computed as

$$H = (x_1H_1 + x_2H_2) + \Delta H \tag{12.47}$$

When solids or gases are dissolved in liquids, the accompanying enthalpy change is usually measured as heats of solution, which is defined as the enthalpy change when one mole of the solute dissolves in the liquid. Thus,

$$\Delta H_S = \frac{\Delta H}{x_1} \tag{12.48}$$

where ΔH_S is the heat of solution per mole of solute (component 1). When the constituents are all liquids and solutions of all proportions are possible, the heat effect is usually termed *heat of mixing*. Figures 12.6 and 12.7 illustrate the two types of presentation of heat of mixing.

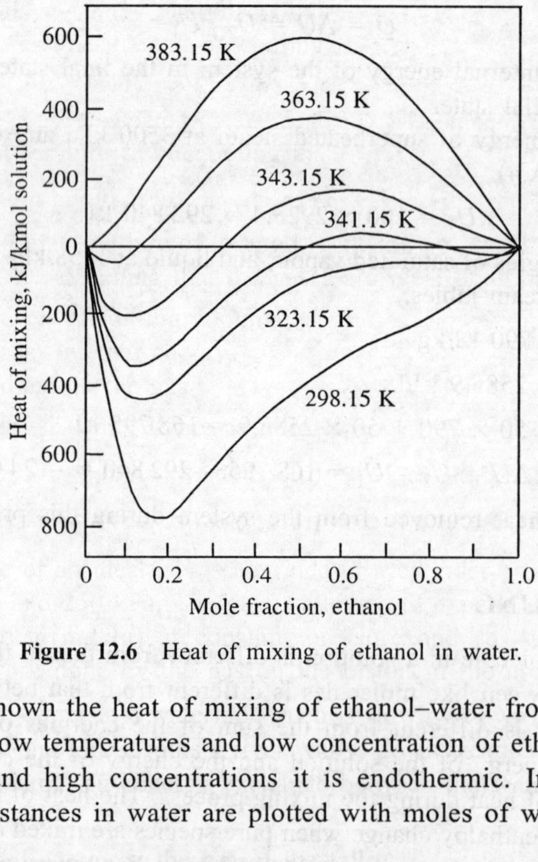

Figure 12.6 Heat of mixing of ethanol in water.

In Figure 12.6 is shown the heat of mixing of ethanol–water from which it is clear that the mixing process at low temperatures and low concentration of ethanol is exothermic and at high temperatures and high concentrations it is endothermic. In Figure 12.7 heats of solution of various substances in water are plotted with moles of water per mole of solute as the abscissa.

Using the heat of mixing at one temperature and heat capacity data of pure species and the solution (Table 12.6), the heat of mixing at any temperature can be calculated. The method of calculation is similar to the one employed for the calculation of standard heat of reaction at any temperature from the values at 298 K.

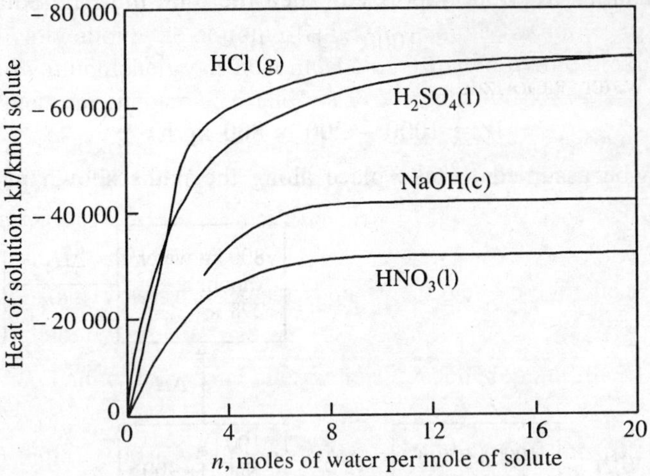

Figure 12.7 Heat of solution at 298 K.

Table 12.6 Heat of solution data in aqueous systems (kJ/mol) at 298 K (n = moles of H_2O/mole of material)

n	Na_2CO_3(s)	NaOH(s)	NH_4NO_3(s)	HNO_3(l)	H_2SO_4(l)	HCl(g)
1				–13.10	–28.09	–26.25
2				–20.10	–41.95	–48.86
4		–34.46	+17.04	–27.00	–54.09	–61.25
8		–41.95	+19.18	–31.15	–64.64	–68.29
15	–32.78	–42.87	+21.19	–32.49	–70.21	–71.05
25	–31.07	–42.83	+23.15	–32.78	–72.35	–72.31
50	–28.43	–42.54	+24.20	–32.78	–73.39	–73.31
100	–26.25	–42.37	+25.00	–32.78	–74.02	–73.90

EXAMPLE 12.25 Caustic soda is concentrated from 10% to 50% in a single effect-evaporator. The feed at 305 K enters at a rate of 1000 kg/h. The concentrated solution leaves the evaporator at 380 K and the vapour leaves at 373.15 K. Determine the heat to be supplied in the evaporator. The mean heat capacity applicable to the solutions are 3.67 kJ/kg K for the feed and 3.34 kJ/kg K for the product. The heat of solution of 10% and 50% NaOH solution at 298 K are, respectively –42.85 kJ/mol and –25.89 kJ/mol. The latent heat of vaporization of water at 298 K is 2442.5 kJ/kg and the mean heat capacity of water vapour is 1.884 kJ/kg K.

Solution *Basis:* 1000 kg/h of 10% NaOH solution

The product is 50% NaOH solution. Let P be the weight of the product. Take an NaOH balance:

$$0.5P = 0.1 \times 1000$$

Therefore,

$$P = 200 \text{ kg/h}$$

Let W be the quantity of water vaporized. Then the total material balance gives

$$1000 = P + W$$

The amount of water vaporized,

$$W = 1000 - 200 = 800 \text{ kg/h}$$

The process can be assumed to take place along the paths shown in Figure 12.8.

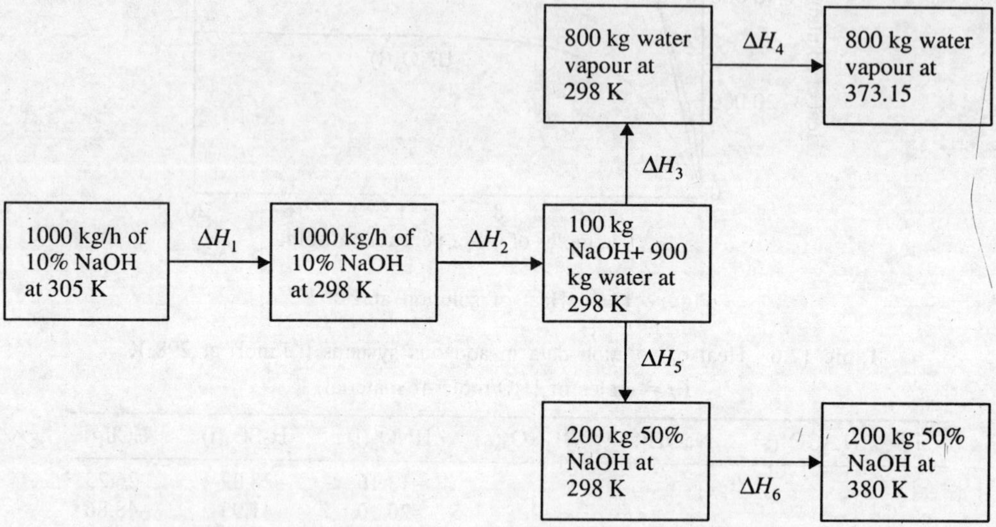

Figure 12.8 Enthalpy balance for the evaporation process in Example 12.25.

Step 1. Cooling 1000 kg/h of 10% solution from 305 K to 298 K.

$$\Delta H_1 = 1000 \times 3.67 \times (298 - 305) = -25\,690 \text{ kJ}$$

Step 2. Separation of the solution into the pure constituents at 298 K. The enthalpy change in this step is the negative of the heat of solution. The heat of solution is – 42.85 kJ/mol. The number of moles of NaOH in the solution is

$$\frac{100}{40} = 2.5 \text{ kmol} = 2500 \text{ mol}$$

$$\Delta H_2 = -(-42.85 \times 2500) = 107\,125 \text{ kJ}$$

Step 3. 800 kg of water at 298 K is converted to water vapour at 298 K.

$$\Delta H_3 = 2442.5 \times 800 = 1\,954\,000 \text{ kJ}$$

Step 4. Water vapour at 298 K is heated to 373.15 K.

$$\Delta H_4 = 800 \times 1.884 \times (373.15 - 298) = 113\,266 \text{ kJ}$$

Step 5. Formation of 200 kg of 50% NaOH solution at 298 K.

$$\Delta H_5 = 2500 \times (-25.89) = 64\,725 \text{ kJ}$$

Step 6. Heating the solution from 298 K to 380 K.

$$\Delta H_6 = 200 \times 3.34 \times (380 - 298) = 54\,776 \text{ kJ}$$

The enthalpy change accompanying the complete process is

$$\begin{aligned}\Delta H &= \Delta H_1 + \Delta H_2 + \Delta H_3 + \Delta H_4 + \Delta H_5 + \Delta H_6 \\ &= -\ 25\,690 + 107\,125 + 1\,954\,000 + 113\,266 - 64\,725 + 54\,776 \\ &= 2\,138\,752 \text{ kJ}\end{aligned}$$

EXAMPLE 12.26 Calculate the mean heat capacity of a 20-mole percent solution of alcohol in water at 298 K given the following:

Heat capacity of water: 4.18×10^3 J/kg K; heat capacity of ethanol: 2.18×10^3 J/kg K; and heat of mixing for 20 mole-percent ethanol–water at 298 K: –758 J/mol; heat of mixing for 20 mole-percent ethanol-water at 323 K: –415 J/mol. Assume that the heat capacities of pure liquids are constant between 298 and 523 K.

Solution The enthalpy change when 0.8 moles of water and 0.2 moles of ethanol both at 323 K are mixed together is given by the heat of mixing at 323 K which is equal to –415 J/mol of solution.

0.8 mol water at 323 K + 0.2 mol ethanol at 323 K → 1.0 mole 20 percent ethanol–water, $\Delta H = -415$ J/mol.

This change can be assumed to be taking place in four steps as detailed in Figure 12.9.

Basis: 1 mol ethanol–water solution at 298 K

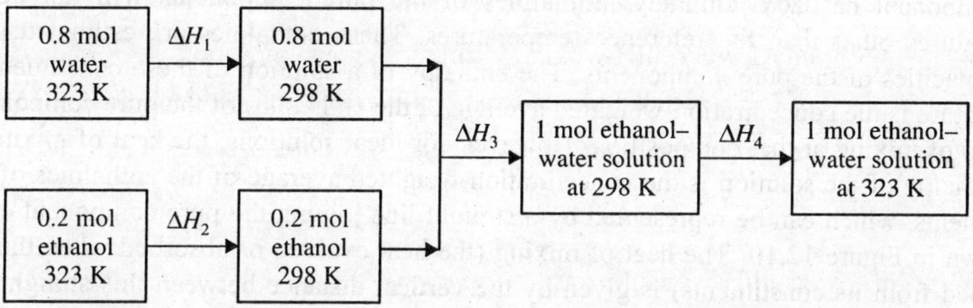

Figure 12.9 Mean heat capacity of ethanol–water solution (Example 12.26).

Step 1. 0.8 moles of water is cooled from 323 K to 298 K. Let ΔH_1 be the enthalpy of cooling. Then

$$\Delta H_1 = 0.8 \times 18.016 \times 4.18 \times (298 - 323) = -1506.1 \text{ J}$$

Step 2. 0.2 mol ethanol cooled from 323 K to 298 K. Let ΔH_2 be the enthalpy of cooling.

$$\Delta H_2 = 0.2 \times 46.048 \times 2.58 \times (298 - 323) = -594.0 \text{ J}$$

Step 3. 0.8 mol water and 0.2 mol ethanol at 298 K are mixed together. The heat of mixing is

$$\Delta H_3 = -758 \text{ J/mol}$$

Step 4. 20 percent ethanol–water solution is heated to 323 K. The enthalpy of heating is

$$\Delta H_4 = C_{P\,m}(323 - 298)$$

where $C_{p\,m}$ is the mean specific heat of solution.

$$\Delta H = \Delta H_1 + \Delta H_2 + \Delta H_3 + \Delta H_4$$

$$-415 = -1506.1 - 594.0 - 758 + C_{P\,m}\,(323 - 298)$$

Thus the mean heat capacity of a 20 percent solution is $C_{P\,m} = 97.72$ J/mol K

12.7.1 Enthalpy–Composition Diagrams

In enthalpy-concentration diagrams, the enthalpy of a binary solution is plotted against the composition with temperature as parameter. The pressure is constant and usually the diagrams at one atmosphere pressure are in common use. The composition is represented by mole fraction or mass fraction of the solute. As shown earlier, the enthalpy of a binary solution is given by Eq. (12.47), i.e.

$$H = (x_1H_1 + x_2H_2) + \Delta H$$

At a given temperature and pressure, enthalpy H depends on the composition, enthalpies of the pure components H_1 and H_2, and the heat of mixing ΔH. Pure components are arbitrarily assigned an absolute enthalpy of zero at a certain reference state. For water, the reference state chosen is the triple point of water (– 273.16 K), same as that used in steam tables. This choice permits the use of enthalpy values read from the steam tables along with the values from the enthalpy-concentration charts without any corrections. The reference temperature at which the enthalpy of the solute is assigned zero value need not be the same as that adopted for water. For solutes such as NaOH, H_2SO_4, etc. 298.15 K is used as the reference temperature at which the component has zero enthalpy. Enthalpies of the pure components will not be zero at temperatures other than the reference temperatures. These enthalpies are estimated using the heat capacities of the pure components. The enthalpy of a solution of these constituents at any temperature is the concentration-weighted average of the enthalpies of the pure components plus the heat of mixing at that composition. However, for ideal solutions, the heat of mixing is zero and enthalpy of the solution is the concentration-weighted average of the enthalpies of the pure components, which can be represented by a straight line joining the pure component enthalpies as shown in Figure 12.10. The heat of mixing (the heat evolved or absorbed when the solution is formed from its constituents) is given by the vertical distance between this straight line and the enthalpy-concentration curve. Figure 12.11 gives the enthalpy-concentration curve for NaOH–water solution.

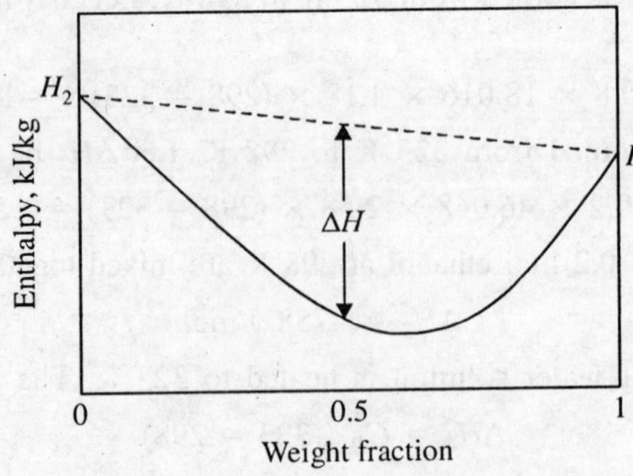

Figure 12.10 Enthalpy of solution versus concentration.

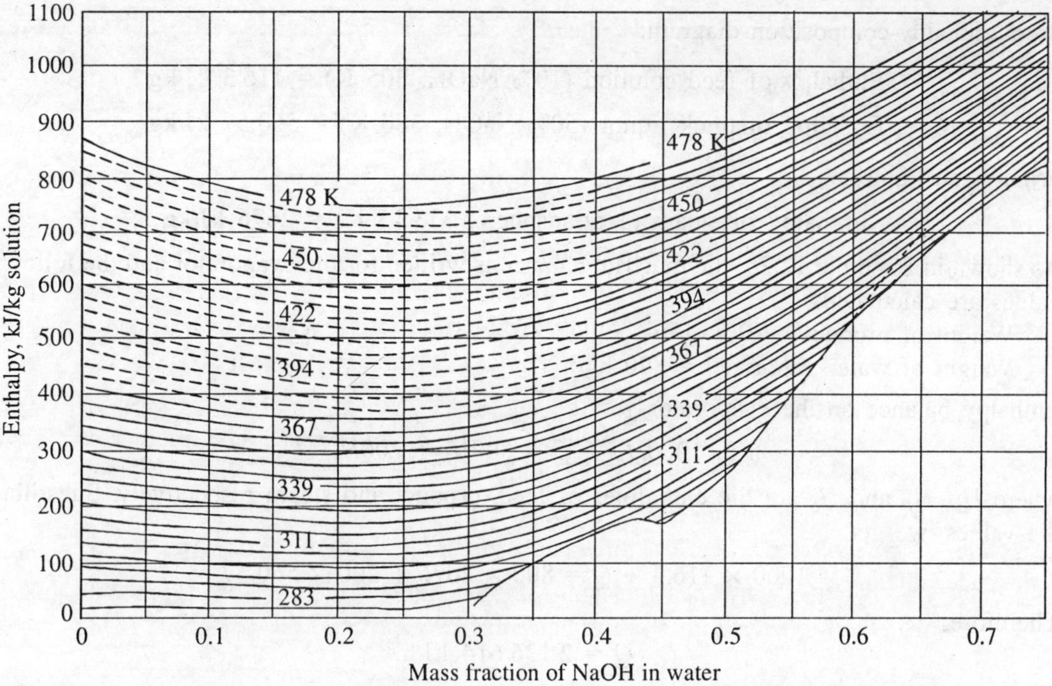

Figure 12.11 Enthalpy-composition diagram for NaOH–water.

EXAMPLE 12.27 Caustic soda is concentrated from 10% to 50% in a single-effect evaporator. The feed at 305 K enters at a rate of 1000 kg/h. The concentrated solution leaves the evaporator at 380 K and the vapour leaves at 373.15 K. Determine the heat to be supplied in the evaporator. Take the necessary data from the enthalpy-concentration diagram and steam tables. Assume that the pressure inside the evaporator is one atmosphere.

Solution

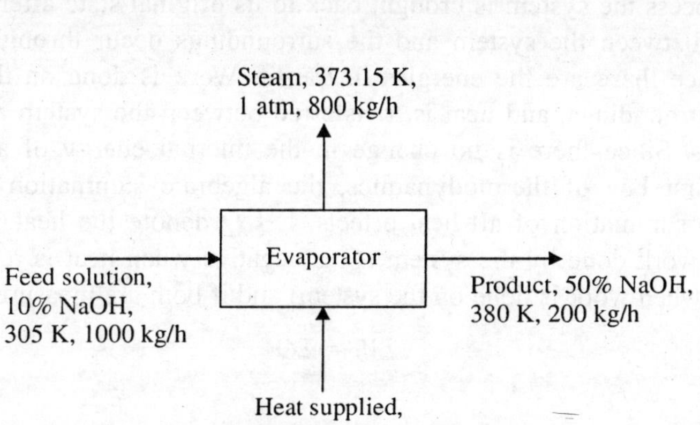

Figure 12.12 Evaporation of NaOH solution (Example 12.27).

From enthalpy-composition diagram:

Enthalpy of feed solution (10% NaOH, 305 K) = 116.3 kJ/kg

Enthalpy of thick liquor (50% NaOH, 380 K) = 560.57 kJ/kg

From steam tables:

Enthalpy of steam (1 atmosphere, 373.15 K) = 2676 kJ/kg

As shown in Example 12.25, the material balance is written for the evaporator and the following values are calculated:

Weight of product = 200 kg/h
Weight of water vaporized = 800 kg/h

Enthalpy balance on the evaporator yields

$$1000H_F + Q = 800H_S + 200H_P$$

where H_F, H_P and H_S are the enthalpies of feed, product and steam respectively. Substituting the values, we get

$$1000 \times 116.3 + Q = 800 \times 2676 + 200 \times 560.57$$

Therefore,

$$Q = 2\,136\,614 \text{ kJ/h.}$$

12.8 ENERGY BALANCE IN CYCLIC PROCESSES

The first law of thermodynamics is a generalization based on our experiences such as the failure to construct a perpetual motion machine, the constancy of mechanical equivalent of heat, etc. It is stated in a number of ways, but essentially it means that *energy is indestructible*. Any loss or gain of energy by the system must be exactly equivalent to the gain or loss of energy of the surroundings.

In a cyclic process the system is brought back to its original state after a series of changes. Energy exchange between the system and the surroundings occur through the mechanism of heat and work since these are the energies in transit. Work is done on the system by forces acting from the surroundings, and heat is transferred between the system and the surroundings during the process. Since there is no change in the internal energy of a system in a cyclic process, by the first law of thermodynamics, the algebraic summation of all work effects exactly equals the summation of all heat effects. Let Q denote the heat added to the system and W denote the work done by the system (Q is negative when heat is rejected by the system and W is negative when work is done on the system) and if both are measured in consistent units,

$$\Sigma W = \Sigma Q$$

12.9 ENERGY BALANCE IN NON-FLOW PROCESSES

12.9.1 First Law of Thermodynamics for Non-flow Process

The first law of thermodynamics requires that the change in the total energy of the system be compensated by an equal but opposite change in the total energy of the surroundings, so that there is no net change in the energy in any process. For a closed system undergoing only changes in the kinetic, potential and internal energies, the total energy change of the system $(\Delta E)_{sys}$ is given by

$$(\Delta E)_{sys} = \Delta(\text{KE}) + \Delta(\text{PE}) + \Delta U \tag{12.49}$$

Δ(KE) and Δ(PE) denote changes in kinetic energy and potential energy respectively, and ΔU is the change in the internal energy. The change in the total energy of the surroundings occurs only through the exchange of heat or work with the system. Then the change in the total energy of the surroundings $(\Delta E)_{sur}$ must be equal to the energy transferred to or from it as heat and work.

$$(\Delta \text{E})_{sur} = -Q + W \tag{12.50}$$

By the first law of thermodynamics,

$$(\Delta E)_{sys} + (\Delta E)_{sur} = 0 \tag{12.51}$$

Substituting Eqs. (12.49) and (12.50) into Eq. (12.51) and rearranging the resulting equation, we get

$$\Delta(\text{KE}) + \Delta(\text{PE}) + \Delta U = Q - W \tag{12.52}$$

Equation (12.52) means that the total energy change in a closed system is equal to the heat added to the system minus the work done by the system. For a steady-state non-flow process in which there are no changes in the kinetic energy and potential energy, the above equation simplifies to

$$\Delta U = Q - W \tag{12.53}$$

EXAMPLE 12.28 A rigid tank holds one kilogram of air at 300 K. The air has an internal energy of 0.25×10^3 kJ with reference to the fixed datum conditions. Heat is added to the air until the internal energy is 0.35×10^3 kJ. Calculate the heat transferred to the air.

Solution Since the tank is rigid, the volume of the gas does not change during heating. Under constant volume, the change in internal energy is equal to the heat supplied. Therefore,

$$Q = \Delta U = U_2 - U_1 = 0.35 \times 10^3 - 0.25 \times 10^3 = 100 \text{ kJ}$$

EXAMPLE 12.29 A system consisting of some fluid is stirred in a tank. The work done on the system by the stirrer is 2.25 hp. The heat generated due to stirring is dissipated to the surroundings. If the heat transferred to the surroundings is 3400 kJ/h, determine the change in internal energy.

Solution The work done on the system can be obtained as:

$$-W = 2.25 \text{ hp} = 2.25 \times 745.7 = 1677.825 \text{ W} = 1677.825 \text{ J/s}$$

Heat transferred to the surroundings is

$$-Q = 3400 \text{ kJ/h} = 3400 \times 10^3/3600 = 944.444 \text{ J/s}$$

By Eq. (12.53),

$$\Delta U = Q - W = -944.444 - (-1677.825) = 733.381 \text{ J/s}$$

Therefore, the internal energy of the system increases by 733.481 J in one second.

EXAMPLE 12.30 Iron filings are contained in a cylinder in an atmosphere of oxygen. It combines with oxygen according to the following reaction.

$$2Fe + \frac{3}{2}O_2 \rightarrow Fe_2O_3$$

The pressure inside the cylinder is maintained at 101 kPa. The temperature is kept constant at 298 K by removing heat. For 2 mol iron reacted, calculate Q, W and ΔU given that 831.08 kJ of heat is liberated in the process.

Solution The heat liberated when 2 mol iron reacts with oxygen is 831.08×10^3 J. Therefore,

$$Q = -831.08 \times 10^3 \text{ J}$$

Neglecting the volume occupied by the solids, the volume change accompanying the process is equal to the change in the volume of oxygen. Assuming the ideal gas behaviour for the gas,

$$P\Delta V = (\Delta n)\, RT$$

where ΔV is the volume change and Δn, the increase in the number of moles accompanying the reaction. Here $\Delta n = -1.5$. Therefore, the work done by the system,

$$W = P\Delta V = (\Delta n)RT = -1.5\, RT$$

$$= -1.5\,(8.314)\, 298 = -3716.4 \text{ J}$$

The change in the internal energy is

$$\Delta U = Q - W = -831.08 \times 10^3 - (-3716.4) = -827.364 \times 10^3 \text{ J}$$

EXAMPLE 12.31 A mixture of saturated steam and saturated water is contained in a rigid tank of volume 0.1 m^3 at a pressure of 2 bar. The liquid occupies 10% of the total volume. How much heat must be added in order that the tank contains only saturated steam? What will be the pressure in the tank? At a pressure of 2 bar, the specific volumes of saturated liquid and saturated vapour are 1.061×10^{-3} m^3/kg and 0.8857 m^3/kg respectively, and the internal energy of saturated liquid and saturated vapour are respectively 504.5 kJ/kg and 2529.5 kJ/kg.

Solution The volume occupied by the gas is 0.09 m^3 and the volume occupied by the liquid is 0.01 m^3. Since the specific volumes are 1.061×10^{-3} m^3/kg and 0.8857 m^3/kg respectively,

$$\text{mass of vapour} = \frac{0.09}{0.8857} = 0.1016 \text{ kg}$$

$$\text{mass of liquid} = \frac{0.01}{1.061 \times 10^{-3}} = 9.425 \text{ kg}$$

The internal energy of the initial mixture $U_1 = U_l + U_g$. Substituting the values,

$$U_1 = 9.425 \times 504.5 + 0.1016 \times 2529.5 = 5012 \text{ kJ}$$

In the final state the entire liquid is vaporized and the tank contains only saturated vapour. That is, a volume of 0.1 m^3 is occupied by a mass of (9.425 + 0.1016 = 9.5266 kg) of vapour. Therefore, the specific volume of the vapour is

$$\frac{0.1}{9.5266} = 0.0105 \text{ m}^3/\text{kg}$$

From steam tables, we can see that this specific volume for saturated vapour corresponds to a pressure of 148.6 bar and the internal energy of saturated vapour under this condition is 2464.6 kJ/kg. Therefore, the total internal energy in the final state is

$$9.5266 \times 2464.6 = 23\,479 \text{ kJ}$$

$$\Delta U = Q - W$$

But $W = 0$; therefore,

$$Q = \Delta U = U_2 - U_1 = 23\,479 - 5012 = 18\,467 \text{ kJ}$$

12.9.2 Energy Changes in Ideal Gas Processes

Constant volume process: For a constant volume process, there is no work of expansion and $dW = 0$. The heat supplied is equal to the product of heat capacity and rise in temperature, i.e.

$$dQ = C_V\, dT$$

Since for differential changes, the first law of thermodynamics is mathematically expressed as

$$dU = dQ - dW \tag{12.54}$$

for constant volume process, the first law reduces to

$$dU = dQ = C_V\, dT \tag{12.55}$$

This equation is applicable to all gases, whether ideal or real, in a constant volume process. Since the internal energy of an ideal gas depends upon the temperature only, the change in internal energy of an ideal gas in any process is

$$dU = C_V\, dT \tag{12.56}$$

Constant pressure process: In a constant pressure process, the heat supplied is the product of heat capacity C_P and the rise in temperature: $dQ = C_P\, dT$. The work involved is the work of expansion as given by $dW = P\, dV$. Equation (12.54) now becomes

$$dU = C_P\, dT - P\, dV \tag{12.57}$$

For a constant pressure process, the change in enthalpy is

$$dH = dU + P\, dV \tag{12.58}$$

Combining Eqs. (12.57) and (12.58), we get

$$dH = C_P\, dT \tag{12.59}$$

Thus, for the process occurring at constant pressure, we have the following general result applicable for all gases whether ideal or not.

$$dH = dQ = C_P\, dT \tag{12.60}$$

Again, as the enthalpy of an ideal gas depends on temperature alone, the change in enthalpy for all processes involving ideal gases is given by

$$dH = C_P\, dT \tag{12.61}$$

Constant temperature process: Internal energy of an ideal gas will change only with change in temperature. In an isothermal process, the change in internal energy and the change in enthalpy of an ideal gas would be zero. Put $dU = 0$ in Eq. (12.54) so that

$$dW = dQ \tag{12.62}$$

The above equation can be integrated to determine the heat and work effects in an isothermal process involving ideal gases.

$$Q = W = \int P\,dV$$

Since $P = RT/V$, the above equation gives

$$Q = W = RT \ln \frac{V_2}{V_1} \tag{12.63}$$

where V_1 and V_2 are the initial and final volumes of the gas in an isothermal process. Let P_1 and P_2 be the pressures in the initial and final states respectively. Then the ideal gas equation leads to the following relation:

$$\frac{V_2}{V_1} = \frac{P_1}{P_2} \tag{12.64}$$

Substituting Eq. (12.64) into Eq. (12.63), the latter can be modified as

$$Q = W = RT \ln \frac{V_2}{V_1} = RT \ln \frac{P_1}{P_2} \tag{12.65}$$

Adiabatic process: In an adiabatic process, there is no heat interaction between the system and the surroundings. dQ is zero in Eq. (12.54) so that we have

$$dU = -dW = -P\,dV \tag{12.66}$$

for an adiabatic reversible process. The pressure and volume in an adiabatic process are related as

$$PV^{\gamma} = \text{Constant}$$

where $\gamma = C_P/C_V$, which is assumed constant for an ideal gas.

Since $C_P - C_V = R$, a constant, it follows that both C_P and C_V are also constant. Also,

$$R = C_V(\gamma - 1) \tag{12.67}$$

Using Eq. (12.56) in Eq. (12.66), we see that

$$dW = P\,dV = -C_V\,dT \tag{12.68}$$

The work done in the adiabatic process is given by Eq. (12.68). Since C_V is constant, Eq. (12.68) can be readily integrated.

$$W = C_V(T_1 - T_2) \tag{12.69}$$

Eliminating C_V from this equation in terms of γ using Eq. (12.67),

$$W = \frac{RT_1 - RT_2}{\gamma - 1} = \frac{P_1V_1 - P_2V_2}{\gamma - 1} \tag{12.70}$$

Equation (12.70) gives the work done in an adiabatic process involving ideal gas. Equation (12.70) can be put into a more useful form as follows:

$$W = \frac{P_1V_1}{\gamma - 1}\left(1 - \frac{P_2V_2}{P_1V_1}\right) = \frac{P_1V_1}{\gamma - 1}\left[1 - \frac{P_2}{P_1}\left(\frac{P_1}{P_2}\right)^{1/\gamma}\right]$$

That is,

$$W = \frac{P_1V_1}{\gamma - 1}\left[1 - \left(\frac{P_2}{P_1}\right)^{\gamma - 1/\gamma}\right] \tag{12.71}$$

The following relationships between temperature and volume, and temperature and pressure in adiabatic processes can be easily derived:

$$\frac{T_2}{T_1} = \left(\frac{V_1}{V_2}\right)^{\gamma - 1} \tag{12.72}$$

$$\frac{T_2}{T_1} = \left(\frac{P_2}{P_1}\right)^{\gamma - 1/\gamma} \tag{12.73}$$

EXAMPLE 12.32 Heat is transferred to 10 kg of air which is initially at 100 kPa and 300 K until its temperature reaches 600 K. Determine the change in internal energy, the change in enthalpy, the heat supplied and the work done in the following processes:

(a) Constant volume process
(b) Constant pressure process

Assume that air is an ideal gas. Take C_P = 29.099 kJ/kmol K and C_V = 20.785 kJ/kmol K.

Solution The number of moles of the gas = 10/29 = 0.3448 kmol.
P_1 = 100 kPa, T_1 = 300 K. Hence,

$$V_1 = nRT_1/P_1 = 0.3448(8.314)\ 300/100 = 8.6\ \text{m}^3$$

(a) V_2 = 8.6 m^3 and T_2 = 600 K.
The change in internal energy:

$$\Delta U = n\int C_V\, dT = (0.3448)20.785(600 - 300) = 2150\ \text{kJ}$$

Heat supplied:

$$Q = \Delta U = 2150\ \text{kJ}$$

Work done:

$$W = Q - \Delta U = 0$$

Change in enthalpy:

$$\Delta H = \Delta U + \Delta(PV) = \Delta U + nR\,\Delta T$$
$$= 2150 + (0.3448)\ 8.314\ (600 - 300) = 3010\ \text{kJ}$$

(b) P_2 = 100 kPa and T_2 = 600 K.

Change in enthalpy:

$$\Delta H = n\int_{300}^{600} C_P\, dT = (0.3448)\, 29.099\, (600 - 300) = 3010 \text{ kJ}$$

Heat supplied:

$$Q = \Delta H = 3010 \text{ kJ}$$

Change in internal energy:

$$\Delta U = \Delta H - \Delta(PV) = \Delta H - nR\,\Delta T$$

$$= 3010 - (0.3448)\, 8.314\, (600 - 300) = 2150 \text{ kJ}$$

Work done:

$$W = Q - \Delta U = 860 \text{ kJ}$$

(*Note:* The work done may also be calculated as

$$W = P(V_2 - V_1)$$

$$V_2 = nRT_2/P = 0.3448\ (8.314)\ 600/100 = 17.2 \text{ m}^3$$

Therefore, $W = P\Delta V = 100\ (17.2 - 8.6) = 860$ kJ)

EXAMPLE 12.33 An ideal gas is undergoing a series of three operations as detailed below:

The gas is heated at constant volume from 300 K and 1 bar to a pressure of 2 bar. It is expanded in a reversible adiabatic process to a pressure of 1 bar. It is cooled at a constant pressure of 1 bar to 300 K. Determine the heat and work effects for each step. Assume C_P = 29.3 kJ/kmol K.

Solution The difference between C_P and C_V of an ideal gas is equal to R, the ideal gas constant. Hence, C_V is given by

$$C_V = C_P - R = 29.3 - 8.314 = 20.986 \text{ kJ/kmol K}$$

Step 1. Volume remains constant. Therefore, the work done is zero and the heat supplied is C_V $(T_2 - T_1)$. Also, $T_2/T_1 = P_2/P_1 = 2$. That is,

$$T_2 = 2 \times 300 = 600 \text{ K}$$

where T_2 is the final temperature attained in Step 1. The heat supplied is

$$20.986 \times (600 - 300) = 6295.8 \text{ kJ/kmol}$$

Step 2. The process is adiabatic. Therefore heat supplied is zero. The work done is given by Eq. (12.69), i.e.

$$W = C_V\,(T_1 - T_2)$$

where T_1 and T_2 are the initial and final temperatures in the adiabatic process which are related to the initial and final pressures by Eq. (12.73), i.e.

$$\frac{T_2}{T_1} = \left(\frac{P_2}{P_1}\right)^{(\gamma-1)/\gamma}$$

Therefore, $$T_2 = T_1\left(\frac{P_2}{P_1}\right)^{(\gamma-1)/\gamma} = 600\times\left(\frac{1}{2}\right)^{(0.4/1.4)} = 492.2\text{ K}$$

So, $W = 20.986\ (600 - 492.2) = 2262.3$ kJ/kmol. Since W is positive, the work is done by the system.

Step 3. The process is isobaric. The heat supplied is calculated as

$$C_P(T_2 - T_1) = 29.3(300 - 492.2) = -5631.5\text{ kJ/kmol}$$

(Heat is given out by the system.)

$$\Delta U = C_V(T_2 - T_1) = 20.896\ (300 - 492.2) = -4016.2\text{ kJ/kmol}$$

The first law states that $\Delta U = Q - W$. Hence,

$$W = Q - \Delta U = -5631.5 + 4016.2 = -1615.3\text{ kJ/kmol}$$

(Work is done on the system.)

EXAMPLE 12.34 Calculate the change in internal energy, the change in enthalpy, the work done and the heat supplied in the following processes:

(a) An ideal gas is expanded from 5 bar to 4 bar isothermally at 600 K.

(b) An ideal gas contained in a vessel of 0.1 m^3 capacity is initially at 1.01325 bar and 298 K. It is heated at constant volume to 400 K. (Assume that C_P = 30 J/mol K.)

Solution

(a) Since the internal energy and enthalpy of ideal gas depend only on temperature, in an isothermal process, they remain constant. $\Delta U = \Delta H = 0$. Equation (12.65) is used to calculate the work done and heat supplied, i.e.

$$Q = W = RT\ln\frac{P_1}{P_2} = 8.314\times 600\times\ln\frac{5}{4} = 1113.13\text{ kJ/kmol}$$

(b) The number of moles of the gas is

$PV/RT = (1.01325 \times 10^5)\ 0.1/(8.314 \times 298) = 4.04$ mol

$$C_V = C_P - R = 30 - 8.314 = 21.686\text{ J/mol K}$$

$$\Delta U = C_V(T_2 - T_1) = 21.686(400 - 298) = 2212\text{ J/mol} = 2212(4.04) = 8936\text{ J}$$

$$\Delta H = C_P(T_2 - T_1) = 30(400 - 298) = 3060\text{ J/mol} = 12{,}362\text{ J}$$

Being a constant volume process, $W = 0$, and $Q = \Delta U + W = \Delta U = 8936$ J.

12.10 ENERGY BALANCE FOR FLOW PROCESSES

The energy balance in steady-state flow processes become important when we consider flow through equipment such as pumps, fans, compressors, turbines, nozzles, heat exchangers, reactors, etc. where energy is transferred or is converted from one form to another. Consider an idealized flow system as shown in Figure 12.13. A fluid is flowing through the apparatus from section 1 to section 2. The velocity, specific volume, pressure and height above the datum level are represented by u, V, P and Z respectively. Suffix 1 indicates conditions at section 1 and suffix 2 the conditions at section 2. Per unit mass of the fluid, heat Q is added by means of the heat exchanger and shaft work W_s is extracted by means of a turbine or any other suitable device.

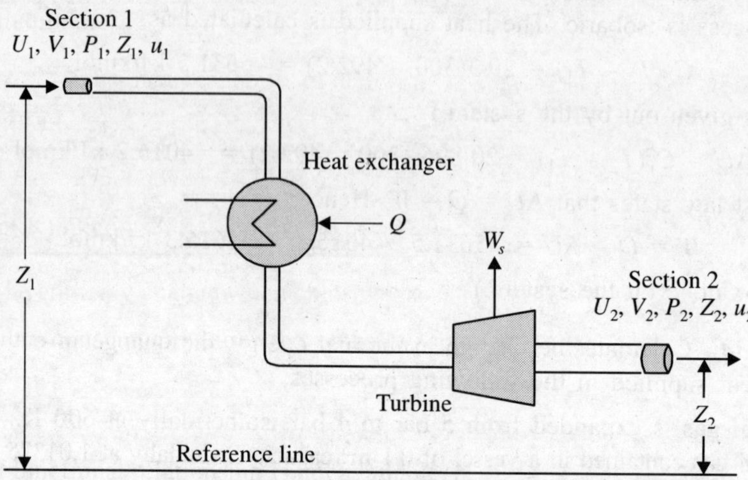

Figure 12.13 Steady-state flow processes.

The total energy of the fluid entering and leaving the system is made up of the internal energy possessed by the molecules, the potential energy due to its elevation above the datum level, the kinetic energy due to the velocity of the fluid and the flow energy which is equal to the product of pressure P and specific volume V of the fluid. Between sections 1 and 2, the change in the total energy of the unit mass of fluid is therefore equal to

$$\Delta U + \Delta(PV) + g\Delta Z + \frac{1}{2}\Delta u^2$$

By the first law, this should be equal to $Q - W_s$, the energy imparted to the fluid externally through heat or work. Therefore,

$$\Delta U + \Delta(PV) + g\Delta Z + \frac{1}{2}\Delta u^2 = Q - W_s \tag{12.74}$$

Substituting Eq. (12.11), i.e. $\Delta H = \Delta U + \Delta(PV)$ into Eq. (12.74), we get

$$\Delta H + g\Delta Z + \frac{1}{2}\Delta u^2 = Q - W_s \tag{12.75}$$

Equation (12.75) is the mathematical statement of the first law of thermodynamics for flow processes and can be used for solving problems involving flow of fluids, power required for pumps and compressors, etc. Equation (12.75) is known as the *total energy balance*. For most applications in thermodynamics, the kinetic energy and potential energy terms in Eq. (12.75) are negligibly small compared with the other terms so that the equation can be written as

$$\Delta H = Q - W \tag{12.76}$$

This equation is analogous to Eq. (12.8) developed for non-flow processes.

12.10.1 Mechanical Energy Balance

The total energy balance considers only the input and output from the system. For fluid mechanics calculations, it is customary to express this equation in the form of mechanical energy balance. The mechanical energy balance takes irreversible effects such as fluid friction also into account.

From the definition of enthalpy of a fluid, we get

$$dH = dU + P\,dV + V\,dP \qquad \text{[from Eq. (12.12)]}$$

From the first law of thermodynamics [Eq. (12.54)],

$$dU = dQ - dW$$

For a reversible process, the work done by the fluid is

$$dW = P\,dV$$

Substituting this into Eq. (12.54) and combining the resulting equation with Eq. (12.12), we get

$$dH = dQ + V\,dP \tag{12.77}$$

Integrate Eq. (12.77) between the entrance and exit points.

$$\Delta H = Q + \int_{P_1}^{P_2} V\,dP \tag{12.78}$$

Utilizing this equation, Eq. (12.75) can be modified to the following form:

$$-W_s = \frac{\Delta u^2}{2} + g\Delta Z + \int_{P_1}^{P_2} V\,dP \tag{12.79}$$

If the changes in kinetic and potential energies are negligible, the shaft work is given by

$$W_s = -\int_{P_1}^{P_2} V\,dP \tag{12.80}$$

The above equations are developed assuming reversible processes. On including the work lost due to the irreversibilities, Eq. (12.79) takes the following form:

$$\frac{\Delta u^2}{2} + g\Delta Z + \int_{P_1}^{P_2} V\,dP + W_s + F = 0 \tag{12.81}$$

Here, F represents the lost work due to friction or mechanical energy dissipated to heat due to irreversibilities in the system. Equation (12.81) is known as the *mechanical energy balance.*

12.10.2 Bernoulli's Equation

Bernoulli's equation is a special form of the mechanical energy balance applicable only to non-viscous incompressible fluids, which do not exchange shaft work with the surroundings. For non-viscous fluids, $F = 0$. The integral in Eq. (12.81) can be evaluated if the functional relationship between pressure and volume is known. For incompressible fluids volume is independent of pressure, and hence

$$\int_{P_1}^{P_2} V\,dP = V\Delta P = \frac{\Delta P}{\rho} \tag{12.82}$$

where ρ is the density of the fluid. Since no shaft work is exchanged, $W_s = 0$. Using these simplifications, Eq. (12.81) can be written as

$$\frac{\Delta u^2}{2} + g\Delta Z + \frac{\Delta P}{\rho} = 0 \tag{12.83}$$

$$\frac{u^2}{2} + gZ + \frac{P}{\rho} = \text{constant} \tag{12.84}$$

Equation (12.84) is known as *Bernoulli's equation.*

EXAMPLE 12.35 Steam at 610 K and 3500 kPa is expanded in a turbine for power generation and then condensed. The steam at a velocity of 60 m/s enters the turbine at an elevation of 7.5 m. The condensate leaves as saturated liquid at 305 K with a velocity of 0.5 m/s at an elevation of 2 m. The work done on the turbine by steam is 800 kJ/kg, and the enthalpy of steam in the turbine is greater than the enthalpy of the condensate by 3000 kJ/kg. Calculate the amount of heat that is removed in the condenser per kg of steam expanded (take g = 9.81 m/s^2).

Solution For a steady-state flow process we can use Eq. (12.75), i.e.

$$\Delta H + g\Delta Z + \frac{1}{2}\Delta u^2 = Q - W_s$$

For 1 kg of steam undergoing the specified change of state, $\Delta H = -3000$ kJ/kg.

$$\frac{\Delta u^2}{2} = \frac{u_2^2 - u_1^2}{2} = \frac{0.5^2 - 60^2}{2} = -1799.875 \text{ J/kg} = -1.8 \text{ kJ/kg}$$

$$g\Delta Z = 9.81 \times (2 - 7.5) = -53.955 \text{ J/kg} = -0.054 \text{ kJ/kg}$$

W_s = 800 kJ/kg

Heat added to the fluid in the system is

$$Q = \Delta H + g\Delta Z + \frac{1}{2}\Delta u^2 + W_s$$

Substituting the values,

$$Q = -3000 - 0.054 - 1.8 + 800 = -2201.85 \text{ kJ/kg}$$

Heat removed from the fluid, $-Q$ = 2201.85 kJ/kg.

EXAMPLE 12.36 Steam flowing at 3 m/s is throttled down from 6000 kPa and 623 K to 1000 kPa by passage through an insulated needle valve. What is the temperature after expansion?

Solution We can show that the throttling process is essentially a constant enthalpy process. In the steady-state energy balance equation

$$\Delta H + g\Delta Z + \frac{1}{2}\Delta u^2 = Q - W_s$$

$g\Delta Z = 0$; $Q = 0$; $W_s = 0$, so that we have

$$\Delta H + \frac{1}{2}\Delta u^2 = 0$$

The velocity of steam increases according to the relation $u_1 v_1 = u_2 v_2$ where u_1 is the velocity and v_1 is the specific volume of steam at inlet, and u_2 and v_2 are the corresponding values at the exit. But the contribution of the kinetic energy term $\frac{1}{2}\Delta u^2$ is negligibly small in comparison with the enthalpy values involved. Therefore, it is justifiable to assume

$$\Delta H = 0$$

for the process, or throttling can be considered as a constant enthalpy process.

Enthalpy of superheated steam at 6000 kPa and 623 K = 3045.8 kJ/kg. Hence the enthalpy of the steam leaving the throttling process is 3045.8 kJ/kg. At 1000 kPa, the temperature of superheated steam for which this enthalpy value corresponds to is 570 K (from steam tables).

EXAMPLE 12.37 Water at 288 K is pumped from a depth of 5 m at the rate of 1.5×10^{-2} m^3/min. The motor for the pump supplies work at the rate of 2 hp. Only about 60% of the rated horsepower is available as work of pumping, and the rest is dissipated as heat. The water passes through a heat exchanger, where it receives heat at the rate of 500 kJ/min and is delivered to a storage tank where the level of water is maintained at an elevation of 50 m above the ground. Heat loss from the whole system is estimated to be at a constant rate of 400 kJ/min. What is the temperature of the water delivered to the tank? Assume that the specific heat of water is constant at 4.2 kJ/kg K (take g = 9.81 m/s^2).

Solution Consider Figure 12.14.

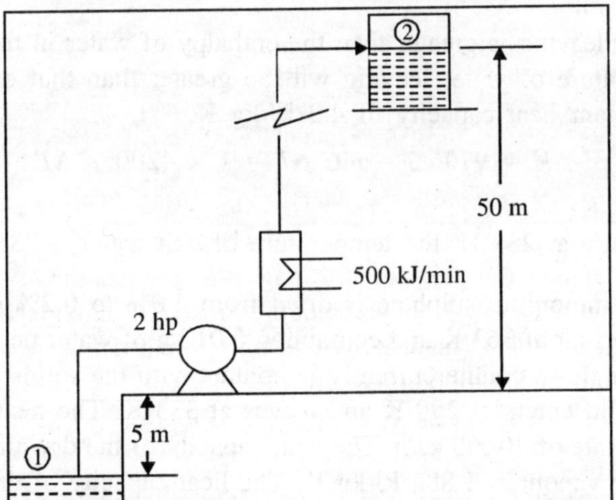

Figure 12.14 Energy balance (Example 12.37).

Point 1 is considered at the surface of water which is at a depth of 5 m from the ground and point 2 at the storage tank. Applying the steady-state energy balance equation,

$$\Delta H + g\Delta Z + \frac{1}{2}\Delta u^2 = Q - W_s$$

The surface at point 1 is the reference plane for estimating the potential energy. $Z_2 = 55$ m and $Z_1 = 0$. Therefore,

$$g\Delta Z = 9.81 \times 55 = 539.55 \text{ J/kg}$$

Neglecting the kinetic energy contribution,

$$\frac{1}{2}\Delta u^2 = 0$$

The flow rate of water = 1.5×10^{-2} m^3/min. Assuming a density of 1000 kg/m^3 for water, the mass flow rate of water = $1.5 \times 10^{-2} \times 1000 = 15$ kg/min. Heat supplied = 500 kJ/min; heat lost = 400 kJ/min. Therefore, the net heat transferred to the system

$$Q = \frac{(500 - 400) \times 1000}{15} = 6666.67 \text{ J/kg}$$

The work done by the pump on the unit mass of water, taking into consideration its efficiency of 60%, is

$$-W_s = \frac{2 \times 745.7 \times 0.60}{15/60} = 3579.36 \text{ J/kg}$$

(*Note:* The work done by the water is positive and the work done by the pump on the water is negative, by the sign convention we have adopted.)

$$\Delta H = Q - W_s - g\Delta Z - \frac{1}{2}\Delta u^2 = 6666.67 + 3579.36 - 539.55 = 9706.5 \text{ J/kg}$$

The enthalpy of water leaving is greater than the enthalpy of water in the well by 9706.5 J/kg. Therefore, the temperature of water leaving will be greater than that of water in the well by ΔT. Assuming a constant heat capacity of 4.2 kJ/kg K,

$$\Delta H = 9706.5 = mC_p\Delta T = 1 \times 4200 \times \Delta T$$

Therefore, $\Delta T = 2.31$ K.
Since water is originally at 288 K, the temperature of exit water = 288 + 2.31 = 290.31 K.

EXAMPLE 12.38 Ammonium sulphate is dried from 4.0% to 0.2% moisture in a counter-current rotary drier. Hot air at 363 K and containing 0.01 kg of water per kg of dry air admitted at one end of the drier flows countercurrently in contact with the solids and leaves at the other end at 305 K. The solid enters at 298 K and leaves at 333 K. The heat lost from the drier is estimated to be at the rate of 40000 kJ/h. The heat capacity of the dry air is 1.005 kJ/kg K. The heat capacity of water vapour is 1.884 kJ/kg K. The heat capacity of dry ammonium sulphate is 1.507 kJ/kg K. The heat capacity of water is 4.2 kJ/kg K. The latent heat of vaporization of water at 273 K is 2502.3 kJ/kg. Estimate the air requirement for the drier for producing ammonium sulphate product at the rate of 1000 kg/h.

Solution *Basis:* 1000 kg/h of dried product

Let S kg be the amount of dry solid in the product stream. Since the product contains 0.2% moisture,

$$S = 1000(1 - 0.002) = 998 \text{ kg}$$

This remains constant in the solid stream as only water is removed from the solid on drying.

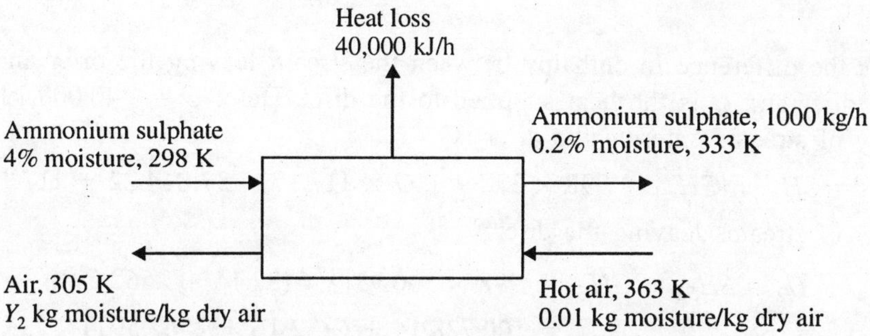

Figure 12.15 Energy balance on rotary dryer (Example 12.38).

Let X_1 be the moisture content in weight ratio in the entering solid and X_2 that in the exit solid. Then

$$X_1 = \frac{4}{96} = 0.0417 \text{ kg water per kg dry solid}$$

$$X_2 = \frac{0.2}{99.8} = 0.002 \text{ kg water per kg dry solid}$$

Let G be the weight of dry air in the air stream passing through the drier which is constant throughout the dryer. Let Y_1 be the humidity in the air entering the drier and Y_2 be the humidity in the air leaving the drier. $Y_1 = 0.01$ kg water per kg dry air.

The enthalpies of the streams entering and leaving the drier can now be calculated. A datum temperature of 273 K is assumed for enthalpy calculation.

Enthalpy of solid entering the drier,

$$H_{s1} = (C_P + X_1C_w)(T_1 - T_0)$$
$$= (1.507 + 0.002 \times 4.2) \times (298 - 273) = 37.89 \text{ kJ/kg of dry solid}$$

Enthalpy of solid leaving the drier,

$$H_{s2} = (C_P + X_2C_w)(T_2 - T_0)$$
$$= (1.507 + 0.0417 \times 4.2) \times (333 - 273) = 100.93 \text{ kJ/kg of dry solid}$$

Enthalpy of wet air can be evaluated as

$$H_G = C_s(T_G - T_0) + Y\lambda_0$$

where C_S is the humid heat of air, T_G is the temperature of air and λ_0 is the latent heat of vaporization of water at T_0.

$$C_s = 1.005 + 1.884Y$$

where Y is the humidity of air. For air entering the drier,

$$H_{G1} = (1.005 + 1.884 \times 0.01)(363 - 273) + 2502.3 \times 0.01 = 117.17 \text{ kJ/kg dry air}$$

For exit air, the enthalpy is

$$H_{G2} = (1.005 + 1.884 \times Y)(305 - 273) + 2502.3 \times Y = 32.16 + 2562.59Y \text{ kJ/kg dry air}$$

Taking an enthalpy balance on the drier,

$$\Delta H = Q \tag{A}$$

where ΔH is the difference in enthalpy between the stream leaving the drier and the streams entering the drier and Q is the heat supplied to the drier. Here, $Q = -40\,000$ kJ/h.

Enthalpy of streams entering the drier is

$$H_1 = SH_{s1} + GH_{G1} = 998 \times 37.89 + G \times 117.17 = 37{,}814.22 + 117.17G$$

Enthalpy of streams leaving the drier

$$H_2 = SH_{S2} + GH_{G2} = 998 \times 100.93 + G(32.16 + 2562.59Y)$$
$$= 100\,728.14 + G(32.16 + 2562.59Y)$$
$$\Delta H = H_2 - H_1 = 629\,13.92 + G(-\,85.01 + 2562.59Y)$$

Substituting this in Eq. (A),

$$62\,913.92 + G(-\,85.01 + 2562.59Y) + 40\,000 = 0$$

or

$$102\,913.92 + G(-85.01 + 2562.59Y) = 0 \tag{B}$$

A moisture balance on the drier gives

$$SX_1 + GY_1 = SX_2 + GY_2$$
$$998(X_1 - X_2) = G(Y - 0.01)$$

or

$$998(0.0417 - 0.002) = G(Y - 0.01)$$

which on simplification gives

$$G(Y - 0.01) = 39.62 \tag{C}$$

Equations (B) and (C) are solved simultaneously to obtain G, the circulation rate of dry air, i.e. $G = 3443$ kg/h.

Since the air supplied contains 0.01 kg of water vapour per kg of dry air, the rate at which wet air is supplied is

$$G(1 + 0.01) = 3477.4 \text{ kg/h}$$

EXAMPLE 12.39 A solution of 10% (weight) acetone in water is subjected to fractional distillation at a rate of 1000 kg/h to produce a distillate containing 90% acetone and a bottom product containing not more than 1% acetone. The feed enters at 340 K, distillate and residue leave the tower at 300 K and 370 K respectively. A reflux ratio (ratio of the weight of reflux to the weight of distillate product) of 8 is employed. The rise in temperature of 30 K is permitted for the cooling water circulated in the condenser employed for condensing the vapours into the distillate product and the reflux. Saturated steam at 276 kPa is available for supplying the heat of vaporization in the reboiler.

Heat losses from the column may be neglected. The heat capacity of acetone is 2.2 kJ/kg K and that of water is 4.2 kJ/kg K. The boiling point of 90% acetone–water solution is 332 K. The latent heat of acetone at 332 K is 620 kJ/kg and that of water is 2500 kJ/kg. The latent heat of steam at 276 kPa is 2730 kJ/kg. Calculate the following:

(a) The cooling water circulation rate
(b) The rate of circulation of steam

Solution *Basis:* 1000 kg/h of feed solution.
Refer to Figure 12.16.

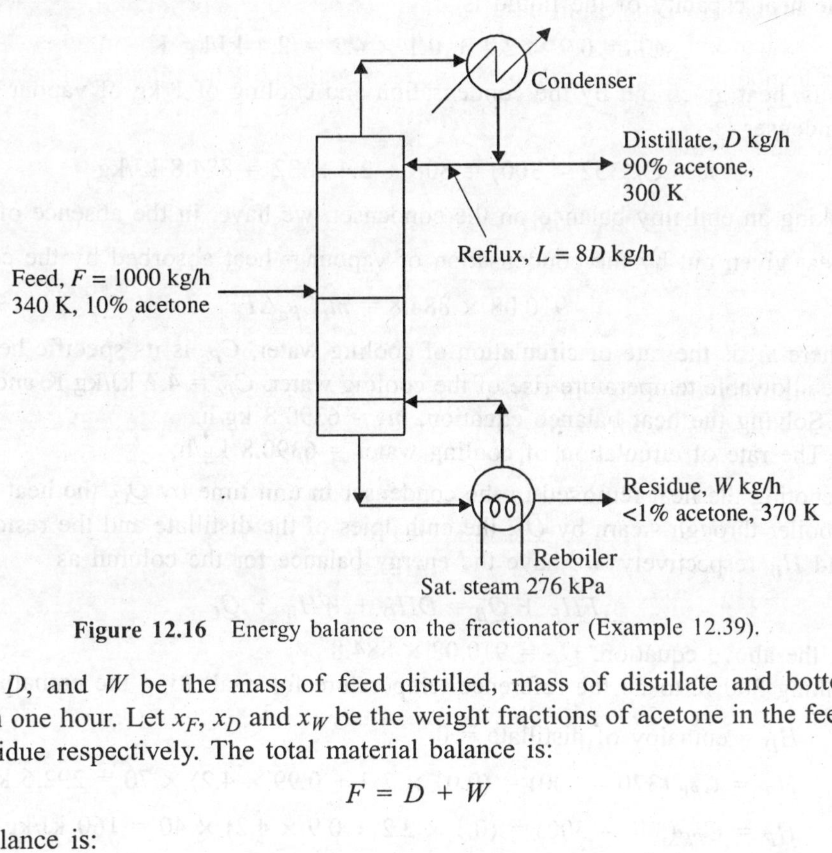

Figure 12.16 Energy balance on the fractionator (Example 12.39).

Let F, D, and W be the mass of feed distilled, mass of distillate and bottom product produced in one hour. Let x_F, x_D and x_W be the weight fractions of acetone in the feed, distillate and the residue respectively. The total material balance is:

$$F = D + W$$

Acetone balance is:

$$Fx_F = Dx_D + Wx_W$$

Here, $F = 1000$, $x_F = 0.10$, $x_D = 0.90$ and $x_W = 0.01$. Substitute these values in the above equations and solve the resulting equations for D and W. Hence, we get $D = 101.12$ kg/h and $W = 898.88$ kg/h.

(a) The reflux ratio

$$R = \frac{L}{D} = 8$$

Therefore,

$$L = 8D = 808.96 \text{ kg/h}$$

Taking a total material balance around the condenser,

$$G = L + D = 808.96 + 101.12 = 910.08 \text{ kg/h}$$

where G is the vapour reaching the condenser from the column. This is the quantity of vapour condensed to saturated liquid at 332 K and further cooled to 300 K to be taken as distillate and reflux.

The latent heat of condensation of the vapour is

$$\lambda = 0.9 \times 620 + 0.1 \times 2500 = 808 \text{ kJ/kg}$$

The heat capacity of the liquid is

$$C_P = 0.9 \times 2.2 + 0.1 \times 4.2 = 2.4 \text{ kJ/kg K}$$

Thus, heat given out by the condensation and cooling of 1 kg of vapour reaching the condenser is

$$\lambda + C_P(332 - 300) = 808 + 2.4 \times 32 = 884.8 \text{ kJ/kg}$$

Taking an enthalpy balance on the condenser, we have, in the absence of heat losses,

heat given out by the condensation of vapour = heat absorbed by the cooling water

$$910.08 \times 884.8 = m_c C_{Pc} \Delta T$$

where m_c is the rate of circulation of cooling water, C_{Pc} is its specific heat and ΔT is the allowable temperature rise of the cooling water. C_{Pc} = 4.2 kJ/kg K and ΔT = 30 K.

Solving the heat balance equation, m_c = 6390.8 kg/h.

The rate of circulation of cooling water = 6390.8 kg/h.

(b) Denoting the heat removed in the condenser in unit time by Q_C, the heat added in the reboiler through steam by Q_B, the enthalpies of the distillate and the residue to be H_D and H_W respectively, we have the energy balance for the column as

$$FH_F + Q_B = DH_D + WH_W + Q_C$$

In the above equation, Q_C = 910.08 × 884.8.

Taking 300 K to be the reference temperature for evaluating the enthalpy,

$$H_D = \text{enthalpy of distillate} = 0$$

$$H_W = C_{PW}(370 - 300) = (0.01 \times 2.2 + 0.99 \times 4.2) \times 70 = 292.6 \text{ kJ/kg}$$

$$H_F = C_{PF}(340 - 300) = (0.1 \times 2.2 + 0.9 \times 4.2) \times 40 = 160 \text{ kJ/kg}$$

Substituting these values in the energy balance equation

$$Q_B = DH_D + WH_W + Q_C - FH_F$$

$$= 0 + 898.88 \times 292.6 + 910.08 \times 884.8 - 1000 \times 160$$

$$= 908\,251.07 \text{ kJ/h}$$

This is the heat that is to be supplied to the reboiler by the condensation of steam. The heat of condensation of steam is 2730 kJ/kg.

Therefore, the amount of steam supplied is

$$\frac{908\,251.07}{2730} = 332.69 \text{ kg/h}$$

EXERCISES

Components of energy balance equations

12.1 The potential energy of a body of mass 20.0 kg is 3.5 kJ. What is the height of the body from the ground? If a body of mass 20 kg is moving at a velocity of 50 m/s, what is its kinetic energy?

12.2 Nitrogen gas, 0.1 cubic metres in volume, is confined in a cylinder at a pressure of 200 kPa and a temperature of 300 K is expanded at constant temperature to double its volume. What is the work done by the gas on the piston? Assume that nitrogen behaves as an ideal gas.

12.3 A man circling the earth in a spaceship weighed 300 N at a location where the local gravitational acceleration was 3.35 m/s^2. Calculate the mass of the man and his weight on the earth where the gravitational acceleration is 9.81 m/s^2.

12.4 A spherical balloon of diameter 0.5 m contains a gas at 1 bar and 300 K. The gas is heated and the balloon is allowed to expand. The pressure inside the balloon is found to vary linearly with the diameter. What would be the work done by the gas when the pressure inside reaches 5 bar.

12.5 A car having a mass of 1200 kg is running at a speed of 60 km/h. What is the kinetic energy of the car in kJ? What is the work to be done to bring the car to a stop.

12.6 A body of mass 50 kg is lifted through a distance of 15 m. What is the work done? If it took 2 min for lifting the mass, calculate the power.

12.7 A 100-kg ball initially at a height of 15 m is dropped to the ground.

(a) Find the initial potential energy and kinetic energy of the ball.
(b) Find the final potential energy and kinetic energy of the ball.
(c) If the initial potential energy is converted to heat and the entire heat is transferred to 100 kg of water, what would be the temperature rise of water?

12.8 A man whose weight is 700 N takes 2.5 minutes for climbing up a staircase. What is the power developed in him, if the staircase is made up of 20 stairs, each 0.18 m in height?

12.9 A balloon filled with hydrogen lifts a mass of 150 kg including its own mass. If the density of hydrogen is 0.09 kg/m^3 at the prevailing conditions, what would be the volume of the balloon?

12.10 Nitrogen gas is confined in a cylinder and the pressure of the gas is maintained by a weight placed on the piston. The mass of the piston and the weight together is 100 kg. The acceleration due to gravity is 9.81 m/s^2 and the atmospheric pressure is 1.01325 bar. Assume that the piston is frictionless. Determine the following:

(a) The force exerted by the atmosphere, the piston and the weight on the gas if the piston is 200 mm in diameter
(b) The pressure of the gas
(c) If the gas is allowed to expand pushing up the piston and the weight by 500 mm, what is the work done by the gas in joules?
(d) What is the change in the potential energy of the piston and the weight after the expansion in part (c)?

12.11 A body of mass 20 kg is dropped from a height of 15 m. What is the potential energy of the body at the time of its release? Assume that the potential energy gets converted into kinetic energy when the body strikes the ground. At what velocity does it strikes the ground?

12.12 Water at a rate of 1×10^{-3} m^3/s is pumped from a storage tank into a tube of 30 mm diameter.
(a) Determine the kinetic energy of water in J/kg.
(b) If the water is discharged into a tank at a height of 15 m, what is the increase in the potential energy of water from that in the original tank?

12.13 A spherical balloon of diameter 0.5 m contains a gas at 1 bar and 300 K. The gas is heated and the balloon is allowed to expand. The pressure inside the balloon is directly proportional to the square of the diameter. What would be the work done by the gas when the pressure inside reaches 5 bar?

12.14 The steam supplied to an engine liberates 5000 J of heat. If the engine efficiency is only 40 percent, to what height a body of mass 10 kg can be lifted using the work output from the engine?

12.15 A vertical cylinder is separated into two compartments by means of a piston, the compartments being filled with two different gases. If the whole cylinder were filled with any one of these gases, it would support a pressure equal to three eighth of the weight of the piston. What would be the position of the piston separating the two gases?

12.16 A balloon which was originally empty is being filled by hydrogen from a cylinder at a constant temperature of 300 K. The atmospheric pressure is 1.01325 bar. What is the work done by the balloon-cylinder system when the balloon attains a spherical shape with a diameter of 6 m?

12.17 Five kilograms of CO_2 gas is contained in a piston-cylinder assembly at a pressure of 7.5 bar and a temperature of 300 K. The piston has a mass of 6000 kg and a surface area of 1 m^2. The friction of the piston on the walls is significant and cannot be ignored. The atmospheric pressure is 1.01325 bar. The latch holding the piston in position is suddenly removed and the gas is allowed to expand. The expansion is arrested when the volume is double the original volume. Determine the work appearing in the surroundings.

12.18 An elevator is initially resting at a height of 7.5 m above the base of the elevator shaft. It is then raised to a level of 75 m above the base when the cable holding it breaks. It falls freely to the base where it is brought to rest by a strong spring. The mass of the elevator is 2500 kg and the acceleration due to gravity is 9.81 m/s^2. Neglecting the effect of friction, calculate

(a) The potential energy of the elevator in its initial position
(b) The potential energy of the elevator in its highest position
(c) The work done in raising the elevator
(d) The kinetic energy and velocity of the elevator just before it strikes the spring
(e) The potential energy of the spring when the elevator rests on it

12.19 Water is being vaporized by passing an electric current of 0.5 A from a 12 V supply through a resistance immersed in saturated water at atmospheric pressure. If the current is passed for 5 minutes and water evaporated is 0.798 g, calculate the change in molar internal energy and molar enthalpy during this process. Assume that the water vapour behaves ideally.

12.20 Liquid CO_2 at 233 K has a vapour pressure of 1.005×10^3 kPa and a specific volume of 0.9×10^{-3} m^3/kg. Assume that CO_2 is a saturated liquid at these conditions and its enthalpy is zero. The latent heat of vaporization of CO_2 is 320.5 kJ/kg and the specific volume of saturated vapour is 38.2×10^{-3} m^3/kg. Calculate the internal energy of saturated liquid and both internal energy and enthalpy of saturated vapour.

12.21 A gas expands from an initial volume of 0.2 m^3 to a final volume of 0.4 m^3 in a reversible steady flow process. During the process the pressure varies as

$$P = 5 \times 10^5\, V + 7 \times 10^4$$

where P is in N/m^2 and V is in m^3. The inlet line is 4 m below the outlet line, and the gas enters with a negligible velocity. The internal energy of the gas decreases by 30 kJ during the process. Determine the heat transferred.

12.22 Liquid water at 373 K and 101.3 kPa has an internal energy of 420 kJ/kg on an arbitrary basis. The specific volume at these conditions is 1.04×10^{-3} m^3/kg. The water is brought to the vapour state at 473 K and 700 kPa. At this condition, its specific volume is 0.3 m^3/kg and its enthalpy is 2844 kJ/kg. Calculate the enthalpy of the liquid and the changes in internal energy and enthalpy accompanying the vaporization process.

12.23 A sample of 10 g of liquid benzene at its boiling point is vaporized using a 12 V, 0.5 A electric supply. The normal boiling point and the latent heat of vaporization of benzene are 353.2 K and 30.8×10^3 kJ/kmol respectively. Determine

(a) The change in internal energy
(b) The time required for complete vaporization

Heat capacities

12.24 The heat capacity of carbon dioxide gas is given by

$$C_P = 26.540 + 42.454 \times 10^{-3}\, T - 14.298 \times 10^{-6}\, T^2$$

where C_P is in J/mole K and T in K.

(a) Find the amount of heat needed to heat one mole from 350 to 450 K.
(b) Find the constants for C_P in cal/mol °C and T in °C.

12.25 Carbon dioxide has a molal heat capacity of $C_P = 26.54 + 42.454 \times 10^{-3}\ T - 14.298 \times 10^{-6}\ T^2$ where C_P is in J/mol K and T in K.

(a) Calculate the heat capacity at 400 K.
(b) Calculate the mean heat capacity between 400 and 300 K.
(c) Calculate the heat requirement for heating 88 kg of the gas from 300 K to 400 K.

12.26 The molal heat capacity of CO_2 gas is given by

$$C_P = 26.540 + 42.454 \times 10^{-3}T - 14.298 \times 10^{-6}\ T^2$$

where C_P is in kJ/kmol K and T in K. Calculate the following:

(a) The mean molal heat capacity between 500 and 1000 K
(b) The heat required to raise the temperature of 200 m³ per hour of CO_2 gas at STP from 500 to 1000 K.

12.27 A 5 kg mass of a gas undergoes a process in which its temperature rises from 10°C to 30°C. The specific heat of the system is

$$C_P = 3.505 + \frac{262.9}{T}$$

where C_P is in kJ/kg K and T is in K. Determine the following:

(a) The heat interaction
(b) The mean heat capacity of the system

12.28 The heat capacity of HCl for the temperature range of 300 to 1500 K is given by

$$C_P = 28.194 + 1.805 \times 10^{-3}T + 1.513 \times 10^{-6}T^2$$

where C_P is in kJ/kmol K and T in K. Calculate the following:

(a) The mean heat capacity at 1300 K with reference to 300 K
(b) The enthalpy change of HCl when it is heated from 300 K to 1300 K

12.29 Determine the specific heat of a gas at 800 K for which C_P varies with temperature as $C_P = a + bT + cT^2$ where a, b and c are constants. The following data are available.

T (K)	300	600	900
C_P (kJ/kmol K)	28.788	30.417	32.033

12.30 The heat capacity of ethylene gas at different temperatures are given below.

T (K)	298	500	700	1000
C_P (J/mol K)	43.6	63.5	78.5	94.5

(a) Fit the data into an equation of the form $C_P = a + bT + cT^2$.
(b) Calculate the heat required to raise the temperature of one mole of ethylene gas from 400 to 1000 K.

12.31 Calculate the amount of heat given off when 5 m³ of SO_3 gas at STP cools from 773 K to 400 K at a constant pressure of 101 kPa. The heat capacity is given as $C_P = a + bT + cT^2$ where a, b and c are constants. The values of the constants are: a = 34.33, $b \times 10^3$ = 42.86 and $c \times 10^6$ = – 14.298. C_P is in kJ/kmol K and T in K.

12.32 The heat capacity of silicon carbide is given by

$$C_P = 37.221 + 1.22 \times 10^{-2}T - 1.189 \times 10^{5}T^{-2}$$

where C_P is in kJ/kmol K and T is in K. Estimate the following:

(a) The mean heat capacity of silicon carbide in the range of 0 to 1000°C

(b) The enthalpy change in silicon carbide in the temperature range between 0 to 1000°C

12.33 A gas mixture contains species A (MW=30) 15%, species B (MW = 45) 45% and species C (MW = 80) 40% by weight. Calculate the quantity of heat necessary to heat one kilomole of the mixture from 300 K to 2700 K. The constants of the molar heat capacity equation $C_P = a + bT + cT^2$ are as follows where C_P is in kJ/kmol K and T is in K.

Species	a	b	$c \times 10^3$
A	25	0.05	−0.01
B	30	0.009	−0.05
C	21	0.08	−0.001

12.34 The gas leaving a pyrites roaster in a sulphuric acid plant has the following composition. SO_2 = 7.00%, O_2 = 10.50%, SO_3 = 0.80%, N_2 = 81.70%. The heat capacity of the gases varies with temperature as given below where C_P is in kJ/kmol K and T is in K.

O_2: $C_P = 25.74 + 12.987 \times 10^{-3}T - 3.864 \times 10^{-6}T^2$

N_2: $C_P = 27.03 + 5.815 \times 10^{-3}T - 0.289 \times 10^{-6}T^2$

SO_3: $C_P = 34.33 + 42.86 \times 10^{-3}T - 14.298 \times 10^{-6}T^2$

SO_2: $C_P = 25.74 + 57.96 \times 10^{-3}T - 38.11 \times 10^{-6}T^2$

Calculate the enthalpy of the gas mixture at 750 K with reference to a base temperature of 298 K.

12.35 The analysis of natural gas showed the following: Methane, 40.00%; Hydrogen, 40%; carbon monoxide, 4.0%; carbon dioxide, 3.0%; oxygen, 2.0% and the rest nitrogen. A 1-kw heater is used to supply heat to the gas so that its temperature increases by 0.2°C. Estimate the flow rate of the gas in cubic metres per minute at 300 K and 100 kPa. The specific heats of the gases in kJ/kmol K are 35.91 for methane, 28.95 for hydrogen, 29.25 for carbon monoxide, 37.31 for carbon dioxide, 29.47 for oxygen and 29.24 for nitrogen.

Enthalpy change of phase changes

12.36 One kg of ice at 273 K and 5 kg of water at 300 K are mixed and kept in an open container till all the ice melts and the temperature of the mixture becomes 295 K. Given that the heat of fusion of ice is 335 kJ/kg, calculate the heat added or removed to or from the water and ice.

12.37 One kilogram of ice at 0°C is heated so that it is completely converted to steam at 150°C and 101.3 kPa. What is the enthalpy change accompanying the process?

The heat of fusion of water at 101.3 kPa and 0°C is 335 kJ/kg and heat of vaporization of water at 101.3 kPa and 100°C is 2256 kJ/kg. The heat capacity equation of liquid water is

$$C_P = 18.296 + 47.212 \times 10^{-2}T - 133.88 \times 10^{-5}T^2 + 1314.2 \times 10^{-9}T^3$$

and the heat capacity equation of water vapour at 101.3 kPa is

$$C_P = 30.475 + 9.652 \times 10^{-3}T + 1.189 \times 10^{-6}T^2$$

where T is in K and specific heat is in (kJ/kmol K).

12.38 The vapour pressure of water at 473 K is 1554.3 kPa. The normal boiling point of water is 373 K. Determine the latent heat of vaporization of water assuming that it is independent of temperature.

12.39 The vapour pressure of ethyl ether at 273 K is 25 kPa and at 293 K is 63.5 kPa. Estimate the latent heat of vaporization of ether in kJ/kg in this temperature range.

12.40 Using the Watson equation, calculate the latent heat of vaporization of *n*-butane at 373 K given that the heat of vaporization at the normal boiling point of 272.7 K is 385 kJ/kg. The critical temperature of *n*-butane is 425 K.

12.41 Using Trouton's rule, find the approximate heat of vaporization of benzene boiling at 80°C.

12.42 The vapour pressure of zinc is given by the Clapeyron equation

$$\log P^S = -\frac{6160}{T} + 8.1$$

where pressure is in mm Hg and temperature is in K. Estimate the heat of vaporization of zinc at the normal boiling point of 1180 K, and compare the result with the value estimated using the Kistyakowsky equation.

12.43 Calculate the enthalpy of zinc vapour at 1500 K and standard atmospheric pressure with reference to the solid at 273 K. The melting point and the boiling point of zinc at atmospheric pressure are 592 K and 1180 K respectively. The mean heat capacity of the solid between 273 K and 592 K = 0.441 kJ/kg K; the mean heat capacity of liquid = 0.458 kJ/kg K; the mean heat capacity of the vapour from 1180 K to 1500 K = 20.874 kJ/kg K; the heat of fusion = 6972 J/g-atom. Use the Kistyakowsky equation.

12.44 Methyl chloride is obtained from the vapour-phase catalytic reaction between methanol and hydrogen chloride. The liquid methanol is vaporized in a separate vaporizer, and the vapours are continuously mixed with HCl gas in equimolar proportions. The mixed gases are preheated in a tube furnace to 200°C before entering the catalytic reactor. The heat capacity of liquid methanol 2.62 kJ/kg K. The mean heat capacity of methanol vapour is 53.47 kJ/kmol K. The mean heat capacity of HCl gas is 41.70 kJ/kmol K. The normal boiling point of methanol is 337.9 K. The heat of vaporization of methanol at 101.3 kPa is 35.3 kJ/mol. On the basis of 400 kg/h of alcohol feed at 25°C, find the following:

(a) The heat duty of methanol vaporizer operating at one standard atmosphere
(b) The heat duty of the tube furnace if the methanol vapour enters saturated at 1 atm, the HCl enters at 30°C, no chemical reaction occurs in the furnace, and heat losses are negligible.

12.45 The following table gives the vapour pressures of benzene and water. Plot the reference temperature vapour pressure curve with the logarithm of vapour pressure of benzene on the *y*-axis and the logarithm of vapour pressure of water on the *x*-axis.

T (K)	288.6	299.3	315.4	333.8	353.3	377.0	415.7	452
P_B^s (kPa)	8.0	13.3	26.7	53.3	101.3	202.7	506.6	1013.0
P_W^s (kPa)	1.8	3.4	8.3	26.1	49.0	116.3	382.5	980.0

(b) Determine the vapour pressure of benzene at 373 K.

(c) Determine the latent heat of vaporization of benzene at 298 K given that the latent heat of vaporization of water at 298 K is 2443 kJ/kg. Compare your result with the experimental value of 434 kJ/kg.

Steam tables

12.46 A steam-jet ejector is used to entrain saturated water vapour at 25 kPa (H = 2618.2 kJ/kg) leaving an evaporator by high pressure saturated steam at 1000 kPa (H = 2778.1 kJ/kg). It is estimated that with every kilogram of high-pressure steam, 0.75 kg of the vapour from the evaporator can be entrained. If the mixed stream leaving the ejector is at 100 kPa what is its temperature?

12.47 A valve on a well-insulated steam pipe carrying saturated steam at 1000 kPa is found leaking. The temperature of the steam escaping from the leak is measured to be 398 K. Determine the quality of steam flowing through the pipe.

The following data are taken from steam tables: the enthalpy of saturated vapour at 1000 kPa =2778 kJ/kg, the enthalpy of saturated liquid at 1000 kPa = 763 kJ/kg and the enthalpy of superheated steam at 398 K and 101.3 kPa = 2726 kJ/kg.

12.48 Wet steam containing 5% by weight of liquid at a pressure of 500 kPa is mixed at a rate of 1 kg/s with superheated steam at 500 kPa and 473 K (H = 2855 kJ/kg) to obtain dry saturated steam at 500 kPa (H_l = 640 kJ/kg; H_V =2749 kJ/kg). Determine the rate of addition of superheated steam if mixing is done adiabatically.

12.49 An evacuated tank is connected to a pipe carrying steam at 1400 kPa and 598 K (enthalpy = 3097 kJ/kg) through a valve. The valve is opened and the tank fills with steam until the pressure is 1400 kPa, and then the valve is closed. Assume that the process is adiabatic and the kinetic and potential energies are negligible. Determine the final temperature of the steam.

12.50 Using steam tables, find the enthalpy change when (a) one kilogram of water is heated from 293.15 to 323.15 K at 101.3 kPa and (b) when one kilogram of water at 293.15 K is heated and vaporized at 200 kPa.

12.51 A closed tank contains 10 kg of water in equilibrium with 2 kg of steam at 349 K. Heat is added until 8 kg of the water is vaporized.

(a) What was the original pressure?
(b) What is the final pressure?
(c) How much heat was added?

12.52 A mixture of saturated steam and saturated water is contained in a rigid tank of volume 0.1 m^3 at a pressure of 2 bar. The liquid occupies 10% of the total volume. How much heat must be added in order that the tank contains only saturated steam? What will be the pressure in the tank?

12.53 A steam boiler is one-fourth filled with liquid water at 101.3 kPa and 373.15 K and three-fourth filled with steam. If it is brought to a pressure of 6817 kPa with all the valves closed, how much heat must be supplied in kJ per kilogram of water in the boiler?

12.54 A vessel of volume 1 m^3 initially contains one percent by volume saturated liquid water and the rest saturated vapour at 1 bar. How much heat is to be supplied so that the vessel gets filled with vapour?

12.55 A steam boiler of volume 2.3 m^3 initially contained 1.7 m^3 liquid water in equilibrium with 0.6 m^3 of vapour at 100 kPa. The boiler is heated keeping the inlet and discharge valves closed. The relief valve is set to lift when the pressure in the boiler reaches 5500 kPa. Determine the amount of heat supplied to the contents in the boiler before the relief valve lifts.

12.56 A cylinder fitted with a piston has a volume of 0.1 m^3 and contains 0.5 kg of steam at 500 kPa. How much heat is to be supplied to bring the temperature of the steam to 823 K, keeping the pressure constant? What is the work done in the process?

Heat of mixing

12.57 Fifty kilograms of NH_4NO_3 is dissolved in 108 kg of water at 298 K. Find (a) the heat absorbed if the process is isothermal and (b) the temperature of the mixture if the dissolution is carried out adiabatically. Take the heat capacity of the mixture to be 3.35 kJ/kg K.

12.58 The heat of formation of LiCl is – 408.610 kJ/mol at 298 K. The heat of solution for 1 mol LiCl in 12 mol water is – 33.614 kJ at 298 K. Calculate the heat of formation of LiCl in 12 mol water at 298 K.

12.59 One hundred kilograms of pure H_2SO_4 is mixed with 110 kg of water at 25°C and 101.3 kPa. Calculate the following:

(a) The heat evolved under isothermal conditions
(b) The final temperature of the mixture if it was formed under adiabatic conditions. Assume the heat capacity of the final mixture is 2.5 kJ/kg K.

12.60 What temperature will be attained when a 20 mole percent ethanol-water mixture is adiabatically formed from the pure liquids at 298 K? The heat of mixing for 20 mole percent ethanol-water at 298 K is – 758 J/mol. The mean heat capacity of a 20 mole percent solution of alcohol in water at 298 K is 97.65 J/mol K.

12.61 A solution of NaOH in water is prepared by diluting a concentrated solution in an agitated, jacketed vessel. 2500 kg of 5% NaOH solution is prepared by diluting 50% NaOH solution. If the concentrated solution is available at 298 K and the solution discharged is also at 298 K, calculate the heat removed by the cooling water.

Heat of solution varies with concentration as:

mol H_2O/mol NaOH	2	4	5	10	∞
ΔH_s, kJ/mol NaOH	−22.9	−34.4	−37.7	−42.5	−42.9

12.62 Water at a rate of 54×10^3 kg/h and $Cu(NO_3)_2{\cdot}6H_2O$ at a rate of 64.8×10^3 kg/h are mixed together in a tank. The solution is then passed through a heat exchanger to bring the temperature to 298 K, same as the components before mixing. Determine the rate of heat transfer in the exchanger. The following data are available. The heat of formation at 298 K of $Cu(NO_3)_2$ is – 302.9 kJ and that of $Cu(NO_3)_2{\cdot}6H_2O$ is – 2110.8 kJ. The heat of solution of $Cu(NO_3)_2{\cdot}nH_2O$ at 298 K is – 47.84 kJ per kmol salt and is independent of n.

12.63 Hydrogen chloride gas at 373 K containing 10% (weight) water vapour is absorbed in 25% hydrochloric acid at 313 K yielding 28% (weight) acid. The enthalpy of HCl gas at 373 K is 2065 J/kg and the enthalpy of the 25% acid at 313 K is 163 J/kg. Calculate the enthalpy of product.

12.64 If pure liquid H_2SO_4 is added to pure water both at 300 K to form a 20 per cent (weight) solution, what is the final temperature of the solution? The heat of solution of sulphuric acid in water is H_2SO_4 (21.8 H_2O) = – 70 × 10^3 kJ/kmol of sulphuric acid. The standard heat of formation of water is – 286 kJ/mol.

12.65 $LiCl{\cdot}H_2O(c)$ is dissolved isothermally in enough water to form a solution containing 5 mol of water per mole of LiCl. What is the heat effect? The following enthalpies of formation are given:

$$LiCl(c) = -409.05 \text{ kJ}, \quad LiCl{\cdot}H_2O(c) = -713.054 \text{ kJ}$$

$$LiCl{\cdot}(5H_2O) = -437.232 \text{ kJ}, \quad H_2O(l) = -286.03 \text{ kJ}$$

12.66 1000 kg/h of $FeSO_4{\cdot}7H_2O$ crystals are to be produced in a Swenson–Walker crystallizer by cooling a saturated solution at 323 K (solubility is 140 parts $FeSO_4{\cdot}7H_2O$ per 100 parts water) to 300 K (solubility is 74 parts $FeSO_4{\cdot}7H_2O$ per 100 parts water). The feed solution has an average specific heat of 2.93 kJ/kg K. The heat of solution of $FeSO_4{\cdot}7H_2O$ at 300 K is –18 400 kJ/kmol. How much heat is evolved during the process?

12.67 Calculate the heat effects when 1.0 kmol of water is added to a solution containing 1.0 kmol sulphuric acid and 3.0 kmol of water. The process is isothermal and occurs at 298 K.

Data: Heat of mixing for $H_2SO_4(3H_2O)$ = – 49 000 kJ per kmol H_2SO_4. Heat of mixing for $H_2SO_4(4H_2O)$ = – 54 100 kJ per kmol H_2SO_4.

12.68 A single-effect evaporator is used to concentrate a 15 percent (weight) solution of LiCl in water to 40 percent. The feed enters the evaporator at 298 K at the rate of 2 kg/s. The normal boiling point of a 40 percent LiCl solution is 405 K and its specific heat is 2.72 kJ/kg K. For what heat transfer rate in kJ/h, should the evaporator be designed?

The heat of solution of LiCl in water per mole of LiCl at 298 are:
ΔH_S for LiCl (13.35 H_2O) = – 33.8 kJ, for LiCl (3.53 H_2O) = – 23.26 kJ, the enthalpy of superheated steam at 405 K = 2740.3 kJ/kg, the enthalpy of water at 298 K = 104.8 kJ/kg, and the molecular weight of LiCl = 42.39.

Energy balance in non-flow processes

12.69 A gas mixture consists of 60% methane and 40% nitrogen. Calculate the heating value of this mixture when it is metered at 300 K and 105 kPa if the partial pressure of water vapour in the gas is 4 kPa. The calorific value of methane gas is 40000 kJ/m³ measured at standard temperature and pressure.

12.70 Sulphur dioxide is circulated as the refrigerant in a small refrigerator. SO_2 gas at a pressure of 5 bar and a temperature of 340 K is to be cooled at a constant volume of 0.142 m³, to 293 K as part of the refrigeration cycle. Calculate

(a) The heat liberated
(b) The work done by the gas on cooling
(c) The final pressure attained on cooling
(d) The change in enthalpy

Sulphur dioxide may be treated as an ideal gas. The specific heat (J/mol K) is found to vary with temperature (K) according to

$$C_P = 25.736 + 5.796 \times 10^{-2}T - 3.8112 \times 10^{-5}T^2 + 8.612 \times 10^{-9}T^3$$

12.71 10 kg of an equimolar mixture of CO and H_2 at 1500 K and 1 bar is cooled at constant pressure to 350 K. Assuming that the gases are ideal determine the following:

(a) The heat given off during cooling,
(b) The change in internal energy,
(c) The work done on the gas

The specific heats in J/mol K are as follows:

$$C_{P,H_2} = 29.086 - 8.3694 \times 10^{-4}T + 2.0130 \times 10^{-6}T^2$$

$$C_{P,CO} = 26.553 + 7.6882 \times 10^{-3}T - 1.1727 \times 10^{-6}T^2$$

12.72 One cubic metre of an ideal gas is confined in a cylinder provided with a frictionless piston at 300 K and 250 kPa. The gas is allowed to expand till its volume is doubled in a process occurring (a) at constant pressure and (b) at constant temperature. Calculate the work done by the gas on the piston in the above processes.

12.73 Consider 1 kg of water in a waterfall which is flowing down from a height of 100 m. At the bottom of the fall, it joins a river whose downstream velocity may be assumed negligible. Neglecting any energy exchange between the water and the surroundings, calculate

(a) The potential energy at the top of the falls
(b) The kinetic energy just before the water strikes the bottom
(c) The change in temperature of the water when it enters the river.

12.74 A car riding downhill at a speed of 20 m/s was applied brake when it was at a height of 30 m vertically above the bottom of a hill. When the car comes to a halt at the bottom of the hill, how much energy as heat must be dissipated by the brakes, if wind and other frictional effects are neglected? The car weighed 1400 kg.

12.75 A system consisting of a gas confined in a cylinder is undergoing the following series of processes before it is brought back to the initial conditions:

Step 1. A constant pressure process when it receives 50 J of work and gives up 25 J of heat.

Step 2. A constant volume process when it receives 75 J of heat.

Step 3. An adiabatic process.

Determine the change in internal energy during each step and the work done during the adiabatic process.

12.76 A 10-g lead ball at 300 K is dropped from a height of 10 m. Calculate (a) the kinetic energy and speed of the ball as it reaches the ground and (b) the temperature rise of the ball if all its kinetic energy is transformed into internal energy as the ball is suddenly stopped after 10 m, given that the specific heat of lead is 125.6 J/kg K and its volume change is negligible.

12.77 (a) Three kmol nitrogen gas (C_P = 29.1 kJ/kmol K and C_V = 20.8 kJ/kmol K) at 350 K is taken in a rigid container and is heated to 550 K. How much heat must be supplied if the mass of the vessel is 100 kg and its specific heat is 0.5 kJ/kg K?

(b) Two kmol nitrogen is taken in a piston/cylinder arrangement at 500 K. Determine the quantity of heat extracted from the gas to cool it to 350 K at constant pressure if the heat capacity of the container is negligible.

12.78 An insulated rigid container contains air at 5 bar and 375 K. The volume of the container is 1 m^3. Air may be assumed to behave as an ideal gas. The specific heat C_V of air is 20.785 kJ/kmol K. The temperature of the air is increased by a rotating paddle wheel. Calculate the work done to rise the temperature of the air to 425 K.

12.79 100 m^3 of CO_2 initially at 423 K and 50 bar is to be isothermally compressed in a frictionless piston and cylinder device to a final pressure of 300 bar. Write a general expression for the energy balance for the gas within the piston and cylinder device as the system and define all terms. Also calculate the heat flow on compression.

12.80 A 14-m^3 storage tank contains 13 m^3 liquid oxygen in equilibrium with its vapour at 100 kPa. The tank is provided with a pressure relief valve which is set at 250 kPa. (a) Determine the mass of total material in the tank and (b) the fraction of the total material that is vapour. (c) When will the pressure relief valve operate if heat leaks into the tank at the rate of 5000 kJ/h? The specific volume of gas and liquid oxygen at 100 kPa are, respectively, 7.15 m^3/kmol and 2.81 × 10^{-2} m^3/kmol. The enthalpies are −133.5 kJ/kg and −116.6 kJ/kg at 100 kPa and 250 kPa respectively.

12.81 One kmole of an ideal gas is originally at 300 K and 1 bar. It is then heated at constant pressure to a temperature of 400 K and compressed isothermally to a volume equal to its initial volume. Assume that C_P = 30 kJ/kmol K. Find ΔU, ΔH, Q and W.

12.82 Argon gas at 300 K and 1.5 Pa is contained in a chamber of volume 2×10^{-3} m^3. A 1000-ohm electric resistance heater is used to heat the gas. The mass of heater is 15 g and its heat capacity is 0.35 kJ/kg K. The gas may be assumed ideal with $C_P = 5R/2$. The heater draws current at 40 V for 5 minutes. Calculate the final gas temperature and pressure at equilibrium.

12.83 Hydrogenation of an edible oil is to be carried out in a reactor at a pressure of 136.8 kPa and a temperature of 453 K. The hydrogen gas at 293 K is heated to this temperature by passing through a coil where it experiences a pressure drop of 300 kPa. Hydrogen behaves like an ideal gas and its heat capacity at constant pressure is 29.4 kJ/kmol. Neglecting kinetic energy effects, determine the heat transfer rate in the coils per 1000 kg of hydrogen.

Energy changes in ideal gas processes

12.84 The heat capacities of nitrogen gas which is assumed as an ideal gas are $C_V = 20.8$ and $C_P = 29.1$ kJ/kmol K. The gas initially at 10 bar and 280 K is undergoing a change of state to the final condition of 1 bar and 340 K. Determine the change in internal energy and the change in enthalpy.

12.85 One cubic metre of an ideal gas is being compressed adiabatically from an initial condition of 1 bar and 310 K to a pressure of 10 bar in an open system. If the efficiency of the process is 80%, how much work is required? Take $\gamma = 1.4$. Compare the result with that if the operation is carried out in a closed system.

12.86 CO_2 is sold commercially in steel containers at 60 bar. The gas is leaking through the outlet valve slowly so that the temperature may be assumed constant at the room temperature of 300 K.

(a) What is the work done in the expansion of 10 kg of this gas from 60 bar to 1 bar?

(b) If the temperature were constant at 290 K, would there be any difference in the work done?

(c) What are the changes in enthalpy in both the above cases?

12.87 An ideal gas is compressed adiabatically from 1.5 bar, 338 K to 9 bar. The process is reversible, and $\gamma = 1.23$ is constant over the entire range of conditions. Calculate the following:

(a) The temperature at the end of compression

(b) The work of compression

(c) The heat transferred

(d) The change in internal energy

(e) The change in enthalpy

12.88 An ideal gas undergoes the following reversible processes:

(a) From an initial state of 343 K and 1 bar, it is compressed adiabatically to 423 K.

(b) It is then cooled to 343 K at constant pressure.

(c) Finally it is expanded to its original state isothermally.

Calculate ΔU, ΔH, W and Q for each step as well as for the entire cycle. Assume $C_V = (3/2)R$.

12.89 Air initially at 389 K and 8 bar is expanded reversibly and isothermally to a pressure such that when it is cooled to 278 K at constant volume, its pressure is 2 bar. Assume air to be an ideal gas with C_P = 29.3 J/mol K. Calculate the work, heat transferred, changes in internal energy and change in enthalpy.

12.90 Assume that the *P-V-T* relationship for nitrogen gas can be approximated by the ideal gas equation $PV = nRT$, where R = 8.314 kJ/kmol K. The heat capacities are C_V = 20.786 kJ/kmol K and C_P = 29.1 kJ/kmol K. Nitrogen which was initially at 1 bar and 280 K is compressed to 5 bar and 280 K by two different reversible processes:

(a) Cooling at constant pressure followed by heating at constant volume
(b) Heating at constant volume followed by cooling at constant pressure

For each of the above paths determine the conditions at the intermediate state and ΔU, ΔH, Q and W.

12.91 Methane is stored in a tank of capacity 5.7 × 10^{-2} m^3 at a pressure of 15 bar and 294 K. The gas is allowed to flow from the tank through a partially opened valve to a gas holder where the pressure is constant at 1.15 bar. When the pressure in the tank dropped to 5 bar, what would be the mass of methane removed under the following conditions:

(a) If the process took place slowly so that the temperature was constant.
(b) If the process took place so rapidly that the heat transferred was negligible.

Methane behaves as an ideal gas with γ = 1.4.

12.92 A rigid non-conducting tank with a volume of 4 m^3 is divided into two equal parts by a membrane. On one side of the membrane the tank contained gas *A* at 5 bar and 350 K and on the other side a gas *B* at 10 bar and 450 K. *A* and *B* are ideal gases with C_V values (5/2)R and (7/2)R respectively. The membrane is suddenly ruptured and the gases get mixed. What are the final temperature and pressure?

12.93 One cubic metre of an ideal gas at 600 K and 20 bar expands to ten times its initial volume as follows:

(a) By a reversible, isothermal process
(b) By a reversible, adiabatic process
(c) By an irreversible, adiabatic process in which the expansion is against a restraining pressure of 1 bar

If C_P = 21 J/mol K, calculate the final temperature, pressure and the work done by the gas.

Energy balance for flow processes

12.94 In a test of water-jacketed air compressor, it was found that the shaft work required to drive the compressor was 150 kJ per kg air compressed. The enthalpy of air leaving the compressor was found to be greater than that of the air entering the compressor by 80 kJ/kg and the heat removed by the cooling water was 100 kJ per kg of air. Determine the amount of energy that must have been dissipated as heat to the atmosphere from the bearings, cylinder walls, etc.

12.95 A steam turbine using steam at 1368 kPa and 645 K and discharging saturated steam at 137 kPa is used to generate power for a certain chemical plant. The turbine acts adiabatically and the feed and discharge velocities may be considered equal. Determine the theoretical horsepower developed by the turbine if it uses 1650 kg of steam per hour. From steam tables, the enthalpy of superheated steam at 1368 kPa and 645 K = 3200 kJ/kg and the enthalpy of saturated steam at 137 kPa = 2690 kJ/kg.

12.96 3600 kg/h of superheated steam at 200 kPa and 673 K enters a turbine with a velocity of 100 m/s. The inlet to the turbine is at an elevation of 10 m and the exit is at an elevation of 3 m. The steam leaves the turbine at a velocity of 150 m/s and is 98% dry at a pressure of 10 kPa. What is the power output of the turbine if the energy loss from it is estimated to be 40000 kJ/h?

12.97 A trial run on a steam turbine power plant gave the following results

Entrance to boiler:	*Exit of turbine:*
Mass flow rate = 3600 kg/h	Velocity of steam = 25 m/s
Enthalpy of water = 850 kJ/kg	Elevation above datum = 0
Elevation above datum = 4.3 m	Enthalpy of steam = 2625 kJ/kg
Velocity of water = 5 m/s	

Determine the power developed by the turbine if the heat added in the boiler is 2100 kJ/s.

12.98 In a small power plant steam at 573 K and 1800 kPa is expanded to 101.3 kPa with 85% quality at a rate of 0.1 kg/s. What is the horse power developed by the turbine?

12.99 A turbine is fed with steam at 78 bar and 698 K (enthalpy = 3213 kJ/kg) at a rate of 1000 kg/h. Saturated steam at 5 bar (H = 2749 kJ/kg) is withdrawn from one point in the turbine at a rate of 250 kg/h. The remaining steam leaves the turbine saturated at 1 bar (H =2676 kJ/kg). Determine the power output from the turbine if it operates adiabatically.

12.100 A boiler receives water at 1800 kPa and 300 K where it is converted to superheated steam at 1800 kPa and 673K. The superheated steam is expanded in an adiabatic turbine and emerges at a pressure of 200 kPa and 98% quality. Per kg of steam calculate the heat transferred, work done and the change in enthalpy for (a) the boiler and (b) the turbine.

12.101 Superheated steam at 2000 kPa and 873 K is expanded in a turbine at a rate of 100 kg/min. The feed line is 75 mm in diameter and the turbine exhausts through a 125 mm diameter line after expanding adiabatically to 25 kPa. What is the horsepower rating of the turbine?

12.102 The work required for compressing a gas from an initial condition of 100 kPa and 300 K to a final pressure of 300 kPa is found to be 280 kJ per kg of the gas. The compressed gas is admitted to a nozzle where its velocity is increased to 700 m/s. If the gas enters the compressor with negligible velocity and leaves the nozzle at 100 kPa and 300 K, what is the heat removed during compression?

12.103 Steam at 700 kPa and 553 K enters a nozzle with negligible velocity and discharges at a pressure of 475 kPa. Determine the following:

(a) The exit velocity

(b) The cross-sectional area at the nozzle exit for a flow rate of 0.5 kg/s

12.104 Water at 368 K is pumped from a storage tank at the rate of 25 m^3/h. The motor for the pump supplies work at the rate of 2 hp. The water passes through a heat exchanger, where it gives up heat at the rate of 42 000 kJ/min and is delivered to a second storage tank at an elevation of 20 m above the first tank. What is the temperature of the water delivered to the second storage tank? Assume that the enthalpy of water is zero at 273 K and the specific heat of water is constant at 4.2 kJ/kg K.

12.105 Water is flowing in a straight horizontal insulated pipe 25 mm ID. There is no device present for adding or removing energy as work. The upstream velocity is 10 m/s. The water flows in a section where the diameter is suddenly increased.

(a) What is the change in enthalpy if the downstream diameter is 50 mm?

(b) What is the maximum enthalpy change for a sudden enlargement in pipe?

12.106 Water at 293 K is being pumped from a constant head tank open to atmosphere to a second tank kept at a constant pressure of 1150 kPa through a 50 mm ID pipe at a rate of 24 m^3/h. The level of water in the second tank is 10 m above that in the first tank. The pump and motor have an overall efficiency of 70 %, and the energy loss in the pipe line is estimated to be 50 J/ kg. Determine the horsepower rating of the pump.

12.107 Water at 200 kPa and 355 K (H = 343.3 kJ/kg) enters a straight horizontal pipe at a velocity of 3 m/s, where it is heated by flue gases from the outside. Steam leaves the system at 100 kPa and 423 K (H = 2776.3 kJ/kg) at a velocity of 200 m/s. How much heat must have been supplied per kg of water flowing?

12.108 Calculate the horsepower rating of the pump required to pump 1000 kg/h of water from 600 kPa and 300 K to 6000 kPa and 310 K. The discharge is 5 m above the suction point.

12.109 Oil flows at a rate of 1000 kg/min from an open reservoir at the top of a hill, 400 m in height to another reservoir at the bottom of the hill. Heat is supplied to the oil on its way at the rate of 1800 kJ/min and work is supplied by a 1-hp pump. Take the mean specific heat of oil to be 3.35 kJ/kg K. Determine the temperature change of the oil.

12.110 A municipal water-supply system delivers water from a storage tank at a rate of 12 cubic metres per minute to consumers 15 m below the tank level utilizing pump work at a rate of 1.5 hp. If the exit velocity of water is 0.5 m/s and its temperature is the same as the temperature of water in the tank, estimate the heat loss from the system. Assume that the water level in the tank is essentially constant.

12.111 A pump is used to transfer a solution of density 1200 kg/m^3 from a mixing vessel to a storage tank through a pipe of diameter 0.08 m at a velocity of 1 m/s. The level difference between the liquid in the mixing vessel and the storage tank is 20 m. Both tanks are open to the atmosphere. Frictional loss is estimated to be 300 W. Determine the pressure increase over the pump. What is the power input to the pump?

12.112 A pump is used to transfer a solution of density 1250 kg/m^3 at a rate of 12 m^3/h from an open storage tank to the top of an absorption tower, which is operated at a pressure of 500 kPa. The pump discharges into the tower through openings equivalent in area to a 25 × 10^{-3} m pipe. The point of discharge is 30 m above the level of solution in the tank. The pump intake is through a pipe of diameter 50 × 10^{-3} m which extends to a depth of 2 m below the level of solution in the tank. The friction head in the suction line is estimated to be 1.5 m of water and that in the discharge line is 10 m of water. If the efficiency of the pump is 70%, what is the power input to the pump? What pressures will be indicated by the pressure gauges at the inlet and exit of the pump?

12.113 Air, assumed to be an ideal gas with molar heat capacity C_P = 30 kJ/kmol K, is flowing through a pipe of diameter 0.15 m at a rate of 0.3 m^3/s at 100 kPa and 300 K before entering a compressor. A cooler removes heat from the compressed gas at a rate of 75.0 kJ/s. The gas at 315 K and 550 kPa is then carried away through a pipe of diameter 0.03 m. What is the power input to the compressor?

12.114 A solution of 10% (weight) acetone in water is subjected to fractional distillation at a rate of 1000 kg/h to produce a distillate containing 99% acetone and a bottom product containing not more than 100 ppm acetone. The feed enters at 308 K, and the distillate and residue leave the tower at 298 K and 373 K respectively. A reflux ratio (ratio of the weight of reflux to the weight of distillate product) of 10 is employed. The rise in temperature by 30 K is permitted for the cooling water circulated in the condenser employed for condensing the vapours into the distillate product and the reflux. Saturated steam at 276 kPa is available for supplying the heat of vaporization in the reboiler. Heat losses from the column may be neglected. The heat capacity of acetone is 2.2 kJ/kg K and that of water is 4.2 kJ/kg K. The boiling point of 99% acetone–water solution is 330 K. The latent heat of acetone at 330 K is 620 kJ/kh and that of water is 2500 kJ/kg. The laten heat of steam at 276 kPa is 2730 kJ/kg. Calculate the following:

(a) The cooling water circulation rate
(b) The rate of circulation of steam

12.115 Air at a rate of 100 kg/h enters a compressor at 100 kPa and 260 K with an enthalpy of 490 kJ/kg where it is compressed to 1000 kPa and 280 K. The air leaves the compressor at 65 m/s with enthalpy 510 kJ/kg. Calculate the power required for the compressor.

12.116 Water flowing upward through a vertical pipe enters a reducer with a velocity of 1 m/s. The diameters at the entrance and exit of the reducer are 0.2 m and 0.1 m respectively. If the pressure at the entrance to the section is 105 kPa, what is the pressure at the exit given that the entrance and exit are 1.5 m apart?

12.117 1000 kg/h of steam at 4500 kPa and 573 K enters the turbine through a 75-mm diameter main. In the turbine steam expands at constant entropy (reversible adiabatic) to 10 kPa pressure. Determine (a) the power produced by the plant in kWh/day if the conversion of mechanical to electrical work is 90% efficient and (b) the size of exhaust main if the steam were to have the same velocity as in the inlet main.

12.118 The feed to a distillation column consists of 10 000 kg/h of a mixture of benzene and toluene containing 45% (mole) benzene at 300 K. On fractionation of this mixture a distillate containing 99% (mole) benzene and a residue consisting of 1.5% (mole) benzene are obtained. The distillate is condensed at 355 K and further cooled to 350 K in the overhead condenser. The residue leaves at 383 K. A reflux ratio of 8 is employed. Saturated steam at 412 K (λ = 2148 kJ/kg) is supplied to the reboiler. The vapour produced in the reboiler is 3.5% (mole) benzene. The temperature rise for cooling water is limited to 30 K. The latent heat of condensation of the distillate vapour is 32.05 kJ/mol. The heat capacity of all liquid streams in the column may be assumed to be 160 kJ/kmol K. Calculate the rate of circulation of cooling water and the steam supply rate in kg/h.

12.119 A steam heated dryer evaporates 1000 kg/h of water from a wet solid material. The dryer is operated at atmospheric pressure. Air enters the dryer at 303 K and 293 K wet bulb and leaves at 365 K containing 0.3 mol water per mol dry air. The solid enters the dryer at 290 K and the enthalpy change in the solid may be neglected. The heat loss from the dryer is estimated to be 5% of the total heat duty of the dryer. Calculate the heat duty of the dryer.

12.120 A steam-jacketed vessel is charged with 200 kg of a feed material (heat capacity 3.5 kJ/kg K) at 295 K to heat it to 373 K. The heating is done by complete condensation of saturated steam at 40 bar (λ = 1714 kJ/kg) in the jacket. The rate of heat loss from the vessel is estimated to be at a rate of 1.5 kJ/s. Determine the mass of steam needed if the charge was heated for one hour.

12.121 10 000 kg/h of a wet solid containing 5% water on a wet basis is to be dried to 0.10% in a counter-current rotary drier. The feed enters the drier at 300 K. Drying air enters the drier at 422 K and leaves at 343 K. The solid leaves the drier at 366 K. The heat capacity of solid may be taken as 1.5 kJ/kg K. The evaporation of the moisture occurs at 307 K. The latent heat of vaporization of water at 307 K is 2420 kJ/kg and the specific heat of water vapour is 1.884 kJ/kg K. Calculate the following:

(a) The amount of heat required to bring the product to the discharge temperature
(b) The amount of heat required to remove the moisture in the solid

12.122 Saturated steam at 500 K circulated at a rate of 100 kg/h is used as the heating medium for preheating the feed to a reactor. Determine the amount of heat transferred to the feed stream in one hour if the condensate leaving is at 425 K. If subcooling is not allowed in the condenser, how much additional steam is to be supplied for the same heat duty?

12.123 A liquid mixture consisting of 30% (mole) ethane and the rest *n*-butane at 280 K and a very high pressure is first heated by passing through a heat exchanger and then throttled to a lower pressure of 1400 kPa. On throttling, partial vaporization of the mixture takes place. The mixture is admitted into a flash chamber maintained at 1400 kPa. The vapour and liquid products are separated here at 328 K. The vapour product contains 60% (mol) ethane and the liquid product contains 17% (mol) ethane. The enthalpy values are: feed at 280 K = – 8500 kJ/kmol, vapour at 328 K = 14 000 kJ/kmol and liquid at 328 K = – 3500 kJ/kg. Find the quantity of heat required for the flash vaporization of 100 kmol of the feed.

12.124 The discharge air from a drier has a dry-bulb temperature of 330 K and a wet-bulb temperature of 317 K (humidity = 0.06 kg of water per kg of dry air, enthalpy 228 kJ/kg dry air). To keep the drying rate within limits and to avoid undesirable effects, a portion of this air is recycled and is mixed with fresh air before being sent to the heater and then to the drier. Eighty percent of the dry air entering the heater is recycled. Fresh air supplied is at 249 K with humidity 0.006 kg of water per kg of dry air (enthalpy of fresh air is 49 kJ/kg). How much heat is required per kilogram of water evaporated?

12.125 Ammonium sulphate is dried from 3.5% moisture to 0.2% moisture in a counter-current rotary drier at a rate of 1000 kg/h of product. Atmospheric air at 298 K, 50% humidity will be heated by passage over steam coils to 363 K before it is admitted into the drier. The air leaves the drier at 305 K. The solids enter the drier at 298 K and is expected to be discharged at 333 K. The heat loss from the drier is 40×10^3 kJ/h. Determine the following:

(a) The air circulation rate, kg/h

(b) The heat duty of the drier, kJ/h

The heat capacity of ammonium sulphate is 1.5 kJ/kg K, the heat capacity of dry air is 1.005 kJ/kg K, the specific heat of water vapour is 1.884 kJ/kg k and the latent heat of vaporization of water at 273 K is 2502 kJ/kg.

12.126 Air, with dry-bulb temperature 311 K and wet-bulb temperature 300 K, is scrubbed with water to remove dust. The water is maintained at 298 K. The air leaving the scrubber is in equilibrium with water. It is then heated to 366 K in an air pre-heater and admitted to an adiabatic rotary drier. The air leaves the drier at 322 K. The material to be dried enters and leaves the drier at 319 K. The material loses moisture at a rate of 0.05 kg of water per kg of product The total product is 1000 kg/h. Using the psychrometric chart, determine the heat supplied to the air pre-heater in kJ/h.

13 Energy Balance Thermochemistry

Thermochemistry plays a very significant role in the design and analysis of chemical processes. Energy balances in processes involving chemical reactions are important in the design of process reactors for carrying out industrial reactions under specified conditions. A chemical engineer should be able to make accurate estimates of energy requirements for carrying out chemical processes and the energy changes involved in various stages of the process. The enthalpy changes involved in chemical reactions, the influence of temperature on the enthalpy changes of chemical reactions and the methods for the evaluation of the heat of reaction are important for process calculations and these are discussed in the following sections.

13.1 HEAT EFFECTS ACCOMPANYING CHEMICAL REACTIONS

Energy changes are involved in the course of a chemical reaction; heat is either absorbed or evolved during the reaction. The reactions in which heat is absorbed are called *endothermic reactions* and those in which heat is evolved are called *exothermic reactions*. Knowledge of the heat effects accompanying chemical reactions and the influence of the operating parameters on these energy changes is essential for the proper design and operation of reaction vessels.

13.1.1 The Standard Heat of Reaction

By heat of reaction we mean the change in enthalpy of the system for the reaction proceeding at constant temperature. It is the difference between the enthalpy of products and the enthalpy of reactants and is denoted by ΔH. Thus, positive values of ΔH indicate increase in the enthalpy and therefore represent endothermic reactions and, negative values of ΔH mean decrease in enthalpy and therefore refer to exothermic reactions.

$$\Delta H_r = \Sigma H_P - \Sigma H_R \tag{13.1}$$

where ΔH_r is the heat of reaction and H_P and H_R are the enthalpies of the products and reactants respectively. To complete the definition of the heat of reaction, it is necessary to specify the conditions of every product and reactant involved in the reaction.

The *standard heat of reaction* is the enthalpy change accompanying a reaction when both the reactants and products are at their standard states at constant temperature T. It is represented by the symbol ΔH_T^0, the superscript '0' indicating that the heat of reaction refers to the standard conditions. By convention, the standard heats of reaction are reported at a temperature of 298 K and are represented by ΔH_{298}^0. The standard state is specified by specifying the pressure, composition and the state of aggregation, or the physical state of the species. By convention, the standard state pressure is one standard atmosphere. The choice of one bar as the standard state pressure is now widely accepted. The species are assumed to be pure components in the standard state. The physical state of the components are: for gases, the pure substances in the ideal gas state at 1 bar; for solids and liquids, the pure substances in the solid or liquid state respectively at 1 bar.

For example, consider the following reaction:

$$CH_4(g) + 2O_2(g) \rightarrow CO_2(g) + 2H_2O(l) \qquad \Delta H_{298}^0 = -890 \text{ kJ}$$

This equation means 1 mole of methane in the form of pure methane gas at 1 bar reacts completely with 2 moles of pure oxygen gas at 1 bar to produce one mole of pure CO_2 gas and 2 moles of pure liquid water at 1 bar, all the substances being at 298 K. Since the heat of reaction is negative, the above equation says that 890 kJ heat is liberated in the reaction. In the reaction

$$2C(s) + O_2(g) \rightarrow 2CO(g) \qquad \Delta H_{298}^0 = -221.2 \text{ kJ}$$

2 mol solid carbon reacts with 1 mol gaseous oxygen both at their standard state of 1 bar giving 2 mol gaseous carbon monoxide also at the standard state of 1 bar, the temperature being kept constant at 298 K. During this reaction 221.2 kJ of heat is liberated.

The heat evolved or absorbed in a chemical reaction corresponds to the stoichiometric numbers of the reacting species as given in the chemical equation. For example, when the above reaction is written as follows, the standard heat of reaction is just half of the value in the previous equation:

$$C(s) + \frac{1}{2}O_2(g) \rightarrow CO(g) \qquad \Delta H_{298}^0 = -110.6 \text{ kJ}$$

In the absence of chemical equation, the reported values of heat of reaction should clearly mention the basis as given below:

$$\Delta H_r^0 = -110.6 \text{ kJ per mol CO}$$

In the case of *incomplete reactions*, the heat of reaction is to be calculated considering only the quantity of reactants actually consumed or the quantity of the products actually formed in the reaction. Suppose that the heat liberated on the combustion of 12 kg of carbon is to be calculated. We have seen that on complete combustion, the heat liberated is 110.6×10^3 kJ. If only 50% carbon is reacted, the heat liberated will be $0.5 \times (110.6 \times 10^3)$ kJ.

EXAMPLE 13.1 A gas mixture analyzing 20% (mol) CO, 30% (mol) H_2 and 50% (mol) N_2 is completely burned in air. The following heat of reaction data are available:

$$CO(g) + \frac{1}{2}O_2(g) \rightarrow CO_2(g) \qquad \Delta H_r^0 = -282.91 \text{ kJ/mol}$$

$$H_2(g) + \frac{1}{2}O_2(g) \rightarrow H_2O(g) \qquad \Delta H_r^0 = -241.83 \text{ kJ/mol}$$

(a) Determine the amount of heat liberated on the complete combustion of 100 mol of the gas mixture.

(b) If only 90% of the CO and 80% of the H_2 react, how much heat is liberated on the combustion of 100 mol of the mixture?

Solution *Basis:* 100 mol gas mixture burned

(a) The heats of reactions are negative, indicating that heat is liberated during the reaction. According to the stoichiometric equation, 1 mol CO on combustion liberates 282.91 kJ of heat. The number of moles of CO present is 20 mol. Therefore, the amount of heat liberated on combustion of CO = 20 × 282.91 kJ = 5658.2. Similarly, the amount of heat liberated on combustion of 30 mol hydrogen = 30 × 241.83 kJ = 7254.9 kJ. Therefore, the total amount of heat liberated is

$$5658.2 + 7254.9 = 12913.1 \text{ kJ}$$

(b) The number of moles of CO reacted is

$$0.9 \times 20 = 18 \text{ mol}$$

The number of moles of hydrogen reacted is

$$0.8 \times 30 = 24 \text{ mol}$$

The total amount of heat liberated is

$$18 \times 282.91 + 24 \times 241.83 = 10{,}896.3 \text{ kJ}$$

13.1.2 The Standard Heat of Combustion

When the reaction under consideration is a combustion reaction, the heat of reaction is known as *heat of combustion*. The heat of combustion of a substance is the heat of reaction when a substance is oxidized with molecular oxygen. The standard heat of combustion at temperature T is the enthalpy change when the substance at its standard state and temperature T, undergoes combustion, yielding products also at their standard state and temperature T. In the following reaction, the standard heat of combustion of liquid acetaldehyde at 298 K, $\Delta H^0_{c,298}$, is −1168.12 kJ/kmol.

$$CH_3CHO\ (l) + \frac{5}{2}O_2(g) \rightarrow 2CO_2(g) + 2H_2O\ (l) \qquad \Delta H^0_{c,298} = -1168.12 \text{ kJ}$$

(*Note:* The heat of combustion values are reported on the basis that water formed is in the liquid state.)

The heat of combustion is usually expressed per mole of the substance reacted. The negative of the heat of combustion of a fuel is frequently referred to as its *heating value*. This is the energy exchanged with the surroundings when the unit mass of fuel is burnt in oxygen. The heat of combustion of a fuel is a negative quantity, whereas the heating value is positive. When water formed during combustion is in the liquid state, the energy liberated will be more than the energy liberated when the combustion products contain water in the vapour state. The heating value in the former is known as *higher heating value* or *gross heating value*, and that in the latter is known as *lower heating value* or *net heating value*. Unless otherwise stated, the heating value reported along with the fuel analysis is the gross heating value since this is the one that is usually determined using a calorimeter.

The gross heating value of coal can be calculated using the Dülong formula:

$$\text{Heating value (kJ/kg of coal)} = 338.2w_C + 1442.8w_{H(net)} + 94.2w_S$$

where w_C, w_S, $w_{H(net)}$ are, respectively, the weight percentage of carbon, sulphur and net hydrogen. The net hydrogen

$$w_{H(net)} = w_H - \frac{w_O}{8}$$

where w_O is the weight percentage of oxygen and w_H is the weight percentage of hydrogen.

13.1.3 The Standard Heat of Formation

The heat of formation is the heat evolved or absorbed in the formation of one mole of a substance from its constituent elements or the change in enthalpy accompanying the formation of one mole of a substance from the constituent elements is termed the *heat of formation*. When the reactants and products are at their standard states, the heat of formation is called *standard heat of formation*. The enthalpy of formation of the elements is taken as zero. The standard heat of formation of carbon monoxide gas at 298 K, $\Delta H^0_{f,298}$, is – 110.6 kJ/mol. The standard heat of formation at 298 K for methyl chloride is – 81.923 kJ/mol. These statements are equivalent to the following equations:

$$C(s) + \frac{1}{2}O_2(g) \rightarrow CO(g) \qquad \Delta H^0_{f,298} = -110.6 \text{ kJ}$$

$$C(s) + \frac{3}{2}H_2(g) + \frac{1}{2}Cl_2(g) \rightarrow CH_3Cl(g) \qquad \Delta H^0_{f,298} = -81.923 \text{ kJ}$$

The standard heat of formation of sulphuric acid is given by the equation

$$H_2(g) + S(s) + 2O_2(g) \rightarrow H_2SO_4(l), \qquad \Delta H^0_{f,298} = -811.86 \text{ kJ}$$

which means that when one mole sulphuric acid is formed from its constituent elements all at 298 K and a pressure of one bar, 811.86 kJ heat is evolved.

Standard heats of formation and combustion data for certain compounds are given in Appendix (Table A.2).

13.1.4 Hess's Law of Constant Heat Summation

Heat of reaction represents the difference between the enthalpy of products and the enthalpy of reactants. Enthalpy as we know, is a state function and the heat of reaction, therefore, depends only on the initial and final states, no matter how this change is carried out. *The net heat evolved or absorbed in a chemical reaction is the same whether the reaction takes place in a single step or in a series of steps.* This is known as *Hess's law.* This law permits us to treat all thermochemical equations as algebraic equations. Using this, we can calculate the heat of formation of a compound from a series of reactions not involving the direct formation of the compound from its elements. For example, the heat of formation of a compound can be calculated if the heat of combustion data of all the species involved in the formation reaction are known. Similarly, if the data on the heat of formation of all the substances taking part in a chemical reaction are available, the heat of reaction may be readily calculated.

Hess's law permits us to estimate accurately the heat of formation of compounds that are difficult to determine experimentally. For example, the heat of formation of $CaCO_3(s)$ can be estimated to be – 1207.8 kJ, given the following heat of reaction data:

$$Ca(s) + \frac{1}{2}O_2(g) \rightarrow CaO(s) \qquad \Delta H^0 = -635.8 \text{ kJ}$$

$$C(s) + O_2(g) \rightarrow CO_2(g) \qquad \Delta H^0 = -393.8 \text{ kJ}$$

$$CaO(s) + CO_2(g) \rightarrow CaCO_3 \qquad \Delta H^0 = -178.2 \text{ kJ}$$

Adding the above three equations together gives

$$Ca(s) + C(s) + \frac{3}{2}O_2(g) \rightarrow CaCO_3(s)$$

$$\Delta H^0 = -635.8 + (-393.8) + (-178.2) = -1207.8 \text{ kJ}$$

Heat of reaction from heat of combustion or heat of formation data: The heat of reaction may be calculated as the difference between the algebraic sum of the heat of combustion of the reactants and the algebraic sum of the heat of combustion of products.

$$\Delta H^0 = \sum_{\text{Reactants}} \Delta H_c^0 - \sum_{\text{Products}} \Delta H_c^0 \tag{13.2}$$

It can also be shown that the standard heat of reaction is the difference between the algebraic sum of the standard heat of formation of products and that of the reactants.

$$\Delta H^0 = \sum_{\text{Products}} \Delta H_f^0 - \sum_{\text{Reactants}} \Delta H_f^0 \tag{13.3}$$

EXAMPLE 13.2 The heat of combustion of methane, carbon and hydrogen are –890.4 kJ/mol, – 393.51 kJ/mol and – 285.84 kJ/mol respectively. Calculate the heat of formation of methane.

Solution The desired reaction is

$$C(s) + 2H_2(g) \rightarrow CH_4(g) \qquad \Delta H_f^0 = ?$$

The heat of reaction (in this case, the heat of formation) can be calculated either by using the Hess's law or by using Eq. (13.2) directly.

Using Hess's law: Since the heat of combustion data are available, the following combustion reactions may be written:

$$CH_4(g) + 2O_2(g) \rightarrow CO_2(g) + 2H_2O(l) \qquad \Delta H^0 = -890.4 \text{ kJ} \qquad \text{(A)}$$

$$C(s) + O_2(g) \rightarrow CO_2(g) \qquad \Delta H^0 = -393.51 \text{ kJ} \qquad \text{(B)}$$

$$H_2(g) + \frac{1}{2}O_2(l) \rightarrow H_2O(l) \qquad \Delta H^0 = -285.84 \text{ kJ} \qquad \text{(C)}$$

Multiply Eq. (C) by 2 and add the result to Eq. (B) and subtract Eq. (A) from the sum. The result is the desired reaction. That is, Eq. (B) + 2 × Eq. (C) – Eq. (A) gives

$$C(s) + 2H_2(g) \rightarrow CH_4(g)$$

The same algebraic operations are to be carried out on the respective heat of reactions. By Hess's law, the heat of desired reaction is

$$\Delta H^0 \text{ (B)} + 2\Delta H^0 \text{ (C)} - \Delta H^0 \text{ (A)} = -393.51 + 2(-285.84) - (-890.4) = -74.79 \text{ kJ}$$

Using Eq. (13.2): Heat of reaction can be calculated from heat of combustion data using Eq. (13.2). The heat of reaction is the sum of the heat of combustion of all the reactants in the desired reaction minus the sum of the heat of combustion of all the products of the desired reaction. Here the reactants are one mole of carbon and two moles hydrogen, and the product is one mole of methane. Therefore, the heat of reaction is

$$1 \times (-393.51) + 2 \times (-285.84) - (-890.4) = -74.79 \text{ kJ}$$

Therefore, the heat of formation of methane is – 74.79 kJ.

EXAMPLE 13.3 Coal having the following analysis on a weight basis has a gross heating value of 29 000 kJ/kg: carbon, 70.0%; hydrogen, 5.5%; nitrogen, 1.5%; sulphur, 3.0%; oxygen: 13.0%; ash, 7.0%. Calculate the net heating value given that the latent heat of vaporization of water is 2370 kJ/kg water.

Solution *Basis:* 1 kg of coal burned

The amount of hydrogen present in the coal is

$$\frac{5.5}{100 \times 2.016} = 27.28 \times 10^{-3} \text{ kmol}$$

The amount of water formed on combustion is

$$27.28 \times 10^{-3} \text{ kmol} = 0.492 \text{ kg}$$

The amount of heat required for vaporization of 0.492 kg of water is

$$0.492 \times 2370 = 1164.9 \text{ kJ}$$

The net heating value of the coal is the heat given out when all the water formed is in the vapour state. Therefore, the net heating value will be less than the gross heating value by the quantity of heat required for vaporization of the water formed on combustion. Thus, the net heating value is

$$29\,000 - 1164.9 = 27\,835.1 \text{ kJ/kg}$$

EXAMPLE 13.4 Use the Dülong formula to calculate the gross and net heating values of coal having the following composition:

C = 75%, H = 5%, N =1%, O = 9%, S = 1%, Moisture = 2.0%, Ash = 9.0%

Take the latent heat of vaporization of water to be 2370 kJ/kg of water.

Solution *Basis:* 1 kg of coal burned
Gross heating value is given by the Dülong formula:

Gross heating value (kJ/kg of coal) = $338.2w_C + 1442.8w_{H(net)} + 94.2w_S$

where w_C, w_S, $w_{H(net)}$ are, respectively, the weight percentages of carbon, sulphur and net hydrogen.

$$w_{H(net)} = w_H - \frac{w_O}{8} = 5 - 9/8 = 3.875$$

Gross heating value = $338.2w_C + 1442.8w_{H(net)} + 94.2w_S$

= $338.2 \times 75 + 1442.8 \times 3.875 + 94.2 \times 1$ = 31050 kJ/kg of coal

Water present in the combustion products = $0.05 \times 18.016/2.016 + 0.02$ = 0.4668 kg/kg of coal

Heat liberated on condensation = 0.4668×2370 = 1106.4 kJ/kg of coal
Net heating value = 31050 – 1106 = 29944 kJ/kg

EXAMPLE 13.5 Calculate the heat of reaction for the esterification of ethyl alcohol with acetic acid if the standard heats of combustion are: ethyl alcohol (l), – 1366.91 kJ/mol; acetic acid (l), – 871.69 kJ/mol; ethyl acetate (l), –2274.48 kJ/mol.

Solution The esterification reaction may be represented as

$$C_2H_5OH\ (l) + CH_3COOH\ (l) \rightarrow C_2H_5COOCH_3\ (l) + H_2O\ (l) \qquad \Delta H^0_{298} = ?$$

Using Eq. (13.2): To calculate heat of reaction from the heat of combustion data Eq. (13.2) may be directly used.

$$\Delta H^0 = \sum_{\text{Reactants}} \Delta H_c^0 - \sum_{\text{Products}} \Delta H_c^0$$

Here,

$$\sum_{\text{Reactants}} \Delta H_c^0 = -1366.91 - 871.69 = -2238.6 \text{ kJ}$$

$$\sum_{\text{Products}} \Delta H_c^0 = -2274.48 + 0 = -2274.48 \text{ kJ}$$

(Note that the heats of combustion of H_2O and CO_2 are zero.)
Substituting these into Eq. (13.2), we get

$$\Delta H^0 = -2238.6 - (-2274.48) = 35.88 \text{ kJ}$$

Using Hess's law: The given data on heat of combustion may be used to write the following chemical equations:

$$C_2H_5OH\,(l) + 3O_2(g) \rightarrow 2CO_2(g) + 3H_2O\,(l) \qquad \Delta H_c^0 = -1366.91 \text{ kJ} \qquad (A)$$

$$CH_3COOH\,(l) + 2O_2(g) \rightarrow 2CO_2(g) + 2H_2O\,(l) \qquad \Delta H_c^0 = -871.691 \text{ kJ} \qquad (B)$$

$$C_2H_5COOCH_3\,(l) + 5O_2(g) \rightarrow 4CO_2(g) + 4H_2O\,(l) \qquad \Delta H_c^0 = -2274.48 \text{ kJ} \qquad (C)$$

Add Eqs. (A) and (B) and subtract Eq. (C) from the result. We get

$$C_2H_5OH\,(l) + CH_3COOH\,(l) \rightarrow C_2H_5COOCH_3\,(l) + H_2O\,(l)$$

By Hess's law the same algebraic operations may be performed on the heat of combustion data as well, with the result that

$$\Delta H_{298}^0 = \Delta H_c^0(A) + \Delta H_c^0(B) + \Delta H_c^0(C) = -1366.91 - 871.69 - (-2274.48) = 35.88 \text{ kJ}$$

EXAMPLE 13.6 The vapour-phase hydration of ethylene to ethanol is represented by:

$$C_2H_4(g) + H_2O\,(g) \rightarrow C_2H_5OH(g)$$

Calculate the standard heat of reaction if the following data are available:

$$2CO_2(g) + 3H_2O\,(l) \rightarrow C_2H_5OH\,(l) + 3O_2(g) \qquad \Delta H_{298}^0 = 1366.91 \text{ kJ}$$

The standard heat of combustion of ethylene at 298 K is –1410.99 kJ/mol and heats of vaporization of water and ethanol are, respectively, 44.04 kJ/mol and 42.37 kJ/mol.

Solution The following equations may be formulated using the given data:

$$2CO_2(g) + 3H_2O\,(l) \rightarrow C_2H_5OH\,(l) + 3O_2(g) \qquad \Delta H_{298}^0 = 1366.91 \text{ kJ} \qquad (A)$$

$$C_2H_4(g) + 3O_2(g) \rightarrow 2CO_2 + 2H_2O\,(l) \qquad \Delta H_{298}^0 = -1410.99 \text{ kJ} \qquad (B)$$

$$H_2O\,(l) \rightarrow H_2O\,(g) \qquad \Delta H_{v,298} = 44.04 \text{ kJ} \qquad (C)$$

$$C_2H_5OH\,(l) \rightarrow C_2H_5OH\,(g) \qquad \Delta H_{v,298} = 42.37 \text{ kJ} \qquad (D)$$

Perform the following arithmetic operations on the above equations:

$$\text{Eq. (A)} + \text{Eq. (B)} + \text{Eq. (D)} - \text{Eq. (C)}$$

The result is

$$C_2H_4(g) + H_2O\,(g) \rightarrow C_2H_5OH\,(g)$$

$$\Delta H_{298}^0 = \Delta H_{298}^0(A) + \Delta H_{298}^0(B) + \Delta H_{298}^0(D) - \Delta H_{298}^0(C) = -45.75 \text{ kJ}$$

EXAMPLE 13.7 Calculate the standard heat of formation of acetylene (C_2H_2) given that the standard heat of combustion of acetylene is – 1299.61 kJ, the standard heat of combustion of carbon is – 393.51 kJ and the standard heat of formation of liquid water is – 285.84 kJ.

Solution We have the following data:

$$C_2H_2(g) + \frac{5}{2}O_2(g) \rightarrow 2CO_2(g) + H_2O(l) \qquad \Delta H^0_{298} = -1299.61 \text{ kJ} \qquad (A)$$

$$C(s) + O_2(g) \rightarrow CO_2(g) \qquad \Delta H^0_{298} = -393.51 \text{ kJ} \qquad (B)$$

$$H_2(g) + \frac{1}{2}O_2(g) \rightarrow H_2O(l) \qquad \Delta H^0_{298} = -285.84 \text{ kJ} \qquad (C)$$

We can apply Hess's law to calculate the heat of the desired formation reaction.

$$2C(s) + H_2(g) \rightarrow C_2H_2(g) \qquad \Delta H^0_{298} = ?$$

Multiply Eq. (B) by 2, add Eq. (C) to it and subtract Eq. (A) from the result. We get

$$2C(s) + H_2(g) \rightarrow C_2H_5(g)$$

$$\Delta H^0_{298} = 2 \times (-393.51) - 285.84 - (-1299.61) = 226.75 \text{ kJ}$$

EXAMPLE 13.8 In a sulphuric acid plant, sulphur dioxide is obtained by the roasting of iron pyrites containing 80.0% FeS_2 and 20% gangue. Iron sulphide reacts with oxygen according to the reaction

$$4FeS_2 + 11O_2 \rightarrow 2Fe_2O_3 + 8SO_2$$

The cinder formed on the combustion analyzes 5.0% FeS_2. Determine the standard heat of reaction per kilogram of ore, given the following standard heat of formation values at 298 K: $FeS_2(s) = -178.02$ kJ/mol, $Fe_2O_3(s) = -822.71$ kJ/mol and $SO_2(g) = -296.9$ kJ/mol.

Solution *Basis:* 100 kg of pyrites charged.

Let x kg of FeS_2 be leaving with the cinder unburned. Then the cinder has the following composition:

Material	*Weight* (kg)
FeS_2	x
Fe_2O_3	$\frac{(80-x)}{119.98}\times\frac{1}{2}\times 159.69 = 0.6655(80-x)$
Gangue	20
Total	$73.24 + 0.3345x$

Since the cinder contains 5.0% FeS_2, we have $\frac{x}{73.24 + 0.3345x} = 0.05$, which gives $x =$ 3.724 kg.

Therefore, the amount of FeS_2 reacted is

$$80 - 3.724 = 76.276 \text{ kg} = \frac{76.276}{119.98} = 0.6357 \text{ kmol} = 635.7 \text{ mol}$$

The heat of reaction for the following reaction can be calculated from the heat of formation data.

$$4FeS_2 + 11O_2 \rightarrow 2Fe_2O_3 + 8SO_2$$

Using Eq. (13.3)

$$\Delta H^0 = \sum_{\text{Products}} \Delta H_f^0 - \sum_{\text{Reactants}} \Delta H_f^0$$

$$= 2 \times (-822.71) + 8 \times (-296.9) - 4 \times (-178.02) - 0 = -3308.54 \text{ kJ}$$

This is the heat of reaction when 4 mol FeS_2 is completely consumed. Here 635.7 mol FeS_2 is consumed. Therefore, the heat of reaction is

$$\frac{-3308.54}{4} \times 635.7 = -5.258 \times 10^5 \text{ kJ}$$

The heat of reaction per 1 kg of pyrites burned is

$$\frac{-5.258 \times 10^5}{100} = -5.258 \times 10^3 \text{ kJ}$$

EXAMPLE 13.9 Calculate the standard heat of formation of liquid methanol, given the standard heat of combustion of liquid methanol is – 726.55 kJ/mol and the standard heat of formation of gaseous CO_2 and liquid water are, respectively, – 393.51 and –285.84 kJ/mol.

Solution

Using Eq. (13.3): The combustion of liquid methanol is

$$CH_3OH\,(l) + \frac{3}{2} O_2\,(g) \rightarrow CO_2\,(g) + 2H_2O\,(l) \qquad \Delta H_{298}^0 = -726.55 \text{ kJ} \qquad \text{(A)}$$

By Eq. (13.3), the heat of reaction is given as

$$\Delta H^0 = \sum_{\text{Products}} \Delta H_f^0 - \sum_{\text{Reactants}} \Delta H_f^0$$

Applying this equation to reaction (A), we can write

$$-726.55 = \Delta H_{f,CO_2}^0 + 2\Delta H_{f,H_2O}^0 - \Delta H_{f,CH_3OH}^0$$

$$= -393.51 + 2 \times (-285.84) - \Delta H_{f,CH_3OH}^0$$

$$= -965.19 - \Delta H_{f,CH_3OH}^0$$

or

$$\Delta H_{f,CH_3OH}^0 = -965.19 + 726.55 = -238.64 \text{ kJ}$$

Using Hess's law: The desired reaction is the formation of liquid methanol from the constituent elements.

$$C\,(s) + 2H_2\,(g) + \frac{1}{2} O_2\,(g) \rightarrow CH_3OH\,(l)$$

The available data may be put into the following forms:

$$CH_3OH\,(l) + \frac{3}{2}O_2(g) \rightarrow CO_2(g) + 2H_2O\,(l) \qquad \Delta H^0_{298} = -726.55 \text{ kJ} \qquad \text{(A)}$$

$$C\,(s) + O_2(g) \rightarrow CO_2(g) \qquad \Delta H^0_{298} = -393.51 \text{ kJ} \qquad \text{(B)}$$

$$H_2\,(g) + \frac{1}{2}O_2(g) \rightarrow H_2O(l) \qquad \Delta H^0_{298} = -285.84 \text{ kJ} \qquad \text{(C)}$$

The stoichiometric equation representing the formation of liquid methanol is obtained by performing the following algebraic operations on these equations.

$$\text{Eq. (B)} + 2 \times \text{Eq. (C)} - \text{Eq. (A)}$$

Therefore, the heat of formation of methanol is

$$-393.51 + 2 \times (-285.84) - (-726.55) = -238.64 \text{ kJ}$$

EXAMPLE 13.10 Calculate the gross heating value and net heating value expressed in kJ/m^3 at STP of a fuel gas having the following analysis:

CO: 21.0%, H_2 = 15.6%, CO_2: 9.0%, CH_4: 2.0%, C_2H_4: 0.4% and N_2: 52.0%

Given that the standard heats of combustion are –282.99 kJ/mol for CO, – 285.84 kJ/mol for hydrogen, – 890.4 kJ/mol for methane and – 1410.99 kJ/mol for ethylene, and the heat of vaporization of water at 298 K is 44.04 kJ/mol.

Solution *Basis:* 100 mol fuel gas

The total heat evolved on combustion of the fuel is

$$21 \times (282.99) + 15.6 \times (285.84) + 2.0 \times (890.4) + 0.4 \times (1410.99) = 12\,747 \text{ kJ}$$

Since 100 mol gas at STP occupies a volume of $100 \times 22.4 \times 10^{-3} = 2.24$ m^3, the gross heating value of the fuel is

$$\frac{12\,747}{2.240} = 5691 \frac{\text{kJ}}{\text{m}^3}$$

On combustion one mol hydrogen gives one mol water, one mol methane gives two mol water and one mol ethylene gives two mol water. The number of moles of water formed on combustion of 100 mol fuel gas = 15.6 + 4.0 + 0.8 = 20.4. The amount of heat required to vaporize the water formed on combustion = 20.4 × 44.04 = 898.42 kJ.

Therefore, the net heating value is

$$5691 - 898.42 = 4792.58 \text{ kJ}$$

EXAMPLE 13.11 Calculate the standard heat of the following reaction at 298 K:

$$C_5H_{12}(g) + 8O_2\,(g) \rightarrow 5CO_2(g) + 6H_2O\,(l)$$

The standard heats of formation are as follows:

$$CO_2\,(g) = -393.51 \text{ kJ},\ H_2O\,(g) = -241.826 \text{ kJ},\ C_5H_{12}(g) = -146.4 \text{ kJ}$$

The latent heat of vaporization of water at 298 K is 43.967 kJ/mol.

Solution The desired reaction can be obtained as the sum of the following two reactions:

$$C_5H_{12}(g) + 8O_2(g) \rightarrow 5CO_2(g) + 6H_2O(g) \quad \Delta H_1$$

$$6H_2O(g) \rightarrow 6H_2O(l) \quad \Delta H_2$$

The heat of reaction $\Delta H^0 = \Delta H_1 + \Delta H_2$

$$\Delta H_1 = \sum_{\text{Products}} \Delta H_f^0 - \sum_{\text{Reactants}} \Delta H_f^0$$

$$= 5 \times \Delta H^0_{f,CO_2} + 6 \times \Delta H^0_{f,H_2O(g)} - \Delta H_{f,C_5H_{12}(g)}$$

$$= 5 \times (-393.51) + 6 \times (-241.826) - (-146.4) = -3272.106 \text{ kJ}$$

$$\Delta H_2 = -6 \times 43.967 = -263.802 \text{ kJ}$$

$$\Delta H^0 = \Delta H_1 + \Delta H_2 = -3272.106 - 263.802 = -3535.908 \text{ kJ}$$

13.1.5 Heat of Reaction at Constant Pressure and Constant Volume

Heat of reaction can be experimentally determined in a constant volume calorimeter. However, the standard heat of reaction requires that the pressure is maintained constant at 1 atm. Thus, it becomes necessary that the heat of reaction obtained under constant volume conditions be converted to the heat of reaction at constant pressure conditions. Since no work is done under constant volume conditions, by the first law of thermodynamics, the heat transferred is equal to the change in internal energy.

$$\Delta U_V = Q_V \tag{13.4}$$

Assume that the reaction is now carried out under constant pressure conditions starting from the same initial conditions as for the constant volume process and the final temperature attained be the same as that attained in the constant volume process. By the first law of thermodynamics for a constant pressure process,

$$\Delta U_P = Q_P - W = Q_P - P\Delta V$$

$$Q_P = \Delta U_P + P\Delta V \tag{13.5}$$

The difference between the heat of reaction at constant pressure and that at constant volume is

$$Q_P - Q_V = \Delta U_P - \Delta U_V + P\Delta V \tag{13.6}$$

For solid and liquid components, the change in internal energy is zero as internal energy is mainly dependent on temperature and the effect of pressure is negligible. If the gases involved in the reaction are assumed to behave ideally, the change in internal energy for the gaseous constituents is zero. $\Delta U_P = \Delta U_V$. Therefore, Eq. (13.6) reduces to

$$Q_P - Q_V = P\Delta V \tag{13.7}$$

Neglecting the change in volume of solids and liquids involved in the reaction, the volume change accompanying the reaction is solely the change in volume of the gaseous constituents. Assuming ideal behaviour of gases,

$$Q_P - Q_V = (\Delta n)RT \tag{13.8}$$

Here Δn is the change in the number of moles of gaseous constituents.

EXAMPLE 13.12 A fuel oil consisting of 10% (weight) hydrogen and 90% (weight) carbon is found to give a heat of combustion of 43 000 kJ/kg, when burned in a constant volume-bomb calorimeter. Calculate the constant-pressure heat of combustion of the oil.

Solution *Basis:* 1 kg of oil burned

The amount of carbon burned:

$$0.9 \text{ kg} = \frac{0.9}{12} = 0.075 \text{ kmol}$$

$$C\,(s) + O_2\,(g) \rightarrow CO_2\,(g)$$

The combustion of carbon produces no change in the volume of gaseous components.

The amount of hydrogen burned:

$$0.1 \text{ kg} = \frac{0.1}{2.016} = 0.0496 \text{ kmol}$$

Change in the number of moles of gaseous components accompanying the combustion of one mole of hydrogen in the liquid state is – ½ mol. Therefore, the total change in the volume of gaseous components accompanying the combustion of 0.0496 kmol hydrogen is

$$(\Delta n)RT = -\frac{1}{2} \times 0.0496 \times 8.314 \times 298 = -61.444 \text{ kJ}$$

Since

$$Q_P - Q_V = (\Delta n)RT$$

$$Q_P = Q_V + (\Delta n)RT = -43\,000 - 61.44 = -43\,061.4 \text{ kJ}$$

Therefore, the constant pressure heat of combustion of the fuel = – 43 061.4 kJ/kg.

13.2 EFFECT OF TEMPERATURE ON STANDARD HEAT OF REACTION

Chemical reaction can be represented as

$$\sum_i \nu_i A_i = 0 \tag{13.9}$$

where ν_i represents the stoichiometric number of the substance taking part in the reaction. The stoichiometric number ν_i is positive for products and negative for the reactants. For example, consider the reaction

$$aA + bB \rightarrow lL + mM$$

Here, *a*, *b*, *l* and *m* are called stoichiometric coefficients. This reaction may be written as

$$lL + mM - aA - bB = 0$$

in which –*a*, –*b*, *l* and *m* are called the stoichiometric numbers. If the standard heat of the reaction at temperature T_1 is known, the standard heat at any other temperature *T* can be determined. Generally, the standard heat of reaction at temperature 298 K (T_1) will be available, and it would be necessary to determine the standard heat at some other temperature (*T*). The specific heat of the components constituting the reaction system should be known as function of temperature. The calculation makes use of the fact that enthalpy is a state property. The change in enthalpy for a process occuring from a given initial state to a given final state is the same whatever be the path along which the process is carried out. The actual reaction occurring at temperature *T*, for which the standard heat of reaction is ΔH_T^0, may be treated as occurring along the three paths as shown in Figure 13.1.

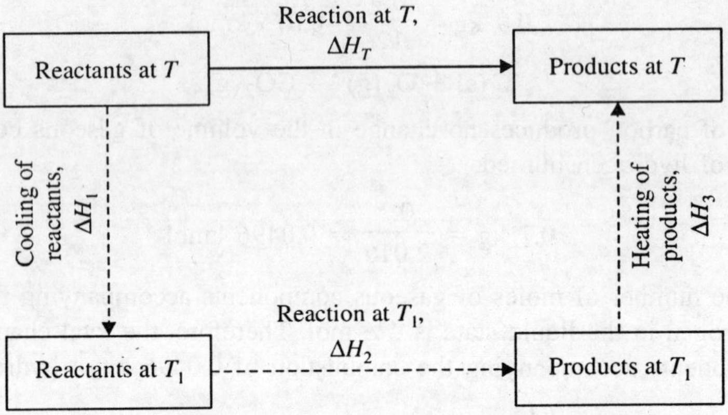

Figure 13.1 Effect of temperature on heat of reaction.

Step 1: From the initial state of temperature *T* the reactants are cooled to temperature T_1. The enthalpy change for this step is

$$\Delta H_1 = \sum_{\text{Reactants}} \int_{T}^{T_1} n_i C_{P,i}\, dT = \sum_{\text{Reactants}} \int_{T_1}^{T} \nu_i C_{P,i}\, dT \tag{13.10}$$

Step 2: The reaction is allowed to occur at temperature T_1. The enthalpy change is

$$\Delta H_2 = \Delta H_{T_1}^0 \tag{13.11}$$

Step 3: The temperature of the products is raised from T_1 to *T* in this step. The enthalpy change is

$$\Delta H_3 = \sum_{\text{Products}} \int_{T_1}^{T} n_i C_{P,i}\, dT = \sum_{\text{Products}} \int_{T_1}^{T} \nu_i C_{P,i}\, dT \tag{13.12}$$

Since the enthalpy change depends on the terminal conditions, the standard heat of reaction at temperature *T* is obtained by adding the preceding three equations.

$$\Delta H_T^0 = \Delta H_1 + \Delta H_2 + \Delta H_3$$

$$\Delta H_T^0 = \sum_{\text{Reactants}} \int_{T_1}^{T} \nu_i C_{P,i}\, dT + \Delta H_{T_1}^0 + \sum_{\text{Products}} \int_{T_1}^{T} \nu_i C_{P,i}\, dT$$

The above result can be written as

$$\Delta H_T^0 = \Delta H_{T_1}^0 + \int_{T_1}^{T} \left(\sum_i \nu_i C_{P,i} \right) dT \tag{13.13}$$

The summation in the above equation is over all species taking part in the reaction. Let the heat capacity of the substances be represented by the following equation.

$$C_P = \alpha + \beta T + \gamma T^2 \tag{13.14}$$

Utilizing Eq. (13.14), Eq. (13.13) can be written as

$$\Delta H_T^0 = \Delta H_{T_1}^0 + \int_{T_1}^{T} \Delta C_P \, dT \tag{13.15}$$

where

$$\Delta C_P = \Delta\alpha + \Delta\beta T + \Delta\gamma T^2 \tag{13.16}$$

and

$$\Delta\alpha = \Sigma v_i \alpha, \quad \Delta\beta = \Sigma v_i \beta, \quad \text{and} \quad \Delta\gamma = \Sigma v_i \gamma \tag{13.17}$$

Equation (13.15) may be expanded as

$$\Delta H_T^0 = \Delta H_{T_1}^0 + \Delta\alpha (T - T_1) + \frac{1}{2}\Delta\beta (T^2 - T_1^2) + \frac{1}{3}\Delta\gamma (T^3 - T_1^3)$$

The constants appearing in the above equation can be grouped together to a single constant $\Delta H'$, so that we have

$$\Delta H_T^0 = \Delta H' + \Delta\alpha T + \frac{\Delta\beta}{2} T^2 + \frac{\Delta\gamma}{3} T^3 \tag{13.18}$$

The constant $\Delta H'$ in Eq. (13.18) can be evaluated if the heat of reaction at a single temperature is known. Equation (13.18) can now be used for the evaluation of the standard heat of reaction at any temperature T.

EXAMPLE 13.13 Ammonia is synthesized according to the following reaction:

$$\frac{1}{2}N_2 + \frac{3}{2}H_2 \rightarrow NH_3 \qquad \Delta H_{298}^0 = -46.191 \text{ kJ}$$

The specific heats of the components are represented by

$$C_P = \alpha + \beta T + \gamma T^2$$

where C_P is in J/mol K and the constants α, β and γ are:

Molecules	α	β	γ
N_2	27.31	5.2335×10^{-3}	-4.1868×10^{-9}
H_2	29.09	-8.374×10^{-4}	2.0139×10^{-6}
NH_3	25.48	36.89×10^{-3}	-6.305×10^{-6}

Determine the heat of reaction at 700 K.

Solution Equation (13.18) can be used to calculate the heat of reaction at any temperature.

$$\Delta H_T^0 = \Delta H' + \Delta\alpha T + \frac{\Delta\beta}{2} T^2 + \frac{\Delta\gamma}{3} T^3$$

$$\Delta\alpha = \alpha_{NH_3} - \frac{1}{2}\alpha_{N_2} - \frac{3}{2}\alpha_{H_2}$$

$$= 25.48 - \frac{1}{2}\times 27.31 - \frac{3}{2}\times 29.09 = -31.81$$

$$\Delta\beta = \beta_{NH_3} - \frac{1}{2}\beta_{N_2} - \frac{3}{2}\beta_{H_2}$$

$$= 36.89\times 10^{-3} - \frac{1}{2}\times 5.2335\times 10^{-3} - \frac{3}{2}\times(-8.374\times 10^{-4})$$

$$= 35.5294 \times 10^{-3}$$

$$\Delta\gamma = \gamma_{NH_3} - \frac{1}{2}\gamma_{N_2} - \frac{3}{2}\gamma_{H_2}$$

$$= -6.305\times 10^{-6} - \frac{1}{2}\times(-4.1868\times 10^{-9}) - \frac{3}{2}\times(2.0139\times 10^{-6})$$

$$= -9.3238 \times 10^{-6}$$

Substituting these values in Eq. (13.18), we get

$$\Delta H_T^0 = \Delta H' - 31.81T + \frac{35.5294\times 10^{-3}}{2}T^2 - \frac{9.3238\times 10^{-6}}{3}T^3$$

At 298 K, $\Delta H_T^0 = -46\,191$ J. Substituting this in the above equation, we get $\Delta H' = -38\,207$ J.

The heat of reaction can now be expressed as a function of temperature as

$$\Delta H_T^0 = -38\,207 - 31.81T + \frac{35.5294\times 10^{-3}}{2}T^2 - \frac{9.3238\times 10^{-6}}{3}T^3$$

At $T = 700$ K,

$$\Delta H_T^0 = -38\,207 - 31.81\times 700 + \frac{35.5294\times 10^{-3}}{2}(700)^2 - \frac{9.3238\times 10^{-6}}{3}(700)^3$$

$$= -52\,835 \text{ J}$$

Therefore, the heat of reaction at 700 K is – 52.835 kJ.

EXAMPLE 13.14 Methanol is synthesized according to the following reaction:

$$CO(g) + 2H_2(g) \rightarrow CH_3OH(g)$$

The standard heats of formation at 298 K are – 110.6 kJ/mol for CO and – 238.64 kJ/mol for methanol(l). The latent heat of vaporization of methanol at 298 K is 37.98 kJ/mol. The specific heats (J/mol K) are given by:

$$C_P(CH_3OH) = 18.382 + 101.564 \times 10^{-3}T - 28.683 \times 10^{-6}T^2$$

$$C_P(CO) = 28.068 + 4.631 \times 10^{-3}T - 2.5773 \times 10^4T^{-2}$$

$$C_P(H_2) = 27.012 + 3.509 \times 10^{-3}T + 6.9006 \times 10^4T^{-2}$$

Calculate the standard heat of reaction at 1073 K.

Solution Enthalpy of cooling the reactants consisting of 1 mol CO and 2 mol hydrogen from 1073 K to 298 K:

$$\Delta H_1 = 1\int_{1073}^{298} (28.068 + 4.631\times10^{-3}T - 2.5773\times10^4 T^{-2})\, dT$$

$$+ 2\int_{1073}^{298} (27.012 + 3.509\times10^{-3}T + 6.9006\times10^4 T^{-2})\, dT$$

$$= \int_{1073}^{298} (82.092 + 11.649\times10^{-3}T + 11.2239\times10^4 T^{-2})\, dT = -70\,082.01 \text{ J}$$

Enthalpy for heating the products from 298 to 1073 K:

$$\Delta H_2 = \int_{298}^{1073} (18.382 + 101.564\times10^{-3}T - 28.683\times10^{-6}T^2)\, dT = 56\,644.78 \text{ J}$$

The standard heat of reaction at 298 K is

$$\Delta H^0_{298} = \sum_{\text{Products}} \Delta H^0_f - \sum_{\text{Reactants}} \Delta H^0_f = \Delta H^0_{f,\text{methanol(g)}} - (\Delta H^0_{f,\text{CO(g)}} + 2\Delta H^0_{f,\text{H}_2\text{O(g)}}) \qquad \text{(A)}$$

The standard heat of formation of methanol (g) can be obtained from the standard heat of formation of methanol (l) and the heat of vaporization.

$$\text{CO(g)} + 2\text{H}_2\text{(g)} \rightarrow \text{CH}_3\text{OH(l)} \quad \Delta H^0_f = -238.64 \text{ kJ}$$

$$\text{CH}_3\text{OH(l)} \rightarrow \text{CH}_3\text{OH(g)} \quad \Delta H^0_v = +37.98 \text{ kJ}$$

Adding these two equations we get the desired reaction for the formation of methanol (g).

$$\text{CO(g)} + 2\text{H}_2\text{(g)} \rightarrow \text{CH}_3\text{OH(g)} \quad \Delta H^0_f = -200.66 \text{ kJ}$$

Substituting the heat of formation values in Eq. (A) we get

$$\Delta H^0_{298} = \Delta H^0_{f,\text{methanol(g)}} - (\Delta H^0_{f,\text{CO(g)}} + 2\Delta H^0_{f,\text{H}_2\text{(g)}})$$

$$= -200.66 - (-110.6 + 0) = -90.06 \text{ kJ}$$

The heat of reaction at 1073 K is now calculated:

$$\Delta H^0_T = \Delta H_1 + \Delta H^0_{298} + \Delta H_2$$

$$= -70.082 - 90.06 + 56.645 = -103.497 \text{ kJ/mol}$$

EXAMPLE 13.15 Pure CO is mixed with 100 percent excess air and completely burned at constant pressure. The reactants are originally at 400 K. Determine the heat added or removed if the products leave at 600 K. The standard heat of reaction at 298 K is –282.99 kJ per mol CO burned. The mean specific heats applicable in the temperature range of this problem are 29.10, 29.70, 29.10, and 41.45 J/mol K respectively for CO, O_2, N_2 and CO_2.

Solution *Basis:* 1 mol CO reacted

$$CO + \frac{1}{2}O_2 \rightarrow CO_2$$

Oxygen theoretically required = 0.5 mol

Oxygen supplied = 1 mol (100 percent excess)

Nitrogen in the air supplied = 1 × 79/21 = 3.76 mol

Let ΔH_1 be the enthalpy of cooling the reactants from 400 K to 298 K. The reactants stream contains 1 mol oxygen, 3.76 mol nitrogen, and 1 mol CO.

$$\Delta H_1 = (1 \times 29.70 + 3.76 \times 29.10 + 1 \times 29.10)(298 - 400)$$
$$= -17.158 \text{ kJ}$$

Let ΔH_2 be the enthalpy of heating the products from 298 K to 600 K. The product stream contains 1 mol CO_2, 0.5 mol oxygen, and 3.76 mol nitrogen.

$$\Delta H_2 = (1 \times 41.45 + 3.76 \times 29.10 + 0.5 \times 29.70)(600 - 298)$$
$$= 50.046 \text{ kJ}$$
$$\Delta H = \Delta H_1 + \Delta H^0_{298} + \Delta H_2 = -17.158 - 282.99 + 50.046$$
$$= -250.102 \text{ kJ}$$

EXAMPLE 13.16 Carbon monoxide reacts with water vapour to form carbon dioxide and hydrogen.

$$CO(g) + H_2O(g) \rightarrow CO_2(g) + H_2(g) \qquad \Delta H^0_{298} = -41.190 \text{ kJ}$$

The reactants are at 298 K. 75 percent of CO is converted in the reaction. The products leave the reaction chamber at 800 K. The mean heat capacities in J/mol K are 30.35 for CO, 45.64 for CO_2, 36.00 for water vapour and 29.30 for hydrogen. Determine the quantity of heat to be added or removed in the reaction chamber per 1000 kg of hydrogen produced.

Solution Assume a reactant mixture consisting of 1 mol CO (g) and 1 mol water vapour. Since the reactants are at 298 K, $\Delta H_1 = 0$.

The products consists of 0.75 mol CO_2, 0.75 mol H_2, 0.25 mol CO and 0.25 mol water vapour. The enthalpy of heating the products from 298 K to 800 K is

$$\Delta H_2 = [0.75(45.64 + 29.30) + 0.25(30.35 + 36.00)](800 - 298)$$
$$= 36\,541.8 \text{ J}$$

The heat of reaction is

$$\Delta H^0_{298} = 0.75 \times (-41.190) = -30.8925 \text{ kJ} = -30\,892.5 \text{ J}$$

The heat of reaction is

$$\Delta H_{800}^0 = \Delta H_{298}^0 + \Delta H_2 = -30\,892.5 + 36\,541.8 = 5649.3 \text{ kJ}$$

This means 5.649 kJ of heat is to be added to the system. The amount of hydrogen produced here is 0.75 mol or

$$0.75 \times 2.016 \times 10^{-3} = 1.512 \times 10^{-3} \text{ kg}$$

Therefore, for 1000 kg of hydrogen produced, the amount of heat to be added is

$$\frac{5.649}{1.512 \times 10^{-3}} \times 1000 = 3.736 \times 10^6 \text{ kJ}$$

EXAMPLE 13.17 A flue gas at 1298 K consisting of CO_2, O_2 and N_2 is passed through a bed of carbon. The following reactions occur and both go to completion:

$$CO_2(g) + C(s) \rightarrow 2CO(g), \quad \Delta H_r^0 \text{ at } 298 \text{ K} = 170 \text{ kJ/mol}$$

$$O_2(g) + 2C(s) \rightarrow 2CO(g), \quad \Delta H_r^0 \text{ at } 298 \text{ K} = -221.2 \text{ kJ/mol}$$

The combustion is adiabatic and the product gases exit at 1298 K. Calculate the required moles of CO_2 per mole of O_2 in the feed stream so that the net heat generated is zero and the bed temperature remains constant at 1298 K. The mean molar heat capacities are 0.02, 0.03, 0.03, and 0.05 kJ/mol K for carbon, oxygen, CO and CO_2 respectively.

Solution Let the flue gas contain x mol CO_2 per mole of oxygen.

The products leaving the bed of carbon contains $2(1 + x)$ mol CO. The amount of nitrogen in the reactants and products being the same and both are at the same temperatures, enthalpy of cooling and enthalpy of heating the nitrogen will numerically be the same and need not be considered in the evaluation of net enthalpy change.

The enthalpy of cooling the reactants (x mol CO_2, 1 mol O_2 and $2 + x$ mol carbon) from 1298 K to 298 K:

$$\Delta H_1 = [0.05x + 0.03 + (2 + x) \times 0.02](298 - 1298)$$

$$= -50x - 30 - 40 - 20x = -70x - 70 \text{ kJ}$$

The enthalpy of heating the products

$$\Delta H_2 = 2(1 + x) \times 0.03 \times (1298 - 298) = 60 + 60x \text{ kJ}$$

The heat of reaction at 298 K

$$\Delta H_{298}^0 = 170x - 221.2$$

Since the net heat generated is zero,

$$\Delta H_1 + \Delta H_{298}^0 + \Delta H_2 = 0$$

$$-70x - 70 + 60 + 60x + 170x - 221.2 = 0$$

Solving this equation, we get $x = 1.445$ mol.

The ratio of moles of CO_2 to oxygen in the feed stream should be 1.445:1 for the net heat generation to be zero.

EXAMPLE 13.18 Formaldehyde can be made by the partial oxidation of natural gas using pure oxygen, natural gas being supplied in large excess.

$$CH_4(g) + O_2(g) \rightarrow CH_2O(g) + H_2O(g) \quad \Delta H^0_{298} = -282.926 \text{ kJ}$$

The following side reaction also occurs.

$$CH_4(g) + 2O_2(g) \rightarrow CO_2(g) + 2H_2O(g) \quad \Delta H^0_{298} = -802.372 \text{ kJ}$$

Methane enters the reactor at 673 K and oxygen at 573 K, and the products leave at 873 K. The Orsat analysis of the flue gas showed CO_2 1.9%, CH_2O 11.7%, O_2 3.8% and CH_4 82.6%. How much heat is removed from the reaction chamber per 1000 kg of formaldehyde produced? The mean heat capacities (J/mol K) for a temperature range from 298 K to T K are given below:

	T = 573 K	T = 673 K	T = 873 K
O_2	30.5		31.9
CH_4		45.9	51.4
CO_2			46.3
CH_2O			47.1
H_2O			36.3

Solution *Basis:* 100 mol flue gas

Refer Figure 13.2.

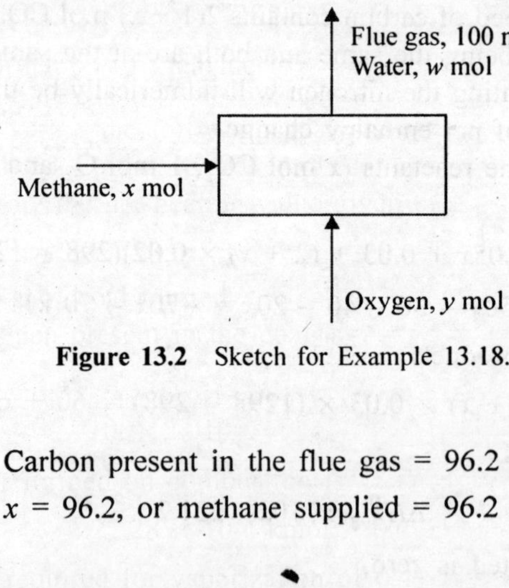

Figure 13.2 Sketch for Example 13.18.

Carbon balance:

$$\text{Carbon present in the flue gas} = 96.2 \text{ mol}$$

$$x = 96.2, \text{ or methane supplied} = 96.2 \text{ mol}$$

Hydrogen balance:

$$\text{Hydrogen in the flue gas} = 11.7 + 2 \times 82.6 = 176.9 \text{ mol}$$

$$2x = 176.9 + w$$

That is,

$$w = 2 \times 96.2 - 176.9 = 15.5 \text{ mol}$$

or the water formed on combustion is 15.5 mol.

Oxygen balance:

Oxygen present in the water = ½ (15.5) = 7.75 mol

Oxygen in the dry flue gas = 1.9 + ½ (11.7) + 3.8 = 11.55 mol

$$y = 7.75 + 11.55 = 19.3 \text{ mol}$$

Thus, the amount of oxygen supplied is 19.3 mol.

The enthalpy change accompanying the cooling of the reactants:

19.3 mol oxygen is cooled from 573 K to 298 K and 96.2 mol methane is cooled from 673 K to 298 K. Therefore,

$$\Delta H_1 = \sum_{\text{Reactants}} \int_{T}^{T_1} n_i C_{P,i}\, dT$$

$$= 19.3 \times 30.5 \times (298 - 573) + 96.2 \times 45.9 \times (298 - 673)$$

$$= -1.817721 \times 10^6 \text{ J} = -1817.721 \text{ kJ}$$

The heat of reaction is obtained as the sum of the heat liberated in the first reaction which forms formaldehyde and the heat liberated in the combustion of methane to carbon dioxide.

$$\Delta H_2 = 11.7 \times (-282.926) + 1.9 \times (-802.372) = -4834.74 \text{ kJ}$$

The products of combustion are CO_2 1.9 mol, CH_2O 11.7 mol, O_2 3.8 mol, CH_4 82.6 mol and water vapour 15.5 mol. Enthalpy change for heating the products to 873 K:

$$\Delta H_3 = \sum_{\text{Products}} \int_{T_1}^{T} n_i C_{P,i}\, dT$$

$$\Delta H_3 = (1.9 \times 46.3 + 11.7 \times 47.1 + 3.8 \times 31.9 + 82.6 \times 51.4 + 15.5 \times 36.3)(873 - 298)$$

$$= 3.2019 \times 10^6 \text{ J} = 3201.9 \text{ kJ}$$

Now the heat evolved or absorbed under the specified conditions is

$$\Delta H_T^0 = \Delta H_1 + \Delta H_2 + \Delta H_3$$

$$= -1817.721 - 4834.74 + 3201.9 = -3450.56 \text{ kJ}$$

The amount of formaldehyde produced is

$$11.7 \text{ mol} = 11.7 \times 30.016 = 351.19 \text{ g}$$

The amount of heat to be removed per 1000 kg of formaldehyde produced is

$$\frac{3450.56}{0.35119} \times 1000 = 9.8055 \times 10^6 \text{ kJ}$$

13.2.1 Temperature of Reaction

If no heat is added to the system from the surroundings or no heat is removed from the system during a reaction, the reaction can be termed an *adiabatic reaction*. Adiabatic conditions can be achieved by insulating the reaction vessel. In the case of combustion or explosions, the heat is generated so rapidly that it is generally impossible to transfer the heat away, and as a consequence, the adiabatic conditions are achieved naturally. The heat liberated during the reaction will be utilized completely to increase the enthalpy of the products. In contrast, in an isothermal reaction either heat is added to or heat is removed from the system to maintain the temperature constant.

When the products get heated up utilizing the heat liberated during the reaction, the temperature attained is termed the temperature of the reaction. In the case of adiabatic reactions, this temperature is the *adiabatic reaction temperature.* When a fuel is burned in air or oxygen under adiabatic conditions the heat of reaction is the heat of combustion and the temperature attained by the system is known as the *adiabatic flame temperature.* The temperature attained by the products of combustion when the unit quantity of a fuel is burned depends on the amount of oxygen supplied as well as whether the oxygen is supplied as air or as pure oxygen. The maximum adiabatic flame temperature is attained when the fuel is burned in theoretically required amount of pure oxygen, and when there is no heat loss to the surroundings. This maximum temperature is also known as the *theoretical flame temperature.* The adiabatic flame temperature attained, when the fuel is burned in air or excess oxygen, is less than the theoretical maximum flame temperature attained in oxygen.

Denoting as before, the enthalpy of cooling the reactants from temperature T_1 to 298 K by ΔH_1 and the enthalpy of heating the products from 298 to T, the adiabatic flame temperature by ΔH_2, we can write

$$\Delta H = \Delta H_1 + \Delta H_{298}^0 + \Delta H_2$$

Here ΔH is the difference between the enthalpy of products and the enthalpy of reactants, or simply, the heat added or removed during the reaction. For an adiabatic reaction, this quantity is zero, and therefore,

$$\Delta H_2 = -\Delta H_1 - \Delta H_{298}^0 \tag{13.19}$$

Equation (13.19) may be used to evaluate the temperature attained in an adiabatic reaction.

EXAMPLE 13.19 Ammonia is produced by the following reaction

$$N_2 + 3H_2 \rightarrow 2NH_3$$

In a commercial process for ammonia production, the feed to an adiabatic reactor contains 1 kmol/s of nitrogen and stoichiometric amounts of hydrogen at 700 K. What is the maximum allowable conversion in the reactor, if the adiabatic temperature rise across the reactor should not exceed 100 K.

Assume the feed and product streams to be ideal gas mixtures. The heat of reaction at 700 K for the above reaction is calculated to be −94.2 kJ/mol. The mean molar heat capacities in the range of 700–800 K are 0.03, 0.0289, and 0.0492 kJ/mol K for nitrogen, hydrogen and ammonia, respectively.

Solution *Basis:* Feed consisting of 1 kmol/s nitrogen and 3 kmol/s hydrogen

Let x be the fraction of nitrogen converted. Since the heat of reaction at 700 K is −94.2 kJ/mol, the heat liberated by the reaction is $-94.2 \times 10^3 x$ kJ. The product stream consists of

$2x$ kmol NH_3, $(1 - x)$ kmol N_2 and $3(1 - x)$ kmol hydrogen. When the temperature of the product stream rises by 100 K, the heat utilized is

$$(1 - x) \times 0.03 \times 10^3 \times 100 + 3(1 - x) \times 0.0289 \times 10^3$$
$$\times 100 + 2x \times 0.0492 \times 10^3 \times 100 \text{ kJ}$$
$$= 11.67 \times 10^3 - 1.83 \times 10^3 x \text{ kJ}$$

Since the reaction is adiabatic, the only heat available for raising the temperature of the products is the heat liberated by the reaction which is $94.2 \times 10^3 x$ kJ. Therefore,

$$11.67 \times 10^3 - 1.83 \times 10^3 x = 94.2 \times 10^3 x$$

On solving this equation, we get $x = 0.1215$. It means that the maximum conversion for nitrogen should be 13.15 % if the temperature rise permitted is 100 K.

EXAMPLE 13.20 Calculate the adiabatic flame temperature for CO when burned with 100 percent excess air when both the reactants are at 373 K. The heat capacities (J/mol K) may be assumed constant at 29.23 for CO, 34.83 for O_2, 33.03 for N_2 and 53.59 for CO_2. The standard heat of combustion at 298 K is −282.99 kJ/mol CO.

Solution *Basis:* 1 mol CO.

$$CO + \frac{1}{2} O_2 \rightarrow CO_2$$

Reactants: $O_2 = 1$ mol; $N_2 = 3.76$ mol; $CO = 1$ mol
Products: $O_2 = 0.5$ mol; $N_2 = 3.76$ mol; $CO_2 = 1$ mol

$$\Delta H_1 = (1 \times 34.83 + 3.76 \times 33.03 + 1 \times 29.23)(298 - 373) = -14\,119 \text{ J}$$

$$\Delta H_2 = (0.5 \times 34.83 + 3.76 \times 33.03 + 1 \times 53.59)(T - 298) = 195.20T - 58\,169$$

where T is the adiabatic flame temperature.
Using Eq. (13.19),

$$195.20T - 58\,169 = 14\,119 + 282\,990$$

Solving this, we get the flame temperature as $T = 1820.1$ K.

EXAMPLE 13.21 Hydrogen gas is burned in an adiabatic reactor with two times the theoretical quantity of air, both air and hydrogen being at 298 K initially. What will be the temperature of the reaction products? The standard heat of formation of gaseous water is −241.826 kJ/mol. The heat capacities (kJ/kmol K) of the gases are given below:

Water vapour: $C_P = 30.475 + 9.652 \times 10^{-3}T + 1.189 \times 10^{-6}T^2$
Nitrogen: $C_P = 27.034 + 5.815 \times 10^{-3}T - 0.2889 \times 10^{-6}T^2$
Oxygen: $C_P = 25.611 + 13.260 \times 10^{-3}T - 4.2077 \times 10^{-6}T^2$

Solution *Basis:* 1 kmol hydrogen burned. The theoretical requirement of oxygen is 0.5 kmol.
The amount of oxygen supplied = 2 × 0.5 = 1 mol

The amount of air supplied = $1 \times \frac{100}{21} = 4.762$ kmol

The amount of nitrogen in the air supplied = 4.762 – 1 = 3.762 kmol

Reactants: Hydrogen = 1 kmol, Air = 4.762 kmol

Products: Water vapour = 1 kmol, Oxygen = 1 – 0.5 = 0.5 kmol, Nitrogen = 3.762 kmol

Let T be the temperature attained.

$$\text{Then}\quad \Delta H_2 = \int_{298}^{T} \begin{bmatrix} 1\times(30.475+9.652\times10^{-3}T+1.189\times10^{-6}T^2) \\ +0.5\times(27.034+5.815\times10^{-3}T-0.2889\times10^{-6}T^2) \\ +3.762\times(25.611+13.260\times10^{-3}T-4.2077\times10^{-6}T^2) \end{bmatrix} dT$$

$$= \int_{298}^{T} (140.34+62.444\times10^{-3}T-14.785\times10^{-6}T^2)\, dT$$

$$= 140.34(T-298)+\frac{62.444\times10^{-3}}{2}(T^2-298^2)-\frac{14.785\times10^{-6}}{3}(T^3-298^3)$$

$$= 140.34T + 31.222\times10^{-3}T^2 - 4.928\times10^{-6}T^2 - 44\,463.54 \text{ kJ}$$

The standard heat of reaction at 298 K, $\Delta H^0_{298} = -241.826\times10^3$ kJ.

Let us use Eq. (13.19), i.e.

$$\Delta H_2 = -\Delta H_1 - \Delta H^0_{298}$$

Since $\Delta H_1 = 0$, we have

$$140.34T + 31.222\times10^{-3}T^2 - 4.928\times10^{-6}T^3 - 44\,463.54 = 241.826\times10^3 \text{ kJ}$$

On solving this equation, we get T = 1609.8 K.

EXERCISES

Standard heat of reaction and Hess's law

13.1 Using Hess's law evaluate the heat of formation of solid $CaCO_3$. The following data are available:

$$Ca(s)+\frac{1}{2}O_2(g)\rightarrow CaO(s) \qquad \Delta H^0_{298} = -635.77 \text{ kJ}$$

$$C(s) + O_2(g) \rightarrow CO_2(g) \qquad \Delta H^0_{298} = -393.51 \text{ kJ}$$

$$CaO(s) + CO_2(g) \rightarrow CaCO_3(s) \qquad \Delta H^0_{298} = -178.15 \text{ kJ}$$

13.2 The heats of combustion of gaseous benzene, gaseous cyclohexane (C_6H_{12}), and hydrogen are respectively –3287.4, –3949.2 and –285.58 kJ. Calculate the heat of reaction for the hydrogenation of benzene to cyclohexane.

13.3 The standard heats of combustion of gaseous ethane and acetaldehyde are – 1559.9 kJ/kmol and – 1192.36 kJ/kmol respectively. The heat of vaporization of water at 298 K is 44.014 kJ/mol. Calculate the heat of oxidation of gaseous ethane to acetaldehyde.

13.4 Find the standard heat of formation of methyl chloride from the following data:

$$CO_2(g) + H_2O(l) + HCl(aq) \rightarrow CH_3Cl(g) + \frac{3}{2}O_2 + aq \quad \Delta H^0 = 765.4 \text{ kJ}$$

$$C(s) + O_2(g) \rightarrow CO_2(g) \quad \Delta H^0 = -393.51 \text{ kJ}$$

$$H_2(g) + \frac{1}{2}O_2(g) \rightarrow H_2O(l) \quad \Delta H^0 = -285.84 \text{ kJ}$$

$$\frac{1}{2}H_2(g) + \frac{1}{2}Cl_2(g) \rightarrow HCl(g) \quad \Delta H^0 = -92.31 \text{ kJ}$$

$$HCl(g) + aq \rightarrow HCl(aq) \quad \Delta H^0 = -75.2 \text{ kJ}$$

13.5 Using Hess's law, calculate the heat of formation of chloroform ($CHCl_3$) with the help of the following data:

$$CHCl_3(g) + \frac{1}{2}O_2(g) + H_2O(l) \rightarrow CO_2(g) + 3HCl(g) \quad \Delta H^0_{298} = -281.67 \text{ kJ}$$

$$H_2(g) + \frac{1}{2}O_2(g) \rightarrow H_2O(l) \quad \Delta H^0_{298} = -285.84 \text{ kJ}$$

$$C(s) + O_2(g) \rightarrow CO_2(g) \quad \Delta H^0_{298} = -393.51 \text{ kJ}$$

$$\frac{1}{2}H_2(g) + \frac{1}{2}Cl_2(g) \rightarrow HCl(g) \quad \Delta H^0_{298} = -92.311 \text{ kJ}$$

13.6 Heats of reactions were measured experimentally for the following reactions as given below:

$$C(s) + O_2(g) \rightarrow CO_2(g) \quad \Delta H_r = -393.51 \text{ kJ/mol}$$

$$CO(g) + \frac{1}{2}O_2(g) \rightarrow CO_2(g) \quad \Delta H_r = -282.99 \text{ kJ/mol}$$

Use the above data to calculate the heat of formation of CO, which is difficult to measure experimentally.

13.7 Calculate the heat of formation of methane gas from the following heat of combustion data:

$$CH_4(g) + 2O_2(g) \rightarrow CO_2(g) + 2H_2O(l) \quad \Delta H^0_{298} = -890.4 \text{ kJ}$$

$$C(s) + O_2(g) \rightarrow CO_2(g) \quad \Delta H^0_{298} = -393.51 \text{ kJ}$$

$$H_2(g) + \frac{1}{2}O_2(g) \rightarrow H_2O(l) \quad \Delta H^0_{298} = -285.84 \text{ kJ}$$

Heat of combustion

13.8 The heats of combustion of solid carbon and gaseous CO are respectively – 393.51 kJ/mol and – 282.99 kJ/mol. Determine the heat of formation of CO.

13.9 Bituminous coal with a calorific value of 20000 kJ/kg is used for generating steam in a boiler. The efficiency of combustion is 75%. Assume that coal contains 67% carbon and 33% ash. (a) How much coal has to be burned to generate 1 MW of energy? (b) How much air is needed if 50% excess air is to be used?

13.10 Calculate the gross and net heating value (kJ/m^3 at STP) of a fuel gas analyzing 80% methane, 10% ethane, 6% propane and 4% butane. The standard heats of combustion are –890.4, –1559.9, –2220, –2878.52 kJ/mol, respectively, for methane, ethane, propane, and butane, and the latent heat of vaporization of water at 298 k is 44.014 kJ/mol.

13.11 Hydrocarbon fuels can be produced from methanol by reactions such as the following:

$$6CH_3OH\,(g) \rightarrow C_6H_{12}(g) + 6H_2O\,(g) \qquad \Delta H^0_{298} = -142.216 \text{ kJ}$$

Compare the standard heat of combustion at 298 K of CH_3OH (g) with the standard heat of combustion at 298 K of C_6H_{12} (g), reaction products in both cases being CO_2(g) and H_2O (g).

Heat of formation

13.12 Calculate the standard heat of reaction at 298 K for the following reaction:

$$4HCl\,(g) + O_2(g) \rightarrow 2H_2O\,(g) + 2Cl_2(g)$$

The standard heats of formation are – 92.311 kJ/mol for HCl (g) and – 241.826 kJ/mol for H_2O (g).

13.13 Calculate the heat of reaction for the following reaction:

$$4NH_3(g) + 5O_2(g) \rightarrow 4NO\,(g) + 6H_2O\,(g)$$

The heat of formation data are as follows:

NH_3 = – 46.191 kJ/mol; NO = 90.374 kJ/mol; H_2O (g) = – 241.826 kJ/mol

13.14 Calculate the standard heat of reaction for the following:

$$2FeS_2\,(s) + \frac{11}{2}O_2\,(g) \rightarrow Fe_2O_3\,(s) + 4SO_2\,(g)$$

The standard heats of formation at 298 K are –178.02 kJ/mol for FeS_2(s), –822.71 kJ/mol for Fe_2O_3 (s) and – 296.9 kJ/mol for SO_2 (g).

13.15 If the standard heat of formation of H_2O (l) is – 285.84 kJ/mol and the heat of vaporization of water is 44.01 kJ/mol, what is the standard heat of formation of H_2O (g)?

13.16 The heat of formation in kJ/mol of C_3H_8(g), CO_2(g) and H_2O(g) are, respectively, –103.85, – 393.51 and –241.826. Estimate the heat of reaction in the standard state for the following:

$$C_3H_8\,(l) + 5O_2\,(g) \rightarrow 3CO_2\,(g) + 4H_2O\,(l)$$

The heats of vaporization at 298 K in kJ/mol are 16.00 and 44.04, respectively, for C_3H_8 and H_2O.

13.17 The standard heat of combustion of benzene at 298 K is – 3267.6 kJ/mol when burned completely to CO_2 and liquid water. The standard heat of combustion of hydrogen to liquid water is – 285.84 kJ/mol and that of carbon to carbon dioxide is – 393.51 kJ/mol. Calculate the standard heat of formation of liquid benzene.

13.18 Calculate the standard heat of formation of 1-butene if the standard heat of combustion is – 2720.4 kJ/mol and the heats of formation of gaseous CO_2 and liquid water are – 393.51 kJ/mol and – 286.03 kJ/mol respectively.

13.19 In the production of ethylene from ethane and propane, the following yields were found:

Reaction	*Percent of reactant proceeding by the reaction*
$C_2H_6 \rightarrow C_2H_4 + H_2$	100
$C_3H_8 \rightarrow C_2H_4 + CH_4$	60
$C_3H_8 \rightarrow C_3H_6 + H_2$	40

Calculate the standard heat of reaction for a mixture consisting of equimolar amounts of ethane and propane, given the following heat of formation in kJ/mol:

C_2H_6 (g): – 84.667, C_3H_8: – 103.85, C_2H_4: 52.283, CH_4: – 74.84, C_3H_6 (g): 20.41

13.20 An iron pyrite ore containing 85.0% FeS_2 and 15.0% gangue is roasted with 200% excess air according to the reaction

$$4FeS_2 + 11O_2 \rightarrow 2Fe_2O_3 + 8SO_2$$

in order to produce SO_2. The cinder analyzed 5.0% FeS_2. Determine the standard heat of reaction per kilogram of ore. The standard heat of formation data (kJ/mol) are as follows: FeS_2(s) = – 177.9, Fe_2O_3(s) = – 822.156, SO_2(g) = – 296.90.

13.21 Calculate the standard heat of reaction for the following:

$$NH_3 + HCl(g) \rightarrow NH_4Cl(s)$$

The standard heats of formation of ammonia, hydrogen chloride and ammonium chloride are – 46.2 KJ, – 92.4 kJ and – 315.6 kJ respectively.

13.22 Calculate the standard heat of the following reaction:

$$2FeO(s) + \frac{1}{2}O_2 \rightarrow Fe_2O_3(s)$$

The standard heat of formation of FeO (s) and Fe_2O_3 (s) are – 268.1 kJ and – 822.7 kJ, respectively.

Heat of reaction at constant pressure and constant volume

13.23 What would be the heat of reaction at constant pressure for the following reaction if the heat evolved under constant volume conditions is 111.760 kJ/mol at 298 K.

$$C + \frac{1}{2}O_2 \rightarrow CO(g)$$

13.24 Calculate the heat of reaction at constant volume at 298 K for the following gas phase reaction:

$$C_2H_4(g) + 2H_2(g) \rightarrow 2CH_4(g)$$

The standard heat of formation at 298 K is 52.283 kJ/mol for gaseous ethylene and – 74.84 kJ/mol for gaseous methane.

13.25 The standard heat of combustion of liquid benzene at 298 K is – 3269.8 kJ/mol. 2.5 g of liquid benzene is placed in a bomb calorimeter at 298 K. The water equivalent of the bomb and surrounding water is 18.4 kJ/K. Determine the rise in temperature of water assuming all water formed on the combustion of benzene is condensed.

Effect of temperature on standard heat of reaction

13.26 Calculate the heat of combustion of methane at 800 K given that the heat of combustion at 298 K is – 890.4 kJ/mol and the mean heat capacity in the temperature range of 298 to 800 K are 41.868 J/mol K, 30.563 J/mol K, 41.449 J/mol K and 34.332 J/mol K for methane, oxygen, CO_2 and water vapour respectively.

13.27 Calculate the constant pressure and constant volume heat requirements of a catalytic reactor oxidizing methanol to formaldehyde at 775 K using air according to the following reaction:

$$CH_3OH(g) + \frac{1}{2}O_2(g) \rightarrow HCHO(g) + H_2O(g) \qquad \Delta H^0_{298} = -156.67 \text{ kJ/mol}$$

The reactants are admitted in stoichiometric proportions, and methanol is completely converted to formaldehyde. The enthalpies of methanol and formaldehyde vapours at 775 K relative to 298 K are 29.84 kJ/mol and 21.44 kJ/mol, respectively. The mean heat capacities of air, nitrogen and water vapour are, respectively, 30.3 kJ/kmol, 30.0 kJ/kmol and 35.76 kJ/kmol.

13.28 CO at 1000 K is burned with 90 percent excess air at 800 K. The products of combustion leave the reaction chamber at 1250 K. Calculate the heat evolved in the reaction chamber per kmol of CO burned. The standard heat of reaction at 298 K is – 282.99 kJ/mol CO. The mean specific heat applicable in the temperature range of this problem are 29.38, 49.91, 33.13 and 31.43 J/mol K for CO, CO_2, O_2 and N_2 respectively.

13.29 Carbon monoxide at 473 K is burned at a rate of 1 kmol/h in a furnace using 100% excess air at 373 K. The combustion is complete and the gases leave at 1273 K. Calculate the heat removed in the furnace. The mean heat capacity values in kJ/kmol K in the temperature range of 298 to 1273 K are 49.91, 33.25, 31.43, 29.29, and 29.38 for CO_2, O_2, N_2, air and CO, respectively. The standard heat of combustion of CO at 298 K is –282.99 kJ/mol.

13.30 Hydrochloric acid is produced from chlorine according to the following reaction:

$$Cl_2 + H_2O \rightarrow 2HCl + \frac{1}{2}O_2$$

The standard heats of formation at 298 K are – 241.826, – 92.311 kJ/mol for water vapour and hydrochloric acid respectively. Determine the amount of heat to be added or to be removed per kmol of acid obtained if the reactants enter at 500 K with a ratio of 3 mol steam per mol chlorine, reaction goes to completion, and the products leave at 500 K. The mean specific heats of chlorine gas, oxygen and HCl may be taken as 35.0, 33.0, 40.0 J/mol K respectively, and enthalpies of saturated water vapour at 298 K and 500 K are 2547 kJ/kg and 2801 kJ/kg respectively.

13.31 Ammonia oxidation occurs as per the reaction

$$4NH_3(g) + 5O_2(g) \rightarrow 4NO(g) + 6H_2O(g) \qquad \Delta H_R^0 = -226 \text{ kJ/mol NO}$$

The reactor is fed at a rate of 100 mol/h ammonia and 200 mol/h oxygen, both at 298 K. The product leaves the reactor at 575 K. Given the following heat capacity data at 575 K, calculate the heat transferred from the reactor: O_2 = 30.8 kJ/kmol K, NO = 30.74 kJ/kmol K and H_2O = 34.8 kJ/kmol K.

13.32 The standard heat of reaction at 298 K for the following reaction is –42.433 kJ.

$$C_2H_4(g) + H_2O(g) \rightarrow C_2H_5OH(g)$$

Calculate the heat of reaction at 400 K given that the constants in the heat capacity equation $C_P = \alpha + \beta T + \gamma T^2$ are as given below: (C_P is in J/mol K and T in K).

	α	β	γ
C_2H_4	11.85	119.75×10^{-3}	-36.53×10^{-6}
H_2O	30.38	9.62×10^{-3}	$+1.19 \times 10^{-6}$
C_2H_5OH	29.27	166.39×10^{-3}	-49.93×10^{-6}

13.33 For the following reaction the standard heat of reaction at 298 K is – 164.987 kJ.

$$CO_2(g) + 4H_2(g) \rightarrow 2H_2O(g) + CH_4(g)$$

The constants in the heat capacity (J/mol K) equation are given below:

	α	β	γ
CO_2	26.75	42.26×10^{-3}	-14.25×10^{-6}
H_2	26.88	4.35×10^{-3}	-0.33×10^{-6}
H_2O	29.16	14.49×10^{-3}	-2.02×10^{-6}
CH_4	13.41	77.03×10^{-3}	-18.74×10^{-6}

Calculate the standard heat of reaction at 773 K.

13.34 For the following gas-phase reaction, determine the heat of reaction if the reactants enter and products leave at 800 K.

$$CO_2 + 4H_2 \rightarrow 2H_2O + CH_4$$

The standard heats of formation are – 393.51 kJ/mol for CO_2, – 241.826 kJ/mol for H_2O (g) and – 74.84 kJ/mol for CH_4. The constants in the heat capacity equation

$$C_P = a + bT + cT^2$$

are given below where C_P is in kJ/kmol K and T in K.

	a	$b \times 10^3$	$c \times 10^6$
Carbon dioxide	26.540	42.454	−14.2979
Hydrogen	29.082	−0.821	1.9917
Methane	13.415	77.079	−18.7569
Water vapour	30.38	9.62	−1.19

13.35 Methane is burned in a furnace with 100% excess air to generate steam in a boiler. Methane and dry air enter the combustion chamber at 500 K and the product leaves the furnace at 1200 K. If the effluent gases contain only CO_2, O_2, H_2O and N_2, how much heat is absorbed by water per kg of methane burned?

The standard heat of combustion of methane at 298 K is –890.4 kJ/mol. The latent heat of vaporization of water at 298 K is 44.014 kJ/mol. The constants in the heat capacity equation $C_P = a + bT + cT^2$ where C_P is in kJ/kmol and T in K are as follows:

	a	$b \times 10^3$	$c \times 10^6$
Water	30.38	9.62	1.19
Carbon dioxide	26.540	42.454	−14.2979
Methane	13.415	77.079	−18.7569
Nitrogen	27.034	5.815	−0.2889
Oxygen	25.611	13.260	−4.2077

13.36 The standard heat of combustion of graphite at 298 K is –393.51 kJ/mol. Determine the heat of combustion at 800 K. The heat capacities in J/mol K are given as follows:

Carbon: $$11.19 + 1.096 \times 10^{-2} T - \frac{4.894 \times 10^5}{T^2}$$

Oxygen: $$34.62 + 1.08 \times 10^{-3} T - \frac{7.859 \times 10^5}{T^2}$$

Carbon dioxide: $$43.29 + 1.147 \times 10^{-2} T - \frac{8.185 \times 10^5}{T^2}$$

13.37 Calculate the heat of the following gas-phase reaction

$$H_2 + \frac{1}{2} O_2 \rightarrow H_2O \qquad \Delta H_{298}^0 = -241.826 \text{ kJ}$$

if the reactants are at 473 K and the product is at 993 K. The specific heats (J/mol K) may be evaluated using the data given below:

$$C_P = \alpha + \beta T + \gamma T^2$$

	α	β	γ
H_2	29.09	-8.374×10^{-4}	2.0139×10^{-6}
O_2	25.74	12.987×10^{-3}	-3.864×10^{-6}
H_2O	30.38	9.621×10^{-3}	-1.185×10^{-6}

13.38 A gaseous mixture at 600 K and 1 bar consisting of methane and steam in the ratio 1 : 2 by mole is sent to a catalytic reformer where the following reaction occurs.

$$CH_4(g) + H_2O(g) \rightarrow CO(g) + 3H_2(g) \qquad \Delta H^0_{298} = 206.284 \text{ kJ}$$

The following side reaction may also occur:

$$CO(g) + H_2O(g) \rightarrow CO_2(g) + H_2(g) \qquad \Delta H^0_{298} = -41.20 \text{ kJ}$$

Methane is completely converted during the reaction and the product stream contains 17.4% (mol) CO. Heat is supplied in the reactor so that the products reach a temperature of 1300 K. The mean heat capacities in the temperature range of 298–600 K are 43.8 kJ/kmol K for methane and 34.8 kJ/kmol K for steam, and the mean heat capacities in the temperature range of 298–1300 K are 31.8 kJ/kmol K for CO, 29.8 kJ/kmol K for hydrogen, 50.1 kJ/kmol K for CO_2 and 38.9 kJ/kmol K for steam, respectively. Calculate the heat requirement for the reactor per kmol methane.

13.39 0.5 mol hydrogen at 300 K, a mixture of 1.5 mol nitrogen and 0.5 mol oxygen at 400 K, and 0.3 mol CO at 500 K are mixed. 50 percent of CO present reacts. If the final temperature of the system raises to 882 K under adiabatic conditions, calculate the standard heat of formation of CO_2, if it is four times the standard heat of formation of CO. The mean heat capacities in J/mol K are 32.5 for O_2, 31.1 for N_2, 29.0 for H_2, 31.0 for CO, and 49.0 for CO_2.

13.40 A gas with the analysis 8.5% SO_2, 2.0% SO_3, 9.5% O_2 and 80% N_2 is admitted to a catalytic converter where SO_2 gets converted to SO_3. The outlet gases contain 82.47% N_2. The reaction occurs at 700 K. Determine the heat evolved per 100 mol gas entering the converter. The data given are: heat of formation of SO_3 at 298 K = – 395.72 kJ/mol, heat of formation of SO_2 at 298 K = – 296.9 kJ/mol, mean heat capacity of SO_3 = 51.6 J/mol K, mean heat capacity of SO_2 = 45.7 J/mol K, mean heat capacity of O_2 = 32.5 J/mol K.

13.41 Calculate the standard heat of reaction at 800 K for the combustion of pentane gas given that the mean heat capacities in J/mol K are 247 for C_5H_{12}, 33.62 for O_2, 52.32 for CO_2, and 38.49 for H_2O. The standard heat of combustion at 298 K is – 3271.67 kJ/mol.

13.42 Sulphur dioxide gas is oxidized in 100% excess air with 80% conversion to sulphur trioxide. The gases enter the converter at 673 K and leave at 723 K. The mean heat capacities are 45, 51, 32 and 31 J/mol K for SO_2, SO_3, O_2, and N_2 respectively, and the standard heat of reaction is – 98.8 kJ/mol. How much heat is absorbed in the heat exchanger of the converter per kilomole of SO_2 introduced?

13.43 In a fermentation process, sucrose ($C_{12}H_{22}O_{11}$) is used as a nutrient and is oxidized to CO_2 gas and H_2O liquid. Calculate the heat of oxidation at 318 K in this fermentation. The standard heat of combustion of sucrose at 298 K = – 5648.8 kJ/mol. Mean heat capacities are 1.25 kJ/kg K for sucrose (s), 4.20 kJ/kg K for H_2O (l), 29.38 J/mol K for O_2 and 37.55 J/mol K for CO_2.

13.44 Pure CO is mixed with 100% excess air and burned. Only 80% of CO burns. The reactants are at 373 K and the products at 573 K. Calculate the amount of heat to be added or removed per kmol of CO fed to the reactor. Standard heats of formation in kJ/mol at 298 K are – 110.6 for CO and – 393.51 for CO_2. The mean molal heat capacity between 298 K and T K in kJ/kmol K are

Gas	T = 373 K	T = 573 K
CO	29.22	30.61
CO_2		43.77
O_2	29.84	30.99
N_2	29.17	29.66

13.45 An internal combustion engine uses octane as fuel. The air and fuel vapour mixture enter the engine at 298 K. 20 percent excess air is supplied. 75 percent of the carbon present in the fuel is converted to CO_2 and the rest to CO, and the combustion products leave the engine at 800 K. Calculate the energy transferred as heat to the engine per kg of fuel burned. The mean heat capacities in J/mol K are 32.5 for O_2, 31.1 for N_2, 31.0 for CO, 50.0 for H_2O and 49.0 for CO_2. The heat of formation of *n*-octane, CO_2, CO and H_2O vapour at 298 K are –208.75 kJ/mol, – 393.51 kJ/mol, – 110.6 kJ/mol and – 241.826 kJ/mol respectively.

13.46 A mixture of methane and steam at 600 K in the ratio 1:2 is sent to the reactor where the following reaction occurs:

$$CH_4(g) + H_2O(g) \rightarrow CO(g) + 3H_2(g) \qquad \Delta H^0_{298} = 205.813 \text{ kJ}$$

The following undesired side reaction also occurs:

$$CO(g) + H_2O(g) \rightarrow CO_2(g) + H_2(g) \qquad \Delta H^0_{298} = -41.198 \text{ kJ}$$

Methane is completely converted and the products contain 15.0% (mole) CO. If the products get heated to 1300 K, determine the quantity of heat supplied per mole of methane. The following mean heat capacity values (J/mol K) are applicable—methane: 43.79, steam: 34.83, CO: 31.82, H_2: 29.81 and CO_2: 50.12.

13.47 One hundred kilograms per hour of dry pyrites at 298 K containing 85% FeS_2 and the rest inert gangue is burned with dry air in a plant for the manufacture of sulphuric acid.

$$4FeS_2 + 11O_2 \rightarrow 2Fe_2O_3 + 8SO_2$$

$$4FeS_2 + 15O_2 \rightarrow 2Fe_2O_3 + 8SO_3$$

The cinder is at 800 K and contains 5% sulphur as FeS_2. Burner gas at 800 K contains 10% SO_2, 7.0% O_2 and 83% N_2 on SO_3-free basis. Standard heats of reaction are – 827.1 kJ for the first reaction and – 1023.7 kJ for the second reaction. The heat capacity of the cinder is 0.75 kJ/kg K. The mean heat capacities in kJ/kmol K are 50.2 for SO_3, 46.9 for SO_2, 31.4 for O_2 and 29.9 for N_2. The pressure is 101.3 kPa. Calculate the following:

(a) The amount of FeS_2 lost in the cinder, kg/h
(b) The amount of burner gas, m^3/h
(c) The amount of heat dissipated in the burner, kJ/h

13.48 Vinyl chloride is manufactured by the pyrolysis of 1,2,dichloroethane at 773 K.

$$C_2H_4Cl_2(g) \rightarrow C_2H_3Cl\,(g) + HCl\,(g) \qquad \Delta H_r^0 = 70.224 \text{ kJ}$$

The conversion is estimated to be 55%. How much heat is to be supplied to produce 500 kg/h of vinyl chloride from a feed of pure liquid dichloroethane at 290 K? The specific heat in kJ/kmol K is represented by the equation $C_P = a + bT + cT^2 + dT^3$ with the following values for the constants:

Component	*a*	$b \times 10^2$	$c \times 10^5$	$d \times 10^9$
Vinyl chloride	5.94	20.16	–15.34	47.65
HCl	30.28	–0.761	1.325	–4.305
Dichloroethane	20.45	23.07	–14.36	33.83

Take the specific heat of liquid dichloroethane to be 116 kJ/kmol K, which is constant and the latent heat of vaporization at 298 K is 34300 kJ/kmol.

Temperature of reaction

13.49 Calculate the maximum flame temperature attained when methane is burned with theoretical air when both fuel and air are at 298 K initially. The mean heat capacities in J/mol K are 62.75 for CO_2, 52.96 for H_2O, 38.67 for O_2 and 37.13 for N_2. The standard heat of combustion of methane at 298 K is – 890.4 kJ/mol.

13.50 A gas mixture consisting of 20% ethane and 80% oxygen at 298 K is burned completely after diluting it with double the volume of CO_2, also at 298 K. The mean heat capacities in J/mol K are 54.56 for CO_2, 43.02 for water vapour and 35.52 for oxygen. The standard heat of reaction is – 1428 kJ/mol of ethane at 298 K. Determine the theoretical flame temperature.

13.51 Calculate the theoretical flame temperature for a gas containing 25% CO and 75% N_2 when burned with 100% excess air when both the reactants are at 298 K. The standard heat of formation are – 393. 51 kJ/mol for CO_2 and – 110.6 kJ/mol for CO. The mean molar specific heats in J/mol K are

T (K)	CO_2	O_2	N_2
800	45.43	31.59	30.31
1000	47.56	32.37	30.64
1200	49.35	33.02	31.22
1400	50.82	33.60	31.77
1600	51.99	34.05	32.25
1800	53.18	34.40	32.65

13.52 Dry methane is burned with dry air. Both are at 298 K initially. The flame temperature is 1573 K. If complete combustion is assumed, how much excess air is used?

$$CH_4 + 2O_2 \rightarrow CO_2 + 2H_2O$$

Standard heat of reaction is – 8.028 × 10^5 J/mole of methane reacted. Mean molal heat capacities of gases between 298 K and 1573 K in J/mol K are: 51.88 for CO_2, 34.01 for O_2, 40.45 for H_2O, and 32.21 for N_2.

13.53 The dehydrogenation of butene is an endothermic reaction:

$$C_4H_8\,(g) \rightarrow C_4H_6\,(g) + H_2\,(g)$$

This reaction is to be carried out adiabatically and at atmospheric pressure, and in order to minimize the temperature drop, the reactor feed will consist of 10 moles of steam per mole of butene. The steam is unreactive. 20% of butene is converted. The feed mixture enters the reactor at 900 K. Calculate the temperature of the products if mean molar heat capacity in kJ/kmol K for the temperature range of 298 to 900 K are 148.7, 131.2, 36.6, and 29.4 for butene, butadiene, steam and hydrogen respectively. The standard heats of formation are – 0.54 kJ/mol for butene and 109.24 kJ/mol for butadiene.

13.54 Calculate the theoretical flame temperature of a gas mixture consisting of 20% CO and 80% N_2 when burned with 100% excess air, both air and gas initially being at 298 K. The standard heat of reaction at 298 K is – 282.99 kJ/mol CO. The heat capacities in J/mol K are given by $C_P = \alpha + \beta T + \gamma T^2$ where the values of the constants are given as follows:

	α	β	γ
CO_2	26.54	42.45×10^{-3}	-14.298×10^{-6}
O_2	25.61	13.26×10^{-3}	-4.208×10^{-6}
N_2	27.03	5.815×10^{-3}	-0.289×10^{-6}

13.55 Calculate the adiabatic reaction temperature for complete combustion of ethyl alcohol with 25% excess air. The alcohol and air are supplied at 298 K, and the water formed is in the vapour state. The standard heat of combustion of alcohol is – 1366.91 kJ/mol, the latent heat of vaporization of water is 44 kJ/mol, and heat capacities in kJ/kmol K where *T* is in K are:

CO_2 : $24.55 + 0.0423T$
H_2O : $31.57 + 0.0069T$
O_2 : $25.90 + 0.0116T$
N_2 : $28.40 + 0.0022T$

13.56 Aluminium metal powder is investigated as a fuel source for jet engines. Calculate the theoretical flame temperature when aluminium metal is burned with 100% excess air. Assume that the air and metal powder enter at 25°C and that combustion is complete. Heat capacities in kJ/kg are 140.26 for Al_2O_3 (s) and 142.35 for Al_2O_3 (l). The heat of fusion of Al_2O_3 is 108.857 kJ/mol. The standard heat of formation of Al_2O_3 (s) is – 1670.91 kJ/mol. The heat capacities of gases in J/mol K are as follows:

$$C_P(O_2) = 30.255 + 4.207 \times 10^{-3}T - 1.8873 \times 10^5 T^{-2}$$

$$C_P(N_2) = 27.270 + 4.930 \times 10^{-3}T - 3.3256 \times 10^4 T^{-2}$$

13.57 The catalytic converter in a sulphuric acid plant is supplied with a gas mixture analyzing 7.8% SO_2, 11.0% O_2 and 81.2% N_2, which enters the converter at a temperature of 675 K where 80% of SO_2 is oxidized to SO_3 by combination with the oxygen present in the gases. Assume that there is no heat losses from the reactor. Calculate the temperature of the gas leaving the converter.

Standard heats of formation at 298 K in kJ/mol are: SO_2 = –296.9 and SO_3 = –395.72. The heat capacities in kJ/kmol K and T in K are as follows:

Gas	*a*	$b \times 10^3$	$c \times 10^6$
Nitrogen	27.034	5.815	– 0.2889
Oxygen	25.611	13.260	– 4.2077
Sulphur dioxide	29.077	41.910	– 15.8847
Sulphur trioxide	31.208	80.093	– 27.7501

13.58 Methane is burned with 20 percent excess air, both methane and air being at 298 K. The standard heat of combustion of methane at 298 K is – 890.4 kJ. The heat capacities in J/mol K are as follows:

$$C_P(CH_4) = 14.150 + 75.499 \times 10^{-3}T - 17.9915 \times 10^{-6}T^2$$

$$C_P(O_2) = 30.255 + 4.207 \times 10^{-3}T - 1.8873 \times 10^5 T^{-2}$$

$$C_P(N_2) = 27.270 + 4.930 \times 10^{-3}T - 3.3256 \times 10^4 T^{-2}$$

$$C_P(CO_2) = 45.369 + 8.688 \times 10^{-3}T - 9.6193 \times 10^5 T^{-2}$$

$$C_P(H_2O) = 28.850 + 12.055 \times 10^{-3}T - 1.006 \times 10^5 T^{-2}$$

Calculate the adiabatic flame temperature attained.

Appendix Tables of Properties

Table A.1 Atomic weights of elements, 2001

Atomic number	*Symbol*	*Name*	*Atomic weight*
89	Ac	Actinium	[227]
13	Al	Aluminium	26.981538(2)
95	Am	Americium	[243]
51	Sb	Antimony	121.760(1)
18	Ar	Argon	39.948(1)
33	As	Arsenic	74.92160(2)
85	At	Astatine	[210]
56	Ba	Barium	137.327(7)
97	Bk	Berkelium	[247]
4	Be	Beryllium	9.012182(3)
83	Bi	Bismuth	208.98038(2)
107	Bh	Bohrium	[264]
5	B	Boron	10.811(7)
35	Br	Bromine	79.904(1)
48	Cd	Cadmium	112.411(8)
55	Cs	Caesium	132.90545(2)
20	Ca	Calcium	40.078(4)
98	Cf	Californium	[251]
6	C	Carbon	12.0107(8)
58	Ce	Cerium	140.116(1)
17	Cl	Chlorine	35.453(2)
24	Cr	Chromium	51.9961(6)
27	Co	Cobalt	58.933200(9)
29	Cu	Copper	63.546(3)
96	Cm	Curium	[247]
110	Ds	Darmstadtium	[281]
105	Db	Dubnium	[262]

(Contd.)

Table A.1 Atomic weights of elements, 2001 (Contd.)

Atomic number	*Symbol*	*Name*	*Atomic weight*
66	Dy	Dysprosium	162.500(1)
99	Es	Einsteinium	[252]
68	Er	Erbium	167.259(3)
63	Eu	Europium	151.964(1)
100	Fm	Fermium	[257]
9	F	Fluorine	18.9984032(5)
87	Fr	Francium	[223]
64	Gd	Gadolinium	157.25(3)
31	Ga	Gallium	69.723(1)
32	Ge	Germanium	72.64(1)
79	Au	Gold	196.96655(2)
72	Hf	Hafnium	178.49(2)
108	Hs	Hassium	[277]
2	He	Helium	4.002602(2)
67	Ho	Holmium	164.93032(2)
1	H	Hydrogen	1.00794(7)
49	In	Indium	114.818(3)
53	I	Iodine	126.90447(3)
77	Ir	Iridium	192.217(3)
26	Fe	Iron	55.845(2)
36	Kr	Krypton	83.798(2)
57	La	Lanthanum	138.9055(2)
103	Lr	Lawrencium	[262]
82	Pb	Lead	207.2(1)
3	Li	Lithium	[6.941(2)]
71	Lu	Lutetium	174.967(1)
12	Mg	Magnesium	24.3050(6)
25	Mn	Manganese	54.938049(9)
109	Mt	Meitnerium	[268]
101	Md	Mendelevium	[258]
80	Hg	Mercury	200.59(2)
42	Mo	Molybdenum	95.94(2)
60	Nd	Neodymium	144.24(3)
10	Ne	Neon	20.1797(6)
93	Np	Neptunium	[237]
28	Ni	Nickel	58.6934(2)
41	Nb	Niobium	92.90638(2)
7	N	Nitrogen	14.0067(2)
102	No	Nobelium	[259]
76	Os	Osmium	190.23(3)
8	O	Oxygen	15.9994(3)
46	Pd	Palladium	106.42(1)
15	P	Phosphorus	30.973761(2)
78	Pt	Platinum	195.078(2)
94	Pu	Plutonium	[244]
84	Po	Polonium	[209]

(Contd.)

Table A.1 Atomic weights of elements, 2001 (Contd.)

Atomic number	*Symbol*	*Name*	*Atomic weight*
19	K	Potassium	39.0983(1)
59	Pr	Praseodymium	140.90765(2)
61	Pm	Promethium	[145]
91	Pa	Protactinium	231.03588(2)
88	Ra	Radium	[226]
86	Rn	Radon	[222]
75	Re	Rhenium	186.207(1)
45	Rh	Rhodium	102.90550(2)
37	Rb	Rubidium	85.4678(3)
44	Ru	Ruthenium	101.07(2)
104	Rf	Rutherfordium	[261]
62	Sm	Samarium	150.36(3)
21	Sc	Scandium	44.955910(8)
106	Sg	Seaborgium	[266]
34	Se	Selenium	78.96(3)
14	Si	Silicon	28.0855(3)
47	Ag	Silver	107.8682(2)
11	Na	Sodium	22.989770(2)
38	Sr	Strontium	87.62(1)
16	S	Sulfur	32.065(5)
73	Ta	Tantalum	180.9479(1)
43	Tc	Technetium	[98]
52	Te	Tellurium	127.60(3)
65	Tb	Terbium	158.92534(2)
81	Tl	Thallium	204.3833(2)
90	Th	Thorium	232.0381(1)
69	Tm	Thulium	168.93421(2)
50	Sn	Tin	118.710(7)
22	Ti	Titanium	47.867(1)
74	W	Tungsten	183.84(1)
112	Uub	Ununbium	[285]
116	Uuh	Ununhexium	
118	Uuo	Ununoctium	
114	Uuq	Ununquadium	[289]
111	Uuu	Unununium	[272]
92	U	Uranium	238.02891(3)
23	V	Vanadium	50.9415(1)
54	Xe	Xenon	131.293(6)
70	Yb	Ytterbium	173.04(3)
39	Y	Yttrium	88.90585(2)
30	Zn	Zinc	65.409(4)
40	Zr	Zirconium	91.224(2)

Note: The value enclosed in brackets, e.g. [289], indicates the mass number of the longest-lived isotope of the element that has no stable nuclides. A number in parentheses for, e.g. 91.224(2) indicates the uncertainty in the last digit of the atomic weight.
Reference: Values from the 2001 table of atomic weights, *Pure Appl. Chem.*, **75**, 1107–1122 (2003).

Table A.2 Standard heat of formation and combustion

Compound	ΔH_f^0 (kJ/mol)	ΔH_c^0 (kJ/mol)
Acetic acid (l)	– 486.2	– 871.69
Acetaldehyde (g)	– 166.4	– 1192.36
Acetylene (g)	226.75	– 1299.61
Ammonia (g)	– 46.191	– 382.58
Benzene (l)	48.66	– 3267.6
n-Butane (l)	– 147.6	– 2855.6
Isobutane (l)	– 158.5	– 2849.0
1-Butene (g)	– 0.54	– 2718.58
1,3-Butadiene (g)	109.24	– 2540.8
Calcium oxide (s)	– 635.77	
Carbon (s, graphite)	0	– 393.51
Carbon dioxide (g)	– 393.51	
Carbon disulphide (l)	87.86	– 1075.2
Carbon monoxide (g)	– 110.6	– 282.99
Carbon tetrachloride (l)	– 139.5	– 352.2
Ethane (g)	– 84.667	– 1559.9
Ethyl acetate (l)	– 442.92	– 2274.48
Ethyl alcohol (l)	– 277.63	– 1366.91
Ethylene (g)	52.283	– 1410.99
Ferric oxide (s)	– 822.71	
Ferrous sulphide (s)	– 178.02	
Formaldehyde (g)	– 115.89	– 563.46
n-Heptane (l)	– 224.4	– 4816.91
n-Hexane (l)	– 198.8	– 4163.1
Hydrogen (g)	0	– 285.84
Hydrogen Chloride (g)	– 92.311	
Hydrogen sulphide (g)	– 20.15	– 562.589
Methane (g)	– 74.79	– 890.4
Methyl alcohol (l)	– 238.64	– 726.55
Methyl chloride (g)	– 81.923	– 766.63
Nitric acid (l)	– 173.23	
Nitric oxide (g)	90.374	
Nitrogen dioxide (g)	33.85	
n-Pentane (g)	– 146.4	– 3536.15
Phosphorous pentoxide (c)	– 1506.0	
Propane (g)	– 103.85	– 2220.0
n-Propyl alcohol (l)	– 255.0	– 2068.6
Sodium carbonate (c)	– 1130	
Sodium chloride (c)	– 411.0	
Sodium nitrate (s)	– 466.68	
Sodium sulphate (s)	– 1384.5	
Sucrose (s)		–5648.8
Sulphur dioxide (g)	– 296.9	
Sulphur trioxide (g)		– 395.18
Sulphuric acid (l)	– 811.32	
Tolune (l)	11.99	– 3909.9
Water (l)	– 285.84	
Water (g)	– 241.826	

Table A.3 Thermodynamic properties of saturated steam

Pressure (kPa)	Temp. (K)	Specific volume ($\times 10^3 \frac{m^3}{kg}$) Sat. liquid	Sat. vapour	Internal energy (kJ/kg) Sat. liquid	Sat. vapour	Enthalpy (kJ/kg) Sat. liquid	Sat. vapour	Entropy (kJ/kg K) Sat. liquid	Sat. vapour
0.6113	273.16	1.000	206140	0.00	2375.30	0.01	2501.40	0.0000	9.1562
1.00	280.13	1.000	129210	29.30	2385.00	29.30	2514.20	0.1059	8.9756
1.50	286.18	1.001	87980	54.71	2393.30	54.71	2525.30	0.1957	8.8279
2.00	290.65	1.001	67000	73.48	2399.50	73.48	2533.50	0.2607	8.7237
2.50	294.23	1.002	54250	88.48	2404.40	88.49	2540.00	0.3120	8.6432
3.00	297.23	1.003	45670	101.04	2408.50	101.05	2545.50	0.3545	8.5776
4.00	302.11	1.004	34800	121.45	2415.20	121.46	2554.40	0.4226	8.4746
5.00	306.03	1.005	28190	137.81	2420.50	137.82	2561.50	0.4764	8.3951
7.50	313.44	1.008	19240	168.78	2430.50	168.79	2574.80	0.5764	8.2515
10.00	318.96	1.010	14670	191.82	2437.90	191.83	2584.70	0.6493	8.1502
15.00	327.12	1.014	10020	225.92	2448.70	225.94	2599.10	0.7549	8.0085
20.00	333.21	1.017	7649	251.38	2456.70	251.40	2609.70	0.8320	7.9085
25.00	338.12	1.020	6204	271.90	2463.10	271.93	2618.20	0.8931	7.8314
30.00	342.25	1.022	5229	289.20	2468.40	289.23	2625.30	0.9439	7.7686
40.00	349.02	1.027	3993	317.53	2477.00	317.58	2636.80	1.0259	7.6700
50.00	354.48	1.030	3240	340.44	2483.90	340.49	2645.90	1.0910	7.5939
75.00	364.93	1.037	2217	384.31	2496.70	384.39	2663.00	1.2130	7.4564
100.00	372.78	1.043	1694.0	417.36	2506.10	417.46	2675.50	1.3026	7.3594
125.00	379.14	1.048	1374.9	444.19	2513.50	444.32	2685.40	1.3740	7.2844
150.00	384.52	1.053	1159.3	466.94	2519.70	467.11	2693.60	1.4336	7.2233
175.00	389.21	1.057	1003.6	486.80	2524.90	486.99	2700.60	1.4849	7.1717
200.00	393.38	1.061	885.7	504.49	2529.50	504.70	2706.70	1.5301	7.1271
225.00	397.15	1.064	793.3	520.47	2533.60	520.72	2712.10	1.5706	7.0878
250.00	400.59	1.067	718.7	535.10	2537.20	535.37	2716.90	1.6072	7.0527
275.00	403.75	1.070	657.3	548.59	2540.50	548.89	2721.30	1.6408	7.0209
300.00	406.70	1.073	605.8	561.15	2543.60	561.47	2725.30	1.6718	6.9919
325.00	409.45	1.076	562.0	572.90	2546.40	573.25	2729.00	1.7006	6.9652
350.00	412.03	1.079	524.3	583.95	2548.90	584.33	2732.40	1.7275	6.9405
375.00	414.47	1.081	491.4	594.40	2551.30	594.81	2735.60	1.7528	6.9175
400.00	416.78	1.084	462.5	604.31	2553.60	604.74	2738.60	1.7766	6.8959
450.00	421.08	1.088	414.0	622.77	2557.60	623.25	2743.90	1.8207	6.8565
500.00	425.01	1.093	374.9	639.68	2561.20	640.23	2748.70	1.8607	6.8213
550.00	428.63	1.097	342.7	655.32	2564.50	655.93	2753.00	1.8973	6.7893
600.00	432.00	1.101	315.7	669.90	2567.40	670.56	2756.80	1.9312	6.7600
650.00	435.16	1.104	292.7	683.56	2570.10	684.28	2760.30	1.9627	6.7331
700.00	438.12	1.108	272.9	696.44	2572.50	697.22	2763.50	1.9922	6.7080
750.00	440.93	1.112	255.6	708.64	2574.70	709.47	2766.40	2.0200	6.6847
800.00	443.58	1.115	240.4	720.22	2576.80	721.11	2769.10	2.0462	6.6628
850.00	446.11	1.118	227.0	731.27	2578.70	732.22	2771.60	2.0710	6.6421
900.00	448.53	1.121	215.0	741.83	2580.50	742.83	2773.90	2.0940	6.6226
950.00	450.84	1.124	204.2	751.95	2582.10	753.02	2776.10	2.1172	6.6041

(Contd.)

Table A.3 Thermodynamic properties of saturated steam (Contd.)

		Specific volume $\left(\times 10^3 \frac{m^3}{kg}\right)$		*Internal energy* (kJ/kg)		*Enthalpy* (kJ/kg)		*Entropy* (kJ/kg K)	
Pressure (kPa)	*Temp.* (K)	*Sat. liquid*	*Sat. vapour*	*Sat. liquid*	*Sat. vapour*	*Sat. liquid*	*Sat. vapour*	*Sat. liquid*	*Sat. vapour*
1000	453.06	1.127	194.4	761.68	2583.60	762.81	2778.10	2.1387	6.5865
1100	457.24	1.133	177.53	780.09	2586.40	781.34	2781.70	2.1792	6.5536
1200	461.14	1.139	163.33	797.29	2588.80	798.65	2784.80	2.2166	6.5233
1300	464.79	1.144	151.25	813.44	2591.00	814.93	2787.60	2.2515	6.4953
1400	468.22	1.149	140.84	828.70	2592.80	830.30	2790.00	2.2842	6.4693
1500	471.47	1.154	131.77	843.16	2594.50	844.89	2792.2	2.3150	6.4448
1750	478.91	1.166	113.49	876.46	2597.80	878.50	2796.40	2.3851	6.3896
2000	485.57	1.177	99.63	906.44	2600.30	908.79	2799.50	2.4474	6.3409
2250	491.60	1.187	88.75	933.83	2602.00	936.49	2801.70	2.5035	6.2972
2500	497.14	1.197	79.98	959.11	2603.10	962.11	2803.10	2.5547	6.2575
3000	507.05	1.217	66.68	1004.78	2604.10	1008.42	2804.20	2.6457	6.1869
3500	515.75	1.235	57.07	1045.43	2603.70	1049.75	2803.40	2.7253	6.1253
4000	523.55	1.252	49.78	1082.31	2602.30	1087.31	2801.40	2.7964	6.0701
5000	537.14	1.286	39.44	1147.81	2597.10	1154.23	2794.30	2.9202	5.9734
6000	548.79	1.319	32.44	1205.44	2589.70	1213.35	2784.30	3.0267	5.8892
7000	559.03	1.351	27.37	1257.55	2580.50	1267.00	2772.10	3.1211	5.8133
8000	568.21	1.384	23.52	1305.57	2569.80	1316.64	2758.00	3.2068	5.7432
9000	576.55	1.418	20.48	1350.51	2557.80	1363.26	2742.10	3.2858	5.6772
10000	584.21	1.452	18.026	1393.04	2544.40	1407.56	2724.70	3.3596	5.6141
11000	591.30	1.489	15.987	1433.70	2529.80	1450.10	2705.60	3.4295	5.5527
12000	597.90	1.527	14.263	1473.00	2513.70	1491.30	2684.90	3.4962	5.4924
13000	604.08	1.567	12.780	1511.10	2496.10	1531.50	2662.20	3.5606	5.4323
14000	609.90	1.611	11.485	1548.60	2476.80	1571.10	2637.60	3.6232	5.3717
15000	615.39	1.658	10.337	1585.60	2455.50	1610.50	2610.50	3.6848	5.3098
16000	620.59	1.711	9.306	1622.70	2431.70	1650.10	2580.60	3.7461	5.2455
17000	625.52	1.770	8.364	1660.20	2405.00	1690.30	2547.20	3.8079	5.1777
18000	630.21	1.840	7.489	1698.90	2374.30	1732.00	2509.10	3.8715	5.1044
19000	634.69	1.924	6.657	1739.90	2338.10	1776.50	2464.50	3.9388	5.0228
20000	638.96	2.036	5.834	1785.60	2293.00	1826.30	2409.70	4.0139	4.9269
21000	643.04	2.207	4.952	1842.10	2230.60	1888.40	2334.60	4.1075	4.8013
22000	646.95	2.742	3.568	1961.90	2087.10	2022.20	2165.60	4.3110	4.5327
22090	647.29	3.155	3.155	2029.60	2029.60	2099.30	2099.30	4.4298	4.4298

Table A.4 Thermodynamic properties of superheated steam

T, K	V	U	H	S	V	U	H	S
	P = 10 kPa (Sat. Temp = 318.96 K)				P = 50 kPa (Sat. Temp = 354.48 K)			
Sat.	14.674	2437.9	2584.7	8.1502	3.240	2483.9	2645.9	7.5939
323.15	14.869	2443.9	2592.6	8.1749				
373.15	17.196	2515.5	2687.5	8.4479	3.418	2511.6	2682.5	7.6947
423.15	19.512	2587.9	2783.0	8.6882	3.889	2585.6	2780.1	7.9401
473.15	21.825	2661.3	2879.5	8.9038	4.356	2659.9	2877.7	8.1580
523.15	24.136	2736.0	2977.3	9.1002	4.820	2735.0	2976.0	8.3556
573.15	26.445	2812.1	3076.5	9.2813	5.284	2811.3	3075.5	8.5373
673.15	31.063	2968.9	3279.6	9.6077	6.209	2968.5	3278.9	8.8642
773.15	35.679	3132.3	3489.1	9.8978	7.134	3132.0	3488.7	9.1546
873.15	40.295	3302.5	3705.4	10.1608	8.057	3302.2	3705.1	9.4178
973.15	44.911	3479.6	3928.7	10.4028	8.981	3479.4	3928.5	9.6599
1073.15	49.526	3663.8	4159.0	10.6281	9.904	3663.6	4158.9	9.8852
1173.15	54.141	3855.0	4396.4	10.8396	10.828	3854.9	4396.3	10.0967
1273.15	58.757	4053.0	4640.6	11.0393	11.751	4052.9	4640.5	10.2964
1373.15	63.372	4257.5	4891.2	11.2287	12.674	4257.4	4891.1	10.4859
1473.15	67.987	4467.9	5147.8	11.4091	13.597	4467.8	5147.7	10.6662
1573.15	72.602	4683.7	5409.7	11.5811	14.521	4683.6	5409.6	10.8382
	P = 100 kPa (Sat. Temp = 372.96 K)				P = 200 kPa (Sat. Temp = 393.38 K)			
Sat.	1.6940	2506.1	2675.5	7.3594	0.8857	2529.5	2706.7	7.1272
373.15	1.6958	2506.7	2676.2	7.3614				
423.15	1.9364	2582.8	2776.4	7.6134	0.9596	2576.9	2768.8	7.2795
473.15	2.1720	2658.1	2875.3	7.8343	1.0803	2654.4	2870.5	7.5066
523.15	2.4060	2733.7	2974.3	8.0333	1.1988	2731.2	2971.0	7.7086
573.15	2.6390	2810.4	3074.3	8.2158	1.3162	2808.6	3071.8	7.8926
673.15	3.1030	2967.9	3278.2	8.5435	1.5493	2966.7	3276.6	8.2218
773.15	3.5650	3131.6	3488.1	8.8342	1.7814	3130.8	3487.1	8.5133
873.15	4.0280	3301.9	3704.7	9.0976	2.0130	3301.4	3704.0	8.7770
973.15	4.4900	3479.2	3928.2	9.3398	2.2440	3478.8	3927.6	9.0194
1073.15	4.9520	3663.5	4158.6	9.5652	2.4750	3663.1	4158.2	9.2449
1173.15	5.4140	3854.8	4396.1	9.7767	2.7060	3854.5	4395.8	9.4566
1273.15	5.8750	4052.8	4640.3	9.9764	2.9370	4052.5	4640.0	9.6563
1373.15	6.3370	4257.3	4891.0	10.1659	3.1680	4257.0	4890.7	9.8458
1473.15	6.7990	4467.7	5147.6	10.3463	3.3990	4467.5	5147.3	10.0262
1573.15	7.2600	4683.5	5409.5	10.5183	3.6300	4683.2	5409.3	10.1982

(Contd.)

Table A.4 Thermodynamic properties of superheated steam (Contd.)

	V	*U*	*H*	*S*	*V*	*U*	*H*	*S*
	P = 300 kPa (Sat. Temp = 406.7 K)				*P* = 400 kPa (Sat. Temp = 416.78 K)			
Sat.	0.6058	2543.6	2725.3	6.9919	0.4625	2553.6	2738.6	6.8959
423.15	0.6339	2570.8	2761.0	7.0778	0.4708	2564.5	2752.8	6.9299
473.15	0.7163	2650.7	2865.6	7.3115	0.5342	2646.8	2860.5	7.1706
523.15	0.7964	2728.7	2967.6	7.5166	0.5951	2726.1	2964.2	7.3789
573.15	0.8753	2806.7	3069.3	7.7022	0.6548	2804.8	3066.8	7.5662
673.15	1.0315	2965.6	3275.0	8.0330	0.7726	2964.4	3273.4	7.8985
773.15	1.1867	3130.0	3486.0	8.3251	0.8893	3129.2	3484.9	8.1913
873.15	1.3414	3300.8	3703.2	8.5892	1.0055	3300.2	3702.4	8.4558
973.15	1.4957	3478.4	3927.1	8.8319	1.1215	3477.9	3926.5	8.6987
1073.15	1.6499	3662.9	4157.8	9.0576	1.2372	3662.4	4157.3	8.9244
1173.15	1.8041	3854.2	4395.4	9.2692	1.3529	3853.9	4395.1	9.1362
1273.15	1.9581	4052.3	4639.7	9.4690	1.4685	4052.0	4639.4	9.3360
1373.15	2.1121	4256.8	4890.4	9.6585	1.5840	4256.5	4890.2	9.5256
1473.15	2.2661	4467.2	5147.1	9.8389	1.6996	4467.0	5146.8	9.7060
1573.15	2.4201	4683.0	5409.0	10.0110	1.8151	4682.8	5408.8	9.8780
	P = 500 kPa (Sat. Temp = 425.01 K)				*P* = 600 kPa (Sat. Temp = 432 K)			
Sat.	0.3749	2561.2	2748.7	6.8213	0.3157	2567.4	2756.8	6.7600
473.15	0.4249	2642.9	2855.4	7.0592	0.3520	2638.9	2850.1	6.9665
523.15	0.4744	2723.5	2960.7	7.2709	0.3938	2720.9	2957.2	7.1816
573.15	0.5226	2802.9	3064.2	7.4599	0.4344	2801.0	3061.6	7.3724
623.15	0.5701	2882.6	3167.7	7.6329	0.4742	2881.2	3165.7	7.5464
673.15	0.6173	2963.2	3271.9	7.7938	0.5137	2962.1	3270.3	7.7079
773.15	0.7109	3128.4	3483.9	8.0873	0.5920	3127.6	3482.8	8.0021
873.15	0.8041	3299.6	3701.7	7.3522	0.6697	3299.1	3700.9	8.2674
973.15	0.8969	3477.5	3925.9	8.5952	0.7472	3477.0	3925.3	8.5107
1073.15	0.9896	3662.1	4156.9	8.8211	0.8245	3661.8	4156.5	8.7367
1173.15	1.0822	3853.6	4394.7	9.0329	0.9017	3853.4	4394.4	8.9486
1273.15	1.1747	4051.8	4639.1	9.2328	0.9788	4051.5	4638.8	9.1485
1373.15	1.2672	4256.3	4889.9	9.4224	1.0559	4256.1	4889.6	9.3381
1473.15	1.3596	4466.8	5146.6	9.6029	1.1330	4466.5	5146.3	9.5185
1573.15	1.4521	4682.5	5408.6	9.7749	1.2101	4682.3	5408.3	9.6906

(Contd.)

Table A.4 Thermodynamic properties of superheated steam (Contd.)

	V	U	H	S	V	U	H	S
	P = 800 kPa (Sat. Temp = 443.58 K)				P = 1000 kPa (Sat. Temp = 453.06 K)			
Sat.	0.2404	2576.8	2769.1	6.6628	0.19444	2583.6	2778.1	6.5865
473.15	0.2608	2630.6	2839.3	6.8158	0.2060	2621.9	2827.9	6.6940
523.15	0.2931	2715.5	2950.0	7.0384	0.2327	2709.9	2942.6	6.9247
573.15	0.3241	2797.2	3056.5	7.2328	0.2579	2793.2	3051.2	7.1229
623.15	0.3544	2878.2	3161.7	7.4089	0.2825	2875.2	3157.7	7.3011
673.15	0.3843	2959.7	3267.1	7.5716	0.3066	2957.3	3263.9	7.4651
773.15	0.4433	3126.0	3480.6	7.8673	0.3541	3124.4	3478.5	7.7622
873.15	0.5018	3297.7	3699.4	8.1333	0.4011	3296.8	3697.9	8.0290
973.15	0.5601	3476.2	3924.2	8.3770	0.4478	3475.3	3923.1	8.2731
1073.15	0.6181	3661.1	4155.6	8.6033	0.4943	3660.4	4154.7	8.4996
1173.15	0.6761	3852.8	4393.7	8.8153	0.5407	3852.2	4392.9	8.7118
1273.15	0.7340	4051.0	4638.2	9.0153	0.5871	4050.5	4637.6	8.9119
1373.15	0.7919	4255.6	4889.1	9.2050	0.6335	4255.1	4888.6	9.1017
1473.15	0.8497	4466.1	5145.9	9.3855	0.6798	4465.6	5145.4	9.2822
1573.15	0.9076	4681.8	5407.9	9.5575	0.7261	4681.3	5407.4	9.4543
	P = 1200 kPa (Sat. Temp = 461.14 K)				P = 1400 kPa (Sat. Temp = 468.22 K)			
Sat.	0.16333	2588.8	2784.8	6.5233	0.14084	2592.8	2790.0	6.4693
473.15	0.16930	2612.8	2815.9	6.5898	0.14302	2603.1	2803.3	6.4975
523.15	0.19234	2704.2	2935.0	6.8294	0.16350	2698.3	2927.2	6.7467
573.15	0.2138	2789.2	3045.8	7.0317	0.18228	2785.2	3040.4	6.9534
623.15	0.2345	2872.2	3153.6	7.2121	0.2003	2869.2	3149.5	7.1360
673.15	0.2548	2954.9	3260.7	7.3774	0.2178	2952.5	3257.5	7.3026
773.15	0.2946	3122.8	3476.3	7.6759	0.2521	3121.1	3474.1	7.6027
873.15	0.3339	3295.6	3696.3	7.9435	0.2860	3294.4	3694.8	7.8710
973.15	0.3729	3474.4	3922.0	8.1881	0.3195	3473.6	3920.8	8.1160
1073.15	0.4118	3659.7	4153.8	8.4148	0.3528	3659.0	4153.0	8.3431
1173.15	0.4505	3851.6	4392.2	8.6272	0.3861	3851.1	4391.5	8.5556
1273.15	0.4892	4050.0	4637.0	8.8274	0.4192	4049.5	4636.4	8.7559
1373.15	0.5278	4254.6	4888.0	9.0172	0.4524	4254.1	4887.5	8.9457
1473.15	0.5665	4465.1	5144.9	9.1977	0.4855	4464.7	5144.4	9.1262
1573.15	0.6051	4680.9	5407.0	9.3698	0.5186	4680.4	5406.5	9.2984

(Contd.)

Table A.4 Thermodynamic properties of superheated steam (Contd.)

	V	U	H	S	V	U	H	S
	P = 1600 kPa (Sat. Temp = 474.56 K)				P = 1800 kPa (Sat. Temp = 480.3 K)			
Sat.	0.12380	2596.0	2794.0	6.4218	0.11042	2598.4	2797.1	6.3794
498.15	0.13287	2644.7	2857.3	6.5518	0.11673	2636.6	2846.7	6.4808
523.15	0.14184	2692.3	2919.2	6.6732	0.12497	2686.0	2911.0	6.6066
573.15	0.15862	2781.1	3034.8	6.8844	0.14021	2776.9	3029.2	6.8226
623.15	0.17456	2866.1	3145.4	7.0694	0.15457	2863.0	3141.2	7.0100
673.15	0.19005	2950.1	3254.2	7.2374	0.16847	2947.7	3250.9	7.1794
773.15	0.2203	3119.5	3472.0	7.5390	0.19550	3117.9	3469.8	7.4825
873.15	0.2500	3293.3	3693.2	7.8080	0.2220	3292.1	3691.7	7.7523
973.15	0.2794	3472.7	3919.7	8.0535	0.2482	3471.8	3918.5	7.9983
1073.15	0.3086	3658.3	4152.1	8.2808	0.2742	3657.6	4151.2	8.2258
1173.15	0.3377	3850.5	4390.8	8.4935	0.3001	3849.9	4390.1	8.4386
1273.15	0.3668	4049.0	4635.8	8.6938	0.3260	4048.5	4635.2	8.6391
1373.15	0.3958	4253.7	4887.0	8.8837	0.3518	4253.2	4886.4	8.8290
1473.15	0.4248	4464.2	5143.9	9.0643	0.3776	4463.7	5143.4	9.0096
1573.15	0.4538	4679.9	5406.0	9.2364	0.4034	4679.5	5405.6	9.1818
	P = 2000 kPa (Sat. Temp = 487.57 K)				P = 2500 kPa (Sat. Temp = 497.14 K)			
Sat.	0.09963	2600.3	2799.5	6.3409	0.07998	2603.1	2803.1	6.2575
498.15	0.10377	2628.3	2835.8	6.4147	0.08027	2605.6	2806.3	6.2639
523.15	0.11144	2679.6	2902.5	6.5453	0.08700	2662.6	2880.1	6.4085
573.15	0.12547	2772.6	3023.5	6.7664	0.09890	2761.6	3008.8	6.6438
623.15	0.13857	2859.8	3137.0	6.9563	0.10976	2851.9	3126.3	6.8403
673.15	0.15120	2945.2	3247.6	7.1271	0.12010	2939.1	3239.3	7.0148
723.15	0.16342	3031.0	3357.8	7.2859	0.13014	3025.5	3350.8	7.1746
773.15	0.17568	3116.2	3467.6	7.4317	0.13998	3112.1	3462.1	7.3234
873.15	0.19960	3290.9	3690.1	7.7024	0.15930	3288.0	3686.3	7.5960
973.15	0.22320	3470.9	3917.4	7.9487	0.17832	3468.7	3914.5	7.8435
1073.15	0.24670	3657.0	4150.3	8.1765	0.19716	3655.3	4148.2	8.0720
1173.15	0.27000	3849.3	4389.4	8.3895	0.21590	3847.9	4387.6	8.2853
1273.15	0.29330	4048.0	4634.6	8.5901	0.2346	4046.7	4633.1	8.4861
1373.15	0.31660	4252.7	4885.9	8.7800	0.2532	4251.5	4884.6	8.6762
1473.15	0.33980	4463.3	5142.9	8.9607	0.2718	4462.1	5141.7	8.8569
1573.15	0.36310	4679.0	5405.1	9.1329	0.2905	4677.8	5404.0	9.0291

(Contd.)

Table A.4 Thermodynamic properties of superheated steam (Contd.)

	V	*U*	*H*	*S*	*V*	*U*	*H*	*S*
	P = 3000 kPa (Sat. Temp = 507.05 K)				*P* = 3500 kPa (Sat. Temp = 515.75 K)			
Sat.	0.06668	2604.1	2804.2	6.1869	0.05707	2603.7	2803.4	6.1253
523.15	0.07058	2644.0	2855.8	6.2872	0.05872	2623.7	2829.2	6.1749
573.15	0.08114	2750.1	2993.5	6.5390	0.06842	2738.0	2977.5	6.4461
623.15	0.09053	2843.7	3115.3	6.7428	0.07678	2835.3	3104.0	6.6579
673.15	0.09936	2932.8	3230.9	6.9212	0.08453	2928.4	3222.3	6.8405
723.15	0.10787	3020.4	3344.0	7.0834	0.09196	3015.3	3337.2	7.0052
773.15	0.11619	3108.0	3456.5	7.2338	0.09918	3103.0	3450.9	7.1572
873.15	0.13243	3285.0	3682.3	7.5085	0.11324	3282.1	3678.4	7.4339
973.15	0.14838	3466.5	3911.7	7.7571	0.12699	3464.3	3908.8	7.6837
1073.15	0.16414	3653.5	4145.9	7.9862	0.14056	3651.8	4143.7	7.9134
1173.15	0.17980	3846.5	4385.9	8.1999	0.15402	3845.0	4384.1	8.1276
1273.15	0.19541	4045.4	4631.6	8.4009	0.16743	4044.1	4630.1	8.3288
1373.15	0.21098	4250.3	4883.3	8.5912	0.18080	4249.2	4881.9	8.5192
1473.15	0.22652	4460.9	5140.5	8.7720	0.19415	4459.8	5139.3	8.7000
1573.15	0.24206	4676.6	5402.8	8.9442	0.20749	4675.5	5401.7	8.8723
	P = 4000 kPa (Sat. Temp = 523.55 K)				*P* = 4500 kPa (Sat. Temp = 530.64 K)			
Sat.	0.04978	2602.3	2801.4	6.0701	0.04406	2600.1	2798.3	6.0198
548.15	0.05457	2667.9	2886.2	6.2285	0.04730	2650.3	2863.2	6.1401
573.15	0.05884	2725.3	2960.7	6.3615	0.05135	2712.0	2943.1	6.2828
623.15	0.06645	2826.7	3092.5	6.5821	0.05840	2817.8	3080.6	6.5131
673.15	0.07341	2919.9	3213.6	6.7690	0.06475	2913.3	3204.7	6.7047
723.15	0.08002	3010.2	3330.3	6.9363	0.07074	3005.0	3323.3	6.8746
773.15	0.08643	3099.5	3445.3	7.0900	0.07651	3095.3	3439.6	7.0301
873.15	0.09885	3279.1	3674.4	7.3688	0.08765	3276.0	3670.5	7.3110
973.15	0.11095	3462.1	3905.9	7.6198	0.09847	3459.9	3903.0	7.5631
1073.15	0.12287	3650.0	4141.5	7.8502	0.10911	3648.3	4139.3	7.7942
1173.15	0.13469	3843.6	4382.3	8.0647	0.11965	3842.2	4380.6	8.0091
1273.15	0.14645	4042.9	4628.7	8.2662	0.13013	4041.6	4627.2	8.2108
1373.15	0.15817	4248.0	4880.6	8.4567	0.14056	4246.8	4879.3	8.4015
1473.15	0.16987	4458.6	5138.1	8.6376	0.15098	4457.5	5136.9	8.5825
1573.15	0.18156	4674.3	5400.5	8.8100	0.16139	4673.1	5399.4	8.7549

(Contd.)

Table A.4 Thermodynamic properties of superheated steam (Contd.)

	V	U	H	S	V	U	H	S
	P = 5000 kPa (Sat. Temp = 537.14 K)				P = 6000 kPa (Sat. Temp = 548.79 K)			
Sat.	0.03944	2597.1	2794.3	5.9734	0.03244	2589.7	2784.3	5.8892
548.15	0.04141	2631.3	2838.3	6.0544				
573.15	0.04532	2698.0	2924.5	6.2084	0.03616	2667.2	2884.2	6.0674
623.15	0.05194	2808.7	3068.4	6.4493	0.04223	2789.6	3043.0	6.3335
673.15	0.05781	2906.6	3195.7	6.6459	0.04739	2892.9	3177.2	6.5408
723.15	0.06330	2999.7	3316.2	6.8186	0.05214	2988.9	3301.8	6.7193
773.15	0.06857	3091.0	3433.8	6.9759	0.05665	3082.2	3422.2	6.8803
823.15	0.07360	3181.0	3549.0	7.1215	0.06101	3174.6	3540.6	7.0288
873.15	0.07869	3273.0	3666.5	7.2589	0.06525	3266.9	3658.4	7.1677
973.15	0.08849	3457.6	3900.1	7.5122	0.07352	3453.1	3894.2	7.4234
1073.15	0.09811	3646.6	4137.1	7.7440	0.08160	3643.1	4132.7	7.6566
1173.15	0.10762	3840.7	4378.8	7.9593	0.08958	3837.8	4375.3	7.8727
1273.15	0.11707	4040.4	4625.7	8.1612	0.09749	4037.8	4622.7	8.0751
1373.15	0.12648	4245.6	4878.0	8.3520	0.10536	4243.3	4875.4	8.2661
1473.15	0.13587	4456.3	5135.7	8.5331	0.11321	4454.0	5133.3	8.4474
1573.15	0.14526	4672.0	5398.2	8.7055	0.12106	4669.6	5396.0	8.6199
	P = 7000 kPa (Sat. Temp = 559.03 K)				P = 8000 kPa (Sat. Temp = 568.21 K)			
Sat.	0.02737	2580.5	2772.1	5.8133	0.02352	2569.8	2758.0	5.7432
573.15	0.02947	2632.2	2838.4	5.9305	0.02426	2590.9	2785.0	5.7906
623.15	0.03524	2769.4	3016.0	6.2283	0.02995	2747.7	2987.3	6.1301
673.15	0.03993	2878.6	3158.1	6.4478	0.03432	2863.8	3138.3	6.3634
723.15	0.04416	2978.0	3287.1	6.6327	0.03817	2966.7	3272.0	6.5551
773.15	0.04814	3073.4	3410.3	6.7975	0.04175	3064.3	3398.3	6.7240
823.15	0.05195	3167.2	3530.9	6.9486	0.04516	3159.8	3521.0	6.8778
873.15	0.05565	3260.7	3650.3	7.0894	0.04845	3254.4	3642.0	7.0206
973.15	0.06283	3448.5	3888.3	7.3476	0.05481	3443.9	3882.4	7.2812
1073.15	0.06981	3639.5	4128.2	7.5822	0.06097	3636.0	4123.8	7.5173
1173.15	0.07669	3835.0	4371.8	7.7991	0.06702	3832.1	4368.3	7.7351
1273.15	0.08350	4035.3	4619.8	8.0020	0.07301	4032.8	4616.9	7.9384
1373.15	0.09027	4240.9	4872.8	8.1933	0.07896	4238.6	4870.3	8.1300
	P = 10000 kPa (Sat. Temp = 584.11 K)				P = 12000 kPa (Sat. Temp = 597.8 K)			
Sat.	0.01803	2544.4	2724.7	5.6141	0.01428	2517.8	2689.2	5.5002
623.15	0.02242	2699.2	2923.4	5.9443	0.01715	2640.6	2846.4	5.7578
673.15	0.02641	2832.4	3096.5	6.2120	0.02108	2801.8	3054.8	6.0810
723.15	0.02975	2943.4	3240.9	6.4190	0.02411	2921.7	3211.0	6.2944
773.15	0.03279	3045.8	3373.7	6.5966	0.02679	3028.2	3349.6	6.4906
823.15	0.03564	3144.6	3500.9	6.7561	0.02925	3128.5	3479.5	6.6532
873.15	0.03837	3241.7	3625.3	6.9029	0.03160	3226.4	3605.7	6.8022
923.15	0.04101	3338.2	3748.2	7.0398	0.03387	3323.8	3730.2	6.9409

(*Contd.*)

Table A.4 Thermodynamic properties of superheated steam (Contd.)

	V	U	H	S	V	U	H	S
	P = 14000 kPa (Sat. Temp = 609.79 K)				P = 15000 kPa (Sat. Temp = 615.28 K)			
Sat.	0.01150	2481.4	2642.4	5.3803	0.01034	2455.5	2610.5	5.3098
623.15	0.01311	2563.4	2746.9	5.5492	0.01147	2520.4	2692.4	5.4421
673.15	0.01723	2764.4	3005.6	5.9513	0.01565	2740.7	2975.5	5.8811
723.15	0.02007	2896.0	3177.0	6.1969	0.01845	2879.5	3156.2	6.1404
773.15	0.02251	3008.7	3323.8	6.3937	0.02080	2996.6	3308.6	6.3443
823.15	0.02472	3112.7	3458.8	6.5627	0.02293	3104.7	3448.6	6.5199
873.15	0.02680	3213.2	3588.5	6.7159	0.02491	3208.6	3582.3	6.6776
923.15	0.02880	3312.5	3715.6	6.8575	0.02680	3310.3	3712.3	6.8224

Specific volume	in m^3/kg
Internal energy	in kJ/kg
Enthalpy	in kJ/kg
Entropy	in kJ/kg K

Objective Type Questions

1. Which of the following is not equivalent to 1 standard atmosphere pressure?
   (a) 1.01325 bar (b) 1.01325 Pa
   (c) 760 mm Hg (d) 1.01325×10^5 N/m^2
2. Pa·s is a unit of
   (a) Pressure (b) Power
   (c) Thermal conductivity (d) Viscosity
3. Pressure head is defined as
   (a) Absolute pressure/density
   (b) Gauge pressure + atmospheric pressure
   (c) Absolute pressure × density
   (d) acceleration due to gravity × density × height of liquid column
4. Thermal conductivity of material A is 40 W/m K, that of material B is 40 kcal/(m h °C) and that of material C is 40 Btu/(ft h °F). Which material is a better conductor of heat in comparison with others?
   (a) Material A (b) Material B
   (c) Material C (d) Both materials A and B
5. Only one of the following statements is correct regarding g_c, the Newton's law conversion factor. Spot the correct statement.
   (a) It is a dimensionless constant.
   (b) It is numerically equal to g, the local acceleration due to gravity.
   (c) Weight of a body is equal to the product of mass and g_c.
   (d) Numerical values of g_c are different for different system of units.
6. One Btu of heat is supplied to one kilogram of water. What will be the temperature rise for water?
   (a) 0.252°C (b) 1°C
   (c) 3.97°C (d) 4.186°C

7. If the temperature of water rises by 2°C when one kJ heat is supplied to it, how much water is being heated?
(a) 0.153 kg (b) 0.263 lb
(c) 200 g (d) 4.186 kg

8. The SI system uses bar as a unit of pressure. Which one of the following is equivalent to 1 bar?
(a) 1.01325 atm (b) 1.01325×10^5 N/m^2
(c) 100 kPa (d) 101.325 kPa

9. If the velocity v of a fluid is related to the pressure drop ΔP and density ρ of the fluid by $v = k\sqrt{\dfrac{\Delta P}{\rho}}$, where k is a constant, what are the units of k in SI system?
(a) m s^{-1} (b) s^{-1}
(c) kg$^{0.5}$ Pa$^{-0.5}$ m$^{-1.5}$ (d) None of the above

10. The dependence of thermal conductivity of a material on temperature is given by an empirical equation $k = ae^{b/T}$ where a and b are constants, and k is in W/m K and T is in K. Spot the true statement from the following set.
(a) a and b have the same dimension as k.
(b) a has the dimension of k and b is dimensionless.
(c) a and b are dimensionless.
(d) a has the dimension of k and b has the dimension of T.

11. How many moles are there in 500 g of oxygen?
(a) 31.25 (b) 15.625
(c) 9.41×10^{24} (d) 22.4×10^3

12. Two bottles A and B are filled with 100 g each of HNO_3 and H_2SO_4 respectively. Which one of the following statements is true?
(a) Bottle A contains more molecules.
(b) Bottle B contains more molecules.
(c) Bottles A and B contain the same number of molecules.
(d) Bottles A and B contain the same number of moles of the liquids.

13. How many kilogram of CS_2 will contain 42 kg of carbon?
(a) 3.5×10^3 (b) 266.45
(c) 76.13 (d) None of the above

14. The molecular formula of an organic compound is $C_{10}H_7Br$ (molecular weight 206.96). The weight percentages of carbon, hydrogen and bromine in the compound are:
(a) 55.56% C, 38.89% H and 5.55% Br
(b) 71.43% C, 25.00% H and 3.57% Br
(c) 58.00% C, 3.40% H and 38.60% Br
(d) 40.83% C, 4.80% H and 54.37% Br

15. A mole of a compound contains
(a) One molecule of the substance (b) One atom of the substance
(c) 6.023×10^{23} molecules (d) 22.4×10^3 molecules

16. Molecular weight of an ideal gas is 44. Which of the following statements is not true about the gas?
(a) 44 g of the gas contains 6.023×10^{23} molecules.
(b) The mass of one molecule is 44 g.
(c) The molar volume of the gas is 22.4×10^{-3} m^3/mol at STP.
(d) The number of molecules in 44 g of the gas is the same as number of atoms in 12 g C.

17. The molar composition of a gas is 10% H_2, 10% O_2, 30% CO_2 and balance H_2O. If 50% H_2O condenses, the final mole percent of H_2 in the gas on a dry basis will be
(a) 10% (b) 5%
(c) 18.18% (d) 20%

18. Six kg of NaCl is mixed with 2 kg of KCl. The atomic weights of Na, K and Cl are respectively, 23, 39.1 and 35.5. What is the mole percent of NaCl in the mixture?
(a) 75 (b) 70.2
(c) 79.3 (d) 81.5

19. A solution of KOH in water has a molarity of 7.0 and a KOH content of 30.21% (weight). What is the density of the solution?
(a) 900 kg/m^3 (b) 1200 kg/m^3
(c) 1300 kg/m^3 (d) 1288 kg/m^3

20. Two effluent streams are mixed. One stream contains 10% salt and the other none. The combined stream contains 2% salt. The ratio of the two streams are
(a) 1 : 4 (b) 1 : 5
(c) 1 : 2 (d) 1 : 8

21. One hundred kilomoles of ammonia–air mixture containing 10% (volume) ammonia is scrubbed with water to recover 80% of ammonia in the entering gas. Air is assumed insoluble in water. What is the mole percent of ammonia in the gas leaving the scrubber?
(a) 0.2 (b) 20
(c) 1.9 (d) 2.2

22. An evaporator while concentrating an aqueous solution from 10% to 40% solids evaporates 30 000 kg of water. The amount of solids handled by the system in kg is
(a) 4000 (b) 9000
(c) 4600 (d) 3000

23. 1000 kg of wet solids are to be dried from 60% to 20% moisture. The mass of water removed in kilograms is
(a) 520 (b) 200
(c) 400 (d) 500

24. Pure oxygen is mixed with air to produce an enriched air containing 50 volume % of O_2. The ratio of moles of air to oxygen is
(a) 1.72 (b) 0.58
(c) 0.5 (d) 0.2

25. One thousand kilograms of a salt solution containing 50% salt is cooled. 400 kg of anhydrous salt is separated out. The solubility of the salt at the lower temperature in kg/100 kg of water is

(a) 80 (b) 50
(c) 40 (d) 20

26. The weight fraction of methanol in an aqueous solution is 0.64. The mole fraction of methanol (x_M) satisfies

(a) $x_M < 0.5$ (b) $x_M = 0.5$
(c) $0.5 < x_M < 0.64$ (d) $x_M \geq 0.5$

27. Pure *A* in the gas phase enters a reactor. 50% of this *A* is converted to *B* through the reaction $A \rightarrow 3B$. The mole fraction of *A* in the exit stream is

(a) 1/2 (b) 1/3
(c) 1/4 (d) 1/5

28. Assume that 50% of water present in a wet material is removed by drying. Which one of the following statements is true?

(a) The mass fraction of water in the solids leaving is half the mass fraction of water in the entering solid.
(b) The mass ratio of water in the solids leaving is half the mass ratio of water in the entering solid.
(c) The mass fraction and the mass ratio of water in the solid leaving the dryer are half of the respective values in the feed.
(d) The weight percent of water in the solids leaving the dryer is 25%.

29. Two solids of same volume, one of steel and the other of stone, are immersed in water. Which one will suffer a greater loss of weight?

(a) Steel (b) Stone
(c) Equal for both (d) Depends on their mass

30. A body floats in water with three-fourths of it submerged. What is the specific gravity of the body?

(a) 0.25 (b) 0.33
(c) 0.75 (d) 1.25

31. A test tube carrying some lead balls sinks to a depth of 0.15 m when placed in water. The same tube sinks only to a depth of 0.10 m when placed in a liquid. What is the specific gravity of the liquid?

(a) 1.5 (b) 0.75
(c) 2.0 (d) 1.33

32. A metallic spherical ball of radius 0.05 m weighs 0.850 kg. What is its density?

(a) 1800.27 kg/m^3 (b) 1623.4 kg/m^3
(c) 1700 kg/m^3 (d) 850.00 kg/m^3

33. A body weighs 1 kg in air and 0.85 kg in water. What would be the weight of the object in a liquid of specific gravity 0.8?

(a) 0.8 kg (b) 1.15 kg
(c) 0.88 kg (d) 0.78 kg

34. Which of the following statements is not true for a mixture of ideal gases?
(a) Total pressure is equal to the sum of the partial pressures.
(b) Partial pressure is equal to the product of mole fraction and the total pressure.
(c) Molar volume is equal to 22.4 m^3/kmol.
(d) Mole % of the components equal the volume %.

35. The specific gravity of a hydrocarbon oil is 0.903 at 288.8 K. What is the corresponding value on the Baumé scale?
(a) 30°Bé (b) 29°Bé
(c) 25°Bé (d) –16°Bé

36. An aqueous solution of 2.45% by weight H_2SO_4 has a specific gravity of 1.011. The composition expressed in normality is
(a) 0.2500 (b) 0.2528
(c) 0.5000 (d) 0.5055

37. A hydrocarbon oil is rated at 30°API. What is its specific gravity at 288.8 K?
(a) 1.235 (b) 0.876
(c) 0.300 (d) 0.675

38. What is the porosity of a solid if its bulk density is 1125 kg/m^3 and the true density is 1500 kg/m^3?
(a) 0.25 (b) 0.75
(c) 1.33 (d) 0.33

39. The molarity of water in pure water is
(a) 1 (b) 18
(c) 55.55 (d) Infinity

40. An aqueous solution of NaCl contains 20% NaCl. The density of the solution is 1160 kg/m^3. 500 mL of water of density 1000 kg/m^3 is added to one litre of the solution. What is the molality of the resulting solution?
(a) 2.779 mol/kg (b) 2.644 mol/L
(c) 2.779 mol/L (d) 2.644 mol/kg

41. A 10% solution of an organic compound in benzene has a molality of 0.62. The molecular weight of the compound is
(a) 62 (b) 179.2
(c) 111.1 (d) 620

42. A sample of well water contains 140 g/m^3 of Ca^{2+} ions and 345 g/m^3 of Na^+ ions. The hardness of the sample of water expressed in terms of equivalent $CaCO_3$ in g/m^3 (assume atomic masses of Ca: 40, Na: 23, C: 12, and O: 16) is
(a) 350 (b) 485
(c) 140 (d) 345

43. The concentration of SO_2 in the flue gases from a boiler is found to be 0.2 kg/m^3 at STP. The concentration of SO_2 in parts per million (by volume) is
(a) 70 000 (b) 20 000
(c) 200 (d) 45 000

44. A body when immersed in liquid A experiences a greater loss in weight compared to that in liquid B. Which one of the following is true?

(a) Liquid A is lighter than liquid B.
(b) Liquid B is lighter than liquid A.
(c) Body is lighter than liquids A and B.
(d) Body is lighter than A, heavier than B.

45. A solution of specific gravity 1.0 consists of 35% A by weight and the remaining B. If the specific gravity of A is 0.7, the specific gravity of B is

(a) 1.25 (b) 1.3
(c) 1.35 (d) 1.2

46. The loss of weight of a body in a liquid depends on

(a) The volume of the body and the specific gravity of the liquid
(b) The mass of the body and the specific gravity of the liquid
(c) The volume and specific gravity of the body
(d) The volume of the body only

47. The analysis of a gas entering the secondary converter of a contact sulphuric acid plant is 4% SO_2, 13% O_2 and 83% N_2 (volume %). In the converter, SO_2 is oxidized to SO_3. The gases leaving the converter contain 0.45% SO_2 in an SO_3-free basis. The percent conversion of SO_2 to SO_3 is

(a) 88.75% (b) 13.75%
(c) 45.00% (d) 89.35%

48. How much potassium chlorate ($KClO_3$) must be taken to produce the same amount of oxygen that will be produced by 25 kg of mercuric oxide (HgO)? (Atomic weight: Hg = 200.59, K = 39.10)

(a) 5.236 kg (b) 14.145 kg
(c) 4.715 kg (d) 44.186 kg

49. If 90 g of iron react with sulphuric acid, how many litres of hydrogen are liberated at STP? (Atomic weight of Fe: 55.85)

(a) 36.12 L (b) 0.0361 L
(c) 90 L (d) 1.62 L

50. A wet paper pulp contains 75% water. After 100 kg of water is removed in a dryer, it is found that the pulp is now containing 30% water. The weight of the original pulp is

(a) 200.36 kg (b) 150.00 kg
(c) 222.20 kg (d) 155.56 kg

51. It is required to make 100 kg of a solution containing 40% salt by mixing solution A containing 25% salt and solution B containing 50% salt. The mass in kilograms of solution A required is

(a) 40 kg (b) 60 kg
(c) 75 kg (d) 25 kg

52. Pure carbon is completely burnt in oxygen. The flue gas analysis is 70% CO_2, 20% CO and 10% O_2. The percent excess oxygen used is
(a) 20 (b) 12.5
(c) 0 (d) 10

53. The gaseous reaction $A \rightarrow 2B + C$ takes place in a constant volume reactor starting with a mixture of 75% A and the rest inert materials. What would be the conversion of A when the pressure is double the original pressure?
(a) 50% (b) 66.7%
(c) 75% (d) 100%

54. The fraction of the total pressure exerted by oxygen in a mixture containing equal weights of methane and oxygen is
(a) 1/3 (b) 1/2
(c) 1/4 (d) 3/4

55. Assuming that CO_2 obeys perfect gas law, calculate the density of CO_2 at 540 K and 202 kPa, and express it as kilograms per cubic metre.
(a) 1 (b) 2
(c) 3 (d) 4

56. 1.2 g-atom of carbon and 1.5 moles of oxygen are reacted to give one mole of carbon dioxide. The limiting reactant is
(a) Carbon (b) Oxygen
(c) Both (d) None

57. Forty five moles of oxygen is supplied for the combustion of 36 g-atom of carbon. The percent excess reactant supplied is
(a) 80% (b) 33.3%
(c) 30% (d) 25%

58. A gaseous reaction $A \rightarrow 2B + C$ takes place isothermally in a constant pressure reactor. Starting with a gaseous mixture containing 50% A and the rest inert materials, the ratio of final to initial volume is found to be 1.8. The percent conversion of A is
(a) 80 (b) 50
(c) 60 (d) 74

59. Methane is completely burnt with air. The maximum possible volume percent of carbon dioxide (on a dry basis) in the flue gas is
(a) 11.7 (b) 21.0
(c) 44.0 (d) 28.0

60. H_2S is produced from the following reaction:

$$FeS + 2HCl \rightarrow FeCl_2 + H_2S$$

120 kg of FeS react with 150 kg of HCl and 0.5 kmol of H_2S has been produced. (Atomic weight, Fe = 55.85). The degree of completion of the reaction is
(a) 12.2% (b) 14.2%
(c) 36.6% (d) None of the above

61. Iron pyrites contains 75% FeS_2 by weight. What is the iron content in the ore?
(a) 46.55%
(b) 35%
(c) 75%
(d) 53.45%

62. The molecular weight of an ideal gas is 40. What will be the volume occupied by 0.20 kg of this gas at STP?
(a) 22.4 L
(b) 22.4 m^3
(c) 112 L
(d) 4.48 m^3

63. CO_2 is contained in a closed rigid container. If it is heated to a temperature twice its original value what will be the pressure of the gas?
(a) Twice the original pressure
(b) Half the original pressure
(c) Same as the initial pressure
(d) Cannot be predicted

64. How many molecules of nitrogen will be present in 28 g of nitrogen at standard temperature and pressure?
(a) 1 molecule
(b) 2 molecules
(c) 6.023×10^{23} molecules
(d) 12.046×10^{23} molecules

65. What is the volume of oxygen at STP required for the combustion of 5 L carbon monoxide at STP?
(a) 56 L
(b) 0.223 L
(c) 2.5 L
(d) 5 L

66. If on combustion, 4 kg of organic compound gives 11 kg of CO_2, the percentage of carbon in the compound will be
(a) 50
(b) 25
(c) 60
(d) 75

67. Forty litres of a mixture of methane and nitrogen requires 24 L of pure oxygen for complete combustion. The percentage of methane in the mixture is
(a) 40
(b) 25
(c) 30
(d) 60

68. 15 kg of oxygen and 15 kg of hydrogen are mixed together. Which one will have greater partial pressure in the resulting mixture?
(a) Hydrogen
(b) Oxygen
(c) Both have same partial pressure
(d) Partial pressure depends on their vapour pressure.

69. 100 m^3 of a gas mixture containing 20% CO_2 is sent to an absorber in which 90% of CO_2 is absorbed. The temperature and pressure of the gas entering are the same as those of the gas leaving. What will be the pure-component volume of CO_2 in the gas leaving?
(a) 2 m^3
(b) 80 m^3
(c) 82 m^3
(d) 100 m^3

70. Six grams of carbon is burnt with an amount of air containing 18 g of oxygen. The product contains 16.5 g of CO_2 and 2.8 g of CO besides other constituents. What is the degree of conversion on the basis of the disappearance of the limiting reactant?

(a) 100% (b) 95%
(c) 75% (d) 20%

71. Nitrogen dioxide associates and forms nitrogen tetroxide:

$$2NO_2 \rightarrow N_2O_4$$

One cubic metre of nitrogen dioxide at 100 kPa and 300 K is taken in a rigid closed container. If 29.5% of the gas associates to form N_2O_4, what will be the final pressure of the gas in the vessel at constant temperature?
(a) 70.5 kPa (b) 736 kPa
(c) 120 kPa (d) 85 kPa

72. How many kilograms of hydrogen can be stored at 325 K and 100 bar in a compressed gas cylinder that normally holds 0.5 kg of hydrogen at 300 K and 130 bar?
(a) 0.665 (b) 0.207
(c) 0.335 (d) 0.5

73. A tank contains oxygen gas at 1.40×10^7 N/m^2. On withdrawal of 140 L of oxygen measured at 1.01×10^5 N/m^2, the pressure in the tank is observed to fall to 1.30×10^5 N/m^2. Assume that there is no change in the temperature of the gas. What is the volume of the tank in m^3?
(a) 1.0 (b) 0.88×10^{-3}
(c) 1.02×10^{-3} (d) 0.860 m^3

74. A mixture of oxygen and sulphur dioxide is at 200 kPa. The average molecular weight of the mixture is 44.8. The partial pressure of oxygen in the mixture is
(a) 89.6 kPa (b) 120 kPa
(c) 101.3 kPa (d) 80 kPa

75. In an air-naphthalene ($C_{10}H_8$) mixture at 100 kPa and 300 K, the partial pressure of naphthalene is found to be 10 kPa. What volume of this mixture will contain 1 kg of naphthalene?
(a) 1.95 m^3 (b) 0.0195 m^3
(c) 3.25 m^3 (d) 249.6 m^3

76. N_2O_4 decomposes to NO_2 at high temperatures. A sample of N_2O_4 weighing 27.6 g, when heated to 300 K at 100 kPa, is found to occupy a volume of 0.012 m^3. Assuming the applicability of the ideal gas law, calculate the percent dissociation of N_2O_4 to NO_2.
(a) 78% (b) 75%
(c) 60% (d) 50%

77. A closed container of volume 0.02 m^3 contains a mixture of neon (atomic weight, 20) and argon (atomic weight, 40) gases, at a temperature of 300 K and a pressure of 1×10^5 N/m^2. The total mass of the mixture is 28 g. The gases may be assumed to be ideal. How many grams of argon are there in the container?
(a) 14 g (b) 24 g
(c) 4 g (d) 10 g

78. A mixture of ethane (C_2H_6) and ethylene (C_2H_4) occupies 6.5×10^{-2} m^3 at 1.0233 bar and 400 K. The mixture reacts completely with 0.2112 kg of O_2 to produce CO_2 and

H_2O. Assume the ideal gas behaviour. What is the volume percent of ethane in the mixture?

(a) 40% (b) 60%
(c) 50% (d) 75%

79. The weight ratio of carbon to hydrogen in a paraffinic hydrocarbon fuel is found to be 5.25. What is the chemical formula for the fuel?

(a) C_7H_{16} (b) C_6H_{14}
(c) C_5H_{12} (d) C_8H_{18}

80. The pressure of a real gas is less than the pressure of an ideal gas because in a real gas,

(a) The molecules have finite size.
(b) The number of molecular collisions is greater.
(c) The kinetic energy of molecules is greater.
(d) Intermolecular forces exist.

81. In the van der Waals equation of state, the term that accounts for intermolecular forces is

(a) $V - b$ (b) a/V^2
(c) RT (d) $(RT)^{-1}$

82. A gas above its critical temperature can be condensed to the liquid state by

(a) Reducing the temperature at constant pressure
(b) Increasing the pressure at constant temperature
(c) By reducing the temperature at constant pressure or by increasing the pressure at constant temperature
(d) Can not be condensed

83. Which one of the following statements regarding critical point of a substance is not true?

(a) It represents the condition of the highest temperature and pressure at which a gas and a liquid can exist in vapour–liquid equilibrium.
(b) It is an idealized concept and can never be attained in practice.
(c) The *PV*-isotherm at critical point exhibits a point of inflection.
(d) Liquid and vapour phases at the critical point have identical properties.

84. One of the following statements regarding compressibility factor *Z* is incorrect. Identify it.

(a) For real gases *Z* is greater than 1.
(b) For ideal gases *Z* is equal to 1.
(c) For real gases the value of *Z* depends on pressure and temperature of the gas.
(d) All gases have approximately the same *Z* when compared at the same reduced temperature and reduced pressure.

85. At 286 K and 139.2 bar, the compressibility factor of methane is found to be 0.8. At approximately what temperature and pressure, nitrogen gas will have the compressibility factor of 0.8? (The critical constants are T_c = 190.7 K, P_c = 45.8 bar for methane and T_c = 126.2 K, P_c = 33.5 bar for nitrogen.)

(a) 189 K, 100 bar (b) 229 K, 111 bar
(c) 286 K, 33.5 bar (d) 84 K, 11.1 bar

86. Which one of the following statements is not true?
(a) Vapour pressure of a liquid is independent of the size of the liquid surface.
(b) Vapour pressure is independent of the size and nature of the molecules.
(c) Vapour pressure increases with temperature.
(d) Surrounding pressure has only a negligible effect on the vapour pressure.

87. A pure gas is above its critical temperature.
(a) It cannot be condensed by cooling at constant pressure.
(b) It cannot be condensed by compressing isothermally.
(c) It cannot be condensed by cooling or by compression.
(d) It can be liquefied either by cooling or by compression.

88. The process occurring by heating a pure substance at a pressure below that corresponding to the triple point is known as
(a) Evaporation (b) Condensation
(c) Sublimation (d) Fusion

89. The saturation temperature of a pure substance is generally known as
(a) The critical temperature (b) The three-phase temperature
(c) The triple point (d) The boiling point

90. The logarithm of the vapour pressure of the substance is plotted as the ordinate against the logarithm of the vapour pressure of the reference at the same temperature as the abscissa in
(a) Hausbrandt chart (b) Dühring plot
(c) Cox chart (d) Psychrometric chart

91. If the Dühring lines are straight parallel lines, what does that indicate?
(a) Boiling point elevation is independent of concentration of solute.
(b) Boiling point elevation is independent of pressure for a given solution.
(c) Boiling point of solution is the same as that of the solvent.
(d) Boiling point of the solution is independent of the pressure.

92. The vapour pressure of an organic liquid is given by the Antoine equation

$$\ln P^S = 14.5463 - \frac{2940.46}{T - 49.19}$$

where P^S is in kPa and T is in K. What is the normal boiling point of the substance?
(a) 373.2 K (b) 1022.8 K
(c) 345.4 K (d) 283.7 K

93. A vessel of volume 1000 m^3 contains air which is saturated with water vapour. The total pressure and temperature are 100 kPa and 293 K respectively. Assuming that the vapour pressure of water at 293 K is 2.34 kPa, the amount of water vapour in kilograms in the vessel is approximately
(a) 17 (b) 20
(c) 25 (d) 34

94. The vapour pressure of water is given by $\ln P^{sat} = A - \dfrac{5000}{T}$, where A is constant, P^{sat} is vapour pressure in atm and temperature in K. The vapour pressure of water in atm at 323 K is approximately

(a) 0.07 (b) 0.11
(c) 0.09 (d) 0.13

95. The saturation temperature of a liquid at 101.3 kPa is 350 K. Which one of the following statements about this liquid is wrong?

(a) The vapour pressure of the liquid at 350 K is 101.3 kPa.
(b) At a temperature of 350 K and pressure of 80 kPa the substance is a superheated vapour.
(c) At a temperature of 300 K and a pressure of 101.3 kPa the substance is a cold liquid.
(d) The vapour and liquid cannot exist in equilibrium at 350 K and 101.3 kPa.

96. Vaporization equilibrium constant of ideal solutions depends on

(a) Temperature alone
(b) Temperature and pressure
(c) Temperature, pressure and composition
(d) Temperature and composition

97. Which one of the following statements is not true in the case of ideal solutions?

(a) There is no volume change on mixing the constituents.
(b) There is no temperature change on mixing.
(c) The total pressure over the solution varies linearly with x, the liquid composition at constant temperature.
(d) The total pressure varies linearly with y, the vapour concentration at constant temperature.

98. The temperature at which a binary solution exists as a heterogeneous mixture consisting of finite quantities of liquid and vapour phases

(a) Is known as its dew-point
(b) Is known as its bubble point
(c) Lies between the bubble point and the dew-point of the mixture
(d) Will be below the boiling points of pure constituents

99. A liquid solution is in equilibrium with its vapour. The concentration of the more volatile component in the vapour is maximum

(a) At the bubble point temperature
(b) At the dew-point temperature
(c) Between the bubble point and the dew-point temperatures
(d) At the normal boiling point of the mixture.

100. An air–water vapour mixture has a dry-bulb temperature of 333 K and a dew-point temperature of 313 K. The total pressure is 101.3 kPa and the vapour pressure of water at 313 K is 7.3 kPa and that at 333 K is 19.91 kPa. The wet bulb temperature T_w for the above mixture would be

(a) less than 313 K (b) 313 K
(c) $313 < T_w < 333$ (d) 333 K

101. The slope of the *P-x* diagram of an ideal binary solution at constant temperature is

(a) Equal to $P_A^S - P_B^S$ (b) Equal to P_A^S
(c) Independent of the vapour pressures (d) Equal to the Henry's law constant.

102. An ideal solution containing 40% *A* and 60% *B* is in equilibrium with its vapour. The vapour pressures of pure liquids at the equilibrium temperature are 80 kPa for *A* and 40 kPa for *B*. The vapour composition is

(a) 80% *A* (b) 67% *A*
(c) 57% *A* (d) 40% *A*

103. In a real solution the partial pressure is

(a) Greater than that in an ideal solution
(b) Less than that in an ideal solution
(c) Greater than or less than that in an ideal solution
(d) Equal to that in an ideal solution

104. An air–water vapour mixture has a dry-bulb temperature of 333 K and a dew-point temperature of 313 K. The total pressure is 101.3 kPa and the vapour pressure of water at 313 K is 7.3 kPa and that at 333 K is 19.91 kPa. The humidity of air expressed as kg of water vapour per kg of dry air is

(a) 0.048 (b) 0.079
(c) 0.122 (d) 0.152

105. In a dilute solution

(a) The solvent obeys Henry's law
(b) The solute obeys Henry's law
(c) The solvent and solute obey Henry's law
(d) The solute obeys Raoult's law

106. Which one of the following statements is incorrect with regard to a solution?

(a) The partial pressure approaches the vapour pressure as mole fraction tends to unity.
(b) The partial pressure is directly proportional to concentration as mole fraction tends to zero.
(c) The volume of the solution is greater than the sum of the volumes of the constituents.
(d) The total pressure over the solution is equal to the sum of the partial pressures.

107. Air, initially at 101.3 kPa and 313 K and with a relative humidity of 50%, is cooled at constant pressure to 303 K. The cooled air has

(a) A higher dew point (b) A higher absolute humidity
(c) A higher relative humidity (d) A higher wet bulb temperature

108. The solubility of gases in liquid at a given partial pressure is

(a) Directly proportional to the Henry's law constant
(b) Inversely proportional to the Henry's law constant

(c) Increases with increasing temperature
(d) Not related to the Henry's law constant

109. The boiling point of a mixture of two immiscible liquids
(a) Is greater than the boiling points of both pure liquids
(b) Lies between the boiling points of the pure liquids
(c) Is less than the boiling point of the more volatile liquid
(d) May be greater or less than the boiling point of the low volatile liquid.

110. The partial pressure over an immiscible solution is
(a) Equal to the product of mole fraction and vapour pressure of pure components
(b) Equal to the vapour pressure of components
(c) Is not related to vapour pressure
(d) Is equal to the product of weight fraction and vapour pressure

111. The three-phase temperature of a system made up of two immiscible liquids
(a) Depends only on the pressure
(b) Depends on the amounts of both liquids constituting the system
(c) Depends on the pressure as well as the amounts of liquids
(d) Is independent of both pressure as well as the amounts present

112. A mixture of two immiscible liquids is in equilibrium with its vapour. The composition of the vapour
(a) Will not change by changing the pressure
(b) Will change by changing the pressure
(c) Will change only if the amounts of the constituent liquids are altered.
(d) Will change by changing the pressure as well as the amounts of liquids

113. Two immiscible liquids *A* and *B* are taken in the mole ratio 1: 3 and heated. At the three-phase equilibrium temperature, the vapour pressures are 32 kPa for *A* and 68 kPa for B. The mole percent of *A* in the vapour is
(a) 13.6% (b) 32%
(c) 68% (d) 47.1%

114. A mixture of two immiscible liquids *A* and *B* is in equilibrium with its vapour at 101.3 kPa. The normal boiling points are: 373 K for *A* and 473 K for *B*. Which one of the following statements can be true regarding the boiling point (T) of the two-phase mixture?
(a) T = 373 K (b) 373 K < T < 473 K
(c) T = 363 K (d) T = 473 K

115. The presence of a non-volatile solute decreases
(a) The boiling point of the solvent
(b) The boiling point and vapour pressure of the solvent
(c) The boiling point and increase the vapour pressure
(d) The vapour pressure

116. If Raoult's law is applicable, the predicted boiling point of the solution will be independent of
(a) The mole fraction of the solvent (b) The mole fraction of the solute
(c) The solute (d) The surrounding pressure

117. The boiling point elevation of an aqueous 1 molal solution of sodium chloride will be greater than that of a 1-molal solution of sucrose in water. This is because of
(a) Ionisation of sodium chloride
(b) The increase in the effective mole fraction of sucrose in the solution
(c) The higher solubility of sucrose in water
(d) The low vapour pressure of sodium chloride

118. Components *A* and *B* form ideal solution. At 350 K, a liquid mixture containing 40% (mole) *A* is in equilibrium with a vapour containing 70% (mole) *A*. If the vapour pressure of *A* at 350 K is 70 kPa, what is the vapour pressure of *B*?
(a) 25 kPa (b) 20 kPa
(c) 40 kPa (d) 12 kPa

119. A vapour containing 70% *A* and 30% *B* is compressed at a constant temperature of 350 K. It is known that *A* and *B* form ideal solutions and at 350 K, the vapour pressures of *A* and *B* are respectively 70 kPa and 30 kPa. At what pressure does condensation begin?
(a) 70 kPa (b) 30 kPa
(c) 58 kPa (d) 50 kPa

120. A vapour containing 70% *A* and 30% *B* is compressed at a constant temperature of 350 K. It is known that *A* and *B* form ideal solutions and at 350 K the vapour pressures of *A* and *B* are respectively 70 kPa and 30 kPa. What is the composition of the first drop of condensate?
(a) 50% *A*, 50% *B* (b) 30% *A*, 70% *B*
(c) 70% *A*, 30% *B* (d) None of the above

121. A liquid mixture of benzene and toluene is in equilibrium with its vapour at 101 kPa and 373 K. The vapour pressures of benzene and toluene are 156 and 63 kPa respectively. Assuming that the system obeys Raoult's law, the mole fraction of benzene in the liquid phase is
(a) 0.65 (b) 0.41
(c) 0.065 (d) 0.04

122. The solubility of oxygen in water at 298 K is found to be 10 ppm for a partial pressure of oxygen of 25 kPa. What is the Henry's law constant for oxygen in water at 298 K?
(a) 2.5 kPa (b) 2.5×10^6 kPa
(c) 4.4×10^6 kPa (d) 1.4×10^6 kPa

123. At a specified temperature the molal humidity at saturation
(a) Depends on pressure but is independent of the properties of the gas
(b) Independent of pressure and properties of the gas
(c) Depends on the properties of the gas but is independent of the pressure
(d) Depends on properties of the gas and the total pressure.

124. At what temperature, the saturation humidity value is infinity?
(a) At the triple point temperature of the liquid
(b) At the critical temperature of the liquid

(c) At the boiling point of the liquid
(d) At the dew-point of the system

125. Assume that benzene is insoluble in water. The normal boiling points of benzene and water are 353.3 K and 373.2 K respectively. At a pressure of 101.3 kPa, the boiling point of a mixture of benzene and water is
(a) 353.3 K
(b) 373.2 K
(c) greater than 353.3 K, less than 373.2 K
(d) less than 353.3 K

126. The vapour pressures of benzene and toluene are 3 atm and 4/3 atm respectively. A liquid feed of 0.4 mol benzene and 0.6 mol toluene is vaporized. Assuming that the products are in equilibrium, the vapour phase mole fraction of benzene is
(a) 0.4 (b) 0.6
(c) 0.8 (d) 0.2

127. Which one of the following statements is true with regards to the percent saturation (PS) and relative saturation (RS)?
(a) PS = RS (b) PS > RS
(c) PS < RS (d) PS and RS are not interrelated.

128. Which of the following statements is true with respect to the dew point of a vapour–gas mixture?
(a) The dew point is independent of the humidity of the gas–vapour system.
(b) The vapour pressure at the dew point is equal to the partial pressure of the vapour in the gas.
(c) When the given gas–vapour mixture is subjected to saturation, the temperature attained is the dew point.
(d) The saturation humidity of the gas vapour system is equal to the saturation humidity at the dew point.

129. What will happen if an air–water vapour sample is cooled below the dew point?
(a) Its humidity will increase. (b) Its percent saturation will increase.
(c) Its percent saturation will decrease. (d) Its humidity will decrease.

130. The wet-bulb temperature of an unsaturated gas–vapour system
(a) Is the same as its dew point (b) Is always greater than the dew point
(c) Is always less than the dew point (d) None of the above.

131. Choose the right statement.
(a) The vapour pressure at the wet-bulb temperature is equal to the existing partial pressure of the vapour in the vapour–gas mixture.
(b) The existing partial pressure of the vapour is less than the vapour pressure at the wet-bulb temperature.
(c) Wet-bulb and dry-bulb temperatures can never be equal.
(d) The lower the wet bulb depression the higher the percent saturation of the vapour–gas mixture.

132. The enthalpy of a gas–vapour mixture can be increased by

(a) Increasing the humidity and decreasing the temperature
(b) Increasing the humidity and temperature
(c) Decreasing the humidity and increasing the temperature
(d) Decreasing both temperature and humidity.

133. Adiabatic saturation temperature of a vapour–gas mixture is

(a) The steady-state temperature attained by the liquid evaporating into the gas
(b) The temperature to which the gas can be cooled at constant pressure without condensation
(c) The temperature attained by the gas when it is humidified to saturation in contact with the liquid at constant temperature
(d) The temperature at which the existing vapour content is sufficient to saturate the gas.

134. An air–water vapour mixture has a dry bulb temperature of 303 K and wet bulb temperature of 293 K. The mixture is heated at constant pressure to 333 K. The wet-bulb temperature will be

(a) Constant at 293 K
(b) Less than 293 K
(c) Greater than 293 K
(d) Equal to 303 K.

135. At certain temperature, the saturation humidity of air–water vapour system is found to be 0.065 kg of water vapour/kg of dry air at 101.3 kPa. What is the vapour pressure of water at this temperature?

(a) 3.9 kPa
(b) 9.6 kPa
(c) 6.2 kPa
(d) 13.1 kPa

136. Air at 318.15 K and 101.3 kPa has a dew point of 298.15 K. What is the partial pressure of water vapour in the air if the vapour pressure of water in kPa is given by the Antoine equation

$$\ln P^S = 16.26205 - \frac{3799.887}{T - 46.854}$$

(a) 6.2 kPa
(b) 9.72 kPa
(c) 3.13 kPa
(d) 4.52 kPa

137. The enthalpy of dry air and saturated air are respectively 60 kJ/kg of dry air and 470 kJ/kg of dry air at 333 K. What is the enthalpy of air if the dry bulb temperature is 333 K and dew point is 310 K? The vapour pressure of water is given by

$$\ln P^S = 16.26205 - \frac{3799.887}{T - 46.854}$$

(a) 265 kJ/kg
(b) 442 kJ/kg
(c) 345 kJ/kg
(d) 170 kJ/kg

138. The adiabatic saturation temperature of air–benzene system at 300 K is found to be 285 K at 101.3 kPa. The vapour pressure of benzene at 285 K is 6.7 kPa. The humid heat of the system may be taken to be constant at 1.155 kJ/kg K and the latent heat of vaporization of benzene at 285 K is 435.4 kJ/kg. The benzene content in the air at 300 K is .... kg of benzene vapour per kg of dry air.

(a) 0.15 (b) 0.20
(c) 0.30 (d) 0.25

139. It is desired to produce air at 308 K and 30% saturated with water vapour from an air supply of 30% saturated air at 298 K. Pressure is constant at 101.3 kPa. Suggest a suitable method for achieving this.

(a) Heating at constant pressure from 298 K to 308 K.
(b) Heating at constant pressure from 298 K to 308 K followed by adiabatic humidification.
(c) Adiabatic humidification followed by heating at constant pressure.
(d) Adiabatic dehumidification.

140. It is required that the air leaving an adiabatic spray chamber contains 0.02 kg of water/kg of dry air and is 30% saturated with water vapour. If the ambient air at 308 K with 30% saturation is preheated before sending into the spray chamber, to what temperature this air need be preheated?

(a) 321 K (b) 342 K
(c) 368 K (d) No preheating is required.

141. The absolute humidity of air at 101.3 kPa is measured to be 0.02 kg of water per kg of dry air. Determine the partial pressure of water vapour in the air?

(a) 1.99 kPa (b) 2.55 kPa
(c) 3.16 kPa (d) 3.87 kPa

142. If the specific heats of air and vapour are 0.2 kJ/kg K and 1.5 kJ/kg K respectively and the humidity is 0.01, the humid heat in kJ/kg K is

(a) 0.31 (b) 0.107
(c) 0.017 (d) 0.215

143. A process flow sheet analysis results in the degrees of freedom having a value −2. Which one of the following steps must be next carried out?

(a) Identify and add two independent new equations from the process model.
(b) Remove two equations that have been wrongly assumed to be independent.
(c) Assign values of two variables in the process.
(d) Assign value to one variable and remove one equation that was wrongly assumed to be independent.

144. A butane isomerisation process produces 70 kmol/h of pure isobutane. A purge stream, removed continuously, contains 85% *n*-butane and 15% impurity (mole%). The feed stream is *n*-butane containing 1% impurity. The flow rate of the purge stream will be:

(a) 3 kmol/h (b) 4 kmol/h
(c) 5 kmol/h (d) 6 kmol/h

145. 5.85 g of NaCl is dissolved in one kilogram of water.

(a) The molarity of the solution is 0.1. (b) The molarity of the solution is 1.0.
(c) The molality of the solution is 0.1. (d) The normality of the solution is 0.1.

146. The average molecular weight of a mixture of oxygen and other gases is calculated to be 23.2 using an incorrect value of 16 for the molecular weight of oxygen whereas the correct value is 29.6. What is the volume percent of oxygen in the mixture?

(a) 40% (b) 60%
(c) 50% (d) 28%

147. A flow sheet is depicted below:

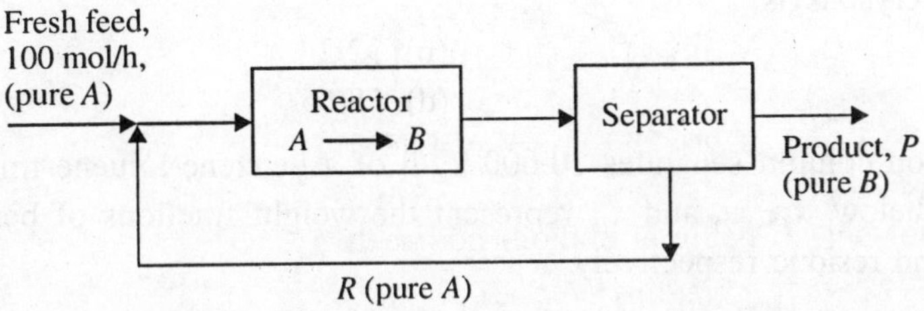

If the single pass (once through) conversion of *A* to *B* is 20%, then the rate of recycle *R* (mol/h) is

(a) 300 (b) 400
(c) 500 (d) 600

148. Ideal solution obeys

(a) Boyle's law (b) Amagats's law
(c) Raoult's law (d) All the above laws

149. With increase in the molecular weight, the vapour pressure of chemically similar liquids at any given temperature

(a) Increases (b) Decreases
(c) Remains unchanged (d) May increase or decrease

150. Assuming that CO_2 obeys the perfect gas law, calculate the density of CO_2 in kilograms per cubic metre at 540 K and 202 kPa.

(a) 1 (b) 2
(c) 3 (d) 4

151. A mixture of ethyl acetate vapour and air has a relative saturation of 50% at 303 K and a total pressure of 100 kPa. If the vapour pressure of ethyl acetate at 303 K is 16 kPa, the percent of air is

(a) 92% (b) 45.7%
(c) 50% (d) 8%

152. A mixture of ethyl acetate vapour and air has a relative saturation of 50% at 303 K and a total pressure of 100 kPa. If the vapour pressure of ethyl acetate at 303 K is 16 kPa, the molal saturation is

(a) 0.080 (b) 0.087
(c) 0.264 (d) 0.029

153. Air at 293 K and 750 mm Hg pressure has a relative humidity of 80%. What is its percent humidity? The vapour pressure of water at 293 K is 17.5 mm Hg.

(a) 80.38 (b) 80
(c) 79.62 (d) 78.51

154. $Na_2SO_4 \cdot 10H_2O$ crystals are formed by cooling 100 kg of 30% by weight aqueous solution of Na_2SO_4. The final concentration of the solute in the solution is 10%. The weight of crystals is

(a) 22.33 (b) 32.2
(c) 45.35 (d) 58.65

155. A distillation column separates 10,000 kg/h of a benzene-toluene mixture as shown in the figure below. x_F, x_D and x_W represent the weight fractions of benzene in the feed, distillate and residue respectively.

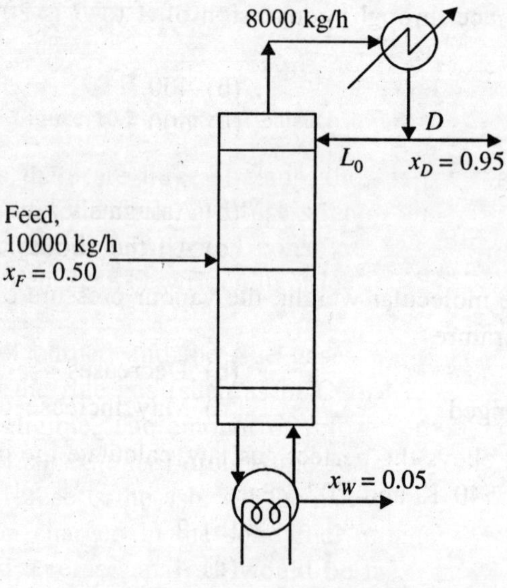

The reflux ratio is

(a) 0.5 (b) 0.6
(c) 1.0 (d) 2.0

156. Purging operation is performed on recycle streams for

(a) Increasing yield (b) Reducing the accumulation of inerts.
(c) Conserving heat (d) Improving efficiency

157. A substance which is at a pressure smaller than the saturation pressure at the temperature in question is in the .... state.

(a) Superheated vapour (b) Subcooled liquid
(c) Saturated vapour (d) Saturated liquid

158. The weather bureau reports a dry-bulb temperature of ambient air as 302 K and a relative humidity of 80%. The barometer reads 100 kPa. The vapour pressure of water at 302 K is 30 mm Hg. The percent humidity of the ambient air is

(a) 80% (b) 30%
(c) 74% (d) 84%

159. Which of the following gives specific gravity on the Baumé scale for liquids heavier than water?

(a) $\dfrac{141.5}{\text{sp gr}} - 131.5$ (b) $\dfrac{140}{\text{sp gr}} - 130$

(c) $145 - \dfrac{145}{\text{sp gr}}$ (d) 200(sp gr – 1.00)

160. If 50 kg of dry solid containing 6% water is obtained by drying 65 kg of wet material, what was the initial moisture content?

(a) 27.7% (b) 77%
(c) 23% (d) 21%

161. There exists a unique solution for a set of material balance equations if

(a) The number of degrees of freedom is positive
(b) The number of degrees of freedom is negative
(c) The number of degrees of freedom is zero
(d) The number of degrees of freedom is non-zero value

162. With increase in pressure, the heat of vaporization of liquids

(a) Decreases (b) Increases
(c) Remain unchanged (d) May increase or decrease

163. The pH value of a solution increases by one unit. The hydrogen ion concentration

(a) Decreases by one unit
(b) Increases by one unit
(c) Becomes 10 times the original value
(d) Decreases to one-tenth of the original value

164. A multiple effect evaporator has a capacity to process 4000 kg of solid caustic soda per day when it is concentrating from 10% to 25% solids. The water evaporated in kilograms per day is

(a) 800 (b) 24 000
(c) 60 000 (d) 48 000

165. An evaporator is fed with 100 kg/h of a solution which contains 10% NaCl, 10% NaOH and the rest water. During evaporation water is removed as vapour and NaCl crystallizes and is settled and removed. The mother liquor contains 50% NaOH and 2% NaCl. How many kilograms of salt are precipitated per hour?

(a) 9.6 (b) 8.0
(c) 9.0 (d) None of the above

166. An evaporator is fed with 100 kg/h of a solution which contains 10% NaCl, 10% NaOH and the rest water. During evaporation, water is removed as vapour and NaCl crystallizes and is settled and removed. The mother liquor contains 50% NaOH and 2% NaCl. What is the weight of concentrated liquor leaving per hour?

(a) 50 kg (b) 25 kg
(c) 20 kg (d) 75 kg

167. A metal recovery unit (MRU) of intake capacity 5000 kg/h treats a liquid product from a plant and recovers 90% of the metal in the pure form. The unrecovered metal and its associated liquid are sent to a disposal unit along with the untreated product from the plant (see figure). Find the flow rate m_6 and the weight fraction of the metal (w_6). The liquid product flow rate is 7500 kg/h of composition 0.1 (weight fraction). Assume steady state.

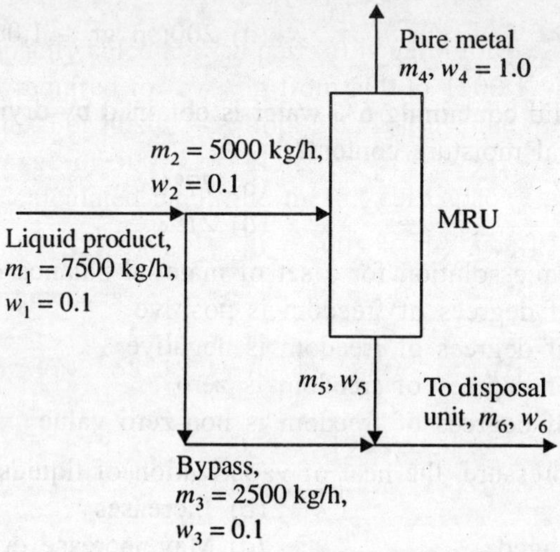

(a) m_6 = 7500 kg/h, w_6 = 0.0 (b) m_6 = 7050 kg/h, w_6 = 0.04255
(c) m_6 = 4500 kg/h, w_6 = 0.1712 (d) m_6 = 5600 kg/h, w_6 = 0.0314

168. A feed stream (S1) at 100 kg/h and containing only *A* mixes with recycle stream *S*5 before entering the reactor (see figure), where the reaction $A \rightarrow B$ takes place. The operation is at steady state. The stream S3 leaving the reactor is separated, without either phase change or composition change, into two streams S4 and S5. If the mass fraction of *B* in S4 is 0.95 and the total flow rate of S5 is 10 kg/h, then the ratio of flow rates of streams S3/S5 and the flow rate of *A* in S3 are, respectively,

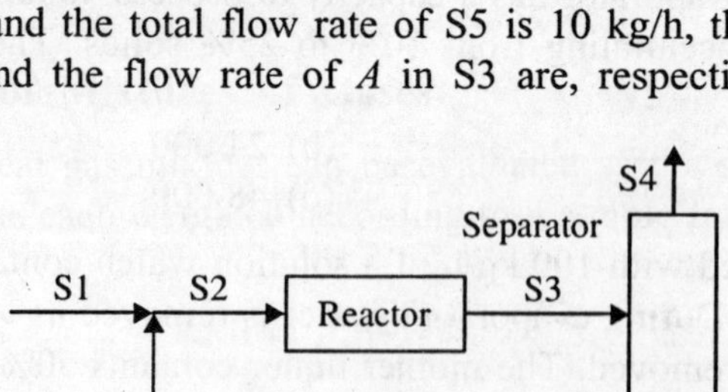

(a) 11 and 110 kg/h (b) 24 and 240 kg/h
(c) 11 and 5.5 kg/h (d) 70 and 330 kg/h

169. Which of the following conditions are to be satisfied for using the Clausius–Clapeyron equation to estimate the vapour pressure?
(a) The vapour behaves as an ideal gas.
(b) The latent heat of vaporisation is constant.
(c) The molar volume of liquid is negligibly small.
(d) All the above.

170. A rigid vessel, containing three moles of nitrogen gas at 303 K is heated to 523 K. Assume the average heat capacity of nitrogen to be $C_P = 29.1$ J/mol K and $C_V = 20.8$ J/mol K. The heat required, neglecting the heat capacity of the vessel, is
(a) 13 728 J (b) 19 206 J
(c) 4576 J (d) 12 712 J

171. One cubic metre of an ideal gas at 500 K and 1000 kPa expands reversibly to 5 times its initial volume in an insulated container. If the specific heat capacity at constant pressure of the gas is 21 J/mol K, the final temperature will be
(a) 35 K (b) 174 K
(c) 274 K (d) 154 K

172. At 373 K, water and methyl cyclohexane both have a vapour pressure of 1.0 atm. The latent heats of vaporization are 40.63 kJ/kmol for water and 31.55 kJ/kmol for cyclohexane. The vapour pressure of water at 423 K is 4.69 atm. The vapour pressure of methylcyclohexane at 423 K is expected to be
(a) Significantly less than 4.69 atm (b) Nearly equal to 4.69 atm
(c) Significantly more than 4.69 atm (d) Indeterminate due to lack of data

173. For the estimation of heat capacity of a solid compound, one can use
(a) Clapeyron equation (b) Gibb's equation
(c) Kopp's rule (d) Trouton's rule

174. An insulated container holds 20 kg of water initially at 298 K. It is stirred by an agitator, which is made to turn by a body weighing 40 kg slowly falling through a height of 4 m. The process is repeated 500 times. The acceleration due to gravity is 9.8 m/s^2. Neglecting the heat capacity of the agitator, the temperature attained by the water is
(a) 313.5 K (b) 307.4 K
(c) 299.8 K (d) 298 K

175. One mole of nitrogen at 8 bar and 600 K is contained in a piston-cylinder assembly. It is brought to 1 bar isothermally against a resisting pressure of 1 bar. The work done in kJ by the gas is
(a) 30.554 (b) 10.373
(c) 4.9884 (d) 4.3649

176. Heat capacity of gas can be approximately expressed as $C_P = 26.693 + 7.365 \times 10^{-3}T$ where C_P is in J/mol K and T is in K. The heat given off by one mole of air when cooled at atmospheric pressure from 773 K to 173 K is
(a) 10.73 kJ (b) 16.15 kJ
(c) 18.11 kJ (d) 18.33 kJ

177. One kilogram of saturated steam at 373 K and 1.01325 bar is contained in a rigid walled vessel. It has a volume of 1.673 m^3. It is cooled to a temperature at which the specific volume of water vapour is 1.789 m^3. The amount of water vapour condensed in kilograms is

(a) 0.0 (b) 0.065
(c) 0.1 (d) 1.0

178. One kilogram of saturated steam at 373 K and 1.01325 bar is contained in a rigid walled vessel. It has a volume of 1.673 m^3. It is cooled to a temperature at which the specific volume of water vapour is 1.789 m^3. If the heat liberated is 147 kJ, the latent heat of condensation in kJ/kg under these conditions is

(a) 40 732 (b) 2676
(c) 2263 (d) 540

179. The heat capacity at constant pressure of a single-component system consisting of liquid and vapour phases in equilibrium is

(a) infinite (b) zero
(c) positive (d) negative

180. For ideal gases, $\Delta H = \int C_P\, dT$ is valid

(a) For a constant volume process
(b) For a constant pressure process
(c) Irrespective of the nature of the process
(d) The statement is never true.

181. The difference between the heat supplied and the work extracted in a steady-flow process, in which the kinetic and potential energy changes are negligible, is equal to

(a) The change in internal energy (b) The change in enthalpy
(c) The change in the work function (d) The change in the Gibbs free energy

182. To obtain the integrated form of the Clausius–Clapeyron equation

$$\ln \frac{P_2}{P_1} = \frac{\Delta H_V}{R}\left(\frac{1}{T_1} - \frac{1}{T_2}\right)$$

from the exact Clapeyron equation, it is assumed that

(a) The volume of the liquid phase is negligible compared to that of the vapour phase
(b) The vapour phase behaves as an ideal gas
(c) The heat of vaporization is independent of temperature
(d) All the above are applicable

183. Which one of the following is not true?

(a) Internal energy is a thermodynamic property of a system.
(b) Internal energy is a state function.
(c) Internal energy of a system does not change in a cyclic process.
(d) Internal energy is independent of temperature and pressure of the system.

184. The heat supplied to a system under constant pressure is equal to
(a) The work done by the system (b) The change in internal energy
(c) The change in enthalpy (d) The change in kinetic and potential energy

185. Which one of the following is true?
(a) Generally $C_P > C_V$ for all substances. (b) For solids and liquids, $C_V > C_P$.
(c) For all substances, $C_P - C_V = R$. (d) $C_P = C_V$ for compressible substances.

186. If the heat capacity is given by $C_P = a + bT$ where C_P is in kJ/kmol K and T is in K
(a) The constants a and b are dimensionless constants
(b) Both a and b have the dimensions kJ/kmol K
(c) The dimension of a is kJ/kmol K and that of b is kJ/kmol
(d) The dimensions of a is kJ/kmol K and that of b is kJ/kmol K^2

187. The Kistyakowsky equation is
(a) Used to estimate the heat capacity of non-polar liquids
(b) Used to estimate the heat capacity of organic compounds
(c) Used to estimate the heat of vaporization of non-polar liquids
(d) Used to estimate the heat of vaporization of polar liquids

188. Trouton's ratio is
(a) The ratio of heat of vaporization to heat of vaporization at normal boiling point
(b) The ratio of heat of vaporization to the normal boiling point
(c) The ratio of heat of vaporization to heat of vaporization at critical temperature
(d) The ratio of heat capacity to the heat capacity at normal boiling point

189. The heat capacity of potassium permanganate is equal to the heat capacity of potassium + the heat capacity of manganese + 4 × (the heat capacity of oxygen). This is a consequence of
(a) Trouton's rule (b) Kopp's rule
(c) Kistayakowsky equation (d) Dulong and Petit rule

190. If the heat capacity of a substance varies linearly with temperature the mean heat capacity of the substance is equal to
(a) The average of the heat capacity at T_1 and heat capacity at T_2
(b) The heat capacity at $(T_1 + T_2)/2$
(c) The ratio of the heat required to raise the temperature from T_1 to T_2 to $(T_1 - T_2)$
(d) All the above

191. Which of the following is valid if dH is the enthalpy change for an ideal gas?
(a) $dH = C_V\, dT$ only for constant volume process
(b) $dH = C_P\, dT$ only for constant pressure process
(c) $dH = dQ = C_P\, dT$ for all processes
(d) $dH = C_P\, dT$ for all processes

192. For processes involving ideal gases only one of the following is true.
(a) $dH = C_V\, dT$ (b) $dU = C_V\, dT$
(c) $dU = Q = C_V\, d\mathrm{T}$ (d) $dH = Q = C_P\, dT$

193. In a cyclic process involving ideal gases, the work extracted is equal to the heat supplied for

(a) Isobaric process (b) Isothermal process
(c) Adiabatic process (d) Isochoric process

194. The absolute temperature of an ideal gas gets doubled in an adiabatic process. If C_P = 1.4 C_V, the initial volume V_1 and final volume V_2 are related as

(a) $V_2 = 5.66\ V_1$ (b) $V_2 = 1.22\ V_1$
(c) $V_2 = 0.82\ V_1$ (d) $V_2 = 0.18\ V_1$

195. Calculate the change in internal energy when one kilomole water is vaporized at a constant temperature of 373 K and a constant pressure of 101.3 kPa. The specific volumes of liquid and vapour under these conditions are 1.04×10^{-3} m^3/kmol and 1.675 m^3/kmol respectively. 1030 kJ heat is added for this change.

(a) 169.6 kJ (b) 1030 kJ
(c) 860.4 kJ (d) 1199.6 kJ

196. Determine the heat capacity of $Na_2SO_4 \cdot 10H_2O$ at room temperature using Kopp's rule. The atomic heat capacities of elements (J/g-atom K) are 26.04 for Na, 22.6 for S, 16.8 for O, and 9.6 for H.

(a) 325.4 (b) 501.9
(c) 65.44 (d) 177.90

197. The higher heating value is obtained when water formed on combustion of a fuel is

(a) In the liquid state (b) In the vapour state
(c) Negligible (d) In the liquid or in the vapour state

198. The calorific value of a fuel is equal to

(a) Heat of formation of the fuel
(b) Heat of combustion of the fuel
(c) Negative of the heat of combustion
(d) Heat of combustion – Heat of vaporization of water

199. The standard state for heat combustion data assumes water in the liquid state at 298 K. The standard heat of combustion is thus numerically equal to

(a) Net heating value (b) Gross heating value
(c) Heat of formation of water (d) None of the above

200. The heats of combustion of yellow phosphorus and red phosphorus are –9.91 kJ and –8.78 kJ respectively. The heat of transition of yellow phosphorus to red phosphorus is

(a) – 18.69 kJ (b) 1.13 kJ
(c) 18.69 kJ (d) – 1.13 kJ

201. In an adiabatic humidification

(a) The temperature and humidity of air increase
(b) Humid heat remains constant
(c) Enthalpy of air remains constant
(d) Humidity increases and temperature remains constant.

202. The difference between heats of reaction at constant pressure and constant volume for the reaction

$$2C_6H_6(l) + 15O_2(g) \rightarrow 12CO_2(g) + 6H_2O(l)$$

at 298 K in kJ is

(a) -7.43 (b) 3.72
(c) -3.72 (d) 7.43

203. The standard heats of combustion of ethylene, ethane and butane in kJ/mol are -1410.99, -1559.9 and -2855.6 respectively. The heat absorbed for the isothermal reaction

$$C_4H_{10}(g) \rightarrow C_2H_4(g) + C_2H_6(g)$$

at 298 K and 1 atmosphere pressure is

(a) 2706.7 kJ (b) -2706.7 kJ
(c) 115.3 kJ (d) -115.3 kJ

204. Given that

$$2C(s) + 2O_2 \rightarrow 2CO_2(g) \qquad \Delta H = -789 \text{ kJ}$$

$$H_2(g) + \frac{1}{2}O_2(g) \rightarrow H_2O(l) \qquad \Delta H = -286 \text{ kJ}$$

$$C_2H_5(g) + \frac{5}{2}O_2(g) \rightarrow 2CO_2(g) + H_2O(l) \qquad \Delta H = -1310 \text{ kJ}$$

the heat of formation of acetylene is

(a) 1802 kJ (b) -1802 KJ
(c) -800 kJ (d) 235 kJ

205. The actual flame temperature is less than adiabatic flame temperature. This is due to
(a) Loss of heat
(b) Incomplete combustion
(c) Energy expended for performing mechanical work and increasing external kinetic and potential energy
(d) All the above

206. The maximum flame temperature is attained
(a) When the fuel and air are in stoichiometric quantities
(b) When fuel is burned with an excess of pure oxygen
(c) When stoichiometric amount of oxygen is used
(d) When excess of air supply is provided

207. An exothermic reaction takes place in an adiabatic reactor. The product temperature .... the reactor feed temperature.
(a) Is always equal to (b) Is always greater than
(c) Is always less than (d) May be greater or less than

208. A batch adiabatic reactor at an initial temperature of 373 K is being used for the reaction $A \rightarrow B$. Assume that the heat of reaction is -1 kJ/mol at 373 K and the heat capacity of both A and B to be constant and equal to 50 J/mol K. The temperature rise after a conversion of 0.5 will be

(a) 5 K (b) 10 K
(c) 20 K (d) 100 K

209. In the hydrodealkylation of toluene to benzene, the following reactions occur.

$$C_7H_8 + H_2 \rightarrow C_6H_6 + CH_4$$
$$2C_6H_6 \Leftrightarrow C_{12}H_{10} + H_2$$

Toluene and hydrogen are fed to a reactor in a molar ratio of 1:5. 80% of the toluene gets converted and the selectivity of benzene (defined as moles of benzene formed per mole of toluene converted) is 90%. The fractional conversion of hydrogen is

(a) 0.16 (b) 0.144
(c) 0.152 (d) 0.136

210. Determine $Q_P - Q_V$ for the following reaction at 298 K

$$C(s) + \frac{1}{2}O_2(g) \rightarrow CO(g) \qquad \Delta H = -110.5 \text{ kJ}$$

where Q_P is the heat of reaction at constant pressure and Q_V is the heat of reaction at constant volume.

(a) 1239 J/mol (b) 2178 J/mol
(c) – 1239 J/mol (d) 4956 J/mol

211. One mole of methane undergoes complete combustion in a stoichiometric amount of air. Both the reactants and products are in the gas phase. The mole fraction of water vapour in the product gas is about

(a) 0.19 (b) 0.33
(c) 0.40 (d) 0.67

212. One mole of methane at 298 K undergoes complete combustion in a stoichiometric amount of air also at 298 K. Both the reactants and products are in the gas phase.

$$CH_4 + 2O_2 \rightarrow CO_2 + 2H_2O \qquad \Delta H^0_{298} = -730 \text{ kJ/mol}$$

If the average specific heat of all the gases/vapours is 40 J/mol K, the maximum temperature rise (K) of the exhaust gases would be approximately

(a) 1225 (b) 1335
(c) 1525 (d) 1735

213. The heat of reaction is

(a) Independent of temperature and pressure
(b) Independent of temperature but changes with pressure
(c) Independent of the number of intermediate steps involved
(d) Independent of the state of aggregation of the reactants and products

214. The adiabatic flame temperature of a fuel in air is .... that in pure oxygen.

(a) Greater than (b) Less than
(c) Equal to (d) Greater than or equal to

215. Steam at 1500 kPa is throttled (i.e., expanded at constant enthalpy) to 100 kPa. If the temperature rises to 423 K on expansion, what was the quality of original steam?
(a) 0.844 (b) 0.992
(c) 0.112 (d) 1.115

216. In the absence of experimental values, ... may be used to calculate the heat capacity of solids
(a) Dulong and Petit law (b) Kopp's rule
(c) the Kistyakowsky equation (d) Trouton's rule

217. At standard conditions,

$$N_2 + 2O_2 \rightarrow 2NO_2 \qquad \Delta H^0 = -90 \text{ kJ/mol}$$

$$NO + \frac{1}{2}O_2 \rightarrow NO_2 \qquad \Delta H^0 = -11 \text{ kJ/mol}$$

The standard heat of formation of NO in kJ/mol is
(a) −78 (b) 78
(c) −34 (d) −101

218. Enthalpy change in a chemical reaction is independent of the steps through which the reaction proceeded. This is a consequence of
(a) Boyle's law (b) Van't Hoff's law
(c) Hess's law (d) LeChatlier's principle

219. When methane is burnt in a stoichiometric quantity of air, the Orsat analysis of flue gas will be
(a) 9.5% CO_2, 19.0% H_2O, 71.5% N_2 (b) 33.3% CO_2, 66.7% H_2O
(c) 21.0% CO_2, 79.0% N_2 (d) 11.7% CO_2, 88.3% N_2

220. Which one of the following represents the Orsat analysis of the flue gas when methane is completely burnt in a theoretical supply of air?
(a) CO_2 5.7%, CO 5.7%, O_2 2.9%, N_2 85.7%
(b) CO_2 9.5%, H_2O 19.0%, N_2 71.5%
(c) 33.3% CO_2, 66.7% N_2
(d) None of the above

221. The proximate analysis of coal gives
(a) Carbon, hydrogen and ash
(b) Volatile matter, moisture, ash and fixed carbon
(c) Carbon, hydrogen, sulphur and nitrogen
(d) Volatile matter, moisture, nitrogen and carbon

222. Trouton's ratio is the ratio of
(a) Latent heat of vaporization to the normal boiling point
(b) The heat supplied to the absolute temperature
(c) The heat capacity of a substance to the heat capacity of water
(d) None of the above

223. Trouton's ratio in SI units is

(a) 21 (b) 87.9
(c) 5 (d) 79.8

224. A sample of natural gas containing 80% methane and the rest nitrogen is burnt with 20% excess air. With 80% of the combustibles producing CO_2 and the remainder going to CO the Orsat analysis in volume percent is

(a) CO_2: 6.26, CO: 1.56, O_2: 3.91, H_2O: 15.66, N_2: 72.60
(b) CO_2: 7.42, CO: 1.86, O_2: 4.64, N_2: 86.02
(c) CO_2: 6.39, CO: 1.60, O_2: 3.99, H_2O: 15.96, N_2: 72.06
(d) CO_2: 7.60, CO: 1.90, O_2: 4.75, N_2: 85.74

225. The heat capacity of air can be approximately expressed as

$$C_P = 26.693 + 7.365 \times 10^{-3}T$$

where C_P is in J/mol K and T is in K. The heat given off by one mole of air when cooled at 1 atmospheric pressure from 500–100°C is

(a) 10.73 kJ (b) 16.15 kJ
(c) 18.11 kJ (d) 18.33 kJ

226. One mole methane undergoes complete combustion in a stoichiometric amount of air. The reaction proceeds as

$$CH_4 + 2O_2 \rightarrow CO_2 + 2H_2O$$

Both the reactants and the products are in gas phase.

$$\Delta H_0 = -730 \text{ kJ/mol of methane}$$

The mole fraction of water vapour in the product gases is about

(a) 0.19 (b) 0.33
(c) 0.40 (d) 0.67

227. One mole methane undergoes complete combustion in a stoichiometric amount of air. The reaction proceeds as

$$CH_4 + 2O_2 \rightarrow CO_2 + 2H_2O$$

Both the reactants and the products are in gas phase.
$\Delta H_0 = -730$ kJ/mol of methane. If the average specific heat of all gases and vapours is 40 J/mol K, the maximum temperature rise of the exhaust gases in K would be approximately equal to

(a) 1225 (b) 1335
(c) 1525 (d) 1735

228. By increasing the air/fuel ratio, the adiabatic flame temperature

(a) increases (b) decreases
(c) remains unchanged (d) change is unpredictable

229. The reaction A(1) + B(g) → C(1) + D(g) is carried out in a reactor followed by separator as shown below:

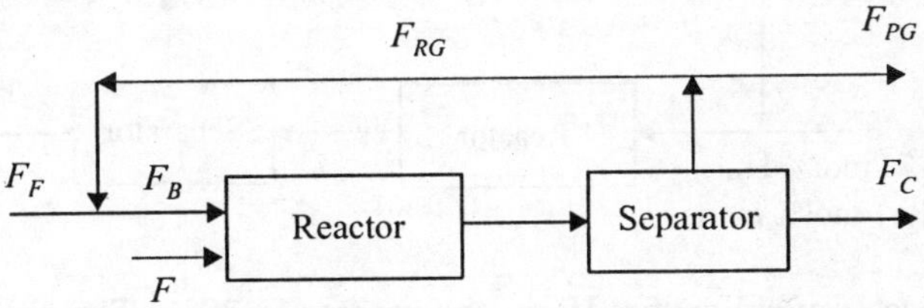

Notation: Molar flow rate of fresh B = F_{FB}; Molar flow rate of A = F_A; Molar flow rate of recycle gas = F_{RG}; Mole fraction of B in recycle gas = Y_{RB}; Molar flow rate of purge gas = F_{PG}; Molar flow rate of C = F_C; F_{FB} = 2 mol/s, F_A = 1 mol/s, F_B/F_A = 5, A is completely converted. If Y_{RB} = 0.3, the ratio of recycle gas to purge gas (F_{RG}/F_{PG}) is

(a) 2 (b) 5
(c) 7 (d) 10

230. The reaction A(1) + B(g) → C(1) + D(g) is carried out in a reactor followed by separator as shown below:

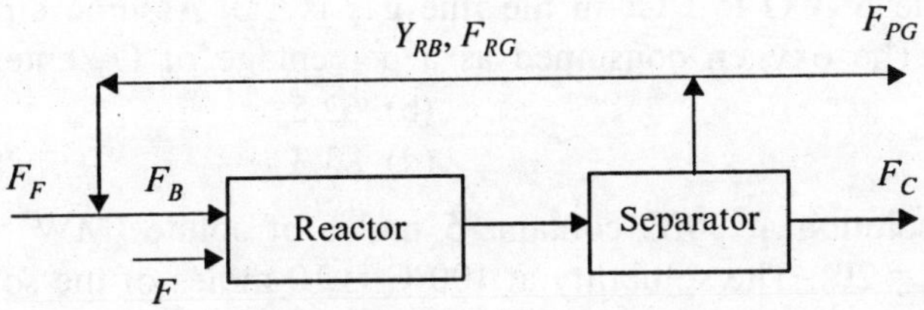

Notation: Molar flow rate of fresh B = F_{FB}; Molar flow rate of A = F_A; Molar flow rate of recycle gas = F_{RG}; Mole fraction of B in recycle gas = Y_{RB}; Molar flow rate of purge gas = F_{PG}; Molar flow rate of C = F_C; F_{FB} = 2 mol/s, F_A = 1 mol/s, F_B/F_A = 5, A is completely converted. If the ratio of recycle gas to purge gas (F_{RG}/F_{PG}) is 4, then F_{RB} is

(a) 3/8 (b) 2/5
(c) 1/2 (d) 3/4

231. One mole of methane is contained in a leak-proof piston cylinder assembly at 8 bar and 1000 K. The gas undergoes isothermal expansion to 4 bar under reversible condition. Methane can be considered as an ideal gas under these conditions. The value of universal gas constant is 8.314 J/mol K. The heat transferred (in kJ) during the process is

(a) 11.52 (b) 5.76
(c) 4.15 (d) 2.38

232. Ammonia is synthesized at 200 bar and 773 K by the reaction $N_2 + 3H_2 \leftrightarrow 2NH_3$. The yield of ammonia is 0.45 mol/mol of fresh feed. Flow sheet for the process (along with available compositions) is shown below:

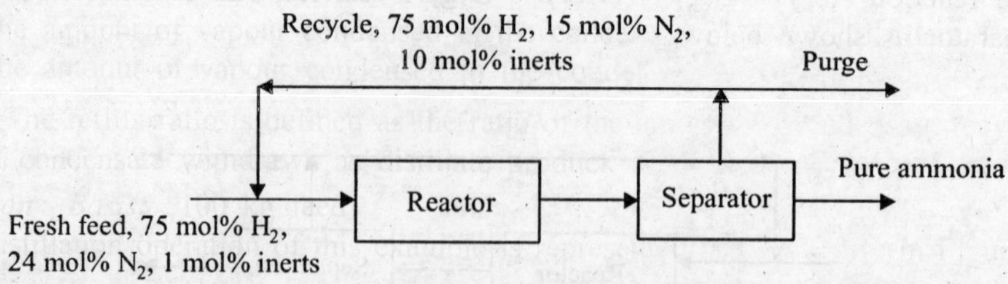

The single pass conversion for H_2 in the reactor is 20%. The amount of H_2 lost in the purge as a percentage of H_2 in the fresh feed is

(a) 10 (b) 20
(c) 45 (d) 55

233. The following combustion reactions occur when methane is burned.

$$CH_4 + 2O_2 \rightarrow CO_2 + 2H_2O$$

$$2CH_4 + 3O_2 \rightarrow 2CO + 4H_2O$$

20% excess air is supplied to the combustor. The conversion of methane is 80% and the molar ratio of CO to CO_2 in the flue gas is 1:3. Assume air to have 80 mol% N_2 and rest O_2. The oxygen consumed as a percentage of O_2 entering the combustor is

(a) 20 (b) 62.5
(c) 80 (d) 83.3

234. A saturated solution at 30°C contains 5 moles of solute (MW = 50) per kilogram of solvent (MW = 20). The solubility at 100°C is 10 moles of the solute per kg of solvent. If 10 kg of the original solution is heated to 100°C, then the weight of the additional solute that can be dissolved in it, is

(a) 0.25 kg (b) 1 kg
(c) 2 kg (d) 3.34 kg

235. The products of combustion of methane in atmospheric air (21% O_2, 79% N_2) have the following composition on dry basis:

$$CO_2 = 10.00\%,\ O_2 = 2.37\%,\ CO = 0.53\%,\ N_2 = 87.1\%$$

The ratio of the moles of CH_4 to moles of O_2 in the feed stream is

(a) 1.05 (b) 0.60
(c) 0.51 (d) 0.45

236. At 25°C and 90% relative humidity, water evaporates from the surface of a lake at the rate of 1 kg/m^2 h. The relative humidity that will lead to an evaporation rate of 3 kg/m^2 h with other conditions remaining the same, is

(a) 30% (b) 50%
(c) 60% (d) 70%

237. Pure water (stream W) is to be obtained from a feed containing 5 weight % salt using a desalination unit as shown below:

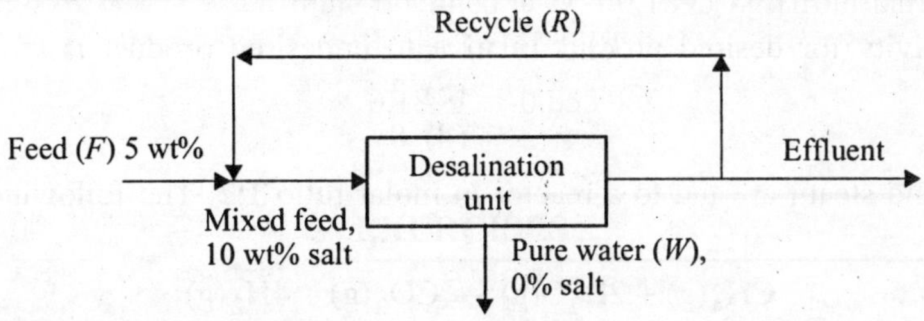

If the overall recovery of pure water (through stream W) is 0.75 kg/kg feed, the recycle ratio (R/F) is

(a) 0.25 (b) 0.5
(c) 0.75 (d) 1.0

238. A 35 weight% Na_2SO_4 solution in water initially at 50°C is fed to a crystallizer at 20°C. The product stream contains hydrated crystals $Na_2SO_4.10H_2O$ in equilibrium with a 20 weight% Na_2SO_4 solution. The molecular weights of Na_2SO_4 and $Na_2SO_4.10H_2O$ are 142 and 322, respectively. The feed rate of 35% solution required to produce 500 kg/h of hydrated crystals is

(a) 403 kg/h (b) 603 kg/h
(c) 803 kg/h (d) 1103 kg/h

239. 600 kg/h of saturated steam at 1 bar (Enthalpy = 2675.4 kJ/kg) is mixed adiabatically with superheated steam at 450°C and 1 bar (Enthalpy = 3382.4 kJ/kg). The product is superheated steam at 350°C and 1 bar (Enthalpy = 3175.6 kJ/kg). The flow rate of the product is

(a) 711 kg/h (b) 1111 kg/h
(c) 1451 kg/h (d) 2051 kg/h

240. Carbon black is produced by the decomposition of CH_4.

$$CH_4(g) \rightarrow C(s) + 2H_2(g)$$

The single pass conversion of CH_4 is 60%. If fresh feed is pure CH_4 and 25% of CH_4 exiting the reactor is recycled, then the molar ratio of fresh feed stream to recycle stream is

(a) 0.9 (b) 9
(c) 10 (d) 90

241. Methane and steam are fed to a reactor in molar ratio 1:2. The following reactions take place:

$$CH_4(g) + 2H_2O(g) \rightarrow CO_2(g) + 4H_2(g)$$

$$CH_4(g) + H_2O(g) \rightarrow CO(g) + 3H_2(g)$$

where CO_2 is the desired product, CO is the undesired product and H_2 is a by-product. The exit stream has the following composition:

Species	CH_4	H_2O	CO_2	H_2	CO
Mol%	4.35	10.88	15.21	67.39	2.17

The selectivity for desired product relative to undesired product is
(a) 2.3 (b) 3.5
(c) 7 (d) 8

242. Methane and steam are fed to a reactor in molar ratio 1:2. The following reactions take place:

$$CH_4(g) + 2H_2O(g) \rightarrow CO_2(g) + 4H_2(g)$$

$$CH_4(g) + H_2O(g) \rightarrow CO(g) + 3H_2(g)$$

where CO_2 is the desired product, CO is the undesired product and H_2 is a by-product. The exit stream has the following composition:

Species	CH_4	H_2O	CO_2	H_2	CO
Mol%	4.35	10.88	15.21	67.39	2.17

The fractional yield of CO_2 is (fractional yield is defined as the ratio of the moles of the desired product formed to that would have been formed if there were no side reactions and the limiting reactant had reacted completely).
(a) 0.7 (b) 0.88
(c) 1 (d) 3.5

243. Methane and steam are fed to a reactor in molar ratio 1:2. The following reactions take place:

$$CH_4(g) + 2H_2O(g) \rightarrow CO_2(g) + 4H_2(g)$$

$$CH_4(g) + H_2O(g) \rightarrow CO(g) + 3H_2(g)$$

where CO_2 is the desired product, CO is the undesired product and H_2 is a by-product. The exit stream has the following composition:

Species	CH_4	H_2O	CO_2	H_2	CO
Mol%	4.35	10.88	15.21	67.39	2.17

The fractional conversion of methane is
(a) 0.4 (b) 0.5
(c) 0.7 (d) 0.8

244. Air (79% N_2 and 21% O_2) is passed over a catalyst at high temperature. Oxygen completely reacts with nitrogen as shown below:

$$\frac{1}{2}N_2 + \frac{1}{2}O_2 \rightarrow NO$$

$$\frac{1}{2}N_2 + O_2 \rightarrow NO_2$$

The molar ratio of NO to NO_2 in the product stream is 2:1. The fractional conversion of nitrogen is

(a) 0.13
(b) 0.20
(c) 0.27
(d) 0.40

245. A simplified flowsheet is shown in the figure for production of ethanol from ethylene. The conversion of ethylene in the reactor is 30% and the scrubber following the reactor completely separates ethylene as top stream and ethanol and water as bottom stream. The last (distillation) column gives an ethanol-water azeotrope (90 mol% ethanol) as the final product and water as the waste. The recycle to purge ratio is 34. The reaction is:

$$C_2H_4(g) + H_2O(g) \rightarrow C_2H_5OH(g)$$

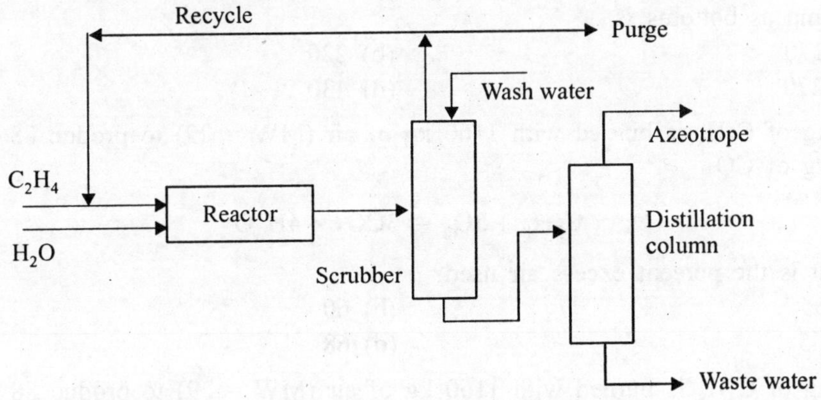

For an azeotrope product rate of 500 mol/h, the recycle gas flow rate in mol/h is

(a) 30
(b) 420
(c) 1020
(d) 1500

246. A simplified flowsheet is shown in the figure for production of ethanol from ethylene. The conversion of ethylene in the reactor is 30% and the scrubber following the reactor completely separates ethylene as top stream and ethanol and water as bottom stream. The last (distillation) column gives an ethanol-water azeotrope (90 mol% ethanol) as the final product and water as the waste. The recycle to purge ratio is 34. The reaction is:

$$C_2H_4(g) + H_2O(g) \rightarrow C_2H_5OH(g)$$

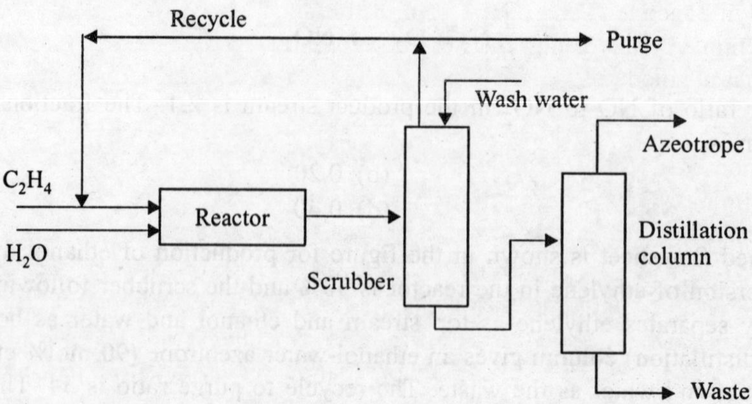

If the fresh H_2O feed to the reactor is 600 mol/h, and wash water for scrubbing is 20% of the condensables coming out of the reactor, the water flow rate from the distillation column as bottoms is

(a) 170 (b) 220
(c) 270 (d) 430

247. 44 kg of C_3H_8 is burned with 1160 kg of air (MW = 29) to produc 88 kg CO_2 and 14 kg of CO.

$$C_3H_8 + 5O_2 \rightarrow 3CO_2 + 4H_2O$$

What is the percent excess air used?

(a) 55 (b) 60
(c) 65 (d) 68

248. 44 kg of C_3H_8 is burned with 1160 kg of air (MW = 29) to produc 88 kg CO_2 and 14 kg of CO.

$$C_3H_8 + 5O_2 \rightarrow 3CO_2 + 4H_2O$$

What is the percent carbon burned?

(a) 63.3 (b) 73.3
(c) 83.3 (d) 93.3

249. Carbon monoxide (CO) is burnt in the presence of 200% excess pure oxygen and the flame temperature achieved is 2298 K. The inlet streams are at 25°C. The standard heat of formation (at 25°C) of CO and CO_2 are – 110 kJ/mol and –390 kJ/mol, respectively. The heat capacities in J/mol K of the components are:

$$C_{pO_2} = 25 + 14 \times 10^{-3} T$$

$$C_{pCO_2} = 25 + 42 \times 10^{-3} T$$

where T is the temperature in K. The heat loss (in kJ) per mole of CO burnt is

(a) 100 (b) 51.6
(c) 34.6 (d) 121.6

250. A reverse osmosis unit treats feed water (F) containing fluoride and its output consists of a permeate stream (P) and a reject stream (R). Let C_F, C_P and C_R denote the fluoride concentration in the feed, permeate and reject streams, respectively. Under steady state conditions, the volumetric flow rate of the reject is 60% of the volumetric flow rate of the inlet stream, and C_F = 2 mg/L and C_P = 0.1 mg/L. The value of C_R in mg/L is

(a) 4.3 (b) 5.3
(c) 3.3 (d) 2.3

251. A reverse osmosis unit treats feed water (F) containing fluoride and its output consists of a permeate stream (P) and a reject stream (R). Let C_F, C_P and C_R denote the fluoride concentration in the feed, permeate and reject streams, respectively. Under steady state conditions, the volumetric flow rate of the reject is 60% of the volumetric flow rate of the inlet stream, and C_F = 2 mg/L and C_P = 0.1 mg/L. A fraction f of the feed is bypassed and mixed with the permeate to obtain treated water having fluoride concentration of 1 mg/L. The flow rate of the reject stream is 60% of the flow rate entering the reverse osmosis unit (after the bypass). The value of f is

(a) 0.26 (b) 0.22
(c) 0.18 (d) 0.30

252. What is the heat required (in kJ up to 1 digit after the decimal point) to raise the temperature of one mole of the solid material from 100°C to 1000°C? The specific heat C_p of the material in J/mol K is expressed as $C_p = 20 + 0.005\,T$ where T is in K. Assume no phase change.

(a) 21.7 (b) 31.4
(c) 42.3 (d) 11.6

253. An ideal gas is initially at a pressure of 0.1 MPa and a total volume of 2 m^3. It is compressed to 1 MPa by a reversible adiabatic process and then cooled at constant pressure to a final volume of 0.2 m^3. The total work done in kJ on the gas for the entire process is (take R = 8.314 J/mol K, $C_p = 2.5R$)

(a) 827 (b) 757
(c) 631 (d) 503

254. Two elemental gases (A and B) are reacting to form liquid (C) in a steady state process as per the reaction A + B → C. The single pass conversion of the reaction is only 20%, and hence, recycle is used. The product is separated completely in pure form. The fresh feed has 49 mol% A and B each along with 2 mol% impurities. The maximum allowable impurities in the recycle stream is 20 mol%. The amount of purge stream (in moles) per 100 moles of the fresh feed is

(a) 5 (b) 15
(c) 20 (d) 10

255. For a pure liquid, the rate of change of vapour pressure with temperature is 0.1 bar/K in the temperature range of 300 K to 350 K. If the boiling point of the liquid at 2 bar is 320 K, the temperature in K at which it boils at 1 bar (up to one decimal place) is

(a) 305 K (b) 310 K
(c) 300 K (d) 330 K

256. The schematic diagram of a steady state process is shown below. The fresh feed (F) to the reactor consists of 96 mol% reactant A, and 4 mol% inert I. The stoichiometry of the reaction is A → C. A part of the reactor effluent is recycled. The molar flow rate of the recycle stream is $0.3F$. The product stream P contains 50 mol% C. The percent conversion of A in the reactor based on A entering the reactor at point 1 in the figure is

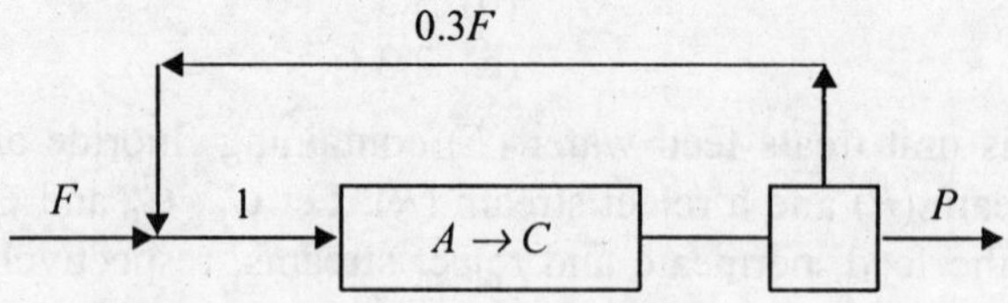

(a) 50.6 (b) 45.5
(c) 60.0 (d) 35.5

257. A liquid mixture of ethanol and water is flowing as inlet stream P into a stream splitter. It is split into two streams, Q and R, as shown in the figure below.

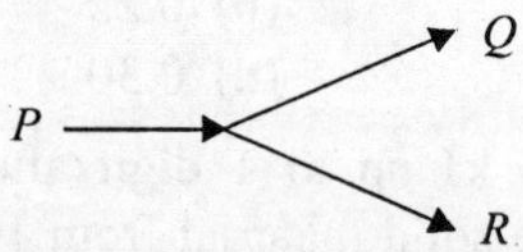

The flow rate of P, containing 30 mass% of ethanol, is 100 kg/h. What is the least number of additional specification(s) required to determine the mass flow rates and compositions (mass%) of the two exit streams?

(a) 0 (b) 1
(c) 2 (d) 3

258. A jacketed stirred tank with a provision for heat removal is used to mix sulphuric acid and water in a steady state flow process. H_2SO_4 (l) enters at a rate of 4 kg/h at 25°C and H_2O (l) enters at a rate of 6 kg/h at 10°C. The following data are available:

Specific heat capacity of water = 4.2 kJ/kg K.
Specific heat capacity of aqueous solution of 40 mass% H_2SO_4 = 2.8 kJ/kg solution K.
Assume the specific heat capacities to be independent of temperature.

Based on reference states of H_2SO_4 (l) and H_2O (l) at 25°C, the heat of mixing for aqueous solution of 40 mass% H_2SO_4 = – 650 kJ/kg H_2SO_4.

If the mixed stream leaves at 40°C, what is the rate of heat removal (in kJ/h)?

(a) 1802 (b) 2558
(c) 5702 (d) 6458

259. A catalytic reforming plant produces hydrogen and benzene from cyclohexane by dehydroaromatisation. In order to increase the production of hydrogen, the owner plans to change the process to steam reforming of the same feedstock that produces hydrogen and carbon dioxide. Stoichiometrically, what is the maximum ratio of pure hydrogen produced in the proposed process to that in the existing process?

(a) 1 (b) 2
(c) 5 (d) 6

ANSWERS TO OBJECTIVE TYPE QUESTIONS

1. (b)	**2.** (d)	**3.** (a)	**4.** (c)	**5.** (d)	**6.** (a)
7. (b)	**8.** (c)	**9.** (d)	**10.** (d)	**11.** (b)	**12.** (a)
13. (b)	**14.** (c)	**15.** (c)	**16.** (b)	**17.** (d)	**18.** (c)
19. (c)	**20.** (a)	**21.** (d)	**22.** (a)	**23.** (d)	**24.** (a)
25. (d)	**26.** (b)	**27.** (c)	**28.** (b)	**29.** (c)	**30.** (c)
31. (a)	**32.** (b)	**33.** (c)	**34.** (c)	**35.** (c)	**36.** (d)
37. (b)	**38.** (a)	**39.** (c)	**40.** (a)	**41.** (b)	**42.** (a)
43. (a)	**44.** (b)	**45.** (b)	**46.** (a)	**47.** (d)	**48.** (c)
49. (a)	**50.** (d)	**51.** (a)	**52.** (c)	**53.** (b)	**54.** (a)
55. (b)	**56.** (a)	**57.** (d)	**58.** (a)	**59.** (a)	**60.** (c)
61. (b)	**62.** (c)	**63.** (a)	**64.** (c)	**65.** (c)	**66.** (d)
67. (c)	**68.** (a)	**69.** (a)	**70.** (b)	**71.** (d)	**72.** (c)
73. (c)	**74.** (b)	**75.** (a)	**76.** (c)	**77.** (b)	**78.** (b)
79. (a)	**80.** (d)	**81.** (b)	**82.** (a)	**83.** (b)	**84.** (a)
85. (a)	**86.** (b)	**87.** (b)	**88.** (c)	**89.** (d)	**90.** (c)
91. (d)	**92.** (c)	**93.** (a)	**94.** (d)	**95.** (d)	**96.** (b)
97. (d)	**98.** (c)	**99.** (a)	**100.** (c)	**101.** (a)	**102.** (c)
103. (c)	**104.** (a)	**105.** (b)	**106.** (c)	**107.** (c)	**108.** (b)
109. (c)	**110.** (b)	**111.** (a)	**112.** (b)	**113.** (b)	**114.** (c)
115. (d)	**116.** (c)	**117.** (a)	**118.** (b)	**119.** (d)	**120.** (a)
121. (b)	**122.** (c)	**123.** (a)	**124.** (c)	**125.** (d)	**126.** (b)
127. (c)	**128.** (b)	**129.** (d)	**130.** (b)	**131.** (b)	**132.** (b)
133. (c)	**134.** (c)	**135.** (b)	**136.** (c)	**137.** (d)	**138.** (a)
139. (c)	**140.** (a)	**141.** (c)	**142.** (d)	**143.** (b)	**144.** (c)
145. (c)	**146.** (a)	**147.** (b)	**148.** (c)	**149.** (b)	**150.** (b)
151. (a)	**152.** (b)	**153.** (c)	**154.** (d)	**155.** (b)	**156.** (b)
157. (a)	**158.** (c)	**159.** (c)	**160.** (a)	**161.** (c)	**162.** (a)
163. (c)	**164.** (b)	**165.** (a)	**166.** (c)	**167.** (b)	**168.** (c)
169. (d)	**170.** (a)	**171.** (b)	**172.** (a)	**173.** (c)	**174.** (b)
175. (b)	**176.** (c)	**177.** (b)	**178.** (c)	**179.** (a)	**180.** (c)
181. (b)	**182.** (d)	**183.** (d)	**184.** (c)	**185.** (a)	**186.** (d)
187. (c)	**188.** (b)	**189.** (b)	**190.** (d)	**191.** (d)	**192.** (b)
193. (b)	**194.** (d)	**195.** (c)	**196.** (b)	**197.** (a)	**198.** (c)
199. (b)	**200.** (d)	**201.** (c)	**202.** (a)	**203.** (c)	**204.** (d)
205. (d)	**206.** (c)	**207.** (b)	**208.** (b)	**209.** (c)	**210.** (a)
211. (a)	**212.** (d)	**213.** (c)	**214.** (b)	**215.** (b)	**216.** (b)
217. (c)	**218.** (c)	**219.** (d)	**220.** (a)	**221.** (b)	**222.** (a)
223. (b)	**224.** (b)	**225.** (c)	**226.** (a)	**227.** (d)	**228.** (c)
229. (b)	**230.** (a)	**231.** (b)	**232.** (a)	**233.** (b)	**234.** (c)
235. (d)	**236.** (d)	**237.** (b)	**238.** (c)	**239.** (c)	**240.** (b)
241. (c)	**242.** (a)	**243.** (d)	**244.** (b)	**245.** (c)	**246.** (d)
247. (d)	**248.** (c)	**249.** (c)	**250.** (c)	**251.** (a)	**252.** (a)
253. (b)	**254.** (d)	**255.** (b)	**256.** (b)	**257.** (c)	**258.** (c)
259. (d)					

Bibliography

Atkins, P.W., *Physical Chemistry*, 4th ed., ELBS-Oxford, Oxford University Press, New Delhi, 1990.

Badger, W.L. and J.T. Banchero, *Introduction to Chemical Engineering*, McGraw-Hill, Tokyo, 1955.

Benedict, M., G. Webb, and L. Lubin, *Art. J. Chem. Phys.*, **8**, 334, 1940.

Bhat, B.I. and S.M. Vora, *Stoichiometry*, Tata McGraw-Hill, New Delhi, 1996.

Callen, H.B., *Thermodynamics*, John Wiley, New York, 1960.

Coulson, J.M., J.F. Richardson and E.K. Sinnot, *Chemical Engineering*, vol 6, Pergamon Press, New York, 1983.

Cox, E.R., *Ind. Eng. Chem.*, **15**, 592, 1923.

Daubert, T.E., *Chemical Engineering Thermodynamics*, McGraw-Hill, New York, 1985.

DePriester, C.L., *Chem. Eng. Progr.*, Symposium Ser., 7, 49, 1953.

Doran, P.M., *Bioprocess Engineering Principles*, Academic Press, 2005.

Geankoplis, C.J., *Transport Processes and Unit Operations*, 4th ed., Prentice-Hall of India, New Delhi, 2005.

Green, D.W. and J.O. Malony (Eds.), Perry's *Chemical Engineer's Handbook*, 6th ed., McGraw-Hill, New York, 1984.

Groggins, P.H., *Unit Processes in Organic Synthesis*, McGraw-Hill, Tokyo, 1958.

Henley, E.J. and H. Bieber, *Chemical Engineering Calculations: Mass and Energy Balances*, McGraw-Hill, New York, 1959.

Himmelblau, D.M., *Basic Principles and Calculations in Chemical Engineering*, 6th ed., Prentice-Hall of India, New Delhi, 1992.

Hougan, O.A., K.M. Watson and R.A. Ragatz, *Chemical Process Principles, Part. I and II*, 2nd ed., Wiley, New York, 1963.

https://en.wikipedia.org/wiki/Molecular_engineering (2015).

King, C.J., *Separation Processes*, McGraw-Hill, New York, 1971.

Kirkwood, J. and I. Oppenheim, *Chemical Thermodynamics*, McGraw-Hill, New York, 1961.

Kyle, B.G., *Chemical and Process Thermodynamics*, 3rd ed., Prentice-Hall of India, New Delhi, 2004.

Lee, B.L. and M.G. Kessler, "A Generalized Thermodynamic Correlation Based on Three-parameter Corresponding States", *AIChE J.*, **21**, 510, 1975.

Lewis, G.N., M. Randall, K.S. Pitzer and L. Brewer, *Thermodynamics*, McGraw-Hill, New York, 1981.

Lewis, W.K., A.H. Radasch and H.C. Lewis, *Industrial Stoichiometry: Chemical Calculations of Manufacturing Processes*, McGraw-Hill, New York, 1954.

Mashelkar R.A. and J.V. Rajan, "Chemical Engineering Developments in India" in *One Hundred Years of Chemical Engineering*, Peppas, N.A. (Ed.), Kluwer Academic Publishers, 1989.

McCabe, W.L. and J.C. Smith, *Unit Operations in Chemical Engineering*, McGraw-Hill, New York, 1976.

Modell, M. and R.C. Reid, *Thermodynamics and Its Applications*, Prentice Hall Inc., New Jersey, 1974.

Narayanan, K.V., *A Textbook of Chemical Engineering Thermodynamics*, Prentice-Hall of India, New Delhi, 2001.

Nelson, L.E. and E.F. Obert, "Generalized Compressibility Charge" *Chem. Eng.*, **61**(7), 203–208, 1954.

Peppas N.A., "The Origins of Academic Chemical Engineering" in *India in One Hundred Years of Chemical Engineering* by Peppas N.A. (Ed.), Kluwer Academic Publishers, 1989.

Pitzer K.S., "The Volumetric and Thermodynamic Properties of Fluids II: Compressibility Factor, Vapour Pressure and Entropy of Vaporization", *J. Am. Chem. Soc.*, **77**, 3427 1955.

Prausnitz, J.M., R.N. Lichtenthaler and E.G. Azevedo, *Molecular Thermodynamics of Fluid Phase Equilibria*, 2nd ed., Prentice Hall, New Jersey, 1986.

Rao, Ch.D. and D.V.S. Murthy, *Process Calculations for Chemical Engineers*, Chemical Engineering Education Development Centre, IIT Madras, 1981.

Rao, M.G. and M. Sittig, *Outlines of Chemical Technology*, Affiliated East-West Press, New Delhi, 1973.

Redlich, O. and J.N.S. Kwong, "On the Thermodynamics of Solutions", *Chem. Rev.*, 44, 233, 1949.

Reid, R.C., J.M. Prausnitz and Pauling B.E., *The Properties of Gases and Liquids*, McGraw-Hill, New York, 1988.

Sandler, S.I., *Chemical and Engineering Thermodynamics*, Wiley, New York, 1977.

Sharma, M.M., "Chemical Engineering Science in India—Needs, Opportunities and Challenges", *Chem. Eng. Sci.*, **42**, 2497, 1987.

Shreve, R.N., *Chemical Process Industries*, McGraw-Hill, New York, 1967.

Shuler and Kargi, "*Bioprocess Engineering: Basic Concepts*", 2nd ed., Prentice-Hall, New Jersey, 2004.

Smith, J.M. and H.C. Van Ness, *Introduction to Chemical Engineering Thermodynamics*, 4th ed., McGraw-Hill, New York, 1987.

Soave, G., "Equilibrium Constants from a Modified Equation of State", *Chem. Engg. Sci.*, **27**, 1197, 1972.

Sonntag, R.E. and G.J. Van Wylen, *Introduction to Thermodynamics*, Wiley, New York, 1971.

Treybal, R.E., *Mass-Transfer Operations*, 3rd ed., McGraw-Hill, New York, 1981.

Tyner, M., *Process Engineering Calculations: Material and Energy Balance*, Ronald Press Co., New York, 1960.

Van Ness, H.C., "Chemical Engineering Education: Will We Ever Get it Right?", *Chem. Eng. Prog.*, p. 18, Jan. 1989.

Vilbrandt, F.C. and C.E. Dryden, *Chemical Engineering Plant Design*, McGraw-Hill, New York, 1959.

Walker, W.H., W.K. Lewis, W.H. McAdams and E.R. Gilliland, *Principles of Chemical Engineering*, McGraw-Hill, New York, 1937.

Wark, K., *Thermodynamics*, 2nd ed., McGraw-Hill, New York, 1971.

Weber, H.E. and H.C. Meissner, *Thermodynamics for Chemical Engineers*, 2nd ed., Wiley, New York, 1957.

Westmoreland, P.R., "Chemical Engineering in the Next 25 Years", *CEP*, pp. 31–41, November, 2008.

Williams, E.T. and R.C. Johnson, *Stoichiometry for Chemical Engineer*, McGraw-Hill, New York, 1958.

Answers to Exercises

Chapter 2

2.1 (a) 10^{-3} Pa s (b) 2.419 lb/(ft h)

2.2 1.0197 kgf/cm^2, 14.5038 psi

2.3 0.2596 W/m K

2.4 (a) 257.95 kcal/(h m^2 °C) (b) 52.832 Btu/(h ft^2 °F)

2.5 iron

2.6 5.69×10^{-2} lb/ft^2.s

2.7 36.81 kg/s

2.8 66.67 kg, 654 N

2.9 18 W

2.10 15.3 m, 12.5 kJ

2.11 (a) 1286.305 N (b) 1.6378 bar (c) 0.5145 kJ (d) 196.2 J

2.12 $V(\text{inch}^3) = 0.0610\ e^{3600t}$

2.13 5.6629×10^{-8} W/(m^2.K^4)

2.14 (a) $\dfrac{\text{Btu (inch)}^{0.25}}{\text{ft}^2\text{(h)(°F)}^{1.25}}$

(b) $q = 1.313 \dfrac{\Delta T^{1.25}}{D^{0.25}}$, where q is in W/m^2, D is in m and ΔT is in K

2.15 a = 3.602 atm L^2/(mol)^2, $b = 4.28 \times 10^{-2}$ L/(mol)

2.16 (a) $k_1 = \dfrac{\text{kmol}}{\text{m}^3(\text{min})(\text{atm})^{2.2}}$; $k_2 = \dfrac{\text{kmol}}{\text{m}^3(\text{min})(\text{atm})^{0.5}}$

(b) $k_1' = 4.795 \times 10^5$, $k_2' = 24.185$

2.17 $\Delta P = 7.5885 \left[\dfrac{\mu^{0.15} H \rho^{0.85} V^{1.85}}{D_P^{1.15}} \right]$

where ΔP is in Pa, μ is in lb/(ft s), H is in ft, ρ is in lb/ft^3, V is in ft/s and D_P is in ft.

2.18 (a) $A = 18.2773$, $B = 3799.887$, $C = -226.296$
(b) $A = 14.331\ 28$, $B = 6839.7966$, $C = -375.0628$

2.19 (a) $C_P = 96.55 + 0.1562T$ (b) $C_P = 33.24 + 0.0373T$

2.20 (a) $C_P = 6.35 + 1.811 \times 10^{-3}T - 2.675 \times 10^{-7}T^2$
(b) $C_P = 6.825 + 1.665 \times 10^{-3}T - 2.675 \times 10^{-7}T^2$
(c) $C_P = 9.495 \times 10^2 + 0.271T - 4.0 \times 10^{-5}T^2$
(d) $C_P = 6.795 + 0.93 \times 10^{-3}T - 8.256 \times 10^{-8}T^2$

2.21 $C_P = 26.5386 + 4.24 \times 10^{-2}T - 1.4295 \times 10^{-5}T^2$

2.22 (a) 3.038 (b) 10.94

2.23 1.8829×10^{-4}

Chapter 3

3.1 267.41 g

3.2 3 mol

3.3 266.45 kg

3.4 72 g

3.5 58% C, 3.4% H and 38.6% Br

3.6 9 kg-equivalents

3.7 57.026

3.8 15.82 g.equivalents

3.9 49

3.10 C_7H_5N

3.11 (a) 125.38 kg (b) 12.685 g

3.12 (a) 684.48 kg (b) 254.10 kg

3.13 (a) 93.65 kg (b) 88.68 kg

3.14 (a) 146.58 g (b) $BaCl_2$: 44.61 g, Na_2SO_4: 30.43 g (c) 25.04 g

3.15 1292.76 kg

3.16 0.41%

3.17 43.74 kg, 22.28 m^3 at STP
3.18 (a) 17.64 m^3 (b) 164.21 kg
3.19 1053.2 kg, 895.2 kg
3.20 0.8651 N
3.21 1.5
3.22 1.076
3.23 7.273
3.24 (a) 420 kg/m^3 (b) 2.24 kmol/h
3.25 18.91°Bé, 30°Tw
3.26 45°Bé
3.27 29.1°Bé , 29.3°API
3.28 0.25
3.29 947 kg/m^3
3.30 (a) 768.37 kg/m^3 (b) 52.46
(c) gasoline = 24.66%, kerosene = 26.82%, gasoil = 28.30% and isopentane = 20.22%
3.31 0.9263
3.32 131.53 kg of Mg, 1136.78 kg of Na, 2132.67 kg of Cl_2
3.33 75.69
3.34 K 44.87%, S 18.4%, O_2 36.73%
3.35 47.23
3.36 Ferrous 44.06%, Ferric 34.43%
3.37 12.38% $NaNO_3$, 87.62% water
3.38 567.04 kg
3.39 90.02%
3.40 0.4959, 0.33
3.41 55.23%
3.42 3.7203 × 10^5 kg
3.43 557.3 mol
3.44 24.57% $SiO_2 \cdot 2CaO$, 43.43% $SiO_2 \cdot 3CaO$
3.45 95.58% (weight) ethanol
3.46 (a) 1.688% (b) 3.48%
3.47 720 kg
3.48 1.0753
3.49 0.2860
3.50 39.75%, 0.6596
3.51 (a) 622.5 (b) 1.2971 (c) 1.213 kg

3.52 (a) 0.4751 (b) 42.77 (c) 33.75

3.53 (a) 40.78% A, 27.18% B, 32.04% C (b) 63.07 (c) 41.38% (d) 0.6242

3.54 16.66%

3.55 30.73

3.56 ₹ 275.5

3.57 (a) 1.557 g-atom Ca, 3.133 g-atom C (b) CaC_2

3.58 (a) 33.27 kg (b) 36.96 kg

3.59 (a) 3.09% (b) 0.0319

3.60 (a) 23.29% O_2, 76.71% N_2 (b) 28.85

3.61 Chlorine 74.03%, Bromine 13.73%, Oxygen 12.24%

3.62 7.7%

3.63 10.43

3.64 1291 kg/m^3

3.65 70 000 ppm

3.66 (a) 179.2 (b) 0.046

3.67 2.78 mol/kg, 2.646 mol/L

3.68 (a) 12.94% (b) 65.00% (c) 5.773 mol/L (d) 8.25 mol/kg solvent

3.69 (a) 29.77% (b) 8.1467 mol/L (c) 8.1467 g eq/L (d) 15.27% LiCl, 84.73% water

3.70 (a) 0.293 (b) 0.0727, (c) 4.35 mol/kg water (d) 8 g eq/L

3.71 (a) 58.59% (b) 9.92 mol/L (c) 21.716 mol/kg water

3.72 (a) KCl = 24%, water = 76% (b) 0.0709 (c) 5.15 mol/L (d) 4.24 mol/kg

3.73 (a) 0.0287 kg (b) 0.26602 mol/L

3.74 (a) Nitrogen (b) 32.76% (c) 30%

3.75 94%, 94.95%

3.76 (a) 90% (b) 100% (c) 75%

3.77 $\dfrac{0.5\varepsilon}{n_0 + 0.5\varepsilon}$

3.78 $\dfrac{4 + 3\varepsilon}{8 + 2\varepsilon}$

3.79 $\dfrac{3\varepsilon_1 + 4\varepsilon_2}{5 + 2\varepsilon_1 + 2\varepsilon_2}$

3.80 $\dfrac{3 - \varepsilon_1 - \varepsilon_2}{5 - 0.5\varepsilon_1}$

3.81 10%, 94.4%, 5.44%

3.82 (a) carbon 240 kg, silica 865 kg (b) 188 kg (c) 1491 kg

3.83 (a) 474.8 kg (b) 8.4% SO_2, 10.11% O_2 and 81.49% N_2

3.84 (a) Ammonia, 17.97% (b) 53.89 kg

3.85 (a) 25.96 m^3 (b) 12.46 m^3 (c) 35.3 m^3
(d) 35.25% SO_3, 23.53% SO_2, 41.18% O_2 (e) 56.47

3.86 (a) 29.5% (b) 14.71 kPa

3.87 (a) 0.4 (b) 0.343 bar

3.88 20%

3.89 (a) 300 m^3 (b) 994.29 m^3 (c) N_2 = 29.32 m^3, H_2 = 87.96 m^3, NH_3 = 58.64 m^3

3.90 (a) 0.292 m^3 (b) 40%

Chapter 4

4.1 1010.79 kg

4.2 91.67 m^3

4.3 7.688 m^3

4.4 (a) 2.167 kg (b) 1.702 kg

4.5 0.3557 m^3/kg; 0.65% greater

4.6 12.67 bar

4.7 5.6686×10^5 N/m^2

4.8 288.8 K

4.9 (a) 0.7153 kg/m^3 (b) 2.8582 kg/m^3

4.10 7.6098×10^4 kg

4.11 78.98 kg

4.12 0.355 kg

4.13 1.0195×10^{-3} m^3

4.14 375 K

4.15 1.119×10^{-3} m^3

4.16 2.48 g, 7.52 g

4.17 (a) 71.73% ethane, 2.01% hydrogen and 26.26% carbon dioxide
(b) 2.8865×10^{-2} mol/L ethane, 1.2027×10^{-2} mol/L hydrogen, 7.2163×10^{-3} mol/L carbon dioxide
(c) 241.82 kg/h

4.18 (a) 35.74 kPa (b) 11.59 m^3

4.19 3.2462 m^3

4.20 (a) 20.14 kPa (b) 4.42 m^3 (c) 2.196 m^3/kg

4.21 C_3H_8

4.22 78%

4.23 air: 4.313×10^5 N/m², methyl alcohol: 1.951×10^5 N/m², and steam: 1.736×10^5 N/m²

4.24 30.25

4.25 28.96, 1.1836 kg/m³

4.26 (a) CO_2 2.12%, N_2 5.39%, CH_4 92.5% (b) 16.64 (c) 0.7424 kg/m³

4.27 20

4.28 (a) 30.53 (b) CO_2 = 20.18, SO_2 = 1.05, CO = 1.83, O_2 = 2.62, N_2 = 74.32
(c) 1.7213 kg/m³ (d) 1.0582

4.29 (a) 19.68 (b) 22.36% CO_2, 53.44% Cl_2, 17.8% N_2, 6.4% H_2

4.30 (a) 34.62 (b) 38.82

4.31 (a) 25.15 (b) 1.21 kg/m³

4.32 (a) 3.0068 kmol (b) 0.3757 kmol, 0.125 (c) 0.7578 kg, 0.007
(d) 6.25 m³ (e) 35.95

4.33 (a) 120 kPa (b) 1.3826 m³ (c) 2 kg/m³

4.34 (a) 250 bar, 750 bar (b) 8.5155 (c) 204.84 kg/m³

4.35 (a) CO: 54.34, H_2: 4.72, CH_4: 0.36, CO_2: 13.50, O_2: 0.36, N_2: 26.72,
(b) 17.93 (c) 0.8

4.36 (a) 38.98
(b) Ethylene: 22.0%, Benzene: 49.05%, Oxygen: 1.07%,
Methane: 6.37%, Ethane: 19.27%, Nitrogen: 2.23% (c) 1.7391

4.37 0.0252 kg

4.38 (a) 0.011, (b) 0.648% NH_3, 99.352% air, (c) 27.8, 28.87

4.39 (a) 121.6 m³ (b) 24.7 kg

4.40 (a) 70.76 m³ (b) 37.1 kg (c) 2.1% HCl, 97.9% Air

4.41 (a) 70.85 m³ (b) 36.98 kg

4.42 (a) 92.02 m³ (b) 1.082 kg

4.43 (a) 46.44 m³ (b) 34.22 m³ (c) 87.72%

4.44 (a) 81.91 m³ (b) 25.92 kg

4.45 (a) 33.36 (b) 30.28 (c) 20.73 kg (d) 60.28%

4.46 (a) 2.778 m³/min (b) 22.727 m³/min

4.47 (a) 91.43 m³/h (b) 15.05 kg/h (c) 91.84%

4.48 (a) 37.41 kg (b) 115.838 kg

4.49 0.3177 m³

4.50 (a) Ammonia, 17.97% (b) 53.89 kg

4.51 (a) 25.96 m³ (b) 12.46 m³ (c) 35.31 m³ (d) 35.29% SO_3, 23.53% SO_2, 41.18% O_2
(e) 56.51

4.52 (a) 29.5% (b) 14.71 kPa

4.53 (a) 70% (b) 0.6 bar

4.54 (a) 195.4 m^3 (b) 36.036 kg

4.55 (a) 64.34% (b) 15.074 m^3 (c) 8.605 m^3

4.56 404

4.57 (a) 4.98% CH_3OH, 3.73% O_2, 51.48% N_2, 19.9% HCHO, 19.9% H_2O
(b) 395.13 m^3

4.58 20%

4.59 30%

4.60 0.333 ethylene, 0.667 ethane

4.61 (a) 11.08% CO_2, 83.35% N_2, 5.57% H_2O (b) 26.98 kg (c) 0.0836 bar

4.62 (a) 1142.58 m^3 (b) 4588.6 m^3/h (c) 259.39 kg, 51.37% HNO_3

Chapter 5

5.1 380.1 K

5.2 601.3 K

5.3 (a) 68.30 bar (b) 51.8 bar

5.4 23.84 bar

5.5 93.8 kPa, 94.23 kPa by ideal gas equation

5.6 9.78 kg

5.7 1.48 kg

5.8 (a) 3.502×10^{-3} m^3/mol, 0.936 (b) 3.5021×10^{-3} m^3/mol, 0.9361

5.9 0.045 m^3, 0.06751 m^3

5.10 0.3206×10^{-3} m^3/mol

5.11 (a) $V = 0.249 \times 10^{-3}$ m^3 (b) 0.23×10^{-3} m^3/mol

5.12 (a) 3×10^{-3} m^3/mol (b) 2.98×10^{-3} m^3/mol

5.13 1.8×10^{-4} m^3/mol, $a = 0.54985$ N m^4/mol^2, $b = 6.41 \times 10^{-5}$ m^3/mol

5.14 7.134×10^{-5} m^3/mol, 1.712×10^{-3} m^3/mol

5.15 3.485×10^{-3} m^3/mol, 0.8862

5.16 (a) 4.157×10^{-5} m^3/mol (b) 6.44×10^{-5} m^3/mol (c) 5.3×10^{-5} m^3/mol

5.17 76.76 bar

5.18 551 K

5.19 0.1255 m^3, 0.1357 m^3 using ideal gas equation

5.20 (a) 3.795×10^{-4} m^3/mol (b) 3.726×10^{-4} m^3/mol

5.21 (a) 3.639×10^{-3} m^3/mol (b) 3.52×10^{-3} m^3/mol (c) 3.514×10^{-3} m^3/mol

5.22 (a) 2.036×10^{-4} m^3/mol (b) 1.433×10^{-4} m^3/mol (c) 1.4575×10^{-4} m^3/mol

5.23 165.2 bar

5.24 $V = 1.426 \times 10^{-4}$ m^3/mol

5.25 0.1196 m^3

5.26 371.8 K

5.27 (a) 3.1178×10^{-4} m^3/mol (b) 2.8635×10^{-4} m^3/mol
(c) 2.847×10^{-4} m^3/mol (Amagat's law) 3.046×10^{-4} m^3/mol (Dalton's law)
(d) 2.962×10^{-4} m^3/mol

5.28 (a) 2.8776 kg/m^3 (b) 2.8769 kg/m^3

Chapter 6

6.1 63.51 kPa and 118.04 kPa

6.2 4.184×10^2 kJ/kg

6.3 2.34 kPa (293 K), 7.63 kPa (313 K), 13.04 kPa (323 K), 20.87 kPa (333 K), 47.95 kPa (353 K)

6.4 0.4653 bar

6.5 234.39 K

6.6 (a) 314.6 K (b) 28.06 kPa

6.7 15.33 kPa

6.8 6.7 kg

6.9 42 kPa, 0.5%

6.10 1.39 kg

6.11 1.29 kg

6.12 27.76 kPa, 0.18%

6.13 350.6 K, 30.55 kPa, 4.5%

6.14 (a) A = 14.655, B = 2209.7, C = 43 (b) 2861 kPa

6.15 370 K

6.16 (a) 181.97 kPa (b) 444.94 kJ/kg

6.17 (a) 27.102 kPa, 1.66% (b) 379.5 K

6.18 27.46 kPa, 3.0% deviation

Chapter 7

7.1 mole fraction of heptane x = 0.853, y = 0.893

7.2 7.47 kPa (benzene), 1.91 kPa (toluene); 76.8% (weight) benzene

7.3 (a) 25.2% A (b) 57.55 kPa

7.4 The liquid contains 67.42% (mol) nitrogen and the vapour contains 90.34% (mol) nitrogen

7.5 92.49% A

7.6 x = 0.5608, y = 0.7344

7.7 (a) 366.82 K (b) 28.05% *n*-hexane

7.8

T (K)	353.1	358	363	368	373	378	383	383.6
x	1.0	0.78	0.581	0.411	0.258	0.130	0.017	0
y	1.0	0.90	0.777	0.632	0.456	0.261	0.039	0

7.9 $\bar{P}_A$ = 8.53 and $\bar{P}_B$ = 2.24 kPa. Weight percent of benzene in vapour = 76.4%

7.10 (a) 39.06 kPa (0, 0), 66.99 kPa (0.2, 0.5335), 94.92 kPa (0.4, 0.7531), 108.88 kPa (0.5, 0.8206), 122.84 kPa (0.6, 0.8728), 150.77 kPa (0.8, 0.9482), 178.7 kPa (1, 1)
(b) 108.9 kPa, 64 kPa

7.11 Between 350.9 K and 357.6 K

7.12 72.32% (Raoult's law)

7.13 94.7 kPa and 77.2 kPa

7.14 54%

7.15 64.86% *A*, 34.00% *B*, 1.14% *C*

7.16 (a) 25 kPa (b) 12.5% *A*, 37.5% *B*, 50.0% *C*

7.17 16.7% *A*, 83.3% *B*

7.18 (a) 26.7% *A*, 26.7% *B*, 46.6% *C* (b) 11.47% *A*, 33.25% *B*, 55.3% *C*

7.19 (a) T = 330.4 K, y = 0.5886; (b) P = 57.63 kPa, y = 0.5908

7.20 f = 0.1982, x = 0.5608, y = 0.7344

7.21 Exhibits positive deviation from Raoult's law. If Raoult's law is obeyed, vapour composition would be 0.413 mole fraction methanol.

7.22 131.24 kPa, 46.2% *A*

7.23 (a) 8.245×10^2 kPa,
(b) 25.1% butane, 46.3% propane, 28.6% ethane,
(c) 6.19×10^2 kPa, 35.17% butane, 64.83% propane

7.24 (a) 342 K (1, 1), 343 K (0.958, 0.983), 348 K (0.726, 0.875), 353 K (0.527, 0.735), 358 K (0.360, 0.581), 363 K (0.210, 0.389), 368 K (0.094, 0.195) 372.4 K (0, 0)

7.25 (a) x and y are given in brackets against P in kPa:
33.79 (0, 0), 43.426 (0.2, 0.3775), 53.062 (0.4, 0.6179), 62.698 (0.6, 0.7844), 72.334 (0.8, 0.9066), 81.79 (1, 1)

(b) x and y are given in brackets against T in K:
311.45 (1, 1), 315 (0.787, 0.902), 319 (0.581, 0.773), 323 (0.405, 0.623), 327 (0.254, 0.449), 331 (0.123, 0.250), 335.33 (0, 0)

7.26 (a) 352.8 K, 405.3 K,
(b) 352.8 (1, 1), 363.2 (0.685, 0.914), 373.2 (0.446, 0.786), 383.2 (0.266, 0.611), 393.2 (0.129, 0.378), 405.3 (0,0)
(c) 369 K, 386.5 K

7.27 (a) x and y are given in brackets against P in kPa:

30.74 (0, 0), 57.58 (0.2, 0.573), 84.42 (0.4, 0.782), 111.26 (0.6, 0.89), 138.1 (0.8, 0.956), 164.94 (1, 1)

(b) x and y are given in brackets against T in K:

352.88 (1, 1), 360.0 (0.7658, 0.9474), 370.0 (0.5160, 0.8512), 380.0 (0.3310, 0.7137), 390.0 (0.1900, 0.5273), 400.0 (0.0797, 0.2805), 408.9 (0, 0)

7.28 (a) 77.3% benzene (b) 57.5% benzene (c) 37.9%

7.29 (a) 89.26 kPa (b) 59.5% benzene, 29.5% toluene and 11.0% xylene (c) 92.80, 85.26

7.30 (a) $x = 0.5095$, $y = 0.8866$ (b) 338.9 K, 75.7% pentane (c) 108.5 kPa, 83.4% pentane

7.31 (a) 355.7 K (b) 0.56 (c) mole fraction of alcohol 0.16

7.32 9.0 kPa

7.33 0.101×10^{-4} kg oxygen per kg water

7.34 3.074×10^{-4} m^3

7.35 44.73 bar

7.36 $H_A = 0.23$ bar, $H_B = 0.217$ bar

7.37 5.0564×10^{-4} mol/kg

7.38 533 bar

7.39 (a) 15.7 kPa (b) 391.4 K (c) 374.1 K (d) 36.68 kPa

7.40 329.7 K, 356.9 K

7.41 1310 kPa, liquid composition: 6.8% ethane, 66.2% propane, 23.8% isobutene and the rest n-butane

7.42 (a) 330 K (b) 340.6 K

(c) Mole % in the liquid and vapour given in brackets: Ethane (0.48, 0.09), Propane (36.3, 17.45), Isobutane (18.18, 18.74), n-Butane (44.98, 63.35), Isopentane (0.13, 0.38)

7.43 500 mol each.

Component	*Mole % in liquid*	*Mole % in vapour*
Methane	3.26	28.78
Ethane	16.30	29.68
Propane	42.10	29.90
Isobutene	16.30	5.70
n-Butane	22.04	5.94

7.44 (a) 930.3 kPa

(b) 337.4 K, Composition of condensate: Ethane (1.8%), Propane (6.2%), Isobutane (17.3%), n-Butane (67.1%), Isopentane (7.50%) 330.2 K, Composition of the liquid and vapour: Ethane (3.22%, 16.81%), Propane (9.31%, 18.7%), Isobutane (19.55%, 18.46%), n-Butane (63.38%, 44.57%), Isopentane (4.54%, 1.46%)

7.45 2205 kPa, Methane: 41.91, Ethane: 20.25, Propane: 21.96, Isobutane: 8.57, *n*-Butane: 7.31 706 kPa, Methane: 0.2, Ethane: 1.91, Propane: 16.14, Isobutane: 30.88, *n*-Butane: 50.88

7.46 861.4 kPa, 2446 kPa

7.47 758 kPa

7.48 (a) 717 kPa (b) Propane: 63.93%, *n*-Butane: 36.07%
(c) Propane: 36.07%, *n*-Butane: 63.93%

7.49 (a) 356 K, 58.8% methanol, 24.1% ethanol, 8.4% propanol, 8.7% butanol,
(b) 373 K, 8.8% methanol, 8.9% ethanol, 13.6% propanol, 68.6% butanol

7.50 (a) 84.2 kPa (b) Ether = 51.26%, Water = 48.74%

7.51 (a) 347.97 K, Pure benzene (b) 342.2 K

7.52 3.35 kg

7.53 (a) 107.3 kPa (b) 66.44%

7.54 342 K, 0.0975 kg

7.55 (a) 342 K (b) 0.7 (c) 0.0989 kg

7.56 (a) 372.3 K (b) 3% aniline

7.57 (a) 295.68 K, 32.67% chlorobenzene (b) 341.1 K, 28.96% chlorobenzene

7.58 (a) 343.24 K (b) 80.15% Chloroform (c) 77.09% Chloroform

7.59 (a) 178.65 kPa (b) 361.14 K

7.60 (a) 327.2 K (b) Pure water (c) 316.7 K

7.61 (a) 369.2 K (b) 4.78% aniline

7.62 (a) 361.6 K, 21.9% toluene, 13.1% ethylbenzene, 65.0% water
(b) 387.8 K, 26.8% toluene, 73.2% ethyl benzene

7.64 (a) 366 K, pure water
(b) 374.9 K, pure toluene, last drop of vapour contains 44.36% toluene

7.65 (a) 388.8 K, pure component 1 (b) 365.5 K, pure component 2

7.66 (a) 78.76 kg (b) 94.18%

7.67 (a) L = 74.6 mol, V = 25.4 mol
(b) Liquid: 0.062 (A), 0.737 (B), 0.201 (C) Vapour: 0.409 (A), 0.591 (B)

7.68 2.842 kPa

7.69 372.47 K

7.70 (a) 8.85 K (b) 7.8 K

Chapter 8

8.1 (a) 16.6% (b) 29.2% (c) 0.1991 (d) 0.4127 (e) 0.4120 kg/m^3

8.2 (a) 3.56% (b) H_2O = 1.49%, CO_2 = 98.51% (c) 0.0369 (d) 0.0151 (e) 0.0257 kg/m^3

8.3 (a) 104.86 kPa (b) 0.0351

8.4 25.71 kg

8.5 Contains 0.946×10^{-2} kg benzene per m^3

8.6 51.93 m^3

8.7 50.99 kg

8.8 4 kg

8.9 18.8%

8.10 262.7 K

8.11 1623 kPa

8.12 43.6%

8.13 (a) 0.1429 (b) 0.3172 (c) 0.1667 (d) 0.4643 (e) 0.7988 (f) 2.2254 (g) 42.67 kg (h) 20.87% (i) 32.17%

8.14 (a) 10.13 kPa (b) 0.2302 (c) 0.111 (d) 0.2990 (e) 70.44% (f) 73.41% (g) 106.2 kg

8.15 (a) 0.3710 (b) 0.7690 (c) 28.4 kPa (d) 85.4% (e) 296.3 K

8.16 (a) 3.6122 kPa (b) 49.65% (c) 0.023 (d) 47.79% (e) 340.2 K

8.17 7.35%

8.18 (a) 0.0537 (b) 82.42% (c) 81.48% (d) 327.45 K

8.19 19349 kJ/m^3

8.20 (a) 271 K (b) 269 K

8.21 0.035

8.22 315.4 K

8.23 (a) 288.9 K (b) 0.0954 m^3

8.24 (a) 0.2281, 0.0846 (b) Benzene 7.8% (*V*), 18.6% (*W*) (c) 49.37%, 45.08%

8.25 (a) 306.8 K (b) 420.9 kPa

8.26 322.9 K

8.27 13.25 kg

8.28 344.22 kPa

8.29 (a) 0.3467 kg (b) 41.73%

8.30 309.7 K, 0.0157 kg/kg of dry air

8.31 0.0117 kg, 0.9857 m^3

8.32 (a) 19.2%, 90.6% (b) 0.67% water, 2.2% water (c) 18.63%, 90.3% (d) 0.487 kg, 1.625 kg (e) 0.484 kg, 1.614 kg

8.33 (a) 2.36% (b) 0.244 (c) 0.384 (d) 298.9 K (e) 236.6 kPa

8.34 (a) 0.377 kg of acetone/kg dry of nitrogen, 0.6727 kg of acetone/kg dry nitrogen (b) 0.2953 kg (c) 28.05 kg (d) 111.3 m^3

8.35 285 kPa

8.36 6498.35 m^3/h

8.37 (a) 78.8% (b) 24.23 m^3/min

8.38 20.5%

8.39 1.65 kg/h

8.40 (a) 0.0191 kmol/kmol of dry air, 0.00494 kmol/kmol of dry air (b) 1.0276 kg (c) 38.10 m^3

8.41 283.03 K

8.42 159.75 kPa

8.43 164.53 m^3/h

8.44 0.0258 kg of water per kg of dry air

8.45 0.0408 kg of water per kg of dry air

8.46 (a) 314.1 K (b) 0.224 kg of benzene per kg of dry air (c) 216.7 kJ

8.48 (a) 299.3 K (b) 304 K

8.49 (a) 279 K (b) 312.3 K

8.51 (a) 6.25% (b) 1.602 kPa (c) 0.16 kg of water/kg of dry air (d) 20.7 kPa (e) 7.73% (f) 287.7 K (g) 1.0238 kJ/kg (h) 300.7 K

8.52 (a) 0.04 (b) 0.0644 (c) 26.5% (d) 1.0036 m^3 (e) 310.2 K (f) 158.05 kJ/kg

8.53 (a) 0.035 kg/kg (b) 0.0564 kmol/kmol (c) 14% (d) 1.0255 m^3/kg (e) 313 K (f) 162.39 kJ/kg

8.54 55.81 kJ

8.55 0.99875 m^3/kg, 153 kJ/kg, 155 kJ/kg

8.56 (a) 0.013 kg/kg of dry air, 0.0209 kmol/kmol of dry air, 1%
(b) 329 K, 0.028 kg/kg dry air, 22.5%

8.57 (a) 0.035 kg/kg of dry air, 0.0564 kmol/kmol of dry air, 30%
(b) 5.407 kPa, 34.4% (c) 310.7 K (d) 1.071 kJ/kg, 0.979 m^3/kg, 150.3 kJ/kg

8.59 0.007 kg

8.60 (a) 299.5 K (b) 0.004 kg

8.61 (a) 298 K (b) 5.55 kg

8.62 (a) 0.04 kg/kg of dry air (b) 313 K (c) 310 K (d) 70%
(e) 0.046 kg/kg of dry air, 318 K (f) 0.598 kg

8.63 (a) 0.01 kg/kg dry air (b) 40% (c) 370 K (d) 68.6 kJ

8.64 0.0085 kg, 41.3 kJ/kg

8.65 (a) 333 K (b) 298 K (c) 31.82 kJ (d) 27.6 kJ

8.66 (a) 0.0322, 50%, 283.2 K (b) 0.1206, 85%, 316.2 K (c) 6.07 kg

8.67 (a) 289 K (b) 0.01 kg/kg of dry air (c) 0.004 kg

8.68 12122 kg/h wet air

8.69 6514.1 m^3/h

8.70 (a) 44.9% (b) 18 912.3 m^3/h

8.71 (a) 3604.6 m^3/h (b) 3355.4 m^3/h

8.72 (a) 305 K (b) 0.2489 kg (c) 8.76 m^3/h

8.73 3.15 kg

8.74 (a) 0.017 kg/kg of dry air, 58%, 295 K (b) 6782.5 kJ (c) 2.706 kg (d) 310 K

8.75 (a) 312 K, 323 K (b) 48% (c) 90.07 m^3

8.76 (a) 0.028 and 0.096 kg/kg of dry air (b) 45.45 kg/h (c) 82612 kJ/h

8.77 (a) 0.0175, 44%; 0.02, 100%; 0.02, 0.9%, 0.038, 45% (b) 2778 kg (c) 2686 m^3

8.78 (a) 295.7 K, 300.7 K (b) 16.69 kg/h (c) 342.7 K

8.79 (a) 318.65 K (b) 70%

8.80 (a) 311 K (b) 2.43 kg (c) 9.11 kg

8.81 (a) 310 K (b) 382.2 K (c) 3393.3 kg/h (d) 2851.9 m^3/h

Chapter 9

9.1 25 kg

9.2 3.457 kg

9.3 Urea 214.4 kg, Phosphoric acid 359 kg, KCl 411.5 kg

9.4 $0.8F - 0.11P = 400$; $B - 0.675P = 0$; $0.2F - 0.215P = 0$; $F = 573.3$ kg, $P = 533.3$ kg, $B = 360$ kg

9.5 6250

9.6 5625 kg

9.7 33.33 kg

9.8 43.48 kg

9.9 Waste acid 419.18 kg; Conc. HNO_3 397.26 kg, Conc. H_2SO_4, 183.56 kg

9.10 0.0244 g/L

9.11 106.28 mol/h

9.12 530.95 kg/h

9.13 0.3184 m^3

9.14 8 : 3 : 9

9.15 1 : 10.64

9.16 ₹ 19.19

9.17 (a) 1.6349×10^{-3} m^3 (b) 853 kg
9.18 (a) ₹ 102105 (b) ₹ 75 958
9.19 3.3322×10^4 kg/h
9.20 9.51 kg
9.21 (a) 2000 kg, 1200 kg (b) 2000 kg/h (c) 800 kg/h
9.22 (a) 25% NaOH (b) 60% NaOH, 2% NaCl, 38% H_2O (c) 4500 kg
9.23 NaCl 62.5%, Water 37.5%
9.24 (a) NaBr 2.68%, $MgCl_2$ 8.04% (b) NaCl 78.95%, NaBr 5.26%, $MgCl_2$ 15.79%
9.25 (a) 35200 kg (b) 4800 kg (c) 10000 kg
9.26 10.526 kg/100 kg of water
9.27 299.7 K
9.28 Na_2SO_4 5.2%, water 94.8%
9.29 $Na_2SO_4 : Na_2CO_3 = 1 : 2.37$
9.30 30% Ca $(NO_3)_2.4H_2O$, 70% $CuSO_4.5H_2O$
9.31 536.86 kg
9.32 (a) Na_2SO_4 = 27.96%, H_2O = 72.04% (b) 0.0147 kg
9.33 39.15 kg crystals
9.34 211 kg
9.35 (a) 294.12 kg (b) 85.47 kg
9.36 (a) 251.26 kg (b) 10.3 kg
9.37 (a) 1857 kg of water, 368 kg of crystals (b) 399 kg of crystals
9.38 (a) 6370 kg (b) 6630 kg
9.39 (a) 17.5 kg (b) 18.61 kg
9.40 700 kg, 94.7%
9.41 (a) 3.58 m^3 (b) 534 kg
9.42 (a) 844 kg (b) 285.69 kg
9.43 10.1%
9.44 (a) 47.42 kg (b) 25.91 kg
9.45 (a) 7607.1 kg (b) 1085.7 kg
9.46 (a) 332.74 kg (b) 16.48 kg $MgSO_4$/100 kg of water
9.47 (a) 607.95 kg (b) 130.85 kg
9.48 (a) 29.63 (b) 329.52 kg (c) 8487.14 kg (d) 102.86 kg
9.49 (a) 19.33 kg (b) 66.3%
9.50 75%

9.51 93.1%

9.52 (a) 588.24 kg (b) 420.59 kg

9.53 96.31%

9.54 (a) 229.17 kg/h (b) 55%

9.55 (a) 5.79% (b) 44.22% (c) 96.03%

9.56 6250

9.57 (a) 42 kg (b) Pulp = 51.72%, Water = 48.28%

9.58 150 kg

9.59 0.4167 kg

9.60 (a) 1.105 kg of water per kg of dry leather
(b) 1.05 kg of water per kg of dry leather (c) 95.24%

9.61 (a) 125 kg, 102.04 kg (b) 97.22%, 2.55%

9.62 (a) 313.15 K, 318.65 K (b) 70% (c) 89.22 m^3

9.63 (a) 318.65 K (b) 70%

9.64 (a) 311 K (b) 2.43 kg (c) 9.11 kg

9.65 5279.5 kg of wet air/h

9.66 (a) 820 kg (b) 105.26 kg (c) 4774.02 kg

9.67 (a) 911.77 kg (b) 508 kg

9.68 (a) 17.74 kg/h (b) 97.99 m^3/h

9.69 30 kg, 70 kg

9.70 50.82 g (upper layer), 49.18 g (lower layer)

9.71 (a) 68.23 kg, 131.77 kg (b) 74.13%

9.72 (a) 166.93 kg, 133.62 kg (b) 31.72 kg

9.73 (a) 40% acid, 60.0% water (b) 90 kg of raffinate, 160 kg of extract (c) 150 kg

9.74 82.05%

9.75 2.2383×10^{-2} kg

9.76 1.318 kg

9.77 36.11 kg

9.78 (a) 1373.93 m^3 (b) N_2 = 10.97%, CH_4 = 89.03%

9.79 (a) 2916.37 kg (b) 75.25%

9.80 75%

9.81 (a) 7.58 kg (b) 986 mol/min (c) 3.41% (d) 14.9 m^3/min

9.82 (a) 157.73 kg (b) 1336.68 kg (c) 221.05 kg (d) 34.9 kg (e) 186.15 kg

9.83 (a) 903.06 kmol/h (b) Benzene: 0.554%, Toluene = 99.446%

9.84 (a) 40 kg (b) 4.44 weight % alcohol

9.85 (a) 37.5 kg (b) 0.6

9.86 462.48 kg

9.87 (a) Feed 44.02%, distillate 97.44%, residue 5.85%
(b) Distillate 41.67 mol, residue 58.33 mol

9.88 (a) 615.84 kg (b) B = 3.2%, T = 64.32%, X = 32.48%

9.89 (a) 78.95 mol; 49.6% propane, 50.4% butane (b) 35.9 mol

9.90 Overhead: Ethylene 7.79%, Ethane 11.68%, Propylene 19.46%, Propane 56.07%, Isobutane 5.0%;
Bottoms: Propane 0.8%, Isobutane 31.91%, *n*-Butane 47.10%, *n*-Pentane 20.19%

9.91 (a) F = 40 kg/s, G = 60 kg/s, H = 20 kg/s
(b) Feed: B = 36.36%, C = 23.64%; D: B = 4.55%, C = 0.45%;
F: B = 86.35%, C = 8.64%

9.92 (a) 92.33% (b) 240.9 mol/s

Chapter 10

10.1 CO_2 = 12.34%, O_2 = 5.97%, H_2O = 13.41%, N_2 = 68.28%

10.2 (a) Ethane: 0.21%, CO_2 = 3.32%, CO = 0.41%, H_2O = 5.6%,
O_2 = 14.41% and N_2 = 76.05%
(b) Ethane: 0.22%, CO_2 = 3.51%, CO = 0.44%, O_2 = 15.27% and N_2 = 80.56%

10.3 49.1%

10.4 (a) 195.4 m^3 (b) 36.036 kg

10.5 (a) 11.08% CO_2, 83.35% N_2, 5.57% H_2O (b) 26.98 kg (c) 0.0836 bar

10.6 0.3374 ethene, 0.6626 ethane

10.7 (a) CO_2 = 9.97%, H_2O = 15.03%, N_2 = 75.0% (b) 0.553 kg (c) 28.09 kg

10.8 CO_2 9.83%, CO 0.52%, O_2 6.57%, N_2 83.08%

10.9 (a) 6.93 m^3 (b) 54.286 m^3 (c) 20.45% CO_2, 1.34% O_2, 78.21% N_2

10.10 (a) CO 1.71%, CO_2 17.5%, O_2 2.19%, N_2 78.6% (b) 177.29 kg

10.11 (a) 6216 m^3/h (b) 10.96% CO_2, 0.61% CO, 85.69% N_2 and 2.74% O_2
(c) 17 138 m^3/h

10.12 (a) 2237.76 kg (b) CO_2 = 8.76%, H_2O = 9.19%, O_2 = 6.68%, N_2 = 75.37%
(c) 28.76

10.13 (a) CO_2 = 6.61%, O_2 = 6.45%, N_2 = 71.88%, H_2O = 15.06%
(b) 2784.93 m^3 (c) 327.56 K

10.14 (a) 16.24% (b) 1216.7 m^3

10.15 0.811

10.16 94.23% C, 5.77% H_2

10.17 319.08 K

10.18 37.0%

10.19 25%

10.20 (a) 24.7% (b) C : H = 1 : 0.309

10.21 (a) Methane = 50.0%, N_2 = 50.0% (b) 7.73%

10.22 (a) 1:7.5 (b) 22.69 kg (c) 0.812 kmol

10.23 CO_2 = 11.33%, SO_2 = 0.33%, CO = 2.12%, O_2 = 3.38%, H_2O = 6.49%, N_2 = 76.35%

10.24 75.3%

10.25 (a) 6.67 kg (b) 0.834%

10.26 (a) 24.6% (b) 37.62 kmol

10.27 (a) 2.188 kg (b) 4.6% (c) 14.26 kg

10.28 (a) 33.33% (b) 8

10.29 (a) 60% CH_4, 40% C_2H_6 (b) 40%

10.30 24.2%

10.31 (a) 19.71% (b) 0.374 (c) 13.72 m^3

10.32 (a) 40% (b) 3:1, C_2H_6 (c) 18.72 m^3

10.33 (a) 12.19% (b) 60.81% (c) 12.3304 mol (d) 1767.14 m^3

10.34 (a) 1.900 kmol (b) 13.78% CO_2, 4.59% CO, 4.25% O_2, 77.38% N_2

10.35 79.36%

10.36 (a) 2.67:1, C_3H_8 (b) 85:1 (c) 77.81 m^3/min

10.37 (a) 10% (b) C_4H_{10} (c) 5716.7 m^3 (d) 3812.8 kg/h

10.38 (a) 12.44% (b) 1:0.2778 (c) 15.66 m^3 (d) 2.72 kg

10.39 26.57%

10.40 19.9%

10.41 (a) 251.61 kg (b) 1.056 (c) 2.28 kmol (d) 34.7%

10.42 (a) 63.42% (b) 41.71 m^3 (c) 13.54 m^3

10.43 (a) 16.61% (b) 11.41 kg (c) 16.4517 m^3 (d) 22.08% C, 77.92% inerts

10.44 (a) 1190.53 kg (b) 43.02 kmol/h
(c) CO_2 13.08%, CO 1.44%, H_2O 6.60%, O_2 4.03%, SO_2 0.09%, N_2 74.76%

10.45 (a) 8.92 kmol (b) 55.3% (c) 381.93 m^3

10.46 (a) 196.154 kmol (b) 170.57 kmol (c) 1.75%

10.47 7.88 kg, 5.12 kg

10.48 (a) 1224.5 kg (b) 10.4%

10.49 (a) 3.72% (b) 40% (c) 10.87 m^3

10.50 (a) 50.2% (b) 1362.6 kg (c) 61.9 kg (d) 1095.7 m^3

10.51 (a) 2.572 kg (b) 1458.6 kg (c) 982.82 m^3 (d) 3.897%
(e) C: 73.8%, H: 5.12%, O: 9.77, S: 1.1, Ash: 10.2%

10.52 (a) 27.14% (b) CO_2 7.71%, CO 1.94%, N_2 83.37%, O_2 6.92% (c) 14.256 kPa

10.53 80.07%

10.54 (a) 60% (b) 14.23 m^3 (c) 8.92 m^3

10.55 95.6%

10.56 (a) 957.08 kg (b) SO_2 = 9.45%, O_2 = 11.55%, N_2 = 79.00%
(c) SO_2 = 0.5%, SO_3 = 9.4%, O_2 = 7.39%, N_2 = 82.71%

10.57 6.34% SO_2, 1.59% SO_3, 10.10% O_2, 81.97% N_2

10.58 89.35%

10.59 17.02%

10.60 2.0276 kg

10.61 2.7886 kg

10.62 312.8 m^3

10.63 (a) 80.789 kg, 2.24% FeS_2, 54.11% Fe_2O_3, 18.09% ZnO, 6.99% SO_3, 18.57% inerts
(b) 7.73%

10.64 (a) 347 m^3 (b) 70.0% (c) 864.04 m^3

10.65 76.8%

10.66 0.2975

10.67 (a) 73.64 kg (b) 1.8501 kmol O_2, 14.8371 kmol N_2, 1.4511 kmol SO_2, 0.007 kmol SO_3
(c) 40.5% (d) 3.47%

10.68 67.6%

10.69 (a) 604.29 kg (b) SO_3 7.21%, SO_2 8.65%, O_2 10.90%, N_2 73.24%
(c) 110.05 kg (d) SO_3 7.01%, SO_2 0.22%, O_2 8.87%, N_2 83.91%
(e) 71.57 kg

10.70 (a) 6.043 kg (b) SO_2 9.13%, SO_3 7.66%, N_2 73.01%, O_2 10.20%
(c) 117.24 kg (d) 87.84%

10.71 (a) 1.2465 kmol (b) 8.8% (c) 182.77 kg

10.72 (a) 1.32×10^5 kg (b) 2.64×10^4 kg

10.73 (a) 9.70% (b) 197.5 kg (c) 95.95 (d) 507.84 kg/h
(e) 0.32% SO_2, 10.30% O_2, 89.38% N_2

10.74 (a) 523.81 kg (b) 0.4334 kg (c) 1765 kg

10.75 (a) $CaCO_3$ 60.31%, $MgCO_3$ 35.13%, inerts 4.56% (b) 9.12% (c) 43.93 kg

10.76 $CaCO_3$ 70%

10.77 (a) 65% $CaCO_3$, 30% $MgCO_3$, 5% inerts
(b) 10.92 kmol (c) 86.2 kg (d) 8.54%

10.78 (a) 40% (b) Air/Fuel: 15.35:1 (c) 99.55 kg (d) 71.57 kg

10.79 (a) Phosphate:sand:charcoal = 1:0.663:0.288 (b) 6.04 kg

10.80 (a) 3.36% (b) 49.886%

10.81 (a) 114.15 kg (b) 718.44 kg (c) 72%

10.82 (a) 187.12 kg
(b) $CaSO_4$ = 70.35%, H_2SO_4 = 15.11%, HF = 1.03%, inerts = 13.51%

10.83 51.58 kg

10.84 298.3 kg

10.85 (a) 13.16% (b) 91.52% (c) 595.62 kg; Al_2O_3 16.05%, impurities 83.95%

10.86 (a) 276 mol (b) 0.7567 kg

10.87 2.8303×10^7 m^3/day

10.88 (a) NO 2%, N_2 78%, O_2 20% (b) 0.7 kPa 0.68 kPa (c) 29.49 kg

10.89 (a) 131.84 kg (b) Na_2SO_4 63.38%, H_2SO_4 34%, H_2O 1.5%, HNO_3 1.12%
(c) 107.91 kg (d) 72.35 kg, 3.42 kg

10.90 (a) 1244.27 m^3 (b) 300.04 kg

10.91 (a) 1142.86 m^3
(b) 51.34 kmol; 1.19% NH_3, 10.58% O_2, 71.42% N_2, 6.72% NO, 10.09% H_2O
(c) 49.79 kmol, 1.22% NH_3, 7.79% O_2, 73.65% N_2, 0.69% NO,
6.24% NO_2, 10.4% H_2O (d) 355.16 kg, 36.75%

10.92 (a) 1:1.2313
(b) HCl 17.98%, Cl_2 26.22%, O_2 6.9%, N_2 42.24%, H_2O 6.66%

10.93 (a) 178.6 m^3 (b) 297.24 m^3 (c) 117.62 kg
(d) 9.15% HCl, 8% O_2, 18.3% Cl_2, 64.55% N_2

10.94 (a) 1.136 kg
(b) HCl = 18.78%, Cl_2 = 27.38%, O_2 = 6.18%, N_2 = 40.70%, H_2O = 6.96%
(c) 324.53 K

10.95 (a) 2020.29 kg of CCl_4, 842.37 kg of S (b) 1660 kg of CC_4, 644.42 kg of S

10.96 (a) CS_2, 20.28% (b) 71.89% (c) 51.98 kg

10.97 (a) 1417.26 kmol, 1079.76 kmol (b) 44 477.92 kg, 44 477.92 kg
(c) HCl = 40.98%, Ethylene = 16.54%, O_2 = 9.89%, N_2 = 32.58%
(d) HCl = 4.1%, Ethylene = 2.36%, O_2 = 1.8%, N_2 = 32.58%,
H_2O = 9.11%, EDC = 50.05%

10.98 90.7%, 91.9%

10.99 (a) 97.49% (b) 94.55% (c) 1123.4 kg (d) 1255.77 kg
(e) HCl 603.07 kg, H_2O 266.18 kg

10.100 247.8 kg

10.101 30 300 kg/h

10.102 404

10.103 (a) 30.1% (b) 4.54:1

10.104 (a) 32.88 kg (b) 63.42 kg (c) 0.52 kg/h

10.105 (a) 4.98% CH_3OH, 3.73% O_2, 51.48% N_2, 19.9% HCHO, 19.9% H_2O
(b) 395.13 m^3

10.106 50%

10.107 (a) 5 mol (b) 50% (c) 86.5 mol (d) 72.57%

10.108 (a) 2677.2 kg (b) 2369.7 kg (c) 27.81:1

10.109 C_2H_4 23.08%, H_2O 64.23%, ether 0.13%, ethanol 2.31%, inerts 10.26%

10.110 (a) 18.94 kg, 86.28 kg
(b) monosulphonate 7.58%, disulphonate 34.51%, H_2SO_4 51.34%, H_2O 6.57%

Chapter 11

11.1 0.1

11.2 55.56%

11.3 0.1667

11.4 0.1837

11.5 (a) 0.3591 (b) 4.22 kg (c) 129.9 m^3

11.6 (a) 375 kg (b) 10% (c) 30%

11.7 (a) 100000 kg/h (b) 6.75%

11.8 0.7093 kg/kg

11.9 (a) 26.56% (b) 6000 kg/h (c) 1816.24 kg/h

11.10 6729.63 kg

11.11 23.44 kg/h

11.12 (a) 1.2 (b) 46.51 (c) 0.973 kg/min

11.13 (a) 3571.43 kg (b) 58 204 kg

11.14 (a) 0.2 (b) 2.25 kg

11.15 (a) 404.21 m^3 (b) 57.86 kg (c) 190.97 kg

11.16 (a) Benzene 2%, toluene 55.34%, xylene 42.67%
(b) 2438.21 kg, 6768.62 kg (c) 96.92% (d) 9752.84 kg

11.17 (a) 7205.93 kg/h (b) 6195.52 kg/h (c) 93.66 kg/h (d) 6.05%

11.18 23 295.37 kg; 10.83% NaCl, 84.88% H_2O, 4.29% KCl

11.19 0.6132

11.20 (a) 926.32 kg/h (b) 818.67 kg (c) 409.33 kg

11.21 (a) 1:6 (b) 75.19 kg (c) 12.53 kg

11.22 (a) 2777.78 kg/h (b) 2177.8 kg/h (c) 17 400 kg/h
(d) 22 177 kg/h, 11.12% NaCl, 2.71% KCl, 86.17% H_2O

11.23 (a) 130.16 kg/h (b) 0.838 kg of dry air/kg of dry air in the fresh air feed

11.24 (a) 4.0953 kg/min (b) 52.21 kg/min (c) 45.44 m^3/min (d) 358.34 kg/min

11.25 (a) 315.8 kg/h (b) 105.26 kg/h (c) 894.74 kg/h

11.26 (a) 0.9067 (b) 6.9% *A*, 12.3% *B*, 6.3% *C*, 74.5% *D*

11.27 (a) 0.9 (b) 30% *A*, 43.33% *B*, 2.22% *C*, 24.45% *D*
(c) A = 13.33%, B = 10.00%, C = 5.00%, D = 71.67%

11.28 (a) 20% (b) 9.09%

11.29 20%

11.30 3.1

11.31 0.6312

11.32 (a) 567.31 kg/h (b) 432.69 kg/h (c) 0.5851

11.33 (a) 2401.3 kg (b) 121 kg (c) 2562.6 kg

11.34 S_1 60 kmol/h C_3H_8; S_2 200 kmol/h C_3H_8; S_3 140 kmol/h C_3H_8, 60 kmol/h C_3H_6, 60 kmol/h H_2; S_4 60 kmol/h H_2; S_5 60 kmol/h C_3H_6; S_6 140 kmol/h C_3H_8

11.35 (a) 300 (b) 1700

11.36 (a) 3:11
(b) Mass%: ethanol 29.22%, acetaldehyde 9.31%, H_2O 3.81%, O_2 10.83%, N_2 46.82%;
Mol%: ethanol 20.69%, acetaldehyde 6.90%, H_2O 6.90%, O_2 11.03%, N_2 54.48%

11.37 77.27%

11.38 (a) 290.35 kg, 91.4% (b) 0.49 (c) 108.9 kmol

11.39 (a) 4 kmol/h (b) 80% (c) 3 kmol/h

11.40 (a) 1.66 kmol, 2.13 kmol (b) 314.86 kg/h (c) 28.21% (d) 37.24 m^3

11.41 (a) 42.86 mol NH_3 (b) 1:2.14

11.42 (a) 227.6 kmol/h to *A*, 190.6 kmol/h to *B* (b) 404.04 kmol/h

11.43 (a) 544 mol
(b) CO 31.27%, CH_3OH 4.82%, CH_4 0.69%, H_2O 0.69%, H_2 62.53%
(c) 28 mol

11.44 4 mol to 100 mol feed

11.45 (a) 177.45 kg/h (b) 186.9 kg/h

11.46 (a) 0.0079 (b) 48.4 mol (c) 97.0%

11.47 (a) 673.74 mol (b) 6.667 mol
(c) CO 20.38%, H_2 76.17%, CH_4 3.0% (d) 31.11 mol

11.48 (a) 1:5.71 (b) 24.13 kmol (c) 111.41 kg (d) 27.24 kg/h

11.49 (a) 860 kmol (b) 4 kmol (c) 9.43%

11.50 (a) 50.08 kmol (b) 5.87 kmol (c) 2.93 kmol

11.51 (a) 0.0157 (b) 0.2 molAr/3 mol N_2–H_2 (c) 382 mol

11.52 (a) 130.92 mol (b) 42.71% H_2, 57.29% CH_4 (c) 127.17 mol

11.53 (a) 63.79 kmol (b) 95.69% (c) 6.18 kmol

Chapter 12

12.1 17.845 m, 25 kJ

12.2 13.86 kJ

12.3 89.55 kg, 878.51 N

12.4 30.6305×10^5 J

12.5 1.667×10^2 kJ

12.6 7.355 kJ, 61.29 W

12.7 (a) 14.715, 0 (b) 0, 14.715 (c) 0.035 K

12.8 16.8 W

12.9 123.46 m^3

12.10 (a) 4.164×10^3 N (b) 1.3254×10^5 N/m_2 (c) 2.082×10^3 J (d) 490.5 J

12.11 17.15 m/s

12.12 (a) 1 J (b) 147.15 J

12.13 2.156×10^5 J

12.14 20.39 m

12.15 3:1

12.16 1.146×10^4 kJ

12.17 6.073×10^4 J

12.18 (a) 183.94 kJ (b) 1839.4 kJ (c) 1655.46 kJ (d) 38.36 m/s (e) 1839.4 kJ

12.19 ΔU = 37.9 kJ/mol; ΔH = 41 kJ/mol

12.20 U_l = – 0.9045 kJ/kg, U_v = 282.11 kJ/kg, H_v = 320.5 kJ/kg

12.21 58.4×10^4 J

12.22 420.1 kJ, 2214 kJ/kg, 2423.9 kJ/kg

12.23 (a) 3.572 kJ (b) 10.97 min

12.24 (a) 4122.2 J (b) $C_P = 8.85394 + 8.2744 \times 10^{-3}T - 3.4150 \times 10^{-6}T^2$

12.25 (a) 41.2339 J/mol K (b) 39.6355 J/mol K (c) 7927 kJ

12.26 (a) 50.0395 kJ/kmol K (b) 22.3251×10^4 kJ

12.27 (a) 440.22 kJ (b) 4.40 kJ/kg K

12.28 (a) 30.7324 kJ/kmol K (b) 30732.4 kJ/kmol

12.29 31.496 kJ/kmol K

12.30 (a) $C_P = 7.7844 + 0.1344T - 4.7696 \times 10^{-5}T^2$ (b) 46 237.6 J

12.31 4525.07 kJ

12.32 46.31 kJ/kmol K, 46 310 kJ/kmol

12.33 1.55×10^{-4} kJ

12.34 14 187 kJ/kmol

12.35 233.55 m^3/min

12.36 322.44 kJ

12.37 3105 kJ

12.38 2223.3 kJ/kg

12.39 4.184×10^2 kJ/kg

12.40 255.95 kJ/kg

12.41 397.7 kJ/kg

12.42 117 895 kJ/kmol, 112 582 kJ/kmol

12.43 8918.5 kJ/kg

12.44 (a) 482 624 kJ/h (b) 178 831 kJ/h

12.45 (a) 181.97 kPa (b) 444.94 kJ/kg

12.46 389.8 K

12.47 2.6% liquid

12.48 0.9948 kg/s

12.49 759 K

12.50 (a) 125.44 kJ/kg (b) 2622.14 kJ/kg

12.51 (a) 40 kPa (b) 223 kPa (c) 18242 kJ

12.52 18 467 kJ, 148.6 bar

12.53 995.96 kJ/kg of water

12.54 195 814 kJ

12.55 1.246×10^6 kJ

12.56 914.86 kJ, 139.35 kJ

12.57 (a) 11 987.5 kJ (b) 275.4 K

12.58 442.224 kJ

12.59 (a) 60 522 kJ (b) 140.3°C

12.60 T = 305.8 K

12.61 62 187.5 kJ

12.62 9.661×10^6 kJ

12.63 250.77 J/kg

12.64 340.62 K

12.65 –10.217 kJ/mol

12.66 178 873 kJ

12.67 5100 kJ removed

12.68 1.2914×10^7 kJ/h

12.69 21 783 kJ

12.70 (a) 38 kJ (b) 0 (c) 4.31 bar (d) – 47.8 kJ

12.71 (a) – 24.0685 kJ (b) – 17.695 kJ (c) 6.374 kJ

12.72 (a) 250 kJ (b) 173.3 kJ

12.73 (a) 981 J (b) 981 J (c) 0.234 K

12.74 6.92×10^5 J

12.75 25 J, 75 J, –100 J, 100 J

12.76 (a) 14 m/s (b) 0.78 K

12.77 (a) 22 480 kJ (b) 8730 kJ

12.78 166.67 kJ

12.79 – 895 879.7 kJ

12.80 (a) 1.48×10^4 kg (b) 3.022×10^{-4} (c) 50 h

12.81 2168.6 kJ, 3000 kJ, –125.32 kJ, 2043.28 kJ

12.82 391.4 K, 1.96 Pa

12.83 2328.7 kJ

12.84 1243.6 kJ, 1742.4 kJ

12.85 407.18 kJ

12.86 (a) 2282 kJ (b) 2206 kJ (c) 0

12.87 (a) 472.5 K (b) – 4863 kJ (c) 0 (d) 4863 kJ (e) 5981 kJ

12.88 (a) 997.68, 1662.8, – 997.68 kJ/kmol (b) – 997.68, – 1662.8, – 665.12 kJ/kmol (c) 0, 0, 1490, 1490 kJ/kmol (d) 0, 0, –172.8, –172.8 kJ/kmol

12.89 3407, 1077, –2329, –3252 kJ/kmol

12.90 (a) 1 bar, 56 K, 4656, 6518, 0, 4656 kJ
(b) 5 bar, 1400 K, 22 380.3, 32 592.0, 0, 22 380.3 kJ

12.91 (a) 0.373 kg (b) 0.304 kg

12.92 418.5 K, 7.65 bar

12.93 (a) 600 K, 2 bar, 4605 kJ (b) 132.9 K, 0.443 bar, 2376 kJ (c) 900 kJ

12.94 70 kJ

12.95 313.46 hp

12.96 722.51 kW

12.97 435.5 hp

12.98 92.9 hp

12.99 144.1 kW

12.100 (a) Q = 3122 kJ/kg, W_s = 0, ΔH = 3122 kJ/kg
(b) Q = 0, W_s = 588.34 kJ/kg, ΔH = –588.34 kJ/kg

12.101 1744 hp

12.102 35 kJ

12.103 (a) 402.64 m/s (b) 5.751×10^{-4} m^2

12.104 344 K

12.105 (a) 46.88 J/kg (b) 50 J/kg

12.106 15.35 hp

12.107 2453 kJ/kg

12.108 2.03 hp

12.109 1.72 K

12.110 30.523 kW

12.111 2.95×10^5 N/m^2, 1481.3 W

12.112 4.316 kW, 86.22 kPa, 966 kPa

12.113 81.51 kW

12.114 (a) 6259 kg/h (b) 378 kg/h

12.115 0.824 hp

12.116 82.785 kPa

12.117 (a) 5725.2 kWh/day (b) 1.099 m

12.118 122 155 kg/h, 7759 kg/h

12.119 3.174×10^6 kJ/h

12.120 35 kg

12.121 (a) 9.4313×10^5 kJ/h (b) 12.3470×10^5 kJ/h

12.122 214 030 kJ/h, 17.26 kg

12.123 1 029 025 kJ

12.124 3314.8 kJ

12.125 (a) 2976.1 kg/h (b) 1.94×10^5 kJ/h

12.126 178 354 kJ/h

Chapter 13

13.1 –1207.43 kJ

13.2 –194.9 kJ/mol

13.3 –323.53 kJ/kmol

13.4 –81.69 kJ

13.5 –102.933 kJ/mol

13.6 –110.52 kJ/mol

13.7 –74.79 kJ

13.8 –110.52 kJ/mol

13.9 (a) 358.2 kg/h (b) 142.86 kmol/h

13.10 49.819×10^3 kJ/m^3 and 45.224×10^3 kJ/m^3 at STP

13.11 $\Delta H_{C_6H_{12}} = 6\Delta H_{CH_3OH} + 142216$

13.12 –114.408 kJ

13.13 –904.7 kJ/mol

13.14 –1654.27 kJ/mol

13.15 –241.83 kJ/mol

13.16 –2204.14 kJ/mol

13.17 49.02 kJ/mol

13.18 2.24 J/mol

13.19 117.715 kJ/mol

13.20 -5.623×10^3 kJ/kg

13.21 –177.0 kJ

13.22 –286.46 kJ

13.23 –110.52 kJ/mol

13.24 –199.5 kJ/mol

13.25 5.7 K

13.26 –886.83 kJ/mol

13.27 155.52 kJ/mol, 158.74 kJ//mol

13.28 2.072×10^5 kJ

13.29 118.5×10^3 kJ/h

13.30 32 525.5 kJ/kmol of acid to be added

13.31 19 662.7 kJ

13.32 –42.62 kJ/mol

13.33 –183.950 kJ

13.34 –185.342 kJ/mol

13.35 23 178 kJ/kg of methane

13.36 –393.83 kJ/mol

13.37 –224.5 kJ

13.38 328 299.8 kJ/kmol of CH_4

13.39 –377.45 kJ/mol

13.40 617.88 kJ

13.41 –3283.43 kJ/mol

13.42 70.35×10^3 kJ/mol

13.43 5641.26 kJ/mol

13.44 -191.88×10^3 kJ/kmol

13.45 28 098 kJ

13.46 320.97 kJ

13.47 (a) 6.91 kg/h (b) 804.3 m^3/h (c) 8646 kJ/h

13.48 2.103×10^6 kJ/h

13.49 2285.3 K

13.50 2079.6 K

13.51 1308 K

13.52 82%

13.53 854 K

13.54 1215 K

13.55 1990 K

13.56 351.1 K

13.57 857 K

13.58 2237 K

Index